Lippincott® Illustrated Reviews: Physiology

Third Edition

Robin R. Preston, PhD

Professor of Physiology

Department of Biomedical Sciences

Noorda College of Osteopathic Medicine

Provo, Utah

Thad E. Wilson, PhD

Professor

Department of Physiology

University of Kentucky College of Medicine

Lexington, Kentucky

.®. Wolters Kluwer

Philadelphia • Baltimore • New York • London
Buenos Aires • Hong Kong • Sydney • Tokyo

Senior Acquisitions Editor: Crystal Taylor
Development Editors: Debbie Bordeaux, Kelly Horvath
Editorial Coordinator: Erin E Hernandez
Marketing Manager: Phyllis Hitner
Production Project Manager: Matthew West
Manager, Graphic Arts & Design: Stephen Druding
Art Director, Illustration: Jennifer Clements
Manufacturing Coordinator: Margie Orzech
Prepress Vendor: TNQ Technologies

3rd edition

Library of Congress Cataloging-in-Publication Data

ISBN-13: 978-1-975196-33-2

Cataloging in Publication data available on request from publisher.

The cover illustration shows an artificially colorized scanning electron microscope image of a tangle of renal glomerular capillaries. The capillaries are ensheathed in a filtration slit diaphragm formed by interdigitating podocyte foot processes. The slit diaphragm is a component of the filtration barrier that ensures fluid filtering into the renal tubule from capillaries is cell- and protein-free. For more details, see Chapter 25.

Image copyright sciencephoto.com.

Dedication

This book is dedicated to our students, who continue to challenge and inspire us.

Acknowledgments

Medical terminology and our understanding of human physiology are in a state of constant flux. The third edition of *Lippincott® Illustrated Reviews: Physiology* incorporates all recent changes and has been expanded to cover new areas that are impacting current medical practice and that are likely to be tested on future licensing examinations.

Producing new editions of textbooks is a lengthy process and relies on a team of talented professionals to bring them to fruition. First and foremost, we acknowledge the contributions of Kelly Horvath, freelance Development Editor, whose keen eye for detail, extensive experience in medical publishing, and deep understanding of how to make content connect with the current generation of readers have had a major impact on the direction of this latest revision. We also would like to thank Jennifer Clements, Art Director (WK), for editing all but a handful of the existing 600 illustrations, plus the 20 or so new ones that appear in the current version of the text. We are also grateful to several other individuals at WK, including Crystal Taylor (Senior Acquisitions Editor) for her continued support of the authors, Lindsey Porambo (Acquisitions Editor), Debbie Bordeaux (Development Editor), and Erin Hernandez (Editorial Coordinator).

We are extremely grateful to Barbara Mroz, MD, for reviewing and editing all the new Case Studies to ensure that a basic scientist's view of patient presentations and underlying conditions lines up with clinical realities.

We would also like to thank the Chair (Alan Daugherty, PhD, DSc) and faculty of the Department of Physiology at the University of Kentucky for their support. We would also like to thank year-one and year-two Course Directors at the University of Kentucky College of Medicine for their insights and shared medical education goals. Finally, we would like to thank all of the faculty, clinicians, and medical and health professions students (a special thanks to Alyssa Curry, Noorda-COM Class of 2026) whose continued feedback drives improvements in this text.

Preface

The third edition of *Lippincott® Illustrated Reviews: Physiology* is a non-fiction novel masquerading as a textbook. It tells a tale of how the body works, how it adapts to changes in its internal and external environment, how it ages, and how derangements in organ function cause disease and death. It follows the organization of the human body, each unit addressing a different organ system and considering its role in the life of the individual. Physiology texts typically take a "macro-to-micro" approach, their descriptions of organs following the history of human physiologic discovery (gross anatomy, microanatomy, cellular biology, and, finally, molecular biology). We begin most units by identifying key organ functions and then showing how cells and tissues are designed to fulfill these functions. Although physiologic design is shaped by natural selection and not by purpose, this teleologic approach can help us understand why cells and organs are structured the way they are. It also helps us understand the rationale behind responses to stress that are advantageous in the short term but ultimately may lead to system failure. Understanding the "why" aids retention and gives future health care providers a powerful tool for anticipating how and understanding why disease processes present clinically in the way that they do.

What does the third edition of* Lippincott® Illustrated Reviews: Physiology *cover?
Physiology is a burgeoning discipline, which no single text can cover exhaustively. We used several guides to help us decide what material to include:

- Topics currently being tested by the U.S. Medical Licensing Examiners (USMLE® Step 1 Content Outline & COMLEX-USA Level 1 master blueprint)
- Learning objectives that many medical school physiology courses cover (American Physiological Society and the Association of Chairs of Departments of Physiology, 2012, and The Physiological Society, 2020)
- Topics covered in *BRS Physiology*, WK's popular board review book

Who should use this book? *Lippincott® Illustrated Reviews: Physiology*, third edition, is intended to help medical students preparing for their licensing examinations, but the material is presented with a clarity and level of detail that also suits it for a primary course text by any health professions (e.g., dentistry, nursing, pharmacy, physical therapy, and physician assistant) as well as serving as a reference and review text for established clinicians.

Format: *Lippincott® Illustrated Reviews: Physiology*, third edition, follows a lecture-note format, with minimal introductions, history, or discussions of ongoing research—the chapters quickly cut to the chase in a narrative form customized for fast assimilation. The subheadings break the presentation into easy-to-absorb paragraphs that are appropriate to skim for review yet sufficiently detailed to instruct a student who may be new to or unsure of a topic. The writing style is engaging yet succinct, rendering complex topics accessible and memorable.

Art: The text is also heavily illustrated with step-by-step guides to help the visual learner and for ease of review by students preparing for examinations. *Lippincott® Illustrated Reviews: Physiology*, third edition, has fully integrated art into the text to seamlessly tell the story of physiology in a completely new way. More than 600 original and energetic full-color line drawings are supplemented by abundant clinical images that together illustrate physiology with a dynamism belying their two-dimensionality. Legends are deliberately minimal to allow the art itself to "speak." Step-by-step dialogue boxes *guide* the viewer through physiologic processes. Medical physiology textbooks typically rely on a set of common time-tested figures to illustrate concepts. *Lippincott® Illustrated Reviews: Physiology*, third edition, takes a slightly more irreverent approach to illustration, so look for irritated octopi, cells that spit nuclei, and hatchet-induced hemorrhages.

Features: *Lippincott® Illustrated Reviews: Physiology*, third edition, incorporates multiple features to facilitate comprehension of material:

- **Real-world examples:** Physiologic concepts are notoriously difficult to grasp, so we have used real-world examples wherever possible to aid comprehension.

- **Clinical Applications:** All chapters include Clinical Applications—many with accompanying clinical images—that show how physiology gone awry can present clinically.

- **Biologic Sex and Aging:** Physiologic differences between the sexes are highlighted, along with a consideration of how normal (healthy) aging impacts all organ systems.

- **Equations boxes:** Real numbers are run through tricky equations for equilibrium potential, alveolar–arterial oxygen difference, and renal clearance and are featured in yellow margin boxes to show students examples they might encounter in practice.

- **Consistency:** Cellular physiology can be overwhelming in its details, especially where transport physiology is concerned. We have kept details in artwork to a bare minimum and consistently used the same colors to denote different ion species throughout the text: Readers will quickly become familiar with visual cues and spend less time reading labels. Easy-to-follow-and-recall flowcharts and concept maps are also widely used.
  - Sodium = red
  - Calcium = indigo
  - Potassium = purple
  - Anions (chloride and bicarbonate) = green
  - Acid = orange

- **Cross-references:** Linking topics across chapters, intravolume cross-references in an easy-to-locate format specify section number to the nearest heading level (e.g., see Chapter 25·III·B). Chapter number and section level are provided at the head of every page for ease of location.

- **Infolinks:** These cross-references among the *Lippincott® Illustrated Reviews* series provide resources for students to delve more deeply into related topics across several disciplines, including biochemistry, pharmacology, microbiology, neuroscience, immunology, and cell and molecular biology. Infolinks in *Lippincott® Illustrated Reviews: Physiology*, third edition, cite chapter and subsection rather than page numbers to facilitate navigation of the e-books.

- **Practice questions:** Each chapter is accompanied by study questions (>160 in total), many new to the third edition of *Lippincott® Illustrated Reviews: Physiology*, that students can use to self-assess their current physiologic knowledge and level of understanding. Questions have been formatted to follow the style used in medical licensing exams. Additional questions can be found online at thePoint.

- **NEW—Case studies:** New to the third edition of *Lippincott® Illustrated Reviews: Physiology* are a set of 15 case studies. Each begins with a clinical vignette describing a patient presentation. Provocative thought questions ask the student to test the extent of their understanding of key concepts. Multiple choice review questions test for a more general grasp of the subject area covered by the case study. Thought questions and review questions are accompanied by fully illustrated explanations. Case studies are aimed at students wishing to better understand the relevance of the underlying physiology for clinical practice and also instructors looking for material around which to create small group discussions.

- *Bonus material:* A companion website on thePoint provides additional resources, including an interactive board-style (i.e., USMLE and COMLEX) question bank, which includes full answer explanations. Students using McGraw-Hill's *First Aid for the USMLE Step 1* and LWW's *BRS Physiology* can also download a map cross-referencing both texts to *Lippincott® Illustrated Reviews: Physiology*, third edition, which allows them to quickly locate a more complete explanation of a difficult concept than that found in either review series. Students will also find animations to clarify important concepts and flash-cards to assist with study and review.

- **Comments?** Our current understanding of physiologic mechanisms is evolving constantly in the light of new research findings. Subsequent editions of *Lippincott® Illustrated Reviews: Physiology* will be updated to take into account new findings and reader feedback. If you have any suggestions for improvement or other comments on content or the *Lippincott® Illustrated Reviews: Physiology* approach, you are welcome to submit them to the publisher directly at http://www.lww.com or contact the authors by e-mail at LIRphysiology@gmail.com.

Contents

Cell and Membrane Physiology

1

I. OVERVIEW

The human body comprises several distinct organs, each of which has a unique role in supporting the life and well-being of the individual. Organs are, in turn, composed of tissues. Tissues are collections of cells specialized to perform specific tasks that are required of the organ. Although cells from any two organs may appear strikingly dissimilar at the microscopic level (e.g., compare the shape of a red blood cell [RBC] with the branching structure of a nerve cell's dendritic tree, as in Fig. 1.1), morphology can be misleading because it masks a set of common principles in design and function that apply to all cells. All cells are enclosed within a membrane that separates the inside of the cell from the outside. This barrier allows the cells to create an internal environment that is optimized to support the biochemical reactions required for normal function. The composition of this internal environment varies little from cell to cell. Most cells also contain an identical set of membrane-bound organelles: **nuclei**, **endoplasmic reticulum (ER)**, **lysosomes**, **Golgi apparatuses**, and **mitochondria**. Specialization of cell and organ function is usually achieved by adding a novel organelle or structure, or by altering the mix of membrane proteins that provide pathways for ions and other solutes to move across the barrier. This chapter reviews some common principles of molecular and cellular function that will serve as a foundation for later discussions of how the various organs contribute to maintaining normal bodily function.

II. CELLULAR ENVIRONMENT

Cells are bathed in an **extracellular fluid (ECF)** that contains ionized sodium (Na^+), potassium (K^+), magnesium (Mg^{2+}), chloride (Cl^-), phosphates (HPO_4^{3-} and $H_2PO_4^{2-}$), bicarbonate (HCO_3^-), glucose, and small amounts of protein (Table 1.1). It also contains around 2 mmol free calcium (Ca^{2+}). Ca^{2+} is essential to life, but many of the biochemical reactions

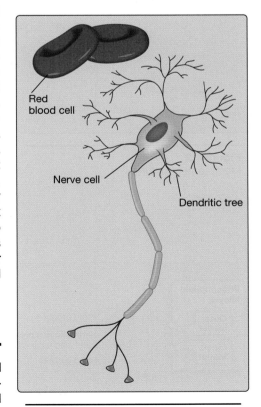

Red blood cell

Nerve cell

Dendritic tree

Figure 1.1.
Differences in cell morphology.

Table 1.1: Extracellular and Intracellular Fluid Composition

Solute	ECF	ICF
Na$^+$	145	12
K$^+$	4	120
Ca^{2+}	2.5	0.0001
Mg^{2+}	1	0.5
Cl$^-$	110	15
HCO$_3^-$	24	12
Phosphates	0.8	0.7
Glucose	5	<1
Proteins (g/dL)	1	30
pH	7.4	7.2

ECF values are from the interstitium. For blood values, see Table 3.1. Values are approximate and represent free concentrations under normal metabolic conditions. All values (except protein concentration and pH) are given in mmol/L.

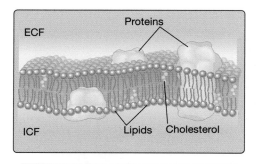

Figure 1.2.
Membrane structure.

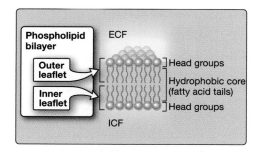

Figure 1.3.
Membrane lipid bilayer.

required of cells can only occur if free Ca^{2+} concentrations are lowered 10,000-fold to around 10^{-7} mol. Thus, cells erect a barrier that is impermeable to ions (the **plasma membrane**) to separate **intracellular fluid** (**[ICF]** or **cytosol**) from ECF and then selectively modify the composition of ICF to facilitate the biochemical reactions that sustain life. ICF is characterized by low Ca^{2+}, Na$^+$, and Cl$^-$ concentrations compared with ECF, whereas the K$^+$ concentration is increased. Cells also contain more free protein than does the ECF, and the pH of ICF is slightly more acidic.

III. MEMBRANE COMPOSITION

Membranes comprise lipid and protein (Fig. 1.2). Lipids form the core of all membranes. Lipids are ideally suited to a barrier function because they are **hydrophobic**: They repel water and anything dissolved in it (**hydrophilic** molecules). Proteins allow cells to interact with and communicate with each other, and they provide pathways that allow water and hydrophilic molecules to cross the lipid core.

A. Lipids

Membranes contain three predominant types of lipids: **phospholipids**, **cholesterol**, and **glycolipids**. All are **amphipathic** in nature, meaning that they have a polar (hydrophilic) region and a nonpolar (hydrophobic) region. The polar region is referred to as the **head group** (Fig. 1.3). The hydrophobic region is usually composed of fatty acid "**tails**" of variable length. Some of these fatty acid chains lack double bonds and are considered "saturated." Introducing a double bond into the tail causes it to kink and creates an "unsaturated" fatty acid (Fig. 1.4). Unsaturated fatty acids increase membrane fluidity. When the membrane is assembled, the lipids naturally gather into a continuous bilayer (Fig. 1.3). The polar head groups gather at the internal and external surfaces where the two layers interface with ICF and ECF, respectively. The hydrophobic tail groups dangle from the head groups toward the center of the bilayer to form the fatty membrane core. Although the two halves of the bilayer are closely apposed, no significant lipid exchange occurs between the two membrane leaflets.

1. **Phospholipids:** Phospholipids are the most common membrane lipid type. Phospholipids comprise a fatty acid tail coupled via glycerol to a head group that contains phosphate and an attached alcohol.

2. **Cholesterol:** Cholesterol is the second most common membrane lipid. It is hydrophobic but contains a polar hydroxyl group that draws it to the bilayer's outer surface, where it nestles between adjacent phospholipids (see Fig. 1.4). Between the hydroxyl group and the hydrocarbon tail is a steroid nucleus. The four steroid carbon rings make it relatively inflexible, so adding cholesterol to a membrane reduces its fluidity and makes it stronger and more rigid. Cholesterol is concentrated in and helps stabilize lipids rafts (e.g., caveolae; see Chapter 13·II·C). Rafts are membrane microdomains that serve as organizing centers for receptors and signaling molecules.

3. **Glycolipids:** The outer leaflet of the bilayer contains glycolipids, a minor but physiologically significant lipid type comprising a fatty

acid tail coupled via sphingosine to a carbohydrate head group. The glycolipids create a carbohydrate cell coat that is involved in cell-to-cell interactions and that conveys antigenicity.

B. Proteins

The membrane's lipid core seals the cell in an envelope across which only lipid-soluble materials, such as O_2, CO_2, and alcohol can cross. Cells exist in an aqueous world, however, and most of the molecules that they need to thrive are hydrophilic and cannot penetrate the lipid core. Thus, the surface (**plasma**) membrane also contains proteins whose function is to help ions and other charged molecules across the lipid barrier. Membrane proteins also allow for intercellular communication and provide cells with sensory information about the external environment. Proteins are grouped based on whether they localize to the membrane surface (**peripheral**) or are **integral** to the lipid bilayer (Fig. 1.5).

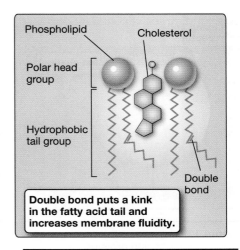

Double bond puts a kink in the fatty acid tail and increases membrane fluidity.

Figure 1.4.
Cholesterol location with the membrane.

1. **Peripheral:** Peripheral proteins are found on the membrane surface, loosely attached to polar head groups or glycolipids. The links are weak, and, thus, these proteins can be washed free using simple salt solutions. Peripheral proteins can be found associated with both intracellular and extracellular plasma membrane surfaces.

   a. **Intracellular:** Proteins that localize to the intracellular surface include many enzymes; regulatory subunits of ion channels, receptors, and transporters; and proteins involved in vesicle trafficking and membrane fusion as well as proteins that tether the membrane to a dense network of fibrils lying just beneath its inner surface. The network is composed of spectrin, actin, ankyrin, and several other molecules that link together to form a **subcortical cytoskeleton** (see Fig. 1.5).

   b. **Extracellular:** Proteins located on the extracellular surface include enzymes, antigens, and adhesion molecules. Many peripheral proteins are attached to the membrane via **glycophosphatidylinositol** ([**GPI**] a glycosylated phospholipid) and are known collectively as **GPI-anchored proteins**.

2. **Integral:** Integral membrane proteins penetrate the lipid core. They are anchored by covalent bonds to surrounding structures and can only be removed by treating the membrane with a detergent. Some

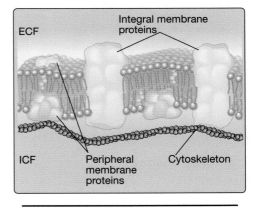

Figure 1.5.
Membrane proteins.

Clinical Application 1.1: Paroxysmal Nocturnal Hemoglobinuria

Paroxysmal nocturnal hemoglobinuria (PNH) is a rare, X-linked disorder caused by a defect in the *PIGA* gene that encodes phosphatidylinositol glycan A (PIGA). PIGA is required for synthesis of the glycophosphatidylinositol anchor used to tether peripheral proteins to the outside of the cell membrane. The gene defect prevents hematopoietic stem cells from expressing proteins that normally protect their progeny (RBCs and platelets) from the immune system. The nighttime appearance of hemoglobin in urine (hemoglobinuria) reflects RBC lysis by immune complement. Patients typically manifest symptoms associated with hemolytic anemia (fatigue, dyspnea). PNH is associated with a significant risk of morbidity due to bone marrow suppression, which leaves affected individuals susceptible to infections and venous thrombosis.

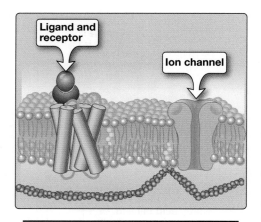

Figure 1.6.
Membrane-spanning proteins.

integral proteins may remain localized to one or the other of the two membrane leaflets without actually traversing its width. Others may weave across the membrane many times (**transmembrane proteins**) as shown in Figure 1.6. Examples include various classes of **ion channels**, **transporters**, and **receptors**.

IV. DIFFUSION

Movement across a membrane requires a motive force. Most substances cross the plasma membrane by **diffusion**, their movement driven by a transmembrane concentration gradient. When the concentration difference across a membrane is unfavorable, however, then the cell must expend energy to force movement "uphill" against the concentration gradient (**active transport**).

A. Simple diffusion

Consider a container filled with water and divided into two compartments by a pure lipid membrane (Fig. 1.7). Blue dye is now dropped into the container at left. Initially, the dye remains concentrated and restricted to its small entry area, but molecules of gas, water, or anything dissolved in water are in constant thermal motion. These movements cause the dye molecules to distribute randomly throughout the entire chamber, and the water eventually becomes a uniform color, albeit lighter than the original drop. The example shown in Figure 1.7 assumes that the dye is unable to cross the membrane, so the chamber on the right remains clear, even though the difference in dye concentrations across the barrier is high. However, if the dye is lipid soluble or is provided with a pathway (a protein) that allows it to cross the barrier, diffusion will carry the molecules into the second chamber, and the entire tank will turn blue (Fig. 1.8).

B. Fick law

The rate at which molecules such as blue dye cross membranes can be determined using a simplified version of the **Fick law**:

$$J = P \times A(C_1 - C_2)$$

where J is diffusion rate (in mmol/s), P is a **permeability coefficient**, A is **membrane surface area** (cm^2), and C_1 and C_2 are dye concentrations (mmol/L) in compartments 1 and 2, respectively.

1. **Permeability coefficient:** The permeability coefficient takes into account a molecule's **diffusion coefficient**, **partition coefficient**, and the **thickness** of the barrier that it must traverse.

   a. **Diffusion coefficient:** Diffusion rates increase when a molecule's velocity increases, which is, in turn, determined by its **diffusion coefficient**. The coefficient is proportional to temperature and inversely proportional to molecular radius and the viscosity of the medium through which it diffuses. In practice, small molecules diffuse quickly through warm water, whereas large molecules diffuse very slowly through cold, viscous solutions.

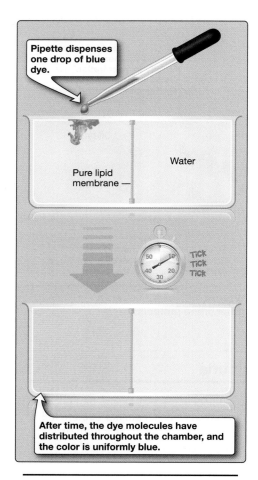

Figure 1.7.
Simple diffusion in water.

b. Partition coefficient: Lipid-soluble molecules, such as fats, alcohols, and some gases (e.g., O_2, CO_2, anesthetics) can cross the membrane by dissolving in its lipid core, and they have a high partition coefficient. Conversely, ions such as Na^+ and Ca^{2+} are repelled by lipids and have a very low partition coefficient. A molecule's partition coefficient is determined by measuring its solubility in oil compared with water.

c. Barrier thickness: Net diffusion rate slows when molecules must traverse thick or layered membranes compared with thin ones. The practical consequences can be seen in the lungs (see Chapter 21·II·C) and fetal placenta (see Chapter 37·III·B), organs designed to maximize diffusional rates by minimizing diffusional distance between two compartments.

2. Surface area: Increasing the surface area available for diffusion also increases the rate of diffusion. This relationship is used to practical advantage in several organs. The lungs comprise 300 million small sacs (**alveoli**) that have a combined surface area of ~80 m^2 that allows for efficient O_2 and CO_2 exchange between blood and the atmosphere (see Chapter 21·II·C). The lining of the small intestine is folded into fingerlike **villi** (Fig. 1.9), and the villi sprout **microvilli** that, together, create a combined surface area of ~200 m^2 (see Chapter 31·II). Surface area amplification allows for efficient absorption of water and nutrients from the gut lumen. Efficient exchange of nutrients and metabolic waste products between blood and tissues is ensured by a vast network of small vessels (**capillaries**) whose combined surface area exceeds 500 m^2 (see Chapter 18·II·C).

3. Concentration gradient: The rate at which molecules diffuse across a membrane is directly proportional to the concentration difference between the two sides of the membrane. In the example shown in Figure 1.8, the concentration gradient (and, thus, the dye diffusion rate) between the two compartments is high initially, but the rate slows and eventually ceases as the gradient dissipates and the two sides equilibrate. Note that thermal motion causes the dye molecules to continue moving back and forth between the two compartments at equilibrium, but net movement between the two is zero. If there were a way of removing dye continually from the chamber on the right, the concentration gradient and diffusion rate would remain high (we would also have to keep adding dye to the left chamber to compensate for movement across the barrier).

O_2 movement between the atmosphere and the pulmonary circulation occurs by simple diffusion, driven by an air–blood O_2 concentration gradient. Blood carries away O_2 as fast as it is absorbed, and breathing movements constantly renew the O_2 content of the lungs, thereby maintaining a favorable concentration gradient across the air–blood interface (see Chapter 22·V).

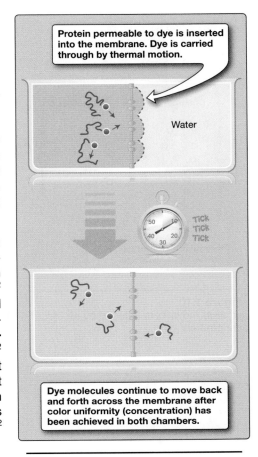

Protein permeable to dye is inserted into the membrane. Dye is carried through by thermal motion.

Water

Dye molecules continue to move back and forth across the membrane after color uniformity (concentration) has been achieved in both chambers.

Figure 1.8.
Diffusion through a lipid bilayer.

Small intestinal wall bears fingerlike villi to increase the surface area available for nutrient absorption.

Villi

Figure 1.9.
Intestinal villi.

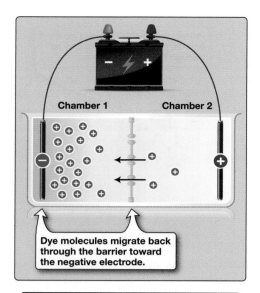

Figure 1.10.
Charge movement induced by an electrical gradient.

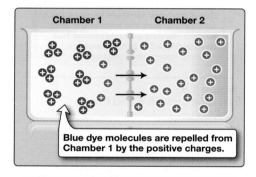

Figure 1.11.
Repellent effects of like charges.

C. Charged molecules

The prior discussion assumes that the dye is uncharged. The same basic principles apply to diffusion of a charged molecule (an **electrolyte**), but electrolytes are also influenced by electrical gradients. Positively charged ions such as Na^+, K^+, Ca^{2+}, and Mg^{2+} (**cations**) and negatively charged ions such as Cl^- and HCO_3^- (**anions**) are attracted to and will move toward their charge opposites. Thus, if the dye in Figure 1.8 carries a positive charge, and an electrical gradient is imposed across the container, dye molecules will move back through the membrane toward the negative electrode (Fig. 1.10). The electrical gradient in Figure 1.10 was generated using a battery, but the same effect can be achieved by adding membrane-impermeant anions to chamber 1. If chamber 1 is filled with cations, the dye molecules will be repelled by the positive (like) charge and will accumulate in chamber 2 (Fig. 1.11). Note that the *electrical* gradient causes the dye *concentration* gradient between the two chambers to reform. Dye molecules will continue migrating from chamber 2 back to chamber 1 until the concentration gradient becomes so large that it equals and opposes the electrical gradient, at which point an **electrochemical equilibrium** has been established. Most cells actively expel Na^+ ions to create an electrical gradient across their membranes (see Chapter 2·II·E). They then use the power of the combined electrical and chemical gradient (**electrochemical gradient**) to move ions and other small molecules (e.g., glucose) across their membranes and for electrical signaling.

V. PORES, CHANNELS, AND CARRIERS

Small, nonpolar molecules (e.g., O_2 and CO_2) diffuse across membranes rapidly, and they require no specialized pathway. Most molecules common to the ICF and ECF are charged, however, meaning that they require assistance from a **pore**, **channel**, or **carrier** protein to pass through the membrane's lipid core.

A. Pores

Pores are integral membrane proteins containing unregulated, water-filled passages that allow ions and other small molecules to cross the membrane. Pores are relatively uncommon in higher organisms because they are always open and can support very high transit rates (Table 1.2). Unregulated holes in the lipid barrier potentially can kill cells by allowing valuable cytoplasmic constituents to escape and

Table 1.2: Approximate Transit Rates for Pores, Channels, and Carriers

Pathway	Example	Molecule(s) Moved	Transit Rate (Number/s)
Pores	AQP1	H_2O	3×10^9
Channels	Na^+ (voltage-gated, Nav)	Na^+	10^8
	Cl^- (voltage-gated, ClC1)	Cl^-	10^6
Carriers	Na^+/K^+-ATPase	Na^+, K^+	3×10^2

Clinical Application 1.2: *Staphylococcus aureus*

Staphylococcus aureus is a skin-borne bacterium that is a leading cause of community- and hospital-acquired, bloodborne "staph" infections (bacteremia). The incidence of such infections has been rising steadily in recent decades and is of growing concern because of the high associated mortality rates and the increasing prevalence of antibiotic-resistant strains such as MRSA (methicillin-resistant *S. aureus*).[1] The bacterium's lethality derives from its production of a toxin (α-hemolysin) that kills red blood cells. The toxin targets the plasma membrane, where toxin monomers assemble into a multimeric protein with an unregulated pore at its center. The pore causes the cell to lose control of its internal environment. Ion gradients dissipate, and the cell swells and then blisters and ruptures, releasing nutrients that are believed to nourish the bacterium as it proliferates.

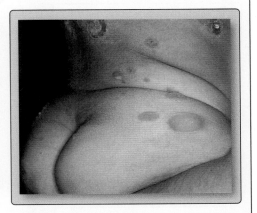

Skin lesions caused by *S. aureus*.

Ca^{2+} to flood into the cell from the ECF. **Aquaporin** (**AQP**) is a ubiquitous water-selective pore. Of the 13 known family members (AQP0–AQP12), 2 are expressed widely throughout the body (AQP1 and AQP3). AQP is found wherever there is a need to move water across membranes. AQPs play a critical role in regulating water recovery from the renal tubule (see Chapter 27·V·C), for example. Because AQP is always open, cells must regulate their water permeability by adding or removing AQP from the membrane.

B. Channels

Ion channels are transmembrane proteins that assemble to create one or more water-filled passages across the membrane. Channels differ from pores in that the permeability pathways are revealed transiently (**channel opening**) in response to a membrane-potential change, ligand binding, or some other stimulus, thereby allowing small ions (e.g., Na^+, K^+, Ca^{2+}, and Cl^-) to enter and traverse the lipid core (Fig. 1.12). Ion movement is driven by simple diffusion and powered by the transmembrane electrochemical gradient. Ions are forced to interact with the channel pore briefly so that their chemical nature and suitability for passage can be established (a **selectivity filter**), but the rate at which ions traverse the membrane via channels can be as high as 10^8 per second (see Table 1.2). All cells express ion channels, and there are numerous types, including the voltage-gated Na^+ channel that mediates nerve action potentials and voltage-gated Ca^{2+} channels that mediate muscle contraction. Ion channels are discussed in detail in Chapter 2.

 [1]For more information on *Staphylococcus aureus* and MRSA, see *LIR Microbiology*, 3e, Chapter 8.

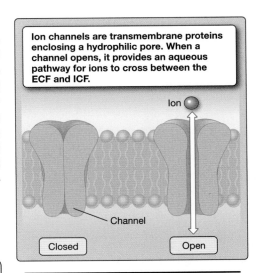

Ion channels are transmembrane proteins enclosing a hydrophilic pore. When a channel opens, it provides an aqueous pathway for ions to cross between the ECF and ICF.

Figure 1.12.
Ion channel opening.

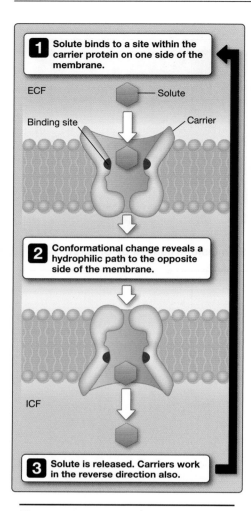

Figure 1.13.
Model for transport by a carrier protein.

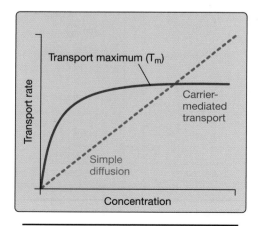

Figure 1.14.
Carrier saturation kinetics.

C. Carriers

Larger solutes, such as monosaccharides and amino acids, are typically assisted across the membrane by carriers. Carriers can be considered enzymes that catalyze movement rather than a biochemical reaction. Translocation involves a binding step, which slows transport rate considerably compared with pores and channels (see Table 1.2). There are three principal carrier modes: **facilitated diffusion**, **primary active transport**, and **secondary active transport**.

1. **Transport kinetics:** Carriers, like enzymes, show substrate specificity, saturation kinetics (Michaelis-Menten kinetics), and susceptibility to competition. A general scheme for carrier-mediated transport envisions a solute-binding step, a change in carrier conformation that reveals a conduit through which the solute may pass, and then release on the opposite side of the membrane (Fig. 1.13). When solute concentrations are low, carrier-mediated transport is more efficient than simple diffusion, but a finite number of solute-binding sites means that a carrier can saturate when substrate concentrations are high (Fig. 1.14). The transport rate at which saturation occurs is known as the **transport maximum** (T_m) and is the functional equivalent of V_{max} that defines maximal reaction velocity catalyzed by an enzyme.[1]

2. **Facilitated diffusion:** The simplest carriers use electrochemical gradients as a motive force (facilitated diffusion) as shown in Figure 1.15A. They simply provide a selective pathway by which organic solutes, such as glucose, organic acids, and urea, can move across the membrane down their electrochemical gradients. The binding step ensures selectivity of passage. Common examples of such carriers include the GLUT family of glucose transporters and the renal tubule urea transporter (see Chapter 27·V·D). The GLUT1 transporter is ubiquitous and provides a principal pathway by which all cells take up glucose. GLUT4 is an insulin-regulated glucose transporter expressed primarily in adipose tissue and muscle (see Chapter 33·III·B·3).

3. **Primary active transport:** Moving a solute uphill against its electrochemical gradient requires energy. **Primary active transporters** are **ATPases** that move or "pump" solutes across membranes by hydrolyzing adenosine triphosphate (ATP) as shown in Figure 1.15B. The three main types of pumps are all related P-type ATPase family members: a **Na⁺/K⁺-ATPase**, a group of **Ca²⁺-ATPases**, and a **H⁺/K⁺-ATPase**.

   a. **Na⁺/K⁺-ATPase:** The Na⁺/K⁺-ATPase (**Na⁺/K⁺ exchanger** or **Na⁺/K⁺ pump**) is common to all cells and uses the energy of a single ATP molecule to transport 3 Na⁺ out of the cell while simultaneously bringing 2 K⁺ back from the ECF. Movement of both ions occurs uphill against their respective electrochemical gradients. The physiologic importance of the Na⁺/K⁺-ATPase cannot be overstated. The Na⁺ and K⁺ gradients

 [1]For further discussion of enzymatic maximal velocity, see *LIR Biochemistry*, 8e, Chapter 5·A·1.

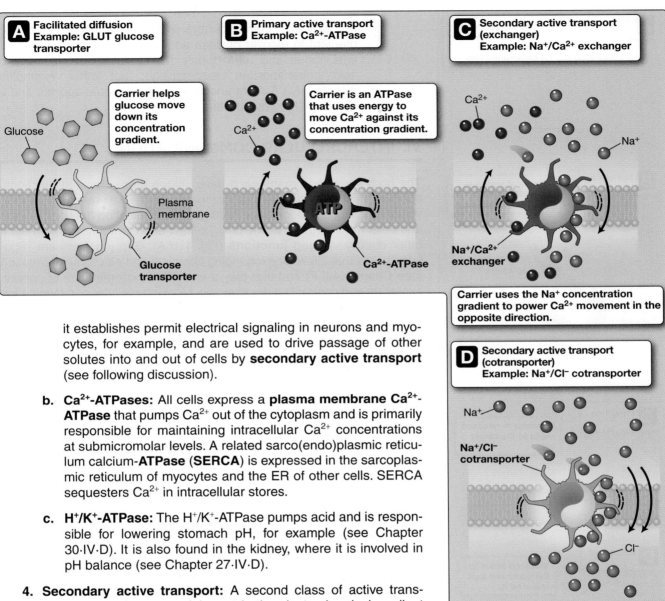

Figure 1.15.
Principal modes of membrane transport.

it establishes permit electrical signaling in neurons and myocytes, for example, and are used to drive passage of other solutes into and out of cells by **secondary active transport** (see following discussion).

   b. **Ca²⁺-ATPases:** All cells express a **plasma membrane Ca²⁺-ATPase** that pumps Ca^{2+} out of the cytoplasm and is primarily responsible for maintaining intracellular Ca^{2+} concentrations at submicromolar levels. A related sarco(endo)plasmic reticulum calcium-**ATPase** (**SERCA**) is expressed in the sarcoplasmic reticulum of myocytes and the ER of other cells. SERCA sequesters Ca^{2+} in intracellular stores.

   c. **H⁺/K⁺-ATPase:** The H⁺/K⁺-ATPase pumps acid and is responsible for lowering stomach pH, for example (see Chapter 30·IV·D). It is also found in the kidney, where it is involved in pH balance (see Chapter 27·IV·D).

4. **Secondary active transport:** A second class of active transporters uses the energy inherent in the electrochemical gradient of one solute to drive uphill movement of a second solute (**secondary active transport**). Such carriers do not hydrolyze ATP directly, although ATP may have been used to create the gradient being harnessed by the secondary transporter. Two transport modes are possible: **countertransport** and **cotransport**.

   a. **Countertransport:** Exchangers (**antiporters**) use the electrochemical gradient of one solute (e.g., Na⁺) to drive flow of a second (e.g., Ca²⁺) in the opposite direction to the first (see Fig. 1.15C). The Na⁺/Ca²⁺ exchanger helps maintain low intracellular Ca^{2+} concentrations by using the inwardly directed Na⁺ gradient to pump Ca^{2+} out of the cell. Other important exchangers include a Na⁺/H⁺ exchanger and a Cl⁻/HCO₃⁻ exchanger.

   b. **Cotransport:** Cotransporters (**symports**) use the electrochemical gradient of one solute to drive flow of a second or even

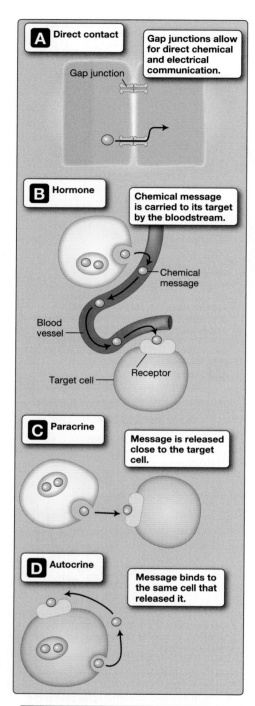

Figure 1.16.
Chemical signaling pathways.

a third solute in the same direction as the first (see Fig. 1.15D). For example, cotransporters use an inwardly directed Na^+ gradient to recover glucose and amino acids from the intestinal lumen and renal tubule (Na^+-glucose and Na^+-amino acid cotransporters, respectively), but other examples include a Na^+/Cl^- cotransporter, a K^+/Cl^- cotransporter, and a $Na^+/K^+/2Cl^-$ cotransporter.

VI. INTERCELLULAR COMMUNICATION

The body's various organs each have unique properties and functions, but they must work closely together to ensure the well-being of the individual as a whole. Cooperation requires communication between organs and cells within organs. Some cells contact and communicate with each other directly via **gap junctions** (Fig. 1.16A). Gap junctions are regulated pores that allow for exchanging chemical and electrical information (see Chapter 4·II·F) and that play a vital role in coordination of cardiac excitation and contraction, for example. Most intercellular communication occurs using chemical signals, which have traditionally been classified according to the distance and route they have to travel to exert a physiologic effect. **Hormones** are chemicals produced by endocrine glands and some nonendocrine tissues that are carried to distant targets by the vasculature (see Fig. 1.16B). Insulin, for example, is released into the circulation by pancreatic islet cells for carriage to muscle, adipose tissue, and the liver. **Paracrines** are released from cells in very close proximity to their target (see Fig. 1.16C). For example, the endothelial cells that line blood vessels release nitric oxide as a way of communicating with the smooth muscle cells that make up the vessel walls (see Chapter 19·II·E·1). Paracrines typically have a very limited signaling range because either they are degraded or are taken up rapidly by neighboring cells. **Autocrine** messengers bind to receptors on the same cell that released them, creating a negative feedback pathway that modulates autocrine release (see Fig. 1.16D). Autocrines, like paracrines, have a very limited signaling range.

VII. INTRACELLULAR SIGNALING

Once a chemical message arrives at its destination, it must be recognized as such by the target cell and then transduced into a form that can modify cell function. Most chemical messengers are charged and cannot permeate the membrane, so recognition must occur at the cell surface. Recognition is accomplished using receptors, which serve as cellular switches. Hormone or neurotransmitter binding trips the switch and elicits a preprogrammed instruction set that culminates in a cellular response. Receptors are typically integral membrane proteins such as **ligand-gated ion channels**, **G protein–coupled receptors** (**GPCRs**), or enzyme-associated **catalytic receptors**. Lipophilic messengers can cross the plasma membrane and are recognized by **intracellular receptors**.

A. Channels

Ligand-gated ion channels facilitate communication between neurons and their target cells, including other neurons (see Chapter 2·VI·B).

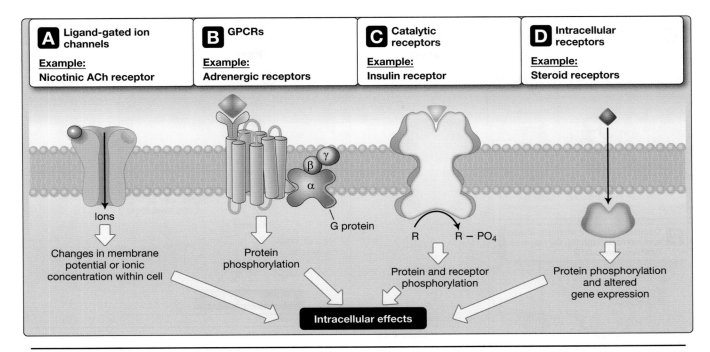

Figure 1.17.
Neurotransmitter and hormone receptors.

For example, the nicotinic acetylcholine (ACh) receptor is a ligand-gated ion channel that allows skeletal muscle cells to respond to excitatory commands from α-motor neurons. Neurotransmitter binding to its receptor causes a conformational change that opens the channel and allows ions such as Na^+, K^+, Ca^{2+}, and Cl^- to flow across the membrane through the pore (Fig. 1.17A). Charge movement across the membrane constitutes an electrical signal that influences target cell activity directly, but channel-mediated Ca^{2+} influx can have additional and potent effects on cell function by activating various Ca^{2+}-dependent signal transduction pathways (see following discussion).

B. G protein–coupled receptors

GPCRs sense and transduce a majority of chemical signals, and the GPCR family is large and diverse (human genome contains >900 GPCR genes). They are found in both neural and nonneural tissues. Common examples include the muscarinic ACh receptor, α- and β-adrenergic receptors, and odorant receptors. GPCRs all share a common structure that includes seven **membrane-spanning regions** that weave back and forth across the membrane (Fig. 1.18). Receptor binding is transduced by a **G protein** (guanosine triphosphate [GTP]-binding protein), which then activates one or more **second messenger** pathways (see Fig. 1.17B). Second messengers include **cyclic 3′5′-adenosine monophosphate (cAMP)**, **cyclic 3′5′-guanosine monophosphate (cGMP)**, and **inositol trisphosphate (IP₃)**. Multistep signal relay pathways allow for profound amplification of receptor-binding events. Thus, one occupied receptor can activate several G

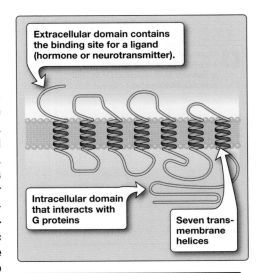

Figure 1.18.
GPCR structure.

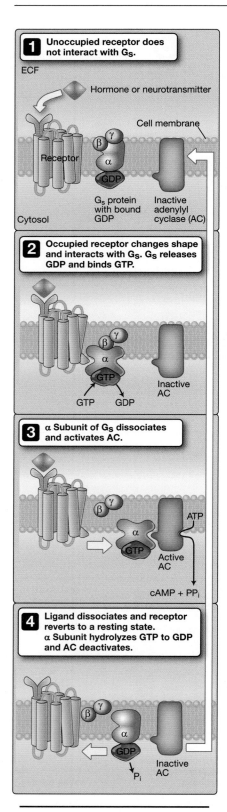

Figure 1.20.
cAMP signaling pathway.
G_s = stimulatory G protein; P_i = inorganic phosphate; PP_i = pyrophosphate.

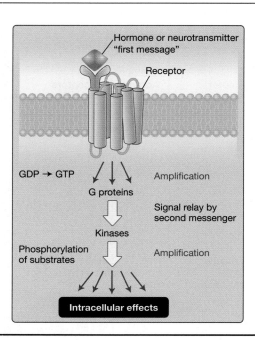

Figure 1.19.
Signal amplification by second messengers.

proteins, each of which can yield multiple second messenger molecules that, in turn, can activate multiple effector pathways (Fig. 1.19).

1. **G proteins:** G proteins are small membrane-associated proteins with GTPase activity. Two types of G protein have been described. Members of the class that associates with hormone and neurotransmitter receptors comprises assemblies of three subunits: α, β, and γ. The GTPase activity resides in the α-subunit (G_α), which is normally bound to guanosine diphosphate (GDP). Receptor binding causes a conformational change that allows it to interact with its G-protein partner. The α-subunit then releases GDP, binds GTP, and dissociates from the protein complex (Fig. 1.20). An occupied receptor can activate many G proteins before the hormone or transmitter dissociates. Active G_α subunits can interact with a variety of second-messenger cascades, the principal ones being the cAMP and IP_3 signaling pathways. The duration of G_α's effects are limited by the protein's intrinsic GTPase activity. The rate of hydrolysis is slow, but, once GTP has been converted to GDP (and inorganic phosphate), the subunit loses its ability to signal. It then redocks with the $G_{\beta\gamma}$ assembly in the surface membrane and awaits a further opportunity to bind to an occupied receptor.

At least 16 different G_α subunits have been described. They can be classed according to their effects on a target pathway. $G_{\alpha s}$ subunits are stimulatory. $G_{\alpha i}$ subunits are inhibitory, meaning that they suppress second messenger formation when active.

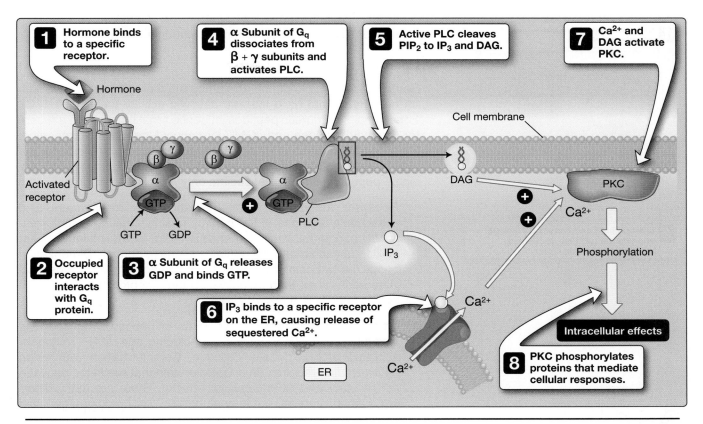

Figure 1.21.
IP$_3$ signaling pathway. G$_q$ = stimulatory G protein, PKC = protein kinase C, PLC = phospholipase C.

2. **cAMP signaling pathway:** cAMP is a second messenger synthesized from ATP by **adenylyl cyclase** (**AC**). AC is regulated by G proteins (see Fig. 1.20). G$_{\alpha s}$ stimulates cAMP formation, whereas G$_{\alpha i}$ inhibits it. cAMP activates **protein kinase A** (**PKA**), which phosphorylates and modifies the function of a variety of intracellular proteins, including enzymes, ion channels, and pumps. The cAMP signaling pathway is capable of tremendous signal amplification, so two checks are in place to limit its effects. Protein phosphatases counter the effects of the kinase by dephosphorylating the target proteins. The effects of AC are countered by a **phosphodiesterase** that converts cAMP to 5′-AMP.

3. **IP$_3$ signaling pathway:** G$_{\alpha q}$ is a G-protein subunit that liberates three different second messengers via activation of phospholipase C (PLC), as shown in Figure 1.21. The messengers are IP$_3$, **diacylglycerol** (**DAG**), and Ca^{2+}. PLC catalyzes the formation of IP$_3$ and DAG from **phosphatidylinositol 4,5-bisphosphate** (**PIP$_2$**), a plasma membrane lipid. DAG remains localized to the membrane, but IP$_3$ is released into the cytoplasm and binds to a Ca^{2+} release channel located in the ER. Ca^{2+} then floods out of the stores and into the cytosol, where it binds to a Ca^{2+}-binding protein such as **calmodulin** (**CaM**), as shown in Figure 1.22. CaM mediates Ca^{2+} activation of many enzymes and other intracellular effectors. Ca^{2+} also coordinates with DAG to activate **protein kinase C** (see Fig. 1.21),

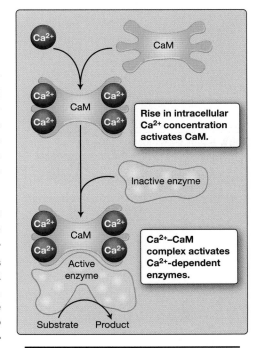

Figure 1.22.
Ca^{2+}-CaM–dependent enzyme activation.

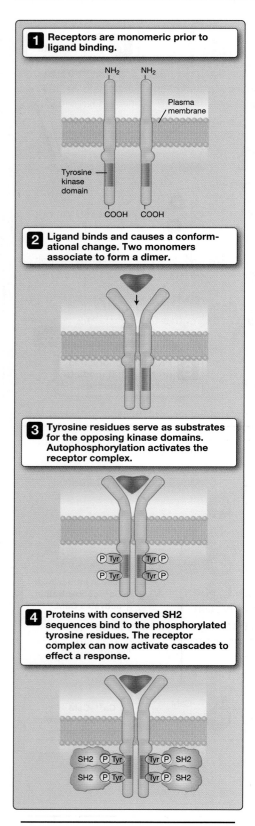

1 Receptors are monomeric prior to ligand binding.

2 Ligand binds and causes a conformational change. Two monomers associate to form a dimer.

3 Tyrosine residues serve as substrates for the opposing kinase domains. Autophosphorylation activates the receptor complex.

4 Proteins with conserved SH2 sequences bind to the phosphorylated tyrosine residues. The receptor complex can now activate cascades to effect a response.

Figure 1.23.
Tyrosine (Tyr) receptor kinase activation.
P = phosphorylation.

which phosphorylates proteins involved in smooth muscle contraction and salivary secretion, for example.

C. Catalytic receptors

Some ligands bind to membrane receptors that either associate with an enzyme or that have intrinsic catalytic activity (see Fig. 1.17C). For example, natriuretic peptides influence renal function via a receptor **guanylyl cyclase** and cGMP formation. Most catalytic receptors are **tyrosine kinases (TRKs)**, the most common example being the insulin receptor. The insulin receptor is tetrameric, but most TRKs are single-peptide chains that associate only after ligand binding.

1. **Receptor activation:** Hormones and other messengers bind extracellularly to one of the peptide chains, causing a conformational change that favors dimerization (Fig. 1.23). The intracellular portion of each monomer contains a kinase domain. Dimerization brings the two catalytic domains into contact, and they phosphorylate each other, thereby activating the receptor complex, which begins signaling.

2. **Intracellular signaling:** Active TRKs influence cell function via several transduction pathways, including the MAP (mitogen-activated protein) kinase cascade. Communication with these pathways first requires an adapter protein that mediates between the receptor and its intracellular effector. There are many different adapter proteins, but they all contain Src homology domains named SH2 and SH3. The SH2 domain recognizes the phosphorylated tyrosine domains on the activated TRK and allows the adapter protein to bind to the signaling complex.

D. Intracellular receptors

A fourth receptor class is located intracellularly and includes receptors for thyroid hormone and a majority of steroid hormones (see Fig. 1.17D). All are transcription factors that influence cell function by binding to DNA and altering gene expression levels. Some of the receptors are cytoplasmic, whereas others are nuclear and may be associated with DNA. The cytoplasmic receptors are normally bound to a "heat shock" protein, which is displaced by the conformational change caused by steroid binding. The occupied receptor then translocates to the nucleus and binds to a **hormone-response element** within the promoter region of the target gene. Nuclear receptors act in a similar way. Once bound, the receptor induces gene transcription, and the product alters cell function.

Chapter Summary

- All cells erect a lipid barrier (**plasma membrane**) to separate the inside of the cell from the outside and then selectively modify the ionic composition in the intracellular environment to facilitate the biochemical reactions that sustain life. **ICF** contains very low concentrations of Ca^{2+} compared with **ECF**. Na^+ concentrations are also lower inside, but K^+ levels are higher.

- The plasma membrane contains three principal lipid types: **phospholipids**, which dominate the structure; **cholesterol**, which adds strength; and **glycolipids**, which mediate interactions with other cells.

- Movement across membranes occurs primarily by **diffusion**. Diffusion rate is dependent on the transmembrane **concentration difference**, molecular size, membrane thickness and surface area, temperature, viscosity of the solution through which the molecule must diffuse, and the molecule's solubility in lipid (**partition coefficient**).

- Integral membrane proteins such as **pores**, **channels**, and **carriers** provide pathways by which hydrophilic molecules cross the lipid barrier.

- Pores are always open and are rare, the principal example being **AQP**, a ubiquitous water channel. Channels are regulated pores that open transiently to allow passage of small ions, such as Na^+, Ca^{2+}, K^+, and Cl^-. Movement through pores and channels occurs by simple diffusion down electrical and chemical concentration gradients (**electrochemical gradient**).

- Carriers selectively bind ions and small organic solutes, carry them across the membrane, and then release them on the opposite side. Carriers operate by two modes of transport: **facilitated diffusion** and **active transport**. Facilitated diffusion moves solutes "downhill" in the direction of the electrochemical gradients (e.g., glucose transport by the GLUT transporter family). Active transport uses energy to move solutes "uphill" from an area containing low-solute concentration to an area of higher concentration.

- **Primary active transporters**, or **pumps**, use ATP to drive solutes uphill against their electrochemical gradient. Pumps include the Na^+/K^+-ATPase that is present in all cells, Ca^{2+}-ATPases, and H^+/K^+-ATPase.

- **Secondary active transporters** move solutes uphill by harnessing the energy inherent in electrochemical gradients for other ions. **Exchangers** move two solutes across the membrane in opposite directions (e.g., Na^+/H^+ exchanger). **Cotransporters** (e.g., $Na^+/K^+/2Cl^-$ cotransporter and the $Na^+/$glucose cotransporter) move two or more solutes in the same direction.

- The plasma membrane also contains receptor proteins that allow cells to communicate with each other using chemical messages. Signaling can occur over long distances via the release of **hormones** (e.g., insulin) into the bloodstream. Cells close to each other communicate using **paracrines** (e.g., nitric oxide). **Autocrines** target the same cell that released them (e.g., norepinephrine).

- **Receptor binding** is transduced in various ways. **Ligand-gated ion channels** transduce binding using changes in membrane potential. Other receptor classes release **G proteins** to activate or inhibit **second messenger** pathways. Many receptors possess intrinsic kinase activity and signal occupancy through protein phosphorylation. Intracellular receptors affect levels of gene expression when a message binds.

- G proteins modulate two major second-messenger cascades. The first involves **cAMP** formation by **adenylyl cyclase**. cAMP acts primarily through regulation of **protein kinase A** and protein phosphorylation.

- Other G proteins activate **phospholipase C** and cause the release of IP_3 and **DAG**. IP_3, in turn, initiates Ca^{2+} release from intracellular stores. Ca^{2+} then binds to CaM and activates Ca^{2+}-dependent transduction pathways. Ca^{2+} and DAG together activate protein kinase C and cause phosphorylation of target proteins.

- Catalytic receptors with intrinsic **tyrosine kinase** activity autophosphorylate when the message binds. This allows them to complex with adapter proteins that initiate signal cascades affecting cell growth and differentiation.

- **Intracellular receptors** translocate to the nucleus and bind to **hormone-response elements** within the promoter region of target genes. Cell function is altered through increased levels of target gene expression.

Study Questions

Choose the ONE best answer.

1.1. An investigator is comparing the composition of intercellular fluid (ICF) and extracellular fluid (ECF) and notes significant differences between the two. Which of the following is most likely to be present at a higher concentration in ICF compared with ECF?

    A. Ca^{2+}

    B. Cl^-

    C. HCO_3^-

    D. K^+

    E. Na^+

Best answer = D. ICF is characterized by its relatively high K^+ concentration (see Table 1.1). ECF contains significantly higher Ca^{2+} (A), Cl^- (B), HCO_3^- (C), and Na^+ (E) concentrations compared with ICF.

1.2. Dye indicators are important physiologic tools used to calculate unknown volumes or concentrations within the body. If a dye is membrane permeable, which of the following changes will most likely increase dye diffusion rate?

    A. Decreasing fluid temperature

    B. Increasing membrane surface area

    C. Increasing membrane thickness

    D. Lowering dye concentration

    E. Lowering dye partition coefficient

Best answer = B. Increasing membrane surface area increases the opportunity for dye to cross the membrane, which increases its diffusion rate (see Section IV·B). Decreasing a fluid's temperature (A) increases its viscosity, which slows diffusion rate, and it takes longer for molecules to diffuse across thick membranes (C) than thin ones. Concentration gradients provide the driving force for diffusion, so lowering dye concentration (D) flattens the gradient and reduces diffusion rate. A partition coefficient is a measure of dye lipid solubility. Molecules with high lipid solubility diffuse through membranes faster than poorly soluble molecules, so lowering dye partition coefficient (E) would decrease diffusion rate.

1.3. A 66-year-old male is treated with the loop diuretic furosemide ($Na^+/K^+/2Cl^-$ cotransport inhibitor) to reduce symptoms associated with congestive heart failure. Which of the following most likely describes this cotransporter's mode of action?

    A. A rise in intracellular K^+ would decrease transport rate.

    B. It is a primary active transporter.

    C. It is an ATPase.

    D. It transports Na^+ against its electrochemical gradient.

    E. It transports Na^+ and K^+ into the cell and 2 Cl^- out of the cell.

Best answer = A. Cotransporters, by definition, move two or more ions in the same direction (see Section V·C). The $Na^+/K^+/2Cl^-$ cotransporter simultaneously carries two anions and two cations across the plasma membrane. Transport rate is sensitive to the transmembrane gradient of all three ions, such that increasing the intracellular concentration of K^+ will slow net uptake. Primary active transporters (B) use ATP (C) to pump ions against their electrochemical gradients. The $Na^+/K^+/2Cl^-$ cotransporter uses the inwardly directed Na^+-gradient (D) to cotransport K^+ and Cl^-. Transporters that move Na^+ in one direction while simultaneously bringing other ions back in the opposite direction (E) are exchangers, not cotransporters.

1.4. A pharmaceutical company develops a candidate drug for use in treating hypertension. The drug inhibits formation of a second messenger in vascular smooth muscle cells. Which of the following most likely identifies this second messenger?

    A. G protein

    B. Guanosine triphosphate

    C. Inositol trisphosphate

    D. Phospholipase C

    E. Protein kinase A

Best answer = C. IP_3 is a second messenger that triggers Ca^{2+} release from intracellular stores in response to hormone or neurotransmitter receptor activation (see Section VII·B). Ca^{2+} is also a second messenger that activates Ca^{2+}-dependent effector pathways. G proteins (A) transduce receptor binding. Guanosine triphosphate (B) binding is required for G-protein activation. Phospholipase C (D) is an enzyme that liberates IP_3 from PIP_2 upon receptor activation. Protein kinase A (E) is a key effector enzyme in the cAMP-signaling pathway.

Membrane Excitability

2

I. OVERVIEW

All cells selectively modify the ionic composition of their internal environment to support the biochemistry of life (see Chapter 1·II) as shown in Figure 2.1. Moving ions into or out of a cell creates a charge imbalance between the intracellular fluid (ICF) and the extracellular fluid (ECF) and thereby allows a voltage difference to form across the surface membrane (**membrane potential**, or V_m). This process creates an electrochemical driving force for diffusion that can be used to move charged solutes across the membrane or that can be modified transiently to create an electrical signal for intercellular communication. For example, nerve cells use changes in V_m (**action potentials [APs]**) to signal to a muscle that it needs to contract. The muscle cell, in turn, uses a change in V_m to initiate Ca^{2+} release from its internal stores. Ca^{2+} release then facilitates actin and myosin interactions, and contraction begins. Neuronal and muscle APs both involve carefully coordinated sequences of **ion channel** events that allow transmembrane passage of ions (e.g., Na^+, Ca^{2+}, and K^+) between ICF and ECF.

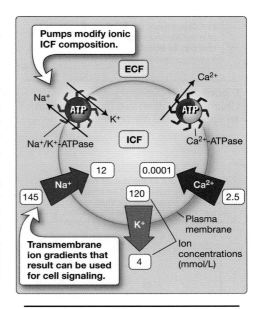

Figure 2.1.
ICF modification by ion transporters.

II. MEMBRANE POTENTIALS

The term "membrane potential" refers to the voltage difference that exists across the plasma membrane. By convention, the ECF is considered to be at zero volts, or electrical "**ground**." Inserting a fine electrode across the surface membrane reveals that the cell interior is negative with respect to ECF by several tens of millivolts. A typical nerve cell has a **resting potential** of −70 mV, for example (Fig. 2.2). V_m is established by membrane-permeant ions traveling down their concentration gradients and generating **diffusion potentials**.

A. Diffusion potentials

Imagine a model cell in which the plasma membrane is composed of pure lipid, the ICF is rich in potassium chloride (KCl, which dissociates into K^+ and Cl^-), and ECF is pure water (Fig. 2.3). Although there is a strong KCl concentration gradient for diffusion across the membrane, the lipid barrier prevents K^+ and Cl^- from leaving the cell, constraining both ions to the ICF. The charges carried by K^+ and Cl^- cancel each other out, so there is no voltage difference between ECF and ICF. If a protein that permits passage of K^+ alone is inserted into the lipid barrier, K^+ is now free to diffuse down its concentration gradient from ICF

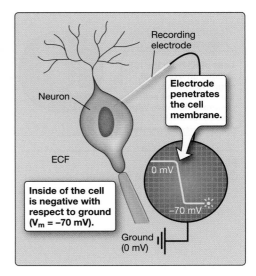

Figure 2.2.
Membrane potential (V_m).

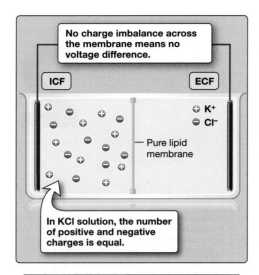

No charge imbalance across the membrane means no voltage difference.

ICF   ECF

⊕ K⁺
⊖ Cl⁻

— Pure lipid membrane

In KCl solution, the number of positive and negative charges is equal.

Figure 2.3.
Charge distribution in a model cell with an impermeable membrane.

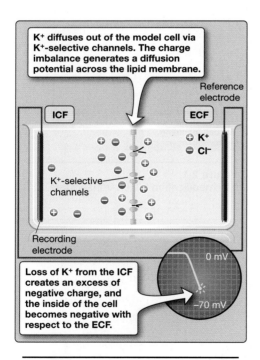

K⁺ diffuses out of the model cell via K⁺-selective channels. The charge imbalance generates a diffusion potential across the lipid membrane.

Reference electrode

ICF   ECF

⊕ K⁺
⊖ Cl⁻

K⁺-selective channels

Recording electrode

0 mV

Loss of K⁺ from the ICF creates an excess of negative charge, and the inside of the cell becomes negative with respect to the ECF.

−70 mV

Figure 2.4.
Origin of a diffusion potential.

to ECF, and the membrane is said to be **semipermeable** (Fig. 2.4). Because potassium ions carry charge, their movement causes a **diffusion potential** to form across the membrane in direct proportion to the magnitude of the concentration gradient. The potential may be significant (tens of millivolts) but involves relatively few ions.

> The **principle of bulk electroneutrality** notes that the number of positive charges in a solution is always balanced by an equal number of negative charges. The ICF and ECF are also subject to this rule, even though all cells create a negative V_m by altering charge distribution between the two compartments. In practice, V_m is established by just a few charges moving in the immediate vicinity of the cell membrane, and their *net effect* on overall charge distribution within the bulk of the ICF and ECF is negligible.

B. Equilibrium potentials

When K⁺ moves down its concentration gradient and crosses the membrane, it leaves a negative charge (i.e., Cl⁻) behind. Net charge magnitude builds in direct proportion to the number of ions leaving the ICF (see Fig. 2.4), but, because opposite charges attract, K⁺ movement slows and eventually stops when the attraction of the negative charges inside the cell precisely counters the outward driving force created by the concentration gradient (**electrochemical equilibrium**). The potential at which equilibrium is established is known as the **equilibrium potential** for K⁺.

1. **Nernst equation:** Equilibrium potentials can be calculated for any membrane-permeant ion assuming that the ion's charge and concentrations on either side of the membrane are known:

   Equation 2.1 $$E_X = \frac{RT}{zF} \ln \frac{[X]_o}{[X]_i}$$

   where E_X is the equilibrium potential for ion X (in mV), T is absolute temperature, z is the valence of the ion, R and F are physical constants (the ideal gas constant and the Faraday constant), and $[X]_o$ and $[X]_i$ are ECF and ICF concentrations of X (in mmol/L), respectively. Equation 2.1 is known as the **Nernst equation**. If T is assumed to be normal human body temperature (37°C), Equation 2.1 can be simplified:

   Equation 2.2 $$E_X = \frac{60}{z} \log_{10} \frac{[X]_o}{[X]_i}$$

> Most of the common inorganic ions (Na⁺, K⁺, Cl⁻, HCO₃⁻) have an electrical valence of 1 (**monovalent**). Ca²⁺ and Mg²⁺ have a valence of 2 (**divalent**).

2. Equilibrium potentials: The ionic compositions of ICF and ECF are maintained within well-defined ranges (see Fig. 2.1; also see Table 1.1). Using known values for concentrations of the common ions, we can use the Nernst equation to predict that, for most cells in the body, $E_K = -90$ mV, $E_{Na} = +61$ mV, and $E_{Ca} = +120$ mV. Intracellular Cl^- concentration can vary considerably, but E_{Cl} usually lies very close to V_m. If any of these ions are provided with a pathway that allows them to diffuse across the plasma membrane, they will drag V_m toward the equilibrium potential for that ion (Fig. 2.5).

C. Resting potential

The plasma membrane of all cells is rich in ion channels that are permeable to one or more of the ions mentioned earlier, and some of these channels are open at rest. Resting V_m (resting potential) thus reflects the sum of the diffusion potentials generated by each of these ions flowing through open channels at rest. V_m can be calculated mathematically as follows:

$$V_m = \frac{g_{Na}}{g_T}E_{Na} + \frac{g_K}{g_T}E_K + \frac{g_{Ca}}{g_T}E_{Ca} + \frac{g_{Cl}}{g_T}E_{Cl}$$

where g_T is total membrane conductance (membrane conductance is the reciprocal of membrane resistance, in Ohms^{-1}); g_{Na}, g_K, g_{Ca}, and g_{Cl} are individual conductances for each of the common ions (Na$^+$, K$^+$, Ca^{2+}, and Cl$^-$, respectively); and E_{Na}, E_K, E_{Ca}, and E_{Cl} are equilibrium potentials for these ions (in mV). V_m can also be calculated using the Goldman-Hodgkin-Katz (GHK) equation, which is similar in form to the Nernst equation (see Equation 2.1). The GHK equation derives V_m using relative membrane permeabilities for each of the ions that contribute to V_m.

> In practice, most cells have a negligible permeability to either Na$^+$ or Ca^{2+} at rest. Cells do have a significant resting K$^+$ conductance, however. Thus, V_m typically rests close to the equilibrium potential for K$^+$ (Fig. 2.6). The approximate value for resting potential in neurons is −70 mV, −90 mV in non-nodal cardiac myocytes, −55 mV in smooth muscle cells, and −40 mV in hepatocytes, for example.

D. Extracellular ion effects

The ionic composition of the ECF is regulated within a fairly narrow range, but significant disturbances can occur through inadequate or excessive ingestion of salts or water. Because resting membrane permeability to Na$^+$ and Ca^{2+} is low, V_m is relatively insensitive to changes in ECF concentration of either ion. V_m is *very* sensitive to changes in extracellular K$^+$ concentration, however, because V_m is closely tied to E_K (see Fig. 2.6). Increasing extracellular K$^+$ concentration (**hyperkalemia**) reduces the electrochemical gradient that drives K$^+$ efflux,

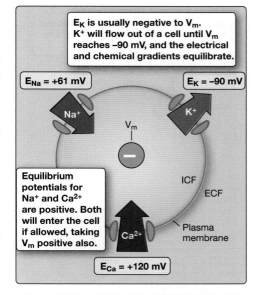

E_K is usually negative to V_m. K$^+$ will flow out of a cell until V_m reaches −90 mV, and the electrical and chemical gradients equilibrate.

$E_{Na} = +61$ mV

$E_K = -90$ mV

Equilibrium potentials for Na$^+$ and Ca^{2+} are positive. Both will enter the cell if allowed, taking V_m positive also.

ICF

ECF

Plasma membrane

$E_{Ca} = +120$ mV

Figure 2.5.
Equilibrium potentials for Na$^+$ (E_{Na}), Ca^{2+} (E_{Ca}), and K$^+$ (E_K). V_m = membrane potential.

Example 2.1

A cell has an intracellular free Mg^{2+} concentration of 0.5 mmol/L and is bathed in a saline solution with a Mg^{2+} composition that approximates plasma (1.0 mmol/L). The saline is held at 37°C. If the cell has a membrane potential (V_m) of −70 mV, and the membrane contains a gated channel that is Mg^{2+} permeable, will Mg^{2+} flow into or out of the cell when the channel opens?

We can use Equation 2.2 to calculate the Mg^{2+} equilibrium potential (E_{Mg}) for:

$$E_{Mg} = \frac{60}{z}\log_{10}\frac{[Mg^{2+}]_o}{[Mg^{2+}]_i}$$

$$= \frac{60}{+2}\text{mV}\log_{10}\frac{1.0 \text{ mmol/L}}{0.5 \text{ mmol/L}}$$

$$= 30 \text{ mV}\log_{10} 2.0$$

$$= 30 \text{ mV} (0.3)$$

$$= 9 \text{ mV}.$$

E_{Mg} tells us that Mg^{2+} will flow into the cell, its positive charges tending to drive V_m toward 9 mV.

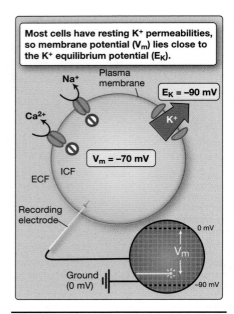

Figure 2.6.
Resting potential origins.

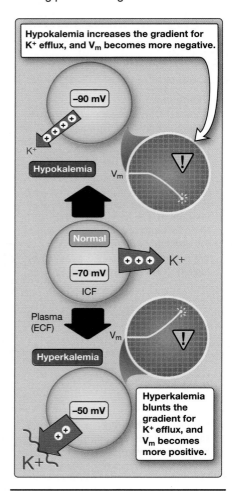

Figure 2.7.
Membrane potential (V_m) dependence on K⁺ concentration in the ECF.

causing the membrane to depolarize (Fig. 2.7). Conversely, lowering the extracellular K⁺ concentration (**hypokalemia**) steepens the gradient, and V_m becomes more negative.

E. Transporter contribution

The Na⁺/K⁺-ATPase that resides in the plasma membrane of all cells drives 3 Na⁺ out of the cell while simultaneously transferring 2 K⁺ from ECF to ICF. The 3-for-2 exchange results in an excess of positive charges being removed from the cell. Because the transporter creates a charge imbalance across the membrane, it is **electrogenic**. The direct contribution of this exchange to V_m is insignificant, however. The main role of the Na⁺/K⁺-ATPase is to maintain a K⁺ concentration gradient across the membrane because the K⁺ gradient ultimately determines V_m via the K⁺ diffusion potential.

III. EXCITATION

Many cell types use changes in V_m and transmembrane ion fluxes as a means of signaling or initiating intracellular events. Sensory cells (e.g., mechanosensors, olfactory receptors, and photoreceptors) transduce sensory stimuli by generating a V_m change called a **receptor potential**. Neurons signal to each other and to effector tissues using APs. Myocytes and secretory cells also use changes in V_m to increase intracellular Ca²⁺ concentration, thereby facilitating contraction and secretion, respectively. All such cells are said to have **excitable membranes**.

A. Terminology

The electrical changes caused by increased membrane permeability to ions do not consider the permeant ion's species (e.g., Na⁺ vs. K⁺ or Cl⁻), only the charge that it carries.

1. **Membrane potential changes:** The inside of a cell at rest is always negative with respect to the ECF. When a positively

---

Clinical Application 2.1: Hypokalemia and Hyperkalemia

Excitable cell function critically depends on maintaining membrane potential within a narrow range, so plasma levels normally range between 3.5 and 5.0 mmol/L. Hypokalemia and hyperkalemia are both commonly encountered clinically, however. Hypokalemia is generally of less concern than hyperkalemia, although some individuals with a rare inherited disorder (hypokalemic periodic paralysis) can experience muscle weakness when plasma K⁺ concentrations dip following a meal, for example. Hyperkalemia is, potentially, a more serious condition. The slow depolarization caused by rising plasma K⁺ levels inactivates Na⁺ channels that are required for muscle excitation, resulting in skeletal muscle weakness or paralysis and cardiac arrhythmias and conduction abnormalities. Hyperkalemia usually results from renal failure and impaired ability to excrete K⁺. Treatment typically requires either diuresis or dialysis to remove excess K⁺ from the body.

charged ion (**cation**) flows into a cell, negative charges (**anions**) are neutralized, and the membrane loses polarization. The influx has **depolarized** the cell, or caused **membrane depolarization**. By convention, depolarization is shown as an upward deflection on a voltage record (Fig. 2.8). Conversely, if a cation leaves the cell, V_m becomes more negative: The efflux **hyperpolarizes** the cell (**membrane hyperpolarization**) and yields a downward deflection on a recording device.

2. **Currents:** When positive charges flow into a cell, they generate an **inward current**. By convention, recording devices are configured so that inward currents cause a downward deflection (see Fig. 2.8). Positive charges leaving the cell cause an **outward current** and an upward deflection.

> Anions and cations are equally effective in changing V_m, but, because anions carry negative charges, their effects are opposite to those of cations. When an anion enters the cell from the ECF, it hyperpolarizes the membrane and yields an outward current. Conversely, anions leaving a cell create an inward current, and the cell depolarizes.

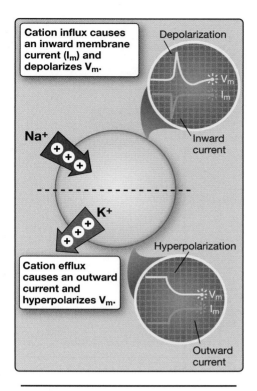

Figure 2.8.
Membrane potential (V_m) changes and ion currents.

B. Action potentials

AP size, shape, and timing may vary widely between the different cell types, but there are several common characteristics, including the existence of a **threshold** for AP formation, **all-or-nothing** behavior, **overshoots**, and **afterpotentials** (Fig. 2.9). The following discussion focuses on a nerve AP whose **upstroke** is mediated by voltage-dependent Na^+ channels, but voltage-dependent Ca^{2+} channels can support APs also (e.g., see Chapter 16·V·B·3).

1. **Threshold potential:** Because APs are explosive membrane events that have consequences (e.g., initiating muscle contraction), they must be triggered with care. V_m normally fluctuates over a range of a few millivolts with changes in extracellular K^+ concentration and other variables, even at rest, but such changes do not trigger spikes. Neurons only fire APs when V_m depolarizes sufficiently to cross the voltage threshold for AP formation (V_{th}), which, in a neuron, is usually around –60 mV. V_{th} corresponds to the voltage needed to open the number of voltage-dependent Na^+ channels required for the response to become self-perpetuating (**regenerative**).

2. **All or nothing:** The voltage-dependent Na^+ channels that mediate APs are typically present in the membrane in high numbers. When V_m crosses threshold, they open to allow a massive inward current, and the membrane depolarizes in a regenerative fashion toward E_{Na} (+61 mV). This "all-or-nothing" behavior can be likened to breaching a dam wall. Once depolarization begins, it does not stop until the ionic flood is complete.

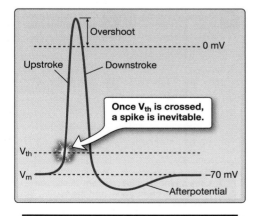

Figure 2.9.
Action potential. V_m = membrane potential; V_{th} = threshold potential.

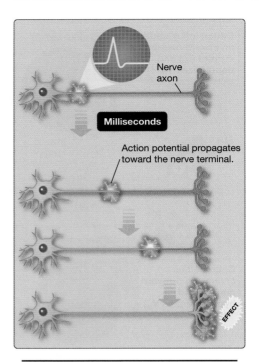

Figure 2.10.
Action potential propagation.

3. **Overshoot:** The AP peak typically does not reach E_{Na}, but it often does "overshoot" the zero-potential line, and the inside of the cell becomes positively charged with respect to the ECF.

4. **Afterpotentials:** APs are transient events. The **downstroke** is caused in part by voltage-dependent K^+ channels that open to allow K^+ efflux, causing V_m to repolarize. In some cells, the AP may be followed by an afterpotential of varying size and polarity. A hyperpolarizing afterpotential takes the membrane negative to resting V_m for a brief period.

C. Action potential propagation

When neurons fire an AP, the electrical event does not instantaneously involve the entire cell but rather the spike begins at one end of the cell and then **propagates** at speeds of up to 120 m/s to the far end (Fig. 2.10). Muscle cells behave similarly, although **conduction velocities** are typically lower in muscle than in nerve (~1 m/s). The advantage to propagation is that it allows a message to be carried unlimited distances. By way of analogy, the travel distance of a written message within a hollow baton thrown to a recipient is limited by the strength of the throw. Pass the baton to a team of relay runners, however, and travel distance is limited only by the number of runners available. In practice, signal propagation allows spinal neurons to communicate with the feet, which are typically a meter away! Neuronal signaling involves several sequential steps, including membrane excitation, AP initiation, signal propagation, and recovery.

1. **Excitation:** APs are typically initiated by sensory receptor potentials or dendritic postsynaptic potentials, for example. These are minor membrane events whose amplitude is graded with input intensity. Their reach is limited, much as throwing a baton is limited by arm muscle strength. The potential spreads passively and instantaneously (**electrotonically**), using the same physical principles as electricity traveling in a wire. Its reach is limited because the local currents created by the receptor potential are short-circuited by **leak channels**, which are found in all excitable membranes (Fig. 2.11). Leak channels (typically K^+ channels) are open at rest, allowing voltage changes to fizzle before they can travel far by "leaking" current across the membrane.

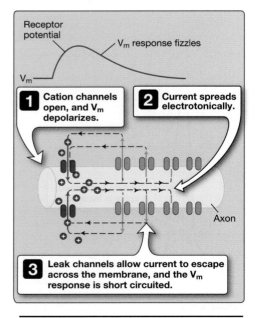

Figure 2.11.
Passive current spread and degradation in a neuron. V_m = membrane potential.

Electrical impulses travel through conductive materials like shock waves. A Newton cradle (i.e., classic desk toy comprising five silver balls suspended side by side within a frame) provides a good visual analogy. When a ball at one end is lifted and released, it impacts its neighbor and imparts its kinetic energy via a shock wave to the ball at the opposing end without disturbing the three intervening balls. The ball rises on its nylon line with little apparent energy loss. Electricity similarly creates shock waves between adjacent metal atoms within a copper wire that are transmitted at close to the speed of light. Electrical currents cause electrons to move also, but they travel at speeds closer to that of cold molasses.

2. **Initiation:** If the receptor potential is sufficiently large to cause V_m in a region of the membrane that contains voltage-dependent Na^+ channels to cross threshold, it will trigger a spike.

3. **Propagation:** Na^+-channel opening allows Na^+ to flow into the cell, driven by the electrochemical gradient for Na^+ and generating an active current (Fig. 2.12). The current then spreads electrotonically and causes a V_m depolarization that extends some distance down the axon. If the distant region contains Na^+ channels and the change in V_m is sufficiently large to cross threshold, the channels in this region open and **regenerate** the signal, much as a relay runner picks up a baton. The cycle of Na^+ influx, electronic spread, and regenerative Na^+-channel opening is repeated (propagates) down the length of the cell.

4. **Recovery:** Na^+ channels inactivate rapidly (within milliseconds) upon membrane depolarization. This temporarily inhibits further excitation and prevents APs endlessly boomeranging back and forth along an axon. Excitation is followed by a period of recovery, during which time ion gradients are renormalized by ion pumps and channels recover from excitation.

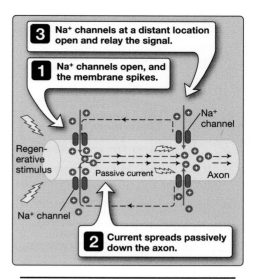

Figure 2.12.
Regenerative signal propagation in a neuron.

All cells have a V_m, but not all are excitable. By definition, nonexcitable cells do not generate APs, but many do show functional changes in V_m. For example, glucose causes the V_m of pancreatic β cells to oscillate in a sustained, rhythmic fashion. The electrical events correlate with insulin release.

D. Currents

APs are gross membrane events reflecting net charge movements through many thousands of individual ion channels. Each channel-opening event generates a "unitary current," the size of which is directly proportional to the number of charges moving through its pore (Fig. 2.13). The sum of individual Na^+ channel–opening events yields a whole-cell Na^+ current. Similarly, the sum of individual K^+ channel events yields a whole-cell K^+ current. Because there are many different ion channel classes with selective permeabilities for all the common inorganic ions, a whole-cell K^+ current, for example, may represent K^+ efflux through two or more discrete K^+-channel types. Such currents can be dissected into their individual components based on their physical properties using voltage-clamp (whole-cell recordings) and patch-clamp (recordings made from small membrane patches) techniques.

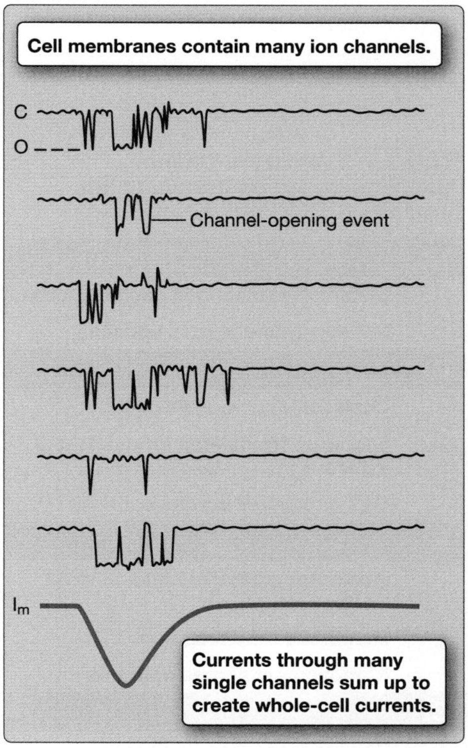

IV. ION CHANNELS

Ion channels are integral membrane proteins containing one or more hydrophilic pores that open transiently to allow ions to cross the membrane. Channels have several distinguishing features that identify them

Figure 2.13.
Single-channel and whole-cell ion currents recorded using patch-clamp and voltage-clamp techniques, respectively. C = closed; I_m = membrane current; O = open.

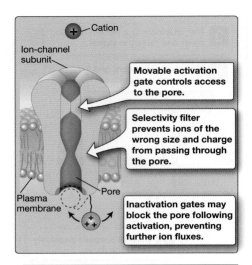

Figure 2.14.
Ion-channel structure.

Table 2.1: Typical Ion-Channel Functions Based on Permeant Ion Species

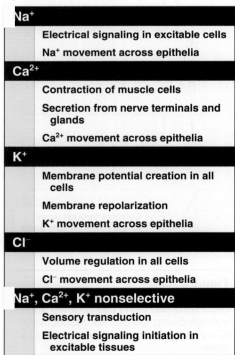

Na⁺		
	Electrical signaling in excitable cells	
	Na⁺ movement across epithelia	
Ca²⁺		
	Contraction of muscle cells	
	Secretion from nerve terminals and glands	
	Ca²⁺ movement across epithelia	
K⁺		
	Membrane potential creation in all cells	
	Membrane repolarization	
	K⁺ movement across epithelia	
Cl⁻		
	Volume regulation in all cells	
	Cl⁻ movement across epithelia	
Na⁺, Ca²⁺, K⁺ nonselective		
	Sensory transduction	
	Electrical signaling initiation in excitable tissues	

as such, including an **activation** mechanism, a **selectivity filter**, and a finite **conductance**. Many channels also inactivate with time during prolonged stimulation.

A. Activation

Ion channels create holes in the lipid barrier separating the ICF from ECF. If they were unregulated, ions would continue to flow across the membrane and collapse their respective concentration gradients, along with V_m. Thus, most channels have activation gates that regulate passage through the pore (Fig. 2.14). When a channel is in its **closed state**, the gate seals the pore, and ions cannot pass. Channel activation (e.g., in response to a change in V_m; see below) initiates a change in protein conformation that opens the gate and allows ions to pass (i.e., the **open state**). Some channels transition between the open and closed states thousands of times per second, with the net open time (or, **open probability**) increasing in direct proportion to the strength of the activating stimulus.

B. Selectivity

Before an ion can traverse the membrane, it must pass through a selectivity filter that determines its suitability for passage (see Fig. 2.14). Selectivity filters reside within the pore and comprise regions where the permeant ion is forced to interact with one or more charged groups that limit passage based on molecular size and charge density. Thus, **Na⁺ channels** only pass Na⁺, **K⁺ channels** are selective for K⁺, **Ca²⁺ channels** are Ca²⁺ selective, and **Cl⁻ channels** pass only Cl⁻ (Table 2.1). Other **nonselective cation** channels and nonselective anion channels may allow passage of two or more different ions. Note that there are many different classes of each type of channel, differentiated by their mode of activation, kinetics, conductance, regulatory mechanisms, tissue specificity, and pharmacology.

> The human genome includes genes for over 400 ion channels. Almost half of these are K⁺ channels. Although they all have selectivity for K⁺ over other ions, the individual members of this large family all have unique properties (e.g., mode of activation, activation kinetics, conductance, regulatory mechanism) and roles to play in membrane physiology (e.g., membrane repolarization, absorption, and secretion).

C. Conductance

When a channel opens, membrane resistance falls because the channel allows current to flow across the lipid barrier. The extent to which resistance falls depends on the number of ions flowing through its pore per unit time, or its **conductance**. A channel's maximum conductance is one of its distinguishing hallmarks and is usually expressed in picosiemens (pS).

D. Inactivation

Some channels possess an "inactivation gate" that is tripped upon activation, causing it to seal the channel pore and prevent further passage of ions (see Fig. 2.14). The timescale of inactivation can vary from milliseconds to tens of seconds, depending on channel class. Regardless, once a channel has been inactivated, it remains unresponsive to new stimuli, no matter how large the activating stimulus might be. Reactivation can only occur once the inactivation gate has been reset, a process that also has a variable timescale depending on channel type.

V. CHANNEL STRUCTURE

There are many ways in which a protein can be configured to create a transmembrane channel. Most mammalian channels follow a similar design principle, however, in which four to six subunits assemble around a central, water-filled pore. Tetrameric channels are the most common form, as shown in Figure 2.15, but many ligand-gated channels are pentameric, and connexin channels are hexameric (see Chapter 4·II·F). Tetrameric channel subunits typically comprise six membrane-spanning domains (S1–S6). The S5 and S6 domains include charged residues that fashion a pore and selectivity filter when the subunits are assembled. Voltage-dependent Na^+ and Ca^{2+} channels are products of a single gene incorporating four subunit-like domains, but most channels are assembled from independent proteins. The advantage to a modular approach to channel design is that it allows for infinite channel diversity. Changes in a single subunit can cause a voltage-gated channel to become a second messenger–gated channel, for example, or change its selectivity or its regulatory mechanism.

VI. CHANNEL TYPES

Channels are usually identified based on their ion selectivity and their **gating** (activation) mechanism. Thus, a "voltage-dependent Na^+ channel" is activated by membrane depolarization and is Na^+ selective. Several different gating mechanisms are known.

A. Voltage gated

Voltage-gated Na^+ channels, Ca^{2+} channels, and K^+ channels all belong to the same channel superfamily related by a common tetrameric structure. The voltage-gated Na^+ channel responsible for the AP upstroke in nerve and muscle cells comprises a single pore-forming α subunit that associates with several smaller regulatory β subunits. The α subunit contains four related subunit-like domains that assemble around a central pore, as discussed in Section V. Each domain includes a highly charged peptide sequence (the S4 region) that functions as a voltage sensor. Membrane depolarization alters the charge distribution between the inner and outer membrane surfaces, and the voltage sensor shifts within the membrane, initiating a conformational change that opens the gate and reveals the channel pore.

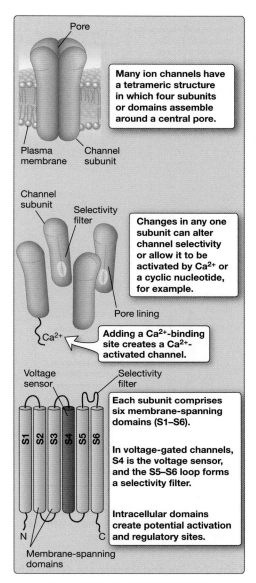

Many ion channels have a tetrameric structure in which four subunits or domains assemble around a central pore.

Changes in any one subunit can alter channel selectivity or allow it to be activated by Ca^{2+} or a cyclic nucleotide, for example.

Adding a Ca^{2+}-binding site creates a Ca^{2+}-activated channel.

Each subunit comprises six membrane-spanning domains (S1–S6).

In voltage-gated channels, S4 is the voltage sensor, and the S5–S6 loop forms a selectivity filter.

Intracellular domains create potential activation and regulatory sites.

Figure 2.15.
Voltage-gated ion-channel structure. N = the N-terminal and C = the C-terminal ends of the peptide sequence.

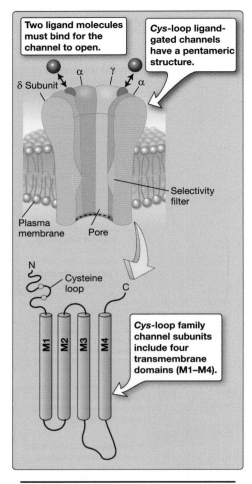

Two ligand molecules must bind for the channel to open.

Cys-loop ligand-gated channels have a pentameric structure.

δ Subunit α γ α

Selectivity filter

Plasma membrane   Pore

N

Cysteine loop

C

M1 M2 M3 M4

Cys-loop family channel subunits include four transmembrane domains (M1–M4).

Figure 2.16.
Cys-loop ligand-gated channel structure.
N = the N-terminal and C = the C-terminal ends of the peptide sequence.

B. Ligand gated

Ligand-gated channels transduce chemical signals and are the principal means by which neurons communicate with their targets. The diversity of the ligand-gated channel family is discussed in more detail in Chapter 6, but the six principal classes can be placed in three groups: ***cys*-loop receptors**, **ionotropic glutamate receptors**, and **ATP receptors**.

1. ***Cys*-loop superfamily:** The *cys*-loop family includes the **nicotinic acetylcholine receptor (nAChR)**, the **5-hydroxytryptamine (5-HT, or serotonin)** receptor, the **GABA (γ-aminobutyric acid)** receptor, and the **glycine** receptor. All family members share a short, highly conserved amino acid sequence that gives the family its name, and all comprise five subunits arranged around a central pore (Fig. 2.16). The nAChR and serotonin receptors are relatively nonspecific cation channels that support a mixed Na^+, K^+, and Ca^{2+} influx upon ligand binding. The resultant membrane depolarization is excitatory. GABA and glycine receptors are anion channels that mediate Cl^- fluxes. These fluxes tend to stabilize V_m around resting potential and thereby inhibit membrane excitation. The nAChR and other family members have two ligand-binding sites that must be occupied simultaneously before the channel opens.

2. **Ionotropic glutamate receptors:** Ionotropic glutamate receptors are common in the central nervous system, where they play a critical role in learning and memory. All are tetrameric structures that support relatively nonselective Na^+ and K^+ fluxes when active. There are three principal groups that are differentiated pharmacologically (see Table 5.2): **AMPA** (α-amino-3-hydroxy-5-methyl-4-isoxazolepropionic acid) receptors, **kainate** receptors, and **NMDA** (*N*-methyl-D-aspartate) receptors.

3. **ATP receptors:** ATP-gated channels are purinergic receptors activated by ATP and support a nonspecific Na^+, K^+, and Ca^{2+} flux

Table 2.2: Second Messenger–Gated Channels

Second Messenger	Permeant Ion	Selected Functions	Typical Locations	Notes
Ca^{2+}	K^+	Membrane repolarization	Most cells	Three classes based on conductance; eight known members of the group
	Cl^-	Membrane repolarization; Exocrine gland secretion	Most cells	Two gene families with varied sub-structures
	Ca^{2+}	Contraction	Sarcoplasmic reticulum	
G protein	K^+	Heart rate control	Heart	Four subunits assembled in hetero-meric complexes
		Membrane repolarization	Nervous system	
	Ca^{2+}	Signaling; regulation	Nervous system	
cAMP, cGMP	Na^+, K^+, Ca^{2+}	Signal transduction	Visual and olfactory systems	Closely related to voltage-gated channels
		Pacemaker	Heart	
		Na^+ absorption	Kidney	
IP_3	Ca^{2+}	Contraction; secretion; transcription; others	Endoplasmic reticulum of all cells	Three related genes; four assembled subunits

when open. They are believed to form trimeric channels *in vivo*. ATP receptors are involved in taste transduction, for example (see Chapter 10·II).

C. Second messenger gated

A third class of channel opens or closes in response to changes in intracellular messenger concentration (Table 2.2). Ca^{2+}-gated channels are ubiquitous, opening any time intracellular Ca^{2+} levels rise, regardless of whether the source of Ca^{2+} is an intracellular store or the ECF via a voltage-gated Ca^{2+} channel. Other channels are activated by G proteins, cyclic nucleotides, IP_3, and several additional messengers.

D. Sensory

Transient receptor-potential channels (**TRPs**) form a large and diverse group of channels that function as cellular sensors transducing temperature, taste, pain, and mechanical stress (cell swelling and shear stress), for example. TRPs are also required for Ca^{2+} and Mg^{2+} reabsorption from the renal tubule (see Chapter 27·III). TRPs are currently the subject of intense study, and many aspects of their behavior *in vivo* have yet to be delineated, but they are known to be tetrameric assemblies like the voltage-gated channels described earlier. Most family members are weakly cation selective, passing Na^+, K^+, and Ca^{2+}, the net result being membrane depolarization when activated. The TRP family comprises six or more structurally distinct groups whose functions are summarized in Table 2.3.

Table 2.3: Sensory Channels

TRPC (Canonical [TRPC1–7])
Ubiquitous
TRPV (Vanilloid [TRPV1–6])
Ca^{2+} selective; transduce noxious heat and "hot" chemicals (e.g., capsaicin [TRPV1]); osmosensory (TRPV4); Ca^{2+} recovery from the renal tubule (TRPV5, TRPV6)
TRPM (Melastatin [TRPM1–8])
Transduce taste (TRPM5), cool sensation and "cold" chemicals (e.g., menthol [TRPM8]); recover Mg^{2+} from renal tubule (TRPM6)
TRPP (Polycystin [TRPP2, 3, 5])
Mutations cause polycystic kidney disease.
TRPML (Mucolipin [TRPML1–3])
Mutations in TRPML1 cause a lysosomal storage disease.
TRPA (Ankyrin [TRPA1])
Mechanosensor; transduces pain associated with cold, inflammatory mediators, and chemical irritants such as mustard oil and wasabi

Chapter Summary

- All cells modify their internal ionic environment using ion pumps (ATPases) that cause chemical concentration gradients to form across their surface membrane. Ions diffusing back down these concentration gradients create **diffusion potentials**. **Membrane potential** represents the sum of the diffusion potentials for all permeant ions (Na^+, K^+, Ca^{2+}, Mg^{2+}, and Cl^-).

- Ions are also influenced by electrical gradients, so their tendency to cross a membrane is governed by the net **electrochemical gradient**. The potential at which the chemical and electrical gradients balance precisely (the **equilibrium potential**) can be calculated using the **Nernst equation**.

- Most cells are impermeable to Na^+ and Ca^{2+} at rest, but the presence of a significant resting K^+ conductance causes **resting potential** to settle at close to the K^+ equilibrium potential. The resting K^+ conductance makes resting potential highly susceptible to changes in extracellular K^+ concentration (**hypokalemia** and **hyperkalemia**).

- Excitable cells use changes in membrane potential (**action potentials** [**APs**], or spikes) to communicate with each other and to trigger cellular events, such as muscle contraction and secretion. APs are caused by the sequential opening and closing of **ion channels**. **Voltage-dependent Na^+-channel** opening facilitates an **inward Na^+ current** to cause **membrane depolarization**. Membrane **repolarization** is brought about (in part) by an **outward K^+ current** through voltage-dependent K^+ channels.

- APs are initiated locally at the site of stimulation and then **propagate** in a self-sustaining, **regenerative** fashion along the length of a cell.

- Most cells express many different ion-channel classes in their surface membrane, which can be distinguished based on their mode of activation (**gating**), ion **selectivity**, **activation** and **inactivation kinetics**, **conductance**, and pharmacology.

- Voltage-dependent Na^+ channels, K^+ channels, Ca^{2+} channels, and Cl^- channels are activated by changes in membrane potential. **Ligand-gated channels** are activated by neurotransmitters, including acetylcholine, GABA, and glutamate. **Second messenger–gated channels** are sensitive to intracellular Ca^{2+}, G proteins, cyclic nucleotides, and IP_3. **TRPs** are cellular sensors, mediating responses to chemicals, hot and cold temperatures, and mechanical stress.

Study Questions

Choose the ONE best answer.

2.1. Serum electrolytes levels are ordered on a 12-year-old male with a GI infection that induced prolonged and severe vomiting episodes. Plasma K^+ concentrations were found to be abnormally low (2 mmol/L). Which of the following would most likely result from mild hypokalemia?

A. K^+-channel activation would yield K^+ influx.
B. K^+ equilibrium potential would shift negative.
C. Na^+ channels would inactivate.
D. Neuronal action potentials would be inhibited.
E. Resting potentials would shift positive.

Best answer = B. Hypokalemia, or reduced extracellular K^+ concentration, enhances the electrochemical gradient favoring K^+ efflux from cells and causes the K^+ equilibrium potential to shift negative (see Section II·B). K^+-channel activation always causes K^+ efflux (A), except in rare instances (e.g., in the inner ear; see Chapter 9·IV·C). Na^+ channels inactivate (C) during membrane depolarization, not hyperpolarization. A negative shift in V_m means that a stronger depolarization would be necessary to take V_m to the threshold for voltage-gated Na^+ channel activation (see Section III·B), but, once reached, an action potential would be initiated (D). Because membrane potential (V_m) is determined largely by the transmembrane K^+ gradient, resting V_m would shift negative (E) also.

2.2. A 35-year-old male carries an epilepsy gene. The gene mutation affects the neuronal voltage-dependent Na^+ channel, causing it to inactivate more slowly (by ~50%). Which of the following would most likely describe the effects of this gene mutation on nerve function?

A. Action potentials would be prolonged.
B. Action potentials would no longer overshoot 0 mV.
C. Action potentials would rise very slowly.
D. Resting potential would settle close to 0 mV.
E. There would be no action potentials.

Best answer = A. The voltage-dependent Na^+ channel is opened by membrane depolarization to yield the upstroke of the neuronal action potential ([AP] see Section VI·A). An inactivation gate closes shortly after activation, blocking passage of Na^+ and allowing membrane potential to return to resting levels. If inactivation were slowed, membrane recovery would be delayed, and the AP would be prolonged. AP overshoot (B) is largely determined by the Na^+ equilibrium potential. Activation and inactivation are separate processes, and, therefore, the rate at which the AP rises (C) should be normal. Resting potential (D) should not be affected by an inactivation defect unless it prevented the channel from closing, causing a sustained Na^+ influx. Inactivation and recovery from inactivation are separate processes, so slowing of inactivation would not prevent APs being triggered (E).

2.3. A researcher is studying the principles underlying the rate of electrical signal propagation along neuronal axons. Propagation rate is found to be limited by the presence of leak channels. Which of the following most likely describes the properties of leak channels?

A. Calcium selective
B. Inactivate rapidly during membrane depolarization
C. Mediate electronic signal propagation
D. Open at resting potential
E. Voltage-gated

Best answer = D. Leak channels are open at resting potential and remain open during membrane depolarization. Leak channels are typically permeable to potassium ions, not calcium ions (A). They do not inactivate (B). Their presence in the membrane short-circuits the spread of local currents created by a receptor potential, for example, thereby preventing electrotonic signal propagation (C; see Section III·C). Leak channels do not need to be activated by membrane potential changes (E).

2.4. An agricultural worker is packing hot chili peppers for transport. He removes his protective mask and becomes incapacitated with a sensation of nasal burning caused by capsaicin from the peppers. Capsaicin is most likely stimulating which of the following receptor types?

A. *Cys*-loop family receptors
B. Ionotropic glutamate receptors
C. Purinergic receptors
D. Transient receptor-potential channels
E. Voltage-gated Na^+ channels

Best answer = D. The transient receptor-potential (TRP) channel family transduces a variety of sensory stimuli, including heat, cold, and osmolality (see Section VI·D). TRPV1 channels are stimulated by capsaicin. *Cys*-loop family (A), glutamate (B), and purinergic receptors (C) are activated by specific ligands (i.e., acetylcholine, L-glutamate, and ATP, respectively). Capsaicin is not an agonist for these receptor classes. Voltage-gated Na^+ channels (E) are activated primarily by membrane depolarization.

Osmosis and Body Fluids

3

I. OVERVIEW

The average human body comprises 50% to 60% water by weight, depending on body composition, sex, and age. The proportion of water in cells is even greater (~80%) as shown in Figure 3.1, the remainder largely comprising proteins. Water is the universal solvent, facilitating molecular interactions, biochemical reactions, and providing a medium that supports molecular movement between cellular and subcellular compartments. The biochemistry of life is highly sensitive to solute concentration, which, in turn, is determined by how much water is contained within a cell. Thus, the **autonomic nervous system (ANS)** closely monitors changes in **total body water (TBW)** and adjusts intake and output pathways (drinking and urine formation, respectively) to maintain water balance (see Chapter 28·II). Although TBW is tightly regulated, water moves freely across cell membranes and among the body's fluid compartments. Loss of water from the cell raises intracellular solute concentrations, thereby interfering with normal cell function. The body does not contain a transporter capable of redistributing water between compartments, so its approach to water management at the cellular and tissue level is to manipulate solute concentrations within intracellular fluid (ICF), interstitial fluid, and plasma. This approach is effective because water is enslaved to solute concentration by **osmosis**.

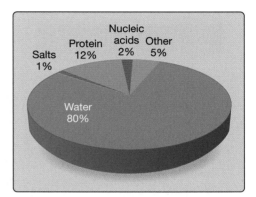

Figure 3.1.
Cellular composition.

II. OSMOSIS

Osmosis describes a process by which water moves passively across a semipermeable membrane, driven by a difference in water concentration between the two sides of the membrane. Pure water has a molarity of >55 mol/L. Although cells do not contain pure water, it is nevertheless a superabundant chemical. The concentration difference required to generate physiologically significant water flow across membranes is very small, so, in practice, it is far more convenient to discuss osmosis in terms of the amount of *pressure* that water is capable of generating as it moves down its concentration gradient. Thus, a chemical concentration gradient becomes an **osmotic pressure gradient**.

A. Osmotic pressure

Osmotic pressure gradients are created when solute molecules displace water, thereby decreasing water concentration. An appar-

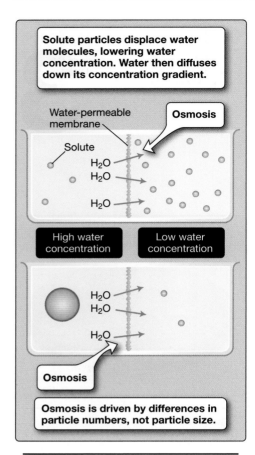

Figure 3.2.
Osmosis.

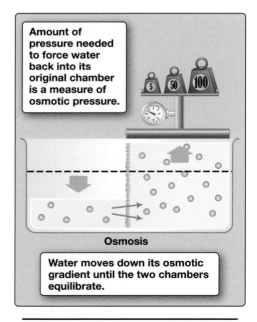

Figure 3.3.
Osmotic pressure.

ent peculiarity of the process is that pressure is determined entirely by solute particle number and is largely independent of the size, mass, chemical nature of the solute, or even its electrical valence. Therefore, two small ions such as Na^+ (23 MW) generate a higher osmotic pressure than a single complex glucose polymer such as starch (MW > 40,000), as shown in Figure 3.2. The osmotic pressure of a solution (π; measured in mm Hg) can be calculated from:

Equation 3.1 $\qquad \pi = nCRT$

where n is the number of particles that a given solute dissociates into when in solution, C is solute concentration (in mmol/L), and R and T are the universal gas constant and absolute temperature, respectively. Osmotic pressure can be measured physically as the amount of pressure required to precisely counter water movement between two solutions with dissimilar solute concentrations (Fig. 3.3).

B. Osmolarity and osmolality

Osmolarity is a measure of a solute's ability to generate osmotic pressure that takes into account how many particles a solute dissociates into when dissolved in water. Glucose does not dissociate in solution, so a 1 mmol/L–glucose solution has an osmolarity of 1 mOsmol. NaCl dissociates into two osmotically active particles in solution (Na^+ and Cl^-); thus, a 1 mmol/L–NaCl solution has an osmolarity of ~2 mOsmol. $MgCl_2$ dissociates into three particles ($Mg^{2+} + 2Cl^-$), and so a 1 mmol/L–$MgCl_2$ solution has an osmolarity of 3 mOsmol.

Osmolality is an almost identical measure to osmolarity but uses water mass in place of volume (i.e., Osmol/kg H_2O). A liter of water has a mass of 1 kg at 4°C, but water volume increases with temperature, which causes osmolarity to fall slightly. Because mass is invariant, Osmol/kg H_2O is the preferred unit for use in discussions of human physiology.

C. Tonicity

Tonicity measures a solute's effect on *cell volume*, the term recognizing that membrane-permeant solutes cause cells to shrink or swell through effects on ICF osmolality.

1. **Nonpermeant solutes:** Sucrose cannot cross the plasma membrane of most cells. Therefore, if a cell is placed in a sucrose solution whose osmolality matches that of the ICF (300 mOsmol/kg H_2O), cell volume will remain unchanged because the solution is **isotonic** (Fig. 3.4, top). Volume changes only occur when an osmotic gradient across the plasma membrane forces water to enter or leave the cell. A 100 mOsmol/kg–H_2O sucrose solution is **hypotonic** compared with the ICF. Water molecules will migrate across the membrane from extracellular fluid (ECF) to ICF following the osmotic gradient, and the cell will swell (see Fig. 3.4, middle). Conversely, a 500 mOsmol/kg–H_2O sucrose solution is **hypertonic**: Water will be drawn out of the cell by osmosis, causing the cell to shrink (see Fig. 3.4, bottom).

> Note that ICF has an osmolality of 285–295 mOsmol/kg H_2O *in vivo*. The value of 300 mOsmol/kg H_2O used in this and the following examples is for ease of illustration only.

2. **Permeant solutes:** Urea is a small (60 MW) organic molecule that, unlike sucrose, readily permeates the membranes of most cells via ubiquitous urea transporters (UTs). Thus, although 300 mOsmol/kg–H_2O urea and 300 mOsmol/kg–H_2O sucrose have identical osmolalities (i.e., they are **isosmotic**), they are *not* isotonic. When a cell is placed in a 300 mOsmol/kg–H_2O urea solution, urea crosses the membrane via UTs and raises ICF osmolality. Water then follows urea by osmosis, and the cell swells. So, a 300 mOsmol/kg–H_2O urea solution is considered to be **hypotonic**.

3. **Mixed solutions:** A solution containing 300 mOsmol/kg–H_2O urea *plus* 300 mOsmol/kg–H_2O sucrose has an osmolarity of 600-mOsmol/kg H_2O and is, thus, **hyperosmotic** relative to the ICF. It is also functionally isotonic, however, because urea rapidly crosses the membrane until the intracellular and extracellular urea concentrations equilibrate at 150-mOsmol/kg H_2O. With solution osmolality on both sides of the membrane now standing at 450-mOsmol/kg H_2O, the driving force for osmosis is zero, and cell volume remains unchanged.

4. **Reflection coefficient:** When calculating the osmotic potential of a solution that bathes a cell, it is necessary to add a reflection coefficient (σ) to Equation 3.1.

$$\pi = \sigma nCRT$$

The reflection coefficient is a measure of the ease with which a solute can traverse the plasma membrane. For highly permeant solutes such as urea, σ approaches 0. The reflection coefficient for nonpermeant solutes (such as sucrose and plasma proteins) approaches 1.0.

D. Water movement between intracellular and extracellular fluids

The plasma membrane's lipid core is hydrophobic, but water enters and exits the cell with relative ease. Some water molecules slip between adjacent membrane phospholipid molecules, whereas others are swept along with solutes in ion channels and transporters. Most cells also express **aquaporins (AQPs)**, which are large tetrameric proteins that form water-specific channels across the surface membrane. AQPs, unlike most ion channels, are always open (see Chapter 1·V·A).

E. Cell volume regulation

ECF solute composition is maintained within fairly narrow limits by the pathways involved in TBW homeostasis (see Chapter 28·II), but ICF osmolality changes constantly with changing activity levels. When cell metabolism increases, for example, nutrients are absorbed, metabolic waste products accumulate, and water moves into the cell by

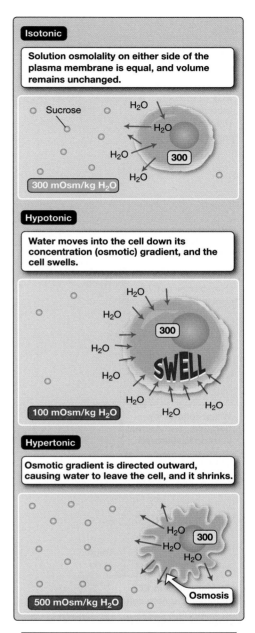

Figure 3.4.
Tonicity. All osmolality values are in mOsmol/kg H_2O.

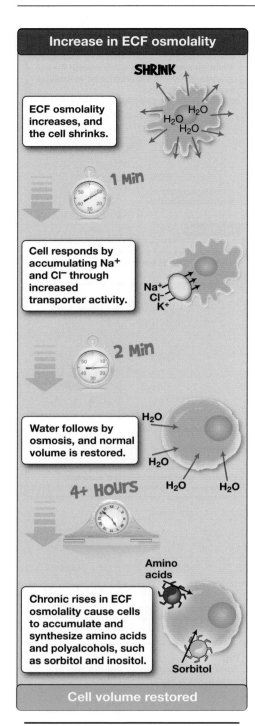

Increase in ECF osmolality

SHRINK

ECF osmolality increases, and the cell shrinks.

H_2O H_2O H_2O

1 Min

Cell responds by accumulating Na^+ and Cl^- through increased transporter activity.

Na^+ Cl^- K^+

2 Min

Water follows by osmosis, and normal volume is restored.

H_2O H_2O H_2O H_2O

4+ Hours

Amino acids

Chronic rises in ECF osmolality cause cells to accumulate and synthesize amino acids and polyalcohols, such as sorbitol and inositol.

Sorbitol

Cell volume restored

Figure 3.5.
Regulatory volume increase.

osmosis, causing it to swell. Cells that exist on the boundary between the internal and external environments (e.g., intestinal and renal epithelial cells) are also subject to acute changes in extracellular osmolality, causing frequent changes in cell volume. The mechanisms by which cells sense and transduce volume changes are not well defined, but they respond to osmotic shrinkage and swelling by enacting a **regulatory volume increase (RVI)** or a **regulatory volume decrease (RVD)**, respectively.

1. **Regulatory volume increase:** When ECF osmolality rises, water is drawn out of the cell by osmosis, and it shrinks. The cell responds with an RVI, which, in the short term, involves accumulation of Na^+ and Cl^- through increased Na^+/H^+ exchanger and $Na^+/K^+/2Cl^-$ cotransporter activity (Fig. 3.5). Na^+ and Cl^- uptake raises ICF osmolality and restores cell volume by osmosis. In the longer term, cells may accumulate small organic molecules, such as betaine (an amino acid), sorbitol, and inositol (two polyalcohols) to maintain increased ICF osmolality and retain volume.

2. **Regulatory volume decrease:** Cell swelling initiates an RVD, which principally involves K^+ and Cl^- efflux via swelling-activated K^+ channels and Cl^- channels. The resulting fall in ICF osmolality causes water loss by osmosis, and cell volume renormalizes. Cells may also release amino acids (principally glutamate, glutamine, and taurine) as a way of reducing their osmolality and volume.

III. BODY FLUID COMPARTMENTS

A 70-kg male contains 42 L of water, or around 60% of total body weight. Females generally have less muscle and more adipose tissue as a percentage of total body mass than do males. Because fat contains less water than muscle, their total water content is correspondingly lower (55%). TBW usually decreases with age in both sexes due to loss of muscle mass (**sarcopenia**) associated with aging.

A. Distribution

Two thirds of TBW is contained within cells (ICF = ~28 L of the 42 L). The remainder (14 L) is divided between the interstitium and blood plasma (Fig. 3.6).

1. **Plasma:** The cardiovascular system comprises the heart and an extensive network of blood vessels that together hold ~5 L of blood, a fluid composed of cells and protein-rich plasma. Approximately 1.5 L of total blood volume is contained within red and white blood cells and is included in the value given for ICF (see Fig. 3.6). Plasma accounts for 3.5 L of ECF volume.

2. **Interstitium:** The remaining 10.5 L of water resides outside the vasculature and occupies spaces between cells (the **interstitium**). Interstitial fluid and plasma have very similar solute compositions because water and small molecules move freely between the two compartments. The main difference between plasma and interstitial fluid is that plasma contains large amounts of proteins, whereas interstitial fluid is relatively protein free.

Clinical Application 3.1: Hyponatremia and Osmotic Demyelination Syndrome

Hyponatremia is a serum Na⁺ concentration of less than 136 mmol/L. Patients who develop hyponatremia usually have an impaired ability to excrete water, often due to an inability to suppress antidiuretic hormone (ADH) secretion. Hyponatremia with appropriate ADH suppression is also seen with advanced renal failure and low dietary sodium intake. Normally, the kidneys can excrete 10–15 L of dilute urine per day and maintain normal serum electrolyte levels, but higher flow rates may exceed their solute reabsorptive capabilities, and hyponatremia ensues. Because Na⁺ is the primary determinant of ECF osmolality, hyponatremia creates an osmotic shift across the plasma membrane of all cells and causes them to swell. Hyponatremic patients may develop severe neurologic symptoms (i.e., lethargy, seizures, coma), which typically only occur with acute and severe hyponatremia (serum sodium concentration <120 mmol/L), and rapid correction with hypertonic saline is necessary in this clinical scenario. Hyponatremia that develops slowly and chronically (more commonly the case) allows time for a regulatory volume decrease, and severe symptoms may be delayed until serum Na⁺ levels fall even further. When hyponatremia has developed slowly and a patient has no neurologic symptoms, correction to normal serum sodium levels must also be undertaken slowly to avoid a treatment complication known as **osmotic demyelination syndrome**

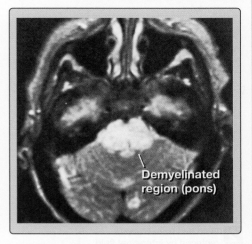

Osmotic demyelination in the pons region of the brain.

(**ODS**). ODS occurs when a too-rapid rise in ECF Na⁺ concentration creates an osmotic gradient that draws water from neurons before they have a chance to adapt, causing cell shrinkage and demyelination (myelin is a lipid-rich layered membrane that electrically insulates axons to enhance their conduction velocity; see Chapter 5·V·A). ODS may manifest as confusion, behavioral changes, quadriplegia, difficulties with speech or swallowing (dysarthria and dysphagia, respectively), or coma. Because these devastating changes may not be reversible, the maximum rate of correction in stable patients with chronic hyponatremia should not exceed ~10 mmol/L in the first 24 hours.

A variable amount of fluid is held behind physiologic barriers that separate it from plasma and interstitial fluid (**transcellular fluid**). This includes cerebrospinal fluid and fluid within the eye (aqueous humor), joints (synovial fluid), bladder (urine), and intestine. Transcellular fluid volume averages between 1 and 2 L and is not considered in calculations of TBW.

B. Restricting water movement

Water moves freely and rapidly across membranes and capillary walls, which creates the possibility of one fluid compartment (e.g., the ICF) becoming hypohydrated or hyperhydrated relative to the other compartments, to the detriment of body function (Fig. 3.7). Thus, the body puts mechanisms in place that independently control the water content and limit net water movement among ICF, ECF, and plasma.

1. **Intracellular fluid:** ICF osmolality typically averages ~275 to 295 mOsmol/kg H₂O, due primarily to K⁺ and its associated anions (Cl⁻, phosphates, and proteins). The ICF's K⁺-rich composition is due to the plasma membrane Na⁺/K⁺-ATPase, which concentrates

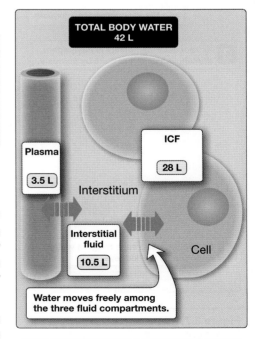

Figure 3.6.
Total body water distribution.

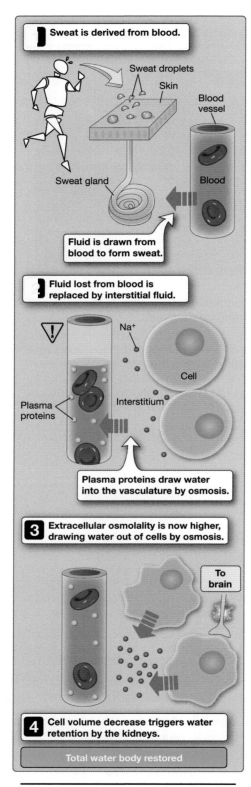

Figure 3.7.
Movement between fluid compartments during dehydration.

K^+ within the ICF and expels Na^+. Net water loss or accumulation from the interstitium is prevented by RVIs and RVDs, respectively, as discussed earlier.

2. **Extracellular fluid:** Plasma and interstitial fluid also have an osmolality of ~275 to 295 mOsmol/kg H_2O, but principal solutes here are Na^+ and its associated anions (Cl^- and HCO_3^-). ECF water content is tightly controlled by centrally located osmoreceptors acting through antidiuretic hormone (ADH). When TBW falls, because of excessive sweating, for example (see Fig. 3.7, panel 1), ECF osmolality rises because its solutes have concentrated. The rise in osmolality draws water from ICF by osmosis (see Fig. 3.7, panel 2) and triggers an RVI in all cells but not before the central osmoreceptors have initiated ADH release from the posterior pituitary as shown in Figure 3.7, panel 3 (also see Chapter 28·II·B). ADH stimulates thirst and enhances AQP expression by the renal tubule epithelium, permitting increased water recovery from urine. As a result, TBW and ECF osmolality are restored to normal (see Fig. 3.7, panel 4). When TBW is too high, AQP expression is suppressed, and the excess water is expelled from the body.

3. **Plasma:** Plasma is the smallest but also the most vital of the three internal fluid compartments. The heart absolutely depends on blood volume to generate pressure and flow through the vasculature (see Chapter 17·IV). Plasma volume must be preserved even if ECF volume is falling due to prolonged sweating or reduced water ingestion, for example. The body cannot regulate plasma volume directly because most small blood vessels (capillaries and venules) are inherently leaky, so plasma and interstitial fluid (the two ECF components) are always in equilibrium with each other. The solution to maintaining adequate plasma volume lies with plasma proteins, such as albumin, which are synthesized by the liver and remain trapped in the vasculature because of their large size. Here, they exert an osmotic potential (**plasma colloid osmotic pressure**) that draws fluid from the interstitium, regardless of changes in bulk ECF osmolality or ECF volume depletion, as shown in Figure 3.7, panel 2 (also see Chapter 18·VII·A).

IV. BODY FLUID pH

H^+ is a common inorganic cation that is similar in many ways to Na^+ and K^+. It is attracted to and binds to anions, and it depolarizes cells when it crosses the plasma membrane. H^+ deserves special consideration and cellular handling because its small atomic size allows it to form strong bonds with proteins. Such interactions alter a protein's internal charge distribution, weaken interactions between adjacent polypeptide chains, and cause conformational changes that may inhibit enzyme function or hormone binding, for example (Fig. 3.8). High H^+ concentrations denature proteins and cause cell degradation. Thus, the pH of fluid in which cells are bathed must be tightly controlled at all times.

Clinical Application 3.2: Electrolytes

ECF is Na^+ rich, but it also contains a number of other charged solutes, or **electrolytes**, the bulk comprising common inorganic ions (K^+, Ca^{2+}, Mg^{2+}, Cl^-, phosphates, and HCO_3^-). All cells are bathed in ECF. Because changes in the concentrations of any of these electrolytes can have significant effects on cell function, serum levels are maintained within a fairly narrow range, principally through modulation of kidney function (see Chapter 28). Blood tests typically include a standard electrolyte panel that measures serum Na^+, K^+, and Cl^- (Table 3.1). Serum Na^+ is measured, in part, to assess kidney function but also because Na^+ is a primary determinant of ECF osmolality and total body water. K^+ is measured because normal cardiac function depends on stable serum K^+ levels.

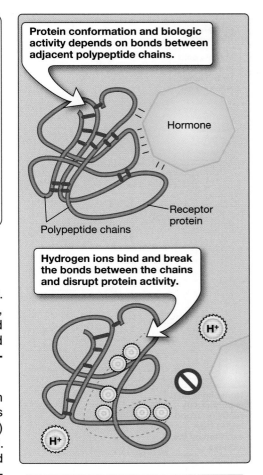

Figure 3.8.
Protein denaturation by acid.

A. Acids

Blood has a pH of 7.4 and seldom varies by more than 0.05 pH units. This corresponds to a H^+ concentration range of 35 to 45 nmol/L, which is impressive given that metabolism of carbohydrates, fats, and proteins pours ~22 mol of acid into the vasculature every day! Acid generated by metabolism comes in two forms: **volatile** and **nonvolatile** (Fig. 3.9).

1. **Volatile:** The vast majority of daily acid output comes in the form of carbonic acid (H_2CO_3), which is created when CO_2 dissolves in water. CO_2 is generated from carbohydrates (such as glucose) during aerobic respiration ($C_6H_{12}O_6 + 6O_2 \rightarrow 6CO_2 + 6H_2O$). Carbonic acid is known as a volatile acid because it is converted back to CO_2 and water in the lungs and then liberated to the atmosphere (see Fig. 3.9).

2. **Nonvolatile:** Metabolism also generates smaller amounts (~70–100 mmol/d) of **nonvolatile** or **fixed acid** that cannot be disposed of via the lungs. Nonvolatile acids include sulfuric, nitric, and phosphoric acids, which are formed during catabolism of amino acids (e.g., cysteine and methionine) and phosphate compounds, oxalic acid, lactic acid, and ketoacids. Nonvolatile acids are excreted in urine (see Fig. 3.9).

3. **Range:** Life can only exist within a relatively narrow pH range (pH 6.8–7.8, corresponding to a H^+ concentration of 16–160 nmol/L), so excreting H^+ in a timely manner is critical for continued survival. A decrease in plasma pH below 7.35 is called **acidemia. Alkalemia** is an increase in plasma pH above 7.45. **Acidosis** and **alkalosis** are more general terms referring to processes that result in acidemia and alkalemia, respectively.

Table 3.1: Serum Electrolytes

Electrolyte	Reference Range (mmol/L)
Na^+	136–145
K^+	3.5–5.0
Ca^{2+}	2.1–2.8
Mg^{2+}	0.75–1.00
Cl^-	95–105
HCO_3^-	22–28
Phosphorus (inorganic)	1.0–1.5

B. Buffer systems

Cells produce acid continually. Their intracellular structures are protected from the deleterious effects of this acid by buffer systems, which immobilize H^+ temporarily and limit its destructive effects until it can be disposed of. The body contains three primary buffer systems: the **bicarbonate buffer system**, **phosphate buffer system**, and proteins.

1. **Bicarbonate:** HCO_3^- is the body's primary defense against acid. HCO_3^- is a base that combines with H^+ to form carbonic acid, H_2CO_3:

$$HCO_3^- + H^+ \leftrightarrows H_2CO_3 \leftrightarrows CO_2 + H_2O$$
carbonic anhydrase

H_2CO_3 can then be broken down to form CO_2 and water, both of which are readily expelled from the body via the lungs and kidneys, respectively. Spontaneous conversion of H_2CO_3 to CO_2 and H_2O occurs too slowly for the HCO_3^- buffer system to be of any practical use, but the reaction becomes essentially instantaneous when catalyzed by carbonic anhydrase (CA). CA is a ubiquitous enzyme expressed by all tissues, reflecting the central importance of the HCO_3^- buffer system.

> There are at least 12 different functional CA isoforms, many of which are expressed in virtually all tissues. CA-II is a ubiquitous cytosolic isoform. CA-I is expressed at high levels in red blood cells (RBCs), whereas CA-III is found primarily in muscle. CA-IV is a membrane-bound isoform that is expressed on the surface of pulmonary and renal epithelia, where it facilitates acid excretion.

2. **Phosphate:** The phosphate buffer system employs hydrogen phosphate to buffer acid, the end product being dihydrogen phosphate:

$$H^+ + HPO_4^{2-} \leftrightarrows H_2PO_4^-$$

HPO_4^{2-} is used to buffer acid in the renal tubule during urinary excretion of nonvolatile acids.

3. **Proteins:** Proteins contain numerous H^+-binding sites and, therefore, make a major contribution to net intracellular and extracellular buffering capacity. One of the most important of these is hemoglobin (Hb), a protein found in RBCs that buffers acid during transit to the lungs and kidney.

C. Acid handling

Most acid is generated intracellularly at sites of active metabolism and then is transported in the vasculature to the lungs and kidneys for disposal. pH is carefully controlled by buffers and pumps at all stages of handling.

1. **Cells:** Intracellular structures are shielded from locally produced acid by intracellular proteins and HCO_3^-. Cells also actively control their internal pH using transporters, although the pathways involved in cellular pH control have not been well delineated.

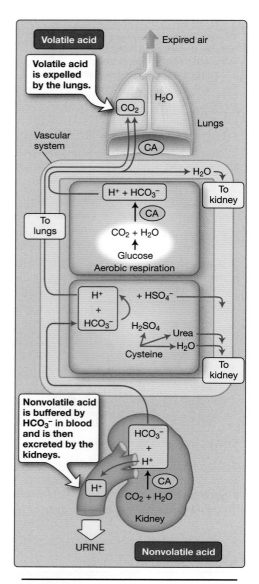

Figure 3.9.
Excretion of volatile and nonvolatile acids. CA = carbonic anhydrase.

a. **Acid:** Most cells express a Na^+/H^+ exchanger to expel acid and can also take up HCO_3^- from the ECF via a Na^+/HCO_3^- exchanger if the need arises (Fig. 3.10A).

b. **Base:** Most cells also express a Cl^-/HCO_3^- exchanger to expel excess base. Alkalosis simultaneously suppresses Na^+/H^+ exchange to help lower intracellular pH (see Fig. 3.10B).

2. **Lungs:** CO_2 produced by cells during aerobic respiration rapidly diffuses across the cell membrane and crosses through the interstitium to the vasculature. RBCs express high levels of CA-I and CA-II, which facilitate conversion of CO_2 and H_2O to HCO_3^- and H^+ (see Chapter 23·III·3). H^+ then binds to Hb for transit to the lungs. Pulmonary epithelia also contain high levels of CA, which facilitates conversion back to CO_2 for transfer to the atmosphere (see Fig. 3.9).

3. **Kidneys:** Nonvolatile acid is secreted into the renal tubule lumen and excreted in urine as shown in Figure 3.9 (also see Chapter 28·V). The urinary epithelia are protected during excretion by buffers and by combining H^+ with ammonia, which the renal tubule synthesizes specifically for this purpose. Nonvolatile acid is generated at distant sites, however, and the cells responsible must be protected from this acid until transport to the kidney can be arranged. Thus, the renal epithelium also expresses high levels of CA-IV, which generates HCO_3^- and releases it to the vasculature for transport to the sites of acid generation (see Fig. 3.9). H^+ that is formed during HCO_3^- synthesis is pumped into the tubule lumen and excreted.

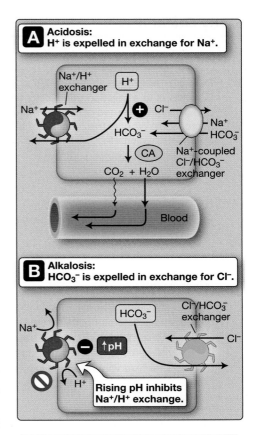

Figure 3.10.
Acid and base handling by cells. CA = carbonic anhydrase.

Chapter Summary

- The human body is composed largely of water that distributes among three principal compartments: **ICF**, **interstitial fluid**, and **plasma**. The latter two compartments together comprise **ECF**. Water movement between these compartments occurs principally by **osmosis**.

- Osmosis is driven by **osmotic pressure gradients** created by local differences in solute particle number. Water moves from regions containing low particle numbers toward regions with high particle numbers, generating osmotic pressure.

- **Osmolarity** and **osmolality** measure a solute's ability to generate osmotic pressure, whereas **tonicity** describes a solution's effect on cell volume.

- Most cells contain channels (**AQPs**) that allow water to move easily between ICF and ECF in response to transmembrane osmolality gradients. Increases in ECF osmolality cause water to leave the cell, and its volume decreases. Cells respond by accumulating solutes (Na^+, Cl^-, and amino acids) to recruit water from the ECF by osmosis (a **regulatory volume increase**). Cell volume increases elicit a **regulatory volume decrease**, involving volume-activated K^+ and Cl^- channel opening and secretion of small organic solutes (amino acids and polyalcohols).

- Regulatory volume changes allow cells to control ICF water content. Kidney function is modulated to control total body Na^+ content, which, in turn, determines how much water is retained by ECF. Plasma proteins determine how much of this ECF is retained by the vasculature.

- All cells rely on **buffer systems** to maintain the pH of ICF and ECF within a narrow range. Acid is produced continually from carbohydrate metabolism and amino acid catabolism. Carbohydrate metabolism yields CO_2, which dissolves in water to form carbonic acid (a **volatile** acid). Amino acid breakdown yields sulfuric and phosphoric acids (**nonvolatile** acids).

- The **bicarbonate buffer system** represents the body's primary defense against acid. The buffer system relies on the ubiquitous enzyme **carbonic anhydrase** to facilitate bicarbonate formation from CO_2 and water. Volatile acid is expelled as CO_2 from the lungs, whereas nonvolatile acid is excreted in urine by the kidneys.

Study Questions

Choose the ONE best answer.

3.1. During a serologic analysis, RBCs were transferred from blood to a solution containing 100 mmol/L $CaCl_2$ and 100 mmol/L urea and then monitored using light microscopy. What is the most likely effect of this transfer on RBC volume?

  A. The cell would shrink by ~50%.
  B. The solution is isosmotic, so no long-term effect.
  C. The solution is isotonic, so no long-term effect.
  D. Transient swelling would occur.
  E. Swelling to the point of lysis would occur.

Best answer = C. $CaCl_2$ dissociates into three particles (1 Ca^{2+} and 2 Cl^-) in water. A 100-mmol/L $CaCl_2$ solution has an osmolality of 300 mOsmol/kg H_2O, which approximates that of RBC ICF. The 100 mmol/L urea brings total osmolality to 400 mOsmol/kg H_2O, but urea would rapidly enter the cell until ICF and ECF equilibrated at ~350 mOsmol/kg H_2O (see Section II·C). The solution is, thus, isotonic but hyperosmotic (B). Cell shrinkage (A) would occur if urea was impermeant, but most cells are highly permeable to urea. Cell swelling (D, E) in this example would only occur if ICF osmolality rose due to active accumulation of the solutes.

3.2. Liver damage may result in decreased synthesis of plasma proteins such as albumin. What is the most likely effect of low plasma albumin on osmosis or fluid distribution?

  A. Interstitial fluid volume would increase.
  B. Plasma colloid osmotic pressure would increase.
  C. Plasma osmolality would decrease.
  D. Plasma osmolality would increase.
  E. Vascular fluid volume would increase.

Best answer = A. Blood contains large amounts of albumin (3.5–5.0 g/dL) that is trapped in the vascular compartment by its large size (see Section III·B). Its function is to help create an osmotic potential (plasma colloid osmotic pressure) that draws ECF into the vasculature. A decrease in plasma albumin concentration would, therefore, allow fluid to leave the vasculature and enter the interstitium. Colloid osmotic pressure is proportional to protein concentration, so liver damage would decrease plasma colloid pressure (B). Plasma osmolality (C, D) does not change significantly with changes in protein concentration. The main determinants of ECF osmolality are ions (e.g., Na^+ and Cl^-) and other solutes (e.g., glucose and urea). Loss of fluid to the interstitium would decrease vascular volume, not increase it (E).

3.3. A 22-year-old female is brought to the emergency department with a headache and altered mental status following a motor vehicle accident. A head CT reveals cerebral edema. Mannitol is administered intravenously to decrease intracranial pressure. Mannitol is a soluble sugar that does not enter cells. Which of the following most likely represents the volume of the water through which the mannitol distributes?

  A. 3.5 L
  B. 10.5 L
  C. 14 L
  D. 28 L
  E. 42 L

Best answer = C. Mannitol cannot cross the plasma membrane, so it is confined to ECF. Total body water is 42 L (E), of which 28 L (D) is located within cells (i.e., ICF). The difference between the two volumes represents ECF, which averages 14 L. ECF can be subdivided into plasma (3.5 L; A) and interstitial fluid (10.5 L; B; see Section III).

3.4. A 95-year-old male with widely metastatic cancer is receiving morphine to help alleviate pain. Brainstem respiratory center function has been depressed as a result, causing hypoventilation. Which of the following would most likely result from reducing ventilation?

  A. Alkalemia
  B. Decreased plasma HCO_3^- levels
  C. Decreased renal HCO_3^- reabsorption
  D. Increased interstitial pH
  E. Increased urinary H^+ excretion

Best answer = E. Metabolism generates large amounts of volatile acid (H_2CO_3) that is excreted via the lungs (see Section IV·A). Hypoventilation allows this acid to accumulate, producing (respiratory) acidemia, not alkalemia (A). The kidneys help compensate by increasing H^+ excretion (E). Accumulation of volatile acid raises plasma HCO_3^- levels (B) because H_2CO_3 dissociates in solution to form HCO_3^- and H^+. Decreasing renal HCO_3^- reabsorption (C) would exacerbate the acidemia through loss of buffer to urine. H^+, along with other small molecules, moves freely between the blood and interstitium. Thus, if blood is acidic, the interstitium will also have a low pH (D).

Epithelial and Connective Tissue

4

I. OVERVIEW

The human body comprises a diverse assemblage of cells that can be placed in one of four groups based on structural and functional similarities: **epithelial tissue**, **nervous tissue**, **muscle tissue**, and **connective tissue**. The four tissue types associate with and work in close cooperation with each other. Epithelial tissue comprises sheets of cells that provide barriers between the internal and external environment, much as the plasma membrane separates cytosol from extracellular fluid (ECF). Skin (**epidermis**) is the most visible example (see Chapter 15), but many unseen internal interfaces are also lined with epithelia (e.g., lungs, gastrointestinal [GI] tract, kidneys, and reproductive organs). Nervous tissue comprises **neurons** and their support cells (**glia**) that provide pathways for communication and coordinate tissue function, as will be discussed in greater detail in Unit II. Muscle tissue is specialized for contraction. The three types of muscle are: **skeletal muscle** (see Chapter 12), **cardiac muscle** (see Chapter 16), and **smooth muscle** (see Chapter 13). Connective tissue is a mix of cells, structural fibers, and **ground substance** that connects and fills the spaces between adjacent cells and gives tissues their strength and form. Bone is a specialized connective tissue that is mineralized to provide strength and to resist compression (see Chapter 14). This chapter considers the structure and varied functions of epithelial tissue (Table 4.1) and connective tissue.

II. EPITHELIA

Epithelia are continuous sheets of cells that line all body surfaces and create barriers that separate the internal and external environments. They help protect us from invasion by microorganisms, and they limit fluid loss to the external environment: They "keep our insides in and the outside out." Epithelia are much more than just barriers, however. Most epithelia have additional specialized secretory or absorptive functions that include sweat formation, food digestion and absorption, and excretion of waste products.

A. Structure

The simplest epithelia comprise a single layer of cells adhered to each other by a variety of **junctional complexes** that impart mechanical strength and create pathways for communication between adjacent cells (Fig. 4.1). The **apical membrane** interfaces with the external

Table 4.1: Epithelial Functions

Function	Examples
Protection	Epidermis; mouth; esophagus; larynx; vagina; anal canal
Excretion	Kidney
Secretion	Intestine; kidney; most glands
Absorption and reabsorption	Intestine; kidney; gallbladder
Lubrication	Intestine; airways; reproductive tracts
Cleansing	Trachea; auditory canal
Sensory	Gustatory, olfactory, and vestibular epithelia; skin
Reproduction	Germinal, uterine, and ovarian epithelia

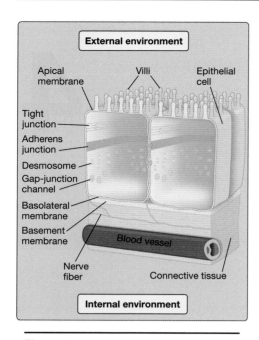

Figure 4.1.
Epithelial structure.

environment (or an internal body cavity), whereas the **basolateral membrane** rests on a **basement membrane** that provides structural support. The basement membrane comprises two fused layers. The **basal lamina** is synthesized by the epithelial cells that it supports and is composed of collagen and associated proteins. The inner layer (**lamina reticularis**, or **reticular lamina**) is formed by underlying connective tissue. Epithelia are avascular, relying on blood vessels lying close to the basement membrane to deliver O_2 and nutrients by diffusion, but they are innervated.

B. Types

Epithelia are classified based on their morphology, which is usually a reflection of their function. There are three types: **simple**, **stratified**, and **glandular epithelia**.

1. **Simple:** Many epithelia are specialized to facilitate exchange of materials between their apical surface and the vasculature. For example, the pulmonary epithelium facilitates gas exchange between the atmosphere and the pulmonary circulation (see Chapter 21·II·C). The renal tubule epithelium reabsorbs fluid from the tubule lumen and returns it to blood (see Chapter 26·II·C), whereas the epithelium that lines the small intestine facilitates transfer of materials between the intestinal lumen and the circulation (see Chapter 31·II). Exchange and transport functions require that the barrier separating the two compartments be minimal, so all of the aforementioned structures are lined with "simple" epithelia. Simple epithelia comprise a single cell layer and can be further subdivided into three groups according to epithelial cell shape.

    a. **Simple squamous epithelium:** Pulmonary alveoli and blood vessels are lined with **simple squamous epithelium**. Squamous epithelial cells are extremely thin to maximize diffusional exchange of gases.

Clinical Application 4.1: Squamous Cell Carcinoma

Squamous cell carcinomas are one of the most common forms of cancer that arise from most epithelia, including the skin, lips, buccal lining, esophagus, lungs, prostate gland, vagina, cervix, and urinary bladder. Cutaneous squamous cell carcinoma is a prevalent skin cancer that typically occurs in sun-exposed skin areas. Squamous cell malignancies are believed to arise from uncontrolled division of epithelial stem cells rather than squamous epithelial cells. Squamous cell cancers usually remain localized and can be treated by Mohs surgery (a specialized dermatologic surgery for skin malignancies), cryotherapy, or surgical excision.

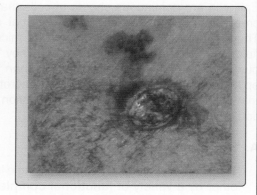

Squamous cell carcinoma.

b. **Simple cuboidal epithelium:** Many segments of renal tubules and glandular ducts are lined with cube-shaped (**simple cuboidal**) epithelial cells. Their shape reflects the fact that they actively transport materials, and, thus, they must accommodate mitochondria to produce the ATP needed to support primary active transporter function.

c. **Simple columnar epithelium: Simple columnar epithelia** comprise sheets of cells that are long and narrow to accommodate large numbers of mitochondria and are found in the distal regions of the renal tubule (see Chapter 27·IV·A) and in the intestines, for example.

2. **Stratified:** Epithelia that are subject to mechanical abrasion are composed of multiple cell layers (Fig. 4.2). These layers are designed to be sacrificed to prevent exposure of the basement membrane and deeper structures. The inner epithelial cell layers are renewed continually, and the damaged outer layers sloughed off. Examples include the skin and the lining of the mouth, esophagus, and vagina. The skin suffers constant exposure to mechanical stress associated with contact with and manipulation of external objects, so the outer layers are reinforced with keratin, a resilient structural protein (see Chapter 15·III·A). **Transitional epithelium** (also known as **urothelium**) is a specialized stratified epithelium that lines the urinary bladder, ureters, and urethra (see Chapter 25·VI). Transitional epithelium comprises cells that readily stretch and change shape (from cuboidal to squamous) without tearing to accommodate volume changes within the structures they line.

3. **Glandular:** Glandular epithelia produce specialized proteinaceous secretions (Fig. 4.3). Glands are formed from columns or tubes of surface epithelial cells that invade the underlying structures to form invaginations. Glandular secretions are then released either via a duct or ductal system onto the epithelial surface (**exocrine glands**) or across the basement membrane into the bloodstream (**endocrine glands**). Endocrine glands include the adrenal glands (which secrete epinephrine), the endocrine pancreas (which secretes insulin and glucagon), and reproductive glands (see Unit VIII). Sweat glands, salivary glands, and mammary glands are all examples of exocrine glands. Exocrine glands are classified based on their method of secretion (e.g., apocrine vs. merocrine) or secretory product (e.g., serous vs. mucous).

a. **Serous:** Serous cells produce a watery secretion containing proteins, typically enzymes. Salivary serous cells produce salivary amylase; gastric chief cells produce pepsinogen (a pepsin precursor); and pancreatic exocrine serous cells produce trypsinogen, chymotrypsinogen, pancreatic lipase, and pancreatic amylase.

b. **Mucous:** Mucous cells secrete **mucus**, a slippery glycoprotein (**mucin**)-rich secretion that lubricates the surface of mucous membranes. Many glands contain a mix of serous and mucous cells that together create an epithelial barrier layer enriched with antibacterial agents, such as lactoferrin, to help ward off infection (e.g., salivary glands) or enriched with HCO_3^- to neutralize acid (e.g., gastric epithelium).

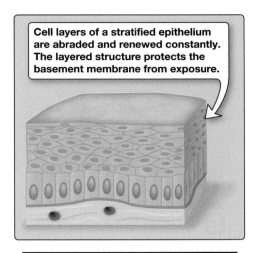

Cell layers of a stratified epithelium are abraded and renewed constantly. The layered structure protects the basement membrane from exposure.

Figure 4.2.
Stratified epithelial structure.

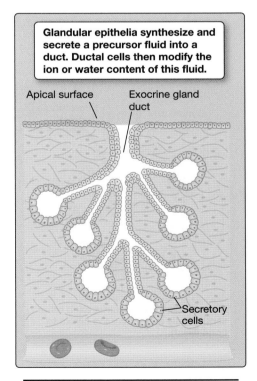

Glandular epithelia synthesize and secrete a precursor fluid into a duct. Ductal cells then modify the ion or water content of this fluid.

Apical surface

Exocrine gland duct

Secretory cells

Figure 4.3.
Glandular epithelial structure.

Figure 4.4.
Apical surface specializations.

C. Apical membrane and surface specializations

Several epithelia support apical modifications that amplify surface area or serve motile or sensory functions, including **villi**, **cilia** (**motile** and **sensory**), and **stereocilia** (Fig. 4.4).

1. **Villi:** Epithelia specialized for high-volume fluid uptake or secretion (e.g., epithelia lining the renal proximal tubule and small intestine) are folded extensively to create fingerlike projections (**villi**) that amplify the surface area available for diffusion and transport (see Fig. 4.4A). The epithelial cells that cover villi may also support **microvilli**, plasma-membrane projections that enhance surface area even further. Villi and microvilli are nonmotile.

2. **Motile cilia:** Epithelia lining the upper airways, brain ventricles, and fallopian tubes are covered with motile cilia. Cilia are hairlike organelles containing a 9 + 2 arrangement of microtubules that run the length of the organelle (Fig. 4.5). Two microtubules are located centrally, and nine microtubule doublets run around the ciliary circumference. Adjacent microtubule doublets are associated with **dynein** (dynein arms are shown in Fig. 4.5), which is a molecular motor (an ATPase). When activated, dynein causes adjacent microtubule doublets to slide against each other sequentially around the ciliary circumference, causing the cilium to bend, or "beat." The synchronized beating of many thousands of cilia causes the mucus (e.g., in the airways; see Chapter 21·II·A) or cerebrospinal fluid ([CSF] see Chapter 6·VII·D) in which they are immersed to move over the epithelial surface (see Fig. 4.4B). Respiratory cilia propel mucus and trapped dust, bacteria, and other inhaled particles upward and away from the blood–gas interface. In the brain, ciliary beating helps circulate CSF.

3. **Sensory cilia:** Epithelial cells lining the renal tubule each sprout a single, nonmotile central cilium that is believed to monitor flow rates through the tubule. The olfactory epithelium also bears nonmotile cilia whose membranes are dense with odorant receptors (see Chapter 10·III·B).

4. **Stereocilia:** The sensory epithelium that forms the lining of the inner ear expresses mechanosensory **stereocilia** that transduce sound waves (organ of Corti; see Chapter 9·IV·A) and detect head motion (vestibular apparatus; see Chapter 9·V·A). Stereocilia are nonmotile epithelial projections more closely related to villi than to true cilia.

D. Basolateral membrane

The membranes of two adjacent epithelial cells come into close apposition just below the apical surface to form **tight junctions** (**zona occludens**) as shown in Figure 4.1. Tight junctions comprise continuous structural bands that link adjacent cells together, much as beverage cans are held together by plastic six-pack rings (Fig. 4.6). Tight junctions effectively seal the apical surface of an epithelium and create a barrier, which, in some epithelia (e.g., distal segments of the renal tubule), is impermeable to water and solutes. Tight junctions also divide the epithelial plasma membrane into two distinct regions (apical and basal) by preventing lateral movement and mixing of membrane proteins. The membrane located on the basal side of the

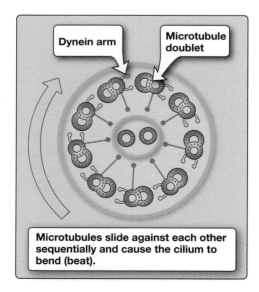

Figure 4.5.
Microtubules within a motile cilium.

tight junction includes the lateral and basal membranes, which are contiguous and together form a functional unit known as the **basolateral membrane**. The basolateral membrane usually contains a different complement of ion channels and transporters from the apical side (e.g., the Na^+/K^+-ATPase is usually restricted to the basolateral membrane) and may be folded to increase the surface area available for transporter proteins (e.g., some portions of the nephron). The basolateral membrane faces the vasculature across an interstitial space.

E. Tight junctions

Tight junctions contain numerous different proteins, the principal ones being **occludin** and **claudin**. Tight junctions serve several important functions: They form molecular "fences," they determine tight-junction "leakiness," and they regulate water and solute flow across epithelia.

1. **Fences:** Tight junctions prevent apical and basolateral membrane proteins from mixing (a fence function), thereby allowing epithelial cells to develop **functional polarity** (Fig. 4.7). The apical membrane becomes specialized for moving material between the external environment and the cell interior, whereas the basolateral membrane moves material between the inside of the cell and the bloodstream (see following discussion).

2. **Leakiness:** Adjacent cells within an epithelium are separated by a narrow space that creates a physical pathway for transepithelial fluid flow (**paracellular pathway**). Tight junctions act as gates that limit paracellular fluid movement and, in so doing, define epithelial leakiness.

   a. **Leaky epithelia:** The tight junctions in a **"leaky"** epithelium (e.g., renal proximal tubule; see Chapter 26·II) are highly permeable and allow solutes and water to pass with relative ease (see Fig. 4.7). Leakiness prevents an epithelium from being able to create strong solute concentration gradients between external and internal surfaces, but leaky epithelia can facilitate significant fluid volume shifts via the paracellular route.

   b. **Tight epithelia:** Tight junctions in **"tight"** epithelia effectively bar paracellular flow of water and solutes and allow the epithelium to become highly selective in what it absorbs or secretes (e.g., nephron distal segments; see Chapter 27·IV) as shown in Figure 4.8. An epithelium's leakiness is defined by its electrical resistance. Because the tight junctions in tight epithelia restrict passage of ions, they have a high resistance to current flow (>50,000 Ohm), whereas leaky epithelia have low resistance (<10 Ohm).

F. Gap junctions

Gap junctions are channels that provide pathways for rapid and direct communication between adjacent cells. Gap junctions also have a structural function. They are found in many areas (including muscle and nervous tissue) but are so abundant in some epithelia (e.g., intestinal epithelia) that they are packed into dense crystalline arrays, each containing thousands of individual channels. Gap-junction channels are formed from hexameric assemblies of gap-junction proteins, or connexins (Cxs), as shown in Figure 4.9.

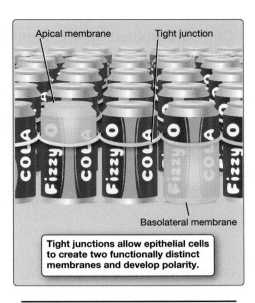

Tight junctions allow epithelial cells to create two functionally distinct membranes and develop polarity.

Figure 4.6.
Epithelium model.

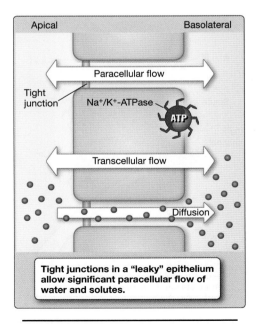

Tight junctions in a "leaky" epithelium allow significant paracellular flow of water and solutes.

Figure 4.7.
Flow across a leaky epithelium.

Clinical Application 4.2: Familial Hypomagnesemia With Hypercalciuria and Nephrocalcinosis

Familial hypomagnesemia with hypercalciuria and nephrocalcinosis (FHHNC) is a rare autosomal recessive disorder characterized by an inability to reabsorb Mg^{2+} from the renal tubule. Plasma Mg^{2+} levels fall as a consequence (hypomagnesemia). The mutation also impairs Ca^{2+} reabsorption, which increases urinary excretion rates (hypercalciuria) and the likelihood of kidney stone formation (nephrocalcinosis). Kidney stones form when urinary Ca^{2+} and Mg^{2+} concentrations are so high that their salts precipitate as crystals, which then aggregate and become lodged within the renal tubule (intrarenal calculi), ureters (ureteral calculi), or bladder. FHHNC is caused by claudin-16 gene mutations (the human genome contains 24 claudin genes). Claudin-16 forms a divalent cation-specific pathway (**paracellin-1**) for Mg^{2+} and Ca^{2+} reabsorption from the thick ascending limb of the loop of Henle. Mutations in claudin-19 can similarly produce renal Mg^{2+} wasting. Affected individuals typically require magnesium supplements and frequent lithotripsy to mechanically fragment kidney stones, which allows them to pass out of the body.

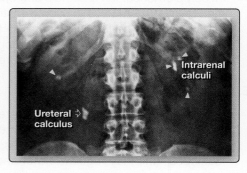

Intrarenal calculi.

1. **Connexins:** Cxs are transmembrane proteins, each containing four membrane-spanning domains. Six Cx subunits assemble around a central pore to form a **connexon hemichannel**. A gap-junction channel forms when hemichannels from two adjacent cells contact each other end to end, align, and form a tight association. Gap junctions have a significant intercellular adhesion function in addition to their role in intercellular communication. The human genome includes genes that encode at least 21 Cx isoforms that assemble into channels with distinct gating properties, selectivities, and regulatory mechanisms (see Clinical Application 4.3).

2. **Gating:** Gap-junction channels are gated by numerous factors, including the potential difference across the junction, membrane-potential (V_m) changes, Ca^{2+}, pH changes, and by phosphorylation. At rest, when the transjunctional potential is 0 mV, gap-junction channels are usually open.

3. **Permeability:** The gap-junction channel pore is sufficiently large to allow passage of ions, water, metabolites, second messengers, and even small proteins of up to around 1,000 MW. Gap junctions allow all cells within an epithelium to communicate with each other both electrically and chemically.

G. Other junctional structures

Two additional structures provide support to the epithelial sheet: **adherens junctions** and **desmosomes** (Fig. 4.10).

1. **Adherens junctions:** All cells in an epithelial sheet are tethered together by bands of protein complexes known as adherens junctions (**zonula adherens**; see Fig. 4.10) that lie just below the tight junction. The complexes straddle two adjacent cells and then link to the cell cytoskeleton.

2. **Desmosomes:** Adjacent cells within an epithelium are also tightly adhered by **desmosomes (macula adherens)**, as shown in Figure 4.10. Desmosomes are small, rounded, membrane specializations

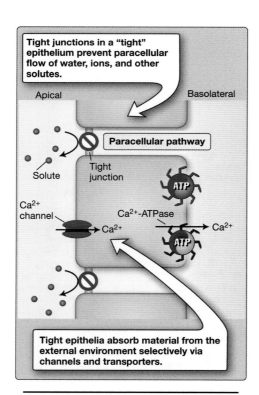

Figure 4.8.
Tight epithelia.

Clinical Application 4.3: Connexin Gene Mutations

The widespread tissue distribution of connexins (Cxs) and their surprisingly high rate of turnover (e.g., the principal Cx in heart muscle has a half-life of ≤2 hours) means that a number of hereditary diseases are associated with Cx gene mutations. Mutations may affect Cx trafficking, their ability to form a functional gap-junction channel, channel conductance, and channel-gating properties.

Neuropathies: Cx32 is expressed in Schwann cells, where it allows communication between tightly packed myelin folds (see Chapter 5·V). Cx32 gene mutation is associated with an X-linked variant of Charcot-Marie-Tooth disease (peripheral neuropathy). Patients present with distal muscle atrophy that results in weakness in the feet and hands, foot deformities, and mild sensory deficits. Cx47 is expressed in oligodendrocytes, which myelinate central neurons. Cx47 gene mutations are associated with Pelizaeus-Merzbacher–like disease, a rare central nervous system disorder caused by myelin deficits. Patients present with nystagmus, speech disorders, spasticity, and gait abnormalities.

Cataracts: The lens of the eye is a flexible disc composed of clear lens fibers enclosed within an epithelium and capsule (see Chapter 8·II·D). The epithelium expresses Cx43, whereas both the epithelium and fibers express Cx50. Mutations in the genes for either Cx result in opaque cataracts, suggesting that gap junctions help maintain lens clarity.

Deafness: Cx26 and Cx30 are involved in K$^+$ recycling within the cochlea (see Chapter 9·IV·C). Mutation in the genes for either Cx results in hearing deficits or profound deafness.

Skin disorders: Cx26, Cx30, and Cx31 are expressed in the skin and are associated with a variety of skin disorders often accompanied by hearing loss. These disorders typically affect keratinocyte differentiation and result in areas of thickened, scaly skin.

Atrial fibrillation: Cardiac atria express Cx40 and Cx43. Cx40 gene mutation has been linked to a hereditary form of atrial fibrillation.

Ectodermal developmental abnormalities: Cx43 is the most widely expressed Cx. Mutations in the Cx43 gene are associated with oculodentodigital dysplasia, a rare but striking developmental disorder. Affected individuals have a long, thin nose; small eyes and teeth; and fusion of the digits (syndactyly).

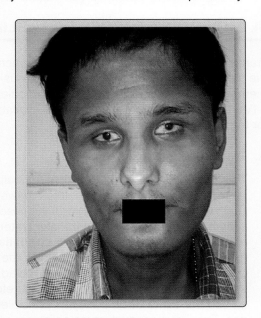
Oculodentodigital dysplasia.

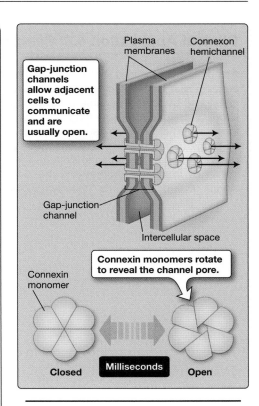
Figure 4.9.
Gap-junction channels.

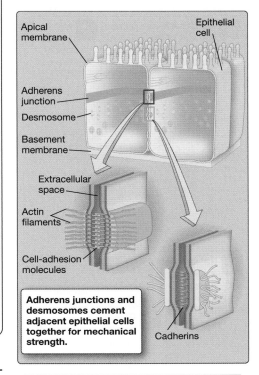
Figure 4.10.
Adherens junction and desmosome structures.

that function much like the spot welds used to join metal body panels to an automobile chassis. Protein complexes link the membrane to the cytoskeleton on the intracellular side, whereas adhesion proteins (**cadherins**) bridge the gap between cells and fuse the

Clinical Application 4.4: Pemphigus Foliaceus

Pemphigus foliaceus is a rare autoimmune disorder that presents as scaly, crusting skin blisters located primarily on the face and scalp, although the chest and back may become involved in later stages. Affected individuals express antibodies to desmoglein 1, an integral membrane protein that forms a part of the desmosomal complex. Symptoms typically are triggered by drugs (e.g., penicillin) and are caused by desmoglein 1 being targeted and degraded by the immune system. Adjacent skin epithelial cells become detached from one another and the skin blisters. The blisters ultimately slough off and leave sores. Treatment includes immunosuppressive therapy.

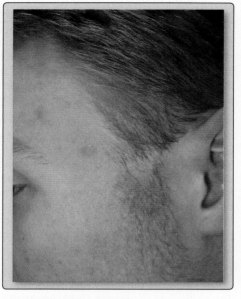

Pemphigus foliaceus.

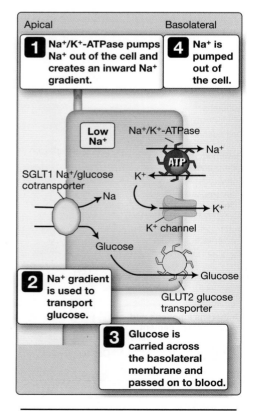

Figure 4.11.
Epithelial transport principles.

two surfaces together. Desmosomes are particularly important for maintaining the integrity of epithelia that are stressed mechanically (e.g., the urinary bladder transitional epithelium).

III. MOVEMENT ACROSS EPITHELIA

Transepithelial flow of water and solutes (vectorial transport) is driven by the same physical forces discussed previously in relation to flow across membranes (i.e., diffusion and carrier-mediated transport; see Section 1·IV). The principal difference is in the availability of a paracellular route for transepithelial transport.

A. Transcellular transport

Transport epithelia (e.g., intestinal epithelia) are specialized to move large volumes of water and solutes between the outside of the body (e.g., GI tract or renal tubule) and the vasculature by way of the interstitium. **Secretory epithelia** transfer water and solutes to the outside of the body, whereas **absorptive epithelia** take up water and solutes from the outside and transfer them to the vasculature. The example that follows considers the steps involved in glucose uptake by the small intestine as an example, but secretory and absorptive epithelia use the same basic transport principles regardless of body location. The first step involves establishing a Na⁺-concentration gradient across the surface membrane. The following steps correspond to the steps in Figure 4.11.

1. **Step 1—Create a sodium gradient:** Transepithelial transport involves work, the energy for which is supplied by ATP. ATP is used to power primary active transport, which, in virtually all instances, involves the ubiquitous Na⁺/K⁺-ATPase located in the basolateral

membrane. The Na$^+$/K$^+$-ATPase takes up K$^+$ and expels Na$^+$, thereby creating an inwardly directed Na$^+$-concentration gradient and an outwardly directed K$^+$-concentration gradient.

2. **Step 2—Glucose uptake:** Gut luminal glucose concentration is usually lower than that of intracellular fluid (ICF), meaning that the sugar must be transported "uphill" against a concentration gradient. Uptake is powered by the inward Na$^+$-concentration gradient using SGLT1 (secondary active transport).

3. **Step 3—Glucose absorption:** Na$^+$/glucose cotransport raises intracellular glucose concentration and creates an outwardly directed concentration gradient that favors glucose movement from the cell to the interstitium. Glucose diffuses across the cell and exits via a GLUT2 transporter in the basolateral membrane. Glucose subsequently diffuses through the interstitium, enters a capillary, and is carried away in the bloodstream.

4. **Step 4—Sodium removal:** Na$^+$ that crossed the apical membrane during glucose transport is removed from the cell by the basolateral Na$^+$/K$^+$-ATPase.

5. **Step 5—Potassium removal:** Not shown in Figure 4.11, the Na$^+$/K$^+$ exchange during Step 4 raises intracellular K$^+$ concentrations. The gradient favoring K$^+$ efflux is already very strong, and excess K$^+$ exits the cell passively via K$^+$ channels. K$^+$ channels are usually present in the basolateral membrane but may also be located apically.

B. Water movement

Water cannot be actively transported across epithelia, but a tried-and-true maxim of transport physiology notes that "**water follows solutes**" (by osmosis). The steps outlined in Section IV·A caused Na$^+$ and glucose to be translocated from the intestinal lumen to the interstitium, which created a transepithelial osmotic gradient used to absorb water. There are two potential routes for water absorption: transcellular and paracellular (Fig. 4.12).

1. **Transcellular flow:** Transcellular water movement only occurs if water is provided with a clear passage through the epithelial cell. In practice, this requires that water channels (aquaporins [AQPs]) be present in both the apical and basolateral membranes. Transport epithelia typically express high AQP levels, which support high volumes of transcellular water uptake (or secretion). The epithelium that lines renal collecting ducts actively regulates its water permeability by modulating apical AQP expression levels (see Chapter 27·V·C). When there is a need to reabsorb water from the tubule lumen, AQPs are recruited to the apical membrane from stores located intracellularly in vesicles. When the body contains water in excess of homeostatic requirements, AQPs are removed from the apical membrane, and the epithelium becomes water impermeable. Although AQPs are the principal route for transcellular water movement, most ion channels and transporters allow a few water molecules to follow along with a permeant ion or organic solute.

2. **Paracellular flow:** Paracellular water flow is also driven by osmotic pressure gradients created by solute transport. The availability of the paracellular route is determined by tight junction leakiness.

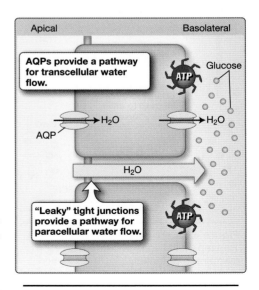

Figure 4.12.
Transepithelial water movement.

Clinical Application 4.5: Oral Rehydration Therapy

Intestinal epithelia are capable of transporting high volumes of watery fluids. In a healthy person, virtually all of the ~10 L of fluids secreted by intestinal epithelia during the digestive phase are subsequently reabsorbed, so that <200 mL/d is lost from the body in stool (see Chapter 31·III·B). The bacterium *Vibrio cholerae* secretes a toxin that increases intestinal epithelial Cl^- permeability and raises intestinal luminal osmolality.[1] Copious amounts of fluid are drawn osmotically across the epithelium as a result. Almost all of the secreted fluid is lost to the external environment, either as a result of vomiting or frequent, watery stools. Death usually occurs as a result of hyponatremia, hypovolemia, and loss of blood pressure. Cholera can be treated, and death prevented, fairly simply, using oral rehydration therapy (ORT). ORT takes advantage of the fact that Na^+ and glucose are rapidly absorbed by the intestinal epithelia (via SGLT1), creating an inwardly directed osmotic gradient that drives water reabsorption. Typical home remedies involve giving patients a solution containing 6 tsp sugar and 1/2 tsp salt (NaCl) per liter of water. The advantage of ORT is that it is highly effective, easy to administer, and cheap, which is of particular public health advantage in developing countries where cholera is endemic, and resources are typically limited.

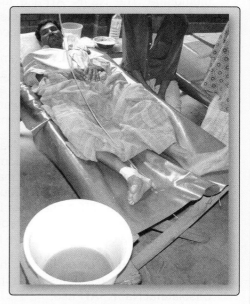

Cholera patients excrete large volumes of watery stool.

C. Solvent drag

Intestinal epithelia secrete and absorb ~10 L of water per day, whereas the renal tubule reabsorbs almost as much on an hourly basis. These secretory and absorptive functions generate high water flow rates, both transcellularly and paracellularly. The resulting water streams carry ions and other small solutes with them, much as a fast-flowing river sweeps along sand and other fine particles. This phenomenon is known as **solvent drag** and can contribute significantly to transepithelial solute movement (e.g., K^+ reabsorption in the renal tubule; see Chapter 26·VIII). The net result of all this solute and water flow is that the secreted or absorbed fluid usually has a composition that is isosmotic relative to the source (**isosmotic flow**).

D. Transepithelial voltage effects

Epithelial cells are located at the interface between two compartments that may have very different chemical compositions. The basolateral membrane faces the inside of the body and is bathed in ECF whose chemical composition is well controlled. The apical membrane is bathed in external fluid whose composition may be indeterminate and variable. Charge differences between the two fluids create a transepithelial voltage difference that influences transport (Fig. 4.13).

Differences in fluid composition between the lumen and ECF create a transepithelial charge difference.

Apical   Basolateral

$V_m = -50$ mV

Charge difference be used to drive absorption of ions.

+3 mV   0 mV

ICF   ECF

Lumen

Figure 4.13.
Transepithelial voltage difference. V_m = membrane potential.

 [1]For more information on the pathogenesis and treatment of cholera, see *LIR Microbiology*, 4e, Chapter 12·VI·B and E.

These voltage differences may amount to <3 mV, but they can significantly affect ion movement.

1. **Transport:** The distal segment of the renal proximal tubule, for example, is positively charged (~3 mV) with respect to ECF. Although the voltage difference is small, it provides a motive force that drives significant amounts of Na^+ out of the tubule toward the interstitium (see Chapter 26·X·B).

2. **Local membrane potential effects:** Epithelial cells, like all cells in the body, establish a V_m across their surface membrane, inside negative. V_m is measured with respect to ECF and is uniform throughout the cell. However, because the apical surface is bathed in a medium of different ionic composition, this can create local differences in V_m. Thus, if V_m is –50 mV and the tubule lumen is +3 mV with respect to ECF, the potential across the apical membrane will be –53 mV relative to the lumen.

IV. CONNECTIVE TISSUE

Connective tissue is the most abundant tissue class that can be found in all areas of the body. There are several different connective tissue types, but they all follow a common organizational principle (Fig. 4.14). Connective tissues are composed of specialized cells, structural proteins, and a fluid-permeated ground substance.

A. Types

The three main types of connective tissue are: **embryonic** (not considered further here), **specialized connective tissue**, and **connective tissue proper**.

1. **Specialized:** Specialized connective tissue includes cartilage, bone (see Chapter 14), hematopoietic tissue and blood (see Chapter 23), lymphatic tissue, and adipose tissue. Cartilage is a flexible connective tissue that cushions bones at sites of articulation and that gives shape to the nose and ears, for example. Lymphatic tissue comprises a system of vessels that drain fluid from the extracellular space (see Chapter 18·VII·C). Adipose tissue is composed largely of adipocytes whose primary function is to store energy in the form of triacylglycerols. Fat conducts heat poorly, so it is layered beneath the skin (subcutaneous fat) to help insulate the body. Adipose tissue deposits can also be associated with internal organs (visceral fat) and in yellow bone marrow.

2. **Proper:** Connective tissue proper forms the **extracellular matrix (ECM)** that occupies the interstitial space. Connective tissue proper can be further subdivided into **loose connective tissue**, a highly pliable form that occupies the space between most cells, **dense connective tissue** (tendons, ligaments, and fibrous fascia and capsules that enclose muscles and organs), and **reticular connective tissue** that forms the scaffolding on which blood vessels, muscle, and the liver is built, for example.

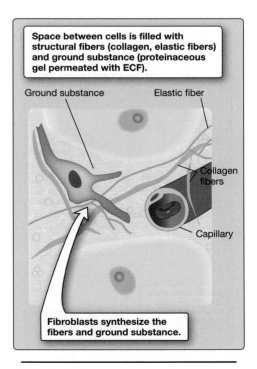

Space between cells is filled with structural fibers (collagen, elastic fibers) and ground substance (proteinaceous gel permeated with ECF).

Ground substance　　　　Elastic fiber

Collagen fibers

Capillary

Fibroblasts synthesize the fibers and ground substance.

Figure 4.14.
Connective tissue.

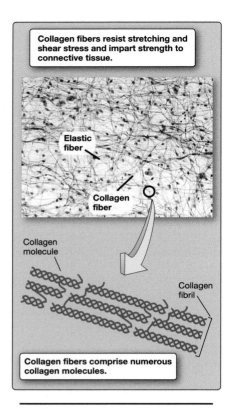

Figure 4.15.
Collagen structure.

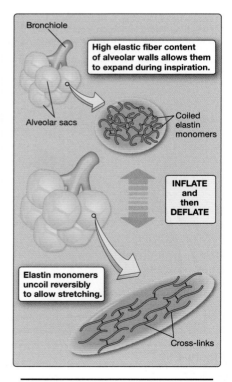

Figure 4.16.
Elastin properties.

B. Extracellular matrix

The ECM is a mix of cells (fibroblasts), structural proteins (collagen and elastic fibers), ground substance, and ECF. The ECM imparts form and strength to tissues and provides pathways for chemical diffusion and for immune system cell migration (e.g., macrophages).

1. **Fibroblasts:** Fibroblasts are motile cells that continually synthesize and secrete structural protein precursors and ground substance. They are essential for ECM maintenance and for wound healing.

2. **Structural proteins:** The ECM is filled with an interconnected structural matrix comprising collagen and elastic fibers.

   a. **Collagen:** Collagen is a tough, fibrous protein that possesses high tensile strength and resistance to shear stress. The body contains 28 different collagen types, but four forms (types I–IV) predominate. Type I collagen is abundant in skin and vascular walls and is bundled to form ligaments, tendons, and bone. Type IV organizes into meshlike networks that make up the basal lamina of epithelia, for example. Collagen molecules are composed of three polypeptide chains braided into a triple helix and then cross-linked extensively for enhanced stress resistance[1] (Fig. 4.15).

   b. **Elastic fibers:** Elastic fibers are composed of elastin and glycoprotein microfibrils (e.g., **fibrillin** and **fibulin**). Elastic fibers stretch like rubber bands when stressed and then recoil and assume their original shape when allowed to relax. Elastic fibers are found in the walls of arteries and veins, which allows them to stretch when intraluminal pressures increase. Elastic fibers also allow lungs to expand during inspiration as well as help reduce stress on teeth during chewing (periodontal fibers). Elastic fibers are synthesized by fibroblasts, which first lay down structural scaffolding made of fibrillin and then deposit tropoelastin monomers onto the scaffolding. Four adjacent elastin monomers are then cross-linked to form an irregular network that comprises the mature elastin molecule (Fig. 4.16).

3. **Ground substance:** Ground substance is a mix of various proteins (principally proteoglycans) and ECF that creates an amorphous gel filling the spaces between cells and structural fibers. The high water content of the gel facilitates chemical diffusion between cells and the vasculature, yet the structural fibers simultaneously impede movement of invading pathogens. Proteoglycans are formed by attaching numerous glycosaminoglycan (GAG) molecules to a core protein, the final structure resembling a bottlebrush or pipe cleaner (Fig. 4.17). GAGs possess a high

 [1]For more information on collagen synthesis and assembly, see *LIR Biochemistry*, 8e, Chapter 4·II.

negative-charge density, which allows them to attract and loosely trap water molecules within the gel. The interstitium contains >10 L of ECF in an average person, representing a substantial buffer volume that helps minimize the impact of changes in total body water on cell and cardiovascular function (see Chapter 3·III and Chapter 18·VIII·C). The gel also cushions underlying tissues and resists compression.

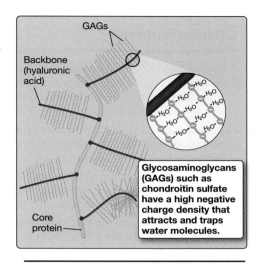

Figure 4.17.
Proteoglycan structure.

Biologic Sex and Aging 4.1: Extracellular Matrix Changes

Physiologic aging is accompanied by profound changes in the structure and composition of the extracellular matrix (ECM). In short, elastin is gradually replaced by collagen, which reduces ECM elasticity and causes stiffening that adversely affects the functioning of all tissues. The consequences are felt most keenly by the heart, vasculature, and lungs (see Biologic Sex and Aging 17.1, 18.1, and 21.1). Three interlinked age-related processes together create a vicious feedforward cycle that make stiffening inevitable: elastin fragmentation, collagen deposition, and cross-linking.

Elastin is formed during fetal development and then we carry these same monomers for life. Elastin is a highly stable protein, but elastin fibers are subject to numerous stressors, including mechanical wear-and-tear (e.g., cardiovascular elastin is stretched >30 billion times during a lifespan), calcification, glycation, and proteolysis that cause it to fragment into tropoelastin monomers and elastin-derived peptides (EDPs). Elastin is not replaced, so tissues inevitably lose elasticity along with their elastin content. Worse still, elastin fragments are biologically active. Both tropoelastin monomers and EDPs bind to elastin receptors whose activation initiates ECM remodeling, recruits fibroblasts, and causes inflammation. Elastin receptor binding also activates matrix metalloproteinases and other elastases that generate yet more elastin fragments.

Lost elastin is replaced by **collagen**, a protein designed for tensile strength rather than elasticity. Calcium mineral deposition in ground substance contributes to loss of elasticity, the net effect being that ECM stiffness increases by greater than a thousandfold over the course of a normal lifespan.

Chronic exposure of elastin and collagen to glucose causes intermolecular **cross-links** to form (**advanced glycation end products [AGEs]**). AGEs accumulate slowly, and long-lived molecules such as elastin are particularly susceptible. Elastin cross-linking makes the protein fragile and prone to fragmentation. Collagen cross-links increase molecular rigidity and prevent it from being degraded, which further potentiates the age-associated increase in collagen-to-elastin ratio.

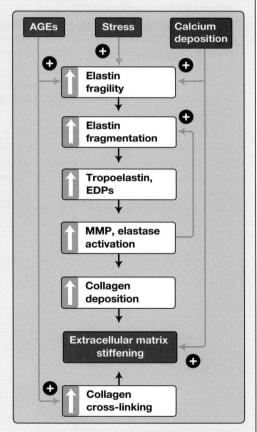

AGEs = advanced glycation end products, EDPs = elastin-derived peptides, MMP = matrix metalloproteinase.

Chapter Summary

- The human body is composed of four tissue types: **epithelial tissue**, **nervous tissue**, **muscle tissue**, and **connective tissue**. Epithelial tissue comprises sheets of tightly packed cells that line all external and internal body surfaces (e.g., skin and the pulmonary and GI linings). Epithelia form **barriers** that protect the body from external threats, but many also have specialized transport functions.

- Epithelia are classed morphologically: **simple epithelia** are composed of a single cell layer (e.g., pulmonary epithelium), **stratified epithelia** (e.g., skin) comprise multiple layers that are sloughed off and renewed, and **glandular epithelia** (endocrine and exocrine glands) are specialized for secretion.

- Epithelia are **polarized** with functionally distinct **apical** and **basal surfaces**. The apical surface faces the external environment, the lumen of a hollow organ, or a body cavity. Apical surfaces may be specialized to include **villi**, **cilia**, and **stereocilia** that amplify surface area, propel mucus layers, or serve a sensory role, respectively.

- The **basolateral membrane** communicates with the body interior via the **interstitium** and the vasculature. It rests on a **basement membrane** that anchors the epithelium to underlying connective tissues.

- Polarization of epithelia is made possible by **tight junctions**, which are structural bands encircling all cells in an epithelium close to their apical surface. The junctions form tight seals with an important barrier function. The junctions also segregate proteins, allowing for specialization of apical and basolateral membrane function.

- Tight-junction permeability is regulated by **claudins**, which determine how much water and solutes cross an epithelium via the space between adjacent cells (**paracellular flow**). The junctions effectively block passage of all water and solutes across a "tight" epithelium. In contrast, "leaky" epithelia secrete and absorb significant amounts of fluid.

- **Gap junctions** are hexameric channels comprising **connexin** monomers that connect adjacent cells and allow all cells in an epithelium to communicate chemically and electrically. **Adherens junctions** and **desmosomes** are junctional structures that provide strength to an epithelium and help prevent it from tearing when stressed mechanically.

- Many epithelia have **secretory** and **absorptive functions**. Transepithelial transport usually occurs **transcellularly** and **paracellularly**, and both routes are regulated. The electrochemical and osmotic driving forces for movement of solutes and ions are established by primary and secondary active transporters (e.g., Na^+/K^+-ATPase and Na^+-coupled transport).

- Connective tissue comprises cells; structural fibers; and an amorphous, fluid-permeated ground substance. Specialized connective tissues include cartilage, bone, and adipose tissue. Connective tissue proper forms the extracellular matrix (ECM).

- The ECM fills the space between all cells, providing mechanical strength and support as well as a loose, gel-like medium that facilitates chemical diffusion and cell migration.

- The ECM is synthesized and maintained by **fibroblasts**. ECM structural proteins include **collagen fibers** for strength and **elastic fibers** to allow stretching. Elastic fibers are composed primarily of **elastin**. Ground substance is a proteinaceous matrix containing large quantities of **proteoglycans**. Proteoglycans have a high negative-charge density that allows them to attract and immobilize ~10 L of **interstitial fluid** in an average person.

Study Questions

Choose the ONE best answer.

4.1. A researcher investigating the properties of intestinal epithelium from a patient with inflammatory bowel disease noted that the diseased areas have a low electrical resistance, whereas the healthy areas have a high resistance. What might be inferred about the properties of the healthy epithelium?

    A. It forms weak transepithelial ionic gradients.
    B. It has a thick basement membrane.
    C. It is specialized for isosmotic transport.
    D. It lacks a basolateral Na^+/K^+-ATPase.
    E. The tight junctions are highly impermeable.

Best answer = E. High electrical resistance is characteristic of a "tight" epithelium, a property conferred in part by the impenetrability of the tight junctions between cells to ions and water (see Section II·E). Tight epithelia are notable for their ability to establish strong osmotic and ionic concentration gradients (A). The areas of inflammation have a low electrical resistance, which makes them "leakier" to ions. Basement membranes (B) do not directly contribute to epithelial electrical resistance. Leaky epithelia are usually specialized for high-volume isosmotic transport (C). All intestinal epithelia express a basolateral Na^+/K^+-ATPase (D).

4.2. A 52-year-old female presents with heart palpitations and lightheadedness. An electrocardiogram shows her to be in atrial fibrillation, which has been linked to increased connexin (Cx) 43 expression. Which of the following most likely describes Cxs normally?

    A. They allow electrical propagation through tissues.
    B. They are found only in the heart.
    C. They are highly ion selective.
    D. They mediate Ca^{2+} influx from the cell exterior.
    E. They open during membrane depolarization.

Best answer = A. Cxs form hexameric assemblies (connexons) with a pore at their center (see Section II·F). Connexons from two adjacent cells fuse to create a gap-junction channel that provides a pathway for electrical and chemical communication between cells. In the heart, they allow waves of contraction to spread across the myocardium (see Chapter 17·III). They have a wide tissue distribution (B). Gap-junction channels are characterized by their wide, nonselective pores (C) that can allow passage of small peptides. Ca^{2+} channels mediate Ca^{2+} influx across surface membranes, not gap-junction channels (D). Connexons are usually open at rest and may close upon depolarization (E).

4.3. Hypokalemia is relatively rare in healthy individuals, but which of the following would most likely favor increased K^+ uptake by a transport epithelium for transfer to the circulation?

    A. Apical glucose cotransport
    B. High interstitial K^+ concentrations
    C. Increased Na^+/K^+-ATPase activity
    D. Increased paracellular water uptake
    E. Lumen-negative potential difference

Best answer = D. The paracellular route is a significant pathway for solute and water movement across many epithelia (see Section III·B). High paracellular water flow rates generate solvent drag, whereby inorganic ions and other solutes are swept along with water. Glucose cotransporters (A) generally couple glucose movement with Na^+, not K^+. High interstitial K^+ concentrations (B) decrease the electrochemical gradient favoring net K^+ uptake. The Na^+/K^+-ATPase increases intracellular K^+ concentrations, which decreases the driving force for apical K^+ uptake (C). Because K^+ is a cation, a negatively charged renal or GI lumen (for example) with respect to blood (E), decreases net uptake.

4.4. Genetic analysis of a young male admitted for an aortic aneurysm reveals a mutation in the *FBN1* gene, which encodes fibrillin. This mutation is most likely to affect the structure of which of the following components of the extracellular matrix?

    A. Collagen
    B. Elastic fiber
    C. Glycosaminoglycan
    D. Ground substance
    E. Matrix metalloproteinase

Best answer = B. Fibrillin is a large glycoprotein and a constituent of microfibrils that form the scaffolding for elastic fiber formation (see Section IV·B). *FBN1* gene mutation can seriously impair microfibril availability and thereby decrease the structural integrity of the extracellular matrix (ECM). Collagen fibers (A) are composed of braided collagen fibrils. Glycosaminoglycans (C) are a constituent of the proteoglycans that make up the bulk of ground substance. Ground substance (D) traps interstitial fluid to form a cushioning gel that fills the space between cells. Matrix metalloproteinases (E) are enzymes involved in ECM turnover.

4.5. A student researcher is interested in the evolutionary history of the different types of membrane permeabilities from prokaryotes through modern eukaryotes. The properties of pores, channels, and carriers common to membranes of living species are compared. Which of the following is most likely to be a feature common to all three permeability types?

A. Active transport
B. G-protein coupling
C. Gating mechanism
D. Selective permeability
E. Voltage sensitivity

Best answer = D. Pores, channels, and carriers all provide a pathway for charged molecules to cross a hydrophobic barrier (i.e., a lipid bilayer). The presence of a nonselective pathway within the plasma membrane would prove lethal to a cell, so all three permeabilities have selectivity filters that limit passage of ions (see Chapter 1·V). Active transport (A) is used by pumps, exchangers, and cotransporters to move ions against a concentration gradient. The three permeability types show great diversity in their gating mechanisms (C). Some are gated by G proteins (B), whereas others are voltage sensitive (E).

4.6. A pharmaceutical company develops a novel drug candidate for use in treating cardiac arrythmia. The drug blocks K^+ channels when membrane potential is closest to the Na^+ equilibrium potential. During which of the following action potential phases is blockage most likely to occur?

A. At threshold
B. Afterpotential
C. Downstroke
D. Overshoot
E. Upstroke

Best answer = D. The Na^+/K^+-ATPase removes Na^+ from the cell interior and establishes a strong, inwardly directed Na^+ gradient used in electrical signaling and secondary active transport. The equilibrium potential for Na^+ (E_{Na}) is around +61 mV as a result (see Chapter 2·II·B). During an action potential (AP) overshoot, membrane potential crosses 0 mV and approaches E_{Na} but usually does not reach it. Threshold (A) is the membrane voltage at which an AP becomes regenerative. During the afterpotential (B), membrane potential is farthest away from E_{Na}. During the downstroke (C) and upstroke (E), membrane potential is moving away from or toward E_{Na}, respectively.

4.7. A 38-year-old male presents to the office with recurrent episodes of unsteadiness, vertigo, and nausea lasting 2–48 hours. Neurologic examination reveals prominent ataxia, dysarthria, and nystagmus. Genetic analysis identifies a mutation in the *CACNA1A* gene affecting the S4 region of a brain-specific voltage-activated Ca^{2+} channel. Which of the following most likely describes the function of the affected region?

A. Acetylcholine binding
B. G-protein binding
C. Inactivation
D. Ion selectivity
E. Voltage sensitivity

Best answer = E. The patient presents with episodic ataxia type 2, a hereditary disease associated with *CACNA1A* gene mutation. The voltage-gated Na^+, Ca^{2+}, and K^+ channel superfamily members share a common tetrameric molecular structure. The S4 transmembrane domain of each subunit is a highly charged voltage sensor that facilitates channel opening on membrane depolarization (see Chapter 2·VI·A). Acetylcholine (A) binds to *cys*-loop receptors, which are ligand-gated channels. Second messenger–gated channels bind G proteins (B). Voltage-gated channel inactivation (C) involves an intracellular domain. Ion selectivity (D) is conferred by a peptide loop between membrane-spanning domains S5 and S6.

4.8. A 32-year-old male is brought to the emergency department with fatigue and dizziness after participating in a marathon on a hot day. Physical examination reveals dry mucous membranes, decreased skin turgor, and elevated heart rate. Laboratory studies reveal elevated blood urea nitrogen (BUN) and hypernatremia. Which of the following most likely describes a mechanism by which cells maintain volume under such conditions?

A. Amino acid release
B. Aquaporin activation
C. Glucose transporter inhibition
D. K^+-channel activation
E. Na^+/H^+-exchanger activation

Best answer = E. The patient is dehydrated because of excessive sweating during the marathon. He is hypernatremic, so extracellular fluid osmolality is expected to be increased. Cells shrink when exposed to such conditions, triggering a regulatory volume increase (RVI). RVIs involve activating Na^+/H^+ exchangers and $Na^+/K^+/2Cl^-$ cotransporters to increase intracellular fluid (ICF) osmolality (see Chapter 3·II·E). Amino acid release (A) occurs during regulatory volume decreases (RVDs). Aquaporin activation (B) would increase water loss from the ICF and cause further shrinkage. Glucose transporters usually facilitate glucose uptake, so their inhibition (C) would not help protect ICF volume. K^+-channel activation (D) and resulting K^+ efflux occurs during an RVD.

Nervous System Organization

5

I. OVERVIEW

To a casual observer, a microscopic pond organism such as *Paramecium* behaves with apparent intent and coordination that suggests the involvement of a sophisticated nervous system. If it bumps into an object, it stops, swims backward, and then moves off in a new direction (Fig. 5.1). This simple behavior minimally requires a sensory system to detect touch, an integrator to process information from the sensor, and a motor pathway to effect a response. *Paramecium* does not possess a nervous system, however. It is a single cell. The human brain contains over a trillion neurons. It has evolved sophisticated structures and networks that allow for self-awareness, creativity, and memory. Yet the human nervous system's basic organizational principles share many similarities with our unicellular cousins. Unicells and neurons both use changes in membrane potential (V_m) to integrate and respond to divergent and, sometimes, conflicting inputs. On an organismal level, humans, like unicells, have sensory systems to inform them about their immediate environment, integrators to process sensory data, and motor systems to produce an appropriate response.

II. NERVOUS SYSTEM

In discussing how the nervous system works, it is useful to define several partially overlapping subdivisions.
- The **central nervous system** (**CNS**) includes the neurons of the brain and spinal cord. The CNS is the nervous system's integrative and decision-making arm.
- The **peripheral nervous system** (**PNS**) collects information from somatic, visceral, and special sensory components and conveys it to the CNS for processing. It then directs motor commands from the CNS to the appropriate targets. Efferent signals target somatic and autonomic motor components. The PNS includes neurons that originate in

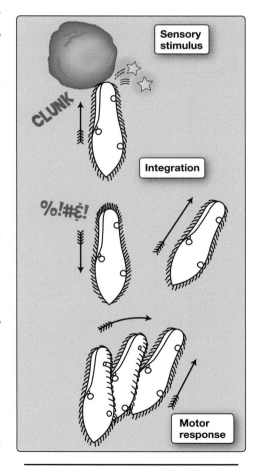

Sensory stimulus

Integration

Motor response

Figure 5.1.
Sensory response in *Paramecium*.

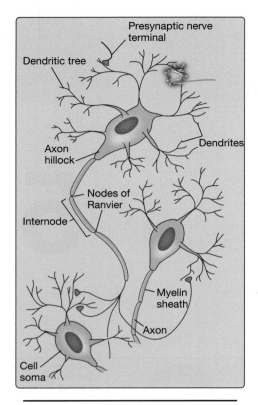

Figure 5.2.
Neuronal anatomy.

the cranium and spinal cord and extend beyond the CNS. Thus, cranial nerves are also considered part of the PNS.

- The autonomic motor and viscerosensory pathways together comprise the **autonomic nervous system** (**ANS**). The ANS can be functionally subdivided into the **sympathetic nervous system** (**SNS**) and **parasympathetic nervous system** (**PSNS**). The ANS acts largely independently of voluntary control. It regulates and coordinates many organ systems and homeostatic reflexes.

- The **enteric nervous system** (**ENS**) is anatomically placed in the PNS and functionally placed in the ANS, with a neuronal network that independently evolved to regulate the gastrointestinal tract.

III. NEURONS

The nervous system comprises a network of **neurons**. Although their shape may vary according to function and location within the body, the basic principles of neuronal design and operation are universal. Their role is to transmit information as rapidly as possible from one area of the body to the next. In the brain, the distance involved may be a few micrometers; in the periphery, it can exceed a meter. Because speed is achieved using electrical signals, a neuron can be thought of as a biologic wire. Unlike a wire, however, a neuron has the ability to integrate incoming signals before transmitting information to a recipient.

A. Anatomy

A neuron can be divided into four anatomic regions: the **cell body**, **dendrites**, an **axon**, and one or more **nerve terminals** (Fig. 5.2).

1. **Cell body:** The cell body (**soma**) houses the nucleus and components required for protein synthesis and other normal housekeeping functions.

2. **Dendrites:** Dendrites are branched projections of the cell body that radiate in multiple directions ("dendrite" is derived from *dendros*, Greek for tree). Some neurons have dense and elaborate dendritic trees, whereas others may be very simple. Dendrites are cellular antennae waiting to receive information from the neural net. Many tens of thousands of nerve terminals may synapse with a single neuron via its dendrites.

3. **Axon:** An axon is designed to relay information at high speed from one end of the neuron to the other. It arises from a swelling of the soma called an **axon hillock**. An axon is long and thin like a wire. It is often wrapped with an insulating material (**myelin**) that enhances signal-transmission rate (see Section B). Myelination begins some distance distal to the axon hillock, leaving a short **initial segment** that is unmyelinated. The **axoplasm** (axonal cytoplasm) is filled with parallel arrays of microtubules and microfilaments. These are partly structural, but they also act like railway tracks in a mineshaft. "Ore carts" (vesicles) filled with neurotransmitters and other materials attach to the tracks and then motor along at relatively high speed (~2 μm/s) from one end of the cell to the other. Movement away from the cell body toward the nerve terminal (**anterograde**

Clinical Application 5.1: Polio

Retrograde transport is believed to be the mechanism by which polio and many other viruses enter the central nervous system from the periphery.[1] Poliovirus is an enterovirus distributed by fecal–oral contact that causes **paralytic poliomyelitis**. After infecting the host, the virus enters and spreads through the nervous system via nerve terminals. After fusing with the surface membrane and entering the axoplasm, the viral capsid (a protein shell) attaches to the retrograde transport machinery and motors to the cell body. Here, it proliferates and, ultimately, destroys the neuron. The result is a flaccid paralysis of the musculature, classically affecting the lower limbs, but it can also cause fatal paralysis of the respiratory musculature. Polio has been largely eradicated in North America and Europe but is endemic in many other regions of the world. Defects in axonal transport are also believed to have a role in precipitating the neuronal death that accompanies Alzheimer, Huntington, and Parkinson diseases and several other adult-onset neurodegenerative diseases.

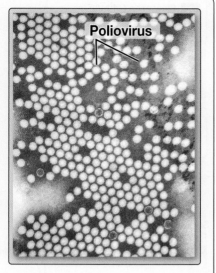

Poliovirus.

transport) is powered by **kinesin**. The return trip (**retrograde transport**) relies on a different molecular motor (**dynein**).

4. **Nerve terminal:** The nerve terminal is specialized to convert an electrical signal into a chemical signal for dispatch to one or more recipients. The junction between the terminal and its target is called a **synapse**. **Presynaptic** and **postsynaptic** cell membranes are separated by a ~30- to 50-nm **synaptic cleft**. Facing the terminal across the cleft may be any of several different postsynaptic effector cells, including myocytes, secretory cells, or even a **dendrite** extending from the cell body of another neuron.

B. Excitability

The speed with which neural nets process and output data is limited by the rate at which signals can be transmitted from one component to the next. Neuronal transmission speed is maximized by using fast **action potentials** (**APs**), by optimizing axonal geometry, and by insulating the axons.

1. **Action potentials:** Axons that convey signals over long distances typically use APs that have a very simple form, and they function as binary digits on the neural net. Neuronal APs are mediated primarily by voltage-gated Na^+ channels, which are very fast activating (Fig. 5.3). When Na^+ channels open, Na^+ flows into the neuron down its electrochemical gradient, and V_m depolarizes rapidly toward the equilibrium potential for Na^+ (see Chapter 2·II·B). Rapid Na^+ channel–opening rate ("gating kinetics") allows electrical signals to propagate at high speeds down an axon's

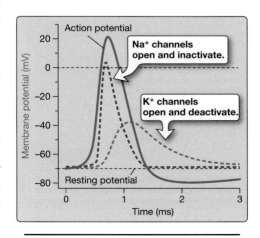

Figure 5.3.
Time course of ion channel events during a neuronal action potential.

 [1]For a discussion of enteroviruses, see *LIR Microbiology*, 4e, Chapter 27·II·A.

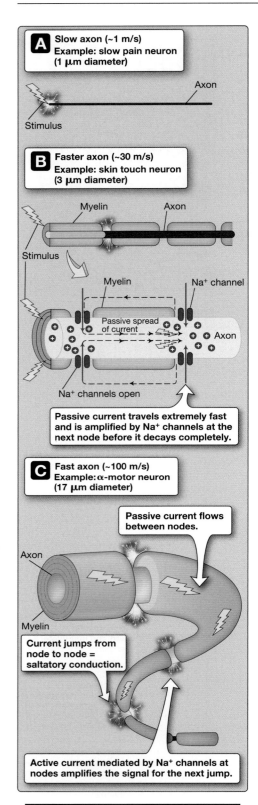

Figure 5.4.
Myelin and diameter effects on axonal conduction velocity.

length. Membrane repolarization occurs largely as a result of Na$^+$-channel inactivation. Voltage-dependent K$^+$ channels activate during a spike also, but their numbers are small, so their contribution to membrane repolarization is limited.

2. **Axon diameter:** The rate at which APs propagate down an axon increases with axon diameter (Fig. 5.4). This is because internal resistance, which is inversely proportional to diameter, determines how far passive current can reach down the axon's length before the signal decays and needs amplifying by an active current (i.e., a Na$^+$ channel–mediated current). The amplification step is slow compared with transmission of passive current, so wide axons transmit information much faster than thin ones.

3. **Insulation:** Passive currents decay with distance because the neuronal membrane contains K$^+$ "leak" channels that bleed current to the extracellular medium (see Fig. 2.11). Leak channels are always open. Current dissipation can be prevented by insulating an axon with myelin to prevent leaks, allowing electrical signals to travel much further before needing to be amplified (see Fig. 5.4B). Myelin is formed by **glial** cells and comprises concentric layers of **sphingomyelin-rich membrane** (see Section V). Insulation increases conduction velocity up to 250-fold.

4. **Saltatory conduction:** An axon's **myelin sheath** is not continuous. Every 1 to 2 mm is a 2- to 3-μm segment of exposed axonal membrane known as a **node of Ranvier**. Nodes are tightly packed with Na$^+$ channels, whereas the **internodal regions** (areas lying hidden beneath the myelin sheath) have virtually no channels. In practice, this means that an AP leapfrogs from one node to the next down the length of the axon, a behavior known as **nodal**, or **saltatory**, **conduction** (see Fig. 5.4C).

C. Classification

CNS neurons are a diverse group of cells, and there are many ways of classifying them. Morphologically, they can be grouped based on the number of **neurites** (processes, such as axons and dendrites) extending from the cell body.

1. **Pseudounipolar: Pseudounipolar neurons** are usually sensory. The cell body gives rise to a single process (the axon) that then splits into two branches. One branch returns sensory information from the periphery (**peripheral branch**), whereas the other branch projects and conveys this information to the CNS (**central branch**).

2. **Bipolar: Bipolar neurons** are usually specialized sensory neurons. They can be found in the retina (see Chapter 8·VII·A) and olfactory epithelium (see Chapter 10·III·B), for example. Their cell body gives rise to two processes. One conveys sensory information from the periphery, and the other (the axon) travels to the CNS.

3. **Multipolar: Multipolar** neurons have a cell body that gives rise to a single axon and numerous dendritic branches. Most CNS neurons are multipolar. They can be further subcharacterized based on the size and complexity of their dendritic tree.

D. Neurons as integrators

Paramecium can integrate multiple sensory signals (e.g., mechanical, chemical, thermal) through their effect on V_m. For example, a noxious signal that depolarizes V_m and increases the tendency to turn might be ignored if an attractant signal indicating nearby food hyperpolarizes the membrane and overrides noxious signal input. A paramecium is not capable of conscious thought, yet it makes a decision that affects behavior based on the summed effect of multiple stimuli on V_m. The dendritic trees of higher cortical neurons receive tens of thousands of competing inputs. The likelihood that neuronal output (spiking) will be modified on the basis of these signals is similarly determined by their net effect on V_m.

1. **Incoming signals:** Neurons hand off information to each other via dendrites. When a presynaptic neuron fires, it releases transmitter into the synaptic cleft. If the neuron is excitatory, transmitter binding to the postsynaptic dendritic membrane causes a transient depolarization known as an **excitatory postsynaptic potential** (**EPSP**), as shown in Figure 5.5A. Inhibitory neurons release transmitters that cause transient hyperpolarizations known as **inhibitory postsynaptic potentials** (**IPSPs**). EPSP and IPSP amplitudes are graded with incoming signal(s) strength.

2. **Filtering:** Much of the information being received by neurons at their dendrites represents sensory "noise." Isolating the strongest and most relevant signals is accomplished using a noise filter that takes advantage of a dendrite's natural electrical properties. Postsynaptic potentials (PSPs) are passive responses that degrade rapidly as they travel toward the cell body (see Fig. 2.11). Degradation is enhanced by a dendrite's inherent electrical leakiness and its lack of myelin. In practice, this means that a small PSP may never reach the cell body. PSPs generated by strong presynaptic activity activate voltage-gated ion currents along the length of the dendrite (see Fig. 2.12). These enhance the signals, thereby increasing their likelihood of reaching the cell body.

3. **Integration:** Signal integration also begins at the dendritic level. PSPs may meet and combine with PSPs arriving from other synapses as they travel toward the soma. This phenomenon is known as **summation** and is reminiscent of the way in which waves (e.g., sound waves and ripples spreading across a pond surface) interfere constructively and destructively. There are two types of summation: **spatial** and **temporal**.

   a. **Spatial summation:** If EPSPs from two different dendrites collide, they combine to create a larger EPSP (see Fig. 5.5B). This is known as spatial summation and applies to IPSPs also. EPSPs and IPSPs can also summate to yield an attenuated membrane response (see Fig. 5.5C).

   b. **Temporal summation:** Two EPSPs (or IPSPs) traveling along a dendrite in rapid succession can also combine to produce a single, larger event. This is known as temporal summation (see Fig. 5.5D).

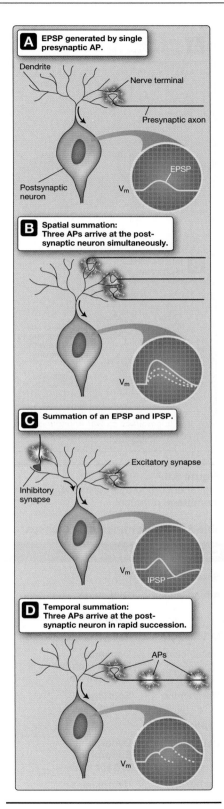

Figure 5.5.
Summation. EPSP and IPSP = excitatory and inhibitory postsynaptic potentials, respectively; V_m = membrane potential.

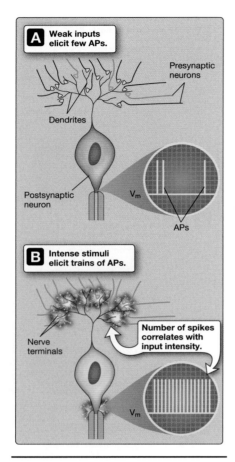

Figure 5.6.
Digital encoding by neurons. V_m = membrane potential.

Table 5.1: Neurotransmitter Classes

Class	Name
Small Molecule	
Amino acid	Glutamate GABA Glycine
Cholinergic	Acetylcholine
Catecholamine	Dopamine Norepinephrine Epinephrine
Monoamine	Serotonin Histamine
Peptides	
Opioid	Dynorphins Endorphins Enkephalins
Tachykinin	Neurokinins Substance P
Enteric	Gastrin-releasing peptide
Others	
Gas	Nitric oxide Carbon monoxide
Purine	ATP

4. **Output:** The net effect of multiple PSPs on V_m determines the likelihood and intensity of neuronal output. If a depolarization is sufficiently strong, it may elicit a train of spikes. Spikes arise from the initial segment (also known as the **spike initiation zone**) and travel down the length of the axon toward the presynaptic terminal.

5. **Encoding:** Because APs are all-or-nothing events, neurons must pass on information about signal strength using digital encoding. Weak stimuli may yield one or two spikes. Strong stimuli elicit spike trains (volleys) that travel in rapid succession down the axon's length. There is tremendous variability in the size, shape, and frequency of spikes generated by different neurons. As a general rule, the number of spikes in a volley reflects incoming stimulus strength (Fig. 5.6).

IV. NEUROTRANSMISSION

Neurons communicate with each other at synapses, specialized regions where two cells come into close apposition with each other. Communication typically occurs chemically via neurotransmitter release and is unidirectional. Although an inherently slow form of communication, placing a neurotransmitter receptor in the signaling pathway allows for a great diversity of responsiveness and unlimited opportunities for regulation.

A. Neurotransmitters

There are two main classes of neurotransmitters: small-molecule transmitters and peptides. A third, lesser group includes gases and other unconventional transmitters such as ATP (Table 5.1). Many tens of neuroactive peptides have also been described, many of which are coreleased along with a small-molecule transmitter.

B. Synaptic vesicles

Neurotransmitters are released into the synaptic cleft from synaptic vesicles. The vesicles are synthesized in the presynaptic cell body and then shuttled by fast axonal transport to the nerve terminal. Here, they are filled with locally produced neurotransmitter for storage and eventual release. Mature vesicles then dock at specialized release sites on the presynaptic membrane and remain there, awaiting the arrival of an AP.

 Peptide transmitters are synthesized and prepackaged in vesicles within the cell body rather than in the nerve terminal.

C. Release

Neurotransmitter release occurs when an AP arrives at the nerve terminal and opens voltage-dependent Ca^{2+} channels in the nerve terminal membrane (Fig. 5.7). Ca^{2+} influx raises local Ca^{2+} concentrations and initiates a Ca^{2+}-dependent secretory event. The details are complex and not fully resolved. The Ca^{2+} signal is sensed by a vesicle-associated Ca^{2+}-binding protein called **synaptotagmin**, which activates a **SNARE**-protein complex that includes **synaptobrevin**, **syntaxin**, and **SNAP-25**. The vesicle then fuses with the presynaptic membrane at an **active zone**, and the contents are emptied into the synaptic cleft. Each vesicle releases a single **quantum** of neurotransmitter ("**quantal signaling**").

> Botulinum toxin and tetanus toxin, two of the most lethal known neurotoxins, both paralyze their victims by targeting the synaptic terminal and disrupting neurotransmitter release. Both toxins degrade SNAPs and SNAREs by virtue of their intrinsic protease activity.

D. Receptors

Once released, a neurotransmitter diffuses across the narrow synaptic cleft and binds to a neurotransmitter receptor expressed on the postsynaptic membrane. Receptors are associated with numerous proteins that anchor them and that regulate activity and expression levels, manifesting as a **postsynaptic density** in micrographs (see Fig. 5.7). Receptors can be classed in at least two ways.

1. **Ionotropic versus metabotropic: Ionotropic receptors** are ion channels that mediate ion fluxes when active (Fig. 5.8). The nicotinic acetylcholine receptor (AChR) is an ionotropic receptor that mediates Na$^+$ influx, for example. **Metabotropic receptors** couple to an intracellular signaling pathway and are usually associated with a G protein. Examples include the muscarinic AChR.

2. **Excitatory versus inhibitory:** Excitatory receptors (e.g., the N-methyl-D-aspartate acid [NMDA] receptor) cause membrane depolarization and increased firing rates when occupied. Conversely, inhibitory receptors (e.g., the glycine receptor) hyperpolarize the membrane and decrease spike frequency.

    Properties of the main neurotransmitter receptor types are summarized in Table 5.2.

E. Signal termination

Signal termination can occur at the receptor level through receptor internalization or desensitization, but, more usually, signaling ends when the transmitter is removed from the synaptic cleft. A neurotransmitter typically suffers one of three fates: degradation, recycling, or diffusion out of the cleft ("spillover"; Table 5.3).

1. **Degradation:** The synaptic cleft usually contains high levels of enzymes that limit signaling by degrading neurotransmitter. For example, cholinergic synapses contain acetylcholinesterase, which degrades ACh.

2. **Recycling:** Many nerves and their support cells (glia) actively take up transmitter from the synaptic cleft and recycle it by repackaging it in synaptic vesicles.

3. **Diffusion:** Transmitter can also diffuse out of the synaptic cleft to affect neighboring neurons. For example, during sympathetic activation, norepinephrine (NE) appears in blood due to "NE spillover." This makes it possible to use plasma NE levels in the

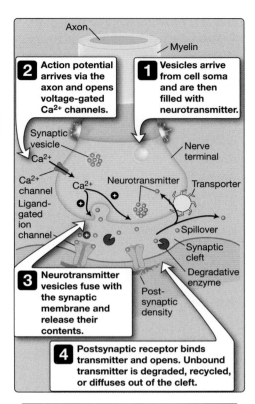

Figure 5.7.
Synaptic vesicle release.

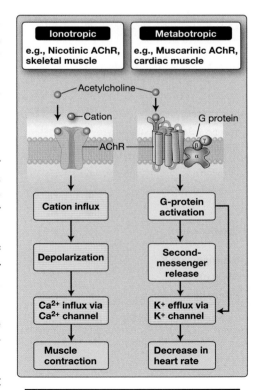

Figure 5.8.
Ionotropic versus metabotropic receptors.
AChR = acetylcholine receptor.

Table 5.2: Neurotransmitter Receptors

Receptor	Type		Transduction	Agonists	Antagonists	Localization
GluN	I		$\uparrow I_{Na}, I_{Ca}$	Glutamate, NMDA	Phencyclidine	CNS
GluA	I		$\uparrow I_{Na}$	Glutamate, AMPA		CNS
GluK	I		$\uparrow I_{Na}$	Glutamate, kainate		CNS
mGluR	M	Group I	$G_q, \uparrow IP_3$	Glutamate		CNS
mGluR	M	Group II, III	$G_i, \downarrow cAMP$	Glutamate		CNS
GABA_A	I		$\uparrow I_{Cl}$	GABA, ibotenate	Bicuculline	CNS
GABA_B	M		$G_i, \uparrow I_K$	GABA, baclofen		ANS
nAChR (nicotinic)	I	Muscle	$\uparrow I_{Na}$	ACh, nicotine	Pancuronium	Skeletal muscle
	I	Ganglion	$\uparrow I_{Na}$	ACh, nicotine	Trimethaphan	ANS
mAChR (muscarinic)	M	M_1	$G_q, \downarrow I_K$	ACh, muscarine	Atropine, diphenhydramine, ipratropium	ANS ganglia
	M	M_2	$G_i, \uparrow I_K \downarrow I_{Ca}$	ACh, muscarine		Heart
	M	M_3	G_q	ACh, pilocarpine		GI glands, eye
	M	M_4	$G_i, \uparrow I_K \downarrow I_{Ca}$	ACh		CNS
Dopamine	M	D_1	$G_s, \uparrow cAMP$	Dopamine		CNS
	M	D_2	$G_i, \downarrow cAMP$	Dopamine	Clozapine	CNS
Adrenergic	M	α_1	$G_q, \uparrow IP_3$	NE, phenylephrine	Prazosin	Vasculature
Epinephrine (Epi), norepinephrine (NE)	M	α_2	$G_i, \downarrow cAMP$	Epi, clonidine	Phentolamine	Heart, vasculature
	M	β_1	$G_s, \uparrow cAMP$	Epi, dobutamine	Propranolol, sotalol	Heart
	M	β_2	$G_s, \uparrow cAMP$	Epi, isoproterenol	Propranolol	Vasculature
Serotonin (5-hydroxytryptamine, or 5-HT)	I	$5-HT_3$	$\uparrow I_{Na}, I_{Ca}$	5-HT	Granisetron	CNS, GI
	M	$5-HT_1$	$G_i, \downarrow cAMP$	5-HT, triptans		CNS, vasculature
	M	$5-HT_2$	$G_q, \uparrow IP_3$	5-HT, lysergic acid	Clozapine	CNS, GI, smooth muscle, vasculature
	M	$5-HT_4$	$G_s, \uparrow cAMP$	5-HT		CNS, GI
	M	$5-HT_5$	$G_i, \downarrow cAMP$	5-HT		CNS
	M	$5-HT_6$	$G_s, \uparrow cAMP$	5-HT		CNS
	M	$5-HT_7$	$G_s, \uparrow cAMP$	5-HT		CNS, ANS
Histamine	M	H_1	$G_q, \uparrow IP_3$	Histamine	Diphenhydramine	CNS, airways, vasculature
	M	H_2	$G_s, \uparrow cAMP$	Histamine	Ranitidine	Heart, stomach
	M	H_3	$G_i, \downarrow cAMP$	Histamine	Ciproxifan	CNS
	M	H_4	$G_i, \downarrow cAMP$	Histamine		Mast cells
Substance P	M	NK1	$G_q, \uparrow IP_3$	Substance P		CNS, pain fibers
Neuropeptide Y	M	Y_{1-2}, Y_{4-5}	$G_i, \downarrow cAMP$	Neuropeptide Y		CNS
Nitric oxide		GC	$\uparrow cGMP$	Nitric oxide		CNS, ANS
Purinergic	I	P2X	$\uparrow I_{Na}, I_{Ca}$	ATP	Suramin	Widespread

ACh = acetylcholine; AMPA = α-amino-3-hydroxy-5-methyl-4-isoxazolepropionic acid; ANS = autonomic nervous system; CNS = central nervous system; G_i, G_q, G_s = G proteins; GC = guanylyl cyclase; GI = gastrointestinal; I = ionotropic; I_{Ca} = Ca^{2+} current; I_{Cl} = Cl^- current; I_K = K^+ current; I_{Na} = Na^+ current; M = metabotropic; NK1 = neurokinin 1; P2X = purinoreceptor.

Table 5.3: Mechanisms of Signal Termination

Neurotransmitter	Fate
Glutamate	Reuptake by glutamate transporters in neurons and glia
GABA	Reuptake by neuronal GABA reuptake transporters and by glia
Acetylcholine	Degraded in synaptic cleft by acetylcholinesterase
Dopamine, norepinephrine, serotonin	Degraded in cleft by catechol-O-methyltransferase, reuptake by Na^+/Cl^--dependent transporters and recycled, or degraded by monoamine oxidases
Histamine	Degraded by histamine methyltransferase and histaminases
Substance P	Internalization of transmitter–receptor complex
Nitric oxide	Oxidized
ATP	Degraded

Table 5.4: Glial Cells and Their Functions

Glial Cell Type	Localization	Function
Peripheral Nervous System		
Schwann cells	Axons	Myelination, phagocytosis
Satellite cells	Ganglia	Regulate chemical environment
Enteric Nervous System		
Enteric glial cells	Ganglia	Various
Central Nervous System ([CNS] Astrocytes)		
Protoplasmic	Gray matter	Nutrient delivery, blood–brain barrier function
Fibrous	White matter	Repair
Müller cells	Retina	Repair
Bergmann glia	Cerebellum	Synaptic plasticity
Oligodendrocytes	White matter (some in gray)	Myelination
Microglia	Throughout CNS	Responses to trauma
Ependymal cells	Ventricles	Regulate exchange between cerebrospinal fluid and brain extracellular fluid

cardiac vein to assess the degree of SNS activation in patients with heart failure. Higher NE levels indicate impaired cardiac function (see Unit IV).

V. NEUROGLIA

Glia (or **neuroglia**) are inexcitable cells that support many aspects of neuronal function. In addition to forming and maintaining myelin, they control local ion concentrations, help recycle neurotransmitters, and provide neurons with nutrients. They are found throughout the PNS and CNS (Table 5.4), where neurons and glia are present in equal numbers.

A. Myelination

Myelin is formed by **Schwann cells** (in the PNS) and **oligodendrocytes** (in the CNS). Oligodendrocytes may simultaneously myelinate the axons of multiple neurons, but Schwann cells remain dedicated to a single axon. Glia form myelin by extending processes that rotate around an axon over 100 times (Fig. 5.9). Cytoplasm is squeezed out from between the membrane layers as the myelin builds, so that the lipid layers become highly compacted. Glial cells remain viable after the layering is complete, with the nucleus and residual cytoplasm occupying the outermost layer.

B. Potassium homeostasis

Renormalization of V_m after neuronal excitation involves K^+ release (see Fig. 5.3). During intense neuronal activity, extracellular K^+ concentration can rise significantly as a result of this release. Because V_m is sensitive to the transmembrane K^+ gradient (see Chapter 2·II·C), K^+ buildup can be detrimental to neuronal function. **Astrocytes** (the

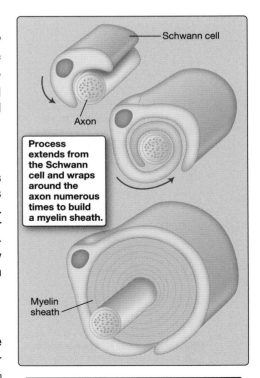

Process extends from the Schwann cell and wraps around the axon numerous times to build a myelin sheath.

Figure 5.9.
Myelin sheath formation.

Clinical Application 5.2: Multiple Sclerosis

Myelin is essential for normal neural communication, so diseases that affect myelin or the cells that produce it have devastating physiologic effects. **Multiple sclerosis (MS)** is a demyelinating disease of central nervous system neurons, and the cause is unknown. Symptoms arise when autoreactive immune cells produce antibodies against one or more sheath components. The myelin swells and degrades, and axonal conduction is interrupted, producing pathologic changes that can be visualized using computed tomography. Patients can present with a variety of neurologic symptoms, including tremors, visual disturbances, autonomic dysfunction, weakness, and fatigue. Some patients present with a relapsing form of the disease, characterized by acute onset of clinical symptoms, followed by a period of remission with full or partial recovery of function. In others, the disease is progressive, and neurologic damage steadily accumulates over a period of 10–20 years. Although immunosuppressive therapies may be helpful in halting symptoms progression during acute episodes, there is no known cure for MS.

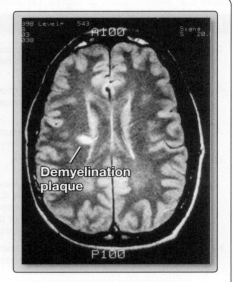

Periventricular frontal lobe demyelination.

predominant glial cell type in the CNS) are inexcitable, but they do possess K⁺ channels and K⁺ transporters that allow them to siphon K⁺ away from active neurons and redistribute it over nonactive regions of the CNS (Fig. 5.10). "**Spatial buffering**" takes advantage of the fact that adjacent astrocytes are tightly coupled via gap junctions (see Chapter 4·II·F) that provide pathways for K⁺ to flow down its concentration gradient from an active zone to a remote site.

C. Neurotransmitter uptake and recycling

Glutamate, γ-aminobutyric acid (GABA), and glycine are widely used CNS neurotransmitters. Neurons that communicate using these

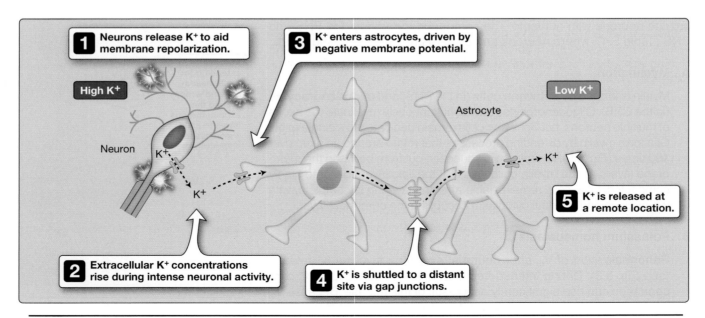

Figure 5.10.
Spatial buffering.

amino acids may spike at such high frequencies that their ability to control synaptic transmitter levels and prevent spillover into adjacent regions is challenged. Intense activity also strains the transmitter synthetic pathways. Glial cells assist with both issues. They extend foot processes that surround the synapse and rapidly take up transmitter from the cleft using high-affinity transport systems (Fig. 5.11). Glutamate is subsequently converted to glutamine by glutamine synthetase and returned to the presynaptic terminal for conversion back to glutamate by glutaminase. Glutamine is also distributed to inhibitory neurons to be used in GABA synthesis. GABAergic neurons have very limited glutamine reserves and are dependent on glia to supply the substrates necessary for continued signaling.

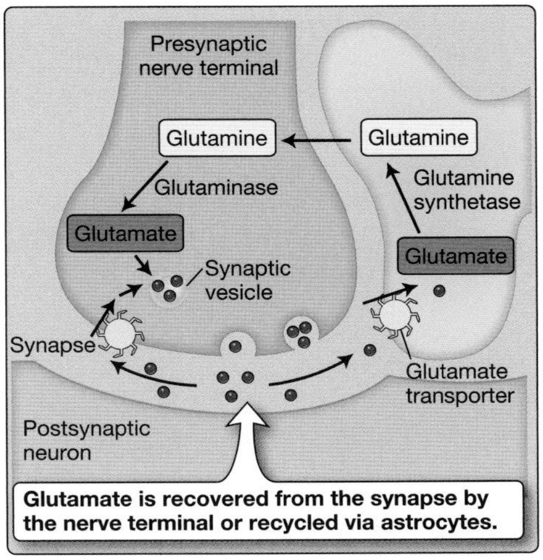

D. Nutrient supply

Neurons are highly dependent on O_2 and glucose for continued activity, the nervous system accounting for 20% of total body usage. Glia play a unique role in ensuring that these needs are met.

1. **Lactate shuttle:** Astrocytes ferry glucose from the vasculature to neurons (see Chapter 20·II). They extend foot processes that surround cerebral capillaries and absorb glucose from blood using transporters. Glucose then diffuses through the glial network via gap junctions. Some glucose is converted to glycogen, and the rest is metabolized to lactic acid. Lactate is then excreted into the extracellular fluid for uptake by surrounding neurons, a process known as the **lactate shuttle**.

2. **Storage:** Neurons have few energy reserves. They rely on astrocytes to maintain a steady lactate supply in the face of changes in neuronal activity or waning blood glucose levels. Astrocytes contain extensive glycogen stores and the necessary pathways to convert them to lactate when needed.

Figure 5.11.
Neurotransmitter recycling by astrocytes.

VI. NERVES

The terms **neuron** and **nerve** are often confused. A *neuron* is a single excitable cell. A *nerve* is a bundle of **nerve fibers** (axons and their supporting cells) that runs through the periphery like a modern telecommunications cable.

A. Conduction velocity

The characteristics of the individual fibers that make up a nerve vary considerably. Some may be thin, unmyelinated, and slow to conduct. Others are thick and myelinated and conduct impulses at high speed. Thick fibers take up more space than thin ones and are expensive to sustain metabolically. They are used only where speed of communication is paramount. In practice, that means the fastest fibers are used for motor reflexes (see Fig. 5.4; also see Table 11.1).

B. Assembly

Individual nerve fibers are wrapped loosely in connective tissue (**endoneurium**), and then several are bundled together to form a **fascicle** (Fig. 5.12). A fascicle is wrapped in yet more connective tissue (**perineurium**), and, finally, several fascicles are gathered together with

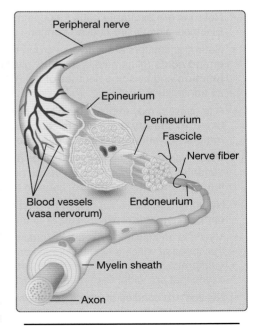

Figure 5.12.
Peripheral nerve anatomy.

blood vessels to form a **nerve**. The nerve is heavily protected with a dense layer of connective tissue (**epineurium**). Peripheral nerves are subjected to considerable mechanical stress associated with locomotion, so the multiple reinforcing layers are essential for protection.

C. Ganglia

The cell bodies of axons that make up a nerve are clustered in swellings called **ganglia**. Ganglia are the sites of connectivity and information relay between neurons. Ganglia may also contain intrinsic circuits that allow for reflex arcs and signal processing.

D. Types

Nerves can be classified according to the type of information that they carry. **Afferent nerves** contain fibers relaying sensory information from the internal and external environment to the CNS. **Efferent nerves** originate in the CNS and contain **somatic** motor neurons that innervate skeletal muscles and autonomic motor neurons. Some peripheral nerves contain a mix of both sensory and motor fibers (**mixed nerves**). This can result in reflex arcs that travel within the same nerve. The vagus nerve, for example, informs the CNS when food is entering the stomach. The CNS responds with a motor command that makes the stomach relax, and it travels via an efferent fiber contained within the vagus nerve (see Chapter 30·IV·B). The arc is called a **vagovagal reflex**.

Chapter Summary

- The **nervous system** comprises the **central nervous system** (**CNS**) and the **peripheral nervous system** (**PNS**). The **autonomic nervous system** (**ANS**) is a functional subdivision of the CNS and PNS. The CNS includes **brain** and **spinal cord** neurons. The PNS includes nerves that carry **sensory** and **motor information** between the CNS and the periphery. The ANS monitors and controls internal organ function.

- Neurons are **excitable cells** that communicate with each other and with target organs using **action potentials** (**APs**). Electrical signals are relayed from one cell to the next at a **synapse** using **chemical neurotransmitters**.

- Neurotransmitters are a diverse group that includes gases and polypeptides. The CNS uses small molecules such as **amino acids** (glutamate, aspartate, glycine, and GABA) as transmitters. **Acetylcholine** is used at the **neuromuscular junction** and by some neurons of the ANS. The ANS and CNS also use **monoamines** (dopamine and norepinephrine) as transmitters.

- A transmitter may excite or inhibit the postsynaptic cell, depending on the nature of the receptor on the postsynaptic membrane. **Excitatory transmitters** cause postsynaptic cell **depolarization** (**excitatory postsynaptic potential**), whereas **inhibitory transmitters** cause **hyperpolarization** and reduced excitability (**inhibitory postsynaptic potential**).

- CNS neurons may receive thousands of synaptic inputs via their dendritic trees. The electrical properties of neuronal dendrites ensure that weak inputs do not propagate. Stronger inputs may **summate** to push membrane potential beyond the **threshold for excitation**, causing the neuron to fire an AP.

- Neuronal function is supported by **glial cells**. Glia may regulate extracellular ion concentrations, aid in uptake and **recycling** of neurotransmitters, supply nutrients, and encase the axons in **myelin**.

- Myelin is an insulating material laid down by **Schwann cells** (PNS) and **oligodendrocytes** (CNS). Myelin increases the speed at which electrical signals propagate down an axon. Myelin is formed from compacted layers of glial surface membrane. Myelinated neurons are used in **reflex arcs** where response timing is critical. **Motor neurons** also have axons that are wider than normal to further increase signal propagation rates.

- **Nerves** are bundles of axons and their supporting cells. Peripheral nerves may convey afferent signals from sensory cells to the CNS, efferent signals from the CNS to effector cells, or a signal mixture.

Study Questions

Choose the ONE best answer.

5.1. Autoimmune diseases such as multiple sclerosis cause neurologic impairment by affecting axon conduction velocity. Which of the following would most likely slow axonal signal propagation to the greatest extent?

A. Decreasing depolarization rate
B. Decreasing leak-channel density
C. Increasing axon diameter
D. Increasing axon length
E. Increasing myelin thickness

Best answer = A. Axonal conduction velocity is dependent on the rate of membrane depolarization during an action potential, which, in turn, is a function of channel gating kinetics (see Section III·B). Increasing, not decreasing, leak-channel density (B) might also be expected to slow axonal conduction velocity. Conduction velocity would be reduced by decreasing (not increasing) axon diameter (C) or reducing (not increasing) myelin thickness (E), which would increase current loss via leak channels. Conduction velocity is independent of axonal length (D).

5.2. The characteristics of postsynaptic potentials (PSPs) allow neurons of the central nervous system to filter out significant incoming signals from background noise on the neural net. Which of the following most likely describes a characteristic of signaling using PSPs?

A. A large PSP may initiate a dendritic Na^+-spike.
B. Inhibitory PSPs do not spatially summate.
C. Inhibitory PSPs take the membrane closer to threshold.
D. PSPs are action potentials.
E. PSPs are graded events.

Best answer = E. Excitatory and inhibitory PSPs are graded events, with the amplitude reflecting incoming signal strength (see Section III·D). Dendrites are extensions of the neuronal cell body and do not support Na^+ action potentials (A), although some may support Ca^{2+}-mediated action potentials. Excitatory and inhibitory PSPs both summate spatially (B) and temporally. Inhibitory PSPs hyperpolarize the membrane and thereby take membrane potential further away from threshold (C). PSPs are graded events, not all-or-nothing action potentials (D).

5.3. Clinical observations of individuals who have misused phencyclidine suggest that the N-methyl-D-aspartate (NMDA) receptor may be involved in the early development of schizophrenia. Which of the following most likely describes the properties of this receptor type?

A. It is a metabotropic glutamate receptor.
B. It is permeable to both Na^+ and Ca^{2+}.
C. Its native ligand is acetylcholine.
D. Its receptor has intrinsic kinase activity.
E. Receptor binding is transduced by a G protein.

Best answer = B. The NMDA receptor is an ionotropic glutamate receptor permeable to both Na^+ and Ca^{2+} (see Section V·D). Receptor activation causes membrane depolarization. Metabotropic receptors are not normally permeable to ions (A). They signal using G-proteins. NMDA is an L-glutamate analog, and NMDA receptors are normally activated by L-glutamate in vivo. Acetylcholine (ACh) normally binds to nicotinic or muscarinic ACh receptors (C). Receptors that have intrinsic kinase activity include insulin receptors (D). NMDA receptor binding is transduced by membrane depolarization, not a G protein (E).

5.4. Epilepsy is a common neurologic disorder characterized by spontaneous episodic neuronal firing and seizures. Research indicates that a glial spatial buffering dysfunction may be involved. Spatial buffering's role most likely includes which of the following?

A. Increasing axonal conduction velocity
B. Limiting K^+ buildup and nerve hyperexcitability
C. Preventing acidification of brain extracellular fluid
D. Synaptic neurotransmitter recycling
E. Transferring nutrients from blood to neurons

Best answer = B. Glia take up K^+ from the neuronal interstitium and transfer it via gap junctions to adjacent cells for disposal at a remote site (or to the circulation; see Section V·B). Neural function is highly sensitive to local K^+ concentrations, and buildup could cause hyperexcitability and inappropriate spiking activity. Axonal conduction velocity (A) is enhanced by myelination, which is also a glial function (see Section V·A). Spatial buffering does not normally play a major role in pH balance (C). Glia also participate in neurotransmitter recycling (D) by synaptic uptake and return to neurons (see Section V·C), but this is not a spatial buffering function. Nutrient transfer via glial cells (E) is referred to as the "lactate shuttle" which is unrelated to spatial buffering (see Section V·D).

6 Central Nervous System

I. OVERVIEW

The central nervous system (CNS) comprises the spinal cord and brain (Fig. 6.1). The **spinal cord** is a thick communications tract that relays sensory and motor signals between the peripheral nervous system (PNS) and the brain. The cord also contains intrinsic circuits that support certain muscle reflexes. The **brain** is a highly sophisticated data processor containing neural circuits that analyze sensory data and then execute appropriate responses via the spinal cord and PNS efferents. Large portions of the brain are devoted to associative functions that integrate information from the various senses and allow us to assign meaning to sounds, associate smells with specific memories, and recognize objects and faces, for example. Associative regions also provide for abstract thinking, language skills, social interactions, and learning and memory. The human body has bilateral symmetry, and the structures of the spinal cord and brain are, for the most part, mirrored on each side of a midline. Sensory and motor information generally crosses the midline at some point in its journey between the brain and the periphery. In practice, this means that the left side of the brain controls the right side of the body and vice versa. For the purposes of discussion, the brain can be divided into four principal areas: **brainstem**, **cerebellum**, **diencephalon**, and **cerebral hemispheres** (**telencephalon**). A full discussion of CNS function is beyond the scope of this book, which focuses on the sensory and motor aspects of CNS function. For more information on higher brain functions, see *Lippincott Illustrated Reviews: Neuroscience, 2e.*

II. SPINAL CORD

The spinal cord lies within the vertebral canal. It extends from the foramen magnum at the base of the skull caudally to the second lumbar vertebra.

A. Segments

The vertebral column consists of a series of stacked vertebrae divided anatomically into five regions: **cervical**, **thoracic**, **lumbar**, **sacral**, and **coccygeal**. The cervical, thoracic, and lumbar vertebrae are separated by intervertebral disks that allow the bones to articulate. The sacral and coccygeal vertebrae are fused to form the **sacrum** and **coccyx**, respectively. The spinal cord can be divided into 31 named segments. Pairs of spinal nerves (one on each side of the body) emerge from corresponding segments (see Fig. 6.1).

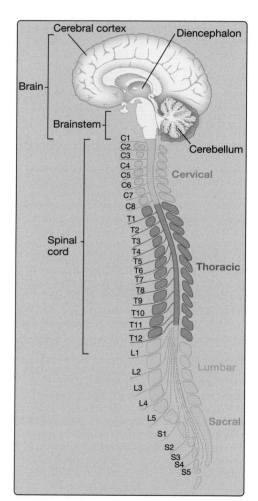

Figure 6.1.
Central nervous system. C1–C8, T1–T12, L1–L5, and S1–S5 are spinal nerves.

Although the spinal cord terminates before it reaches the sacrum, spinal nerves continue caudally within the vertebral canal until they reach an appropriate exit level.

> **Rostral** and **caudal** are anatomical terms meaning "beak" (or mouth) and "tail," respectively. They are commonly used to indicate direction of information flow in the CNS.

B. Nerves

Spinal nerves are a component of the PNS. The nerves contain **sensory afferent** and **motor efferent** fibers (i.e., they are **mixed nerves**) that generally serve tissues on the same level as the nerves. Thus, nerves emerging from the cervical region (C2) control head and neck movements, whereas sacral nerves (S2 and S3) project to the bladder and large intestine.

1. **Sensory:** Somatic and autonomic sensory fibers travel to the spinal cord via peripheral nerves (Fig. 6.2). They relay sensations of pain, temperature, and touch from the skin; proprioceptive signals from muscle and joint receptors; and sensory signals from numerous visceral receptors. Multiple peripheral nerves come together to form the **posterior root** of a spinal nerve and enter the vertebral canal via an **intervertebral foramen**. The cell bodies of these nerves cluster within a prominent **spinal ganglion** located within the foramen. The posterior root then divides into a number of **rootlets** and joins the spinal cord. Sensory nerves travel rostrally to synapse within nuclei en route to the brain. Branches of sensory afferents may also synapse directly with motor neurons or on interneurons that synapse with motor neurons, which makes local spinal cord–mediated reflexes possible (see Chapter 11·III).

2. **Motor:** Motor efferents from the brain travel caudally and synapse with peripheral motor nerves within the spinal cord. These nerves include both somatic and autonomic motor efferents. They leave the spinal cord via **anterior rootlets**, which join to form an **anterior root** and then travel out to the periphery alongside sensory fibers in spinal nerves.

C. Tracts

The spinal cord interior is roughly organized into a butterfly-shaped central area of gray matter surrounded by white matter (Fig. 6.3). The white matter contains bundles of nerve fibers with common origins and destinations that relay information between the PNS and the brain. Sensory nerve fibers from the periphery travel rostrally to the brain in discrete **ascending tracts**. **Descending tracts** carry bundles of motor efferents from the CNS en route to the periphery. The tracts are named according to their origin and destination. For example, the **spinothalamic tract** carries pain fibers from the spine upward to the thalamus. The **corticospinal tract** carries motor fibers from the cortex downward to the spine. The tracts (also known as **fasciculi**) are

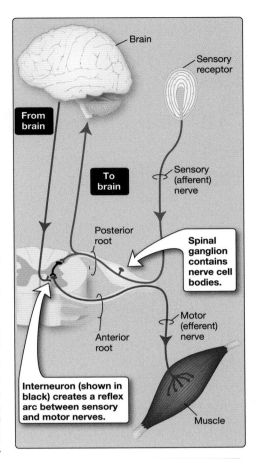

Figure 6.2.
Sensory and motor pathways.

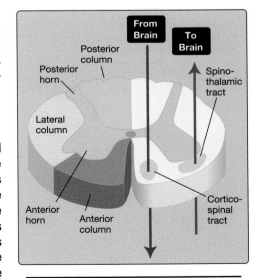

Figure 6.3.
Spinal cord organization.

grouped in **posterior**, **lateral**, and **anterior columns** (also known as **funiculi**). The "wings" of the gray butterflies are divided into **posterior** and **anterior horns** and act as synaptic relay stations for information flow between neurons. They contain neuronal cell bodies, which may be clustered in functionally related groups, or **nuclei**. The gray matter on either side of the cord is connected by **commissures** containing bundles of fibers that allow for information flow across the midline.

> CNS tissue typically appears white or gray. White matter gets its color from myelin, being largely composed of myelinated nerve axons. Gray matter is composed of cell bodies, dendrites, and unmyelinated axons.

III. BRAINSTEM

All sensory and motor information flowing to and from the brain passes through the brainstem (Fig. 6.4). The brainstem contains several important nuclei that act as relay stations for information flow between brain and periphery. Many of the 12 **cranial nerves** (**CNs**) originate from nuclei within the brainstem also (Fig. 6.5). The CNs provide sensory and motor innervation to the head and neck and include nerves that mediate vision, hearing, smell, and taste, among many other functions. Intrinsic circuits within the brainstem create control centers that allow for reflex responses to sensory data. The location and functions of these centers are discussed in more detail in Chapter 7. The brainstem can be subdivided anatomically into three areas:

- **Medulla:** The medulla contains autonomic nuclei involved in the control of respiration and blood pressure and in coordination of swallowing, vomiting, coughing, and sneezing reflexes.
- **Pons:** The pons helps control respiration and is involved in audition.
- **Midbrain:** The midbrain contains areas involved in controlling eye movements.

IV. CEREBELLUM

The cerebellum fine-tunes motor controls and facilitates smooth execution of learned motor sequences (see Chapter 11·IV·C). Cerebellar function requires massive integrative and computational capabilities, which is why this small area comprising ~10% of total brain mass contains more neurons than the rest of the brain combined! The cerebellum is attached to the brainstem by three **peduncles** that contain thick afferent and efferent nerve fiber bundles. The cerebellum receives sensory data from muscles, tendons, joints, skin, and the visual and vestibular systems and inputs from all regions of the CNS involved in motor control. It also sends signals back to most of these areas and modifies their output (Fig. 6.6). Integration of sensory data with motor commands is achieved using feedback and feedforward circuits that include the **Purkinje cell**, a neuronal type renowned for its immense dendritic tree.

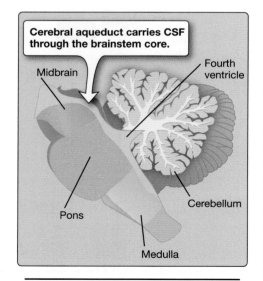

Cerebral aqueduct carries CSF through the brainstem core.

Midbrain

Fourth ventricle

Pons

Cerebellum

Medulla

Figure 6.4.
Brainstem organization.

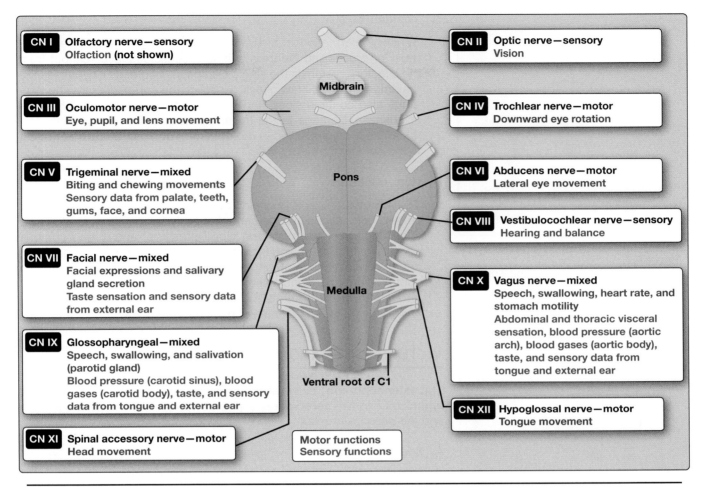

Figure 6.5.
Cranial nerve (CN) functions. C1 = first cervical vertebra.

The dendrites are sites of information flow from hundreds of thousands of presynaptic neurons. The cerebellar circuits allow movements to be finessed with reference to incoming sensory data, even as they are being executed.

Cerebellar injury does not cause paralysis, but it does have profound motor effects (**ataxia**, or an inability to coordinate muscle activity). Patients with cerebellar damage walk with a staggering gait that mimics alcohol intoxication. They may also have slurred speech and difficulties with swallowing and eye movement.

V. DIENCEPHALON

The **diencephalon** and **telencephalon** together make up the forebrain. The diencephalon contains two major structures: the **thalamus** and **hypothalamus** (Fig. 6.7).

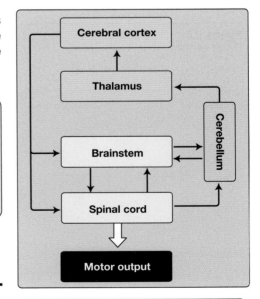

Figure 6.6.
Functional relationships among central nervous system components.

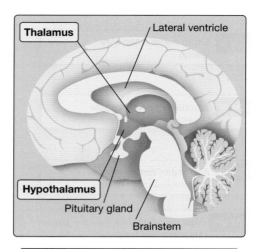

Figure 6.7.
Thalamus and hypothalamus locations.

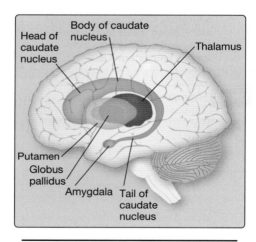

Figure 6.8.
Basal ganglia.

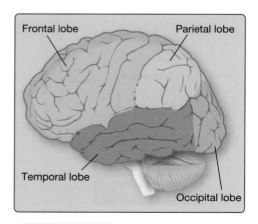

Figure 6.9.
Cortical lobes.

A. Thalamus

Sensory information from the periphery passes through the **thalamus** for processing before reaching a conscious level. Output from the olfactory system is the single exception, insofar as it bypasses the thalamus and feeds raw olfactory data to the cortex directly. The thalamus also controls sleep and wakefulness and is required for consciousness. Damage to the thalamus can result in deep coma. The thalamus is also involved in motor control and has areas that project to the cortical motor regions.

B. Hypothalamus

The hypothalamus is a major autonomic nervous system control center. Its functions include control of body temperature, food intake, thirst and water balance, and blood pressure, and it also controls aggression and rage. The hypothalamus exerts control through direct neural connections to autonomic centers in the brainstem, but it also controls the endocrine system. Endocrine control occurs directly through hormonal synthesis and release and indirectly by secreting hormones that affect release of pituitary hormones.

VI. TELENCEPHALON

The **telencephalon**, or **cerebrum**, is the seat of human intellect. It is organized into two cerebral hemispheres comprising the **basal ganglia** and the **cerebral cortex**.

A. Basal ganglia

The basal ganglia are a group of functionally related nuclei (Fig. 6.8) that work closely with the cerebral cortex and thalamus to effect motor control. Their function is discussed at length in Chapter 11. Major structures within the basal ganglia include the **caudate nucleus** and **putamen** (together forming the **striatum**) and the **globus pallidus**.

B. Cerebral cortex

The cerebral cortex is involved in conscious thought, awareness, language, and learning and memory.

1. **Anatomy:** The cortex comprises a sheet of neural tissue organized in six layers that is folded to accommodate the 15 to 20 billion ($1.5–2.0 \times 10^{10}$) neurons contained within. The folds (**gyri**) are separated by **sulci** (grooves). Deep **fissures** separate the cortex into four lobes: frontal, parietal, occipital, and temporal (Fig. 6.9). The lobes contain discrete areas that can be distinguished on a cytoarchitectural basis and that correlate with regions of specialized function.

2. **Function:** The cortex can be functionally divided into three general areas that stretch across both hemispheres: **sensory**, **motor**, and **associative**.

   a. **Sensory:** Sensory regions process information from the sensory organs (see Chapters 8–10). **Primary sensory regions** receive and process information directly from the thalamus.

Spatial information is preserved as data flows from the senses to the sensory areas and then accurately maps onto the cortex (**topographic mapping**). Thus, the pattern of light falling on the retina is faithfully replicated in the pattern of excitation within the primary visual cortex.

b. **Motor:** Motor areas are involved with planning and executing motor commands. **Primary motor areas** execute movements. Axons from these areas project to the spinal cord, where they synapse with and excite motor neurons. **Supplementary motor areas** are involved with planning and fine control of such movements (see Chapter 11).

c. **Associative:** Most cortical neurons are involved in associative functions. Each cortical sensory region feeds information to a corresponding association area. Here, patterns of color, light, and shade are recognized as a human face, for example, or a series of notes can be recognized as coming from a songbird. Other associative areas integrate sensory information from other parts of the brain to allow for higher mental functions. These include abstract thinking, acquisition of language, musical and mathematical skills, and the ability to engage in social interactions.

VII. CEREBROSPINAL FLUID

Because the CNS has a central role in all aspects of life, its neurons are provided with multiple layers of protection and support.

A. Protective layers

The role of the blood–brain barrier in protecting CNS neurons against bloodborne chemicals is discussed in Chapter 20. The CNS is also enclosed within five protective layers, including three membranes (**meninges**); a layer of cerebrospinal fluid (CSF); and an outer layer of bone, the cranium (Fig. 6.10). The meninges comprise the **pia mater**, the **arachnoid mater**, and the **dura mater**.

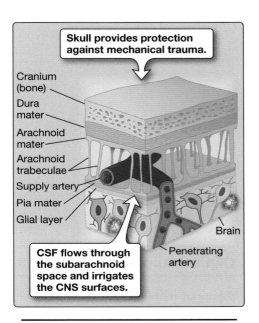

Figure 6.10.
Layers that protect the brain and provide a pathway for CSF flow.

Biologic Sex and Aging 6.1: Central Nervous System

Males and females show differences in the organization of CNS areas related to reproductive physiology and behavior. CNS sexual differentiation begins during the first trimester with an androgen surge in males. Surges also occur during the last trimester and at birth. The result is that specific areas of the brain are reorganized through selective loss of neurons, dendrite density, and synaptic plasticity. For example, females show neuronal population differences in areas of the hypothalamus controlling hormonal changes during the menstrual cycle.

Changes in cognitive function are one of the more recognizable effects of aging, although this may be more commonly a reflection of underlying pathology (such as dementia) rather than healthy physiologic aging. Significant changes in CNS function are apparent after age 65 years and accelerate after age 70 years. Brain volume decreases by 70 cm^3 per decade after age 65 years; most of this loss occurring in the frontal and temporal lobes. Among surviving neurons, the complexity of dendritic trees decreases, along with synaptic density. Numerous changes in neurotransmitter and receptor levels occur throughout the CNS. Cerebral blood flow decreases by up to 20%, which impacts brain O_2 and glucose metabolism. The principal effects on cognitive function are a slow decline in problem-solving abilities (executive functions), learning abilities, attention span, and the ability to retrieve long-term memory, especially as related to word finding.

1. **Pia mater:** The entire surface of the brain and spinal cord is tightly adhered to a thin, fibrous membrane called the **pia mater** (Latin for "dutiful mother"). The pia's cerebral portion is held in place by a continuous layer of astrocyte foot processes.

2. **Arachnoid mater:** The arachnoid mater comprises a layered epithelial membrane loosely connected to the pia by **trabeculae**, small structural supports that give the arachnoid mater a cobwebbed appearance (*arachnoid* comes from the Greek for "spider"). The trabeculae create a **subarachnoid space** through which CSF flows unhindered over the brain's surface. The layer of CSF has multiple functions (see later discussion), including cushioning the brain against trauma.

3. **Dura mater:** The "tough mother" is a thick, leathery membrane comprising two layers. An inner, **meningeal** layer is firmly attached to the arachnoid mater and covers the entire surface of both brain and spinal cord. The cranium is lined by a **periosteal** layer. The two layers separate in places to create an **intracranial venous sinus** that drains blood and CSF from the brain and channels it to the circulation.

B. Functions

CSF is a highly purified, sterile, colorless fluid nearly devoid of proteins that surrounds and bathes the tissues of the CNS. It has four main functions: providing buoyancy, absorbing shock, permitting limited intracranial volume changes, and maintaining homeostasis.

Clinical Application 6.1: Bacterial Meningitis

Bacterial meningitis is a life-threatening disease caused by bacterial infection of the CSF and meningeal inflammation.[1] It is a leading cause of infectious death worldwide. The most common community-acquired causes of meningitis are *Streptococcus pneumoniae* (~70%) and *Neisseria meningitidis* (12%), whereas hospital-acquired cases are usually caused by *Staphylococcus*. Infection is caused by bacteria crossing the blood–brain barrier and establishing colonies in the CSF, affecting both the brain and spinal cord. Symptoms are usually rapid in onset and include a triad comprising a severe headache, nuchal (neck) rigidity, and a change in mental status. Most patients also present with a high fever. Nuchal rigidity is caused by pain and muscle spasm when attempting to flex or turn the head, reflecting meningeal inflammation in the cervical region. Immediate treatment to reduce swelling and address the infection usually leads to full recovery.

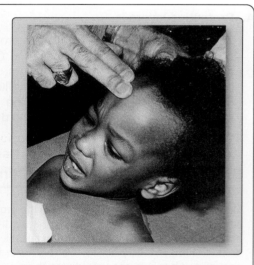

Patient lifts her shoulders rather than flexing her neck when her head is elevated (nuchal rigidity).

---

 [1]For a more complete discussion of bacterial meningitis, see *LIR Microbiology*, 4e, Chapter 11·III·D·1 and Chapter 32 *Neisseria* species.

1. **Buoyancy:** The brain has a relatively high specific gravity compared with CSF, but submerging the brain in CSF reduces its effective weight from ~1500 g to ~50 g. This helps prevent cerebral tissues from being compressed by gravity against the skull. Compression impedes blood flow through the cerebral vasculature and causes ischemia.

2. **Shock absorption:** CSF surrounds the brain on all sides and envelops it in a liquid cushion. Cushioning reduces the chance of mechanical trauma to the brain when the skull is impacted or impacts an object at speed.

3. **Volume changes:** During periods of intense activity, neurons and glia tend to swell due to accumulation of metabolites and other osmotically active materials. CSF allows water to shift from CSF to cells, thus avoiding any gross change in CNS volume. Because the CNS is constrained by bone on all sides, volume changes may compress the cerebral vasculature and cause ischemia (see Chapter 20·II·D).

4. **Homeostasis:** Cell membrane potential (V_m) and neuronal excitability is highly sensitive to changes in extracellular K^+ concentration. Plasma K^+ concentrations can rise by >40% even under normal conditions (normal plasma K^+ concentration = 3.5–5.0 mmol/L), changes that are unacceptable for a V_m-dependent organ such as the CNS. CSF K^+ concentrations are tightly maintained at a relatively low level (2.8–3.2 mmol/L), thereby insulating neurons from large swings in plasma concentration. CSF is also free of potentially neuroactive compounds (such as glutamate and glycine) that constantly circulate in blood. CSF thus provides the CNS with a stable, rarified extracellular environment that is renewed constantly to prevent buildup of neuronal waste products, transmitters, and ions.

C. Choroid plexus

CSF is formed by the **choroid plexus**, a specialized epithelium that lines four fluid-filled **ventricles** within the brain's core (Fig. 6.11).

1. **Ventricles:** The brain contains four ventricles connected by **foramina** that allow CSF to flow caudally to the spinal cord and through its central canal. Two **lateral ventricles** are the largest of the four. They are symmetrical C shapes and are located at the center of the two cerebral hemispheres. They connect with the **third ventricle** via two interventricular channels called the **foramina of Monro.** The third ventricle lies on the midline at the level of the thalamus and hypothalamus. It connects with the **fourth ventricle** via the **cerebral aqueduct (of Sylvius).** The fourth ventricle is located within the brainstem. The caudal end communicates with the spinal cord's central canal. The ventricle also provides a pathway for CSF to flow into the subarachnoid space via three openings. The **foramen of Magendie** is located at the midline. Two **foramina of Luschka** are located laterally.

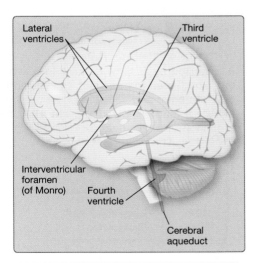

Figure 6.11.
Location of the CSF-filled ventricles and cerebral aqueduct.

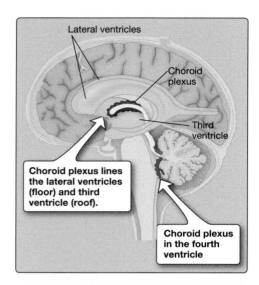

Figure 6.12.
Choroid plexuses.

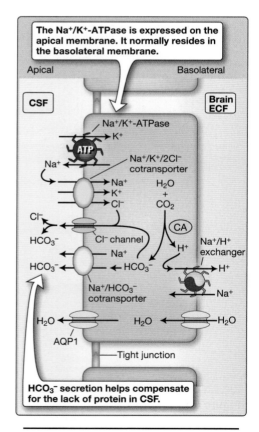

Figure 6.13.
Cerebrospinal fluid formation. CA = carbonic anhydrase.

2. **Location:** The choroid plexuses are localized to specific regions of the ventricles (Fig. 6.12). They line the floor of the lateral ventricles and continue through the interventricular channels to line the roof of the third ventricle. In the fourth ventricle, choroid plexus occupies a small portion of the roof.

3. **Structure:** The ventricles and the spinal cord's central canal are lined with **ependymal epithelium**. In the region of choroid plexus, the ependyma gives way to a ciliated **choroid epithelium**, which is responsible for secreting CSF. Choroid epithelial cells contain large numbers of mitochondria, and their apical surfaces are amplified by microvilli, features that are characteristic of an epithelium specialized for high-capacity ion and water transport. The epithelium rests on a basal lamina, which separates it from the vasculature below, and adjacent cells are all coupled by tight junctions. Choroid epithelial activity is supported by a **vascular plexus** comprising a dense network of arteries, capillaries, and veins. The capillaries are large and leaky, and their walls contain fenestrations to facilitate fluid filtration from blood.

D. **Cerebrospinal fluid formation**

About 30% of total CSF production can be attributed to the brain parenchyma. The remaining 70% is produced by the choroid plexus.

1. **Composition:** The basal side of the choroid epithelium is bathed in plasma filtrate, but tight junctions between adjacent epithelial cells create an effective barrier to exchange of ions and other solutes between blood and CSF. The differences between CSF and plasma are notable in several respects (Table 6.1):

   - CSF contains next to no proteins, making it highly reliant on HCO_3^- for pH buffering.
   - Na^+ and Cl^- levels are higher, which compensates osmotically for the lack of protein.
   - K^+ concentration is lower.

   The necessary gradients for CSF formation are established at the apical (luminal) surface of the choroid epithelium (Fig. 6.13).

Table 6.1: Plasma and Cerebrospinal Fluid Composition

Solute	Plasma	CSF
Na^+	140	149
K^+	4	3
Ca^{2+}	2.5	1.2
Mg^{2+}	1	1.1
Cl^-	100	125
Glucose	5	3
Proteins (g/dL)	7	0.03
pH	7.4	7.3

Values are approximate and represent free concentrations under normal metabolic conditions. All values (except protein concentration and pH) are given in mmol/L.

2. **Cations:** The choroid epithelium is highly unusual in that the Na^+/K^+-ATPase is located on the apical membrane rather than the basolateral membrane. The pump drives Na^+ into the CSF. The Na^+/K^+-ATPase simultaneously removes K^+ from CSF. More K^+ may be absorbed by an apical $Na^+/K^+/2Cl^-$ cotransporter, using the energy of the Na^+ gradient that favors Na^+ reentry into the epithelial cell.

3. **Anions:** HCO_3^- is generated by carbonic anhydrase activity. For every HCO_3^- molecule generated, a H^+ is liberated also. This is released to the vasculature via a Na^+/H^+ exchanger in the basolateral membrane. HCO_3^- is likely secreted into the ventricle via anion (Cl^-) channels and Na^+/HCO_3^- cotransporters. Cl^- is concentrated within the cells by anion exchangers in the basolateral membrane and then flows across the apical membrane via Cl^- channels.

4. **Water:** Water follows an osmotic gradient generated by secretion of Na^+, HCO_3^-, and Cl^-. Aquaporins (AQP1) provide a pathway for movement.

E. Flow

CSF is produced at prodigious rates (~500 mL/d), flushing the ventricles and CNS surfaces once every 7 to 8 hours. The choroid plexuses have a mass of only ~2 g. Their ability to generate so much CSF is made possible both by blood flow that is higher than what most other tissues receive (10 times that supplying neurons) and by the enhanced surface area for secretion created by the villi and microvilli. High CSF flow rates ensure that byproducts of neuronal activity (inorganic ions, acids, and transmitters spilling over from synapses) are removed in a timely manner before they can build to levels that might interfere with CNS function.

1. **Pathways:** CSF secretion by the choroid plexus increases intraventricular pressure by a few millimeters of H_2O, sufficient to drive CSF flow through the ventricles, through the foramina in the fourth ventricle, and into the subarachnoid space (Fig. 6.14). CSF then percolates through the space and flows over the CNS surfaces, eventually joining venous blood contained within the intracranial sinus. CSF enters the sinuses via **arachnoid villi**, which may be organized into large clumps called **arachnoid granulations**. CSF is transported across the villi in giant vesicles, creating a one-way valve that prevents backflow from sinus to subarachnoid space if CSF pressure drops.

2. **Exchange between extracellular fluids:** CSF and brain extracellular fluid (ECF) are separated in the ventricles by ependymal cells and in other regions by the pia and a supporting layer of astrocytic foot processes. Although the pia and astrocyte layers are continuous, the junctions between adjacent cells are leaky, allowing free exchange of materials between CSF and ECF. This allows neuronal and glial waste products to diffuse out of the ECF and be carried away by the CSF.

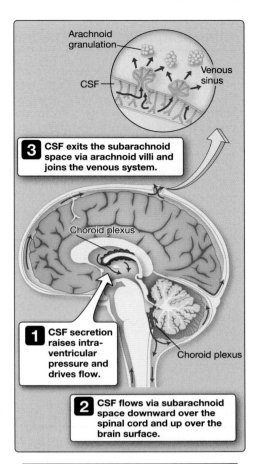

Figure 6.14.
Pathways for CSF flow over brain and spinal cord surfaces.

Clinical Application 6.2: Lumbar Puncture

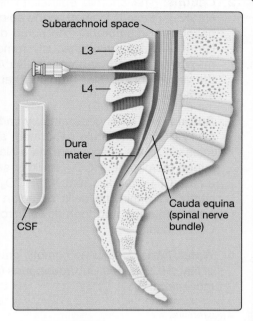

Cerebrospinal fluid (CSF) pressure is normally within a range of 60–200 mm H_2O (~4.5–14.7 mm Hg), but it can rise dramatically when the subarachnoid villi become clogged with bacteria or blood cells. **Lumbar puncture (LP)** offers an opportunity both to measure CSF pressure ("opening pressure") and retrieve fluid samples to test for the presence of white or red blood cells, which might indicate bacterial meningitis or subarachnoid hemorrhage, respectively. LP involves inserting a long, thin (spinal) needle through the dura mater into the subarachnoid space. Fluid is withdrawn from the subarachnoid space in the lumbar region, which is below where the spinal cord terminates. Up to 40 mL of CSF can be withdrawn safely for cytologic analysis and culturing. The risk of complications following LP is relatively low, with headache and backache being the most common. Headache is caused by CSF leakage from the puncture site, although the etiology of the pain is uncertain. Headache is typically mild and resolves spontaneously within 24 hours.

Lumbar puncture.

Chapter Summary

- The **central nervous system** (**CNS**) comprises the **spinal cord** and **brain.** The spinal cord contains bundles of nerve fibers organized into **tracts** that relay information between the **peripheral nervous system** (**PNS**) and the brain. **Ascending tracts** relay **sensory** information from the PNS to the brain, whereas **descending tracts** convey **motor** commands to the PNS. The spinal cord also contains intrinsic circuits that facilitate local reflex arcs that do not require input from the brain.

- PNS neurons enter and leave the spinal cord via 31 pairs of **spinal nerves. Posterior roots** of these nerves contain **sensory afferent** fibers, whereas **anterior roots** contain **motor efferents.**

- All information flowing between the CNS and PNS must pass through the brainstem, which contains the **medulla**, **pons**, and **midbrain.** These areas contain **autonomic nuclei** involved in control of respiration, blood pressure, and upper gastrointestinal tract reflexes. The brainstem is associated with 10 **cranial nerves** that innervate the head and neck.

- The **cerebellum** facilitates fine motor control. It integrates sensory information from muscles, joints, and the visual and vestibular systems, and fine-tunes motor commands in anticipation of and during movements.

- The **diencephalon** comprises the **thalamus** and **hypothalamus.** The thalamus processes sensory information, whereas the hypothalamus is an autonomic nervous system control center.

- The **telencephalon** comprises the **basal ganglia**, which are involved in motor control, and the **cerebral cortex.** The cerebral cortex contains sensory, motor, and associative areas and is the seat of higher functioning.

- The CNS is protected by five layers, comprising the **pia mater**, a layer of **cerebrospinal fluid (CSF)**, **arachnoid mater**, **dura mater**, and **bone.**

- CSF is a colorless, protein-free fluid produced by the **choroid plexus**, a secretory epithelium located within the brain **ventricles.** CSF flows through the ventricles under pressure and then over the surface of both the brain and spinal cord. It drains into venous sinuses located within the dura mater.

- CSF also acts as a liquid cushion that protects the brain from mechanical trauma and helps distribute its weight evenly within the cranium. CSF is produced at relatively high rates, flushing the ventricles and CNS surfaces and carrying away accumulated waste products.

Study Questions

Choose the ONE best answer.

6.1. A 35-year-old male comes to the office reporting that he is unable to raise his left eyebrow and his mouth is drooping on the left side. Physical examination confirms unilateral facial palsy. The patient's condition is most likely due to infection affecting which of the following nerves?

A. Cranial nerve II
B. Cranial nerve III
C. Cranial nerve VII
D. Cranial nerve IX
E. Cranial nerve X

Best answer = C. The patient presents with facial nerve (cranial nerve [CN] VII) palsy (see Fig. 6.5). Also known as Bell palsy, the condition is commonly caused by an infection such as Lyme disease. CN II (A) is the optic nerve that conveys visual sensory information from the eye to the brain. Conditions affecting the optic nerve impair vision. CN III (B) is the oculomotor nerve, a motor nerve controlling the eye, pupil, and lens. CN IX (D) and CN X (E) are the glossopharyngeal and vagus nerves, respectively. These are mixed nerves with wide-ranging functions. Infections of either would not be restricted to facial muscle palsy.

6.2. A 55-year-old male is brought to the emergency department with neck stiffness and what he describes as "the worst headache of my life." Head CT reveals hemorrhage in the space between two of the brain's protective layers through which CSF normally flows and that contains an arterial blood supply. The site of hemorrhage is most likely between which of the following structures?

A. Arachnoid mater and pia mater
B. Meningeal dura and arachnoid mater
C. Periosteal and meningeal layers of the dura mater
D. Pia mater and brain surface
E. Skull and periosteal layer of the dura mater

Best answer = A. The patient presents with subarachnoid hemorrhage (see Fig. 6.10). The arachnoid mater comprises a layered epithelial membrane that extends arachnoid trabeculae. These are thin fibers that attach to the pia mater and give the subarachnoid space a spiderweb appearance. CSF normally flows through this space and irrigates the brain surface. Arterial supply vessels are also housed in this space. Rupture of these vessels because of trauma or aneurysm leads to subarachnoid hemorrhage. The meningeal dura is firmly attached to the arachnoid mater (B). The two layers of the dura (C) separate in places to create a sinus that drains CSF and venous blood. The pia mater is tightly adhered to the brain surface (D), and the skull and periosteal dura (E) are tightly adhered to one another.

6.3. Cerebrospinal fluid loss reduces buoyancy, allowing the brain to sag and triggering a "low CSF pressure headache." In addition to buoyancy, what additional protective feature does CSF most likely provide?

A. It acts as a shock absorber during mechanical trauma.
B. It contains mucin to lubricate the brain.
C. It drains along the olfactory nerve to moisturize the olfactory epithelium.
D. It forms a cohesive film between brain and cranium.
E. It is K^+ free to enhance neuronal K^+ efflux.

Best answer = A. Cerebrospinal fluid surrounds the brain on all sides and thereby envelops it in a liquid cushion (see Section VII·B). This helps lessen the chances of brain trauma when the skull impacts or is impacted by another object. CSF is protein free (i.e., no mucin; B). CSF drains into the venous system via an intracranial sinus, not the olfactory epithelium (C). CSF forms a protective cushion between brain and bone; it does not adhere the brain to the skull (D). CSF contains ~3 mmol/L K^+ (E), slightly less than that of plasma.

6.4. Cerebrospinal fluid is secreted by the heavily vascularized choroid plexus. CSF and cerebral blood come into close proximity but do not mix. Which of the following most likely describes the relationships between CSF and blood?

A. Aquaporins provide a pathway for water to move between blood and CSF.
B. Blood and CSF have a similar composition.
C. CSF is glucose-free.
D. Na^+ is pumped into CSF by an apical Na^+/H^+ exchanger.
E. Tight junctions between adjacent epithelial cells are leaky to facilitate exchange of materials between CSF and blood.

Best answer = A. Choroid plexus secretes CSF at a rate of ~500 mL/d. The water content of CSF is derived from blood and made possible by aquaporins in the basolateral and apical membranes of the choroid epithelium (see Section VII·D). Water moves from blood to CSF by osmosis. Unlike blood, CSF is acellular and contains scant proteins (B), but glucose is present at a concentration of around 3 mmol/L (C). Na^+ is pumped into CSF by an apical Na^+/K^+-ATPase, not an apical Na^+/H^+ exchanger (D). Tight junctions between adjacent choroid epithelial cells are impermeable to water and solutes, which prevents paracellular exchange of materials between blood and CSF (E).

7 Autonomic Nervous System

I. OVERVIEW

Cells erect a barrier around themselves (the plasma membrane) to create and maintain an internal environment optimized to suit their metabolic needs. Likewise, the body is covered with skin to establish an internal environment whose temperature, pH, and electrolyte levels are optimized for tissue function. Maintaining a stable internal environment (i.e., **homeostasis**) is the responsibility of the **autonomic nervous system** (**ANS**). The ANS is organized similarly to the somatic nervous system and uses many of the same neural pathways. Internal sensory receptors gather information about **blood pressure** (**baroreceptors**), **blood chemistry** (**chemoreceptors**), and **body temperature** (**thermoreceptors**) and relay it to autonomic control centers in the brain. The control centers contain neural circuits that compare incoming sensory data with internal preset values. If comparators detect any deviation from the presets, they adjust the function of one or more organs to maintain homeostasis. The principal organs of homeostasis include the skin, liver, lungs, heart, and kidneys (Fig. 7.1). The ANS modulates organ function via two distinct effector pathways: the **sympathetic nervous system** (**SNS**) and the **parasympathetic nervous system** (**PSNS**). The actions of the SNS and PSNS often appear antagonistic, but, in practice, they work in close cooperation with each other.

II. HOMEOSTASIS

The term "homeostasis" refers to a state of physiologic equilibrium or the processes that sustain such an equilibrium. An individual must maintain homeostatic control over numerous vital parameters to survive and thrive, including arterial P_{O_2}, blood pressure, and extracellular fluid osmolality (see Fig. 7.1). Losing control over one or more of these parameters manifests as illness and usually prompts a patient to seek medical attention. It is a clinician's task to identify the underlying cause of the imbalance and intervene to help restore homeostasis.

A. Mechanisms

Homeostatic control pathways are seen at both the cellular and organismal level, and they all include at least three basic components that typically form a negative feedback control system (Fig. 7.2). A sensory component (e.g., a receptor protein) detects and relays information about the parameter that is subject to

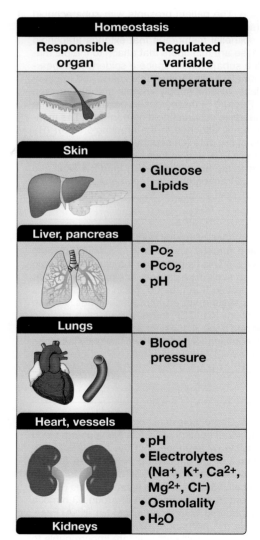

Homeostasis	
Responsible organ	**Regulated variable**
Skin	• Temperature
Liver, pancreas	• Glucose • Lipids
Lungs	• P_{O_2} • P_{CO_2} • pH
Heart, vessels	• Blood pressure
Kidneys	• pH • Electrolytes (Na^+, K^+, Ca^{2+}, Mg^{2+}, Cl^-) • Osmolality • H_2O

Figure 7.1.
Principal homeostatic organs.

homeostatic control, an integrator (e.g., a neural circuit) that compares incoming sensory data with a system preset value, and an effector component capable of changing the regulated variable (e.g., an ion pump or excretory organ). For example, a rise in arterial P_{CO_2} is sensed by chemoreceptors that feed information to a respiratory control center in the brainstem. The control center responds by increasing respiration rate to expel the excess CO_2. Conversely, a decrease in P_{CO_2} reduces respiration rate. Homeostasis may also involve a behavioral component. Behavior drives intake of salt (NaCl), water, and other nutrients and, for example, impels the individual to turn on air conditioning or shed clothing if body temperature is too high.

B. Redundancy

Homeostasis at the organismal level typically involves multiple layered and often hierarchical control pathways, with the number of layers reflecting the relative importance of the parameter under control. Blood pressure, for example, is controlled by numerous local and central regulatory pathways. Layering creates redundancy, but it also ensures that if one pathway fails, one or more redundant pathways can assert control to ensure continued homeostasis. Layering also allows a very fine degree of homeostatic control.

C. Functional reserve

Organ systems responsible for homeostasis typically have considerable **functional reserve**. For example, normal quiet breathing uses only ~10% of total lung capacity, and cardiac output at rest is ~20% of maximal attainable values. Reserves allow the lungs to maintain arterial P_{O_2} and the heart to maintain blood pressure at optimal levels even as body activity level and demand for O_2 and blood flow increases (e.g., during exercise). Functional reserve also allows for progressive decreases in functional capacity, such as occurs with age and disease (see Chapter 40·II·A).

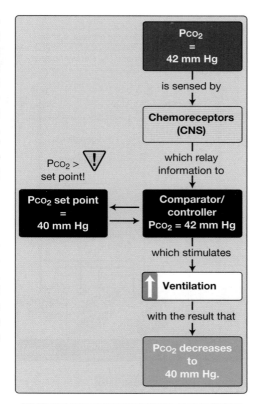

Figure 7.2.
Negative feedback control of P_{CO_2}.

III. ORGANIZATION

The ANS (also known as the **visceral nervous system** or **neurovegetative system**) is responsible for maintaining numerous vital parameters. Homeostasis must continue when we sleep or our conscious minds are focused on a task, so the ANS operates subconsciously and largely independently of voluntary control. Exceptions include the voluntary interruption of breathing to allow for talking, for example. The ANS is organized along similar principles as those of the somatic motor system. Sensory information is relayed via afferent nerves to the central nervous system (CNS) for processing. Adjustments to organ function are signaled via nerve efferents. The main differences between the two systems relate to efferent arm organization. The ANS employs a two-step pathway in which efferent signals are relayed through ganglia (Fig. 7.3).

A. Afferent pathways

ANS sensory afferents relay information from receptors that monitor many aspects of body function, including blood pressure

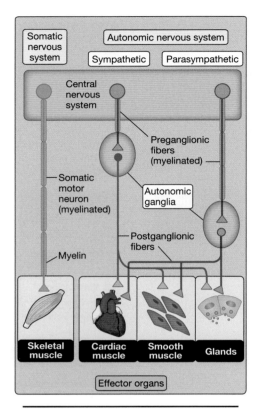

Figure 7.3.
Somatic and autonomic nervous system efferent pathways.

(baroreceptors); blood chemistry, that is, glucose levels, pH, P_{O_2}, and P_{CO_2} (chemoreceptors); skin temperature (thermoreceptors); and mechanical distension of the lungs, bladder, and gastrointestinal (GI) system (mechanoreceptors). Sensory afferent fibers often travel in the same nerves as do autonomic and somatic efferents. Autonomic nerves also contain nociceptive fibers, which provide for visceral pain sensation.

B. Efferent pathways

In the somatic motor system, motor neuron cell bodies originate within the CNS (see Fig. 7.3). In the ANS, the cell bodies of motor efferents are contained within ganglia, which lie outside the CNS (Fig. 7.4; also see Fig. 7.3).

1. **Autonomic ganglia:** Ganglia comprise clusters of nerve cell bodies and their dendritic trees. Commands originating in the CNS are carried to ganglia by myelinated **preganglionic neurons**. Unmyelinated **postganglionic neurons** relay the commands to the target tissues.

    a. **Sympathetic:** Sympathetic ganglia are located close to the spinal cord, so sympathetic preganglionic neurons are relatively short. Postganglionic neurons are relatively long, reflecting the distance between the ganglia and the target cells. There are two types of sympathetic ganglia. **Paravertebral ganglia** are arranged in two parallel **sympathetic chains** located to either side of the vertebral column. The ganglia within the chains are linked by neurons that run longitudinally, which allows signals to be relayed vertically within the chains as well as peripherally. **Prevertebral ganglia** are located in the abdominal cavity.

    b. **Parasympathetic: Parasympathetic ganglia** are located in the periphery near or within the target organ. Thus, parasympathetic preganglionic neurons are much longer than the postganglionic neurons.

2. **Sympathetic efferents:** The cell bodies of sympathetic preganglionic neurons are located in nuclei contained within upper regions of the spinal cord (T1–L3). Neurons located rostrally regulate the upper regions of the body, including the eye, whereas caudal neurons control the function of lower organs, such as the bladder and genitals. Preganglionic neurons leave the spinal cord via a ventral root, enter a nearby paravertebral ganglion, and then terminate in one of several possible locations:

    *   Within the paravertebral ganglion
    *   Within a more distant sympathetic chain ganglion
    *   Within a prevertebral ganglion, a more distal ganglion, or the adrenal medulla

3. **Parasympathetic efferents:** Preganglionic neurons of the PSNS originate in brainstem nuclei or in the sacral region of the spinal cord (S2–S4). Their axons leave the CNS via cranial or pelvic splanchnic nerves, respectively, and terminate within remote ganglia located close to or within the walls of their target organs.

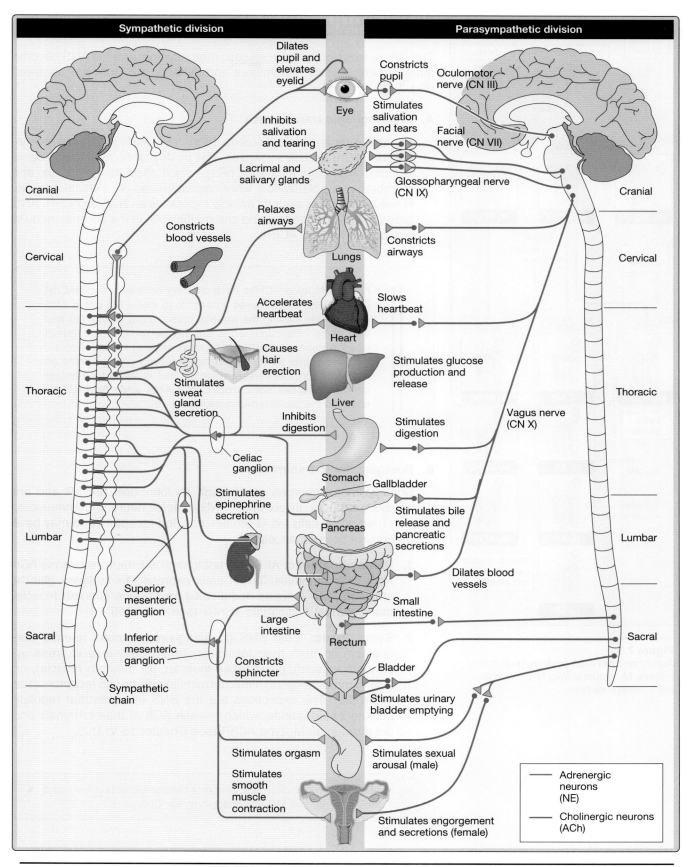

Figure 7.4.
Autonomic nervous system organization. NE = norepinephrine.

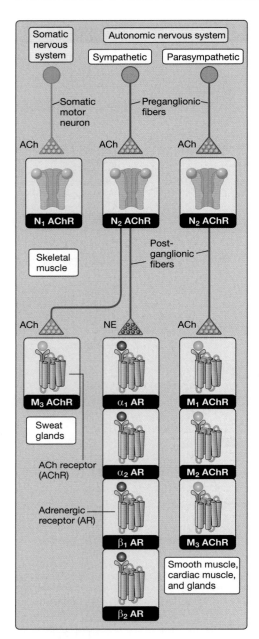

Figure 7.5.
Autonomic nervous system neurotransmitters. M = muscarinic, N = nicotinic, NE = norepinephrine.

IV. NEUROTRANSMISSION

The differences between the somatic motor system and the ANS become more apparent when transmitters and synaptic structure are reviewed (Fig. 7.5).

A. Preganglionic transmitters

All ANS preganglionic neurons (SNS and PSNS) release acetylcholine (ACh) at their synapses. The postsynaptic membrane bears nicotinic ACh receptors (nAChRs), which mediate Na^+ influx and membrane depolarization when activated, as in skeletal muscle. However, whereas skeletal muscle expresses an N_1-type AChR, ANS preganglionic cell bodies and chromaffin cells in the adrenal medulla express an N_2-type AChR.

> N_1- and N_2-type AChRs have different sensitivities to nAChR antagonists, which makes it possible to inhibit the entire ANS output while leaving the skeletal musculature unaffected and vice versa.[1] Pancuronium is an N_1-type receptor antagonist used in general anesthesia to relax skeletal muscle and aid intubation prior to surgery. It has relatively minor effects on ANS function. Conversely, trimethaphan is an N_2-type antagonist that blocks both arms of the ANS while having little effect on the skeletal musculature.

B. Postganglionic transmitters

Somatic motor neurons act through an ionotropic nAChR and are *always* excitatory. In contrast, ANS effector neurons communicate with their target cells via G protein–coupled receptors and may have an array of consequences.

1. **Parasympathetic:** All PSNS postganglionic neurons release ACh from their terminals. Target cells express M_1- (salivary glands, stomach), M_2- (cardiac nodal cells), or M_3-type (smooth muscle, many glands) muscarinic AChRs (see Table 5.2).

2. **Sympathetic:** Most SNS postganglionic neurons release norepinephrine from their terminals. Target cells may express α_1- (smooth muscle); β_1- (cardiac muscle); β_2- (smooth muscle); or, less commonly, α_2- (synaptic terminals) adrenergic receptors (see Table 5.2). The exceptions are the SNS efferents that regulate eccrine sweat glands, which release ACh at their terminals and act through an M_3-type AChR (see Chapter 15·VI·D·2).

 [1] For a more complete discussion of cholinergic antagonists and their actions, see *LIR Pharmacology*, 8e, Chapter 5.

C. Postganglionic synapses

Somatic motor nerves terminate at highly organized neuromuscular junctions. The site of synaptic contact between an ANS neuron and its target cell is very different. Many postganglionic nerve axons exhibit a string of beadlike varicosities (swellings) in the region of their target cells (Fig. 7.6). Each represents a site of transmitter synthesis, storage, and release, functioning as a nerve terminal.

V. EFFECTOR ORGANS

The somatic motor system innervates skeletal musculature. The ANS innervates all other organs. Most visceral organs are innervated by both arms of the ANS. Although the two divisions typically have opposite effects on organ function, they usually work in a complementary rather than antagonistic fashion. Thus, when sympathetic activity increases, output from the parasympathetic division is withdrawn and vice versa. The principal targets and effects of ANS control are summarized in Figures 7.1 and 7.4.

VI. BRAINSTEM

ANS output can be influenced by many higher brain regions, but the main areas involved in autonomic control include the brainstem, the hypothalamus, and the limbic system. The relationship between these areas is shown in Figure 7.7. The brainstem is the primary ANS control center and can maintain most autonomic functions for several years even after clinical brain death has occurred (see Chapter 40·II·C). The brainstem comprises nerve tracts and nuclei. The nerve tracts convey information between the CNS and the periphery. Nuclei are clusters of nerve cell bodies, many of which are involved in autonomic control.

A. Preganglionic nuclei

Preganglionic nuclei are the CNS equivalents of ganglia, comprising clusters of nerve cell bodies at the head of one or more cranial nerves (CNs). Nuclei usually also contain interneurons that create simple negative feedback circuits between afferent and efferent nerve activity. Such circuits mediate many autonomic reflexes, such as reflex slowing of heart rate when blood pressure is too high and receptive relaxation of the stomach when it fills with food (see Clinical Application 7.1). The brainstem contains several important PSNS preganglionic nuclei, including the **Edinger-Westphal nucleus**, **superior** and **inferior salivatory nuclei**, the **dorsal motor nucleus of vagus**, and the **nucleus ambiguus** (Fig. 7.8). The nucleus ambiguus contains both glossopharyngeal (CN IX) and vagal (CN X) efferents that innervate the pharynx, larynx, and part of the esophagus. The nucleus helps coordinate breathing

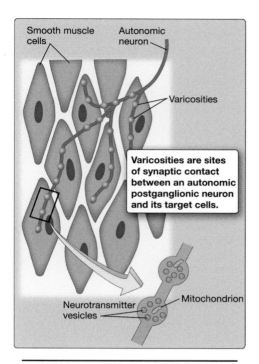

Figure 7.6.
Autonomic nerve varicosities.

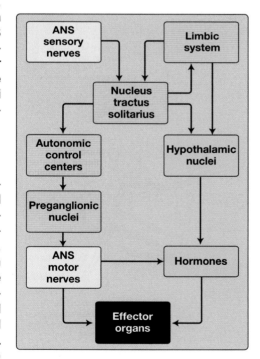

Figure 7.7.
Autonomic control centers.

Clinical Application 7.1: Autonomic Dysfunction

Disruption of autonomic pathways can result in specific functional deficits or more generalized loss of homeostatic function, depending on the nature of the underlying pathology. **Horner syndrome** is caused by disruption of the sympathetic pathway that raises the eyelid, controls pupil diameter, and regulates facial sweat gland activity. The result is unilateral **ptosis** (drooping eyelid), **miosis** (inability to increase pupil diameter), and local **anhidrosis** (inability to sweat). More generalized autonomic dysfunctions are common among patients on maintenance dialysis and those with diabetes whose glucose levels are poorly controlled (diabetic autonomic neuropathy [DAN]). DAN can manifest as an inability to control blood pressure following a meal (postprandial hypotension) or upon standing (postural hypotension), GI motility disorders (difficulty swallowing and constipation), or bladder dysfunction, among other symptoms.

Tests designed to assess autonomic function include monitoring cardiac responses during changes in posture (tilt-table test, used to assess baroreflexes), hand immersion in ice water (cold pressor test, designed to test sympathetic responses to cold-induced pain), and the Valsalva maneuver. The Valsalva maneuver involves forced expiration against a resistance, designed to cause intrathoracic pressures to rise to 40 mm Hg for 10 to 20 seconds. The pressure increase prevents venous blood from entering the thorax, so cardiac filling is impeded, and arterial pressure falls. In a healthy individual, a fall in arterial pressure is sensed by arterial baroreceptors, initiating a reflex increase in heart rate that is mediated by sympathetic efferents traveling in the vagus nerve. Patients with DAN may have impaired baroreceptor or vagal nerve function and fail to respond to a Valsalva maneuver with the expected tachycardia.

and swallowing reflexes, and it also contains vagal cardioinhibitory preganglionic fibers.

B. Nucleus tractus solitarius

The nucleus tractus solitarius is a nerve tract running the length of the medulla through the center of the solitary nucleus (see Fig. 7.8) that coordinates many autonomic functions and reflexes. It receives sensory data from most visceral regions via the vagus and glossopharyngeal nerves (CNs IX and X) and then relays this information to the hypothalamus. It also contains intrinsic circuits that facilitate local (brainstem) reflexes controlling respiration rate and blood pressure, for example.

C. Reticular formation

The reticular formation comprises a collection of brainstem nuclei with diverse functions, including control of blood pressure and respiration (as well as sleep, pain, motor control, etc.). It receives sensory data from the glossopharyngeal and vagal nerves and helps integrate it with effector commands from higher autonomic control centers located in the limbic system and hypothalamus.

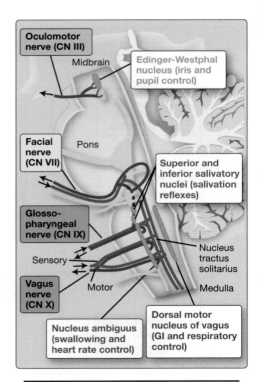

Figure 7.8.
Principal autonomic brainstem nuclei.

D. Control centers

Brainstem areas that have related functions are considered control centers, even if separated spatially. Brainstem control centers include the **respiratory center**, **cardiovascular control center**, and **micturition center** (Table 7.1).

VII. HYPOTHALAMUS

The hypothalamus establishes the set point for many internal parameters, including body temperature (37°C), mean arterial pressure (~95 mm Hg), and extracellular fluid osmolality (~290 mOsmol/kg H_2O). Its influence extends to virtually all of the body's internal systems, belying its tiny size (~4 cm^3, or ~0.3% of brain volume). The ability to establish a set point or narrow operating range minimally requires that the hypothalamus be provided with a way of monitoring the parameters it controls and a way of communicating with the organs that maintain them.

A. Organization

The hypothalamus is located below the thalamus at the base of the brain. It contains several distinct nuclei summarized in Figure 7.9. Note that although some of these nuclei have clearly defined functions, others are organized into functional groups or areas that work cooperatively. In addition to controlling autonomic functions, the hypothalamus can stimulate many behavioral responses, including those associated with sexual drive, hunger, and thirst.

Table 7.1: Brainstem Control Centers

Respiratory Center
Receives sensory information from chemoreceptors that monitor arterial Po_2, Pco_2, and pH
Controls respiration rate through outputs to the diaphragm and respiratory muscles (see Chapter 24·II)

Cardiovascular Center
Receives sensory information from peripheral baroreceptors and chemoreceptors
Controls blood pressure through modulation of cardiac output and vascular tone (see Chapter 19·III·B)

Micturition Center
Monitors urinary bladder distension.
Facilitates urinary bladder emptying by relaxing the urethral sphincter and contracting the bladder (see Chapter 25·VI·D)

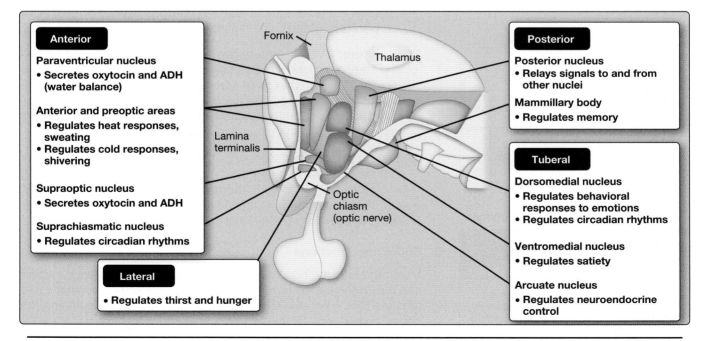

Figure 7.9.
Hypothalamic nuclei. ADH = antidiuretic hormone.

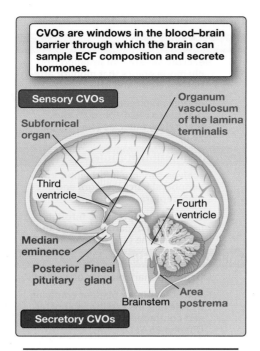

CVOs are windows in the blood–brain barrier through which the brain can sample ECF composition and secrete hormones.

Sensory CVOs

Subfornical organ

Organum vasculosum of the lamina terminalis

Third ventricle

Fourth ventricle

Median eminence

Posterior pituitary

Pineal gland

Brainstem

Area postrema

Secretory CVOs

Figure 7.10.
Circumventricular organs (CVOs).

Table 7.2: Posterior Pituitary Hormones

Pituitary Hormone Released	Pituitary Hormone Target (Effects)
Oxytocin	Uterus (contraction), mammary glands (lactation)
Antidiuretic hormone	Renal tubule (water reabsorption)

B. Neural pathways

The hypothalamus receives reciprocal innervation from many areas, as might be expected of such a key integrative organ. The major pathways for information flow occur between the hypothalamus and the brainstem as well as the hypothalamus and the limbic system (see Fig. 7.7).

C. Circumventricular organs

Any organ tasked with maintaining homeostasis must be able to monitor the parameters it controls. In the case of the hypothalamus, this includes circulating ions, metabolites, and hormones. The hypothalamus is a part of the brain, however, meaning that it is isolated from most such factors by the blood–brain barrier (BBB). Although it does receive feedback from peripheral receptors, they provide limited information. Therefore, the hypothalamus is provided with windows in the BBB through which it can make observations about blood composition directly. These windows are called **circumventricular organs** (**CVOs**).

1. **Location:** The brain contains six CVOs (Fig. 7.10). CVOs comprise specialized brain regions where the BBB is interrupted to allow cerebral neurons to interact with extracellular fluid and the circulation directly. Some CVOs are sensory, whereas others are secretory.

   a. **Sensory:** Sensory CVOs include the **subfornical organ** and the **organum vasculosum of the lamina terminalis**, both associated with the hypothalamus. The **area postrema** is a brainstem CVO.

   b. **Secretory:** Secretory CVOs include the **median eminence** (part of the hypothalamus), the **neurohypophysis** (**posterior pituitary gland**), and the **pineal gland**.

2. **Structure:** CVOs are interfaces between the brain and the periphery. They are highly vascularized, and blood flows through these regions slowly to maximize time available for exchanging materials between the blood and brain. Also, CVO capillaries are fenestrated and leaky, which facilitates movement of ions and smaller proteins between the blood and interstitium.

3. **Sensory functions:** Sensory CVOs contain neuronal cell bodies that are sensitive to numerous bloodborne factors (e.g., Na^+, Ca^{2+}, angiotensin II, antidiuretic hormone [ADH], natriuretic peptides, sex hormones, and feeding and satiety signals). Their axons project to hypothalamic areas that control corresponding variables.

D. Endocrine functions

Most organs in the body are dually regulated by the nervous system and the endocrine system. The hypothalamus's key homeostatic role requires that it be able to influence both systems. It modulates the neural component via nerve tracts and peripheral nerves. It exerts endocrine control using hormones (summarized in Tables 7.2 and 7.3) that are released via the pituitary gland.

Table 7.3: Anterior Pituitary Hormones

Hypothalamic Hormone	Pituitary Target Cell	Pituitary Hormone	Target Organ (Effects)
Corticotropin-releasing hormone	Corticotrope	Adrenocorticotropic hormone	Adrenal cortex (stress responses)
Thyrotropin-releasing hormone	Thyrotrope	Thyroid-stimulating hormone (TSH)	Thyroid gland (thyroxine release, metabolism)
Growth hormone–releasing hormone	Somatotrope	Growth hormone	Widespread (anabolic)
Somatostatin (release inhibitor)	Somatotrope	Growth hormone	Widespread
Somatostatin (release inhibitor)	Thyrotrope	TSH	Thyroid gland
Gonadotropin-releasing hormone	Gonadotrope	Luteinizing hormone	Gonads (androgen production)
Dopamine (release inhibitor)	Lactotrope	Prolactin	Mammary glands (milk production and letdown)
Gonadotropin-releasing hormone	Gonadotrope	Follicle-stimulating hormone	Gonads (follicle maturation, spermatogenesis)

1. **Endocrine axes:** The hypothalamus, pituitary, and a dependent endocrine gland together form a unified control system known as an **endocrine axis**. Most endocrine systems are organized into such axes. The advantage is that it allows for both fine and gross control of hormone output. For example, the **hypothalamic–pituitary–adrenal axis** regulates **cortisol** secretion from the **adrenal cortex** (see Fig. 34.5). The hypothalamus produces **corticotropin-releasing hormone** (**CRH**), which stimulates **adrenocorticotropic hormone** (**ACTH**) release from the anterior pituitary. ACTH stimulates cortisol production by the adrenal cortex. Cortisol exerts negative feedback control on both ACTH production by the anterior pituitary, and both ACTH and cortisol inhibit CRH synthesis by the hypothalamus.

2. **Pituitary gland:** The pituitary gland (also known as the **hypophysis**) projects from the hypothalamus at the base of the brain and nestles in a bony cavity called the **sella turcica** (Latin for "Turkish saddle"). The hypothalamus and pituitary are connected by the pituitary (or **hypophyseal**) stalk, which contains bundles of neurosecretory axons. The pituitary contains two lobes (Fig. 7.11). Although they lie next to each other within a common gland, they have very different embryologic origins and cellular compositions.

   a. **Anterior lobe:** The **anterior lobe** (**adenohypophysis**) has epithelial origins. It comprises a collection of glandular tissues that synthesize and store hormones (see Table 7.3). Hormone release is regulated by the hypothalamus using hormone-releasing or release-inhibiting hormones, which travel from the hypothalamus to the anterior pituitary via the **hypophyseal portal system**.

   b. **Hypophyseal portal system:** The hypophyseal portal system directs blood from the hypothalamus to the anterior lobe of the pituitary gland (see Fig. 7.11). This unusual serial vascular arrangement is used to carry peptide hormones synthesized by hypothalamic **parvocellular** (small cell) **neurosecretory cells** to the anterior pituitary, where they stimulate or inhibit pituitary hormone release.

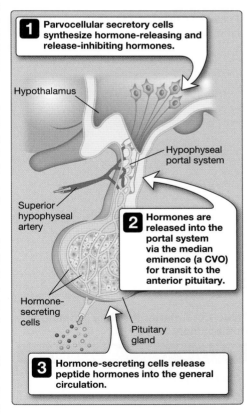

1 Parvocellular secretory cells synthesize hormone-releasing and release-inhibiting hormones.

Hypothalamus

Hypophyseal portal system

Superior hypophyseal artery

2 Hormones are released into the portal system via the median eminence (a CVO) for transit to the anterior pituitary.

Hormone-secreting cells

Pituitary gland

3 Hormone-secreting cells release peptide hormones into the general circulation.

Figure 7.11.
Anterior pituitary. CVO = circumventricular organ.

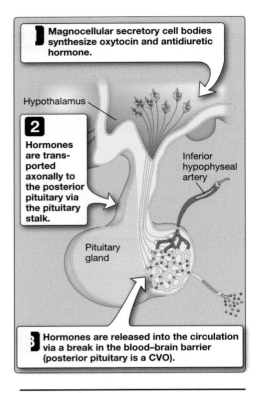

Figure 7.12.
Posterior pituitary. CVO = circumventricular organ.

Table 7.4: Anterior Pituitary Tropic Cell Composition

Cell Type	% of Total
Corticotropes	15–20
Gonadotropes	10
Lactotropes	15–20
Somatotropes	50
Thyrotropes	5

Hypothalamic hormones are synthesized in neurosecretory cell bodies and then transported down their axons to terminals located in the median eminence. The median eminence is a CVO that sits at the head of the pituitary stalk and its portal system. Given an appropriate stimulus, the hormones are released from the nerve terminals into the portal system and carried to the capillaries supplying the anterior lobe's hormone-secreting cells.

c. **Posterior lobe:** The **posterior lobe** (**neurohypophysis**) is neural tissue. Axons from **magnocellular** (large cell) **neurosecretory cells** in the supraoptic and paraventricular nuclei extend down the pituitary stalk and terminate within a CVO located in the posterior lobe (Fig. 7.12). Magnocellular cell bodies synthesize oxytocin (OT) and ADH, two related peptide hormones (see Table 7.2). The hormones are transported to the nerve terminals via the pituitary stalk and stored in secretory granules (**Herring bodies**) awaiting release. The posterior pituitary is highly vascular and the capillaries fenestrated. When peptides are released, they enter the general circulation directly.

3. **Anterior pituitary hormones:** The anterior pituitary comprises five endocrine cell types (Table 7.4). The hormones they produce can be placed in one of three structurally related groups.

a. **Adrenocorticotropic hormone: ACTH** (**corticotropin**) is synthesized by **corticotropes** as a preprohormone, that is, preproopiomelanocortin (prePOMC). Removing the signal sequence yields POMC, a 241–amino acid–residue peptide containing ACTH (39 amino acids), melanocyte-stimulating hormone (MSH), and β endorphin (an endogenous opioid). Corticotropes lack the enzymes necessary to generate MSH or β endorphin, however, so they release ACTH alone.

b. **Glycoprotein hormones: Thyroid-stimulating hormone** (**TSH**), **follicle-stimulating hormone** (**FSH**), and **luteinizing hormone** (**LH**) are related glycoproteins. All three hormones are heterodimers that share a common α subunit called **α-glycoprotein subunit** (**α-GSU**) and a hormone-specific β subunit. TSH is synthesized in **thyrotropes** and comprises an α-GSU–β-TSH dimer. FSH and LH are released by **gonadotropes** and comprise α-GSU–β-FSH and α-GSU–β-LH dimers, respectively. **Human chorionic gonadotropin** (**hCG**) is a related placental hormone comprising an α-GSU–β-hCG heterodimer.

c. **Growth hormone and prolactin: Growth hormone** (**GH**) and **prolactin** are related polypeptides synthesized by **somatotropes** and **lactotropes**, respectively. A related hormone, **human placental lactogen**, is synthesized by the fetal placenta. GH is a single-chain, 191 amino acid–residue polypeptide synthesized and released in several different isoforms. Prolactin, a 199 amino acid–residue polypeptide, is the only anterior pituitary hormone whose release is under tonic inhibition by the hypothalamus (via **dopamine**).

4. **Posterior pituitary hormones:** OT and ADH are near-identical nonapeptide hormones (they differ at only two amino acid positions) with a common evolutionary ancestor (Fig. 7.13). They are both synthesized as preprohormones that contain a signal peptide, the hormone, a **neurophysin**, and a glycoprotein. The signal peptide and glycoproteins are removed to form prohormones during processing and packaging in the Golgi apparatus. **Prooxyphysin** comprises OT and neurophysin I, whereas **propressophysin** comprises ADH and neurophysin II. The hormones are separated by proteolysis from their respective neurophysins after packaging in neurosecretory vesicles and fast axonal transport to the posterior pituitary. The neurophysins (and glycoproteins) are coreleased along with hormone but have no known physiologic function.

The structural similarities between OT and ADH cause some functional crossover when circulating hormone levels are sufficiently high. Thus, OT can have mild antidiuretic effects, whereas ADH can cause milk ejection in lactating women.

E. Clock functions

Most bodily functions, including body temperature, blood pressure, and digestion, have daily ("**circadian**," derived from the Latin *circa dies*) rhythms. All cells appear capable of generating such self-sustaining rhythms. The hypothalamus synchronizes these rhythms and entrains them to a circadian cycle established by a master clock. Entrainment allows the body's various physiologic functions to be modified to anticipate coming nightfall or daybreak and optimized to coincide with a sleep–wake cycle. The master clock is located in the **suprachiasmatic nucleus (SCN)**. It synchronizes bodily functions in part through the endocrine system, with the **pineal gland** acting as a neuroendocrine intermediary.

1. **Molecular clocks:** Although many cortical regions contain circuits that establish seasonal and other rhythms, the master clock responsible for circadian rhythms resides in the SCN (see Fig. 7.9). The molecular cogs that make the clock run comprise two sets of genes and are locked in a negative feedback control system (Fig. 7.14). The *CLOCK* and *BMAL1* gene products are transcription factors that heterodimerize and increase transcription of *PER* and *CRY* genes. The *PER* and *CRY* products dimerize and inhibit the enhancer action of the CLOCK–BMAL1 dimer. The transcription–translation cycle oscillates over a period of ~24 hours.

2. **Setting the time:** Although the master clock oscillates with an inherent periodicity of around 24 hours, the clock is reset daily to entrain it to the light–dark cycle. The clock is set by light falling on a small subset of intrinsically photosensitive retinal ganglion cells (~1%–3% of total). These cells express **melanopsin**, a photopigment that allows them to detect and respond to light. Signals from these cells reach the hypothalamus via afferents traveling in the **retinohypothalamic tract** of the optic nerve (Fig. 7.15).

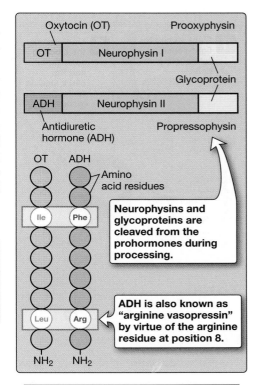

Figure 7.13.
Structural similarities between posterior pituitary hormones.

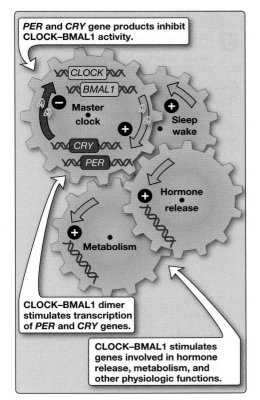

Figure 7.14.
Master clock.

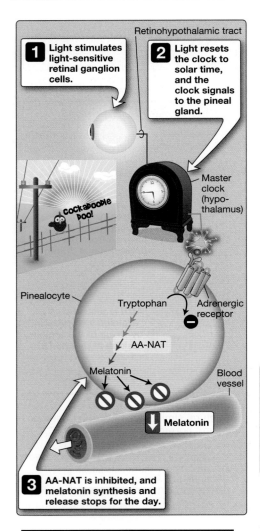

Figure 7.15.
Effects of light on melatonin release. AA-NAT = arylalkylamine *N*-acetyltransferase.

3. **Pineal gland:** The SCN synchronizes body functions in part through manipulation of endocrine axes using the pineal gland as an intermediary. The pineal gland is a small (~8 mm) pinecone-shaped (hence the name) gland located at the midline near the posterior wall of the third ventricle (see Fig. 7.10). It comprises **pinealocytes** and glial support cells that are similar to **pituicytes** (pituitary glial cells). The SCN communicates with the pineal gland via neural connections to the brainstem and spinal cord, and from there via sympathetic connections to the superior cervical ganglion and pineal gland. The pineal gland is a secretory CVO that allows melatonin to be released into the circulation directly.

4. **Melatonin:** Melatonin is an indoleamine (*N*-acetyl-5-methoxy-tryptamine) synthesized from tryptophan that regulates the sleep–wake cycle. The synthetic pathway includes arylalkylamine *N*-acetyltransferase (AA-NAT), which is regulated by the SCN via adrenergic inputs from the SNS. When light falls on the retina, sympathetic pathways from the SCN to the pineal gland are activated, and AA-NAT activity is inhibited (see Fig. 7.15). Melatonin synthesis and secretion fall as a result and do not resume until sundown (Fig. 7.16). Melatonin promotes sleep.

Individuals with **Smith-Magenis syndrome** (developmental disorder) have an inverted melatonin secretory response to light. Melatonin levels peak during the daytime and fall at night. These patients have neurobehavioral problems and sleep disturbances, underscoring melatonin's importance in timing CNS function.

VIII. LIMBIC SYSTEM

The limbic system comprises a collection of functionally related nuclei encircling the brainstem (**hippocampus**, **cingulate cortex**, and **anterior thalamic nuclei**) that strongly influence autonomic activity via connections to the hypothalamus. Many of these nuclei control emotions and motivational drives. These connections explain how emotions such as rage, aggression, fear, and stress can exert such profound physiologic effects. Everyone is familiar with the sensations associated with fright: a rapid, pounding heartbeat (increased heart rate and myocardial contractility); rapid breathing (respiratory center); cold, sweaty palms (sympathetic activation of eccrine sweat glands); and hairs standing erect on the back of the neck (piloerection).

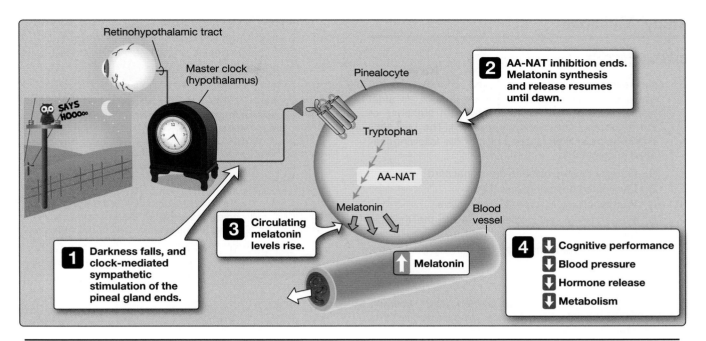

Figure 7.16.
Melatonin effects on physiologic function. AA-NAT = arylalkylamine *N*-acetyltransferase.

Chapter Summary

- The central nervous system (CNS) comprises the somatic nervous system and **autonomic nervous system** (**ANS**). The somatic nervous system controls skeletal musculature, whereas the ANS controls visceral organ function. The ANS's primary function is to maintain internal **homeostasis**.

- The ANS operates subconsciously and largely independently of voluntary control. The ANS incorporates two functionally distinct effector pathways (**sympathetic** and **parasympathetic**) that act cooperatively and in a reciprocal fashion to ensure homeostasis.

- The ANS receives sensory information from receptors located throughout the body that monitor blood pressure, chemistry, and body temperature. This information is used to modify effector function via local reflexes or higher (central) autonomic control centers.

- Effector commands are relayed from autonomic control centers via **ganglia** that lie outside of the CNS. Sympathetic ganglia lie close to the spinal cord, whereas parasympathetic ganglia are located close to or within the walls of their target organs. All preganglionic neurons and parasympathetic effectors release **ACh** at their terminals. Most sympathetic postganglionic motor neurons are adrenergic and release **norepinephrine** at target organs.

- Principal autonomic control centers include the **brainstem** and **hypothalamus**.

- The brainstem contains multiple autonomic control nuclei and control centers. The **nucleus tractus solitarius** and **reticular formation** help integrate autonomic sensory information with effector commands from the hypothalamus and **limbic system**.

- The hypothalamus establishes the set point and operational range for many vital internal parameters. It exerts homeostatic control through modification of brainstem control pathways and hormonally via the **pituitary gland**.

- The pituitary gland has two lobes: one comprising epithelial glandular tissue (**anterior lobe**), the other neural tissue (**posterior lobe**). Two breaches in the blood–brain barrier (**circumventricular organs** [**CVOs**]) allow pituitary hormones to be deposited into the general circulation.

- The hypothalamus regulates release of six peptide (tropic) hormones from the anterior lobe into the circulation using **release-stimulating** or **release-inhibiting hormones**. These hypothalamic hormones reach the pituitary via the **hypophyseal portal system**. Two additional hormones are released from hypothalamic nerve terminals located in the posterior pituitary.

- Sensory CVOs located within the brain allow the hypothalamus to sample the chemistry of ECF and make adjustments to organ function as necessary to maintain homeostasis.

- The hypothalamus is also the location of the master clock that entrains most organs to a **circadian rhythm**. The master clock resides in the **suprachiasmatic nucleus**, which exerts control both through direct neural connections to organs and through endocrine control. Entrainment of endocrine organs is mediated by the **pineal gland** and **melatonin** release.

Study Questions

Choose the ONE best answer.

7.1. Autonomic dysfunction is believed to underlie some cases of cardiac rhythm disturbances such as atrial fibrillation. The heart is served by both branches of the autonomic nervous system (ANS). The two ANS branches most likely share which of the following features in common?

   A. Ganglia located close to the spinal cord
   B. Myelinated postganglionic fibers
   C. Postganglionic N_2-type ACh receptors
   D. Preganglionic cell bodies located in the thoracic spinal cord
   E. Use of norepinephrine as the primary postganglionic transmitter

Best answer = C. The sympathetic (SNS) and parasympathetic (PSNS) branches of the ANS both use ACh as a preganglionic neurotransmitter. ACh binds to nicotinic type 2 (N_2)-ACh receptors on postganglionic fibers (see Section IV). SNS ganglia are located close to the spinal cord, whereas PSNS ganglia usually lie close to the target organ (A). Preganglionic ANS fibers are myelinated, but postganglionic fibers are not (B). Some SNS preganglionic cell bodies are located in the thoracic region of the spinal cord, but PSNS cell bodies are either in the brainstem or the sacral region (D). The SNS primarily uses norepinephrine as a postganglionic neurotransmitter, whereas the PSNS uses ACh (E).

7.2. A 38-year-old female is nauseated after receiving cyclophosphamide, an anticancer drug administered to treat breast cancer. Drug-induced nausea is mediated by the area postrema, a sensory circumventricular organ (CVO). Which of the following most likely describes CVO function?

   A. Aldosterone and thyroxine are released via CVOs.
   B. CVO sensory processes extend across the blood–brain barrier (BBB).
   C. CVOs allow blood and CSF to mix.
   D. The central chemoreceptor that monitors P_{CO_2} is a CVO.
   E. The hypothalamus monitors plasma composition via CVOs.

Best answer = E. The hypothalamus uses CVOs to monitor plasma composition, which facilitates homeostatic control of Na^+, water, and other body parameters (see Section VII·C). Aldosterone and thyroxine are released from the adrenal and thyroid glands (A), respectively, and they do not contain CVOs. CVO sensory neurons do not penetrate the capillary wall and extend across the BBB (B). The BBB is interrupted in CVOs, and the capillaries are leaky, allowing fluid to filter from blood for sensing by CVO neurons. Although CVO capillaries are leaky, blood remains contained in the vasculature by the capillary walls, which prevents blood and brain ECF or CSF from mixing (C). Central chemoreceptors are located behind the BBB, unlike CVOs (D).

7.3. A 32-year-old male presents to the office with decreased libido and a milky nipple discharge (galactorrhea). Laboratory studies suggest the presence of a prolactinoma. The hormone responsible for the patient's symptoms is most likely closely related to which of the following hormones?

   A. Adrenocorticotrophic hormone
   B. Follicle-stimulating hormone
   C. Growth hormone
   D. Oxytocin
   E. Thyroid-stimulating hormone

Best answer = C. Prolactinomas secrete prolactin, an anterior pituitary hormone that normally stimulates mammary gland milk synthesis and letdown in postpartum females. Prolactin is produced by lactrotropes (see Section VII·D). Prolactin is a peptide hormone that is structurally related to growth hormone, which is a product of somatotropes. Adrenocorticotropic hormone ([ACTH] A) is a product of the *POMC* gene that also encodes melanocyte-stimulating hormone. ACTH is secreted by corticotropes. Follicle-stimulating hormone (B) and thyroid-stimulating hormone (E) are closely related heterodimeric glycoprotein hormones produced by gonadotropes and thyrotropes, respectively. Oxytocin (D) is a product of the posterior pituitary and is closely related to antidiuretic hormone.

Vision

8

I. OVERVIEW

The ability to detect light is common to most organisms, including bacteria, reflecting the importance of the visual sense. Designs for visual organs have arisen multiple times over the course of evolutionary history, and many remain extant. In humans, photoreception is the purview of the eyes. Each eye comprises a sheet of photoreceptive cells (**retina**) housed within an optical apparatus (Fig. 8.1). The optics project a spatially accurate representation of the visual field onto the photoreceptors, much as a camera lens projects an image onto photographic film or a photosensor array. The simplest cameras use a pinhole as an aperture, which projects an inverted image of the subject onto film. An eye functions similarly, but aperture size (the **pupil**) is variable to control the amount of light falling on the photoreceptors. The inclusion of a variable-focus lens ensures that the projected image stays sharp when aperture size changes. The retina, located at the back of the eye, contains two types of photoreceptor cells. **Cones** are optimized to function in daylight and provide data that can be used to construct a color image. **Rods** are optimized to collect data under minimal lighting conditions, but the data is sufficient only to construct a monochromatic image.

II. EYE STRUCTURE

The eye is a roughly spherical organ enclosed within a thick layer of connective tissue (**sclera**) that is usually white (see Fig. 8.1). The sclera is protective and creates attachment points for three pairs of skeletal (**extraocular**) muscles that adjust the direction of gaze, stabilize gaze during head movement, and track moving objects. Because the photoreceptors are located at the back of the eye, photons entering the eye must travel through multiple layers and compartments before they can be detected.

A. Cornea

Light enters the eye via the **cornea**, which is continuous with the sclera. The cornea comprises several thin, transparent layers delimited by specialized epithelia. The middle layers are composed of collagen fibers along with supportive **keratinocytes** and an extensive sensory nerve supply. Blood vessels would interfere with light transmission, so the cornea is avascular.

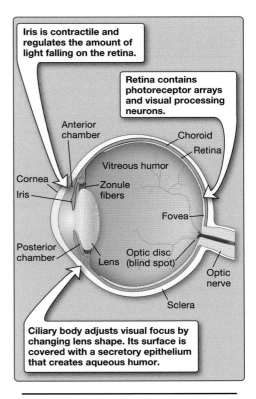

Figure 8.1.
Eye structure.

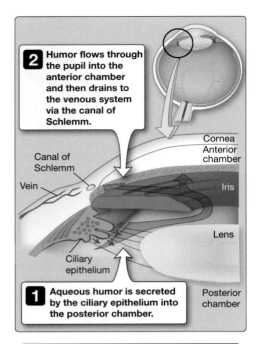

2 Humor flows through the pupil into the anterior chamber and then drains to the venous system via the canal of Schlemm.

Canal of Schlemm

Vein

Cornea
Anterior chamber

Iris

Lens

Ciliary epithelium

1 Aqueous humor is secreted by the ciliary epithelium into the posterior chamber.

Posterior chamber

Figure 8.2.
Aqueous humor secretion and flow.

B. Anterior chamber

The anterior chamber is filled with **aqueous humor**, a watery plasma derivative. It is secreted into the **posterior chamber** by a specialized **ciliary epithelium** that covers the **ciliary body**. It then flows through the pupil into the anterior chamber and drains via the **trabecular meshwork** through the canal of Schlemm to the venous system. Humor is produced continuously to deliver nutrients to the cornea and to create a positive pressure of 10 to 20 mm Hg that stabilizes corneal curvature and optical properties (Fig. 8.2).

C. Iris

The iris is a pigmented, fibrous sheet with an aperture (pupil) at its center that regulates how much light enters the eye. Pupil diameter is determined by two smooth muscle groups under autonomic control. Rings of sphincter muscles controlled by postganglionic parasympathetic fibers from the ciliary ganglion decrease pupil diameter when they contract (**miosis**) as shown in Figure 8.3. A second group of radial muscles controlled by postganglionic sympathetic fibers originating in the superior cervical ganglion widens the pupil (**mydriasis**). Changes in pupil diameter are reflex responses (**pupillary light reflex**) to the amount of light falling on specialized photosensitive ganglion cells located in the retina. Signals from these cells travel

Clinical Application 8.1: Glaucoma

Glaucoma is an optic neuropathy that is the second most common cause of blindness worldwide and a leading cause of blindness among African Americans. Glaucoma commonly occurs when the pathway that allows aqueous humor to pass through the pupil and then drain via the canal of Schlemm is obstructed. Humor production continues unabated, causing intraocular pressure (IOP) to rise. Once IOP exceeds 30 mm Hg, axons traveling in the optic nerve may be damaged irreversibly. Patients typically remain asymptomatic, with their condition being discovered incidentally during a routine ophthalmic examination. Vision loss occurs peripherally during the initial stages. Because central vision is preserved, patients tend not to notice their deficit until retinal damage is extensive. Ophthalmic examination often shows the optic disc to have taken on a hollowed out or "cupped" appearance due to blood vessel displacement, a finding diagnostic of glaucoma. The cup-to-disc ratio increases as the disease progresses. Treatment includes reducing IOP using β-adrenergic antagonists (e.g., timolol) to decrease aqueous humor production, for example,[1] and surgical intervention to correct the cause of obstruction.

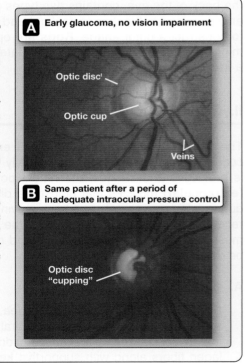

A Early glaucoma, no vision impairment

Optic disc

Optic cup

Veins

B Same patient after a period of inadequate intraocular pressure control

Optic disc "cupping"

[1]For further discussion of drugs used to treat glaucoma, see *LIR Pharmacology*, 8e, Chapter 7·III·B·1.

via the optic nerve to nuclei in the midbrain and then to the Edinger-Westphal nuclei (see Fig. 7.8). Here, they trigger a reflex increase in parasympathetic activity via the oculomotor nerve (cranial nerve [CN] III), and the pupil constricts. Pupillary constriction reduces the amount of light entering the eye and helps prevent photoreceptor saturation. Saturation is undesirable in that it functionally blinds an individual. When light levels are low, reflexive pupillary dilation increases the amount of light reaching the retina. The pupillary reflexes elicit identical muscle responses in both eyes, even though light levels may be changing in one eye only.

> Pupil diameter always reflects a balance between tonic sympathetic and parasympathetic nerve activity. Thus, when atropine (an acetylcholine-receptor antagonist) is applied topically to the cornea during an ophthalmic examination, the pupil dilates because the balance between sympathetic and parasympathetic influence has been shifted in favor of the sympathetic nervous system.

D. Lens

The **lens** is a transparent, ellipsoid disk suspended in the light path by radial bands of connective tissue fibers (**zonule** fibers), attached to the **ciliary body**. The ciliary body is contractile and functions to modify lens shape and adjust its focus. The lens is composed of long, thin cells (**lens fibers**) that are arranged in tightly packed, concentric layers, much like the layers of an onion. The cells are dense with **crystallins**, proteins that give the lens its transparency and determine its optical properties. The lens is enclosed within a capsule composed of connective tissue and an epithelial layer.

E. Vitreous humor

Vitreous humor is a gelatinous substance composed largely of water and proteins. It is maintained under slight positive pressure to hold the retina against the sclera.

F. Retina

When light reaches the retina, it must penetrate multiple layers of neurons and their supporting structures before it can be detected by photoreceptors. The neuronal layers are transparent, so light loss during passage is minimal. The retina contains two specialized regions. The **optic disc** is a small area where the photoreceptor array is interrupted to allow blood vessels and axons from the retinal neurons to exit the eye, creating a **blind spot** (Fig. 8.4). Nearby, in the center of the field of vision, is a circular area called the **macula lutea**. At its center is a small (<1-mm diameter) pit called the **fovea**. The neuronal layers separate here to allow light to fall directly on photoreceptors, creating an area of maximal visual acuity.

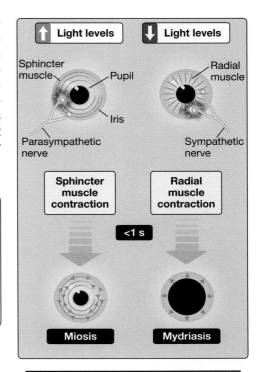

Figure 8.3.
Regulation of pupil diameter.

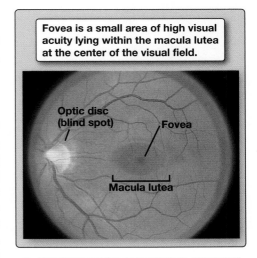

Figure 8.4.
Retinal landmarks.

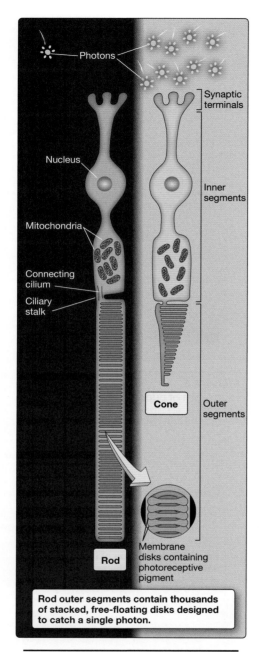

Figure 8.5.
Photoreceptor structure.

III. PHOTORECEPTORS

Retinal photoreceptors are arranged in highly regular arrays so that spatial information can be extracted from photoreceptor excitation patterns. The retina contains two types of photoreceptors that share a similar cellular structure.

A. Types

Rods are specialized to detect single photons of light. They cannot differentiate color, but they can generate an image under low light conditions and thereby facilitate **scotopic vision** (derived from the Greek for "darkness," *skotos*). **Cones** function optimally in daylight and mediate **photopic**, or color, vision.

B. Organization

Photoreceptors are long, thin, excitable cells (Fig. 8.5). At the center is a cell body that encloses the nucleus. The cell body extends in one direction to form a short axon that branches into several presynaptic structures. The opposite end of the cell is long and cylindrical and divided into two segments. The **inner segment** contains all the other organelles required for normal cell function, including numerous mitochondria. The inner segment gives rise to a cilium (**connecting cilium**) that is grossly modified to house the phototransduction machinery. This compartment, the **outer segment**, is connected to the inner segment by a short **ciliary stalk**.

C. Disk membranes

The bloated sensory cilium that comprises the rod outer segment is packed with >1,000 discrete, flattened, membranous disks stacked like dinner plates alongside the ciliary axoneme. Cones contain similar but fewer stacks that are infoldings of the surface membrane. The rod stacks are designed to capture a single photon as it traverses the eye's photosensitive layer. To make this a reality, the disk membrane is so densely packed with photosensory pigment molecules that there is little room left over for lipid!

D. Distribution

The retina lines the inside surface of the eye, covering roughly 75% (~11 cm^2) of its total surface area. Photoreceptors are densely packed within the sheet, with rods outnumbering cones ~20-fold (~130 million rods vs. 7 million cones). Although both rods and cones are found throughout the retina, their distribution is unequal.

1. **Rods:** Rods dominate the peripheral retina, which optimizes these areas for night vision.

2. **Cones:** Cones are concentrated in the central retina, which imparts this area with a high degree of visual acuity. At its center is the fovea, which contains cones alone (see Figs. 8.1 and 8.4). The fovea's lack of rods means that it cannot participate in night vision.

IV. PHOTOSENSOR

The ability to capture the energy of a single photon requires a **chromophore**, a molecule that absorbs certain light wavelengths while reflecting or transmitting others. This property gives the molecule color. The chromophore used in the eye is **retinal**, which can exist in several different conformations. The 11-*cis* conformation is very unstable and, when hit by a photon, immediately flips into a more stable all-*trans* configuration. Transition is rapid (femtoseconds), which makes it an ideal photoreceptive pigment. The task of detecting and reporting the conformational change falls on **opsin**, which is a G protein–coupled receptor. Opsin covalently binds 11-*cis* retinal in the same way that a hormone receptor binds its ligand. The receptor and chromophore combine to create a visual pigment called **rhodopsin**, which has a reddish purple color. When retinal absorbs a photon and transitions, it triggers a change in opsin conformation to generate **metarhodopsin II**. This event initiates a signal cascade that ultimately converts photonic energy into an electrical signal.

V. PHOTOSENSORY TRANSDUCTION

Phototransduction is highly unusual in that stimulus detection causes receptor *hyper*polarization rather than the *de*polarization typical of other sensory systems.

A. Dark current

The outer-segment membrane contains a nonspecific cation channel gated by cGMP, as shown in Figure 8.6. A constitutively active guanylyl cyclase (GC) maintains high intracellular cGMP levels in the dark, and the channel is always open. Na^+ and small amounts of Ca^{2+} flow into the photoreceptor, creating an inward **dark current**. K^+ leak channels in the inner segment allow K^+ to escape the cell and help offset the current, but membrane potential (V_m) still rests at a relatively shallow −40 mV.

B. Transduction

When a photon hits retinal, rhodopsin contorts and activates **transducin**, which is a G protein (G_T; Fig. 8.7). When activated, the G_T α subunit dissociates and activates a membrane-associated phosphodiesterase (PDE). PDE hydrolyzes cGMP to GMP, and intracellular cGMP levels fall. The cation channel deactivates and closes as a result, and the dark current terminates. The K^+ channel in the inner segment remains open, however, which causes V_m to drift negative. The V_m change signals that light has been detected. Although the transduction cascade is slow (tens to hundreds of milliseconds) compared with photon detection, it does provide for tremendous signal amplification that allows the eye to register single photons.

C. Signal termination

The amplification cascade is so powerful that photoreceptors rely on multiple negative feedback mechanisms to limit and terminate signaling in a timely manner (Fig. 8.8).

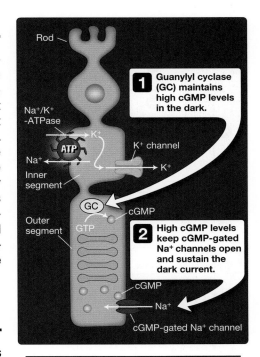

Figure 8.6.
Dark current origins.

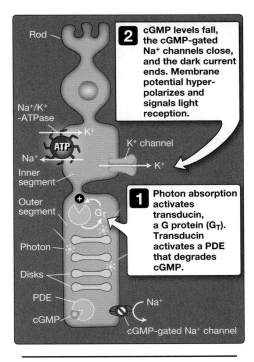

Figure 8.7.
Phototransduction under low light conditions. PDE = phosphodiesterase.

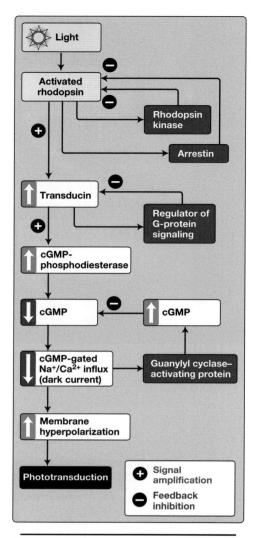

Figure 8.8.
Photosensory transduction pathway and mechanisms for limiting and terminating signaling.

1. **Opsin inactivation:** The active form of opsin is a substrate for rhodopsin kinase. Opsin has multiple phosphorylation sites, and each successive phosphate transfer further reduces its ability to interact with G_T. Phosphorylation also makes the receptor a favorable target for **arrestin** binding. **Arrestin** is a small protein whose sole function is to block interaction between opsin and transducin, preventing further signaling.

2. **Transducin deactivation:** Photoreceptors also contain **regulator of G-protein signaling 9 (RGS9)** that enhances the G_T α subunit's GTPase activity and thereby speeds its deactivation.

3. **Guanylyl cyclase activation:** The dark current is mediated, in part, by Ca^{2+} influx. Light stops this influx, and intracellular Ca^{2+} concentrations fall. This is sensed by one or more Ca^{2+}-dependent GC-activating proteins, which respond by stimulating GC activity, which, in turn, counteracts the effects of PDE. This pathway is important for helping photoreceptors adapt to light levels that saturate the signaling pathway and also helps restore the dark current once signaling ends.

D. Desensitization

Prolonged exposure to bright light desensitizes photoreceptors. Desensitization is partly an extension of the opsin inactivation process described earlier. Rhodopsin kinase phosphorylates opsin at multiple sites, which increases arrestin-binding affinity and blocks further opsin–transducin interactions. With time, transducin translocates from the outer segment to the inner segment, effectively breaking the first crucial link in the phototransduction chain and preventing further signaling.

E. Retinal recycling

Retinal is released from opsin shortly after activation, and the pigment turns yellow (**bleaching**). It is then converted to **retinol**, also known as **vitamin A**. Vitamin A is converted to 11-*cis* retinal, which binds to opsin and restores the visual pigment.

> Vitamin A is essential for synthesis of visual pigments. Inadequate dietary intake results in night blindness, characterized by an inability to see in low light due to impaired rod cell function. The condition can be reversed within hours by administering vitamin A.

VI. COLOR VISION

Night vision is monochromatic because rods contain only one visual pigment. They are designed to register small amounts of light, not provide information about its quality. Distinguishing colors requires two or more pigments that signal maximally at different wavelengths. Color vision employs three cone-cell types, each containing a different visual pigment (Fig. 8.9). All use 11-*cis* retinal as a chromophore; however, the opsins differ in their primary sequence, which shifts pigment sensitivity to different wavelengths on the visible spectrum. **S cones** respond maximally to short wavelengths (violet-blue: ~420 nm), **M cones** to medium wavelengths (green-yellow: ~530 nm), and **L cones** to long wavelengths (yellow-red: ~560 nm). Overlap in pigment absorption spectra means that all three cone types respond to most visible light frequencies, but the intensity of their responses differs according to how close the stimulus is to the cone's optimal range. The brain then extrapolates colors from the data streams emerging from the retina.

VII. VISUAL PROCESSING

The photoreceptor array is capable of generating immense quantities of visual information.[1] This information is relayed to the central nervous system visual centers via the optic nerve, whose exit through the retina creates a "blind spot" (see Figs. 8.1 and 8.4). If the optic nerve carried raw data, its diameter would have to be increased over 100-fold to accommodate the required number of axons, and blind-spot size would increase proportionately. Thus, raw visual sensory data is processed extensively before leaving the retina to compress it and minimize the impact of the optic nerve on sensory array continuity.

A. Retina structure

The retina is a highly organized structure comprising layers of cells (photoreceptors) that generate photosensory data, process visual signals (**bipolar cells**, **horizontal cells**, **ganglion cells**, and **amacrine cells**), or support neuronal activity (**pigment epithelium** and neuroglia) as shown in Figure 8.10.

1. **Pigment epithelium:** The innermost retinal layer is a darkly pigmented epithelium that absorbs stray photons that might otherwise interfere with imaging and decrease visual acuity. The color is imparted by numerous **melanin** pigment granules. The pigment epithelium also supplies photoreceptors with nutrients, is involved with retinal recycling, and assists with photoreceptor turnover. Photoreceptor membranes are subject to constant damage by photons and are therefore turned over continually. The entire rod stack is replaced once every ~10 days.

2. **Neuroglia:** Because the retina is an extension of the brain, the photoreceptors and all associated excitable cells receive support from glia. **Müller cells**, which are a retina-specific glial subtype, occupy the spaces between neurons and form a barrier (**inner delimiting membrane**) that separates the retina from vitreous humor.

 [1]For more information on processing and interpreting visual signals, see *LIR Neuroscience*, 2e, Chapter 15.

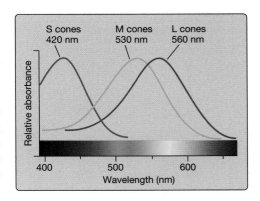

Figure 8.9.
Cone spectral sensitivities.

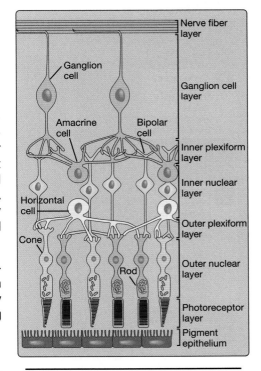

Figure 8.10.
Principal retinal cell layers and interconnections.

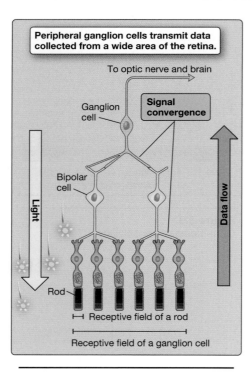

Peripheral ganglion cells transmit data collected from a wide area of the retina.

To optic nerve and brain

Ganglion cell

Signal convergence

Bipolar cell

Light

Data flow

Rod

⊢—⊣ Receptive field of a rod

Receptive field of a ganglion cell

Figure 8.11.
Photosensory data flow in the retina.

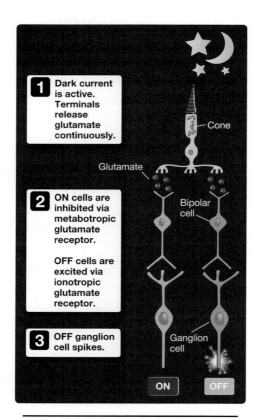

1 Dark current is active. Terminals release glutamate continuously.

Cone

Glutamate

2 ON cells are inhibited via metabotropic glutamate receptor.

OFF cells are excited via ionotropic glutamate receptor.

Bipolar cell

3 OFF ganglion cell spikes.

Ganglion cell

ON   OFF

Figure 8.12.
ON and OFF bipolar cell activity in the dark.

B. Photoreceptor output

The dark current keeps photoreceptors depolarized and excited, and their presynaptic terminals are specialized to release transmitter (glutamate) continuously. Light reduces the dark current to cause a hyperpolarization that is graded with stimulus intensity. Photosensory excitation is signaled first to bipolar cells, which is where signal convergence begins (Fig. 8.11).

1. **Rods:** Rods function under minimal light conditions where the goal is simply capturing an image, and visual acuity is a lesser concern. Bipolar cells, therefore, collect and pool inputs from several rods to increase the likelihood of unitary events being transmitted to the cortex. The portion of the visual field serviced by a single rod represents its **receptive field**. The receptive field of bipolar cells is larger than a single rod's because it incorporates the receptive fields of multiple photoreceptors (see Fig. 8.11).

2. **Cones:** Cones function optimally in daylight. Because all cones are active in daylight, pooling inputs from multiple photoreceptors is not required to form an image, and pooling inputs would degrade sensory data quality. So, cones typically synapse with dedicated bipolar cells that preserve spatial information, thereby increasing visual acuity.

C. Bipolar cell output

Bipolar cells execute the first data-processing step. Individual cells generally receive inputs from either a group of rods or from a single cone but not both. These dedicated relationships preserve the integrity of the data streams from rods and cones. Bipolar cells are unusual in that they do not generate action potentials, but, rather, their output is graded. There are at least 10 different bipolar cell types, but they can be divided into two principal groups: "**ON**" (or, **on-center**) and "**OFF**" (**off-center**) cells. Rods only synapse with ON cells. Most cones synapse with at least one of each.

1. **ON cells:** ON cells employ a metabotropic glutamate receptor (mGluR6) to transduce photoreceptor signals. The glutamate receptor is coupled via an inhibitory G protein to a nonselective cation channel, but, because the relationship is inhibitory, the channel is prevented from opening in the dark (Fig. 8.12). When photoreceptors upstream are illuminated, the channel is freed of its inhibitory influence, and the bipolar cell depolarizes in a graded manner (Fig. 8.13). These bipolar cells are known as **sign-inverting** because receptor *hyperpolarization* causes bipolar cell *depolarization*. They are excited when light is turned ON.

2. **OFF cells:** OFF cells express an ionotropic glutamate receptor. Glutamate binding causes a sustained inward current and membrane depolarization mirroring that of the photoreceptors (see Fig. 8.12). When the photoreceptor is stimulated, its membrane hyperpolarizes, and synaptic glutamate release is inhibited. The bipolar cell that the photoreceptor synapses with also hyperpolarizes (see Fig. 8.13). When the light is turned OFF, the photoreceptor depolarizes, and synaptic glutamate release resumes, causing bipolar cell excitation and signaling.

D. Ganglion cell output

The vertical data streams running from the photoreceptors to the outer layer of the retina are preserved by the ganglion cells. Ganglion cells are one of the few retinal cell types to generate action potentials used to digitally encode visual information for transmission to the brain. ON ganglion cells respond to light with a volley of action potentials (see Fig. 8.13), whereas OFF ganglion cells discharge when the light is turned off (see Fig. 8.12).

E. Horizontal and amacrine cells

Horizontal and amacrine cells manipulate the sensory data streams as they progress through the retinal layers. Horizontal cells extend their processes laterally within the outer plexiform layer, which allows them to synapse with and collate information from photoreceptors within a wide receptive field. There are several different types of horizontal cell. Their output is usually inhibitory, using either γ-aminobutyric acid or glycine as a neurotransmitter, influencing both photoreceptor and bipolar cell signaling. Their net effect is to increase the contrast between signals received from light and dark areas of the retina. The role of amacrine cells is less well understood.

VIII. VISUAL PATHWAYS

Ganglion cell axons gather to form the optic nerves (CN II), one per eye, that convey visual signals from the retina to the brain. The optic nerves meet and merge immediately in front of the pituitary gland at the **optic chiasm** (Fig. 8.14). Here, nasal retinal fibers cross the midline and join temporal fibers from the contralateral eye to form **optic tracts** that converge on the **lateral geniculate nucleus** of the thalamus. In practice, this crossing over (decussation) ensures that sensory data from the right visual fields of both eyes is transmitted to the left side of the brain, and vice versa. The data streams are then transmitted from the thalamus via **optic radiations** to the primary visual cortex (occipital lobe) for further analysis and processing.

IX. OPTICAL PROPERTIES

Light entering a room through a window does not form a perfectly focused image of the outside view on the opposite wall. Similarly, light entering the eye does not form a sharp image on the retina unless the rays are manipulated during passage through the eye.

A. Optical principles

Light rays normally travel in parallel lines. The speed at which they travel depends on the medium they are passing through. Their velocity is slowed in air compared with a vacuum and slowed further during passage through water or a transparent solid such as glass. The ratio of light velocity in a vacuum to velocity in a different medium is known as the **refractive index**. Air has a refractive index of about 1.0003, water has an index of 1.33, and the eye's lens and cornea are closer to 1.4. When light transitions at an angle to a medium of different refractive index, it bends. How much it bends is dependent on the difference in refractive index of the two materials and also the angle of attack.

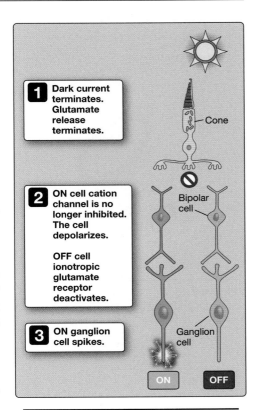

1 Dark current terminates. Glutamate release terminates.

2 ON cell cation channel is no longer inhibited. The cell depolarizes.

OFF cell ionotropic glutamate receptor deactivates.

3 ON ganglion cell spikes.

Cone

Bipolar cell

Ganglion cell

ON   OFF

Figure 8.13.
ON and OFF bipolar cell responses to light.

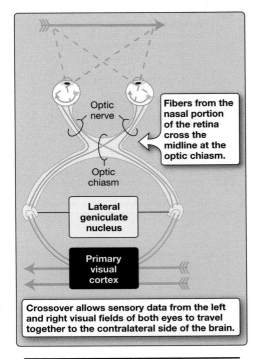

Optic nerve

Fibers from the nasal portion of the retina cross the midline at the optic chiasm.

Optic chiasm

Lateral geniculate nucleus

Primary visual cortex

Crossover allows sensory data from the left and right visual fields of both eyes to travel together to the contralateral side of the brain.

Figure 8.14.
Pathways for visual information flow to the brain.

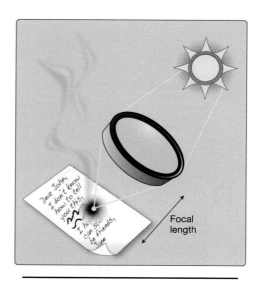

Figure 8.15.
Lens focal distance.

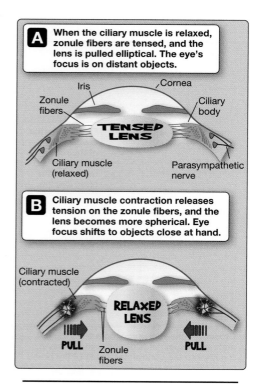

Figure 8.16.
Lens accommodation.

B. Simple lenses

Simple convex lenses (e.g., in a magnifying glass) have two curved surfaces. Their curvature forces parallel light rays to bend toward a central focal point. Thus, a magnifying glass held above a piece of paper in bright sunlight creates a spot of intense white light on the paper (Fig. 8.15). When the glass is held at a distance that equals the **focal length** of the lens, the spot becomes a tightly focused pinhead, and the sun's energy causes the paper to smolder and burn. **Focal power** is the inverse of focal length. Powerful glass lenses are able to bring objects into tight focus within a relatively short distance compared with a weak lens, and this is achieved by increasing the curvature of the surfaces. Focal power is measured in diopters (D) and calculated as the reciprocal of focal length, in meters. A 1D lens brings an object into focus 1 meter from the lens. A 10D lens focuses at 0.1 meter. The human eye has a maximum power of about 59D.

C. Focal power

Light is forced to change direction several times during its passage from air to the retina. It is refracted at the interfaces between air and cornea, cornea and aqueous humor, aqueous humor and lens, and lens and vitreous humor.

The greatest degree of refraction occurs when light transitions from air to cornea, meaning that the cornea is the primary determinant of the eye's ability to bring a distant object into focus.

D. Accommodation

A simple lens has a fixed focal length. Its surfaces can be shaped to image distant objects or objects close at hand but not both simultaneously. More complex imaging devices, such as telescopes, adjust focus by using two or more simple lenses whose positions relative to each other are varied. The eye is also capable of adjusting focus, but it does so by changing lens shape. This is known as the power of **accommodation**. The lens of a young and healthy individual can adjust focal power by up to 14D in a third of a second.

1. **Resting eye:** A human lens freed from all other influences assumes a near-spherical shape due to the capsule's natural elasticity. The lens is suspended in the light path by numerous radial zonule fibers attached to the ciliary body. In a resting eye, the zonule fibers are tensed by surrounding structures, and the lens stretches and flattens into an elliptical shape (Fig. 8.16A). Thus, a resting eye focuses on distant objects.

2. **Accommodation mechanism:** Lens shape and focus is determined by ciliary muscle fibers under parasympathetic control (oculomotor nerve, CN III). When the eye is required to focus on close objects, the ciliary muscle ring is excited and contracts around the lens. This sphincter-like movement releases the tension on the zonule fibers and allows the lens to assume a more rounded shape, and focus shifts accordingly (see Fig. 8.16B). The lens capsule stiffens with age, decreasing the power of accommodation and increasing the reliance on corrective reading lenses for individuals over age 40 years (**presbyopia**, derived from the Greek for "old eye").

Clinical Application 8.2: Refractive Disorders

An individual is assumed to have normal vision (**emmetropia**) if light from a distant object (>6 m away) traverses the eye optics and forms a focused image on the retina when the ciliary muscle is relaxed. Deviation from normal is very common. **Myopia**, or **nearsightedness**, refers to a condition in which the lens projects an image in front of the retina. It most commonly results from an eye that is longer than normal but may also be caused by a lens or cornea that is optically more powerful than normal. Placing a lens with concave surfaces in front of the eye can adjust the plane of focus and restore visual acuity. **Hyperopia**, or **farsightedness**, is caused by an eye that is too short or a lens or cornea that projects an image behind the retina. The cause is usually of genetic origin. **Astigmatism** is a common visual defect in which irregularities in the focal power of the cornea or lens cause portions of the projected image to be blurred. Corrective lenses can restore visual acuity. **Laser-assisted in situ keratomileusis (LASIK)** surgery provides an alternative to wearing corrective lenses as a means of correcting myopic vision. LASIK surgery involves lifting a corneal flap and then reshaping the cornea using an excimer laser to restore the curvature necessary to focus on distant objects.

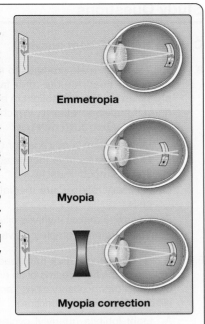

Emmetropia

Myopia

Myopia correction

Emmetropia and myopia.

Chapter Summary

- The eye is a visual sensory organ comprising an optical apparatus (**variable aperture** and **lens**) that projects an image of external objects onto a photosensor array. The array makes it possible to capture a digitized representation of the image for transmission to the **primary visual cortex**.

- Aperture (**pupil**) size is modulated to control how much light enters the eye (**pupillary light reflex**). Pupil size is determined by the iris, which contains two smooth muscle groups under autonomic (**oculomotor nerve**) control. **Sphincter muscles** decrease pupil size when they contract (**miosis**), whereas radial muscles dilate the pupil (**mydriasis**). The autonomic nervous system also regulates lens shape using **ciliary muscles** to allow the eye to focus on near objects (**accommodation**).

- The retina is a thin, layered, pigmented membrane containing a **photoreceptor array**, **signal processing neurons**, and glia (**Müller cells**) lining the back of an eye. Axons from the visual processing neurons (**ganglion cells**) exit the eye via the **optic nerve**, creating a retinal **blind spot**.

- The retina contains two types of photoreceptors. **Rods** are specialized to create monochromatic images in dim light. **Cones** produce color images in daylight. Both photoreceptor types transduce photon detection using modified **sensory cilia** containing stacks of membranes packed with photosensitive pigments.

- Rods are found at highest densities in the peripheral retina; cones are concentrated in the center of the visual field. The **fovea** is a small area at the visual field's center containing only cones and is where the neuronal layers part to allow light to fall on photoreceptors directly. These modifications create an area of high visual acuity.

- Photosensory transduction occurs when photons are absorbed by and trigger a conformational change in a **chromophore (retinal)**. Retinal is associated with a G protein–coupled receptor (**opsin**), together constituting a **visual pigment (rhodopsin)**.

- Different opsin isoforms attune the visual pigment to particular wavelengths optimized for nighttime or color vision. Color vision relies on cones containing one of three such pigments.

- Light falling on rhodopsin activates a G protein (**transducin**). The α subunit activates a phosphodiesterase, and resting cGMP levels then fall. The decrease deactivates a cGMP-dependent inward **dark current**, and the photoreceptor membrane hyperpolarizes. This membrane potential change constitutes a **receptor potential**.

- Photosensory signals are relayed to the cortex via a series of neurons (**bipolar** and **ganglion cells**) that begin processing the visual data while maintaining the integrity of the spatial information they contain.

Study Questions

Choose the ONE best answer.

8.1. A 32-year-old male presents to the emergency department with head trauma after falling from a ladder. A physician shines a flashlight in each eye and observes normal pupillary reflexes. Which of the following most likely describes such reflexes?

A. Light causes cone receptor depolarization.
B. Miosis results from increased sympathetic stimulation of smooth muscle.
C. Pupillary reflexes are mediated by retinal ganglion cells.
D. Reflexive miosis involves ciliary muscle.
E. They are an example of a vagovagal reflex.

Best answer = C. The pupillary light reflex is triggered by light falling on photosensitive retinal ganglion cells (see Section II·C). Light falling on cones hyperpolarizes the receptor cells (A) through a decrease in Na^+ influx through cyclic nucleotide–activated channels. The pupil constricts reflexively through parasympathetic stimulation of iris sphincter muscles, not by sympathetic stimulation of smooth muscle (B) or the ciliary muscle (D). The pupillary light reflex is mediated by the optic nerve and oculomotor nerve, not by the vagus nerve (E).

8.2. A driver traveling a dark rural road at night is temporarily blinded by the high beams of an oncoming vehicle. Which of the following observations most likely describes the blinded driver's retinal function?

A. Light inhibits guanylyl cyclase–activating proteins in rods.
B. Temporary blindness is caused by Na^+ channel internalization.
C. The channel that mediates night vision is a TRP channel.
D. The high beams cause blindness through rod depolarization.
E. Vision recovery involves rhodopsin dephosphorylation.

Best answer = E. Rhodopsin activation by light initiates a signal cascade that causes rod signaling, but it also initiates pathways that limit signaling (see Section V·C). These include rhodopsin phosphorylation by rhodopsin kinase, so recovery necessarily involves rhodopsin dephosphorylation. Guanylyl cyclase–activating proteins are stimulated, not inhibited, by light (A). Photoreceptor desensitization during light exposure involves opsin phosphorylation by rhodopsin kinase, not Na^+ channel internalization (B). Rod cell stimulation is mediated by a cGMP-dependent Na^+ channel, not a TRP channel (C). TRP channels support Ca^{2+}/Na^+ fluxes that depolarize the membrane. Rod cell activation and desensitization involves membrane hyperpolarization, not depolarization (D).

8.3. A 62-year-old female with a history of temporal arteritis suffers sudden monocular vision loss caused by retinal artery occlusion and subsequent ganglion cell ischemia. Which of the following statements most likely describes how these retinal ganglion cells function?

A. Light always causes cell depolarization.
B. They are dedicated to single rods.
C. They assist photoreceptor recycling.
D. They generate action potentials.
E. They signal via the oculomotor nerve.

Best answer = D. Ganglion cells transmit visual information to the brain using action potentials, their axons forming the ocular nerve (the oculomotor nerve controls eye movement). Most other cells in the retina respond to light with graded potentials rather than action potentials (A, D; see Section VII). Ganglion cells activate when light is turned on or off, depending on where light falls on the retina relative to their receptive field. Ganglion cells collate data from groups of photoreceptors (not single rods, B), which gives them a wide receptive field. Pigment cells, not ganglion cells, aid photoreceptor recycling (C).

8.4. Which of the following would be most likely to cause the eye to shift its focus from far objects to near objects?

A. Atropine application
B. Ciliary muscle relaxation
C. Oculomotor nerve stimulation
D. Sphincter muscle contraction
E. Tensing of zonule fibers

Best answer = C. Stimulation of the oculomotor nerve causes ciliary muscle contraction, which releases the tension on zonule fibers. When tensed, zonule fibers pull the lens into an elliptical shape, which allows it to focus on distant objects. Fiber relaxation allows the lens to become rounder, allowing it to focus on near objects (see Section IX·D). Atropine application (A) would block parasympathetic stimulation of ciliary muscle, and it would remain relaxed. Ciliary muscle contraction, not relaxation (B), is required to cause lens rounding and focusing on near objects. Sphincter muscle contraction (D) causes pupil diameter to decrease (see Section II·C). Zonule fibers relax (E) when focusing on near objects.

Hearing and Balance

9

I. OVERVIEW

Modern-day aquatic vertebrates possess lateral-line sensory systems that detect vibrations and movements in their watery surrounds. Lateral lines comprise lines of pits running down both sides of the body. Movements are transduced by clusters of sensory hair cells embedded in a gelatinous dome that protrudes from each lateral-line pit. When the dome is displaced by local water currents or vibrations, the embedded hairs are displaced also, generating a receptor potential in the hair cell body. Although humans did not retain lateral-line organs during evolution, the hair-cell transduction system works so well that it was adapted for use in the inner ear. The inner ear contains two contiguous, hair cell–based sensory systems. The auditory system uses hair cells to transduce vibrations generated by sound waves. The vestibular system uses hair cells to transduce head movements.

II. SOUND

Sounds are atmospheric pressure waves created by moving objects. For example, striking a metal gong causes its metal surface to vibrate (Fig. 9.1). The gong alternately compresses and then decompresses the surrounding air to create a pressure wave that propagates outward at a speed of 343 m/s. We perceive these pressure waves as sounds, the wave's frequency reflecting its pitch. The ability to transduce sound waves (**audition**) allows us to detect objects at a distance and, therefore, has clear survival advantages. If an approaching object represents a threat (e.g., a predator or speeding truck), advance warning of its approach allows time for evasive maneuvers. The ability to hear vocalizations allowed for the development of oral communication and speech. The ability to perceive sound requires that the pressure waves be converted into an electrical signal, a process that occurs within the **inner ear** and relies on **sensory hair cells**.

III. AUDITORY SYSTEM

Designing a system that transduces sound is relatively easy because sound waves vibrate membranes. For example, a dog's bark creates vibrations in the wall of an empty soda can or milk jug that can easily be sensed by mechanoreceptors in the fingertips. Sounds are usually very

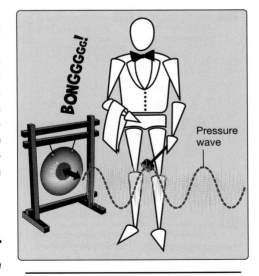

Figure 9.1.
Sounds are pressure waves that travel through air.

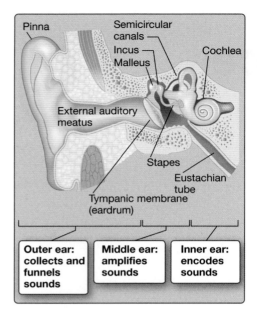

Figure 9.2.
Ear anatomy.

Outer ear: collects and funnels sounds

Middle ear: amplifies sounds

Inner ear: encodes sounds

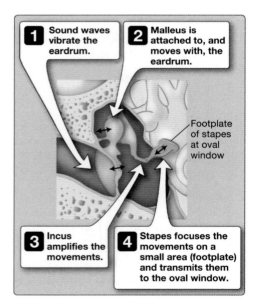

1 Sound waves vibrate the eardrum.

2 Malleus is attached to, and moves with, the eardrum.

Footplate of stapes at oval window

3 Incus amplifies the movements.

4 Stapes focuses the movements on a small area (footplate) and transmits them to the oval window.

Figure 9.3.
Role of ossicles in impedance matching.

complex, however, comprising a series of changing frequencies. The ear decodes such sounds using an array of sensory hair cells embedded within a membrane designed to resonate at different frequencies along its length. It combines this with an amplifier to create a remarkably sensitive acoustic analyzer.

A. Structure

The auditory system, or ear, can be divided into three main anatomical components: the **outer**, **middle**, and **inner ear** (Fig. 9.2).

1. **Outer ear:** The **pinna** collects and focuses sounds. Sounds are channeled into the ear canal (**external auditory meatus**), which allows them to pass through the skull's temporal bone. The canal ends blindly at the **tympanic membrane (eardrum)**, which vibrates in response to sound.

2. **Middle ear:** The **middle ear** is an air-filled chamber lying between the eardrum and the inner ear. It connects with the nasopharynx via the **eustachian tube**, which drains fluids and allows the pressure across the eardrum's two surfaces to equalize. Eardrum vibrations are transmitted to the inner ear by an articulating lever system comprising three, small, fragile bones called **ossicles** (Fig. 9.3). The bones are known as the **malleus** (hammer), **incus** (anvil), and **stapes** (stirrup), their names roughly reflecting their shapes. The malleus is attached to the eardrum's inner surface and transmits vibrations to the incus. The incus transmits them to the stapes. The stapes' footplate inserts into and is firmly attached to the **oval window** of the inner ear.

3. **Inner ear:** The inner ear contains a convoluted series of fluid-filled chambers and tubes (**membranous labyrinth**). The structures are encased within bone (**bony labyrinth**) with a thin layer of **perilymph** trapped between bone and membranes. The labyrinth has two sensory functions. The **auditory portion** is called the **cochlea**. The **vestibular portion** contributes to our sense of balance (see Section V). It comprises the **otolith** organs (**utricle** and **saccule**) and three **semicircular canals**.

B. Impedance matching

The inner ear is filled with fluid that has a high inertia and is difficult to move compared with air. The middle ear's function is, thus, to harness the sound wave's inherent energy and transmit it to the inner ear with sufficient force to overcome the inertia of the fluid contents. This process is called **impedance matching**.

1. **Mechanism:** The ossicles form a lever system that amplifies eardrum movements by ~30% (see Fig. 9.3). It also focuses the movements on the stapes' footplate, whose surface area is ~17 times smaller than the eardrum. Amplification and focusing combined increase force per unit area ~22-fold, which is sufficient to overcome cochlear fluid inertia.

2. **Damping:** Lever system flexibility is modulated to reduce sound amplitude under certain circumstances. The malleus and stapes are attached to two tiny muscles under autonomic control (Fig. 9.4). The **tensor tympani** anchors the malleus to the wall of the middle ear and is innervated by the trigeminal nerve (cranial nerve [CN] V). The stapes is anchored by the **stapedius**, which is innervated by the facial nerve (CN VII). When the two muscles contract, the ossicular chain becomes more rigid, and sound transmission is attenuated. The **attenuation reflex** can be triggered by loud sounds but is probably designed to dampen the sound of our own voices when talking.

C. Cochlea

The cochlea is a long (~3-cm), tapered tube containing three fluid-filled chambers that run the length of the tube. The tube is coiled like a snail shell *in vivo*, but the functional architecture is easier to understand when considered uncoiled (Fig. 9.5). The three chambers are the **scala vestibuli**, **scala media**, and **scala tympani**.

1. **Scala vestibuli and scala tympani:** The upper and lower chambers are both filled with **perilymph** (a fluid approximating plasma) and are physically connected by a small opening (the **helicotrema**) at the cochlear apex.

2. **Scala media:** The center chamber is separated from the scala vestibuli by the **Reissner membrane** (or **vestibular membrane**) and from the scala tympani by the **basilar membrane**. The scala media terminates short of the cochlea apex and is sealed off from the other two chambers. It is filled with **endolymph**, a K^+-rich fluid produced by the **stria vascularis**, a specialized epithelium lining one wall of the chamber (see Fig. 9.5). The scala media contains the **organ of Corti**, which is the auditory sensory organ.

IV. AUDITORY TRANSDUCTION

Sound waves enter the cochlea via the oval window, which forms the basal end of the scala vestibuli (Fig. 9.6). Stapes motion sets up a pressure wave in the perilymph that runs down the chamber's length to the apex, passes through the helicotrema, and then pulses back down the scala tympani to the cochlear base. Here, it encounters the **round window**, a thin membrane located between the inner and middle ear. The membrane vibrates back and forth in reverse phase with the wave generated by stapes movement. The stapes would not be able to displace the oval window and set the perilymph in motion if the round window did not exist because the cochlear chamber walls are otherwise rigidly encased in bone. The scala media, which approximates a fluid-filled sac suspended between the two chambers, is buffeted by the pressure wave as it pulses back and forth. Thus, although the sound wave never enters the scala media directly, the entire structure wobbles, much as a waterbed responds when pushed down on hard at one corner. This buffeting is sensed by and stimulates the organ of Corti.

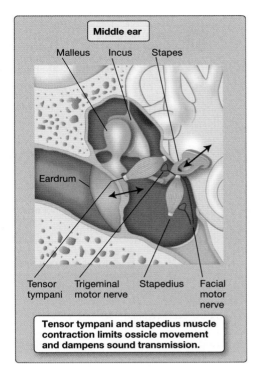

Figure 9.4.
Attenuation reflex.

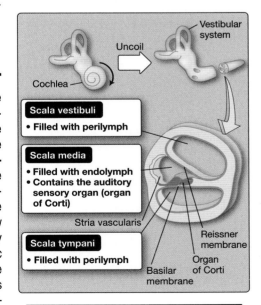

Figure 9.5.
Cochlear chambers.

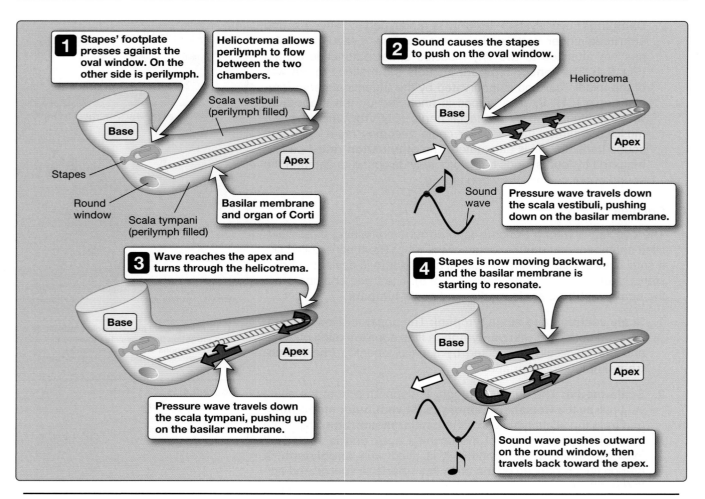

Figure 9.6.
Sound wave passage through the cochlear chambers.

A. Organ of Corti

The organ of Corti comprises a sheet of auditory receptor cells (hair cells) and their associated structures, all of which rest on the basilar membrane (Fig. 9.7).

1. **Hair cell types:** The hair cells are arranged in rows down the length of the cochlea. Two types of hair cell (**inner** and **outer**) can be distinguished based on their location, innervation, and function.

   a. **Inner:** Inner hair cells ([IHCs] ~3,500 total) form a single row toward the center of the cochlea. They are densely innervated by sensory neurons (up to 20 per cell), whose axons make up the bulk of the cochlear nerve (part of the vestibulocochlear nerve, or CN VIII). IHCs are the ear's primary sound transducers.

   b. **Outer:** There are an additional three rows of outer hair cells (OHCs). Although they number around 20,000 in total, their contribution to auditory nerve output is only ~5%. OHCs amplify and fine-tune auditory signals.

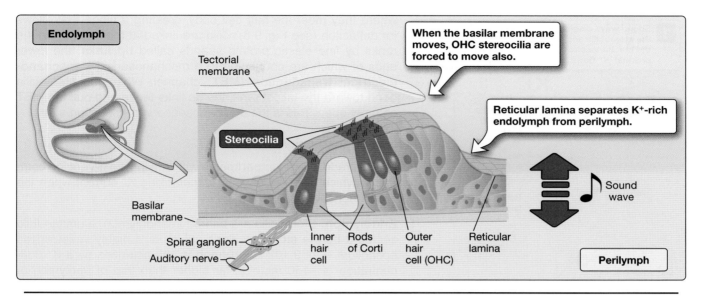

Figure 9.7.
Organ of Corti.

2. **Reticular lamina:** Hair cells are covered with a stiff, membranous **reticular lamina** (see Chapter 4·II·A) that is anchored to the basilar membrane by struts (**rods of Corti**). The reticular lamina both provides structural support for the hair cells and also forms a barrier to ion movement between endolymph and perilymph.

3. **Tectorial membrane:** Hair cells bear ~100 sensory stereocilia at their apical surface. They protrude through the reticular lamina into endolymph, a K⁺-rich fluid whose unusual ion composition is critical for generating an auditory receptor potential (discussed in the following sections). The tips of OHC stereocilia embed in a gelatinous **tectorial membrane**, which lies over the cells like a blanket. When the basilar membrane is buffeted by sound, the hair cells are dragged back and forth beneath the blanket, and the stereocilia are forced to bend (see Fig. 9.7).

B. Hair cell function

Hair cells convert stereociliary motion into an electrical signal that can be relayed to the central nervous system (CNS) for decoding. Stereociliary movement is sensed at the molecular level by mechanoreceptive ion channels.

1. **Hair cell structure:** Hair cells are polarized, nonneuronal sensory cells. The apical side bears several rows of stereocilia, which are stepped in height to form ranks (Fig. 9.8). The basal side synapses with one or more sensory afferent neurons, which the hair cell communicates with using an excitatory neurotransmitter (glutamate) when stimulated appropriately.

2. **Stereocilia:** At birth, hair cells contain a true cilium (**kinocilium**) that may help establish stereociliary orientation. The kinocilium is not involved in auditory transduction, degenerating shortly after birth. **Stereocilia** are actin filled and rigid. They taper at their base

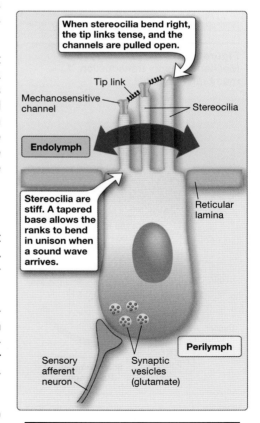

Figure 9.8.
Role of stereocilia in mechanosensory transduction.

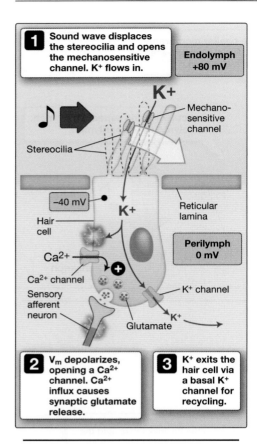

Figure 9.9.
Mechanosensory transduction. V_m = membrane potential.

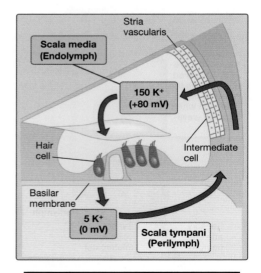

Figure 9.10.
K^+ recycling. K^+ concentrations are given in mmol/L. V_m = membrane potential.

where they meet the hair cell body, creating a hinge that allows for deflection (see Fig. 9.8). Cilia are linked at their tips down the ranks by fine elastic protein strands called **tip links**, the lower ends of which are connected to a mechanosensitive **mechano-electrical transduction** (**MET**) channel. When the stereocilia bend toward the tallest rank, the tip links tense and pull the MET channel open. This is the mechanosensory transduction step.

C. Mechanotransduction

Hair cells straddle two compartments with strikingly different ion compositions, which favor an unusual *depolarizing* K^+ current when the MET channel opens (Fig. 9.9).

1. **Endolymph:** Stereocilia are bathed in endolymph, a unique fluid secreted by the stria vascularis, which is a highly vascularized epithelium (Fig. 9.10). Endolymph is characterized by a K^+ concentration of ~150 mmol/L, far higher than that of perilymph or extracellular fluid ([ECF] ~4 mmol/L). The inside of the hair cell relative to endolymph is −120 mV, creating a strong electrochemical gradient for K^+ influx across the stereociliary membrane.

2. **Perilymph:** The basolateral side of the hair cell is bathed in perilymph. This fluid is high in Na^+ and low in K^+, much like ECF. Perilymph is considered to be at 0 mV. Hair cell membrane potential (V_m) relative to perilymph is −40 mV. The voltage difference between endolymph and perilymph, which approximates 80 mV, is the **endocochlear potential**.

3. **Receptor currents:** The mechanosensitive MET channel is relatively nonselective for cations. When it opens, K^+ flows into the cell and causes a receptor depolarization (see Fig. 9.9). This stands in stark contrast to the usual effects of K^+-channel opening on V_m. Most cells are bathed in low-K^+ ECF, so K^+-channel opening causes K^+ efflux and hyperpolarization.

4. **Potassium recycling:** K^+ exits hair cells via K^+ channels in the basolateral membrane (see Fig. 9.9) and is returned to the endolymph via the stria vascularis (see Fig. 9.10).

Clinical Application 9.1: Congenital Hearing Loss

Congenital hearing loss (HL) results from mutations in any of the many genes required for auditory transduction, signaling, and processing. The most common form of HL results from a recessive mutation in the *GJB2* gene, which encodes connexin-26. Connexins are proteins that form gap-junction channels between adjacent cells in many tissues, including those of the stria vascularis (see Chapter 4·II·F). HL can also result from mutations in the *KCNJ10* gene. The *KCNJ10* gene encodes an ATP-gated K^+ channel expressed in intermediate cells of the stria vascularis (see Fig. 9.10). These cells are responsible for maintaining high endolymph K^+ concentrations. *KCNJ10* mutation interrupts K^+ recycling, collapses the endocochlear potential required for auditory transduction, and causes profound deafness.

D. Converting sounds to electrical signals

When sound displaces the oval window inward, fluid pressure within the scala vestibuli rises. The cochlea is encased within bone, which is incompressible, but the Reissner membrane is thin and flexible. Thus, the pressure increase displaces the membrane downward. Pressure within the scala media now rises, but endolymph is also incompressible, so the pressure increase causes the basilar membrane to flex downward and impinge on the scala tympani. Within the organ of Corti, basilar membrane displacement causes endolymph to flow beneath the tectorial membrane. It also shifts the position of the tectorial membrane relative to the hair cells. These movements cause stereociliary bending and hair-cell excitation. However, the methods IHCs and OHCs use to sense and respond to these movements differ significantly.

1. **Inner hair cells:** IHC stereocilia are not tethered, so they move freely with endolymph flow. Bending occurs in response to the fluid currents caused by basilar membrane displacement. In the absence of sound, stereocilia tip links have a resting tension that keeps some MET channels open. The resultant current flow triggers an occasional action potential (AP) in the auditory nerve (Fig. 9.11A). When the basilar membrane bows downward in response to sound, endolymph currents cause the stereocilia to bend away from the tallest rank. This bending movement relieves the resting tip link tension and allows any open MET channels to close. The resting hair cell current and auditory nerve activity ceases (see Fig. 9.10B). When pressure in the scala vestibuli decreases, stereocilia are displaced toward the tallest rank. The tip links tense, the MET channels open, and the auditory nerve firing rate increases in proportion to pressure wave intensity (see Fig. 9.11C). Note that stereociliary ranks of all IHCs are aligned so that they face in the same direction. This alignment is important because it allows for a unified IHC response to basilar membrane movement.

2. **Outer hair cells:** OHC stereocilia tips are firmly embedded in the tectorial membrane, so they bend in response to shifts in tectorial membrane position relative to the hair cells. Although IHC excitation generates an auditory signal, OHC depolarization, by contrast, causes the hair cell to contract and shorten, pulling down on the tectorial membrane. Conversely, membrane hyperpolarization causes cell lengthening. These OHC shape changes amplify sound-induced basilar membrane movements, creating a **cochlear amplifier** that fine-tunes frequency discrimination by the IHCs. Contraction upon depolarization (electromotility) is mediated by **prestin**, a potential-dependent molecular motor. Prestin is abundant in the lateral walls of OHCs.

> The cochlear amplifier is capable of generating sounds (**otoacoustic emissions**) loud enough to be heard by a bystander! Emissions may occur spontaneously or in response to an applied auditory stimulus. The phenomenon provides a basis for noninvasive testing for hearing defects in newborns and young children. Damage to the inner ear eliminates the emissions.

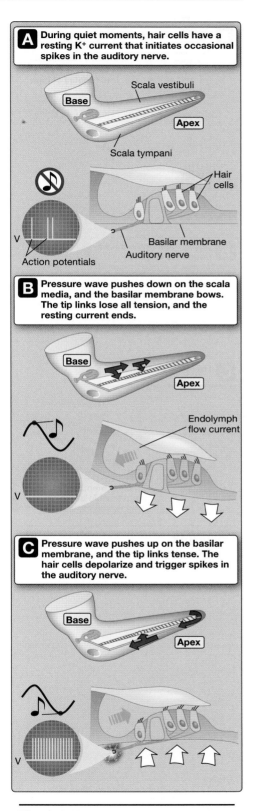

A During quiet moments, hair cells have a resting K+ current that initiates occasional spikes in the auditory nerve.

Scala vestibuli
Base
Apex
Scala tympani
Hair cells
Basilar membrane
Auditory nerve
V
Action potentials

B Pressure wave pushes down on the scala media, and the basilar membrane bows. The tip links lose all tension, and the resting current ends.

Base
Apex
Endolymph flow current
V

C Pressure wave pushes up on the basilar membrane, and the tip links tense. The hair cells depolarize and trigger spikes in the auditory nerve.

Base
Apex
V

Figure 9.11.
Sound-wave effects on hair cells within the organ of Corti. V = voltage.

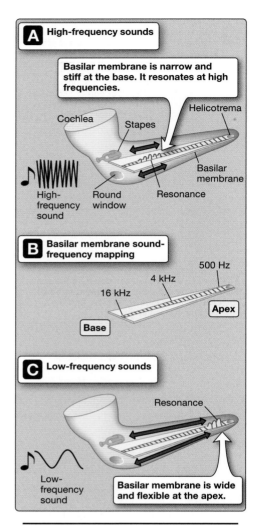

Figure 9.12.
Auditory encoding.

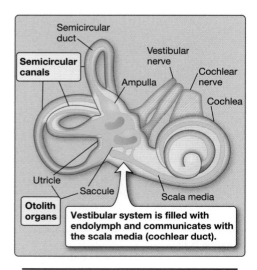

Figure 9.13.
Vestibular system and cochlea.

E. Auditory encoding

Most sounds are a complex mix of tones of variable intensity. The loudness of a sound correlates with the pressure wave's amplitude. Sound intensities are measured on a logarithmic scale in decibels (dB). Wave frequency (measured in **Hertz** [Hz]) determines whether it is perceived as being a low or high note (i.e., its pitch).

1. **Hearing range:** Normal conversation is held at ~60 dB. Leaves rustle at ~10 dB. Sounds of ~120 dB cause discomfort, and anything louder is perceived as being painful and can cause acute auditory damage. The ability to discriminate sound frequencies above 2,000 Hz decreases with age due to loss of hair cells at the basal end of the cochlea (**presbycusis**, Greek for "old hearing"), but most young people can hear sounds within a range of 20 to 20,000 Hz.

2. **Frequencies:** The cochlea is organized much like a stringed instrument, such as a piano or harp. At one end, the strings are short and taut, and they resonate at high frequencies. At the other end, the strings are long and produce low-frequency notes. In the cochlea, the basilar membrane serves a similar function to that of strings. The membrane runs the length of the cochlea and is tapered (Fig. 9.12). At the base, the membrane is relatively narrow (~0.1 mm) and stiff. Stereocilia here tend to be short and stiff also. In practice, this means that the base resonates, and hair cells are displaced maximally, by high-frequency sounds (see Fig. 9.12A and B). The basilar membrane widens to ~0.5 mm and becomes ~100-fold more flexible toward the apex, and the stereocilia become longer and more flexible. The apex is tuned to low frequencies (see Fig. 9.12B and C). This provides a way of breaking down a complex sound into its individual frequencies and relaying information about their relative timing and intensity to the CNS. This acoustic analysis is known as **auditory place coding**.

F. Auditory pathways

The sensory afferents innervating hair cells at the cochlear apex activate at low-frequency sounds, whereas afferents from the base activate at high frequencies (see Fig. 9.11), which allows for spatial mapping of sound frequencies (**tonotopy**). Map integrity is preserved during relay back to the auditory cortex for processing and interpretation. Auditory signals are relayed from the cochlea to the CNS via the spiral ganglion, which is located within the bony **modiolus** that forms the center of the cochlear spiral. The cochlear nerve joins the vestibular nerve to become the vestibulocochlear nerve (CN VIII), which projects to cochlear nuclei in the brainstem medulla. Many fibers then cross the midline and travel to the **inferior colliculus** of the midbrain, an area involved in auditory integration. Information is then relayed to the primary auditory cortex and associated areas involved in speech (**Wernicke area**) for interpretation.

V. BALANCE

The ability to stay upright is a feat we rarely think about until the system is stressed by illness or age or when we place ourselves in situations that force us to pay attention (e.g., walking down a steep trail or stepping into an unstable boat). Maintaining balance requires sensory input from numerous areas that constantly update motor control systems about body position. The most important of these is the vestibular system, a specialized sensory organ that provides rapid and sensitive information about changes in head position. The vestibular system allows us to correct body position before we fall and helps maintain a stable retinal image during head movements.

A. Vestibular system

The vestibular system is a part of the inner ear. The principal components are the otolith organs and the semicircular canals (Fig. 9.13). Like the cochlea, the system comprises membranous, labyrinthine structures encased in bone and bathed in perilymph. The interior, which is continuous with the scala media, contains endolymph. Sensory transduction relies on mechanosensory hair cells, which, unlike those of the cochlea, retain their kinocilium.

1. **Otolith organs:** The otolith organs comprise two chambers near the center of the inner ear. The **utricle** detects linear acceleration and deceleration. The **saccule** is oriented to detect movements caused by vertical acceleration (e.g., riding an elevator). Both also respond to changes in head angle.

2. **Semicircular canals:** The semicircular canals detect angular head rotation. As the name suggests, the canals comprise semicircular tubes with a swelling (**ampulla**) at their base. The vestibular system comprises three such canals (**anterior**, **posterior**, and **horizontal**) that are oriented perpendicular to one another to allow detection of movements in any of three dimensions. These are similar to flight gauges indicating pitch, yaw, and roll in an airplane.

B. Otolith organ function

Each otolith organ contains a sensory epithelium (**macula**) comprising innumerable hair cells and their supporting structures (Fig. 9.14). The kinocilium and stereocilia that project from the hair cell apical surface embed in a gelatinous **otolithic membrane** studded with calcium carbonate crystals. These are the **otoliths** (Greek for "ear stones") that give the organs their name. Their purpose is to add inertial mass to the membrane.

1. **Mechanotransduction:** When the head tilts forward or backward, the otolithic membrane moves under the influence of gravity, and the embedded sensory cilia bend (Fig. 9.15). Similar movements occur when the head accelerates or decelerates suddenly. The mechanotransduction step is identical to that described in Section IV·C. When the stereocilia move toward the kinocilium, tip links tense and open a mechanosensitive channel in the ciliary tip, the membrane depolarizes, and the sensory nerve afferents fire APs. When the stereociliary assembly moves away from the kinocilium, the membrane hyperpolarizes, and vestibular nerve activity decreases.

2. **Hair cell orientation:** Hair cells within both otolith organs are oriented relative to an otolithic membrane ridge (**striola**). The ridge curves across the width of the macula, meaning that hair cell orientation shifts with the curve also. This allows the hair cells to sense movement in any direction (Fig. 9.16). Hair cell orientation also

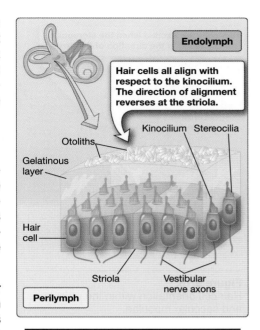

Figure 9.14.
Sensory macula from the utricle.

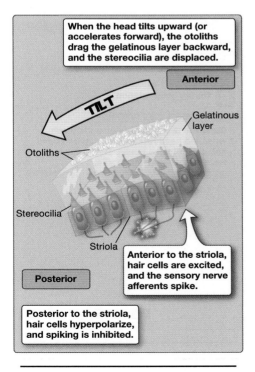

Figure 9.15.
Utricular macular response to head tilt.

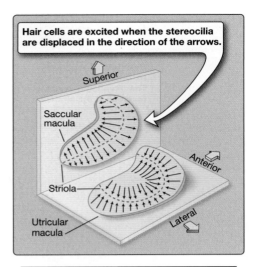

Figure 9.16.
Hair cell orientation within the utricular and saccular maculae.

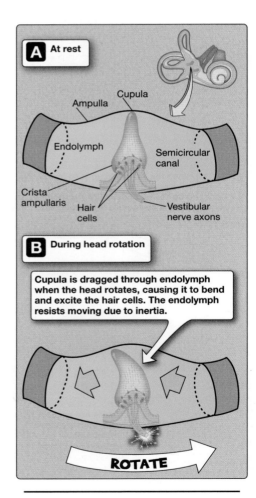

Figure 9.17.
Transduction of rotational head movements by the semicircular canals.

reverses at the striola (see Figs. 9.14 and 9.16), which ensures that linear head movements always activate at least some hair cells within both otolith organs, which further enhances sensory discrimination.

C. Semicircular canal function

The sensory epithelium within semicircular canals is pushed up into a crest (**ampullary crest**, or **crista ampullaris**) covered with sensory hair cells as shown in Figure 9.17. Their cilia are embedded in a gelatinous flap called the **cupula** that occludes the ampulla. The canals are filled with endolymph. When the head moves, the walls of the canal slip past the endolymph, which is held back by inertia. Within the ampulla, head movement drags the cupula through the endolymph, causing it to deflect. This action tugs on the hair cells and generates a receptor potential, the polarity of which correlates with direction of movement. If the head rotates continually, the endolymph eventually achieves the same rotational speed and the receptor potential wanes. Sudden deceleration then initiates a new response. Note that the vestibular system comprises three such canals oriented at right angles to each other so as to detect rotational movements in any direction. The structures of the inner ear are also mirrored on each side of the head. Responses from all six canals and four otolith organs are integrated by brainstem vestibular nuclei.

Float an ice cube in a large measuring cup filled with water and take note of its position relative to the spout or handle. If you now rotate in place while holding the cup, you will note that inertia prevents the contents of the cup from moving even though the cup walls are rotating around it. A similar phenomenon happens in the semicircular canals.

D. Vestibular nuclei

The brainstem contains a group of vestibular nuclei that form an integrative control center responsible for balance. Vestibular sensory nerve signals reach the nuclei via the vestibular nerve, vestibular ganglion, and vestibulocochlear nerve (CN VIII). Vestibular sensory afferents also project to the cerebellum. The vestibular nuclei also receive sensory data from the eyes and somatic proprioceptors located in muscles and joints (see Chapter 11·II). This information is integrated and then used to execute reflex movements of the eye, head, and the muscles involved in postural control.

E. Vestibuloocular reflex

Changes in head position affect eye position also, which is problematic because moving images lack acuity. Camera movement similarly blurs a photographic image. One important vestibular function is, thus, to inform ocular motor control centers about the direction of head movement so that eye position may be adjusted to maintain a stable retinal image even as the head is moving (**vestibuloocular reflex [VOR]**). Figure 9.18 considers what happens when the head is turned to the left as an example, but similar principles govern responses to head movements in any direction. Turning the head to

Clinical Application 9.2: Vestibular Dysfunction

A normal sense of balance is so important that severe cases of vestibular dysfunction can be disabling. Even mild cases can bring on highly disturbing sensations of **vertigo** and nausea. Vertigo is a sense of spinning in space, or that a room is spinning around a person even when stationary. **Benign paroxysmal positional vertigo** (**BPPV**) is the most common form of vestibular dysfunction. The symptoms include dizziness, lightheadedness, and vertigo. It is usually brought on by rolling over in bed or when getting out of bed. Tilting the head to look upward may also precipitate an attack. BPPV is caused by otoliths that have detached from the otolithic membrane and made their way into one of the semicircular canals. They then stimulate the hair cells inappropriately when the head is moved in a particular direction. BPPV usually resolves spontaneously, although episodes may be recurrent. Sequential manipulation of head position by a trained therapist to work the otoliths out of the canals may afford a more permanent solution. Episodes of dizziness and vertigo can also be caused by infection or inflammation of the inner ear (**labyrinthitis**) and are usually accompanied by hearing loss. Appropriate treatment can lead to full recovery of both hearing and balance. **Ménière disease** is an idiopathic inner ear disorder believed to result from inadequate drainage of endolymph from the inner ear.

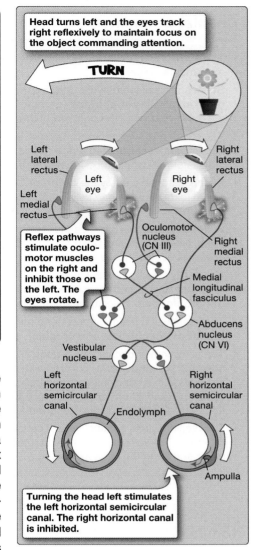

Figure 9.18.
Vestibuloocular reflex.

the left stimulates hair cells in the horizontal semicircular canal on the left side of the head and inhibits output from the horizontal canal on the right. Sensory signals are relayed via the vestibular nuclei to the contralateral **abducens nucleus**, a CN nucleus (CN VI) located in the brainstem. Excitatory outputs from this nucleus travel directly via the abducens nerve to the right lateral rectus eye muscle, one of six muscles that control eye movement. They also travel via the medial longitudinal fasciculus to the oculomotor nucleus (CN III), from where they excite the left medial rectus muscle. Inhibitory pathways simultaneously suppress left lateral rectus and right medial rectus muscle contraction, and the eyes track right, a movement that is exactly equal and opposite to changes in head position. The retinal image remains centered as a result. The pathways between vestibular system and eyes are very short, so the reflex is extremely fast.

Clinical Application 9.3: Caloric Reflex Testing

The vestibuloocular reflex (VOR) provides a way for assessing functionality of the vestibular system, brainstem, and oculomotor pathways. During caloric reflex testing, a supine patient's head is inclined 30°, and the external auditory meatus is then irrigated with ~50 mL of ice water from a syringe. The cold water sets up a temperature gradient across the horizontal canal. Endolymph in the section of the canal closest to the external ear becomes denser as it cools, and it sinks under the influence of gravity. This sinking motion displaces endolymph in sections farthest from the cooled ear, causing fluid movement within the canal that mimics the effects of turning the head away from the stimulated ear. The VOR causes both eyes to track slowly toward the cooled ear and then rapidly reset their focus toward the opposite side (this rapid resetting motion is called **nystagmus**). Irrigating the ear with warm water induces eye turning away from the stimulated ear. The caloric reflex test is a useful means of assessing brainstem function in comatose patients. The VOR is typically abnormal or absent in patients who have suffered brainstem hemorrhage or infarction.

Chapter Summary

- The **auditory** and **vestibular systems** both use **mechanosensory hair cells** to transduce sound waves and head movements, respectively.
- Sounds are collected by the **outer ear** and channeled via the **tympanic membrane** into the middle ear for amplification by an articulating lever system comprising three **ossicles** (**malleus**, **incus**, and **stapes**). The stapes transfers sound via the **oval window** to the **cochlea** of the **inner ear** for auditory transduction.
- The cochlea is a long, coiled, tapered tube, containing three fluid-filled tubular chambers. Two contiguous outer chambers (**scala vestibuli** and **scala tympani**) are filled with **perilymph** and provide a pathway for sounds to travel through the system. A central tube (**scala media**) is filled with K^+-rich **endolymph** and contains the auditory sensor.
- Sound is transduced by the **organ of Corti**, which contains four rows of sensory hair cells. **Stereocilia** project from the apical surface of the hair cells into the central chamber. Stereocilia of outer hair cells embed in an overlying **tectorial membrane**. The basolateral surface of the cells rests on the **basilar membrane**.
- When sound waves displace the basilar membrane, the stereocilia bend. Adjacent stereocilia are connected by **tip links**, which tense and open mechanoreceptor channels in the stereociliary membrane during passage of sounds. The channels allow K^+ influx from the endolymph, producing a depolarizing receptor potential.
- The basilar membrane is tapered: the base is narrow and stiff and resonates at high frequencies, whereas the apex is wide and flexible and resonates at low frequencies. This allows complex sounds to be broken down to create a **spatial** (**tonotopic**) **map**. Spatial information is preserved during data transmission to the auditory cortex.
- The vestibular system detects head movements. It comprises three orthogonally arranged **semicircular canals** and two **otolith organs** (**utricle** and **saccule**), all of which are filled with endolymph.
- The semicircular canals detect **head rotation** in any plane. A swelling at the base of each canal (**ampulla**) contains a ridge (**crista ampullaris**) covered with sensory hair cells, each bearing a kinocilium and several stereocilia. The ciliary tips embed in a gelatinous **cupula** that protrudes into the endolymph. Head rotation forces endolymph against the cupula, causing it (and its embedded sensory cilia) to bend. The sensory cilia contain K^+-permeant mechanosensory channels that open during bending to generate a depolarizing receptor potential.
- The otolith organs contain lines of sensory hair cells whose cilia are embedded in a gelatinous membrane encrusted with **otoliths** (calcium carbonate crystals). Shifts in head position or sudden acceleration cause the heavy membrane to move and bend the sensory cilia.

Study Questions

Choose the ONE best answer.

9.1. A child with congenital hearing loss is diagnosed with round window atresia (absence of a round window) following CT imaging studies. Atresia most likely impairs hearing by which of the following mechanisms?

A. It impairs impedance matching.
B. It prevents ossicular chain movement.
C. It prevents perilymph movement.
D. It stiffens the basilar membrane base.
E. Pressure across the eardrum cannot equalize.

Best answer = C. The round window allows perilymph to move within the cochlear chambers when the oval window is displaced by the stapes (see Section IV). The cochlea is encased in bone, which prevents chamber expansion when the oval window is displaced. Therefore, in the absence of a round window, perilymph movement and basilar membrane flexion cannot occur. Impedance matching (A) is a function of the ossicular chain whose movement (B) would not be affected by atresia. Basilar membrane stiffening (D) would affect frequency discrimination but should not cause hearing loss. Air pressure equalization (E) relies on the eustachian tube.

9.2. A rare inherited disorder that prevents synthesis of tip-link proteins has been observed in animal models. Gene expression would most likely be expected to have which of the following effects on auditory transduction?

A. Endocochlear potential would collapse.
B. Hair cells would lose sensory function.
C. K^+ recycling would be inhibited.
D. Only vestibular function would be impaired.
E. Stereocilia would not be displaced by sound.

Best answer = B. Sounds are transduced by hair cells, which are excited when tip links between adjacent stereocilia tense, and a K^+-permeant mechanoelectrical transduction (MET) channel opens (see Section IV·C). The tip links tense when stereocilia are displaced by sound waves passing through the cochlea. Auditory and vestibular hairs cells would be affected similarly (D). The endocochlear potential (A) and K^+ recycling (C) rely on the stria vascularis to concentrate K^+ within endolymph, which should not be affected by a tip-link disorder. If the tip links were missing, the stereocilia would still be displaced by sound (E), but the hair cells would be unable to generate a receptor potential.

9.3. A hearing screening test on a 1-month-old reveals bilateral hearing loss. Genetic analysis confirms a congenital condition affecting development of the inner ear that results in cochlear aplasia and the absence of an organ of Corti. Which of the following most likely describes the properties of the organ of Corti?

A. Inner hair cells are sound amplifiers.
B. Stereocilia do not bend toward the kinocilium.
C. The apex is attuned to high-frequency sounds.
D. The basilar membrane is wider at the apex.
E. The scala media is filled with perilymph.

Best answer = D. The basilar membrane resonates at different frequencies along its length (see Section IV·D). The membrane is wider at the apex and resonates at low frequencies. The outer hair cells are believed to help amplify these signals, not the inner hair cells (A). Auditory nerve signals are dominated by output from inner hair cells, which signal when stereocilia bend toward the kinocilium (B). The basilar membrane base and the hair cells it supports are attuned to high frequencies, not the apex (C). The scala media is filled with endolymph, not perilymph (E).

9.4. The right ear of a comatose patient is irrigated with cold water to assess vestibuloocular reflex (VOR) function. Which of the following statements most likely describes the VOR or its components?

A. Ear cooling causes receptor-mediated K^+ influx.
B. The horizontal semicircular canal detects vertical motion.
C. The VOR is initiated by otolith displacement.
D. The VOR is mediated by thermosensory nerves.
E. VOR vestibular nuclei are located in the thalamus.

Best answer = A. The temperature gradient created by irrigating the ear canal with cold water causes endolymph to move within the horizontal semicircular canal (see Section V·E). Movement is transduced by mechanosensory channels on sensory hair cells, which open to allow K^+ influx and depolarization. The horizontal canal normally detects rotational head movements in a horizontal, not vertical (B), plane. Otoliths are normally found in the otolith organs, not the semicircular canals (C). The VOR does not involve thermoreceptors (D). VOR vestibular nuclei are located in the brainstem, not the thalamus (E).

10 Taste and Smell

I. OVERVIEW

The gustatory and olfactory systems are probably the oldest of the senses in evolutionary terms. Both systems allow us to detect chemicals in the external environment and are usually grouped together. In practice, however, they represent two very different sensory modalities that complement but cannot replace each other. Taste receptors are modified epithelial cells, whereas olfactory receptors are neurons. Taste allows us to differentiate between very basic flavors, such as sweet vs. salty or savory vs. sour. Taste is closely linked with appetite and cravings, such as a need to ingest salt (NaCl) or something sweet, and it is also protective. Bitter taste often helps us avoid ingesting toxins, whereas the taste of acid (sour) often indicates food decay. Olfaction allows us to detect and identify thousands of unique chemicals, including pheromones.

Table 10.1: Taste Receptor Type

Taste	Perception	Taste Bud Cell Type
Salty	Pleasant	Uncharacterized
Sweet	Pleasant	II
Umami	Pleasant	II
Bitter	Aversive	II
Sour	Aversive	III

II. TASTE

The five basic tastes are **salty**, **sweet**, **umami**, **bitter**, and **sour** (Table 10.1). *Umami* ("good taste" in Japanese) is epitomized by the taste of monosodium glutamate (MSG), which imparts a savory, meaty flavor to food. The taste of fat may constitute a sixth basic taste, but the transduction mechanisms are not fully delineated. In addition to taste buds, a variety of sensory neurons (e.g., mechanosensors and thermosensors) contribute to the gustatory sense. The chemical sensations that mimic hot (e.g., the burning sensation associated with chili peppers) and cold (e.g., menthol) are not tastes but, rather, are mediated by somatosensory pathways located in the oral cavity or nasal passage (see Chapter 15·VII·B).

A. Taste buds

Taste receptor cells are usually clustered in taste buds, which are distributed throughout the oral cavity. Lingual taste buds are organized like garlic bulbs (Fig. 10.1), each containing 50 to 100 elongated neuroepithelial cell "cloves." Adjacent cells are connected apically by tight junctions. Some cells extend microvilli into a taste bud's small central pore, which provides a way for oral fluids (saliva) and their dissolved tastants to enter the taste bud and be sensed. Taste buds contain four main taste bud cell types: type I are glia-like, type II are taste receptor cells, type III are neuron-like presynaptic cells, and type IV comprise a heterogenous mix of progenitor cells.

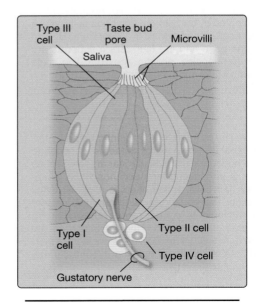

Figure 10.1.
Taste bud organization.

1. **Type I cells: Type I** cells account for ~50% of the total number of cells in a taste bud. They are inexcitable cells with **glia-like** properties. They extend membrane processes that surround other cells within the taste bud and may help regulate extracellular K$^+$ concentrations during excitation. Type I cells also express nucleoside triphosphate diphosphohydrolyse-2 (NTPDase2) on their surface. NTPDase2 is an ecto-ATPase involved in gustatory signaling (Fig. 10.2).

2. **Type II cells: Type II** cells are excitable **sensory receptor cells**. Their membranes contain one of two taste receptor types. Taste receptors are G protein–coupled receptors (GPCRs) that mediate sweet, umami, and bitter tastes. They do not respond to salty or sour tastants. Individual type II cells are tastant specific. Some express type 1 taste receptors (TAS1Rs), which respond to sweet and umami tastants. A separate population expresses type 2 taste receptors (TAS2Rs), which facilitate bitter taste sensation.

3. **Type III cells: Type III** cells make up less than 20% of the total number of cells in a taste bud. They are the only cell type to form a conventional (i.e., structurally familiar) synapse with a sensory nerve. Type III cells are suggested to mediate sour taste, but they also express an ecto-5′-nucleotidase (NT5E) that helps modulate taste bud sensory signaling.

4. **Type IV (basal) cells:** Taste bud cells are continually being damaged and replaced, such that an entire taste bud turns over every 10 to 12 days. Type IV cells comprise a heterogeneous mix of progenitor cells that divide rapidly and can differentiate into type I, II, or III cells as the need arises. Their numbers are maintained by basal keratinocytes, which constitute taste bud stem cells.

B. Sweet, umami, and bitter tastes

Sweet, umami, and bitter tastes are transduced by type II cells. The three tastant classes are perceived as either being pleasant and signaling the presence of food, or noxious and indicative of a potential toxin.

1. **Receptors:** Type II cells express GPCRs that respond to specific tastants.

   a. **Sweet:** Sweet tastes are associated with mono- and disaccharides, such as glucose and sucrose. Sugars are a primary energy source; therefore, the ability to recognize their presence in food has clear evolutionary advantages. Sugars are sensed by a TAS1R. The receptor is a heterodimer of two GPCRs (TAS1R2+3), products of the *TAS1R2* and *TAS1R3* genes.

   b. **Umami:** The umami taste is elicited by glutamate, which imparts a rich, savory taste to meats, mushrooms, cheese, fish, and many vegetables. Umami taste sensation involves a TAS1R and two metabotropic glutamate receptors (mGluRs).

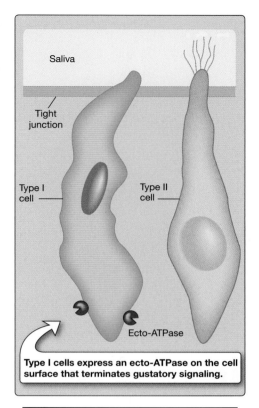

Saliva

Tight junction

Type I cell

Type II cell

Ecto-ATPase

Type I cells express an ecto-ATPase on the cell surface that terminates gustatory signaling.

Figure 10.2.
Type I receptor cell taste transduction.

> The extracellular domain of the sweet taste receptor is large and shaped like a Venus flytrap. The domain contains binding sites for sugars, but it also recognizes certain proteins as sweet (e.g., monellin). This feature has facilitated the development of a wide range of synthetic peptide sweeteners, including aspartame.

i. **Type 1 taste receptors:** The umami TAS1R comprises a TAS1R1+3 GPCR heterodimer. Receptor responsiveness to glutamate is markedly potentiated by GMP and inosine monophosphate (IMP), both of which are common in umami foods.

ii. **Metabotropic glutamate receptors:** Umami taste also involves two taste-specific mGluRs: mGluR1 and mGluR4. These mGluRs have highly truncated extracellular ligand-binding domains compared with the brain versions, but both retain glutamate sensitivity.

c. **Bitter:** Many plants, fungi, and some animals produce toxins as a natural defense mechanism. Evolution has helped guide our choice of food by associating many such poisons with a bitter taste. Most bitter tastants are detected by GPCRs. Toxins are such a diverse group that recognizing them as such requires specific receptor proteins. Type II cells that sense bitter stimuli express subsets of at least 25 TAS2R variants. TAS2Rs are monomeric GPCRs, and some are highly tastant specific, whereas others have broad specificity.

> Quinine is a bitter-tasting toxin with antimalarial properties extracted from cinchona tree bark. It blocks most classes of K^+ channel, including those found in the heart, causing nonspecific membrane depolarization. Side effects include cardiac dysrhythmias (e.g., long QT syndrome and torsades de pointes), which has limited its use as a therapeutic agent.

2. **Transduction pathway:** Tastant binding to its TASR activates **gustducin**, a sense-specific G protein related to transducin (see Chapter 8·V·B) that signals receptor occupancy through ATP release into the taste bud interstitium (Fig. 10.3).

a. **Activation:** The gustducin $G_{\beta\gamma}$ subunit activates phospholipase Cβ_2 (PLCβ_2) and initiates IP$_3$-induced Ca^{2+} release from intracellular stores. Ca^{2+} then activates a Ca^{2+}-dependent Na^+ influx via TRPM5 (a transient receptor-potential channel; see Chapter 2·VI·D), and the receptor cell depolarizes. Voltage-gated Na^+ influx amplifies this receptor potential and initiates trains of action potentials.

b. **ATP release:** Receptor cell depolarization opens a calcium homeostasis modulator 1/3 (CALHM1/3) channel in the

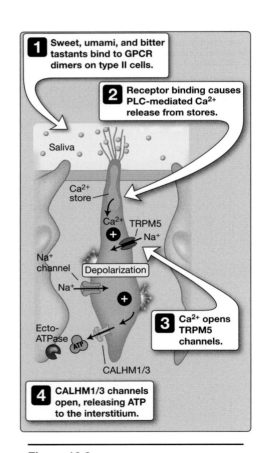

Figure 10.3.
Type II receptor cell taste transduction. CALHM1/3 = calcium homeostasis modulator 1/3 channel; PLC = phospholipase C; TRPM5 = transient receptor-potential M5 channel.

receptor cell membrane. CALHM1/3 channels are related to the connexins that form gap junctions between cells (see Chapter 4·II·F). CALHM1/3 channels are oligomeric proteins with weak ion selectivity that, when open, allow ATP to exit the cell. ATP then diffuses through the interstitium to type III receptor cells, which are innervated. Types II and III receptor cells and afferent nerve fibers all express ATP-sensitive purinergic receptors (see Section E).

> CALHM1 was first identified as being a hippocampal protein linked to an increased risk of developing late-onset Alzheimer disease (AD). Its precise role in the etiology of AD is the subject of ongoing investigations.

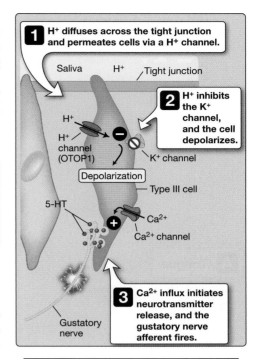

Figure 10.4.
Type III receptor cell taste transduction.

C. Sour

Sourness is the taste of acid (H^+). Common dietary examples include acetate (vinegar), citrate (lemons), and lactate (sour milk). Sour taste is mediated by type III cells. H^+ enters cells via otopetrin 1 (OTOP1) H^+ channels and depolarizes the cell directly, but, once inside the cell, H^+ also reduces K^+ efflux by inhibiting K^+ channels (Fig. 10.4). Inhibition further amplifies the depolarization caused by H^+ entry. If sufficiently large, the H^+-induced receptor potential activates voltage-dependent Na^+ channels in the cell membrane and triggers a spike. Voltage-dependent Ca^{2+} channels then open to allow Ca^{2+} influx and transmitter release (5-hydroxytryptamine [5-HT], also known as serotonin) at a synapse with a sensory afferent neuron.

D. Salt

Salt taste is transduced by taste bud cells that have yet to be characterized. The cells express epithelial Na^+ channels (ENaCs). Na^+ influx via ENaC depolarizes the cells and initiates an action potential that results in ATP efflux from the cells via CALHM1/3 channels, as seen during activation of type II cells.

E. Chemosensory signal integration

Food contains a mix of different tastants, and taste bud output usually represents an integrated response to simultaneous stimulation of all tastant-responsive cell classes. Type III cells are the only taste bud cell type to communicate with a gustatory afferent nerve via a conventional synapse, but the three taste receptor cell classes together fashion an additional, albeit highly unconventional, synapse for relaying sensory signals to the central nervous system (CNS).

1. **Signaling:** Tastant binding by a type II cell initiates ATP release into the taste bud interstitium, which functions as a synaptic cleft. ATP binds to and stimulates the sensory afferent nerve directly. ATP also stimulates type III cells to secrete 5-HT onto the sensory afferent.

2. **Signal termination:** Types I and III cells both express enzymes on their surface that degrade ATP and help terminate signaling.

Table 10.2: Taste Bud Cell Regulation

Compound	Originating Cell	Target Cell	Effect
ATP	II	III	↑ 5-HT
	II	II	↑ ATP
	II	Gustatory neuron	↑ APs
Serotonin (5-HT)	III	Gustatory neuron	↑ APs
	III	II	↓ ATP
GABA	III	II	↓ ATP
Acetylcholine	II	II	↑ ATP
Cholecystokinin	II	II (bitter)	↑ ATP
Neuropeptide Y	II	II (sweet, umami)	↓ ATP
Glucagon-like peptide-1	II (sweet, umami)	Gustatory neuron	↑ APs

APs = action potentials.

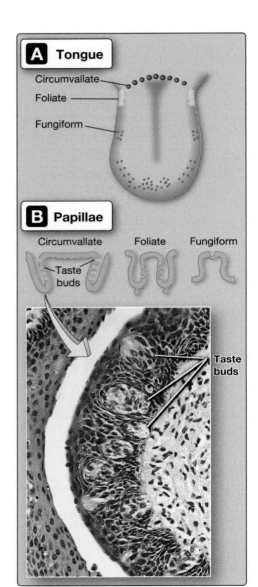

Figure 10.5.
Taste bud distribution on the tongue surface.

Type I cells express NTPDase2, whereas type III cells express NT5E.

3. **Taste bud regulation:** The output of all three taste cell types is modulated through autocrine, paracrine, and endocrine pathways that enhance type II cell output, intensify certain tastes, or alter taste preference.

   a. **Type II cell output:** ATP signaling by type II cells is modulated via autocrine and paracrine signaling pathways involving ATP, 5-HT, γ-aminobutyric acid (GABA), and acetylcholine (ACh), among other neurotransmitters (Table 10.2). These pathways ensure that the amount of ATP released following tastant binding is sufficient to stimulate but insufficient to desensitize the gustatory neuron.

   b. **Taste intensity:** Cholecystokinin, neuropeptide Y, and glucagon-like peptide 1 are involved in autocrine and paracrine signaling pathways that ensure that bitter tastants are perceived in the presence of sweet and umami stimuli (see Table 10.2). This reflects the importance of bitter taste in serving as a warning signal that a food may contain toxins.

   c. **Taste preference:** Type I cells express oxytocin receptors that may be involved in salt appetite. Type II and III cells express galanin receptors that increase fat consumption when stimulated. Leptin receptors present on type II cells enhance sweet tastants.

F. **Taste bud distribution**

   Taste buds are distributed throughout the oral cavity, although the highest concentrations are located on the dorsal surface of the tongue. Lingual taste buds reside on surface projections called **papillae**. Three types of papillae can be distinguished based on shape and taste bud density: **fungiform**, **foliate**, and **circumvallate** (Fig. 10.5).

   1. **Fungiform papillae:** The anterior portions of the tongue bear fungiform papillae. Each thumblike projection carries a few taste buds at its tip.

2. **Foliate papillae:** The posterior lateral edge of the tongue bears ridges called foliate papillae. The sides of the papillae are studded with hundreds of taste buds.

3. **Circumvallate papillae:** The largest concentration of taste buds is found on buttonlike circumvallate papillae. They are located in a line across the back of the tongue.

G. Neural pathways

Taste buds are innervated by three different cranial nerves (CNs). The tongue's anterior portions and the palate are innervated by the chorda tympani, a branch of the facial nerve (CN VII). Taste buds on the posterior tongue signal via the glossopharyngeal nerve (CN IX), whereas the vagus (CN X) innervates the pharynx and larynx. All three nerves relay information via the tractus solitarius to a gustatory area within the solitary nucleus (brainstem). Secondary fibers carry gustatory information to the thalamus and primary gustatory cortex.

III. SMELL

The human sense of smell is not as well developed as that in many animals, but, even so, the human olfactory system is capable of distinguishing hundreds of thousands of different odors. Olfactory sensitivity relies on several hundred unique receptors, each encoded by a different gene.

A. Receptors

Odorants are airborne chemicals that are inhaled and carried through the nasal passages during normal breathing or by intentional sniffing. Odors are detected and transduced by chemoreceptors that belong to the GPCR superfamily. The human genome contains ~900 different olfactory receptor genes or pseudogenes, of which ~390 are functionally expressed.

B. Receptor cells

Odorant receptors are expressed on cilia that project from a sheet of sensory neurons contained within the main olfactory epithelium that lines the roof of the nasal cavity (Fig. 10.6). The sensory

 [1]More information on these drugs can be found in *LIR Pharmacology*, 8e, Chapter 35·II·A–C.

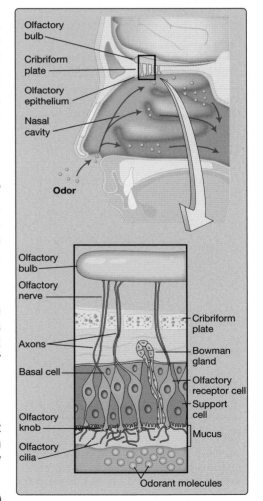

Figure 10.6.
Olfactory epithelium.

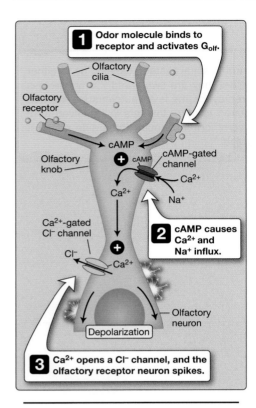

1 Odor molecule binds to receptor and activates G$_{olf}$.

Olfactory cilia

Olfactory receptor

cAMP

Olfactory knob

cAMP-gated channel

Ca^{2+}

Na$^+$

Ca^{2+}-gated Cl$^-$ channel

Cl$^-$

Ca^{2+}

Depolarization

Olfactory neuron

2 cAMP causes Ca^{2+} and Na$^+$ influx.

3 Ca^{2+} opens a Cl$^-$ channel, and the olfactory receptor neuron spikes.

Figure 10.7.
Olfactory transduction. G$_{olf}$ = olfactory-specific G protein.

neurons are bipolar. The apical portion of the cell body gives rise to a single dendrite that extends toward the epithelial surface and then terminates in a swelling (**olfactory knob**). Each knob supports 10 to 30 long, nonmotile, sensory cilia, which project into a thin layer of watery mucus. The mucus traps passing odorant molecules and allows them to be detected by odorant receptors.

Our sense of smell is largely reliant on the main olfactory epithelium, but accessory olfactory regions contribute. These include a rudimentary vomeronasal organ that detects phero-mones, and the septal organ of Masera located on the nasal septum.

C. Transduction

Odorant binding to its receptor activates an olfactory-specific G protein (G$_{olf}$) and initiates an adenylyl cyclase–mediated rise in intracellular cAMP concentration (Fig. 10.7). A cAMP-gated ion channel opens as a result, allowing Na$^+$ and Ca^{2+} influx. The Ca^{2+} flux, in turn, opens a TMEM16B (anoctamin 2) Ca^{2+}-dependent Cl$^-$ channel, and the combined depolarizing effect of cation influx and anion efflux on membrane potential may be sufficient to trigger a spike in the olfactory neuron.

D. Olfactory epithelium

Olfactory receptor neurons have a lifespan of 30–60 days and then are replaced. Receptor neurons are formed from olfactory epithelial **basal cells**, which are neuroblast stem cells. The epithelium also contains **supporting cells**, which have a glia-like function. The mucus that flows over the epithelium is secreted by **Bowman glands**. Olfactory mucus contains odorant-binding proteins that help ferry hydrophobic odorants to the olfactory receptors. Mucus also contains lactoferrin, lysozyme, and various immunoglobulins that help ensure that pathogens do not gain access to the CNS via olfactory nerves. The olfactory epithelium is one of the few regions of the body where CNS nerves directly interface with the external environment.

E. Neural pathways

Olfactory receptor cells are primary sensory neurons that project directly to the olfactory bulb, which is an extension of the forebrain.

Clinical Application 10.2: Anosmia

Although an inability to detect odors caused by food spoilage can increase the likelihood of food poisoning, loss of the sense of smell (**anosmia**) is not life threatening. Anosmia does significantly impact the quality of life, however. Anosmia markedly impairs food enjoyment and often causes appetite and weight loss, depression, and withdrawal from social events that involve food. Hyposmia commonly occurs during aging and as a result of upper respiratory tract infections. Neurodegenerative diseases (Parkinson and Alzheimer disease) can also impair sense of smell. Anosmia may often follow head trauma as a result of damage to the olfactory cortex or shearing of the olfactory nerves as they pass through the cribriform plate.

Biologic Sex and Aging 10.1: Taste and Smell

The senses of taste and smell both decline with age. Children are exquisitely sensitive to bitter taste, which poses problems for pharmaceutical companies, whose products are often perceived as being bitter and are rejected by children. Taste thresholds increase noticeably after age 70 years.

Olfactory senses are more sensitive to age-dependent decline. More than 50% of individuals age 65 years and older have increased olfactory thresholds, reflecting decreased olfactory sensory neuron numbers and olfactory bulb atrophy, among many other detrimental changes in olfactory organ structure and function.

Axons from the sensory neurons form bundles and then pass through foramina in the cribriform plate of the ethmoid bone. Axons travel in the olfactory nerve (CN I) to the glomerulus of the olfactory bulb, where they synapse. Neurons originating in the olfactory bulb project to several brain regions, including the olfactory cortex, thalamus, and hypothalamus. Individual olfactory neurons express a single receptor gene. Because the number of odorants a person can distinguish exceeds receptor gene number by several orders of magnitude, the receptors must recognize specific chemical groups of multiple odorant molecules rather than responding to just one odorant. The brain then extrapolates a unique odor signature based on relative output intensity from each of the different receptor types within the olfactory receptor array.

Chapter Summary

- **Gustation** and **olfaction** allow us to detect chemicals in food and in inhaled air. Taste is used largely to decide whether or not to swallow ingested food. Smell allows for an appreciation of food as well as detection of **pheromones**.

- The five basic tastes are **salty**, **sweet**, **umami** (savory), **bitter**, and **sour**.

- **Taste receptor cells** reside in **taste buds** found throughout the oral cavity. Type I cells are **glia-like** cells that help terminate gustatory signaling.

- Type II cells express tastant-specific GPCRs that detect sweet, umami, and bitter tastes. The GPCRs act through **gustducin**, a taste cell–specific **G protein**. The activated $G_{\beta\gamma}$ subunit elicits intracellular Ca^{2+} release and depolarization. Calcium homeostasis modulator 1/3 channels in the surface membrane then open and allow ATP to diffuse out of the cell. ATP stimulates sensory afferent neurons both directly and indirectly by modulating type III cell output.

- Type III cells transduce the sour taste of acid. H^+ permeates a H^+ channel and causes receptor depolarization. Type III cells are innervated by gustatory nerves and signal excitation to the afferents via 5-HT (serotonin) release.

- **Odors** are detected by a thin **olfactory epithelium** located in the roof of the nasal cavity. **Olfactory receptors** are GPCRs expressed on the surface of cilia that project into a mucus layer lining the olfactory epithelium.

- Olfactory receptor cells are bipolar central neurons that relay information to the **olfactory bulb**. Olfactory receptor binding triggers a cAMP-mediated signaling cascade that leads to cAMP-dependent Ca^{2+} influx and Ca^{2+}-dependent Cl^- efflux causing nerve excitation.

Study Questions

Choose the ONE best answer.

10.1. A 23-year-old female with monosodium glutamate (MSG) symptom complex experiences nausea, palpitations, and diaphoresis after eating food containing MSG, a food additive that enhances umami taste. Which of the following most likely describes the MSG sensory transduction mechanism?

A. It is sensed by type I taste receptor cells.
B. MSG activates type III receptor cells.
C. MSG binds to a domain on "sweet" receptors.
D. The MSG receptor is a Na⁺ channel.
E. Umami cells release ATP.

Best answer = E. MSG binds to a G protein–coupled receptor (GPCR) on umami-specific type II cells (see Section II·A·2). Receptor binding initiates Ca²⁺ release from intracellular stores, which causes membrane depolarization and opening of calcium homeostasis modulator (CALHM1/3) channels. CALHM1/3 allows ATP to diffuse out of the cell and stimulate a gustatory nerve. MSG may have lesser, indirect effects on type I (A) and type III cells (B) but is detected primarily by type II cells. Type II cells are tastant-specific; sweet receptors do not mediate umami taste (C). The MSG receptor is a GPCR, not a Na⁺ channel (D).

10.2 A 45-year-old male with a history of major depressive disorder is prescribed a selective serotonin reuptake inhibitor (SSRI). During a follow-up visit to the office, he notes that his favorite cocktail, which contains quinine, tastes more bitter since he began taking the medication. The SSRI has most likely affected the functioning of which of the following gustatory cell types?

A. Gustatory nerve
B. Type I cells
C. Type II cells
D. Type III cells
E. Type IV cells

Best answer = D. Type III cells communicate taste sensation to sensory afferent neurons using serotonin, also known as 5-hydroxytryptamine (5-HT). SSRIs decrease taste thresholds by increasing taste bud synaptic 5-HT levels. Sensory afferent neurons communicate taste receptor activation to the CNS via the gustatory nerve (A). Type I cells (B) are glia-like and support taste receptor cell function. Type II cells (C) detect bitter stimuli but communicate with the sensory afferents via Type III cells. Type IV cells (E) are taste bud stem cells.

10.3. A 32-year-old male presents with anosmia (loss of sense of smell) following accidental inhalation of a volatile chemical at work. Which of the following statements most likely describes olfactory neuron function?

A. Olfaction is mediated by guanylyl cyclase.
B. The patient's sense of taste is probably intact.
C. Their axons form cranial nerve (CN) II.
D. They do not generate action potentials.
E. They do not regenerate, so anosmia is permanent.

Best answer = B. Taste and smell are mediated by two different types of sensory cell. Taste receptors are epithelial cells, whereas olfactory receptors are primary sensory neurons (see Section III·B). Olfactory receptor binding causes a change in adenylyl cyclase activity rather than guanylyl cyclase activity (A). Olfactory nerve axons form CN I, the olfactory nerve; CN II (C) is the optic nerve. If sufficiently intense, an olfactory stimulus will cause the sensory neuron to fire an action potential (D). Olfactory neurons are turned over and replaced every 30–60 days, so the sensory deficit is probably temporary, not permanent (E).

10.4. One of the commonly reported early symptoms of COVID-19 infection is loss of taste and/or smell. Although taste and smell are very different sensory modalities, which of the following most likely describes a feature that is common to the gustatory and olfactory sensory pathways?

A. Anoctamin 2 Ca²⁺-gated Cl⁻ channel
B. cAMP-gated cation channel
C. CALHM1/3 ATP channel
D. G protein–coupled receptor
E. TRPM5 Ca²⁺ channel

Best answer = D. Type II taste receptors and olfactory neurons both use G protein–coupled receptors (GPCRs) to detect chemical stimuli. Taste receptor binding is transduced by gustducin, and olfactory receptor binding is transduced by G$_{olf}$. These are sense-specific G proteins (see Sections II·B and III·C). The anoctamin 2 Ca²⁺-gated Cl⁻ channel (A; also known as TMEM16B) is specific to the olfaction transduction pathway, whereas the CALHM1/3 ATP channel (C) and the TRPM5 Ca²⁺ channel (E) are components of the taste transduction pathway found in type II receptors. Olfactory neurons employ a cAMP-gated cation channel (B) to signal odorant detection.

Motor Control Systems

11

I. OVERVIEW

The neural pathways that control muscle activity in humans developed during early evolutionary history to facilitate directed locomotion. Coordination of muscle groups that move the limbs was accomplished initially using simple neural feedback loops, but, as body complexity and the difficulty of the tasks that it was required to perform increased, so too the muscle control systems evolved. The human body devotes a large percentage of the nervous system to motor control. The simple neural feedback loops were retained during evolution and now function as muscle reflexes, but these pathways have been supplemented with successively higher layers of control (Fig. 11.1). Motor regions of the cerebral cortex decide when movements are necessary. Basal ganglia compile motor sequences based on learned experience and then relay these sequences through the thalamus for execution by the primary motor cortex. Control sequences ensure that pairs of muscle groups whose actions typically oppose one another (e.g., **extensors** and **flexors**, **abductors** and **adductors**, and **external** and **internal rotators**) contract and relax in a coordinated fashion to effect smooth limb movements. Motor commands are refined even as they are being executed by the cerebellum, which receives streams of sensory data from muscles, joints, skin, eyes, and the vestibular system. The cerebellum allows the cortex to compensate for unexpected changes in terrain, posture, and limb position.

II. MUSCLE SENSORY SYSTEMS

Complex motor behaviors (such as walking) require closely coordinated sequences of muscular contractions whose timing and strength is modified constantly during changes in body position and weight distribution. Such coordination is not possible unless the central nervous system (CNS) is informed about limb movements relative to the torso, made possible by the sense of **kinesthesia**. Kinesthesia is a form of **proprioception** and one of the **somatic senses**. Kinesthesia relies primarily on two sensory systems that sense muscle length (**muscle spindles**) and tension (**Golgi tendon organs** [GTOs]). Muscle requires high blood flow during contraction, the flow increase being mediated, in part, by reflexes involving mechano- and metaboreceptors (chemoreceptors that are responsive to local metabolite levels).

A. Muscle spindles

Skeletal muscle comprises two fiber types. **Extrafusal fibers** (from the Latin *fusus* for "spindle") generate the force needed to move

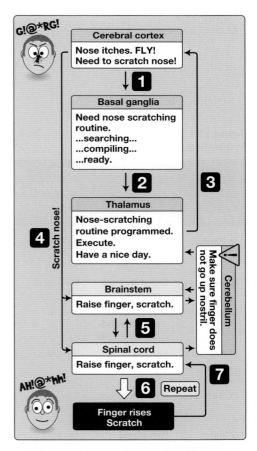

Figure 11.1.
Layering of motor control pathways.

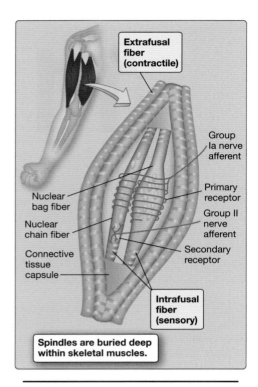

Figure 11.2.
Muscle spindle.

Table 11.1: Muscle Nerve Fiber Properties

Class	Innervation	Conduction Velocity (m/s)
Ia Sensory	Muscle spindle (primary endings)	80–120
Ib Sensory	Golgi tendon organs	80–120
II Sensory	Muscle spindle (secondary endings)	35–75
III Sensory	Muscle connective tissue	2.5–30
IV Sensory	Muscle connective tissue	0.5–2.5
α Motor	Extrafusal muscle fibers	80–120
γ Motor	Intrafusal muscle fibers	15–30

bones. **Intrafusal fibers** are sensory, monitoring muscle length and changes in length. Intrafusal fibers are contained within discrete sensory structures (spindles) that are distributed randomly throughout the muscle body (Fig. 11.2).

1. **Structure:** Muscle spindles contain up to 12 intrafusal fibers enclosed within a connective tissue capsule. Each intrafusal fiber comprises a noncontractile portion centered between two weakly contractile regions. The spindles nestle between contractile fibers and are anchored at either end so that contractile and sensory fibers move as one unit. Spindles contain two types of intrafusal fiber: **nuclear bag fibers** and **nuclear chain fibers** (see Fig. 11.2). Nuclear bag fibers swell centrally to form a "bag" containing numerous clustered nuclei. Nuclear chain fibers are thinner and more numerous than nuclear bag fibers. Their nuclei form a chain down the fiber's length.

2. **Sensory transduction:** Muscle spindles signal via two types of sensory nerve afferents (**group Ia** and **group II**). Both classes have wide, myelinated axons to maximize signal conduction velocity (Table 11.1; also see Chapter 5·III·B). When a muscle is stretched (e.g., by limb extension), the intrafusal fibers are stretched also, distorting the nerves that wrap around them. Stretching activates mechanosensitive cation channels, resulting in depolarization and increased afferent nerve firing frequency.

   a. **Group Ia:** Group Ia fibers coil around the central (equatorial) regions of both nuclear bag and nuclear chain fibers to form **primary muscle spindle receptors**. They yield a **dynamic response** to stretching (Fig. 11.3). Group Ia afferents show maximal firing rates when the muscle fibers (and nerve endings) are being stretched actively. Firing rate decreases when the muscle reaches and maintains a new length.

   b. **Group II:** Group II fiber endings are located at the ends of nuclear chain fibers and some nuclear bag fibers (see Fig. 11.2). They form **secondary muscle spindle receptors** that yield a **static response** to stretching. Their output is proportional to muscle length, and the nerve fibers continue firing at increased rates if the muscle is held at the new length (see Fig. 11.3).

3. **Regulation:** Intrafusal fibers are contractile, but they do not contribute significantly to muscle force development. Instead, the contractile portions serve only to shorten the fiber during muscle excitation and keep the central, sensory portion taut even as the muscle contracts. Maintaining tension allows the intrafusal fibers to continue functioning as stretch sensors throughout the contraction. Intrafusal fibers are innervated by **γ-motor neurons**, which conduct more slowly than the α-motor neurons that stimulate extrafusal muscle contraction (see Table 11.1). The α- and γ-motor neurons fire simultaneously, so that the spindle shortens along with the body of the muscle when the muscle contracts. The combination of muscle spindles and their associated γ-motor neurons constitutes a **fusimotor system**.

B. Golgi tendon organs

Each end of a skeletal muscle is attached to a tendon that typically tethers it to a bone. The musculotendinous junction contains

GTOs, sensory organs that monitor changes in muscle tension when stretched passively or when a muscle contracts (Fig. 11.4).

1. **Structure:** GTOs are situated at the junction between a skeletal muscle and a tendon. In this musculotendinous junction, collagen fibers and **group Ib sensory nerve endings** are interwoven. Group Ib nerve afferents are myelinated to increase signal conduction rates (see Table 11.1).

2. **Sensory transduction:** When a muscle is stretched or contracts, the associated GTO is stretched also. The collagen fibers within the GTO lengthen and compress the nerve endings that weave between them. Compression opens mechanosensitive channels in the nerve endings, causing depolarization and increasing nerve-firing rates.

C. Groups III and IV muscle afferents

Group III and IV muscle afferents are sensory fibers with free nerve endings that associate with connective tissue surrounding muscle fibers, blood and lymph vessels, and nerves. They inform the CNS of muscle activity, stimulating a reflex increase in cardiovascular and ventilatory activity needed to increase O_2 delivery to active muscle (**exercise pressor reflex**). They also limit voluntary muscle activation and contribute to fatigue during high-intensity exercise when muscle is stressed.

1. **Group III:** Group III muscle afferents are thinly myelinated nerve fibers with a conduction velocity of 2.5–30 m/s. They are mainly sensitive to mechanical stimuli and activate when a muscle contracts and develops force.

2. **Group IV:** Group IV muscle afferents are unmyelinated with a conduction velocity of 0.5–2.5 m/s. They are mainly sensitive to metabolites and other chemicals that accumulate within muscle during contraction, so their activation is delayed compared with group III afferents.

D. Joints and skin sensors

Although joints between bones would intuitively seem to be ideal sites for locating receptors that report on limb position, in practice, the role of joint receptors in kinesthesia is minimal. Slow-adapting Ruffini endings in skin do have an important role, however (see Chapter 15·VII·A). Skin that covers joints is stretched whenever a limb or digit is retracted, causing Ruffini endings to fire. The importance of sensory data from Ruffini endings is increased in the fingers, where layering of the various muscles and tendons required for execution of fine movements may impede acquisition of sensory information from spindles and GTOs.

III. SPINAL CORD REFLEXES

Walking with an upright gait is, quite literally, a delicate balancing act. A misplaced foot or slight irregularity of terrain can easily upset the balance and precipitate a fall. Avoiding such mishaps requires that a fall be anticipated and an appropriate correction to gait be made with minimal delay. Motor sensory and control neurons are specialized to conduct signals at up to 120 m/s, representing some of the fastest nerve cells in the body

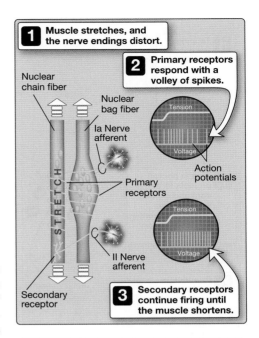

Figure 11.3.
Intrafusal fiber responses to stretch.

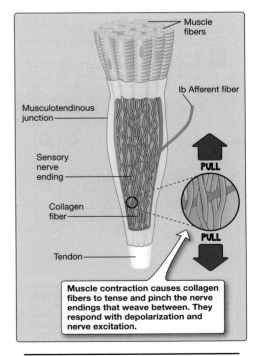

Figure 11.4.
Golgi tendon organ.

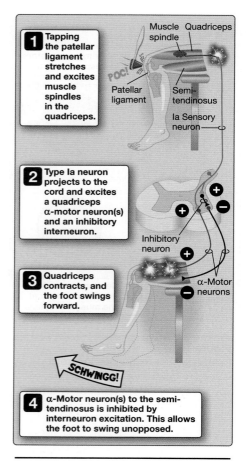

Figure 11.5.
Myotatic reflex.

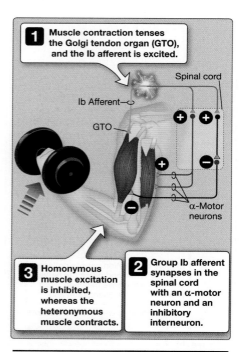

Figure 11.6.
Inverse myotatic reflex (Golgi tendon reflex).

(see Table 11.1). This ensures that sensory information is relayed to the CNS and compensatory commands executed in the shortest time possible. Reaction times are enhanced further by using local reflexes mediated by the spinal cord to make many routine adjustments to gait. The neurons involved are relatively short so as to further minimize signal transmission and processing times.

A. Reflex arcs

Reflex arcs are simple neuronal circuits in which a sensory stimulus initiates a motor response directly. Classic examples include withdrawal reflexes triggered by touching a hot stove or stepping on a sharp object. Such arcs are often mediated by the spinal cord, where a sensory neuron synapses with and activates a motor neuron. More complex arcs involve synapses with multiple neurons, at least one of which may be inhibitory. The spinal cord mediates a number of important reflex arcs, including the **myotatic reflex**, the **inverse myotatic reflex**, and the **flexion reflex**.

B. Myotatic

A myotatic reflex (also known as a **stretch reflex** or **deep-tendon reflex**) is initiated by stretching a muscle and causes contraction of the same ("homonymous") muscle. Reflex contraction of the thigh (quadriceps) muscles caused by tapping the **patellar ligament** is a familiar example (Fig. 11.5).

1. **Response:** Tapping the patellar ligament stretches the quadriceps and activates spindles buried within. Sensory signals are carried by Ia nerve afferents to the spinal cord, where they synapse with and excite α-motor neurons innervating the same muscle. The muscle contracts reflexively, the leg extends, and the foot jerks forward. The myotatic reflex is designed to resist inappropriate changes in muscle length and is important for maintaining posture.

2. **Reciprocal innervation:** Forward foot movement stretches the hamstring muscles at the back of the thigh and stimulates their spindles also. This might be expected to initiate a second reflex that opposes the actions of the first, but the arc is interrupted by an Ia inhibitory spinal interneuron. The Ia interneuron is activated by the same Ia afferent signal that caused the quadriceps to contract. The interneuron synapses with and inhibits the α-motor neurons that innervate the hamstring muscles (e.g., semitendinosus), thus allowing the leg to extend without resistance. This circuitry is referred to as **reciprocal innervation** and is used commonly in situations in which two or more sets of muscles oppose each other around a joint (e.g., flexors and extensors).

C. Inverse myotatic

The inverse myotatic reflex, also known as the **Golgi tendon reflex**, activates whenever a muscle is tensioned and GTOs are stretched (Fig. 11.6). Group Ib afferents from GTOs synapse with Ib inhibitory interneurons on entering the spinal cord. When activated, they inhibit α-motor output to the homonymous muscle. Excitatory interneurons simultaneously activate α-motor output to the heteronymous muscle. The Golgi tendon reflex protects the muscle from being stretched excessively. It is also believed to be important for fine motor control and for maintaining posture, acting synergistically with the myotatic reflex above.

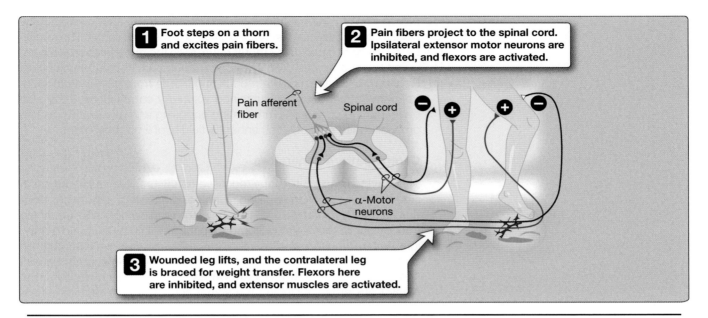

Figure 11.7.
Flexion and crossed-extension reflexes.

D. Flexion and crossed extension

Stepping on a thorn or other injurious object precipitates two urgent actions. The first withdraws the foot from the source of pain (leg flexion). The second braces the opposing limb so that weight can be transferred while still maintaining balance. This complex motion is mediated by **flexion** and **crossed-extension reflexes** (Fig. 11.7). Similar reflexes can be induced in the upper limbs. The action sequence can be broken down into three stages: stimulus sensation, wounded (ipsilateral) limb flexion, and then extension of the opposing (contralateral) limb.

1. **Sensation:** Flexion and crossed-extension reflexes are usually initiated as a response to a noxious, painful stimulus. Pain fibers project to and synapse with interneurons in the spinal cord.

2. **Flexion:** The sensory afferents synapse on the ipsilateral side with excitatory motor neurons that innervate flexor muscles. Extensor muscles are inhibited simultaneously, and the limb retracts from the pain source.

3. **Crossed extension:** Sensory fibers also cross the spinal cord anterior fissure and synapse with motor neurons controlling contralateral limb movement. Extensors are excited and contract, whereas flexors are inhibited and relax. This is known as a **crossed-extension reflex** and braces the contralateral limb for the sudden weight transfer caused by raising the wounded limb.

E. Renshaw cells

Intensely painful stimuli trigger a volley of spikes in the pain afferents that potentially could cause dependent flexor muscles to become tetanic if the reflex circuit were unregulated. Regulation comes in the form of **Renshaw cells**, which are a special class of spinal inhibitory interneuron excited by α-motor neuron collaterals (Fig. 11.8). Renshaw cells fire whenever a muscle receives a command to contract, but they

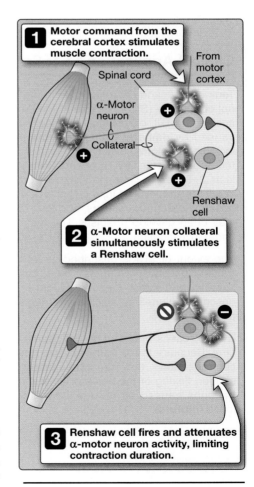

Figure 11.8.
Renshaw cells.

project back to and inhibit the same α-motor neuron that excited them. Renshaw cells can cause inhibition lasting tens of seconds. Their activity level is tied to that of the motor neuron, so the more intense the command to contract, the greater the degree of α-motor neuron inhibition. Renshaw cells also receive modulating inputs from higher motor control centers that allow for sustained voluntary contractions. They also project to motor neurons innervating opposing and associated muscle groups. These relationships enhance fluidity of limb movement.

F. Central pattern generators

Motor systems execute many repetitive behaviors, such as those associated with locomotion (walking, running, swimming), grooming, bladder control, ejaculation, eating (chewing, swallowing), and breathing (movement of the chest wall and diaphragm). These rhythmic behaviors do not require conscious thought, although they can be modified by higher control centers. They are established by neuronal circuits known as **central pattern generators** (**CPGs**), which are found in many CNS areas, including the spinal cord. The only requirement is an excitable cell or cell cluster with intrinsic pacemaker activity (e.g., two neurons that sequentially excite each other) and dependent circuits of interconnected neurons that control motor neurons. Walking, for example, involves a repetitive set of motor commands that move one leg forward, shift weight, and then extend the opposite leg. The motor commands and movements are sequential and predictable. Speeding or slowing the pace requires simple adjustments to CPG timing.

Clinical Application 11.1: Spinal Cord Injury

Tens of thousands of individuals in the United States suffer traumatic spinal cord injury (SCI) every year, most often as a result of a motor vehicle accident, a fall, or violence. Injuries to the spinal cord proper are usually caused by damage to the vertebral column or supporting ligaments, including fractures, dislocation, or disruption or herniation of an intervertebral disk. Acute SCI is often followed by a period of **spinal shock** that lasts 2–6 weeks, characterized by a complete loss of physiologic function caudal to the level of injury. This includes flaccid paralysis of all muscles, absence of tendon reflexes, and loss of bladder and bowel control. Males may often develop priapism. The mechanisms underlying shock are still under investigation.

The extent of SCI is described as being **complete** or **incomplete**. Cord transection causes complete SCI, characterized by a complete loss of sensory and motor function caudal to the trauma site. Incomplete SCI describes injuries in which some degree of sensory or motor function is preserved.

Even in cases of complete SCI, some reflex pathways may recover with time. Because these pathways are now disconnected from higher motor control centers, they can cause inappropriate movements. For example, sudden flexing of the ankle or wrist may trigger prolonged rhythmic contractions caused by unregulated Golgi tendon organ–mediated reflex loops (**clonus**).

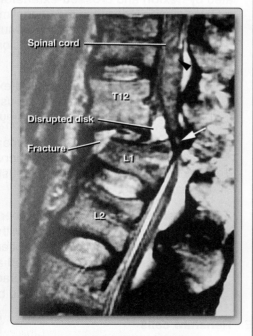

Spinal cord compression (*white arrow*) and hemorrhage (*black arrowhead*) caused by L1 vertebral body fracture and dislocation.

IV. HIGHER CONTROL CENTERS

Spinal cord reflexes and CPGs establish stereotyped behaviors, but planning and recalling learned movements requires higher levels of control. These are added in layers, with each successive layer providing for more sophisticated and finer degrees of motor control (see Fig. 11.1).[1] As is the case for most organ systems, the greatest insights into function come from observing what happens when these pathways are damaged.

A. Cerebral cortex

The cerebral cortex is responsible for planning voluntary motor commands. Many different cortical regions are involved in coordinating motor activities, but the most important are in area 4, the **primary motor cortex**, and area 6, which contains the **premotor cortex** and the **supplementary motor area** (Fig. 11.9).

1. **Primary motor cortex:** The primary motor cortex sends motor fibers via the **corticospinal tract** to the spinal interneurons that ultimately cause muscle contraction. Commands are executed only after extensive processing by the cerebellum and basal ganglia. Their execution also takes into account information being received simultaneously from various skin and muscle proprioceptors.

2. **Premotor cortex:** The premotor cortex may be responsible for planning movements based on visual and other sensory cues.

3. **Supplementary motor area:** The supplementary motor area retrieves and coordinates memorized motor sequences such as those required for playing the piano.

B. Basal ganglia

The cortex makes decisions about when to move and what tasks need to be accomplished, but execution requires careful planning about the timing of contractile events, the distance that limbs and digits need to move, and the force that needs to be applied. Thus, the sequence of movements required to apply fine paint strokes to a watercolor portrait is very different from those required to apply exterior paint in broad strokes to the side of a house. The task of planning and executing motor commands falls on the basal ganglia. These areas are not absolutely required for motor function, but movements become grossly distorted and erratic if they are damaged.

1. **Structure:** The basal ganglia comprise a group of large nuclei located at the base of the cortex close to the thalamus (Fig. 11.10). They work together as a functional unit. The nuclei receive motor commands from the cortex, pass them through a series of feedback loops, and then relay them to the thalamus for return to, and execution by, the primary motor cortex.

    a. **Striatum:** The **striatum** (or **neostriatum**) comprises two nuclei: the **putamen** and the **caudate nucleus**. The striatum is the gateway through which commands from the cortex

[1]A comprehensive description of the anatomical pathways, structures, and mechanisms involved is beyond the purview of this text, but these are considered in greater detail in *LIR Neuroscience*, 2e.

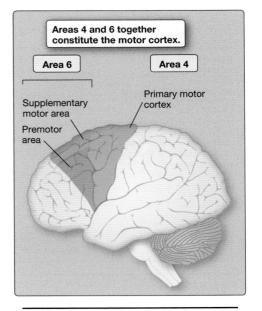

Areas 4 and 6 together constitute the motor cortex.

Area 6

Area 4

Supplementary motor area

Premotor area

Primary motor cortex

Figure 11.9.
Motor cortex organization.

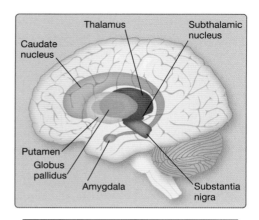

Figure 11.10.
Basal ganglia.

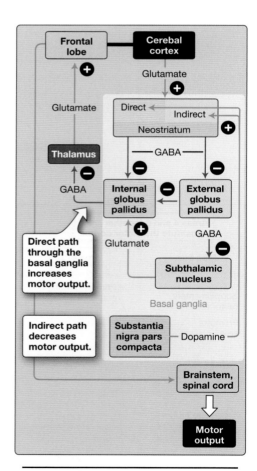

Figure 11.11.
Motor function relationships between the basal ganglia.

enter the nuclear complex. The striatum is dominated by γ-aminobutyric acid (GABA)ergic neurons, and its output is largely inhibitory.

b. **Globus pallidus:** The **globus pallidus** ([**GP**] also known as the **palladium**) can be divided into two regions (**internal [IGP]** and **external [EGP]**) based on function. The palladium is composed of inhibitory GABAergic neurons.

c. **Substantia nigra:** The **substantia nigra** (Latin for "black substance") is filled with melanin, a dark pigment that serves as a substrate for **dopamine** formation. Functionally, it can be divided into two areas: the **pars reticulata** and the **pars compacta**. Both regions contain inhibitory neurons. The pars reticulata is primarily GABAergic, whereas the pars compacta contains dopaminergic neurons. Because the pars reticulata and IGP often function together and have a similar anatomical structure, they are often considered as a single functional unit.

d. **Subthalamic nucleus:** The **subthalamic nucleus** (part of the subthalamus) is a key link in a basal-nuclei feedback circuit and is the only primarily excitatory (glutamatergic) center within the basal ganglia.

2. **Feedback circuits:** The motor cortex communicates its intent to the striatum. There are two pathways for information flow from the striatum through the basal nuclei: a **direct path** and an **indirect path**. Both end at the thalamus, which is tonically active and stimulates cortical areas that ultimately control the musculature (Fig. 11.11).

a. **Direct path:** When the striatum is activated, it inhibits IGP–pars reticulata complex output. These two nuclei are tonically active normally, and their output suppresses the thalamus's tonic output to the motor cortex. Thus, activating the striatum allows the thalamus to stimulate the motor cortex. In practice, the direct path increases motor activity.

b. **Indirect path:** A second pathway involves the EGP and the subthalamic nucleus. Exciting the striatum prevents the EGP from signaling. The EGP normally inhibits the subthalamic nucleus, which would otherwise be tonically active, thus increasing the activity of the IGP. The IGP inhibits the thalamus and prevents it from exciting the motor cortex. In practice, the indirect pathway decreases motor activity.

c. **Biasing output:** When the striatum receives a motor command, the direct and indirect pathways are activated simultaneously, and their effects on the IGP are conflicting and balanced. Any influence that changes this balance might be used to regulate motor output. The substantia nigra pars compacta can potentially have a major influence over motor output because it sends dopaminergic axons back to two areas of the striatum. When active, these neurons increase the activity of the direct pathway via an excitatory dopaminergic (D_1) receptor while simultaneously suppressing the indirect pathway via a dopamine D_2 receptor (see Table 5.2). Both effects favor increased motor activity.

Clinical Application 11.2: Motor Disorders

Tremors are the most common form of movement disorder. A tremor is a rhythmic body movement reflecting an imbalance between the actions of two antagonistic muscle groups. All individuals have a 10–12 Hz **physiologic tremor** that is usually not evident but may be exaggerated by physical stress; hunger; caffeine; and many classes of drugs affecting dopaminergic, adrenergic, and cholinergic neurotransmission.

Essential tremor is a common neurologic disorder that causes an action tremor. It typically affects the arms and hands but can involve the head, voice, and legs. The etiology is unknown, but incidence increases with age and in individuals with a family history. Essential tremor usually improves with alcohol ingestion.

Diseases affecting the basal ganglia cause significant movement disorders. The balance that exists between the direct and indirect pathways from the striatum is delicate. Therefore, disrupting even a single circuit component can have devastating motor consequences. These can include a slowing of movement (bradykinesia) or a complete loss of motor control (akinesia), rigidity due to increased muscle tone (hypertonia), and involuntary writhing movements at rest (dyskinesia). The best-studied motor disorders are Parkinson disease (PD) and Huntington disease (HD).

Parkinson disease is a motor disorder characterized by arm and "pill-rolling" hand rest tremors; increased muscle tone and limb rigidity; bradykinesia; and, in the later stages, postural instability. Patients also have a slow, shuffling gait. These symptoms reflect the death of large numbers of neurons within the pars compacta (>60%), which interrupts the dopaminergic feedback loop between the pars compacta and striatum. Directed movement becomes difficult, and the inherent conflicts between direct and indirect pathways become evident. Treatment options currently include drugs that raise dopamine levels, either by providing a substrate for dopamine formation (L-dopa) or by inhibiting its breakdown (monoamine oxidase inhibitors; see Table 5.3).

Huntington disease is an autosomal-dominant condition affecting huntingtin, a ubiquitous intracellular protein whose normal function is not understood fully. HD is caused by accumulation of CAG trinucleotide repeats within the *HTT* gene. The mutant gene product contains a long tract of repeat glutamine residues. Striatal neurons that normally inhibit motor output via the indirect pathway are destroyed due to abnormal huntingtin accumulations, removing the normal constraints on the direct pathway. Early disease symptoms include chorea (from the Greek for "dance"), characterized by involuntary limb muscle contractions that produce abrupt jerking and writhing motions. HD ultimately involves most brain regions, causing severe psychiatric disturbances and dementia. There is no treatment option, and death usually follows diagnosis by ~20 years.

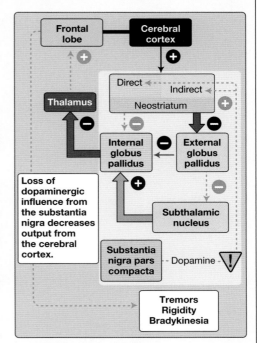

Parkinson disease.

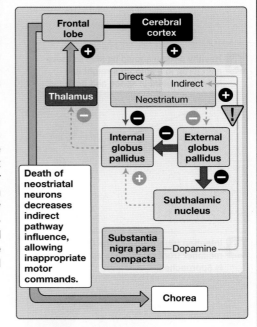

Huntington disease.

C. Cerebellum

The cerebellum is not essential for locomotion, but it is intimately involved in motor control. It verifies that instructions issued by the cortex are executed as intended and makes corrections as necessary.

1. **Function:** The full extent of cerebellar involvement in motor control is not known, but principal functions include fine-tuning and smooth execution of movements.

   a. **Fine-tuning:** The cerebellum receives extensive sensory information about body and head position, muscle contractility and length, and tactile information from the skin. It then compares this information with the motor commands that were issued by the higher centers and makes fine motor adjustments as necessary. This prevents a finger from overshooting its target when reaching out to flip a light switch, for example.

   b. **Sequencing:** Activities such as playing the piano involve finger movements executed so fast that there is insufficient time for sensory information to be relayed back to the CNS for processing and feedback. Such activities are only possible because the cerebellum anticipates when a particular movement should end and then executes a command that arrests it at a precise moment in time. It simultaneously anticipates and executes a command that ensures a smooth transition to the next motion.

   c. **Motor learning:** The cerebellum is able to anticipate and execute motor commands because it stores and constantly updates information about the correct timing of commands required for complex motor sequences.

2. **Cerebellar dysfunction:** The cerebellum finesses motor commands but is not absolutely required for locomotion. Cerebellar lesions cause varying degrees of coordination loss, depending on location and severity. Common symptoms include ataxia and intention tremors.

   a. **Ataxia: Ataxia** refers to a general lack of muscular coordination. Gait may become slow, wide, and swaying. The conscious brain is now forced to think about body position, but the time lag between receipt of proprioceptive information and execution of motor commands means that limbs usually overreach and miss their intended targets. The brain then executes a poorly controlled compensatory movement, resulting in a behavior pattern known as **dysmetria**.

   b. **Intention tremors:** Because the conscious brain must guide and continually update movements, simple tasks, such as reaching for an object, become slow, and the path taken to the target meanders from side to side (an **intention tremor**). Intention tremors are readily discerned using a simple finger-to-nose test (Fig. 11.12).

D. Brainstem

Simple neuronal circuits in the spinal cord produce stereotypical behaviors that facilitate walking and other rhythmic movements. The brainstem sets these pathways in motion and coordinates them with reference to sensory information received from the eyes and vestibular

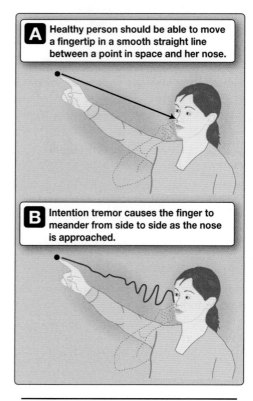

Figure 11.12.
Intention tremor.

system. It also controls eye movement to stabilize visual images during head and body movement. The brainstem contains four important motor control areas: the **superior colliculus (tectum)**, **red nucleus**, **vestibular nuclei**, and **reticular formation** (Fig. 11.13).

1. **Superior colliculus:** The superior colliculus controls head and neck movements with reference to visual information. Fibers from this area project to the cervical spine via the **tectospinal tract**.

2. **Red nucleus:** The **red nucleus** is located in the midbrain. It controls flexor muscles in the upper limbs via the **rubrospinal tract**.

3. **Vestibular nuclei:** There are four **vestibular nuclei**: one in the pons (**superior vestibular nucleus**) and three in the medulla (the **medial**, **lateral**, and **inferior vestibular nuclei**). They receive and integrate information from the inner ear about head and body motion. Output from these regions controls eye movement via the **oculomotor nerve** (cranial nerve III). They also help coordinate head, neck, and body movements via the **vestibulospinal tracts**. The medial vestibulospinal tract arises in the medial vestibular nucleus, which helps stabilize the head during body movements. The lateral vestibulospinal tract projects to all levels of the spinal cord, where it stimulates extensor muscle contraction and inhibits flexors to help control posture during body movements.

4. **Reticular formation:** The **reticular formation** is also involved in many complex motor behaviors. Fibers arising from this area project via the **reticulospinal tract** to all levels of the spinal cord. They influence the activity of both α- and γ-motor neurons to facilitate voluntary body movements originating in the cortex and initiated via the corticospinal tract.

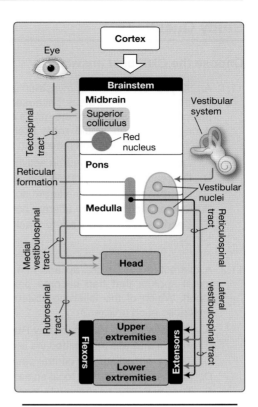

Figure 11.13.
Brainstem motor control centers.

Chapter Summary

- **Skeletal muscles** facilitate locomotion and manipulation of the external environment. Executing complex movements, such as walking, requires multiple levels of coordination, involving the **spinal cord**, **brainstem**, **cerebellum**, **basal ganglia**, and motor areas of the **cerebral cortex**.

- Motor control centers receive sensory data from specialized myofibrils contained within skeletal muscles (**intrafusal fibers**) and from musculotendinous junctions (**Golgi tendon organs [GTOs]**). Intrafusal fibers relay information about muscle length and changes in length. GTOs are tension sensors.

- The spinal cord contains **central pattern generators** that sustain rhythmic limb movements (e.g., during walking). Simple spinal circuits allow for rapid reflex responses to noxious stimuli and unanticipated changes in muscle length.

- The **myotatic reflex** causes muscles to contract when stretched while simultaneously inhibiting opposing muscles to allow free limb movement. The **inverse myotatic reflex** limits muscle contraction and simultaneously activates an opposing muscle. **Flexion** and **crossed-extension reflexes** prepare opposing limbs to brace for transfer of weight (e.g., when stepping on a sharp object).

- Decisions about how and when to move begin in the cortex. Principal motor areas of the cortex include the **primary motor cortex**, the **premotor cortex**, and the **supplementary motor area**.

- Timing and sequencing motor commands is the responsibility of the **basal ganglia**. Motor commands are subjected to a series of feedback loops that hone the sequences and ensure accurate and smooth movements.

- The **cerebellum** fine-tunes movements during execution, referencing information being received from proprioceptors and other sensory systems.

- The **brainstem** executes motor commands and helps coordinate movements with reference to sensory data being relayed from the eyes and vestibular system.

Study Questions

Choose the ONE best answer.

11.1. An 83-year-old male with myasthenia gravis is unable to eat foods such as steak because of rapid skeletal muscle fatigue. Studies of the man's chewing muscles during contraction would most likely have revealed which of the following compared with normal?

    A. Decreased α-motor neuron activity
    B. Decreased γ-motor neuron activity
    C. Decreased Ia sensory afferent activity
    D. Decreased Ib sensory afferent activity
    E. Decreased II sensory afferent activity

Best answer = D. The antibodies produced in patients with myasthenia gravis destroy the nicotinic acetylcholine receptor, interfering with normal excitation and force development (see Clinical Application 12.2). Muscle tension development during contraction is sensed by Golgi tendon organs, which signal via group Ib sensory afferents (see Section II·A). Contraction is initiated by α-motor neurons (A), which might be expected to signal normally, as would the γ-motor neurons (B) that initiate intrafusal fiber contraction. Group Ia (C) and group II afferents (E) relay sensory information from muscle spindles when a muscle is stretched.

11.2. A distracted cook picks up and immediately drops a metal spatula that had become painfully hot to the touch. Which of the following statements most likely describes such reflexes?

    A. Pain stimuli are transduced by Ruffini endings.
    B. Pain stimuli are transmitted via α-motor neurons.
    C. They are mediated by central pattern generators.
    D. They are mediated by local spinal circuits.
    E. They would be unaffected by demyelination.

Best answer = D. Reflexive movements triggered by painful stimuli are mediated by spinal reflex circuits (or "arcs"; see Section III·A). The short signal path and fibers adapted for high conduction velocity decrease reaction time. Painful stimuli are mediated by pain receptors. Ruffini endings sense mechanical stimulation of the skin (see Chapter 15·VII·A; also see Section II·C), not pain (A). Pain stimuli are transmitted to the spinal cord via sensory afferent fibers, not motor neurons (B). Central pattern generators are involved in establishing rhythmic movements (C; see Section III·F). Spinal reflex sensory and motor neurons are both myelinated, so both arms are susceptible to demyelinating disease (E).

11.3. A 75-year-old female with a history of stroke comes to the physician for a follow-up examination. The physician conducts a neurologic examination that includes striking the patellar tendon with a reflex hammer, causing the patient's foot to kick forward. The reflex is most likely initiated by which of the following structures?

    A. Ib Sensory afferent
    B. Central pattern generator
    C. Intrafusal fiber
    D. α Motor neuron
    E. Renshaw cell

Best answer = C. Intrafusal fibers are found within muscle spindles, which are sensory organs. They monitor changes in muscle length (see Section II·A). Tapping the patellar tendon stretches the quadriceps muscle and activates the spindles buried within. The muscle contracts reflexively, causing the foot to kick forward (a myotatic reflex; see Section II·B). Muscle spindles are innervated by group Ia and group II, not group Ib (A), sensory afferent fibers. Central pattern generators (B) are neural circuits that generate rhythmic movements, such as those engaged during walking or running. α Motor neurons (D) initiate muscle contraction. Renshaw cells (E) are spinal inhibitory neurons that limit contractions.

11.4. A 23-year-old male engaged in strength training lifts progressively heavier weights above his head. On the final lift, he is unable sustain muscle contraction, and he drops the weights. Which of the following structures most likely caused his arm muscles to relax and drop the weights?

    A. Extrafusal fiber
    B. Golgi tendon organ
    C. γ Motor neuron
    D. Muscle spindle receptor
    E. Ruffini ending

Best answer = B. Golgi tendon organs (GTOs) are located at the junction between a skeletal muscle fiber and a tendon (see Section II·B). They monitor muscle tension during contraction. When a GTO activates, it inhibits homonymous α motor neuron activity, causing muscle relaxation (see Section II·C). Extrafusal fibers (A) generate the force needed to lift weights. γ Motor neurons (C) stimulate contraction of intrafusal fibers. Muscle spindle receptors (D) monitor changes in muscle length and mediate myotatic reflexes. Ruffini endings (E) are cutaneous receptors that monitor skin stretching during movement of limbs or digits.

Skeletal Muscle

12

I. OVERVIEW

The various body organs are housed within compartments (thorax and abdomen) that are framed and carried by a bony skeleton (bone structure and function is discussed in Chapter 14). Bones also define the limbs, which are used to manipulate objects and for locomotion. Bone movement is facilitated by skeletal muscles, which are usually arranged in antagonistic pairs to create articulating levers. The arm, for example, contains three long bones that form a lever that hinges at the elbow. A pair of antagonistic muscle groups allows for forearm **extension** (triceps) and **flexion** (biceps) as shown in Figure 12.1, but a full range of arm motions uses ~40 more. Skeletal muscles allow the body to sustain itself by hunting and gathering food, but they are also used for nonlocomotory activities, such as maintaining an erect posture and expanding the lungs. Skeletal muscle accounts for ~40% of total body mass in an average person, but another 10% of body mass comprises cardiac muscle and smooth muscle. The three muscle types use common molecular principles to generate force. Contraction is initiated by a rise in intracellular free Ca^{2+} concentrations, which facilitates interaction between actin and myosin filaments through binding to and activation of a Ca^{2+}-dependent regulatory complex. The two filaments then slide against each other at the expense of ATP to generate force. Sliding causes myofilaments and muscle cells (**myocytes**) to contract and shorten. Although the mechanism by which force is generated is similar in the three muscle types, there are significant differences in their organization and in the way they initiate and control contractile force. These differences reflect their unique roles within the human body and will be explored in more detail in Chapter 13 and Unit IV. Skeletal muscle consumes significant amounts of energy (ATP) when maximally active and releases an equally significant amount of heat as a byproduct. It is the integument's (skin's) responsibility to dissipate this heat by transferring it to the external environment (see Chapter 38).

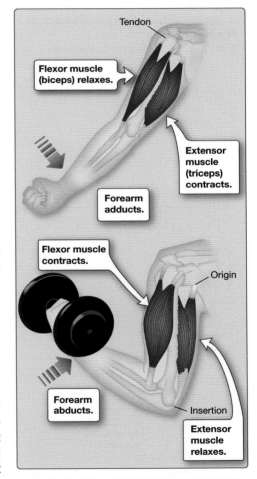

Figure 12.1.
Antagonism between the muscles that extend and flex the forearm.

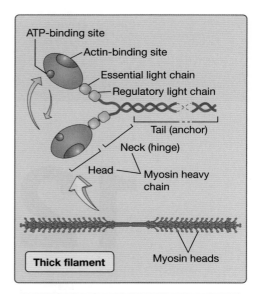

Figure 12.2.
Myosin and thick filament structure.

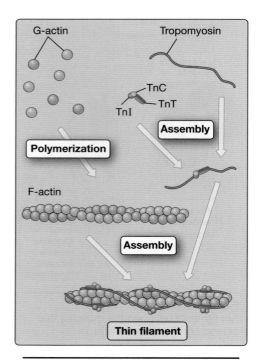

Figure 12.3.
Thin filament assembly. TnC = troponin C; TnI = troponin I; TnT = troponin T.

II. STRUCTURE

Skeletal muscle and cardiac muscle are organized into contractile subunits called **sarcomeres**, which contain orderly arrays of interdigitating thick and thin filaments. Thick filaments are composed largely of **myosin**. Thin filaments are composed largely of **actin**. During muscle contraction, myosin attaches to a thin filament and pulls on it, causing the two filaments to slide past each other.

A. Myosin

Myosin is a large hexameric protein (MW = 52 kDa). The main body is composed of two **myosin heavy chains**, each of which is a golf club–shaped polypeptide with a head; neck; and a long, coiled tail (Fig. 12.2). The head possesses ATPase activity and contains the actin-interaction site. The neck acts as a hinge that allows the head to pivot and pull during contraction. The tail anchors the protein within a larger filamentous assembly. Each head region associates with two light chains: one **regulatory** and one **essential light chain**. The tails of two heavy chains intertwine to form a coiled coil, and then ~100 such assemblies are bundled like golf clubs crammed in a bag, with the heads protruding in various different directions. Two such bundles are joined tail to tail to form a 1.6-μm long **thick filament**.

B. Actin

Actin forms a molecular "rope" that myosin pulls on during contraction. Actin is synthesized as a globular protein (**G-actin**) and then polymerized to form **F-actin** (Fig. 12.3), which contains two beadlike polymers braided in a helical conformation. Actin expresses myosin-binding sites, but these must remain hidden until a signal to contract is received or else the muscle fiber might lock into a rigid state (**rigor**). Access to the binding site is controlled by **tropomyosin** and **troponin** (**Tn**), two regulatory proteins that lie in the grooves of the actin helix near the binding site. Actin, tropomyosin, and Tn together constitute a 1.0-μm long **thin filament**.

1. **Tropomyosin:** Tropomyosin comprises two identical filamentous subunits that are entwined to form a helix. Tropomyosin molecules lay end to end along the actin filament, concealing the myosin-binding sites beneath.

2. **Troponin:** Tn is a Ca^{2+}-sensitive protein that uncovers the myosin-binding site when intracellular Ca^{2+} concentrations rise. Tn is an assembly of three different proteins: **TnC**, **TnI**, and **TnT**.

TnC is a Ca^{2+}-binding protein that senses a rise in intracellular Ca^{2+}. Two of its four Ca^{2+}-binding sites are occupied normally, which allows Tn to attach to the thin filament. When intracellular Ca^{2+} rises, the two vacant binding sites fill, and Tn changes conformation, pulling tropomyosin deeper into the actin-filament groove. This motion uncovers the myosin-binding site, and myosin binds instantly. TnI helps inhibit actin/myosin interaction until the appropriate time. It also masks myosin-binding sites. TnT tethers the Tn complex to tropomyosin.

C. Sarcomere

During contraction, myosin head groups reach along the actin filament from one binding site to the next and then pull. This causes the thin filament to slide past the thick filament, and the muscle shortens. A contracting muscle harnesses the energy from millions of these small movements, but, in order for shortening to occur, the actin ropes must be tethered end to end and ultimately to the ends of the muscle fiber. The two filaments must also be held tightly against each other within a framework of structural proteins that ensure actin–myosin interaction. The result is a contractile unit called a **sarcomere** (Fig. 12.4).

> Many muscle descriptive terms use the "sarco-" prefix, which is derived from the Greek for "flesh," *sarx*. Flesh refers to the soft tissues that cover bone and are composed primarily of muscle and fat.

1. **Sarcomeric structure:** A sarcomere is delineated by two **Z disks**, proteinaceous plates that anchor arrays of thin filaments (see Fig. 12.4A). The filaments protrude from the disks like bristles from a hairbrush. Thick filaments insert between the thin filaments, so that each thick filament is surrounded by, and may pull simultaneously on, six actin ropes (Fig. 12.5). The thick filaments are 60% longer than thin filaments. This allows them to insert into an identical array of thin filaments that extend from the Z disk, which delineates the far end of the sarcomere. This region of overlap is where interactions, or **crossbridges**, between the two types of filament will occur. Overlap between thick and thin filaments produces distinct banding patterns when viewed under polarized light (see Fig. 12.4B) that are repeated across the width of a muscle to give it a **striated** appearance.

2. **Structural proteins:** Numerous specialized cytoskeletal proteins rigidly constrain the thick and thin filaments within the framework of a sarcomere and aid in its assembly and maintenance.

   a. **Actinin:** α-**Actinin** binds the ends of thin filaments to the Z disks (Z disks appear as Z lines in micrographs; see Fig. 12.4B).

   b. **Titin:** **Titin** is a massive (>3-million MW) protein. One end is attached to a Z disk, the other to the thick filaments. It forms a spring that limits how much the sarcomere can be stretched and returns the sarcomere to its pre-contraction state during muscle relaxation. It also centers the thick filaments within the sarcomere.

   c. **Dystrophin:** **Dystrophin** is a large (427,000-MW) protein associated with Z disks. It helps anchor the contractile array to the cytoskeleton and surface membrane. It also aligns the Z disk with disks in adjacent myofibrils and muscle fibers.

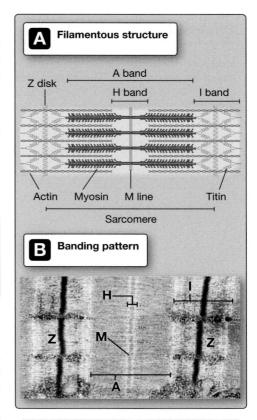

Figure 12.4.
Muscle sarcomere.

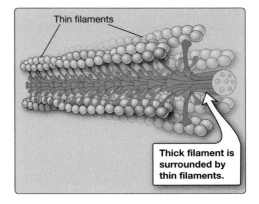

Figure 12.5.
Spatial relationship between thick and thin filaments in muscle.

Clinical Application 12.1: Muscular Dystrophy

Muscular dystrophy (**MD**) refers to a heterogeneous group of inherited disorders that cause muscle wasting. The most common is **Duchenne muscular dystrophy**, which results from a recessive X-linked mutation in the dystrophin (*DMD*) gene. Loss of dystrophin function prevents the cytoskeleton and its embedded contractile machinery from attaching to the sarcolemma, and the muscle fiber necroses as a result, causing gradual muscle wasting. MD affects all voluntary muscles. Patients usually do not survive beyond their early 30s, eventually succumbing to respiratory muscle failure.

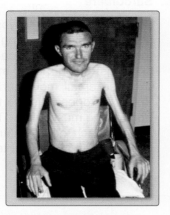

Muscle wasting in a patient with muscular dystrophy.

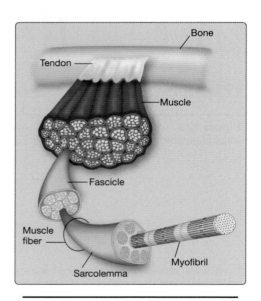

Figure 12.6.
Skeletal muscle structure.

d. **Nebulin: Nebulin** associates with and extends the length of an actin filament. It is believed to act as a molecular ruler that determines thin-filament length during assembly.

D. Skeletal muscle

The structure of the sarcomere is replicated many times down the length of a muscle to build up a **myofibril** (Fig. 12.6). Many hundreds or even thousands of myofibrils are then bundled and wrapped in a **sarcolemma** sheath, which comprises plasma membrane covered by a thin extracellular coat containing numerous collagen fibers to give it strength. The result is a **muscle fiber**. Multiple muscle fibers are bundled and wrapped in connective tissue to form a **fascicle**, and fascicles, in turn, are bundled to form a muscle. A **tendon fiber** is fused to the ends of individual muscle fibers to provide a mechanical link between muscle and bone. Tendon fibers are made of collagen and are well able to withstand the tension generated by muscle contraction. The fibers gather in interlaced, parallel bundles to form tendons, which adhere to bone. The two ends of a muscle are referred to anatomically as its **insertion** and **origin** (see Fig. 12.1). The insertion attaches to the bone that moves when the muscle contracts and is usually more distal than its stationary origin.

E. Sarcoplasm

Sarcoplasm is muscle cell cytoplasm and is rich in Mg^{2+}, phosphates, and glycogen granules. It also contains high levels of **myoglobin**, an oxygen-binding protein that is related to hemoglobin. Sarcoplasm is also dense with mitochondria, which lie tightly alongside the myofibrils to supply the large quantities of ATP needed to fuel contraction.

F. Membrane systems

If a skeletal muscle is to work efficiently, its sarcomeres must contract in unison. The stimulus for contraction is a rise in sarcoplasmic Ca^{2+} levels. Ensuring that every sarcomere receives a signal simultaneously requires two membrane specializations: one external (**transverse [T] tubules**), the other internal (**sarcoplasmic reticulum [SR]**) (Fig. 12.7).

1. **Transverse tubules: T tubules** are radial tubular invaginations of the sarcolemma that extend deep into a muscle fiber's core, branching repeatedly en route and contacting every myofibril. The tubules align with the ends of the thick filaments, two per sarcomere, and the myofibrils are aligned by their Z disks, so that tubules have a straight run through the fiber. T tubules carry signals (action potentials [APs]) from the cell surface to the Ca^{2+} stores.

2. **Sarcoplasmic reticulum:** The SR comprises an extensive system of membranous sacs that wrap around the myofibrils. The role of the sacs is to douse the myofibrils with Ca^{2+} and thereby initiate contraction. The SR scavenges Ca^{2+} from the sarcoplasm using sarco(endo)plasmic reticulum calcium ATPase (SERCA) Ca^{2+} pumps when the muscle is relaxed and then store it temporarily in association with **calsequestrin**, a Ca^{2+}-binding protein that acts like a Ca^{2+} sponge. T tubules communicate with the SR at a junctional **triad**. A triad comprises a T tubule plus SR **terminal cisternae** from two abutting sarcomeres. Triads are the location of specialized ion channels that have a key role in **excitation–contraction coupling** (see Section III).

G. Neuromuscular junction

All skeletal muscles are under voluntary control. Decisions to initiate contractions originate in the cerebral cortex and are then delivered by an α-motor neuron that interfaces with the muscle at a **neuromuscular junction** (**NMJ**), as shown in Figure 12.8A. Motor neurons are flattened in this region to form a **motor endplate** (see Fig. 12.8B). Motor nerve terminals are filled with synaptic vesicles that release **acetylcholine** (**ACh**) into the synapse in a Ca^{2+}-dependent manner upon excitation. The sarcolemma is deeply folded in the region of the NMJ, and the crests of these folds are studded with

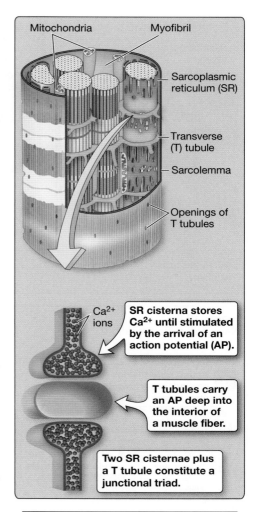

Mitochondria — Myofibril

Sarcoplasmic reticulum (SR)

Transverse (T) tubule

Sarcolemma

Openings of T tubules

Ca^{2+} ions — SR cisterna stores Ca^{2+} until stimulated by the arrival of an action potential (AP).

T tubules carry an AP deep into the interior of a muscle fiber.

Two SR cisternae plus a T tubule constitute a junctional triad.

Figure 12.7.
Skeletal muscle membrane systems.

Clinical Application 12.2: Myasthenia Gravis

The autoimmune condition **myasthenia gravis** (**MG**) is the most common disorder affecting neuromuscular transmission. Patients develop circulating antibodies against nicotinic ACh receptors that interfere with normal signaling at the neuromuscular junction. Patients with a more limited form of the disorder, ocular MG, have muscle weakness of the eyelids and extraocular muscles. Generalized MG affects ocular muscles plus bulbar, limb, and respiratory muscles. Although severe respiratory muscle weakness may cause life-threatening respiratory failure, termed a "myasthenic crisis," most MG patients experience milder episodes of weakness that improve with rest. Treatments include anticholinesterase inhibitors (e.g., pyridostigmine), immunosuppressive drugs, immunomodulating therapies (e.g., plasmapheresis), and surgery (thymectomy). The latter may help because many MG patients have either thymic hyperplasia (~60%) or thymomas (15%) and exhibit symptomatic improvement with thymectomy.

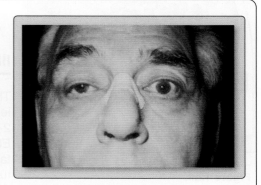

Right eyelid droop caused by myasthenia gravis.

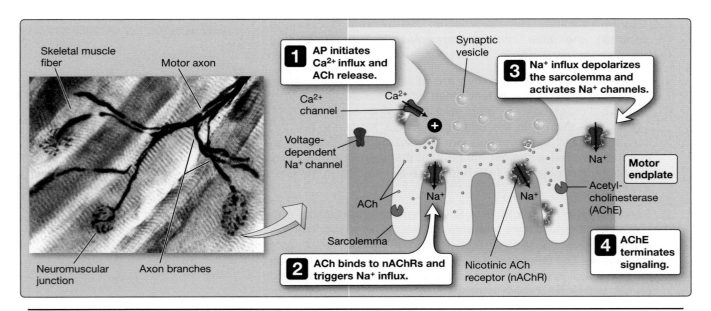

Figure 12.8.
Neuromuscular excitation.

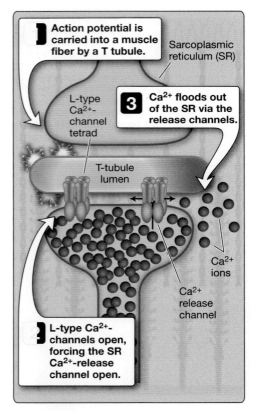

Figure 12.9.
Triad structure and excitation. T tubule = transverse tubule.

nicotinic ACh receptors (nAChRs). ACh binding to the nAChRs initiates contraction.

1. **Nicotinic acetylcholine receptors:** nAChRs are ion channels that open in response to binding of two ACh molecules to mediate a joint Na^+ and K^+ flux across the sarcolemma. Na^+ influx dominates the exchange, and the membrane depolarizes, a response known as a **motor endplate potential**.

2. **Sodium channels:** The sides of the clefts created by SR folding are dense with voltage-dependent Na^+ channels. The local depolarization induced by the nAChR opens the Na^+ channels and initiates an AP that spreads from the clefts over the surface of the muscle fiber.

3. **Signal termination:** The synaptic cleft contains acetylcholinesterase (AChE), an enzyme that rapidly cleaves ACh and terminates signaling.

III. EXCITATION–CONTRACTION COUPLING

The AP triggered by ACh release at the NMJ propagates over the sarcolemma in a similar manner to that described for nerve cells (see Chapter 2·III·C). APs are electrical events, whereas contraction is mechanical. Electromechanical transduction occurs by a process known as excitation–contraction coupling, which begins when an AP enters the T tubules and encounters a triad.

A. Triad role

The T-tubule membrane in the region of a triad is filled with **L-type Ca^{2+} channels (dihydropyridine receptors)** arranged in quadruplets (**tetrads**), as shown in Figure 12.9. They are voltage sensitive, so when the AP arrives, they open and Ca^{2+} flows into the sarcoplasm from the cell exterior (i.e., the tubule lumen). The amount of Ca^{2+} entering the myocyte via these channels is minute compared with the amount needed for contraction, but the conformational change that accompanies channel opening forces open **Ca^{2+}-release channels** in the SR (also known as **ryanodine receptors [RyRs]**). Ca^{2+} then floods out of the stores and saturates the contractile apparatus.

B. Crossbridge cycling

Emptying of the SR Ca^{2+} stores causes sarcoplasmic Ca^{2+} concentrations to rise from 0.1 μmol/L to ~10 μmol/L within a fraction of a second. Tn binds Ca^{2+} and pulls tropomyosin deeper into the groove of the actin filament. With myosin-binding sites now exposed, crossbridge cycling and contraction begins. The following numbered steps correspond to the numbered steps in Figure 12.10.

1. **Myosin binding:** The myosin head immediately latches onto the actin filament, and the two filaments immobilize in a **rigor state**. This step is brief in an active muscle but becomes permanent when ATP is unavailable.

2. **ATP binding:** ATP binds to the myosin head and decreases its actin affinity, thereby releasing it.

3. **ATP hydrolysis:** ATP hydrolysis releases energy that causes the myosin molecule to pivot at the neck hinge, and the head swings forward by ~10 nm. ATP hydrolysis reverses the affinity change that occurred during the previous step, and myosin immediately rebinds actin at this new position. The head group is now tensed for action like a cocked trigger, the potential energy from ATP hydrolysis stored primarily in the neck region.

4. **Power stroke:** Inorganic phosphate dissociates, and the trigger is released, initiating a conformational change that pulls the head back to its original position. The head remains attached to actin during this time, so that the entire actin rope is pulled past the thick filament by a distance of ~10 nm.

5. **ADP release:** ADP dissociates from myosin, and the crossbridge cycle returns to a rigor state. The cycle then repeats, causing the myosin head to ratchet in the actin rope ~10 nm at a time.

Clinical Application 12.3: Rigor Mortis

When the body dies, ATP reserves are depleted rapidly, and the pumps that maintain ion gradients across the membrane stop working. Intracellular Ca^{2+} concentrations rise as a result, causing actin and myosin to bind and lock into a state of **rigor mortis**. Although time of onset can be highly variable depending on ambient temperatures, rigor mortis typically develops within 2–6 hours of death. The rigor state persists for 1–2 days until the myocytes deteriorate, and digestive enzymes released during lysosomal lysis finally break the crossbridges.

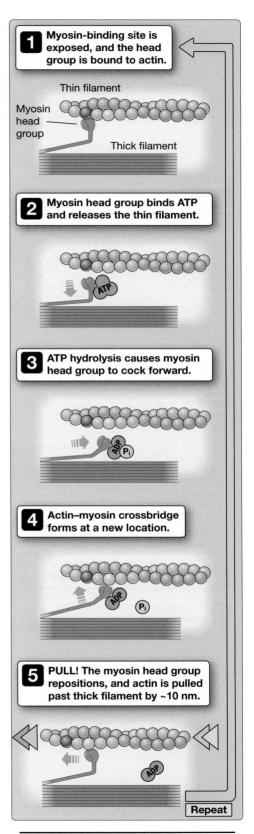

1 Myosin-binding site is exposed, and the head group is bound to actin.

Thin filament
Myosin head group
Thick filament

2 Myosin head group binds ATP and releases the thin filament.

ATP

3 ATP hydrolysis causes myosin head group to cock forward.

ADP Pi

4 Actin–myosin crossbridge forms at a new location.

ADP Pi

5 PULL! The myosin head group repositions, and actin is pulled past thick filament by ~10 nm.

ADP

Repeat

Figure 12.10.
Actin–myosin crossbridge cycle.

C. Relaxation

Once the α-motor neuron stops firing, the muscle is free to relax, but relaxation requires that the crossbridge cycle be broken. The Ca^{2+} channels and release channels close immediately once the motor neuron stops firing, and Ca^{2+}-ATPases in the SR and sarcolemma then rapidly clear the sarcoplasm of Ca^{2+} by returning it to the stores or pumping it out of the cell. Falling sarcoplasmic free Ca^{2+} concentrations cause TnC to release its two labile calcium ions, and tropomyosin slides back into place over the myosin-binding sites. Actin and myosin can no longer interact, crossbridge cycling ceases, and the muscle relaxes.

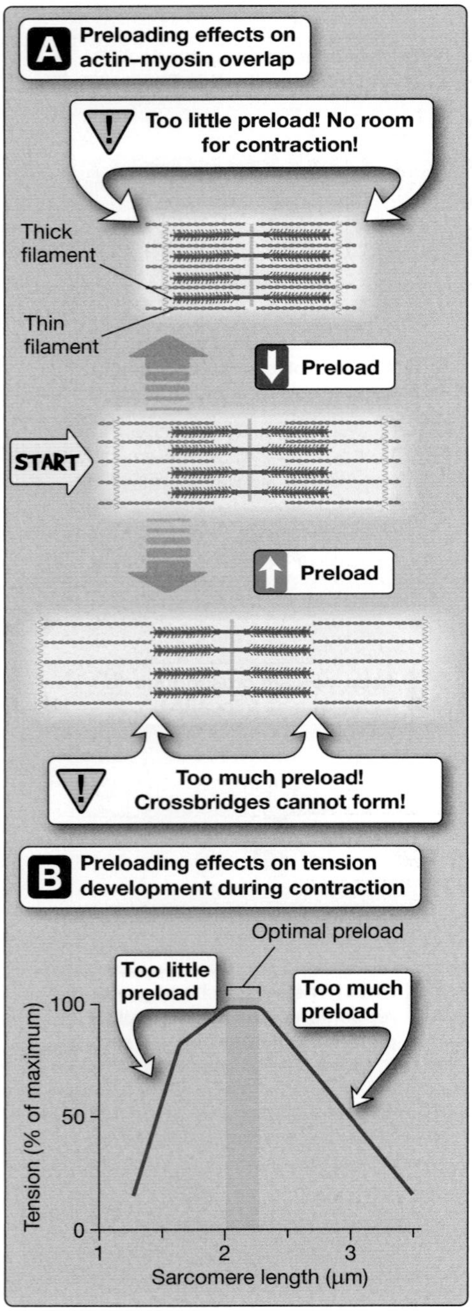

IV. MECHANICS

A skeletal muscle is a complex assembly of contractile and elastic elements. Each crossbridge cycle generates a unit of force that is transferred to a tendon and used to move a bone, lift a load, or tense the diaphragm, for example. Some of the force generated during contraction is wasted in tensioning contractile elements, much as a rubber band must be tensed before it can be used to lift an attached weight (**passive tension**). An understanding of how a skeletal muscle performs *in vivo* begins at the sarcomeric level.

A. Preload

Actin filaments are firmly anchored at either end of the sarcomere by Z disks. Myosin tugging on thin filaments draws the two Z disks closer together, thereby shortening the sarcomere, and the muscle contracts. Shortening continues until the thick filaments run into the Z disks and then stops. If Ca^{2+} release occurs when the thick filaments are already hard up against the Z disks, further shortening cannot occur (Fig. 12.11A). Stretching a sarcomere can also prevent contraction if it physically separates the thick and thin filaments and thus prevents them from interacting. Muscle performance peaks when the potential for crossbridge formation reaches a maximum (see Fig. 12.11B). Stretching a muscle to optimize actin and myosin interaction is known as **preloading**. Sarcomere length is optimal at ~2.2 μm in resting skeletal muscle, but, in cardiac muscle, increasing preload will further increase force production (see Chapter 17·IV·D).

B. Afterload

Although muscle contraction involves innumerable crossbridge cycles, the amount of force they generate is finite. As everyone knows from personal experience, some loads are too heavy to lift. The load that a muscle is asked to move determines how much **tension** a muscle must develop in order to lift it and is known as the **afterload**. Afterload also determines how fast the muscle contracts during a lift. Minimal loads allow the muscle to contract at a maximal velocity, whereas very heavy loads are lifted slowly (Fig. 12.12).

Functionally, there two types of contraction. **Isometric** contraction occurs when a muscle cannot shorten because of an excessive afterload, even though crossbridge cycling continues throughout. A muscle undergoing **isotonic** contraction shortens but maintains a constant tension. In practice, most contractions are a mix of both types.

A Preloading effects on actin–myosin overlap

Too little preload! No room for contraction!

Thick filament

Thin filament

START

↓ Preload

↑ Preload

Too much preload! Crossbridges cannot form!

B Preloading effects on tension development during contraction

Optimal preload

Too little preload

Too much preload

Tension (% of maximum)

100

50

0

1   2   3

Sarcomere length (μm)

Figure 12.11.
Preload effects on force development by skeletal muscle.

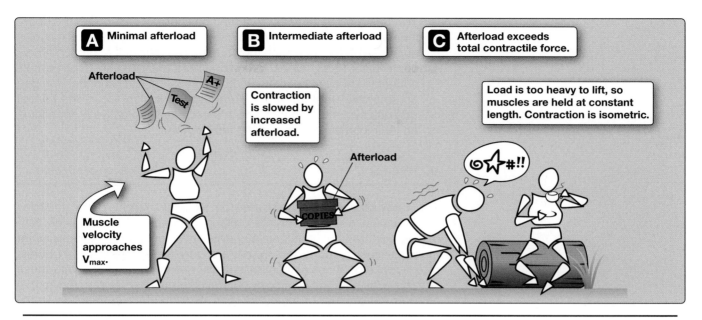

Figure 12.12.
Afterload effects on skeletal muscle shortening velocity. V_{max} = maximal velocity.

C. Stimulation frequency

Motor neurons usually issue commands using a volley of APs rather than just one. The number and frequency of APs within a volley determines how much force a muscle develops during contraction (Fig. 12.13).

1. **Twitch responses:** A single AP triggers a minimal response called a **twitch contraction** (see Fig. 12.13A). Twitches are brief, and tension development is minimal. At the cellular level, a single AP triggers a brief burst of Ca^{2+} from the SR that is cleared from the sarcoplasm almost before the contraction has begun.

2. **Frequency summation:** Full contraction requires additional APs. Electrical events are very fast compared with mechanical events, so a second spike can be delivered within a few milliseconds of the first, even as Ca^{2+} is pouring out of the SR. The second Ca^{2+} burst adds to the first, and Ca^{2+} concentration rises further. A third spike increases Ca^{2+} to an even greater extent. Because Ca^{2+} concentration equates with crossbridge cycling and contraction, muscle tension rises in parallel with Ca^{2+}, an effect known as **frequency summation** (see Fig. 12.13B).

3. **Tetanus:** A maximal contraction requires a steady barrage of spikes. Sarcoplasmic Ca^{2+} levels stay high, and the muscle never has a chance to relax. Tension attains and stays at maximal levels, a condition known as **tetanus** (see Fig. 12.13C).

D. Recruitment

Frequency summation is an imprecise method of controlling muscle tension. Some tasks require a finer degree of control, and this is made possible by dividing muscles into **motor units**. A motor unit

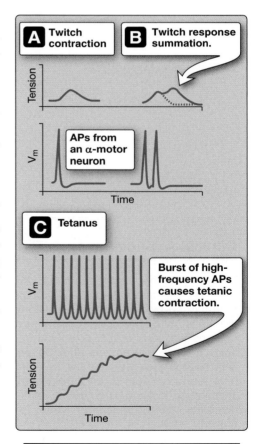

Figure 12.13.
Skeletal muscle summation and tetany. V_m = membrane potential.

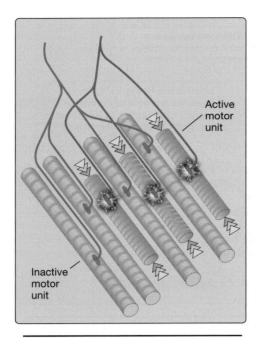

Figure 12.14.
Skeletal muscle motor units.

includes a group of fibers that may be spatially separate but that are innervated by the same α-motor neuron (Fig. 12.14). Skeletal muscles comprise up to 2,000 such motor units. The multiunit approach allows some units to be tetanic even as others are relaxing. This permits contractile force to be maintained at a constant level and then ramped up smoothly (**multiple-fiber summation**). It also allows the central nervous system to choose which muscle units to activate depending on their intrinsic speed and the task at hand.

V. TYPES

The body places many demands on skeletal muscle. Some muscles must sustain a contraction for prolonged periods (e.g., muscles controlling posture), but response latency is not a concern. Conversely, other muscles must respond rapidly but are only ever used for short bursts of activity (e.g., muscles controlling eye movement). To deal with these varied needs, the body employs several different classes of myocyte that differ in the amount of force they can develop, their contraction velocity, how quickly they fatigue, and their metabolism. Traditionally, muscle fibers have been categorized as being either fast or slow (Table 12.1). In practice, most muscles are a mix of slow and fast fiber types.

A. Slow twitch (type I)

Slow-twitch fibers express a myosin heavy chain isoform (MHC-I) that contracts slowly, but these fibers do not readily fatigue. They are used in maintaining posture, for example. Energy is derived primarily from oxidative metabolism, facilitated by a rich vascular system, high levels of oxidative enzymes, and an abundance of myoglobin and mitochondria. The myoglobin makes the slow-twitch fibers appear red.

B. Fast twitch (type II)

Fast-twitch fibers express MHC-II isoforms specialized for rapid movement, but they fatigue more easily. They rely more heavily on glycolysis as a source of ATP than do slow-twitch fibers. Fast-twitch

---

Clinical Application 12.4: Trismus ("Lockjaw")

The soil microbe *Clostridium tetani* produces tetanus toxin. The bacterium typically enters the body via a cut or deep puncture wound and is then carried by retrograde axonal transport via peripheral neurons to the spinal cord. The spinal cord is home to **Renshaw cells**, inhibitory neurons that form a vital component of a feedback loop that normally limits muscle contractions (see Chapter 11·III·E). The toxin interferes with neurotransmitter release by Renshaw cells and thereby allows muscles to contract in an unregulated manner to the point of tetany. Trismus, or "lockjaw," is a classic, early symptom of tetanus toxicity and gives the disease its familiar name. The resulting contractions may be powerful enough to fracture bones. All skeletal muscles are affected, including respiratory muscles, which accounts for the high incidence of

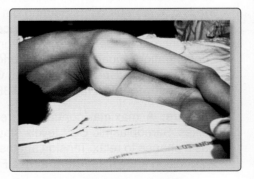

Tetanus-induced opisthotonos.

fatality associated with this disease. The patient shown here exhibits opisthotonos, caused by tetanic contraction of back muscles.

Biologic Sex and Aging 12.1: Skeletal Muscle

Differences between male and female musculature appear after puberty. After age 13 years, males develop increased muscle mass and strength compared with females, particularly in the upper body (upper body muscle mass is ~30% greater in males, ~25% greater in the lower body). These differences are primarily due to the influence of testosterone, causing increased muscle fiber diameter. In males, type IIa fibers appear to dominate, whereas in females, type I fibers outnumber type IIa in muscles such as the quadriceps vastus lateralis. This leads to faster contraction speeds and greater susceptibility to fatigue in the type IIa predominant group, whereas greater oxidative capacity and fatigue resistance occur in muscles that are type I predominant.

Normal aging is accompanied by a 10% per decade decline in muscle mass (**sarcopenia**) and performance, beginning after age 30 years, and resulting in a 30%–50% mass decline and 60% drop in muscle strength (**dynapenia**) by age 80 years. The greater loss in strength compared to mass indicates an age-related change in muscle quality. Males and females are affected similarly, but females begin with less muscle mass than males and are therefore more vulnerable to the aging effects. Upper body muscle mass is affected to a greater extent than lower body muscle mass, although this may be due to disuse rather than healthy aging. The decline is due primarily to decreased myofibril diameter, although myocyte numbers decrease also. Type II fibers appear to be affected more than type I fibers. The decreases in strength are due to both brain (motor cortex, premotor cortex, and cerebellum) and motor neuron changes. In later years, motor neuron numbers decline also. Denervated muscle groups are reinnervated by surviving motor neurons, leading to extensive motor unit remodeling with age. Age-related changes in excitation–contraction coupling include decreased sensitivity to motor stimuli; decreases in dihydropyridine receptor numbers, leading to decreased Ca^{2+} release; and decreased contractile force. Surprisingly, although the progressive decline in muscle function cannot be halted, individuals retain the ability to increase muscle mass by ~200% with exercise training, even in the ninth decade.

fibers are a diverse group that can be further subdivided into **type IIa** and **type IIx** fibers based on myosin isoform.

1. **Type IIa:** Type IIa fibers express MHC-IIa, a product of the *MYH2* gene. These fibers resemble slow-twitch fibers in that they rely primarily on oxidative metabolism, supported by high numbers of mitochondria and myoglobin (which gives them their red color), yet they also have well-developed glycolytic pathways. These properties allow for a wide range of activity types.

2. **Type IIx:** Type IIx fibers express an MHC-IIb isoform (*MYH1* gene). Type IIx fibers are specialized for high-speed contractions of the type used in rapid sprints. They are primarily glycolytic and fatigue easily.

VI. PARALYSIS

The NMJ is a critical component of the motor control pathway because it functions as an on/off switch. If a motor command is interrupted, the muscle is paralyzed. Nature has devised several potent toxins that target the NMJ to help predators immobilize their prey. The pharmaceutical industry has similarly developed many drugs that act at the NMJ. Two primary targets for both toxins and drugs are the nAChR and AChE.

A. Agents that affect nicotinic acetylcholine receptors

nAChR signaling can be blocked both by agonists and antagonists, although the mechanism of action is different.

1. **Agonists:** Succinylcholine comprises two linked ACh molecules that readily binds to the nAChR and initiates contraction.

Table 12.1: Muscle Fiber Types and Properties

	Slow	Fast
Synonyms	Red	White
Myosin ATPase activity	Slow	Fast
Fatigue resistance capacity	High	Low
Oxidative capacity	High	Low
Glycolytic capacity	Low	High
Myoglobin content	High	Low
Mitochondrial volume	High	Low
Capillary density	High	Low

Succinylcholine is resistant to degradation by AChE, however, which allows it to persist in the synaptic cleft and continue stimulating the receptor. The result is nAChR desensitization and Na⁺-channel inactivation. In practice, this means that signaling is blocked after an initial contraction. Succinylcholine is often used clinically to paralyze tracheal muscles to permit endotracheal tube insertion. Its effects are reasonably short lived because it is degraded by a plasma cholinesterase.

2. **Antagonists:** Two natural toxins bind tightly to the nAChR and prevent ACh from interacting with the receptor. **α-Bungarotoxin** is a potent nAChR antagonist found in venom from elapid snakes (e.g., banded krait and cobra). **Curare** is synthesized by some plants in the Amazonian region of South America, the active ingredient in which is d-tubocurarine. Native tribal hunters apply curare to arrowheads to help paralyze and fell prey. Anesthesiologists use similar drugs (muscle relaxants) to limit movement during some surgical procedures.

B. **Agents that affect acetylcholinesterase activity**

Physostigmine is a natural alkaloid that reversibly inhibits AChE, which allows synaptic ACh levels to rise and paralyze the muscle through prolonged stimulation. Synthetic physostigmine derivatives are used to treat the symptoms of the autoimmune disease **myasthenia gravis** (see Clinical Application 12.2).

Chapter Summary

- Most **skeletal muscles** attach to and work in conjunction with bone. Together, they create a system of articulating levers that allow the body to move and manipulate objects.

- The tensile force that allows a muscle to contract is generated at the expense of ATP by **myosin head groups**. They cause muscle shortening by binding to and pulling on **actin filaments**.

- Actin–myosin interactions are initiated by Ca^{2+} release via **troponin**, a Ca^{2+}-sensitive protein that unmasks myosin-binding sites on the actin filament by repositioning **tropomyosin**.

- Actin and myosin monomers are assembled into **thin** and **thick filaments**, respectively. Overlap between these filaments within a sarcomere gives the banding pattern characteristic of **striated muscle**.

- Skeletal muscle is under voluntary control. A command to contract is relayed to the muscle by an **α-motor neuron**. The neuron releases **ACh** at the neuromuscular junction, which binds to a **nicotinic ACh receptor** on the postsynaptic membrane. Binding initiates an **action potential (AP)** that spreads over the surface of the muscle fiber.

- The AP is distributed to individual sarcomeres by **transverse tubules**. **Excitation–contraction coupling** begins at **triads**, specialized junctions within the tubules where the **sarcolemma** and **sarcoplasmic reticulum (SR)** are closely apposed.

- The SR contains high concentrations of Ca^{2+} bound to **calsequestrin**. Opening of **Ca^{2+}-release channels** in the SR membrane allows the stores to empty out into the sarcoplasm. The resulting rise in sarcoplasmic Ca^{2+} concentration initiates **crossbridge cycling** and **contraction**.

- The force that a muscle generates is dependent on sarcomere length (**preload**). Maximal force generation occurs when overlap between thick and thin filaments is optimal. Excessive or insufficient preloading reduces the ability to generate force.

- Maximal contraction velocity occurs when a muscle is unloaded. Increasing **afterload** decreases contraction velocity.

- Single APs produce **twitch responses** and minimal force development. Increasing action-potential frequency increases force by **summation**. Constant stimulation results in **tetanus**.

- Most skeletal muscles are a mix of **slow-twitch** and **fast-twitch** fibers. Slow-twitch fibers contract slowly but can produce a sustained contraction. Fast-twitch fibers are capable of rapid response and high contractile force but **fatigue** quickly.

- Interfering with one or more steps in excitation–contraction coupling can cause muscle **paralysis**. Pharmaceutical drugs and natural toxins usually paralyze muscle by interrupting signals at the neuromuscular junction.

Study Questions

Choose the ONE best answer.

12.1. Genetic analysis of patients with an autosomal recessive form of limb-girdle muscular dystrophy reveals mutations affecting a variety of sarcomeric proteins. One such mutation is in a gene encoding a protein that returns the sarcomere to a resting length during relaxation of striated muscle. The affected gene most likely encodes which of the following proteins?

A. α-Actinin
B. Dystrophin
C. Nebulin
D. Titin
E. Z disk

Best answer = D. Titin is a massive, thick filament–associated structural protein that limits sarcomere length when a muscle is stretched, restores the sarcomere to its resting length during relaxation, and centers the thick filaments within the sarcomere (see Section II·C). Thin filament–associated proteins do not act as springs but, rather, provide structural integrity. For example, α-actinin (A) binds the ends of thin filaments to Z disks (E), structural plates that serve as attachment points for thin filaments; dystrophin (B) anchors the contractile array within the cytoskeletal framework; and nebulin (C), which extends the length of the actin filament, is believed to establish thin-filament length.

12.2. A 22-year-old female receives botulinum toxin type A (a cholinergic presynaptic release inhibitor) injections to treat palmar hyperhidrosis (excess sweating). Her grasp is weakened by the treatments, most likely through a decrease in the synaptic levels of what substance?

A. Acetylcholine
B. Acetylcholinesterase
C. Calsequestrin
D. Myoglobin
E. Nicotinic receptors

Best answer = A. Botulinum toxin is a protease that prevents exocytosis and release of neurotransmitters from nerve terminals (see Chapter 5·IV·C). Botulinum toxin type A is commonly used clinically to inhibit ACh release at the neuromuscular junction, which reduces synaptic ACh levels. It reduces eccrine sweat gland activity through a similar mechanism. Acetylcholinesterase (B), which normally degrades ACh and terminates neuromuscular signaling (see Section II·G), would not be affected by the toxin. Calsequestrin (C) is a Ca^{2+}-binding protein found in the sarcoplasmic reticulum, whereas myoglobin (D) is a sarcoplasmic O_2 storage and transport molecule. Neither is directly involved neuromuscular transmission. ACh binds to postsynaptic nicotinic ACh receptors (E) once released.

12.3. A 17-year-old male undergoes anesthesia for an emergency appendectomy. The procedure was completed without complications, but, while recovering from anesthesia, he develops muscle rigidity, tachycardia, and a rapid temperature rise to above 39°C. Subsequent genetic analysis reveals that he has a mutation in a gene that encodes the ryanodine receptor. The mutation has most likely caused which of the following events following anesthesia?

A. Acetylcholine-receptor binding
B. Ca^{2+} release from the sarcoplasmic reticulum
C. L-type Ca^{2+}-channel activation
D. T-tubule depolarization
E. Troponin C activation

Best answer = B. The ryanodine receptor is a Ca^{2+}-release channel that allows stored Ca^{2+} to flood out of the sarcoplasmic reticulum (SR) during excitation–contraction coupling (see Section III·A). The patient has malignant hyperthermia triggered by inhaled general anesthetics, most often caused by mutations in the ryanodine receptor gene. Excessive Ca^{2+} release stimulates inappropriate muscle contraction, ATP use, and heat generation. ACh-receptor binding (A) results in Na^+ influx at the neuromuscular junction. L-type Ca^{2+} channels (C) are dihydropyridine receptors located in the T-tubule membrane. Ryanodine receptors are not L-type Ca^{2+} channels. Troponin C (E) is a Ca^{2+} binding protein that initiates contraction following Ca^{2+} release from the SR.

12.4. An investigator is comparing the properties of fast-twitch and slow-twitch muscle fibers. During one study, afterload is varied to determine the effects on maximal contraction velocity. Which of the following experimental parameters was most likely varied during this study?

A. External Ca^{2+} levels
B. Muscle length
C. Number of motor units activated
D. Stimulation frequency
E. Weight being lifted

Best answer = E. Afterload equates with the amount of tension that develops in muscle during contraction to lift or move a weight (see Section IV·B). An experiment that involves varying afterload, therefore, most likely involves varying the amount of weight that a muscle fiber must lift. External Ca^{2+} levels (A) determine the size of Ca^{2+} influx during excitation, which is a determinant of maximal force development and contractility. Muscle length (B) prior to contraction establishes the preload, not the afterload. The number of motor units being activated (C) and stimulation frequency (D) vary in response to the afterload, not vice versa.

13 Smooth Muscle

I. OVERVIEW

Skeletal muscle is designed to contract rapidly to facilitate activities such as running, jumping, and throwing and is under voluntary control. However, the human body performs innumerable involuntary activities that require a muscle type capable of contracting rapidly to regulate, for example, the amount of light falling on the retina (i.e., muscles controlling pupil diameter), but also being able to sustain a contraction for an hour or more (e.g., internal anal sphincter). These and many other physiologic functions are the responsibility of smooth muscle. Smooth muscle's most common task involves adjusting the luminal diameter of hollow structures. For example, smooth muscle is layered within the walls of blood vessels, airways, and intestines (Fig. 13.1). In blood vessels and airways, smooth muscle contractile state regulates blood and air flow, respectively. In the gastrointestinal (GI) tract, sequenced smooth muscle contractions move food and digestion products along its length. Some of smooth muscle's many functions are summarized in Table 13.1. The varied demands made of smooth muscle have necessitated many organ-specific adaptations to its structure and function that make generalizations difficult. Also, whereas skeletal muscle is activated by acetylcholine (ACh) released from a somatic nerve terminal, smooth muscle may be activated by chemical, electrical, or mechanical stimuli, and the contractile state may be modulated by hundreds of ligands (neurotransmitters, hormones, and paracrines). Many details of smooth muscle structure and function remain under active investigation, but its core function is to generate force, accomplished using the same basic principles as in skeletal muscle. Contraction begins when intracellular Ca^{2+} concentrations rise, and force is generated when head groups extending from a **myosin** thick filament bind to and pull on **actin** thin filaments.

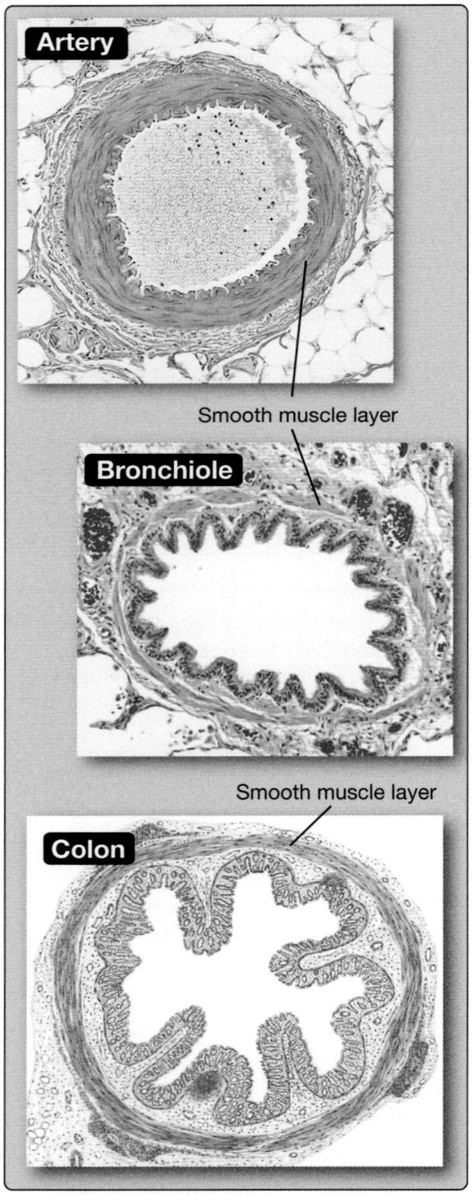

II. STRUCTURE

Smooth muscle, like striated muscle, develops force through actin–myosin interactions. Myosin concentrations are much lower in smooth muscle compared with other muscle types, but actin levels are similar.

A. Contractile units

Individual smooth muscle cells contain small contractile units of closely juxtaposed thick and thin filaments (**minisarcomeres**). Because these sarcomeres are not aligned by adjacent Z disks as they are in skeletal and cardiac muscle, there are no visible striations under polarized light. The lack of striations gives "smooth muscle" its name.

Figure 13.1.
Arterial, airway, and colonic smooth muscle.

Table 13.1: Smooth Muscle Functions

Location	Function
Blood vessels	Control diameter, regulate blood flow
Lung airways	Control diameter, regulate air flow
GI tract	Propulsion, segmentation, sphincter tone
Urinary system	Propels urine through ureter, bladder tone, internal sphincter tone
Male reproductive tract	Secretion, propels semen
Female reproductive tract	Propulsion (fallopian tube), childbirth (uterine myometrium)
Eye	Control of pupil diameter (iris muscle) and lens shape (ciliary muscle)
Kidney	Glomerular regulation (mesangial cells)
Skin	Hair erection (pili muscles)

1. **Thick filaments:** Smooth muscle myosin has a tertiary structure similar to that of striated muscle, but its amino acid sequence is distinct. Smooth muscle myosin is hexameric, like that of other muscle types, containing two **myosin heavy chains** (**MHCs**) that include head and neck regions and two pairs of **myosin light chains** (**MLCs**), a 17-kDa **essential light chain** and a 20-kDa **regulatory light chain** (**MLC$_{20}$**). The myosin heads contain ATPase activity and an actin-binding site. There are fast and slow isoforms of both heavy and light chains. Thick filaments are ~1.6 μm long.

2. **Thin filaments:** Thin filaments average 4.5 μm in length and may contain two different **smooth muscle actin** (**SMA**) isoforms: α SMA is dominant in vascular smooth muscle, whereas γ SMA is largely restricted to the GI tract. Thin filaments in striated muscle are associated with troponin, which imparts Ca^{2+} dependence to contraction. Smooth muscle does not contain troponin, but SMA does associate with two Ca^{2+}-dependent myosin ATPase inhibitors, **caldesmon** and **calponin**, which may serve a similar function to troponin. Their properties have been studied extensively *in vitro*, but their roles *in vivo* are debated.

B. Assembly

For a myocyte to exert a useful force, its contractile units must align roughly with the direction in which they are pulling. Individual smooth muscle contractile units are anchored by α-actinin–rich **dense bodies** (**desmosomes**) found scattered throughout the sarcoplasm (Fig. 13.2), which are believed to be the functional equivalent of Z disks in striated muscle (see Chapter 12·II·C·1). Contractile arrays are aligned within the cytoskeletal framework by a network of intermediate filaments comprising **vimentin** and **desmin**, the relative amounts of which depend on smooth muscle type. The arrays are tethered to the sarcolemma by **focal adhesions** (also known as **dense plaques**; see Fig. 13.2), which are complex assemblies of actin-binding proteins, integrins, and ~80 other proteins that link the cytoskeleton to adjacent cells and to the extracellular matrix. Focal adhesions help direct force transmission during contraction.

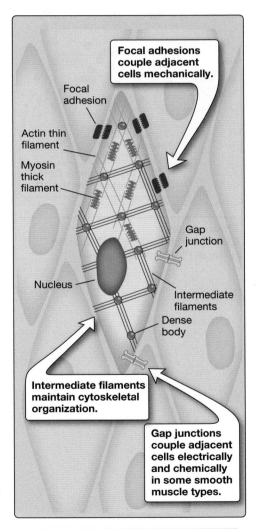

Figure 13.2.
Smooth muscle organization.

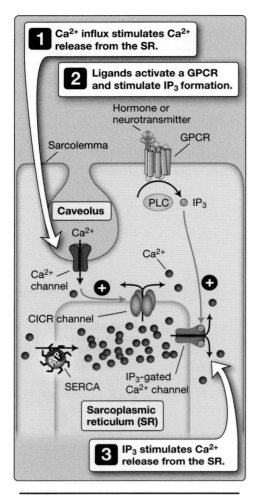

Figure 13.3.
Ca^{2+} sources. CICR = Ca^{2+}-induced Ca^{2+} release, PLC = phospholipase C.

C. Membrane systems

Contraction in all muscle types is initiated when intracellular Ca^{2+} concentration rises above rest (0.1 μmol/L). In striated muscle, this involves Ca^{2+} influx across the sarcolemma via **L-type Ca^{2+}-channel** tetrads concentrated in transverse (T) tubules and **Ca^{2+} release** from the **sarcoplasmic reticulum** ([**SR**] see Chapter 12·III). Smooth muscle contractions are similarly triggered by Ca^{2+} influx across the sarcolemma and Ca^{2+} release from the SR.

1. **Sarcolemma:** Smooth muscle cells do not contain T tubules, but they do show linear arrays of **caveolae** that may have a related function (Fig. 13.3). Caveolae are 50- to 100-nm, flask-shaped sarcolemmal pockets that form narrow junctions (15–30 nm) with the underlying SR. Caveolae contain concentrations of **lipid rafts**, which are small cholesterol- and sphingolipid-enriched lipid platforms. Lipid rafts are dense with signaling molecules, including **G protein–coupled receptors** (**GPCRs**) and their various downstream intermediaries. They also contain clusters of channels, including L-type Ca^{2+} channels.

2. **Sarcoplasmic reticulum:** Smooth muscle cells contain extensive tubular networks of SR that concentrate Ca^{2+} to ~1 mM and store it for release upon contraction. Smooth muscle SR contains two types of Ca^{2+}-release channels. Both are tetrameric structures with a pore at their center. **Ca^{2+}-induced Ca^{2+} release** (**CICR**) channels are opened by Ca^{2+} entering the myocyte via voltage-dependent Ca^{2+} channels in the sarcolemma. Smooth muscle SR also contains an **IP_3-gated Ca^{2+} channel**. IP_3 is a second messenger that communicates binding of one or more chemical signals, including many hormones and neurotransmitters (e.g., norepinephrine and ACh), to cell surface GPCRs (see Chapter 1·VII·B).

D. Neuromuscular junction

Smooth muscle is regulated by the **autonomic nervous system** (**ANS**). Depending on function and location within the body, smooth muscle may receive inputs from the **sympathetic**, **parasympathetic**, and **enteric nervous systems**. The ANS neuromuscular junction is less developed than in skeletal muscle, but pre- and postsynaptic structures are arranged similarly. ANS efferents may contact multiple smooth muscle cells via a series of **varicosities** (swellings) spaced along the length of an axon, each of which is the site of a neuromuscular junction (see Fig. 7.6).

III. EXCITATION–CONTRACTION COUPLING

The number of competing signals that vie for control of smooth muscle function is immense (hundreds). Some of these signals may promote contraction and others relaxation, but, ultimately, most converge on Ca^{2+}, just as in other muscle types. The main difference between smooth muscle and skeletal muscle is that the Ca^{2+}-sensitive step has been transferred from the thin filament (via troponin and tropomyosin) to the thick filament (via myosin phosphorylation).

A. Calcium source

Contraction begins when sarcoplasmic Ca^{2+} concentrations rise. Three potential sources of Ca^{2+} in smooth muscle are Ca^{2+} influx across the sarcolemma, CICR from the SR, and IP_3-mediated Ca^{2+} release from the SR. The relative contribution of these pathways to the net rise in intracellular Ca^{2+} concentration varies according to smooth muscle type, location, and innervation.

1. **Calcium influx:** Smooth muscle cells express at least two types of Ca^{2+} channels that mediate Ca^{2+} influx: L-type Ca^{2+} channels and **receptor-operated Ca^{2+} channels (ROCCs)**.

    a. **L-type Ca^{2+} channels:** L-type Ca^{2+} channels are voltage gated, opening in response to membrane depolarization. Membrane depolarization may occur as a result of membrane stretching and mechanoreceptor activation (e.g., in vascular smooth muscle, where stretch activates a myogenic response; see Chapter 19·II·A), arrival of a wave of excitation from adjacent cells via **gap junctions**, or ROCC activation.

    b. **Receptor-operated Ca^{2+} channels:** Most smooth muscle cells express one or more ROCCs. For example, visceral smooth muscle expresses muscarinic ROCCs, whereas vascular smooth muscle expresses adrenergic ROCCs. Although ROCC-mediated Ca^{2+} fluxes are relatively minor and generally insufficient to support contraction on their own, they do depolarize cells to potentiate (or initiate) Ca^{2+} influx and contraction via L-type Ca^{2+} channels.

2. **Calcium-induced calcium release:** CICR channels are clustered within the SR. Fluxes through individual channels can be visualized as "**Ca^{2+} sparks**" in fluorescent imaging studies. CICR amplifies small changes in sarcoplasmic Ca^{2+} concentration, causing intracellular levels to rise to levels that can initiate or potentiate muscle contraction or activate other Ca^{2+}-sensitive channels or regulatory molecules. Thus, in some smooth muscle types, CICR may potentiate contraction, whereas, in others, Ca^{2+} release relaxes the muscle by opening Ca^{2+}-dependent K^+ channels that hyperpolarize the membrane and decrease L-type Ca^{2+} channel open probability.

3. **IP_3-mediated calcium release:** Most smooth muscle types express a rich variety of GPCRs that initiate contraction through IP_3 formation. When intracellular IP_3 concentrations rise, the SR IP_3-gated Ca^{2+} channel opens, and the Ca^{2+} stores are released (see Fig. 13.3). IP_3-mediated Ca^{2+} release and smooth muscle contraction can (and frequently does) occur independently of an action potential (AP) or other membrane potential (V_m) change and is known as **pharmacomechanical coupling**. The IP_3 pathway is a primary means of initiating contraction in smooth muscle. Other GPCRs may modulate contraction by potentiating or inhibiting this pathway.

[1]For a discussion of antispasmodics, see *LIR Pharmacology*, 8e, Chapter 5·II.

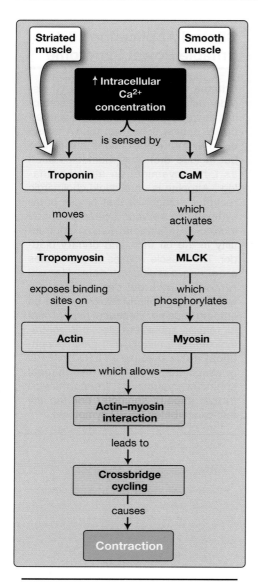

Figure 13.4.
Differences between events facilitating crossbridge cycling in striated and smooth muscle. MLCK = myosin light-chain kinase.

B. Contraction

In striated muscle, rising intracellular Ca^{2+} concentrations prompt troponin to move tropomyosin away from myosin-binding sites on the actin filament (see Chapter 12·III·B). Striated muscle myosin has a high intrinsic ATPase activity, and, with the binding sites exposed, contraction proceeds rapidly. In smooth muscle, the ATPase remains inactive until the MLC_{20} regulatory light chain is phosphorylated. A rise in sarcoplasmic Ca^{2+} is sensed by **calmodulin (CaM)**, which then activates **myosin light-chain kinase (MLCK)**, as shown in Figure 13.4. MLCK phosphorylates MLC_{20}, and, in the presence of ATP, the myosin head group is now able to reach forward and attach to a binding site on the actin thin filament. Crossbridge cycling then proceeds by essentially the same mechanism as described for striated muscle (see Chapter 12·III·B).

C. Relaxation

Muscle relaxation occurs when sarcoplasmic Ca^{2+} concentrations renormalize. Because smooth muscle contraction involves a kinase and a phosphorylation step, relaxation requires a phosphatase.

1. **Calcium renormalization:** When excitatory signaling ends, Ca^{2+} is expelled from the cell by Ca^{2+}-ATPases and Na^+/Ca^{2+} exchangers and returned to the SR by a **sarco(endo)plasmic reticulum calcium ATPase (SERCA)** (Fig. 13.5). MLCK then deactivates.

2. **Store refill:** The sarcolemmal Ca^{2+} transporters and SERCA rapidly renormalize sarcoplasmic Ca^{2+} concentrations. Because variable amounts of Ca^{2+} may be extruded from the cell, SR stores may need to be topped off with Ca^{2+} from the outside of the cell via a **store-operated Ca^{2+} entry (SOCE)**. Refill occurs in specialized regions where the SR comes into contact with the sarcolemma. Store depletion is sensed by **STIM1**, causing it to change conformation and interact with **Orai1** in the sarcolemma. Orai1 monomers gather to form a hexameric **Ca^{2+} release–activated Ca^{2+} channel (CRAC)** that opens to support a sustained Ca^{2+} influx for uptake by the SR (see Fig. 13.5).

3. **Dephosphorylation:** When MLC_{20} is phosphorylated, crossbridge cycling continues for as long as ATP is available to power contraction. Therefore, relaxation of smooth muscle is dependent on **myosin light-chain phosphatase (MLCP)**, which is constitutively active and always undoing the work of MLCK. MLCP is a protein trimer comprising a 37-kDa protein phosphatase catalytic domain (PP1c), a 110- to 130-kDa myosin-binding subunit (myosin phosphatase targeting subunit 1 [MYPT1]), and a 20-kDa subunit. When sarcoplasmic Ca^{2+} concentrations fall and MLCK deactivates, myosin phosphatase quickly strips MLC_{20} of phosphate groups, and the muscle relaxes.

D. Latchbridge formation

If myosin is dephosphorylated when still attached to actin, it locks or "latches" in place. Latchbridges have an intrinsically slow cycling rate that allows blood vessels, sphincters, and hollow organs to sustain contractions for prolonged periods with minimal ATP use (~1% of the amount required by skeletal muscle to achieve the same effect). Because latchbridges are contractile events, they can only occur if intracellular Ca^{2+} levels remain minimally elevated above background. Although many details of latchbridge formation, termination, and regulation have yet to be elucidated, the ability to form latchbridges is a unique and fundamental property of smooth muscle.

E. Regulation

In striated muscle, force regulation usually occurs through control of intracellular Ca^{2+} concentration. In smooth muscle, force is regulated via changes in MLC_{20} phosphorylation state. Force development is dependent on crossbridge formation, which can only occur when MLC_{20} is phosphorylated. Because MLC_{20} phosphorylation state is dependent on Ca^{2+}, MLCK, and MLCP, there are multiple potential control points. Three principal regulatory pathways involve **ROCK** (Rho-associated coiled-coil containing protein kinase [**Rho-kinase**]); **protein kinase C (PKC)**; and, in some tissues, **protein kinase G (PKG)**, as shown in Figure 13.6.

1. **Rho-kinase:** ROCK is a serine–threonine protein kinase regulated by RhoA, a G protein. RhoA is activated indirectly following GPCR binding by, for example, norepinephrine, angiotensin II, endothelin, or any of a number of other ligands. ROCK has several targets, including MYPT1. Phosphorylation inhibits MLCP and thereby promotes contraction. ROCK also has MLCK-like activity and stimulates contraction through direct effects on MLC_{20}. Note that this pathway acts independently of any changes in sarcoplasmic Ca^{2+} concentration.

2. **Protein kinase C:** Agonists that activate phospholipase C and promote contraction via IP_3 simultaneously activate PKC. PKC also phosphorylates many proteins that regulate contraction, including **CPI-17** (PKC-potentiated inhibitor protein of 17 kDa; see Fig. 13.6). CPI-17 is an endogenous protein that binds to the PP1c MLCP catalytic domain and inhibits it, thereby promoting contraction. ROCK also phosphorylates CPI-17 *in vitro*, but the possible significance for smooth muscle function *in vivo* is unclear.

> ROCK- and PKC-mediated contractility increases account for a phenomenon known as "**Ca^{2+} sensitization.**" Ca^{2+} sensitization refers to an observed shift in the Ca^{2+} dependence of contractility toward lower intracellular free Ca^{2+} values following hormone or transmitter binding.

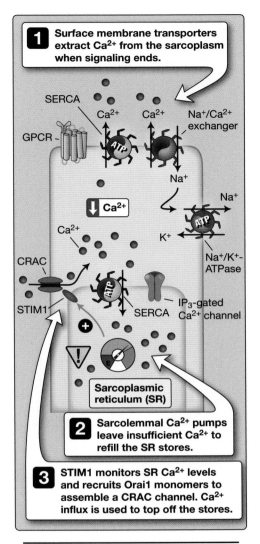

1 Surface membrane transporters extract Ca^{2+} from the sarcoplasm when signaling ends.

2 Sarcolemmal Ca^{2+} pumps leave insufficient Ca^{2+} to refill the SR stores.

3 STIM1 monitors SR Ca^{2+} levels and recruits Orai1 monomers to assemble a CRAC channel. Ca^{2+} influx is used to top off the stores.

Figure 13.5.
Ca^{2+} handling during smooth muscle relaxation. CRAC = Ca^{2+} release–activated Ca^{2+} channel.

3. **Protein kinase G:** PKG is activated by cGMP, which is the product of guanylyl cyclase (GC). In vascular smooth muscle, nitric oxide (NO) activates a soluble GC, which stimulates the activity of a PKG isoform common to smooth muscle (PKG1α; see Fig. 13.6). PKG-1α binds to a carboxyterminal leucine zipper domain on MYPT1, activating it and promoting smooth muscle relaxation. Expression of the PKG-sensitive MLCP may be limited to vascular smooth muscle because NO has no effect on most other smooth muscle types.

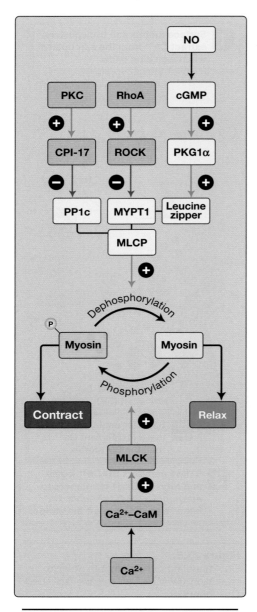

Figure 13.6.
Pathways regulating smooth muscle
contraction. MLCK = myosin light-chain
kinase, MLCP = myosin light-chain
kinase phosphatase, MYPT1 = MLCP
myosin-binding subunit, NO = nitric
oxide, PKC = protein kinase C, PKG-1α =
protein kinase G1α, PP1c = MLCP
protein phosphatase subunit, ROCK =
Rho-kinase.

Clinical Application 13.2: Erectile Dysfunction

Penile erection during male sexual arousal results from increased blood
flow and engorgement of three sinusoidal tissue cavities (**corpora cav-
ernosa** and **corpus spongiosum**) within the penis. Blood flow to these
cavities is regulated in part by cholinergic nerves that signal via nitric
oxide (NO) release. NO relaxes intracavernosal trabeculae, which comprise
smooth muscle cells lined by endothelium. Relaxation allows the erectile
tissue to fill with blood in preparation for sexual activity. NO is a short-lived
signaling molecule that exerts its effects through activation of guanylyl
cyclase, cGMP formation, and phosphorylation of proteins regulated by
protein kinase G. Males who have difficulty achieving or sustaining an
erection have benefited from the availability of drugs such as sildenafil and
tadalafil, which are phosphodiesterase (PDE) inhibitors.[1] PDEs degrade
cGMP and terminate signaling. Inhibiting PDEs increases local NO levels,
allowing an erection to be sustained.

IV. CYTOSKELETAL REMODELING

In addition to the traditional contractile unit (minisarcomere), the cytoskeleton also plays a role in smooth muscle force production. This force
production is minor compared to that of the minisarcomere, but two benefits of cytoskeletal changes are adapting to changes in length and altering rigidity.

A. Length adaptation

Smooth muscle regulates wall tone in many hollow organs whose
lumen size is variable. For example, urinary bladder luminal volume increases from ~6 to ~500 mL when full, yet the detrusor
muscle within its walls can contract and expel urine at any point
during the filling cycle. During bladder filling, the muscle minisarcomeres might be expected to stretch to their physiologic limits.
Stretching striated muscle reduces the degree of actin–myosin
overlap and limits contractility, but smooth muscle is unique in that
it is able to adapt to this stress and retains full contractility even at
an increased length. Length adaptation most likely involves cytoskeletal reorganization.

B. Alternate actin and myosin isoforms

Smooth muscle contains multiple actin and myosin isoforms, only
a subset of which are incorporated into minisarcomeres. During
agonist binding, alternate isoforms of myosin can interact with actin
and the extracellular matrix. This interaction can develop a small
amount of force and provide the minisarcomeres a more ridged
structure to pull on.

 [1]For a discussion of phosphodiesterase inhibitors, see *LIR
Pharmacology*, 8e, Chapter 10·XI·C.

V. TYPES

Smooth muscle is a diverse tissue type that can be difficult to classify. One of the more helpful classifications is based on function. **Phasic smooth muscle** contracts transiently when stimulated. Examples include muscles that make up the walls of the GI tract (stomach, small intestine, large intestine) and urogenital tract (ureters, urinary bladder, vas deferens, fallopian tube, uterus). **Tonic smooth muscle** is capable of sustained contractions, a feature often used to maintain a constant muscular tone. Examples include larger blood vessels and GI sphincters (e.g., lower esophageal and pyloric sphincters). Most organs usually contain a blend of phasic and tonic smooth muscle tailored to specific functional needs, as in the case of the GI tract above (i.e., walls vs. sphincters).

A. Phasic

Phasic smooth muscle functions as a single unit, with all myocytes within the unit being interconnected by gap junctions (Fig. 13.7). Changes in the V_m of one cell spread via gap junctions until it eventually affects all cells within the unit. Skeletal muscle contractions are all-or-nothing events, but smooth muscle may respond to depolarization with a graded contraction; an AP and a stronger contraction; or show no response, depending on location. Phasic smooth muscle typically contracts and reacts several times faster than tonic smooth muscle. This is made possible by expression of a fast-contractile gene program. The program includes fast isoforms of MLC, MHC, and increased expression of MLCK and MYPT1. The fast MLC and MHC isoforms increase muscle shortening velocity but are typically 10- to 20-fold slower than skeletal muscle myosin isoforms. Smooth muscle APs may be initiated by nerves or pacemakers.

1. **Action potentials:** Smooth muscle APs are characteristically slow and their form and time course highly variable. The upstroke is mediated by L-type Ca^{2+} channels, as is the upstroke of cardiac nodal cell "slow" AP (see Chapter 16·V·B). Ca^{2+} influx simultaneously initiates contraction and opens large-conductance Ca^{2+}-activated K^+ channels (BK channels). Ca^{2+}-channel inactivation and BK-mediated K^+ efflux together help repolarize the membrane and yield the AP downstroke. BK activity may also be modulated to increase or decrease the likelihood of AP formation. In vascular smooth muscle, for example, NO increases BK activity, causing membrane hyperpolarization, relaxation, and decreased likelihood of AP formation.

2. **Pacemakers:** Some cells are capable of generating spontaneous V_m changes. In the GI tract, for example, interstitial cells of Cajal are pacemakers that generate electrical "slow waves" with a periodicity of ~3 to 20 cycles per minute, depending on location. These waves spread via gap junctions to adjacent smooth muscle cells and then through the entire muscle body. A wave of contraction may then follow in its wake.

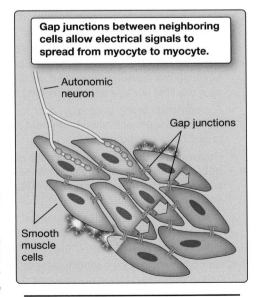

Gap junctions between neighboring cells allow electrical signals to spread from myocyte to myocyte.

Autonomic neuron

Gap junctions

Smooth muscle cells

Figure 13.7.
Contractility control in phasic smooth muscle.

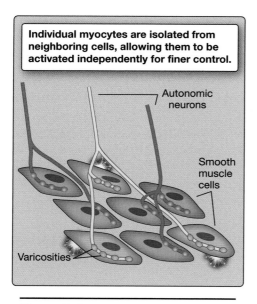

Individual myocytes are isolated from neighboring cells, allowing them to be activated independently for finer control.

Autonomic neurons

Smooth muscle cells

Varicosities

Figure 13.8.
Contractility control in tonic smooth muscle.

B. Tonic

Tonic smooth muscle resembles skeletal muscle in that individual myocytes or groups of myocytes function independently of their neighbors (Fig. 13.8). This feature allows for fine control over movements, which is advantageous for precise control of pupillary diameter and eye lens shape, for example. Multiunit control also allows force to be ramped up through recruitment, much as skeletal muscle force is controlled using combinations of discrete motor units (see Chapter 12·IV·D). Tonic smooth muscle typically does not generate APs; indeed, contraction may occur in the absence of any V_m change through pathways described in Section III. Tonic smooth muscle typically expresses the slower isoforms of MLC, MHC, and associated regulatory enzymes, which are well suited for sustained force production.

Chapter Summary

- **Smooth muscle** serves many diverse functions in all areas of the body. It is found in the walls of many hollow organs, including **blood vessels**, **airways**, and **intestines**, as well as those of the **urogenital tract**. Smooth muscle contracts slowly compared with striated muscle but is able to maintain a steady tone.

- Smooth muscle contraction involves interaction between actin and myosin arranged in minisarcomeres. Minisarcomeres are linked by **dense bodies** and attach to the cytoskeleton at **focal adhesions**.

- Contraction is usually initiated when intracellular Ca^{2+} concentrations rise. Ca^{2+} may enter cells via Ca^{2+} channels or be released from Ca^{2+} stores in the **sarcoplasmic reticulum** (**SR**). Ca^{2+} store release occurs either in response to Ca^{2+} influx from the outside of the cell (Ca^{2+}-induced Ca^{2+} release) or a rise in intracellular IP_3 concentration. The latter occurs as a result of surface receptor binding and acts via an **IP_3-gated Ca^{2+} channel** in the SR.

- Crossbridge cycling is controlled through myosin phosphorylation. Rising intracellular Ca^{2+} concentrations cause Ca^{2+}–CaM-dependent activation of **myosin light-chain kinase** (**MLCK**). MLCK phosphorylates the myosin head group, and crossbridge cycling begins.

- Relaxation of smooth muscle occurs when Ca^{2+} levels fall and MLCK deactivates. **Myosin light-chain phosphatase** (**MLCP**) then dephosphorylates myosin and allows relaxation to occur. When Ca^{2+} levels are barely above baseline, smooth muscle may enter a **latch state** in which muscle tone is maintained for prolonged periods with minimal energy use.

- Smooth muscle contractile state usually represents a balance between MLCK and MLCP activity, and external ligands modulate contractility by manipulating this balance. MLCP is regulated by pathways that include **protein kinase C** and **Rho-kinase**, both of which potentiate contraction, and protein kinase G, which facilitates relaxation.

- Smooth muscle shows **length adaptation**, a process that allows the length–tension relationship to shift in parallel with hollow organ expansion or contraction. Length adaptation may involve cytoskeletal remodeling.

- Smooth muscle can be broadly classified as **phasic** or **tonic**. Phasic smooth muscle functions as a single unit. Adjacent myocytes are connected via gap junctions, which allows for waves of excitation to propagate from one cell to the next. Myocytes in tonic smooth muscle are controlled independently of each other.

Study Questions

Choose the ONE best answer.

13.1. A pharmaceutical company is interested in developing a drug to treat irritable bowel syndrome. One potential drug candidate inhibits channels that are gated by membrane potential change and that cluster in GI smooth muscle caveolae. Which of the following channel classes is the candidate drug most likely targeting?

A. Ca^{2+}-induced Ca^{2+}-release channels
B. Ca^{2+} release–activated Ca^{2+} channels
C. IP_3-gated Ca^{2+} channels
D. L-type Ca^{2+} channels
E. Receptor-operated Ca^{2+} channels

Best answer = D. L-type Ca^{2+} channels are voltage gated, opening in response to membrane depolarization. They are found in many cell types, including smooth muscle, where they are concentrated within plasma membrane pockets called caveolae (see Section II·C). Ca^{2+}-induced Ca^{2-}-release channels (A) and IP_3-gated Ca^{2+} channels (C) are located on the sarcoplasmic reticulum (SR) membrane and mediate Ca^{2+} store release. Ca^{2+} release–activated Ca^{2+} channels (B) are used to top off SR Ca^{2+} stores during muscle relaxation (see Section III·C). Channel opening is controlled by a SR Ca^{2+} sensor (STIM1). Receptor-operated Ca^{2+} channels (D) open when a ligand binds to the associated receptor.

13.2. A 34-year-old female with a history of hypertension comes to the office for a routine prenatal examination. Her pulse is 88/min and blood pressure is 162/108 mm Hg. She is prescribed hydralazine, a drug that lowers blood pressure by relaxing vascular smooth muscle. Which of the following most likely describes the mechanism of action of this drug?

A. Activates myosin light-chain kinase
B. Activates RhoA
C. Inhibits IP_3-gated Ca^{2+} release
D. Inhibits protein kinase G
E. Stimulates protein kinase C

Best answer = C. Hydralazine is a direct-acting vasodilator that inhibits Ca^{2+}-induced Ca^{2+} release in smooth muscle. Ca^{2+} release stimulates myosin light-chain kinase (MLCK) activity (A) and initiates contraction (see Section III·B). Activating RhoA (B) causes phosphorylation and inhibition of myosin light-chain phosphatase (MLCP), resulting in contraction. Stimulating protein kinase C (E) has a similar effect. Protein kinase G normally relaxes smooth muscle, so inhibiting this enzyme (D) would cause contraction (see Section III·E).

13.3. A 20-year-old female comes to the office with a persistent headache. She has a history of muscle weakness and muscle pain. Physical examination reveals permanent miosis (pupillary constriction). Genetic analysis reveals an autosomal dominant gain-of-function mutation in the *STIM1* gene. Which of the following most likely describes the consequences of such a mutation for smooth muscle?

A. Decreased IP_3-gated Ca^{2+} release
B. Decreased Orai1 activity
C. Decreased SERCA activity
D. Increased sarcolemmal Ca^{2+} fluxes
E. Smooth muscle relaxation

Best answer = D. STIM1 monitors Ca^{2+} levels within the sarcoplasmic reticulum (SR). When levels are low, STIM1 causes Ca^{2+} influx across the plasmalemma via Orai1, which constitutes a Ca^{2+} release–activated Ca^{2+} channel. Ca^{2+} influx is used to replenish the SR Ca^{2+} stores (see Section III·C). The patient has Stormorken syndrome. *STIM1* gain-of-function mutations increase, not decrease (B), sarcolemmal Ca^{2+} fluxes via Orai1. The symptoms are wide-ranging, including permanent pupillary constriction. Ca^{2+} influx is used to replenish the SR stores, so IP_3-gated Ca^{2+} release (A) from the SR would be increased. SERCA activity would be increased, not decreased (C), to help counter the increased Ca^{2+} load. Increased sarcolemmal Ca^{2+} fluxes (D) cause smooth muscle contraction, not relaxation (E), including the iris sphincter muscle.

13.4. An investigator is comparing properties of striated and smooth muscle. Which of the following is most likely a feature of smooth muscle that is not seen in striated muscle?

A. Action potentials
B. Ca^{2+}-induced Ca^{2+} release
C. Latchbridges
D. T tubules
E. Troponin C

Best answer = C. Latchbridges are a unique feature of smooth muscle used to sustain contractions for prolonged periods with minimal ATP use. For example, latchbridges allow sphincters to maintain a high resting tone for hours (see Section III·D). Latchbridges form when myosin is dephosphorylated while still attached to actin. Action potentials (A) are common to all muscle types. Ca^{2+}-induced Ca^{2+} release (B) is a feature of smooth and cardiac muscle. T tubules (D) are a feature of striated muscle, as is troponin C (E), a Ca^{2+}-binding protein similar to calmodulin.

14 Bone

I. OVERVIEW

The brain and many other soft body tissues are enclosed within a protective framework made of bone. Bone also fashions limbs that, together with skeletal muscles, facilitate locomotion and allow objects to be manipulated. The 206 bones that make up the human **skeleton** work in conjunction with cartilage, ligaments, tendons, and skeletal muscle. These tissues help maintain skeletal structure and control bone movement. Bone is a connective tissue (see Chapter 4·IV·A) that has been mineralized to give it high resilience to stress and trauma. The mineral component of bone persists long after the body has died and the soft tissues have decomposed, but bones are not lifeless structures. Bone is honeycombed with tunnels and cavities that are teeming with cells and that provide channels through which blood vessels and nerve fibers permeate the matrix. Many bones also contain a central chamber filled with marrow that manufactures blood cells and stores fat (Fig. 14.1). Bone is a highly dynamic tissue that is continually being broken down (**catabolic remodeling**) and reformed (**anabolic remodeling**), such that the skeleton is turned over once every decade. The balance between anabolic and catabolic remodeling determines bone mass (Fig. 14.2). Anabolic remodeling is an adaptive response to mechanical stress. Catabolic remodeling occurs when plasma calcium or phosphate concentrations fall below optimal. Bone contains immense repositories of both minerals (>99% and 80% of body total, respectively) that can be mobilized and made available for use elsewhere in the body when a need arises. Finally, recent evidence suggests that bone is an important endocrine organ. Osteoblasts, which are bone-forming cells, secrete **osteocalcin**, which is involved in regulating body metabolism and stress responses.

II. FORMATION

Bone structure reflects two competing needs. Bones must be strong to protect encased organs from mechanical trauma and to support body weight during locomotion. Bone's resilience is afforded in part by minerals, which makes bones heavy. On the other hand, bones must be sufficiently light to allow for rapid responses to external threats (e.g., from predators). Light bones are prone to fracture, however. Bone breakage is a potentially lethal event (through predation, hemorrhage, or circulatory shock) and, therefore, must be avoided at all costs. Thus, bone design represents a compromise between strength and weight. In practice, bones contain just enough minerals to withstand normal mechanical stress limits plus an added safety margin.

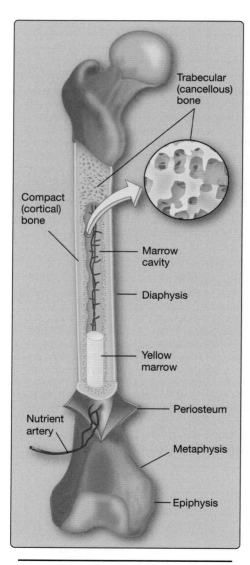

Figure 14.1.
Bone structure.

Trabecular (cancellous) bone

Compact (cortical) bone

Marrow cavity

Diaphysis

Yellow marrow

Nutrient artery

Periosteum

Metaphysis

Epiphysis

Limb bones typically fracture when subjected to stresses that cause deformation by 3–4 times that experienced during normal physiologic activity. If stress levels increase chronically (e.g., during weight training), bone remodels anabolically to compensate for the added stress (the Wolff law) and to reestablish normal safety margins.

Most of a bone's mineral content and strength is concentrated in a thin, highly compacted outer layer. This construction is similar to the hollow steel tubing used to fashion the legs of chairs and tables. Bone's interior is not air filled but is composed of a light, porous matrix. Resistance to compression and mechanical shear is accomplished using a resilient mix of minerals and proteins (Table 14.1).

A. Mineral component

Resistance to compression is achieved using thin, tabular **hydroxyapatite** crystals ~50 nm long. Hydroxyapatite is a mineral comprising calcium and phosphate (Ca_{10} [PO_4]$_6$ [OH]$_2$) that occurs naturally in stalagmites and mineral crusts.

B. Protein scaffolding

Hydroxyapatite crystals do not readily compress, but they do shear. Taking advantage of their natural properties requires that they be cemented within collagen fibers (Fig. 14.3). The cement is made from mucopolysaccharides and is rich in Mg^{2+} and Na^+ (~25% of total body Na^+ is contained within bone). Collagen is the principal component of tendons, tissues notable for their flexibility and high resistance to tensile (stretching) and shear stress. When the collagen fibers with their embedded hydroxyapatite crystals are cemented together by **ground substance** (see Chapter 4·IV·B·3), they create a material that can support heavy loads and resist mechanical impact yet is flexible enough to torque and bend without fracturing. Similar construction techniques are used to make adobe (a mix of straw and mud) and reinforced concrete (a mix of steel rods or "rebar" and cement).

C. Assembly

Bone is formed by **osteoblasts**, which are related to fibroblasts (see Chapter 4·IV·B). Bone formation can be divided into four steps: collagen deposition; secretion of ground substance; crystal seeding; and, finally, mineralization and maturation.

1. **Collagen deposition:** Bone formation begins with the creation of **collagen** scaffolding. Collagen molecules comprise chains of tropocollagen subunits, each composed of three polypeptide chains braided into a helix. Monomers join end to end and then spontaneously assemble into **collagen fibrils**, which are the cellular equivalent of steel rebar in concrete. Monomers within the fibrils are crosslinked extensively and staggered like bricks in a wall for additional stability and strength.

2. **Ground substance:** Ground substance is an amorphous, gel-like matrix of glycosaminoglycans, proteoglycans, glycoproteins, salts, and water that fills the space between cells in all tissues.

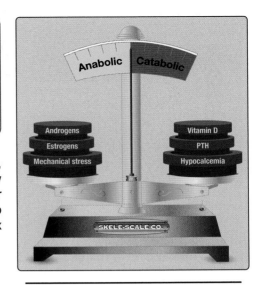

Figure 14.2.
Balance between anabolic and catabolic bone remodeling. PTH = parathyroid hormone.

Table 14.1: Bone Composition

Component (% by Weight)	Composition
Organic (30%)	**Cells (~2%)**
	Type I collagen (~93%)
	Ground substance (~5%)
Inorganic (70%)	**Ca^{2+} and PO_4^{3-} crystals**

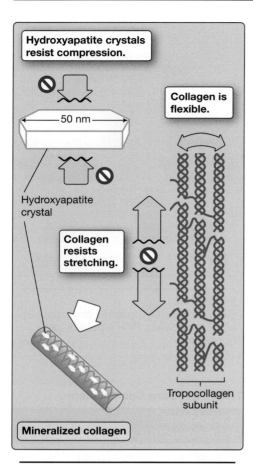

Figure 14.3.
Mineralized collagen formation.

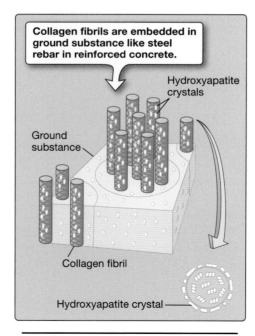

Figure 14.4.
Osteoid.

Clinical Application 14.1: Osteogenesis Imperfecta

Osteogenesis imperfecta, or "brittle bone disease," results from a group of inherited defects in two type I collagen genes. The most common is a point mutation that replaces glycine with a bulkier amino acid at a site where the three strands of the helix normally come close together. The result is a molecular "bleb" that interferes with normal fibril formation and packing of hydroxyapatite crystals. Bones formed with such collagen are weak and prone to fracture.[1]

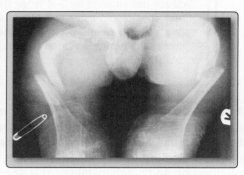

Fractures in an infant with osteogenesis imperfecta.

Bone ground substance differs from that found in other tissue in that it is saturated with Ca^{2+} and PO_4^{3-}. The combination of collagen fibrils and ground substance is called **osteoid** (Fig. 14.4).

3. **Crystal seeding:** Osteoblasts continue secreting Ca^{2+} and PO_4^{3-} into ground substance until it becomes supersaturated, at which point the minerals begin precipitating as calcium phosphate crystals. Osteoblasts also secrete seed crystals that are cemented to collagen fibrils to provide nucleation sites for continued growth. Cementing involves **osteonectin**, which is one of many osteoblast secretory products. Osteonectin is a glycoprotein containing binding sites for both Ca^{2+} and collagen. The collagen fibrils slowly become encrusted with amorphous mineral deposits over time.

4. **Mineralization and maturation:** In succeeding months, bone undergoes mineralization and maturation to yield a mineralized matrix. The calcium phosphate crystals are remodeled by osteoblasts to form mature hydroxyapatite. The tabular hydroxyapatite crystals are oriented with their long axis parallel to the collagen fibrils and then tethered to the fibrils by proteoglycans. The fibrils become extensively interlinked to give bone a tensile strength approaching that of structural steel.

D. Immature

Bone deposition is, by nature, a very slow process, which presents a problem when bone is damaged and must be repaired. Breaking a proximal phalanx in a finger is painful and inconvenient, but breaking a major limb bone (e.g., a femur) is more serious because it impairs motion. In the wild, breaking a leg can leave an animal extremely vulnerable to predation, so deposition of new bone usually occurs with a view to speed rather than strength. The result is **woven bone**, which, although weaker than the mature form, allows breaks to heal in weeks rather than months (Fig. 14.5A). In time, woven bone is

[1]For additional discussion of osteogenesis imperfecta, see *LIR Biochemistry*, 8e, Chapter 4·II·2.

Clinical Application 14.2: Paget Disease

Paget disease is the second most common bone disorder after osteoporosis. The underlying causes are unknown. Paget disease manifests as inappropriately high levels of osteoclast activity, often affecting solitary bones. Osteoclasts normally digest bone during catabolic remodeling. Osteoblasts compensate for the resulting bone loss by laying down new, woven bone, but the rate of turnover in affected bones is so high that it never has time to mature and strengthen. Most Paget disease patients remain asymptomatic. Others present clinically with bone deformities, arthritis, bone pain, symptoms caused by peripheral nerve compression, and increased incidence of fractures.

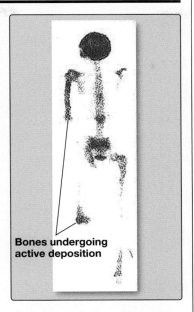

Bones undergoing active deposition

Bone scan of a patient with Paget disease.

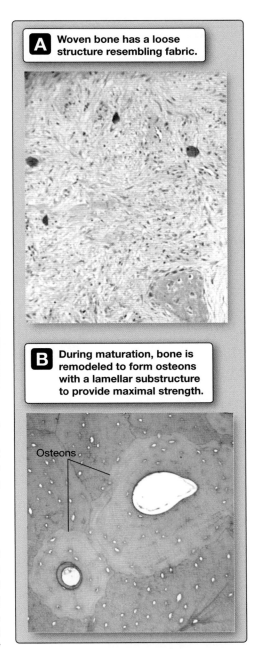

A Woven bone has a loose structure resembling fabric.

B During maturation, bone is remodeled to form osteons with a lamellar substructure to provide maximal strength.

Osteons

Figure 14.5.
Woven and lamellar bone.

replaced through remodeling with the mature **lamellar** form (see Fig. 14.5B). Woven bone has the appearance of fabric when viewed in section, reflecting the fact that osteoblasts deposit collagen fibers at random within osteoid. The random orientation provides resistance to stress in all directions and is therefore an excellent all-purpose patch for a broken bone. Woven bone is also found at bone growth plates.

E. Mature

Bone takes up to 3 years to mature fully. During this time, collagen fibers are aligned with predominant stress lines to provide maximal strength (see Fig. 14.5B). Lamellar bone is laid down in 10 to 30 concentric rings that form cylinders ~200 µm wide and a few millimeters long (~1–3 mm) known as **osteons**, or **Haversian systems** (Fig. 14.6). At the center of each cylinder is a **Haversian canal** that provides a thoroughfare for blood vessels and nerve fibers. Two types of lamellar bone can be distinguished based on density and porosity, **compact bone** and **trabecular bone**.

1. **Compact:** Compact bone is extremely dense and is configured for strength to withstand stress associated with loading, torque, and impact. Also known as **cortical** bone, it is found at the periphery of all bones.

2. **Trabecular:** Trabecular bone (**cancellous** or **spongy bone**) lines the central bone marrow cavity. It has a lacy, porous structure that imparts a very high surface area. When plasma Ca^{2+} levels fall, trabecular bone is the first to be resorbed to release its mineral content. When Ca^{2+} and PO_4^{3-} levels renormalize, the bone is rebuilt. Trabecular bone also provides critical mechanical support, particularly in the vertebrae.

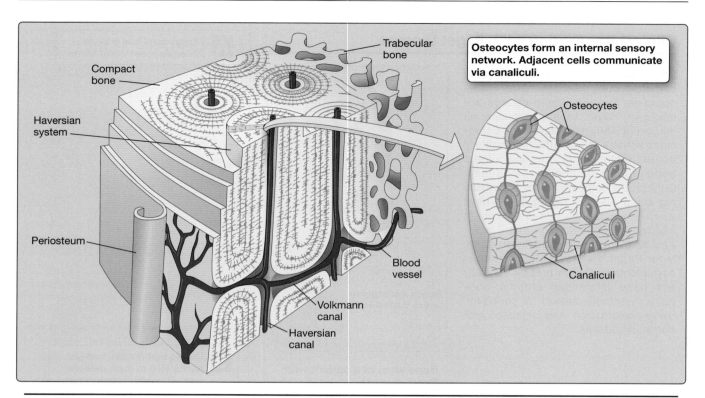

Figure 14.6.
Haversian systems in compact bone.

F. Vasculature

Bone is a living tissue that must be supplied with blood. Blood vessels reach bone by way of the **periosteum**, a fibrous membrane that covers the nonarticulating surfaces of bones and serves as an attachment point for blood vessels and nerves. Supply arteries penetrate the bone cortex and terminate in the medulla. Smaller arterial vessels course through the marrow cavity and then reenter bone to supply the cells within. Vessels travel longitudinally in **Haversian canals** and outward toward the cortex via **Volkmann canals** (see Fig. 14.6).

G. Stress-monitoring system

When bone formation is complete, osteoblasts either undergo apoptosis, or they persist as either **osteocytes** or quiescent **bone-lining cells** (**BLCs**). Together, these two cell types form a vast sensory network that monitors bone stress and integrity.

1. **Osteocytes:** Some osteoblasts become entombed in the osteoid of their own making during bone formation and persist, potentially for 25 years or more, as osteocytes. Osteocytes account for >90% of cells within the adult skeleton. Osteocytes reside in small (~10–20 μm) cavities called **lacunae**, where they remain to monitor bone stress levels associated with bone loading and to signal the need for remodeling if microfractures appear (see Fig. 14.6). Each cell extends 50 to 100 long, thin processes (**dendrites**) in all directions through microscopic channels (**canaliculi**) that permeate the bone matrix (Fig. 14.7). Canaliculi are filled with bone extracellular fluid (ECF) and allow osteocytes to communicate

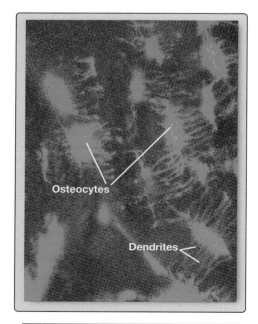

Figure 14.7.
Osteocytes visualized with a fluorescent stain.

with BLCs (via paracrines such as sclerostin, and via gap junctions at sites of cellular contact).

2. **Bone-lining cells:** BLCs form a monolayer covering every bone surface. They communicate with and provide an interface between osteocytes and the bone exterior.

3. **Nutrient supply:** Canaliculi are too small to carry blood vessels (<0.5 μm diameter), but they do provide pathways for bone ECF flow that carries O_2 and nutrients to the osteocytes. Flow is driven by hydrostatic pressure originating in the vasculature. Osteocytes may detect bone stress levels by monitoring changes in bone ECF pressure and flow rates caused by bone deformation.

III. ANATOMY AND GROWTH

The bones that comprise a human skeleton come in several different shapes and sizes. They are usually classified based on their shape.

A. Classification

Bones can be divided into five groups based on anatomy: long, short, flat, irregular, and sesamoid. Long bones are found in arms (humerus, radius, ulna) and legs (femur, tibia, fibula). The bone shaft (**diaphysis**) is a long, thin tube of cortical bone with trabecular bone at the center (see Fig. 14.1). They typically widen (**metaphysis**) toward the end (**epiphysis**) to form a site of articulation with another bone. The ends are broadened to spread the load, and they are filled with trabecular bone that acts as a shock absorber during locomotion. In children, the region between the metaphysis and epiphysis contains a growth plate (physis or **epiphyseal plate**) that is the active site of new bone formation.

B. Growth

Bone growth (widening and lengthening) during fetal development and throughout childhood is effected by **chondrocytes**. Chondrocytes are derived from the same mesenchymal stem cell line that gives rise to osteoblasts and are arranged in 10 to 20 columns within a growth plate. Chondrocytes nearest the ends of the bone divide rapidly (Fig. 14.8). Further down the column, the chondrocytes enlarge and push the ends apart. Chondrocytes also secrete cartilage, which becomes mineralized by osteoblasts and forms a template for further ossification. Older chondrocytes eventually undergo apoptosis, leaving spaces within the matrix that are invaded by nerves, blood vessels, and additional osteoblasts, which complete the task of bone maturation. When skeletal growth is complete, the growth plate dwindles, and the two epiphyses are united with the shaft (**epiphyseal closure**).

C. Bone marrow

Bone marrow is a soft tissue located within the center of some bones. Bone marrow is the source of virtually all blood cells, including red cells, most white cells, and platelets. Blood cells are formed from hematopoietic multipotent stem cells. Marrow is also the location of mesenchymal stem cells, which give rise to osteoblasts and

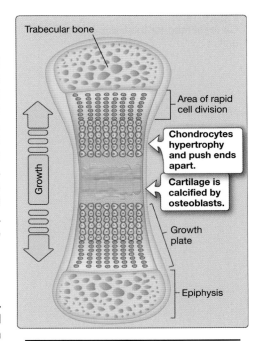

Figure 14.8.
Bone growth.

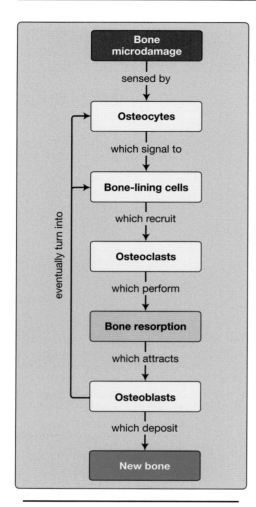

Figure 14.9.
Concept map for the basic multicellular unit.

chondrocytes, among others. Bone marrow in neonates is red with hematopoietic cells, but, in adults, bone marrow is yellowed due to fat deposition. By age 30 years, marrow is 40% adipose tissue and up to 70% fat by age 70 years. Yellow marrow can be converted back to red marrow when increased blood cell production is needed.

IV. REMODELING

Bone is a dynamic tissue constantly being turned over at a rate of about 20% per year in young adults. Remodeling involves four bone cell types that, together, comprise a **basic multicellular unit** (**BMU**). Osteocytes signal the need for remodeling, BLCs facilitate and coordinate remodeling, **osteoclasts** digest old bone, and osteoblasts lay down new bone (Fig. 14.9). BMUs are roving demolition and construction crews replacing bone constantly for the life of an individual. About 1,000,000 BMUs are active at any point in time.

A. Causes

Three main forces drive bone remodeling: mechanical stress, microdamage, and Ca^{2+} and PO_4^{3-} homeostatic needs.

1. **Mechanical stress:** Many bones are subject to repeated mechanical stress associated with lifting and carrying weight. For example, the arm bones form a lever system powered by skeletal muscles. When the muscles contract to lift a weight, the levers are stressed. Bones are designed to withstand such stresses within a normal physiologic range, but a muscle that is exercised repeatedly grows stronger and increases the stress on the levers. Thus, bones are also designed to sense mechanical stress and lay down additional bone mass through anabolic remodeling to compensate if necessary. Conversely, when the stress on bones is reduced, they lose mass (catabolic remodeling).

> The stroke (playing) arm of professional tennis players is subjected to years of repeated mechanical stress. The bones of the forearm respond with anabolic remodeling, which increases bone density, diameter, and length. Individuals who have been freed from gravity and its associated mechanical stresses lose bone mass through catabolic remodeling. Astronauts who remain in space for prolonged periods exercise daily to help offset the effects of bone unloading, but they still lose lower leg and pelvic bone mass at a rate of 1%–1.5% per month.

2. **Microdamage:** Bones constantly develop microscopic damage from mechanical stress, either acutely or as the result of normal actions that are performed repeatedly. The organic component of bone also deteriorates with time, which increases the likelihood of microfissures and microscopic cracks forming. Because fissures and cracks can ultimately result in fracture, damaged areas are replaced with new bone through anabolic remodeling.

Clinical Application 14.3: Remodeling Disorders

In the absence of external influences, such as mechanical stress or disease, total bone mass remains stable despite the continual anabolic and catabolic remodeling. The exact matching of bone deposition to bone resorption requires tight functional **coupling** between osteoblasts and osteoclasts. A disturbance in the balance between the activities of the two cell types causes either loss of bone mass or abnormal bone deposition.

Osteoporosis is a common term for a group of disorders in which the balance between bone resorption and formation is tipped in favor of osteoclasts and catabolic remodeling. The osteoblasts fail to keep up with osteoclast activity, and the bone becomes increasingly porous and fragile as a result. Fracture is common in patients with osteoporosis. Bone loss resulting from catabolic remodeling associated with aging (type II osteoporosis) is common in both women and men, but postmenopausal women are at particular risk. Estrogen limits osteoclast activity, so when circulating levels of this hormone fall after menopause, osteoclasts become increasingly active, and bone is resorbed faster than it is replaced. Osteoporosis affects all bones, but the effects are most dramatic on trabecular bone, which is the primary site of remodeling in healthy individuals. Excessive resorption thins all trabeculae and truncates many of them, which destroys the template required for renewed bone deposition. It also seriously compromises their mechanical functions and greatly increases the likelihood of fracture. Treatment options for both men and women include oral bisphosphonates, which inhibit bone mineral breakdown, and a recombinant form of parathyroid hormone that stimulates osteoblast activity and bone deposition.[1]

Osteopetrosis, or "stone bone," results from a heterogeneous group of rare inherited disorders that impair osteoclast function. The most common form (~60%) results from a mutation in a subunit of the V-ATPase that secretes acid from the ruffled border of an osteoclast onto the bone surface, but other mutations affect genes encoding Cl^- channels, intracellular H^+ pumps, and RANK (receptor activator of nuclear factor κB); the role of these proteins in normal osteoclast function is discussed in Section IV·C. The mutations tip the bone resorption–deposition balance in favor of the osteoblasts, resulting in bones that are dense yet brittle. Affected individuals may show skeletal deformities, an increased likelihood of fractures, and secondary effects related to incursion of bone into the marrow space and the vascular and nerve supply to the bone matrix.

Changes in bone remodeling rate can indicate disease, so several **bone turnover marker** assays have been developed to help monitor bone health. Bone deposition markers include bone-specific alkaline phosphatase (ALP) and a terminal procollagen fragment released during collagen deposition. ALP is expressed on the surface of osteoblasts. Both markers can be detected in serum during anabolic remodeling. Catabolic remodeling is associated with increased urinary levels of N-telopeptide crosslinks and serum levels of C-telopeptide crosslinks. Both are collagen digestion fragments released during osteoclast resorption of bone.

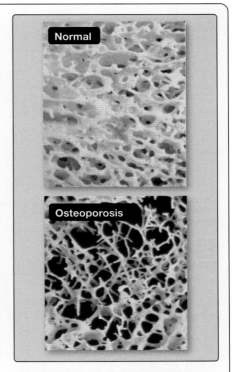

Comparison of normal and osteoporotic trabecular bone.

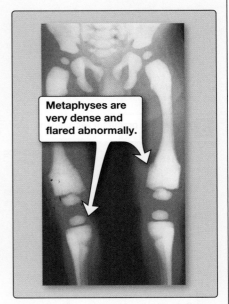

Metaphyses are very dense and flared abnormally.

Abnormal bone density in an infant with osteopetrosis.

[1]For a discussion of agents used to treat osteoporosis, see *LIR Pharmacology*, 8e, Chapter 27·IV.

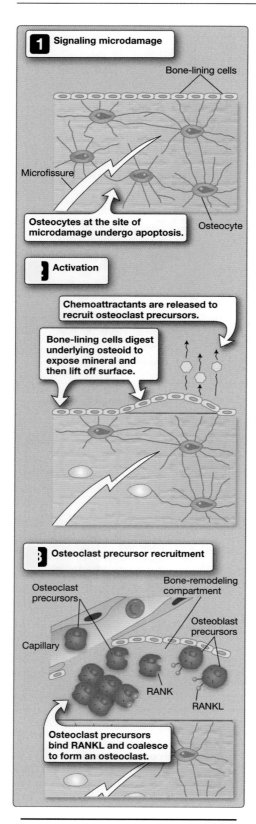

Figure 14.10.
Bone remodeling sequence. RANK =
receptor activator of nuclear factor κB;
RANKL = RANK ligand.

(Continued)

3. **Hormones:** Bone contains immense reservoirs of Ca^{2+} and PO_4^{3-} that can be mobilized and circulated if plasma levels fall. The balance between bone resorption and deposition is controlled by parathyroid hormone (PTH) and vitamin D. The pathways involved are discussed in Chapter 35. Bone mass is also affected by estrogen, androgens, thyroid hormone, and corticosteroids.

B. Signaling

Bone remodeling involves extensive signaling between the various cellular participants. Few of the pathways or signals involved have been characterized fully, although there are many candidates. When microdamage occurs, osteocytes at the fissure site undergo apoptosis and release ATP, which constitutes a signal that remodeling is needed. BLCs then initiate the remodeling sequence (Fig. 14.10; step 1).

C. Remodeling sequence

The bone remodeling sequence lasts ~200 days and can be divided into four phases: activation, resorption, reversal, and formation.

1. **Activation:** During the activation phase, BLCs recruit osteoclast precursors to the remodeling site, expose underlying mineral, and then form a canopy over the BMU worksite (see Fig. 14.10; step 2).

    a. **Osteoclast precursors:** Bone is resorbed by osteoclasts, a blood cell line related to macrophages ("-clast" is derived from the Greek word *klastos*, meaning "broken"). Osteoclasts are large, multinucleate, phagocytic cells formed by fusion of many hematopoietic precursors. The precursors are called to action from the vasculature by chemokines such as macrophage colony–stimulating factor.

    b. **Mineral exposure:** Osteoclasts digest osteoid slowly, so BLCs assist by releasing matrix metalloproteinases (MMPs) and other enzymes onto the bone surface to expose the mineral.

    c. **Canopy:** BLCs then lift off the bone surface as a single sheet and form an epithelial-like canopy over the worksite. This canopy creates a **bone-remodeling compartment (BRC)** whose environment can be optimized for remodeling. The canopy becomes highly vascularized, which allows for recruitment of osteoclast and osteoblast precursors from the vasculature and from marrow (see Fig. 14.10; step 3).

2. **Resorption:** The resorption phase takes about 2 weeks to complete once initiated. Osteoclast precursors fuse to form mature osteocytes, and then digestion and resorption begins.

    a. **Fusion:** Osteoclast precursors express a receptor on their surface that is related to the tumor necrosis factor receptor (**RANK** [receptor activator of nuclear factor κB]). When precursors arrive at the BRC, they encounter osteoblast precursors (bone marrow stromal cells), which express **RANKL** (RANK ligand) on their surface (see Fig. 14.10; step 3). Contact between the two sets of precursors allows

RANK–RANKL binding, and an intracellular cascade is then initiated within the osteoclast precursors that culminates in the synthesis of a variety of fusion proteins. Four or five osteoclast precursors then fuse to form large, multinucleate osteoclasts.

b. **Digestion:** Osteoclasts adhere to the exposed bone matrix and then polarize. The apical membrane establishes a **sealing zone** around the periphery to isolate the demolition site. Apical membrane surface area then increases to form a **ruffled border**, which pumps acid onto the bone surface using a vacuolar-type H^+-ATPase (see Fig. 14.10; step 4). Lysosomes fuse with the apical membrane and empty their contents onto the surface also. Constituents include acid, cathepsin K (a protease), and MMPs to digest collagen and other osteoid connective tissue components. Acid degrades hydroxyapatite and releases Ca^{2+} and HPO_4^{3-} for transfer to the circulation. Osteoclasts usually create a simple pit on the surface of trabecular bone, but in cortical bone, they work deep below the surface, creating tunnels that run longitudinal to the long axis of bone for several millimeters. These tunnels are eventually replaced with a new Haversian system.

3. **Reversal:** Once the resorption phase is complete, bone digestion stops, and the worksite is refilled. The reversal phase involves several concurrent steps, including osteoclast apoptosis, surface cleaning, and recruitment of osteoblast precursors.

a. **Apoptosis:** Osteoblasts determine when sufficient bone has been removed, at which point they release **osteoprotegerin** (**OPG**). OPG is a decoy receptor that binds to and masks RANKL on the osteoblast surface, thereby preventing activation of additional osteoclast precursors. The osteoclasts detach and undergo apoptosis.

b. **Cleaning:** Mononuclear cells arrive at the worksite and clean it with proteases in preparation for new bone deposition.

c. **Osteoblast precursor gathering:** Bone resorption liberates numerous growth factors from the matrix, including insulin-like growth factor 1 (IGF-1). The growth factors recruit osteoblast precursors from blood and bone marrow. The growth factors are incorporated into the bone matrix by osteoblasts during bone formation. Precursors arriving at the worksite migrate into the BRC and differentiate into osteoblasts (see Fig. 14.10; step 5).

4. **Formation:** Bone regrowth is very slow, and it takes several months for the cavity to be refilled. The steps involved in osteoid deposition and mineralization are outlined in Section II·C. Once the cavity has been filled with osteoid, the osteoblasts cease work and either die, differentiate into BLCs, or remain in place as osteocytes (see Fig. 14.10; step 6). Osteocytes inhibit further bone deposition by releasing **sclerostin** into the canalicular system. Sclerostin is a Wnt receptor antagonist that inhibits osteoblast differentiation. Wnt is a growth factor.

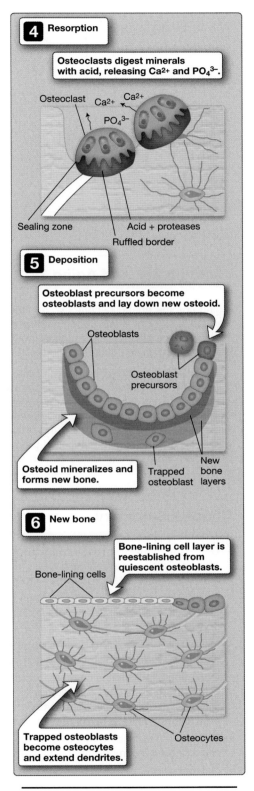

Figure 14.10.
(Continued)

D. Regulation

The anabolic–catabolic remodeling balance is primarily controlled by osteocytes and sclerostin gene (*SOST*) expression levels. The pathways involved are complex and remain to be delineated fully. Local regulatory factors include nitric oxide (NO), prostaglandins, and IGF-1. NO is believed to signal increased bone stress. The resorption–formation cycle is also influenced by hormones involved in Ca^{2+} and PO_4^{3-} homeostasis. Plasma Ca^{2+} and PO_4^{3-} levels are regulated through the concerted actions of vitamin D, calcitonin, and PTH (see Chapter 35). Vitamin D and calcitonin have minimal direct effects on bone remodeling. Chronic elevation in plasma PTH levels causes bone resorption, thereby increasing Ca^{2+} and PO_4^{3-} availability. PTH binds to osteoblasts and stimulates release of factors that recruit and activate osteoclast precursors.

Biologic Sex and Aging 14.1: Bone Mass

Aging tips the anabolic–catabolic remodeling balance in favor of reduced bone mass. Bone loses ~5% of total mass per decade, the cortex thins, trabecular number declines, the marrow cavity enlarges and fills with adipose tissue, and fracture risk increases significantly. At the microscopic level, although osteoclast numbers remain unchanged during aging, osteoblast and osteocyte numbers dwindle, their empty lacunae being filled with bone.

Estrogens and androgens are both required for normal skeletal development and maintenance of a stable skeletal mass. Circulating levels of these hormones decline with age in both men and women, and both sexes may develop osteoporosis in later years (see Clinical Application 14.3). However, because males generally achieve a greater bone mass during development than females, age-related effects have less of an impact on their skeletal integrity. These sex hormones stimulate osteoprotegerin synthesis and reduce RANK (receptor activator of nuclear factor κB) ligand (RANKL) expression by osteoblasts, which limits osteoclast precursor activation and bone resorption. Estrogen also directly increases osteoclast apoptosis, which further favors bone mass retention.

Chapter Summary

- Bone is a connective tissue comprising **mineralized collagen fibers** cemented within an **amorphous ground substance**. Calcium phosphate crystals (predominantly hydroxyapatite) give bone strength, whereas collagen imparts the flexibility and resistance to tensile stress that allows bones to torque and bend without fracturing.

- Bone is formed by **osteoblasts**. Bone is created by embedding collagen fibers within a matrix supersaturated with Ca^{2+} and PO_4^{3-}. Osteoblasts seed the fibrils with **hydroxyapatite crystals**, which then become nucleation points for further crystal growth.

- Newly formed ("**woven**") bone is disorganized and takes several years to mature. Woven bone is gradually replaced with a **lamellar** form in which the collagen fibers are realigned along predominant stress lines to maximize strength.

- Once bone formation is complete, osteoblasts either undergo apoptosis or persist as **osteocytes** and quiescent **bone-lining cells (BLCs)**. Osteocytes reside in small cavities (**lacunae**) located throughout the bone matrix, whereas BLCs cover the surface. Osteocytes and BLCs communicate via **dendrites**, together forming a **sensory network** that monitors bone stress levels and integrity.

- Remodeling is initiated by BLCs, which recruit **osteoclast precursors** to a worksite and then raise a canopy over the site to create a compartment whose microenvironment can be optimized for remodeling.

- Osteoclast precursors fuse to become multinucleate **osteoclasts**. Osteoclast formation is initiated by osteoblast precursors via the **RANK** (receptor activator of nuclear factor κB)–**RANKL** (RANK ligand) signaling pathway.

- Osteoclasts digest bone using acids and proteases. The eroded cavity is then cleaned by mononuclear cells, and new bone is laid down by osteoblasts.

- The remodeling cycle is regulated primarily by **parathyroid hormone (PTH)**, a key Ca^{2+} and PO_4^{3-} homeostatic hormone. PTH stimulates bone resorption when circulating Ca^{2+} levels are low.

Study Questions

Choose the ONE best answer.

14.1. A 3-year-old male is brought to the emergency department with a suspected broken leg. This is his fourth fracture since birth. Physical examination shows blue-tinged sclera, which confirms a diagnosis of brittle bone disease (osteogenesis imperfecta [OI]). OI is most likely caused by mutations that affect which of the following stages in bone formation?

A. Collagen assembly
B. Ground substance deposition
C. Hydroxyapatite crystal seeding
D. Maturation
E. Mineralization

Best answer = A. OI is caused by inherited defects in type I collagen genes that impair normal collagen fibril formation. Fibril strength is compromised as a result. Collagen functions much like steel rebar in concrete; it imparts tensile strength (see Section II·C and Clinical Application 14.1). OI is characterized by brittle bones that break easily and blue-tinged sclera. Ground substance (B) is a gel-like matrix of proteins and minerals. Hydroxyapatite crystals (C) are seeded by osteoblasts. Bone maturation (D) and mineralization (E) increases bone strength over a period of months.

14.2. An 86-year-old female with a history of osteoporosis comes to the office for dual-energy x-ray absorptiometry (DEXA scan) of her spine, hip, and lower arm. DEXA scans assess bone density and mineral content, which is a measure of skeletal health and strength. The scan is most likely detecting which of the following?

A. Calcium oxalate
B. Creatinine
C. Glycosaminoglycan
D. Hydroxyapatite
E. Urate

Best answer = D. Hydroxyapatite is a crystalline mineral containing Ca^{2+} and PO_4^{3-} (see Section II·A). Hydroxyapatite crystals are cemented within collagen fibers and then bundles of mineralized fibers embedded in ground substance to create a material that has a high resistance to compression and tensile stress. Calcium oxalate (A), creatinine (B), and urate (E) are found at high concentrations in urine. When sufficiently concentrated, they form crystals that may be observed in urine sediments. Glycosaminoglycan (C) is a mucopolysaccharide found in ground substance, which fills the spaces between all cells, including bone.

14.3. A pharmaceutical company develops a monoclonal antibody against RANKL (receptor activator of nuclear factor κB ligand) to reduce the effects of osteoporosis in postmenopausal females. The antibody most likely binds to the surface of which of the following cell types?

A. Bone-lining cell
B. Hematopoietic precursor
C. Osteoblast precursor
D. Osteoclast precursor
E. Osteocyte

Best answer = C. Bone resorption and remodeling involves several cell types that work together within a bone-remodeling compartment (see Section IV·C). Osteoblast precursors express RANKL on their surface. RANKL binds to RANK, a receptor expressed on the surface of osteoclast precursors (D), which are hematopoietic precursors (B), causing several such cells to fuse and form large multinucleated osteoclasts. Osteoclasts digest bone and release its mineral content for return to blood. Bone-lining cells (A) are found on the bone surface and form a canopy over the worksite when remodeling is needed. Osteocytes (E) are cells embedded in the bone matrix that monitor stress and integrity.

14.4. A 21-year-old female begins exercise training to increase upper body strength. The training regime includes repeated pull-ups that stress the arm bones and induce microfractures. This microdamage most likely causes apoptosis of which of the following cell types?

A. Bone-lining cells
B. Osteoblasts
C. Osteoclasts
D. Osteocytes
E. Platelets

Best answer = D. Osteocytes are found in small lacunae throughout the bone matrix (see Section II·G). They communicate with adjacent osteocytes and bone-lining cells covering the bone surface using an extensive network of dendrites. Their main function is to monitor bone stress and integrity. When microfractures appear, the osteocytes undergo apoptosis, signaling a need for remodeling. Bone-lining cells (A) form a bone remodeling compartment. Osteoblasts (B) lay down new bone. Osteoclasts (C) digest damaged bone. Platelets (E) participate in blood coagulation.

15 Skin

I. OVERVIEW

Individual cells erect a membrane around their periphery to create a barrier between the extracellular and intracellular environments, which allows them to regulate their cellular contents. The body similarly encloses its tissues within skin, a multilayered barrier separating internal organs from the outside world that makes homeostasis possible. The barrier protects underlying tissues from mechanical insults, chemicals, microbial pathogens, and ultraviolet (UV) radiation and helps the body retain vital fluids. Skin has several additional, nonbarrier functions. Its role in thermoregulation is discussed in Chapter 38. Skin contains exocrine glands that secrete various fluids onto its surface, and it facilitates vitamin D synthesis (see Chapter 35·IV·D). Furthermore, skin actively participates in immune surveillance and contains nerves and receptors that allow us to sense the external environment and manipulate objects. Skin (the **integument**) and its associated structures (hairs, nails, and glands) form the **integumentary system**.

II. ANATOMY

Skin comprises two anatomic layers: a superficial, sacrificial **epidermis** and a deeper, vascularized **dermis** (Fig. 15.1). Their combined thickness varies with body location. The soles of the feet and palms of the hand are subject to constant mechanical abrasion, so skin here is thickest (~5 mm). Other regions are covered with thin skin (~0.5-mm thickness). Skin is one of the body's largest organs, accounting for 15% to 20% of total body mass. Functionally, it is helpful to distinguish between hairy skin (**nonglabrous**) and smooth skin (**glabrous** skin; from *glaber*, Latin for "bald" or "hairless").

A. Hairy skin

Most of the body is covered with hairy skin, which plays an important role in thermoregulation. Hairy skin also contains receptors that sense touch, pressure, temperature, and pain, although in fewer numbers than found in glabrous skin.

B. Glabrous skin

Body areas covered with glabrous skin include lips, soles, palms, and fingertips. The epidermis of glabrous skin is much thicker to resist abrasion and friction force. The palms of the hands and soles of the feet are contact points for grasping and locomotion. Hair here would interfere with motor functions and decrease the ability to discern the

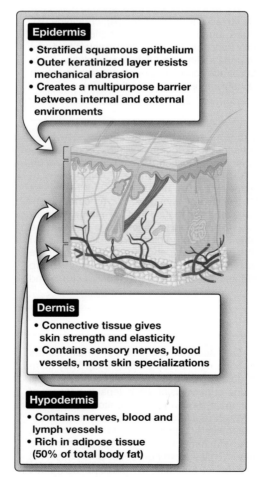

Epidermis
- **Stratified squamous epithelium**
- **Outer keratinized layer resists mechanical abrasion**
- **Creates a multipurpose barrier between internal and external environments**

Dermis
- **Connective tissue gives skin strength and elasticity**
- **Contains sensory nerves, blood vessels, most skin specializations**

Hypodermis
- **Contains nerves, blood and lymph vessels**
- **Rich in adipose tissue (50% of total body fat)**

Figure 15.1.
Skin structure.

texture and temperature of surfaces. Glabrous skin contains a high density of sensory fibers that enhance touch sensitivity. Glabrous skin does not play a large role in thermoregulation, but blood flow is regulated. Palmar erythema is common during pregnancy (caused by estrogen-induced vasodilation) and may also be noted in patients with chronic alcoholic liver disease (Fig. 15.2).

III. EPIDERMIS

The external environment is inherently hostile. It contains many elements capable of causing tissue damage, so the body erects a barrier to protect itself. This barrier is known as the **epidermis**, a reinforced skin layer located at the interface with the external world.

A. Structure

The epidermis is a stratified squamous epithelium that is shed and renewed continuously. It lacks blood vessels, obtaining nutrients by diffusion from the deeper layers. It is composed mainly of **keratinocytes**, but it also contains **melanocytes**, **Langerhans cells**, and **Merkel cells** (Fig. 15.3).

1. **Keratinocytes:** Keratinocytes are the most abundant epidermal cell type. They are formed by division of progenitor cells located in the **stratum basale**. Newly formed cells are pushed upward through the **stratum spinosum**, differentiating as they progress. During differentiation, they synthesize large quantities of **keratins** (a group of >50 related proteins) that assemble into filaments (tonofilaments). Once the keratinocytes reach the surface (**stratum corneum**), they undergo apoptosis, die, and flatten to form **corneocytes**. Transition involves breakdown of the nucleus and other organelles and assembly of the tonofilaments into densely packed keratin tonofibrils. Corneocytes have a 15-nm thick cell envelope composed of insoluble structural proteins, principally keratin and loricrin. The proteins are extensively crosslinked for strength. Corneocytes are mechanically detached from and sloughed off the body surface after ~14 days (a process called **desquamation** or **exfoliation**).

2. **Melanocytes:** Melanocytes produce **melanin**, a photoprotective pigment synthesized in Golgi-derived organelles called **premelanosomes**. Melanin is then passed to keratinocytes and adjacent cells by melanosomes through a process called **pigment donation**. Melanin biosynthesis is regulated by α-melanocyte–stimulating hormone binding to melanocortin receptors. The amount and type of melanin determines skin hue.

3. **Langerhans cells:** Langerhans cells are antigen-presenting cells integral to adaptive immune responses. They ingest foreign matter, digest it, and then present fragments of the material on the cell surface. Presentation allows other immune cells to recognize the fragments. Langerhans cells are also important in delayed (type IV) hypersensitivity reactions.

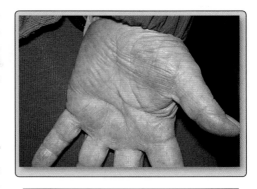

Figure 15.2.
Palmar erythema.

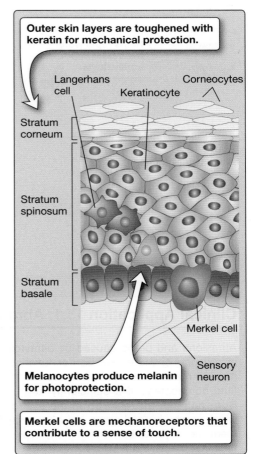

Outer skin layers are toughened with keratin for mechanical protection.

Langerhans cell · Keratinocyte · Corneocytes

Stratum corneum

Stratum spinosum

Stratum basale

Merkel cell

Sensory neuron

Melanocytes produce melanin for photoprotection.

Merkel cells are mechanoreceptors that contribute to a sense of touch.

Figure 15.3.
Epidermal structure.

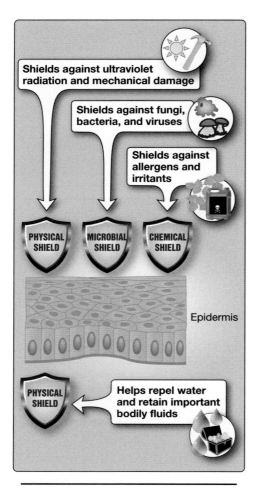

Figure 15.4.
Skin barrier functions.

4. **Merkel cells:** Merkel cells are slow-adapting mechanoreceptors located in skin areas that have a high tactile sensitivity, and they can also be found at the base of hair follicles. Merkel cells associate with a nerve terminal to form a Merkel disk receptor (see Section VII).

B. Barrier functions

The epidermis forms a barrier that both protects tissues from damage and minimizes evaporative water loss. The epidermis has four principal shield functions: **physical**, **photoprotective**, **antimicrobial**, and **water resistive** (Fig. 15.4).

1. **Physical:** The keratinized superficial layers of the epidermis create a tough, multilayered, physical barrier (see Fig. 15.3). Barrier strength is imparted by the corneocyte cell envelope and strong intercellular junctional complexes (desmosomes, adherens junctions, and tight junctions). The densely packed keratin tonofibrils help resist mechanical abrasion and mild penetrating insults that inevitably occur during physical contact with solid objects. The barrier also resists chemicals and prevents tissue exposure to toxins and allergens.

2. **Photoprotective:** UV light can be highly deleterious to biologic tissues because it breaks chemical bonds, thereby disrupting DNA and protein structure. Melanocytes synthesize and donate melanin to adjacent cells specifically to create a photoprotective barrier that absorbs UV radiation and dissipates it harmlessly as heat (see Fig. 15.3). Repeated exposure to the sun or other UV radiation sources stimulates melanocyte proliferation and melanin production, thus increasing the level of photoprotection and causing the skin to darken.

3. **Antimicrobial:** The superficial keratinized layers provide a physical barrier to microbes, but sebaceous glands and sweat glands also secrete a variety of antimicrobial peptides (AMPs) and immunoglobulins onto the epidermal surface. AMPs include dermcidin, cathelicidins, and defensins. If the barrier is breached, Langerhans cells and other immune system components provide a rapid response to microbial invasion, warding off infection until the barrier can be repaired.

Clinical Application 15.1: Abrasions and Burns

Abrasions and burns caused by ultraviolet radiation, heat, and fire can impair one or more skin barrier functions. For example, burn patients have increased transepithelial water loss from the skin that can challenge fluid homeostasis if a sufficiently large body surface area is involved. Burn patients are also at an increased risk of developing infections, which is why burned areas are often wrapped with bandages to supplement the physical barrier, and why topical antibiotics are applied as an antimicrobial shield.

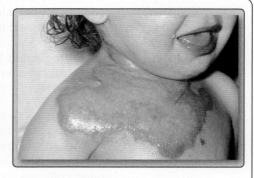

Scalding burn.

4. **Water resistive:** Keratinocytes synthesize and secrete a lipid mixture (ceramides, cholesterol, and free fatty acids) onto their surface, which spreads out to form a 5-nm thick water-repellent layer. The lipid layer is bonded to the underlying cell envelope to prevent it from dissipating. The layer functions much like car wax. It causes water to bead on the epithelial surface, preventing solutes from being washed out of the layers below. The waxy shield also helps minimize evaporative water loss from underlying tissues.

IV. DERMIS

The dermis supports and maintains the epidermis, which does not have its own blood supply. It contains immune cells that react to pathogens that may have breached the epidermal barrier. It also contributes to skin strength and elasticity.

A. Structure

The dermis comprises a meshwork of connective tissue. Type 1 collagen fibers provide structural support to the skin, whereas elastin fibers provide elasticity (see Chapter 4·IV·B·2). Within this matrix are nerve roots and sensory receptors, the cutaneous vasculature, and most skin specializations.

B. Cellular components

The principal cellular components of the dermis include mast cells, macrophages and dermal dendritic cells, and fibroblasts. **Mast cells** are involved in immune and inflammatory responses. Activated mast cells release histamine, prostaglandins, leukotrienes, cytokines, and chemokines (Fig. 15.5), which increase cutaneous blood flow and capillary permeability. **Macrophages** are phagocytic and aid in

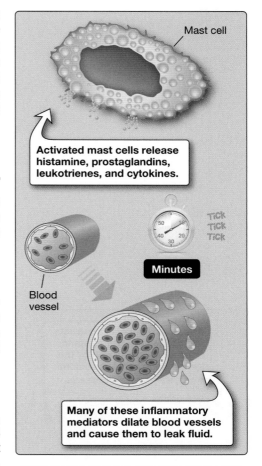

Figure 15.5.
Mast cell activation effects.

Clinical Application 15.2: Triple Response

Wheal, erythema, and flare (known as a "**triple response**") describe the classic reaction to abrasion or histamine-releasing stimuli. First, an erythemic (red) spot develops that spreads outward for a few millimeters, reaching maximal size in about 1 minute. Second, a brighter flush spreads slowly in an irregular flare around the origin. Third, an edemic wheal forms over the site of insult. Mast-cell histamine release can account for the triple response, mediating vasodilation and fluid extrusion into the interstitial space and stimulating nerve endings to give the sensation of itch. The triple response to histamine is often used as a positive control for a skin prick allergy test. Histamine is pricked into the skin followed by a row of other potential allergens such as pet dander, dust mites, and pollens.

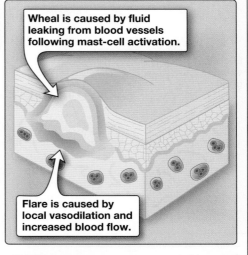

Triple response.

several immune-related responses. **Dermal dendrocytes** are antigen-presenting cells similar to the epidermal Langerhans cells. They are integral to cutaneous adaptive immunity responses. **Fibroblasts** synthesize and degrade fibrous and nonfibrous connective tissue proteins and are involved in wound healing and scarring.

V. HYPODERMIS

The hypodermis lies beneath the dermis (see Fig. 15.1). It contains subcutaneous fat, blood and lymph vessels, and nerves. Around 50% of total body fat is located within the hypodermis on average. Thus, caliper measures of skin-fold thickness can be used to estimate peripheral fat stores. The hypodermis cushions the skin, allows it to slide over underlying structures, and anchors skin to the tissues below.

VI. SPECIALIZED SKIN STRUCTURES

Skin contains several specialized structures, including hairs, nails, sebaceous glands, and sweat glands.

A. Hair

Hairs are constructed from three layers of fused keratinized cells (Fig. 15.6). The outermost protective layer (**cuticle**) is colorless. A middle layer (**cortex**) imparts strength and contains two types of melanin, the relative proportions of which give hair its natural color. Larger hairs also contain an inner **medulla**. The portion that protrudes beyond the epidermis is known as the **hair shaft**. The shaft emerges from a **hair follicle**, a specialized skin structure containing the hair bulb, keratinocytes, and associated glands.

> Chemotherapy, which is used to treat some forms of cancer, targets rapidly dividing cells. The treatment is nondiscriminatory, however. It also affects rapidly proliferating keratinocytes, causing hair thinning and loss.

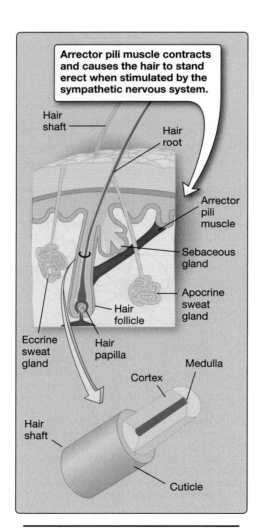

Figure 15.6.
Hair follicle and associated structures.

1. **Hair cycle:** The hair follicle cycle includes growth, rest, regression, and shedding phases. Growth-phase duration determines hair length, which can vary with body location. In the scalp, for example, 85% to 90% of hairs are growing, 10% to 15% are in rest phase, and 1% are shedding.

2. **Associated structures:** The base of each hair is attached to a piloerector muscle under sympathetic control. When excited, piloerector muscles cause hair erection and produce "goose bumps." Follicles also contain a sensory nerve fiber network (**root plexus**) that provides information about touch, pressure, and pain (described in Section VII). Follicles may also be associated with sebaceous and apocrine sweat glands.

B. Nails

Nails are hard, scaly epidermal extensions that shield the posterior fingertips. The nail (known as the **nail plate**) is a hardened keratinized structure that mechanically protects the underlying skin (**nail bed**). Nail keratin contains a high number of disulfide bonds, which give it its strength and rigidity. Skin keratin is softer, reflecting fewer disulfide bonds.

C. Sebaceous glands

Sebaceous glands comprise keratinocytes and **sebocytes**, which produce **sebum**. Sebum is a lipid-rich secretion that is released into the hair follicle when mature sebocytes undergo apoptosis, lyse, and spill their cellular contents into the gland duct. Sebum coats the hair shaft and wicks up onto the epidermal surface. Sebum's lipid content helps seal in moisture and prevents skin desiccation. Sebum also has antioxidant and antimicrobial properties. Sebum secretion is continuous, but gland output is modulated by sex hormones. Androgens and growth hormone increase secretion rate, whereas estrogens reduce secretion rate.

D. Sweat glands

Sweat glands secrete serous fluids onto the epidermal surface (see Fig. 15.6). There are two classes of sweat glands: eccrine and apocrine.

1. **Eccrine:** Eccrine sweat glands produce secretions for evaporation, skin hydration, and protection and to increase friction of glabrous surfaces such as the hands. They are numerous and widespread, the greatest concentrations being on the palms and soles. The glands secrete fluid directly onto the skin surface, where it evaporates and thereby transfers body heat to the environment. The skin is capable of producing copious amounts of sweat (normally ~1.0 L/hr, but heat-acclimated individuals may produce >3 L/hr). Such high levels of fluid loss from the body can compromise cardiovascular function.

   a. **Structure:** Eccrine sweat glands are 3- to 5-mm long tubes comprising a 50- to 100-μm diameter **secretory coil**, a straight **duct**, and a spiral opening onto the body surface (**acrosyringium**; Fig. 15.7).

      i. **Secretory coil:** The secretory coil contains histologically distinct **dark cells**, **clear cells**, and **myoepithelial cells**. Dark cells are dense with secretory granules and produce AMPs. Clear cells produce a sweat precursor fluid consisting of a protein-free plasma filtrate. Sympathetic nervous system postganglionic cholinergic nerves initiate sweating (Fig. 15.8). Acetylcholine (ACh) binds to muscarinic M_3 receptors, which raises cytosolic Ca^{2+} and activates protein kinase C via the IP_3 signaling pathway. Sweat formation involves increased basolateral Cl^- uptake by a $Na^+/K^+/2Cl^-$ cotransporter (NKCC1). Cl^- crosses the apical membrane via a Ca^{2+}-activated Cl^- channel such as TMEM16A (anoctamin 1), and Na^+ follows Cl^- into the lumen paracellularly, driven by the accumulating negative charges. The combined increase in Na^+ and Cl^- in the coil

---

Clinical Application 15.3: Acne

Acne is a common adolescent disorder that may also occur in adults. Closed comedo acne pimples (whiteheads) occur when skin cells block a hair follicle's external opening. Even though sebum is now trapped within the follicle, production continues unabated. Bacterial colonization of the accumulating sebum causes local inflammation. Drugs that reduce sebum secretion (e.g., retinoids) help control acne occurrence and spreading.

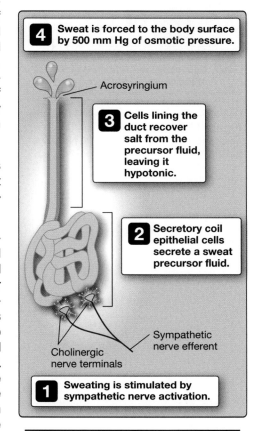

4 Sweat is forced to the body surface by 500 mm Hg of osmotic pressure.

Acrosyringium

3 Cells lining the duct recover salt from the precursor fluid, leaving it hypotonic.

2 Secretory coil epithelial cells secrete a sweat precursor fluid.

Sympathetic nerve efferent

Cholinergic nerve terminals

1 Sweating is stimulated by sympathetic nerve activation.

Figure 15.7.
Eccrine sweat gland activation.

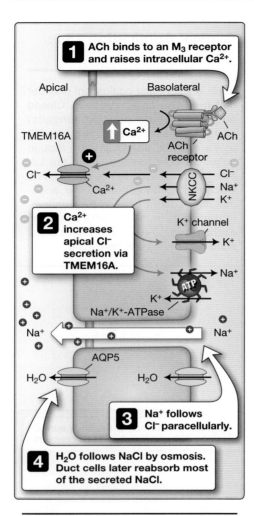

Figure 15.8.
Sweat formation by clear cells. NKCC = Na$^+$/K$^+$/2Cl$^-$ cotransporter type 1.

lumen osmotically draws water into the lumen via aquaporin (AQP) 5. The osmotic pressure that develops (which can exceed 500 mm Hg) propels the fluid to the skin surface. Cholinergic stimulation also causes myoepithelial cell contraction, which braces the coil during pressure development.

ii. **Duct:** Ductal cells reabsorb Na$^+$ and Cl$^-$ from the precursor fluid. Na$^+$ is reabsorbed across the apical membrane via ENaC, an epithelial sodium channel, and Cl$^-$ is reabsorbed via the cystic fibrosis transmembrane regulator (CFTR). The basolateral Na$^+$/K$^+$-ATPase transfers Na$^+$ to the interstitium, and Cl$^-$ leaves via a Cl$^-$ channel. The ductal epithelium has a low water permeability, so NaCl reabsorption leaves sweat hypotonic.

> Cystic fibrosis (CF) is associated with CFTR mutations that prevent ductal Cl$^-$ reabsorption. Sweat normally contains <30 mmol/L Cl$^-$, but levels are significantly higher in CF patients (≥60 mmol/L). A sweat chloride test is routinely used to screen newborns for CF.

b. **Secretions:** Besides AMPs, which contribute to antimicrobial barrier function, secretions aid skin hydration. Stratum corneum water content is normally ~30%. If water content drops below ~10%, the epidermis becomes prone to cracking, which creates a breach in the protective barrier. Sweat contains natural moisturizing factors, including urea and lactate. These, along with amino acids released from apoptotic keratinocytes and sebum from sebaceous glands, help maintain skin hydration.

2. **Apocrine:** Apocrine sweat glands aid in skin hydration; protection; and, possibly, aspects of social display through pheromones. They are largely restricted to the axillae and perineum. The glands activate in response to emotional stimuli and are regulated by adrenergic compounds. Secretions are directed into the hair follicle and are more viscous, containing cell particulates and constituents from the hair follicle. Bacterial action on these secretions can produce odors, which is the reason why underarm deodorants were developed. Pathologically foul-smelling sweat in these areas is known as **bromhidrosis**.

VII. CUTANEOUS NERVES

The cutaneous senses are a part of the **somatosensory system**. Every square millimeter of skin represents an opportunity to interact with and analyze the external environment, and it is therefore dense with sensory nerve fibers. Not only do these nerves provide us with the sense of touch, but they also sense pain (**nociception**), itch (**pruritoception**), and temperature (**thermoreception**; see Chapter 38·II·A·2).

A. Touch

Physical contact can take many forms. Sometimes it might be a light touch, such as dragging a feather across the skin. Other times it might be the intense pressure of holding the handle of a grocery bag full of cartons and cans. The ability to sense such disparate stimuli requires mechanoreceptors that are attuned to varying aspects of stimulus intensity, frequency, and duration. Their depth below the skin surface partly determines the size of their **receptive field**. A receptive field defines the area that a sensory receptor monitors. Receptors that collect stimuli over a wide receptive field have an increased chance of recording events but are unable to locate the source of the stimulus precisely. Receptors with small receptive fields can pinpoint the source of the stimulus with a high degree of accuracy and are usually clustered in large numbers to ensure adequate coverage over a wide surface area.

1. **Tactile receptors:** Skin contains several different types of mechanoreceptors that transduce tactile stimuli (Fig. 15.9). Transduction occurs when a sensory nerve ending is deformed. The endings may be bare or encased in accessory structures that modify their sensitivity and responsiveness to different types of stimuli. Glabrous skin contains rapidly adapting **Pacini** and **Meissner corpuscles**. Hairy skin contains slowly adapting **Merkel disks** and **Ruffini endings** as well as rapidly adapting hair sensory plexus neurons.

   a. **Pacini corpuscles:** Pacini corpuscles are rapidly adapting mechanoreceptors ~1 mm long that sense high-frequency vibrations in glabrous skin (Fig. 15.10A). They reside deep in the skin, and their receptive field is wide. The corpuscles consist of a sensory nerve ending buried beneath numerous layers of fibrous tissue with gelatinous fluid between, so that they resemble an onion in cross section. The entire structure is then wrapped in a connective tissue capsule. The gelatinous layers cushion the nerve so that only transient stimuli are able to deform and excite the nerve membrane. The afferent nerves are myelinated for most of their length, which allows for rapid sensory signal relay.

   b. **Meissner corpuscles:** Meissner corpuscles also adapt rapidly (see Fig. 15.10B). They are exquisitely sensitive to touch and low-frequency vibrations, producing a fluttering sensation. They are smaller than Pacini corpuscles, but their construction is similar, in that a sensory nerve ending meanders between stacked layers of flattened support cells, all enclosed within a capsule. They also have a similar skin distribution to Pacini corpuscles.

   c. **Ruffini endings:** These afferents are slowly adapting receptors (Fig. 15.11A) located in the deeper layers of the skin. The nerve endings branch and weave between bundles of collagen fibers to form a long, thin, spindle-shaped structure. The fibers are enclosed within a connective tissue capsule firmly tethered to surrounding tissues. When skin is stretched, the capsule and structures within distort also.

   d. **Merkel disks:** Merkel disks are also slowly adapting receptors (see Fig. 15.11B) that respond best to low-frequency

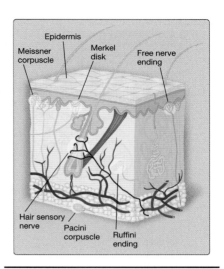

Figure 15.9.
Cutaneous tactile receptors.

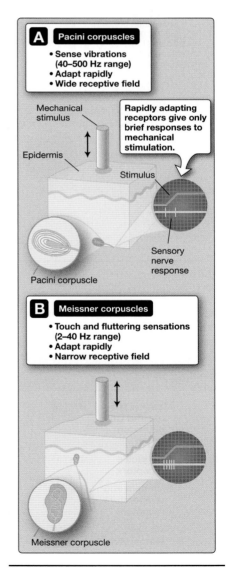

A Pacini corpuscles
- Sense vibrations (40–500 Hz range)
- Adapt rapidly
- Wide receptive field

Mechanical stimulus

Rapidly adapting receptors give only brief responses to mechanical stimulation.

Epidermis

Stimulus

Pacini corpuscle

Sensory nerve response

B Meissner corpuscles
- Touch and fluttering sensations (2–40 Hz range)
- Adapt rapidly
- Narrow receptive field

Meissner corpuscle

Figure 15.10.
Rapidly adapting tactile receptors.

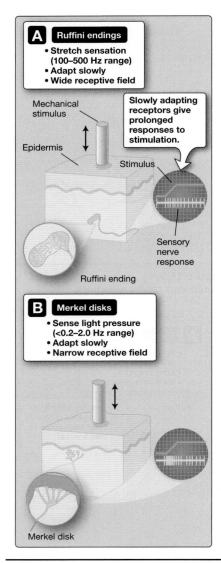

Figure 15.11.
Slowly adapting tactile receptors.

Table 15.1: Classification of Sensory Afferent Nerve Fibers

Nerve Class	Receptor Type	Conduction Velocity (m/s)
Aα	Muscle spindle and Golgi tendon organ	80–120[*]
Aβ	Skin tactile receptors	33–75[*]
Aδ	Skin pain and temperature receptors	2.5–30[*]
C	Skin pain, temperature, and itch receptors	0.5–2.0

[*]Myelinated.

stimulation and light touch. They lie just below the skin surface, which gives them a very narrow receptive field. Fingertips are endowed with very large numbers of Merkel disks, which allows for fine discrimination of form and texture.

e. Hairs: The sensory nerve plexus that wraps around the hair follicle allows a hair to function as a mechanosensor (see Fig. 15.9). When the hair bends, the nerve ending distorts and signals. To understand just how sensitive hairs are to mechanical stimulation, use a pen tip or pin to deflect a single arm hair—you will be impressed!

f. Free nerve endings: Free sensory nerve endings can be found throughout the skin. They contribute to touch, pain, itch, and temperature sensation.

2. **Mechanosensory transduction:** Deformation of a tactile nerve ending opens a Na^+ channel, causing a depolarizing receptor potential (also known as a **generator potential**). Merkel disks transduce touch using a mechanosensitive non-specific cation channel (Piezo2).

3. **Sensory nerve fibers:** Sensory nerve afferents are classified according to how fast they relay signals to the central nervous system ([CNS] Table 15.1). All tactile sensory afferents are myelinated (type Aβ) and conduct at relatively high velocity. These signals then travel via the spinal cord to the CNS for processing (see Chapter 6·II). The area perceived by a particular mechanoreceptor is dependent on receptor type and body area. The hands are more discriminatory than the upper arms, for example (Fig. 15.12).

B. Pain

Mechanical and thermal stimuli that are innocuous or even pleasurable at low intensities can cause significant cellular damage at higher levels. Pain's function is to alert the CNS of local damage and to initiate a motor reflex that causes the body to either avoid or pull back from the source of stimulation (see Chapter 11·III·D).

1. **Nociception:** Several nociceptor types transduce painful stimuli into a membrane potential change.

   a. Mechanical: Mechanical nociceptors respond to intense pressure or mechanical deformation of the skin. They also respond to sharp objects that stab or cut the skin. These receptors are likely very high threshold mechanoreceptors that only respond to mechanical stimuli when they reach noxious levels.

   b. Thermal: Temperature extremes (freezing cold and burning heat) cause tissue damage. Cold stimuli become noxious at ~20°C, with intensity of perceived pain increasing linearly to ~0°C. Cold responses are also sensitive to the rate of cooling, with rapid cooling producing more intense responses. The threshold for noxious heat sensation is ~43°C.

   c. Chemical: Because nociceptive fibers are free nerve endings, they are accessible to chemicals that cross the epidermal barrier or that are released by damaged tissues (Table 15.2). Capsaicin, the active ingredient in hot chili peppers, produces a burning sensation via activation of nociceptors when applied topically.

    d. **Polymodal:** A subpopulation of nociceptors is sensitive to two or more stimuli and is known as **polymodal**.

2. **Nociceptive stimulus transduction:** The precise mechanisms by which nociceptive stimuli are sensed and signaled are not understood fully, but transduction of many noxious stimuli involves multiple transient receptor-potential channel (TRP) family members (see Chapter 2·VI·D).

    a. **Receptors:** Heat activates TRPV1, as does capsaicin. Skin cooling activates TRPM8. Activating either TRP results in Na^+ and Ca^{2+} influx and excitation. Hydrogen ions excite nociceptive neurons by permeating an acid-sensing ENaC family member. Other channels may be involved in pain sensation also.

    b. **Nociceptive fibers:** Nociceptor activation is relayed to the CNS by fast (myelinated) Aδ fibers and slower C fibers. The Aδ fibers mediate sensations of sharp, intense, pricking pain (**first pain**), followed by a more prolonged dull, throbbing, burning pain associated with C-fiber activation (**second pain**).

> C fibers are particularly sensitive to lidocaine, a local anesthetic that is applied topically to relieve skin itching and pain. It blocks the voltage-gated Na^+ channel that mediates the nerve action potential. Lidocaine is also commonly injected to anesthetize teeth prior to dental surgery or is combined with prilocaine (a related Na^+-channel blocker) in an ointment. Lidocaine is sometimes combined with a vasoconstrictor to reduce local blood flow and thereby reduce drug washout effects. Local anesthesia is prolonged as a result.

3. **Sensitization:** Tissue damage initiates a chain of events that sensitizes surrounding afferent nerve endings to innocuous stimuli, such that they are now perceived as painful (**hyperalgesia**). Sensitization initially remains localized to the site of damage (**primary hyperalgesia**) but spreads within minutes to involve surrounding areas (**secondary hyperalgesia**). Sensitization follows the progress of swelling and inflammation and involves many of the common inflammatory mediators. Its effects may persist for months after recovery from the initial injury.

C. Itch

Pruritus (derived from *prurire*, Latin for "itch") appears designed to trigger reflex scratching or rubbing to remove an insect or other irritant. The sensation is mediated by two populations of C-type nerve fibers. One fiber type responds optimally to histamine, whereas the

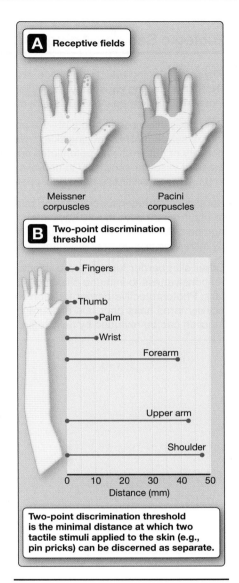

Figure 15.12.
Receptive fields of two receptor types in the hand and sensory discrimination along the arm.

Table 15.2: Nociceptor-Activating Chemicals

Source	Chemical
Mast cells	Histamine
Mast cells, traumatized skin cells	Prostaglandins
Stressed skin cells	K^+
	Bradykinin
	H^+
Sensory afferents	Substance P
Cholinergic efferents	ACh

Biologic Sex and Aging 15.1: Skin

Physiologic aging effects on tissue are most evident from changes in skin appearance and slowed response to trauma. (Note: Photoaging caused by ultraviolet radiation is a separate pathologic process. The effects of photoaging are most obvious in light-skinned individuals with chronic sun exposure.)

Skin loses thickness at a rate of ~7% per decade. Thinning is largely due to changes in extracellular matrix composition and loss of subdermal fat, causing skin to wrinkle and sag. There is an accompanying decrease in skin vascularity, which impairs delivery of nutrients and potentiates age-related changes in skin function. The loss of fat and vascularity, together with decreased vascular sensitivity to neural commands and a 70% decrease in sweat gland output by age 70 years, significantly impairs responsiveness to thermal challenge.

Keratinocyte progenitor cell division rate decreases by ~50% by age 70 years, which causes stratum corneum renewal time to rise from 20 days to >30 days in older adults. Wound repair times are affected similarly. A decrease in sensory receptor numbers (Meissner and Pacini corpuscles, pain receptors) with age decreases skin sensitivity to external stimuli and thereby increases likelihood of mechanical trauma and skin damage.

Other skin barrier functions are affected similarly. A 50% decrease in Langerhans cell numbers by age 70 years impairs responsiveness to immune challenge. Decreased lipid secretion rates increase water loss from the skin, and skin hydration falls. Photoprotective barrier function weakens in parallel with declining melanocyte numbers. Decreased hair melanocyte activity turns hairs gray. It is encouraging to note, however, that many of these age-related changes in skin structure and function can be reversed by daily topical application of all-*trans*-retinoic acid (tretinoin).

other (nonhistamine) type is activated by a wide range of pruritogens, such as prostaglandins, interleukins, proteases, and ACh. Itch sensations can be suppressed by painful stimuli (such as scratching) and by antihistamines and may be potentiated by analgesics. The mechanisms by which painful and pruritic sensations interact are not understood fully.

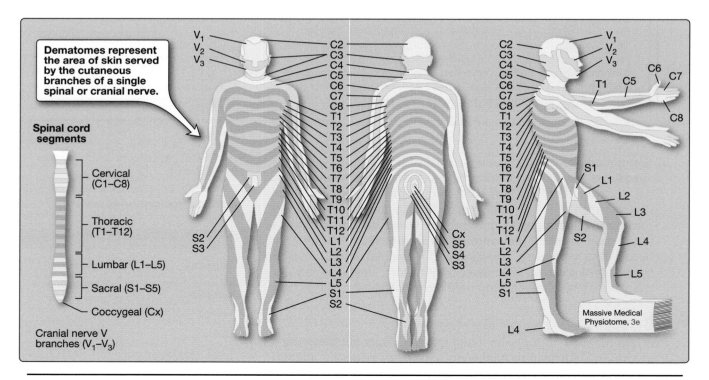

Figure 15.13.
Dermatome maps.

D. Dermatomes

Sensory information from skin receptors is relayed to the CNS via afferent nerves. The nerves have a limited area of coverage, which can be mapped onto the body surface as a series of discrete bands called **dermatomes** (Fig. 15.13). Each band corresponds to a single spinal segment. There is overlap in coverage between the bands, so cutting a single pair of posterior nerve roots does not result in complete sensory loss in the corresponding dermatome. Pain that localizes to a particular dermatome may be helpful in identifying a site of spinal cord injury.

Clinical Application 15.4: Shingles

Varicella-zoster virus causes chickenpox, the symptoms of which include a vesicular skin rash. Viruses from ruptured vesicles infect skin sensory nerve endings and then migrate backward toward the nerve cell body, where they may lie dormant for decades. Upon reactivation, the viruses infect other neurons within the dorsal root ganglion and then spread to the skin. Reactivation causes deep stabbing, throbbing, or burning pains and extreme skin sensitivity, followed by a skin rash and vesicular eruption. The rash characteristically delineates the dermatome served by the infected ganglion, as shown. Shingles usually affects older adults (>60 years), although symptoms may appear at any age. The disease usually resolves spontaneously within a week, but antivirals and analgesics may be helpful in attenuating pain intensity and duration.

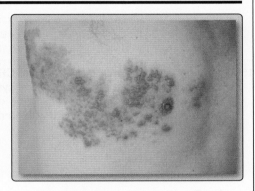

Rash and vesicles with a dermatomal distribution.

Chapter Summary

- The **epidermis** provides most skin barrier functions. It is keratinized to physically resist abrasion, and **lipid layers** are water resistive. **Melanin** absorbs ultraviolet radiation and is photoprotective. Secreted **antimicrobials** retard bacterial colonization.

- The **dermis** houses most specialized skin structures and **mast cells**, which are involved in local inflammatory responses.

- **Hair follicles** are associated with sebaceous and apocrine sweat glands. Hair growth is cyclic, involving active growth, regression, and then shedding.

- **Sebaceous glands** produce **sebum**, a lipid-rich secretion that helps skin retain moisture. **Eccrine sweat glands** are thermoregulatory. An isotonic precursor fluid formed by the secretory coil is modified by ductal cells to produce a hypotonic solution that is secreted onto the epidermal surface. **Apocrine sweat glands** produce pheromones.

- The **cutaneous senses** are a part of the **somatosensory system**, which monitors events occurring within or at the surface of the body.

- **Tactile receptors** are activated by touch, the sensations being transduced by mechanical deformation of a sensory neuron. **Pacini corpuscles** and **Ruffini endings** respond best to high-frequency vibration or stretch. **Meissner corpuscles** and **Merkel disks** are more sensitive to pressure and low-frequency stimulation.

- Sensory afferents from tactile receptors are myelinated and conduct impulses at high speed.

- Pain receptors activate in response to intense and noxious mechanical, thermal, or chemical stimuli.

Study Questions

Choose the ONE best answer.

15.1. An 84-year-old male comes to the office reporting that his fingers are dry, and the skin is cracked and bleeding. He notes that his symptoms are worse during the winter months. Which of the following cell types most likely protects the patient's dermis from microbial invasion?

A. Corneocytes
B. Dendrocytes
C. Keratinocytes
D. Melanocytes
E. Sebocytes

Best answer = B. Skin cracks breach the epidermal barrier and expose the dermis to the external environment. The dermis contains mast cells, macrophages, and dermal dendritic cells that help defend the body against microbial invaders. Dermal dendrocytes are antigen-presenting cells involved in cutaneous adaptive immunity (see Section IV·B). The epidermis forms the primary barrier between body interior and the external environment. Corneocytes (A) are derived from apoptotic keratinocytes (C). Both are enriched in keratin to help the epidermis resist abrasion. Epidermal melanocytes (D) synthesize melanin that protects the skin from the deleterious effects of UV irradiation. Sebocytes (E) are found in sebaceous glands. They secrete sebum into hair follicle to create a water-resistant film on the epidermal surface.

15.2. A 4-year-old male with cystic fibrosis (CF) has a history of respiratory and gastrointestinal symptoms. His sweat composition could most likely be described as which of the following, compared with normal?

A. Copious
B. Hypertonic
C. Hypotonic
D. Isotonic
E. Scant

Best answer = D. CFTR is an ATP-binding cassette transporter that functions as a Cl^- channel in many epithelia. CFTR defects prevent Cl^- secretion by respiratory and GI epithelia. Secretion creates an osmotic gradient used to draw water onto the apical surface, so CF patients typically form mucus that is thick and difficult to expel from the lungs, for example. In sweat glands, CFTR is used to reabsorb Cl^- from the ductal lumen during passage of sweat precursor fluid to the skin surface, causing sweat to become hypotonic (C; see Section VI·D·2). In CF patients, a CFTR defect prevents Cl^- (and Na^+) reabsorption, so their sweat is close to being isotonic, not hypertonic (B). CFTR defects do not cause major changes in sweat volume (A, E).

15.3. A 26-year-old female comes to the office reporting that she sweats excessively, especially under her armpits. She has tried several different antiperspirant brands, but none have been effective. Her hyperhidrosis is most likely to be successfully treated with an antagonist to which of the following receptor types?

A. α-Adrenergic
B. β-Adrenergic
C. Muscarinic
D. Nicotinic
E. Purinergic

Best answer = C. Sweat gland secretory coils are controlled by the sympathetic nervous system, but the nerve terminals are cholinergic rather than adrenergic (see Section VI·D). When active, the nerve terminals release ACh into the synaptic cleft which binds to an M_3 muscarinic receptor. Primary axillary hyperhidrosis can be successfully treated with oral anticholinergics. α-Adrenergic (A) and β-adrenergic receptors (B) bind norepinephrine and epinephrine. Nicotinic ACh receptors (D) are found at the neuromuscular junction. Purinergic receptors (E) bind adenosine or nucleotides such as ATP.

15.4. A 42-year-old jackhammer operator presents to the office with decreased high-frequency vibration sensitivity in the glabrous skin of the hands. Which receptor is most likely being affected?

A. Free nerve endings
B. Hair sensory fibers
C. Merkel disks
D. Pacini corpuscles
E. Ruffini endings

Best answer = D. Pacini corpuscles are rapidly adapting tactile receptors responsible for sensing vibrations in the 40–500 Hz range (see Section VII·A). Glabrous skin contains free nerve endings (A), but these are less sensitive to vibration than Pacini corpuscles. Glabrous skin does not have hair and, therefore, no hair sensory fibers (B). Merkel disks (C) sense light skin pressure. Ruffini endings (E) are slowly adapting tactile receptors that sense skin stretching, not vibrations.

Cardiac Excitation

16

I. OVERVIEW

Single-celled organisms (e.g., *Paramecium*; see Fig. 5.1) are enclosed by a thin plasma membrane, across which gases such as O_2 and CO_2 can readily diffuse. This means that a paramecium's O_2 needs can be satisfied by uptake from the surrounding fluid. The distance over which O_2 can diffuse through multicellular structures is limited, however (<100 μm), so more complex organisms evolved extensive pipework assemblies to carry oxygenated fluid to within a few microns of each and every cell. In animals such as *Homo sapiens*, the pipework assemblies comprise blood vessels, and the motive force for oxygenated fluid (blood) flow through these vessels is hydrostatic pressure generated by the heart. Blood vessels, blood, and the heart together comprise the cardiovascular system (Fig. 16.1). The heart is composed of cardiac muscle, which is layered to fashion four inner chambers. When the muscle contracts, chamber diameter decreases, and blood is forced from chamber to chamber and then out through the vasculature. The heart's unique design and function means that contraction of its various parts must be sequenced for output efficiency. Sequencing is produced by a wave of excitation whose rate of progress is modulated to allow time for blood to move between its chambers. Movement of this wave through time and space creates electrical gradients within the surrounding tissues that can be detected and recorded at the body surface as an electrocardiogram (ECG). Because the timing and pattern of excitation varies little from one individual to the next, an ECG can be used to detect abnormalities in the excitation pathway and in the overall structure of the heart.

II. CARDIOVASCULAR CIRCUITRY

The cardiovascular system is a closed loop through which blood cycles continuously throughout the life of the individual. The loop incorporates two structurally and functionally distinct circulations. The **systemic circulation** supplies all bodily tissues with O_2 and other essential nutrients,

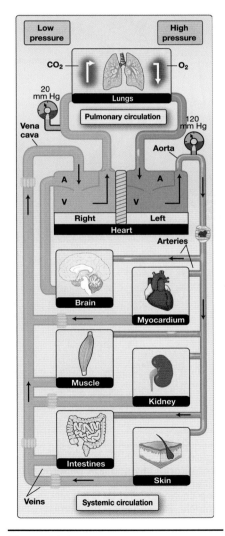

Figure 16.1.
Cardiovascular circuitry. A = atrium,
V = ventricle.

189

carries away metabolic waste products (including heat), and distributes chemical messages and cells of the immune system. The **pulmonary circulation** conveys the contents of the vasculature through the lungs for exchange of CO_2 and O_2.

A. Heart structure

Blood is propelled around the vasculature by two muscular pumps, one on each side of the heart (Fig. 16.2A). Each pump contains two chambers: one **atrium** and one **ventricle**. The left heart pumps blood through the **aorta** to the organs of the systemic circulation (Fig. 16.2B). It returns to the heart via the **vena cavae**. The right side of the heart perfuses the pulmonary circulation. Blood exits the right ventricle (RV) via the **pulmonary artery**, passes through the lungs, and then enters the left heart via the **pulmonary vein**.

B. Cardiac chambers

Atria and ventricles serve different functions, which are reflected in the amount of cardiac muscle contained within their walls.

1. **Atria:** Atria act as holding tanks for blood collected from the venous system during ventricular contraction. Accumulated blood is transferred to the ventricles by atrial contraction at the beginning of each cardiac cycle. Because minimal amounts of pressure are required to push blood into the ventricles when they are relaxed, atria contain relatively small amounts of muscle, and their walls are thin.

2. **Ventricles:** Ventricles drive blood at high pressure through vast networks of vessels, made possible by chamber walls thick with cardiac muscle. The left ventricle (LV) typically generates peak pressures of 120 mm Hg (Fig. 16.2C). The RV pumps blood through a system of relatively low-resistance vessels; therefore, its walls are less muscular than those in the LV. The RV generates peak pressures of about 25 mm Hg.

C. Valves

One-way valves situated between atria and ventricles (**atrioventricular [AV] valves**) and between ventricles and their outlets (**semilunar valves**) help ensure that flow around the cardiovascular system is unidirectional (see Fig. 16.2B).

1. **Atrioventricular:** The **tricuspid** (right side) and **mitral** (left side) valves allow blood to pass from atrium to ventricle and close when ventricular contraction begins. **Chordae tendineae** are filaments attached to the edges of the valve leaflets. The chordae work in conjunction with **papillary muscles** to brace the valves and prevent them from being everted by high pressures generated within the ventricles during contraction (see Fig. 16.2A).

2. **Semilunar:** The **pulmonary** (right side) and **aortic** (left side) valves prevent backflow from the arterial system into the ventricles. The semilunar valves are subject to high shear stress associated with high-velocity ventricular outflow, so their leaflets are thicker and more resilient than AV valve leaflets.

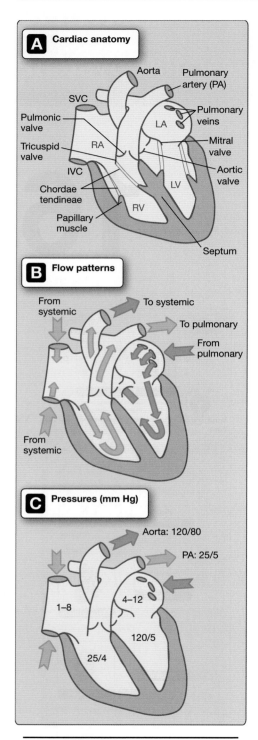

Figure 16.2.
Cardiac anatomy and flow patterns. IVC and SVC = inferior and superior vena cavae, LA and LV = left atrium and ventricle, RA and RV = right atrium and ventricle.

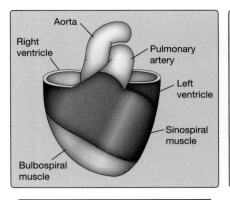

Figure 16.3.
Bands of cardiac muscle.

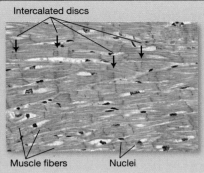

Figure 16.4.
Cardiac muscle structure.

III. CARDIAC MUSCLE

The heart's four chambers are fashioned from wide bands of cardiac muscle (Fig. 16.3). Cardiac muscle is a striated muscle that shares much in common with skeletal muscle. Their sarcomeres are organized similarly, so cardiac muscle and skeletal muscle give similar banding patterns when viewed under polarized light (Fig. 16.4). They exhibit key differences, however, because the tasks required of the two muscles are unique. Cardiac muscle is rich in mitochondria that confer enhanced aerobic capacity and resistance to fatigue. Skeletal muscle fibers contract independently of each other, each responding to individualized commands from the motor cortex. By contrast, cardiac muscle functions independently of somatic motor control, and the fibers are extensively interconnected so as to form functional units.

A. Communication pathways

Skeletal muscle fibers are largely under voluntary control, contracting when stimulated by an α-motor neuron. Cardiac muscle is *regulated* by the nervous system, but a contractile stimulus normally originates within the **sinoatrial (SA) node**, an area of the myocardium specialized to function as a **pacemaker**. Pacemaker cells periodically generate electrical signals that spread from myocyte to myocyte until every fiber within the organ is involved (Fig. 16.5), made possible by gap junctions and extensive cellular branching.

1. **Gap junctions:** Dense arrays of gap junctions couple adjacent myocytes and provide pathways for direct electrical and chemical communication (see Chapter 4·II·F). High junctional density ensures that electrical signals propagate rapidly from myocyte to myocyte. The junctional arrays are located at specialized sarcolemmal contact regions known as **intercalated discs** (Fig. 16.6), which also contain structural elements (**desmosomes** and **fascia adherens**) that fuse the cells together mechanically and allow them to withstand the tension generated during muscle contraction.

2. **Cellular branching:** When a command to contract is issued by the pacemaker, it must spread rapidly throughout the entire heart. Whereas skeletal muscle fibers are long, thin, and unbranching, cardiac myocytes branch extensively. The combination of branching

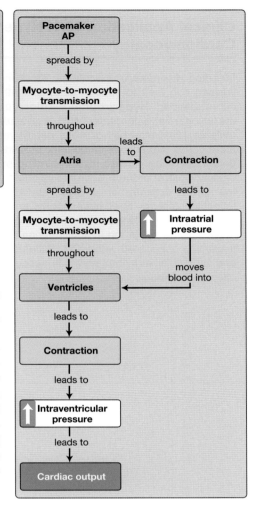

Figure 16.5.
Concept map for cardiac function.

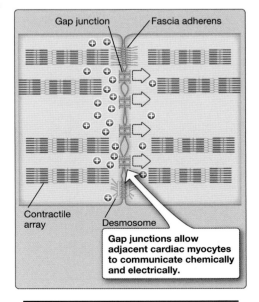

Figure 16.6.
Intercalated disc structure.

Clinical Application 16.1: Hypertrophic Cardiomyopathy

Inherited disorders of sarcomeric function that cause skeletal muscle wasting can lead to sudden cardiac death (SCD) when cardiac muscle is affected. **Hypertrophic cardiomyopathy (HCM)** is inappropriate myocardial enlargement that is a leading cause of SCD in young athletes. More than half of these deaths result from mutations in 15 or more genes that encode sarcomeric proteins. The majority localize to the genes encoding cardiac myosin-binding protein C (*MYBPC3*) and the cardiac isoform of the myosin heavy chain (*MYH7*). HCM alleles disrupt the normal organization of myocytes within the myocardium, causing myocyte disarray, fiber hypertrophy, and cellular structure distortion (compare the micrograph with Fig. 16.4). Extensive deposition of connective tissue within the interstitium contributes to gross ventricular wall thickening associated with HCM. Patients with HCM are prone to atrial and ventricular arrhythmias, which accounts for the high risk of SCD in affected individuals.

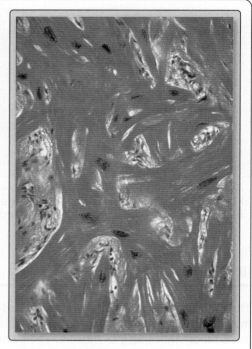

Myocyte disorganization.

cells and gap junctions creates a vast interconnected cellular network that functions as a single unit (**functional syncytium**).

IV. SEQUENCING CONTRACTIONS

If the heart is to function effectively as a pump, its chambers must contract in an orderly sequence. Sequencing is accomplished using a wave of depolarization that spreads from myocyte to myocyte via gap junctions until it envelops the entire heart. Once a wave of depolarization is initiated, it involves and engulfs adjacent myocytes with the inevitability of a line of toppling dominoes (Fig. 16.7). The wave of depolarization that drives myocardial contraction and pump cycling normally originates in the **SA node** (Fig. 16.8).

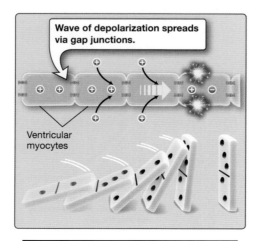

Wave of depolarization spreads via gap junctions.

Ventricular myocytes

Figure 16.7.
Signal propagation in the myocardium.

The current record for number of dominoes toppled in one cascade is just shy of 4.5 million, a record set in the Netherlands on Domino Day 2009. It took a team of 89 builders 2 months to set up! Because a misplaced foot or finger could easily trigger the entire array, it had to be designed with breaks that were completed immediately before the competition began. Because the myocardium does not have such safety features, every cell in the heart has the potential to become the primary pacemaker should its membrane become unstable. "Ectopic" pacemakers are responsible for common mishaps leading to premature ventricular contractions, which are typically one-off events in young, healthy hearts. In aging hearts, ectopic pacemakers may become established, posing a constant and potentially lethal threat to cardiac function.

A. Sinoatrial node

The cardiac pacemaker is located high in the wall of the right atrium near the superior vena cava (see Fig. 16.8; step 1). SA nodal cells are specialized atrial myocytes that have lost most of their contractile fibers and Ca^{2+} stores and, therefore, do not participate in pressure development. Nodal cells have unstable membrane potentials (V_m) that drift slowly positive over time. When V_m crosses **threshold potential** ([V_{th}] see Chapter 2·III·B·1), an action potential (AP) is initiated. The AP spreads regeneratively across both atria and ultimately involves the ventricles, with contraction following in its wake. The rate at which nodal cells depolarize and initiate spikes is modulated by both arms of the **autonomic nervous system** (**ANS**) as a way of controlling heart rate (HR). The **sympathetic nervous system** (**SNS**) increases HR, whereas the **parasympathetic nervous system** (**PSNS**) decreases it.

B. Atria

Nodal cells are linked electrically via gap junctions to surrounding atrial myocytes. Once initiated, the wave of depolarization spreads outward in all directions with a conduction velocity of ~1 m/s, taking about 100 ms to engulf both atria (see Fig. 16.8; step 2).

C. Atrioventricular node

The spreading wave of depolarization is arrested before it can reach the ventricles by a plate of cartilage and fibrous material located at the AV junction. The plate provides structural support for the heart valves, but it also acts as an electrical insulator. By halting the wave, it allows time for the rapidly moving electrical events to be transduced into slower mechanical events and for blood to move from the atria to the ventricles. The wave of excitation is not allowed to fizzle completely, however, because there is an electrical bridge between the atria and ventricles. At the entrance to this bridge is the AV node, a patch of noncontractile cardiomyocytes that are specialized to conduct signals slowly (0.01–0.05 m/s). It takes about 80 ms for the electrical "spark" to traverse the AV node, just long enough for blood to be pushed by atrial contraction through the AV valves.

D. His–Purkinje system

Once the wave of excitation migrates through the AV node, the ventricular walls must be stimulated to contract in a sequence that squeezes blood upward toward the outlets: **septum → apex → free walls → base**. This is made possible using tracts of tissue comprising myocytes specialized to deliver the wave of depolarization at high speeds to the different regions of the ventricles. The pathway to the ventricles begins with the **common bundle of His**, a tract of specialized myocytes that emerges from the AV node and then sweeps downward into the interventricular septum (see Fig. 16.8; step 3). Here, it separates into left and right bundle branches, which then branch again to deliver the excitation signal to all regions of the LV and RV, respectively. High-speed **Purkinje fibers** (conduction velocity ~2–4 m/s) carry the wave of depolarization to the contractile myocytes (see Fig. 16.8; step 4).

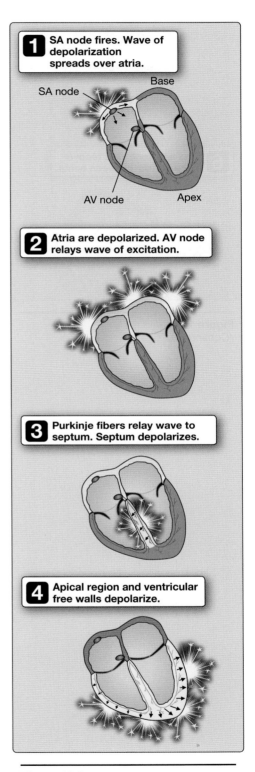

1 SA node fires. Wave of depolarization spreads over atria.

SA node Base

AV node Apex

2 Atria are depolarized. AV node relays wave of excitation.

3 Purkinje fibers relay wave to septum. Septum depolarizes.

4 Apical region and ventricular free walls depolarize.

Figure 16.8.
Cardiac excitation cycle.
AV = atrioventricular, SA = sinoatrial.
(Continued)

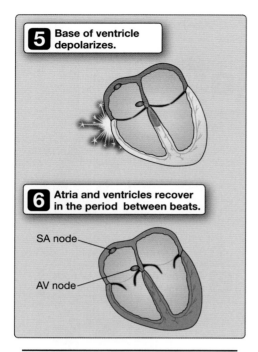

5 Base of ventricle depolarizes.

6 Atria and ventricles recover in the period between beats.

SA node

AV node

Figure 16.8.
(Continued)

The His–Purkinje system is often compared to a system of interstate highways that allow vehicles to travel at high speed through a city's heart to a remote location. The slower myocyte-to-myocyte conduction pathway is equivalent to taking back streets to the same location, a route that is generally much slower.

E. Ventricles

Ventricular myocytes are similar to atrial myocytes, conducting the wave of depolarization from cell to cell via gap junctions at 1 m/s. Excitation of both ventricles is essentially complete within 100 ms (see Fig. 16.8; step 5), although the slower mechanical events will take another 300 ms to complete.

V. ELECTROPHYSIOLOGY

The mechanism by which the wave of excitation is accelerated or delayed during its travel through the various regions of the heart is elegant in its simplicity. The heart is, in essence, a large, sculpted muscle. The rate at which different myocytes within this muscle conduct an electrical signal is dependent on how fast they depolarize, which, in turn, is governed by the relative mix of ion channels contained within the sarcolemma.

A. Ion channels and currents

All cardiac myocytes are excitable cells regardless of their location or role. They all express a Na^+/K^+-ATPase that generates and maintains Na^+ and K^+ gradients across the membrane and establishes V_m. Ion-selective channels then open and close to manipulate V_m. Cardiac function is dependent on four principal ion currents, a **Na^+ current** (I_{Na}), a **Ca^{2+} current** (I_{Ca}), a **K^+ current** (I_K), and a **pacemaker current** (I_f) that is common to those myocytes capable of automaticity (Fig. 16.9). Underlying these currents are numerous individual ion-channel classes (Table 16.1).

1. **Sodium current:** I_{Na} is mediated by the same voltage-dependent Na^+ channel that gives the rapid upstroke to the neuronal AP (see Fig. 5.3). It opens in response to membrane depolarization and is extremely fast activating (0.1–0.2 ms). The resulting Na^+ influx drives V_m to zero and several tens of millivolts positive (**overshoot**). I_{Na} then rapidly inactivates (~2 ms) and stays inactivated until V_m returns to −90 mV. I_{Na} is responsible for the upstroke of the AP in atrial and ventricular myocytes and the Purkinje system. It is not seen in nodal cells, however.

2. **Calcium current:** I_{Ca} is observed in all classes of cardiac myocyte, including nodal cells. It is mediated principally by L-type Ca^{2+} channels. L-type Ca^{2+} channels are voltage gated but activate more slowly than Na^+ channels (~1 ms). Once open, they support a Ca^{2+} influx that both depolarizes the cell and initiates contraction. Ca^{2+} channels also inactivate but slowly (~20 ms).

3. **Potassium current:** Cardiac tissue contains several different classes of K⁺ channels that help repolarize myocytes following excitation, regulate HR, or are cardioprotective (see Table 16.1).

   a. **Membrane repolarization:** Three different classes of K⁺ channel activate during cardiac excitation. The first mediates a minor transient outward K⁺ current (I_{to}) that activates rapidly upon depolarization and then inactivates. Membrane repolarization is largely the responsibility of a voltage-gated K⁺ current that activates very slowly (~100 ms) and after a considerable delay following membrane depolarization. At least two classes of K⁺ channel underlie I_K: the first yields an early and relatively rapid K⁺ current (I_{KR}), whereas the second supports a slower K⁺ current (I_{KS}). I_K does not inactivate.

   b. **Heart rate:** Nodal cells express a G protein–regulated K⁺ current ($I_{K,ACh}$) that slows HR in response to PSNS activation (see Section V·C·3). The channel is gated by G-protein βγ subunits following acetylcholine (ACh) binding to a muscarinic ACh receptor (a G protein–coupled receptor [GPCR]). Receptor activation hyperpolarizes the membrane and decreases the rate of AP formation.

   c. **Cardioprotection:** Cardiac myocytes also express a K⁺ channel that activates when ATP levels fall. The resulting current reduces excitability and is believed to protect myocytes from the toxic effects of Ca^{2+} overload during ischemia, for example.

4. **Pacemaker current:** Nodal cells and Purkinje fibers express a hyperpolarization-activated, cyclic nucleotide–dependent, non-specific (HCN) cation channel that mediates a **"funny current"** (I_f) when active. The current is so named because of its peculiar properties. The channel is activated by hyperpolarization and supports simultaneous K⁺ efflux and Na⁺ influx. Na⁺ dominates the exchange, and the membrane depolarizes. I_f is a pacemaker current, discussed in more detail in the following section.

B. Action potentials

The rate at which myocytes conduct electrical signals is dependent on the mix of ion channels involved. Excitation of contractile myocytes is dominated by voltage-gated Na⁺ channels. Nodal-cell excitation is dominated by L-type Ca^{2+} channels. The consequences are most apparent in the shape of APs recorded from the two myocyte classes. Atrial and ventricular myocytes express a fast AP that rises rapidly during I_{Na} activation. Nodal cells express slow APs that rely on I_{Ca} to provide the upstroke.

1. **Fast:** A fast AP has five distinct phases (Fig. 16.10). The currents that underlie the different phases overlap in time, but pharmaceuticals used to treat dysrhythmias and other cardiac disorders are often grouped according to which AP phase they affect primarily and so are identified here this way.

   a. **Phase 0:** Phase 0, the AP upstroke, is caused by I_{Na}. The sarcolemma of atrial and ventricular myocytes is rich in Na⁺

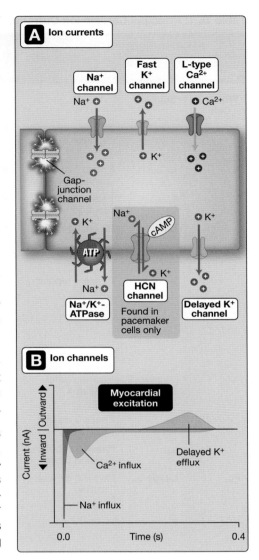

Figure 16.9.
Principal ion channels and currents in cardiac myocytes. HCN = hyperpolarization-activated, cyclic nucleotide–dependent Na⁺ channel.

Table 16.1: Principal Cardiac Ion Currents

Current		Function	Protein	Mode of Activation	Gene	Inhibitor	Inherited Cardiac Diseases
Whole Cell	Subtype						
I$_{Na}$		AP upstroke (phase 0)	Na$_v$1.5	Depolarization	SCN5A	Tetrodotoxin, lidocaine	LQTS, Bruga-da syndrome, SSS
I$_K$	I$_{to, fast}$	Phase 1 repolar-ization	K$_v$4.2/ K$_v$4.3	Depolarization	KCND2/ KCND3	Tedisamil	
	I$_{to, slow}$	Phase 1 repolar-ization	K$_v$1.4	Depolarization	KCNA4		
	I$_{KR}$	Phase 3 repolar-ization	K$_v$11.1	Depolarization, rapid delayed rectifier	KCNH2	Sotalol, E-4031, quinidine, dofetilide, Ba^{2+}, Cs$^+$, TEA	Romano-Ward syndrome, LQTS, SQTS, FAF
	I$_{KS}$	Phase 3 repolar-ization	K$_v$7.1 + minK	Depolarization, slow delayed rectifier	KCNQ1 + KCNE1		Romano-Ward syndrome, LQTS, FAF, SQTS, Jervell, and Lange-Nielsen syndrome
	I$_{K1}$	Phase 3 repolar-ization (late), phase 4	Kir2.1	Depolarization, delayed rectifier	KCNJ2	Ba^{2+}	Andersen-Tawil syndrome, LQTS, SQTS, FAF
	I$_{K1}$	Phase 3 repolar-ization	Kir2.2	Depolarization, delayed rectifier	KCNJ12	Ba^{2+}	
	I$_{Kur}$	Phase 3 repolar-ization (atria)	K$_v$1.5	Depolarization, delayed rectifier	KCNA5	Vernakalant	FAF
I$_{Ca}$	L-type I$_{Ca}$	Phase 2 depolar-ization	Ca$_v$1.2	Depolarization	CACNA1C	Dihydropyri-dines, phenyl-alkylamines, benzothiaze-pines, divalent cations	Timothy syn-drome, BS
	T-type I$_{Ca}$	Phase 2 depolar-ization	Ca$_v$3.2	Depolarization	CACNA1H		
I$_f$		Pacemaker	HCN4	Hyperpolarization, cyclic nucleo-tides	HCN4	Ivabradine	SSS, BS
I$_K$	I$_{K,ACh}$	Heart rate control	Kir3.4	G protein	KCNJ5		
I$_K$	I$_{K,ATP}$	Protection	Kir6.2	Intracellular ATP levels	KCNJ11		

FAF = familial atrial fibrillation, I$_{Ca}$ = Ca^{2+} current, I$_K$ = K$^+$ current, I$_{K,ACh}$ = G protein–regulated K$^+$ current, I$_{K,ATP}$ = ATP-sensitive K$^+$ channel, I$_{KR}$ = rapid K$^+$ current, I$_{KS}$ = slower K$^+$ current, I$_{Kur}$ = ultrarapid K$^+$ current, I$_{to}$ = transient outward K$^+$ current, LQTS = long QT syndrome, SQTS = short QT syndrome, SSS = sick sinus syndrome, TEA = tetraethylammonium.

channels, and they open rapidly once the wave of excitation arrives. The result is a massive Na$^+$ influx that drives V$_m$ toward the equilibrium potential for Na$^+$ (+60 mV).

b. **Phase 1: Phase 1** reflects Na$^+$-channel inactivation, which brings V$_m$ closer to 0 mV. Phase 1 repolarization is aided by I$_{to}$.

c. **Phase 2:** The AP plateau is maintained by I_{Ca}. I_{Ca} inactivates during the AP, but a few Ca^{2+} channels remain open to prolong the plateau and ensure that Ca^{2+} release and contraction completes before excitation terminates.

d. **Phase 3:** Membrane repolarization is mediated by delayed activation of I_K.

e. **Phase 4:** The interval between APs is used to return Ca^{2+} to the intracellular stores and to pump Na^+ out of the cell in exchange for K^+. Return to a resting V_m (-90 mV) also allows Na^+ and Ca^{2+} channels to recover from their inactivated state, a process that takes several tens of milliseconds.

2. **Very fast:** Purkinje cells are designed to conduct the wave of excitation at high speed. Their membranes contain more Na^+ channels and fewer Ca^{2+} channels than ventricular myocytes, meaning that phase 0 more closely follows I_{Na}. It is the rate of phase 0 depolarization that determines conduction velocity. Purkinje cells are also three to four times thicker than ventricular myocytes, which allows for faster conduction velocities (see Chapter 5·III·B·2).

3. **Slow:** Nodal cells express slow APs dominated by I_{Ca} (Fig. 16.11). The primary reason is that nodal cells have a resting V_m that is significantly more positive than contractile myocytes (-65 mV vs. -90 mV). Na^+ channels are inactivated and cannot be opened at -65 mV, forcing nodal cells to rely on the slower L-type Ca^{2+} channels to provide the AP upstroke.

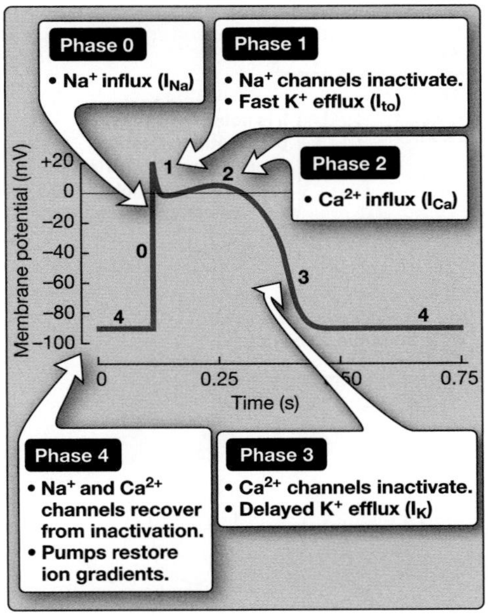

Figure 16.10.
Fast action potential. I_{Ca} = Ca^{2+} current, I_K = K^+ current, I_{Na} = Na^+ current, I_{to} = transient outward current.

Nodal cell membranes contain functional Na^+ channels that will support a small Na^+ current if they are allowed to recover from inactivation. Recovery involves holding V_m at -90 mV under controlled conditions.

a. **Phase 0:** The upstroke of a slow AP is driven by I_{Ca}, which activates by an order of magnitude more slowly than I_{Na}. Slow APs propagate very slowly as a result.

b. **Phase 3:** Phase 3 repolarization is mediated by I_K.

c. **Phase 4:** Phase 4 corresponds to a period of recovery, but nodal cells are notable in that phase 4 drifts slowly positive with time. This drift is caused by I_f and is the key to automaticity and pacemaker function.

C. Pacemakers

Pacemaker ability is conferred on cells by HCN. When open, HCN channels cause V_m to slide gradually toward the threshold for AP formation. The cAMP dependence of this channel also provides the ANS with a way of regulating the rate of phase 4 depolarization, which, in turn, regulates HR. When intracellular cAMP levels rise, HCN open probability increases, and V_m depolarizes at an accelerated rate.

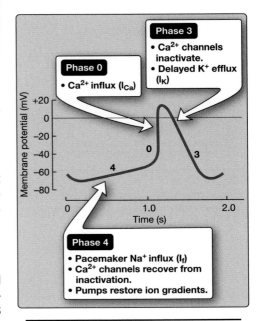

Figure 16.11.
Slow action potential. I_{Ca} = Ca^{2+} current; I_f = funny current; I_K = K^+ current.

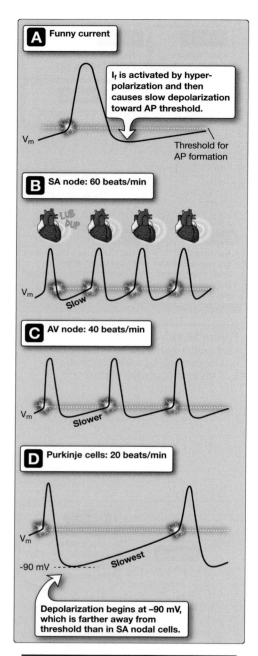

Figure 16.12.
Pacemakers. AV = atrioventricular,
I_f = pacemaker current, SA = sinoatrial,
V_m = membrane potential.

Falling cAMP levels decrease HCN opening, and the rate of phase 4 depolarization slows. Because maintaining a regular heartbeat is critical for survival, three different cell types are granted the ability to function as pacemakers: SA nodal cells, AV nodal cells, and Purkinje fibers.

1. **Funny current:** HCN is activated by the hyperpolarization that occurs at the end of phase 3 (Fig. 16.12A). The ensuing depolarization usually takes several hundred milliseconds to reach V_{th}, at which point a spike and a new wave of excitation is initiated. The spike inactivates HCN until the end of phase 3, and then the cycle repeats.

2. **Other pacemakers:** The SA node is the heart's primary pacemaker. It has an intrinsic rate of ~100 beats/min, but HR is usually lower because the PSNS reduces HR when the need for cardiac output (CO) is low (see Fig. 16.12B). Should the SA node be damaged and fall silent, the AV node takes over as pacemaker. The AV node is normally subservient to the SA node because its intrinsic rate is 40 beats/min. It takes ~1.5 seconds for AV nodal phase 4 to reach V_{th} (see Fig. 16.12C), but the new wave of excitation originating in the SA node typically arrives well before this time (the tendency of a fast pacemaker to inhibit a slower one is known as **overdrive suppression**). Purkinje cells are tertiary pacemakers. Their intrinsic rate is very low (~20 beats/min), partly because V_m is about 25 mV more negative in Purkinje cells than nodal cells; therefore, it takes much longer for V_m to reach and cross threshold from this more negative level (see Fig. 16.12D).

3. **Regulation:** Because HR is a primary determinant of CO, the SA node is heavily regulated by the ANS. The SNS increases HR by releasing norepinephrine onto β_1-adrenergic receptors on nodal cells. These are GPCRs that increase adenylyl cyclase (AC) activity and intracellular cAMP concentration. cAMP binds to and increases HCN open probability and accelerates the rate of phase 4 depolarization (Fig. 16.13). HR thus increases (**positive chronotropy**). PSNS terminals release ACh onto nodal cells. ACh binds to muscarinic type-2 receptors, which are also GPCRs that depress AC activity and decrease cAMP formation. The rate of phase 4 depolarization slows, and HR decreases (**negative chronotropy**).

4. **Other currents:** The rate of phase 4 depolarization is also influenced by a I_{Ca} and $I_{K,Ach}$, both of which are regulated by the ANS (see Fig. 16.13).

   a. **Calcium current:** Catecholamine-induced increases in intracellular cAMP concentration increase Ca^{2+} channel activity via protein kinase A (PKA)-dependent phosphorylation. This contributes to the increased rate of phase 4 depolarization in SA nodal cells, and it also moves V_{th} closer to V_m. PSNS activation decreases AC and PKA activity and, thus, moves I_{Ca} away from V_{th}. The ANS has similar effects on Ca^{2+} channels in the AV node, but here, they manifest as a change in conduction velocity (**dromotropy**).

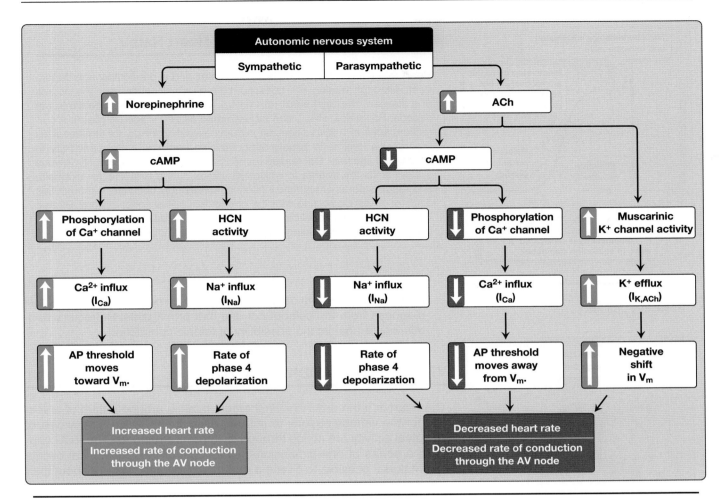

Figure 16.13.
Autonomic regulation of nodal cells. AV = atrioventricular; HCN = hyperpolarization activated, cyclic nucleotide–dependent Na$^+$ channel; I_{Ca} = Ca^{2+} current; $I_{K,ACh}$ = G protein-regulated K$^+$ current; I_{Na} = Na$^+$ current; V_m = membrane potential.

b. **Potassium current:** The PSNS has an additional level of control through $I_{K,ACh}$, causing nodal cell membranes to become more negative when activated. Because phase 4 begins at a more hyperpolarized level, it takes longer to reach V_{th}, and HR slows.

D. Refractory periods and arrhythmias

Because every cell in the myocardium is electrically connected via gap junctions, the heart is vulnerable to pacemakers located within the contractile portions of the myocardium (**ectopic pacemakers**). These may have intrinsic rates that are so fast that the heart is unable to function as a pump. Fortunately, I_{Na} inactivates during depolarization, lessening this possibility by creating an **effective refractory period** (**ERP**), a time during which a myocyte is insensitive to new waves of excitation (Fig. 16.14). Once the current begins recovering from inactivation, the myocyte progresses to a **relative refractory**

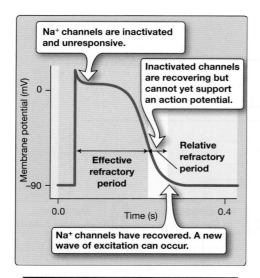

Figure 16.14.
Refractory periods.

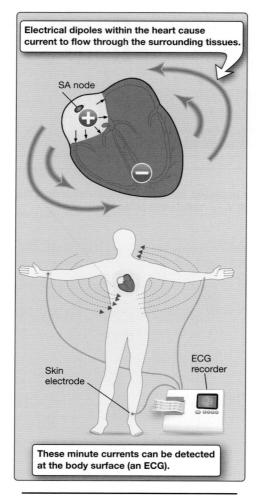

Figure 16.15.
Electrocardiography. SA = sinoatrial.

---

period (**RRP**), a time during which it may be possible to elicit a small response from the cell but not one that propagates. Refractory periods insure against the possibility of tetanic contraction.

VI. ELECTROCARDIOGRAPHY

The heart is a three-dimensional organ. It takes around a third of a second for the various regions to activate fully, during which time waves of electrical activity are racing through its internal structures. The ECG captures a series of one-dimensional snapshots of these electrical events that makes it possible to create a remarkably detailed picture about their timing, direction, and the mass of tissues involved.

A. Theory

An ECG records extracellular potentials using electrodes that adhere to the body surface. The potentials are generated by current flowing through surrounding tissues from depolarized areas of the heart to polarized regions (**electrical dipole**) as shown in Figure 16.15. Current intensity is directly proportional to size of the dipole. Three ECG electrodes are arranged in a triangle (**Einthoven triangle**) around the heart and connected to an ECG recorder. The recorder systematically compares voltage differences between pairs of electrodes and generates a moving paper record. These comparisons, which are facilitated by rapid switching within the ECG recorder, are known as **ECG leads** (Fig. 16.16). Each ECG lead has a positive and a negative pole corresponding to a skin electrode. The two general types of leads are limb leads and precordial leads.

1. **Bipolar limb leads:** The three **bipolar limb leads** compare voltage differences between each of the three ECG electrodes (see Fig. 16.16A). Lead I records voltage differences between the right and left shoulders, lead II compares the right shoulder and the left foot, and lead III compares the left shoulder and left foot. By convention, the left shoulder is designated the positive pole of lead I, whereas the foot is designated the positive pole of leads II and III.

2. **Augmented limb leads:** Three **unipolar leads** compare voltage differences between skin electrodes and a common reference point (**central terminal**) held close to zero potential (see Fig. 16.16B). Leads **aVL**, **aVR**, and **aVF** measure voltage differences between this point and the left shoulder, right shoulder, and foot, respectively. The skin electrodes are considered the positive pole in each case.

3. **Precordial leads: Precordial** (or **chest**) leads compare voltage differences between the common reference point and six additional skin electrodes placed in a line directly above the heart (V_1-V_6).

B. Electrocardiograph

All ECG recordings are standardized so that their interpretation becomes a simple matter of pattern recognition to a trained eye. By convention, when a wave of depolarization is moving through the heart toward the positive pole of a lead, it causes an upward (positive) deflection on the ECG. Movement toward the negative pole causes a downward (negative) deflection. Depolarization of a large muscle mass generates a larger dipole than a smaller mass, so it generates a larger deflection on the record.

C. Normal electrocardiogram

A typical ECG recording comprises five waves, P through T, that correspond to the sequential excitation and recovery of the different regions of the heart (Fig. 16.17).

1. **P wave:** The myocardium rests between beats, and the ECG recording rests at the **isoelectric line**. Excitation begins with the SA node, but the current that it generates is too small to record at the body surface. The wave of depolarization then spreads across the atria, registering as the **P wave**. When both atria are depolarized fully, the record returns to baseline. A normal P wave has a duration of 80 to 100 ms.

2. **QRS complex:** The P wave is followed by a brief period of quiet during which the wave of excitation moves slowly through the AV node and crosses from the atria to the ventricles via the bundle of His. This progression does not register on the recording. Ventricular depolarization produces the **QRS complex**. The three components reflect excitation of the interventricular septum (**Q wave**), the apex and the free walls (**R wave**), and finally the regions near the base (**S wave**). The recording returns to baseline when the entire ventricular myocardium is depolarized, roughly coinciding with phase 2 of the ventricular AP. The QRS complex lasts 60 to 100 ms.

3. **T wave:** Ventricular repolarization registers on the ECG recording as the **T wave**. On rare occasions, the T wave may be followed by a small **U wave**, thought to represent papillary muscle repolarization.

The time intervals between the waves are also named and can provide important insights into cardiac function (Table 16.2).

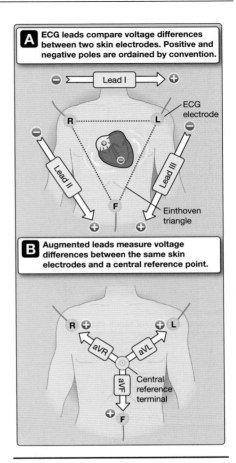

Figure 16.16.
ECG limb leads.

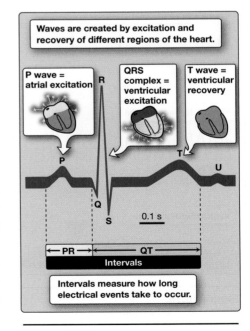

Figure 16.17.
ECG waves and intervals.

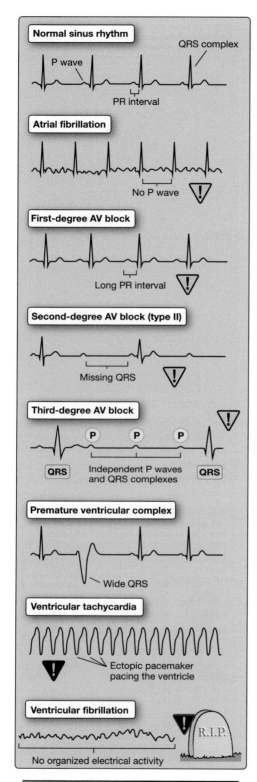

Figure 16.18.
Normal and abnormal cardiac rhythms.
AV = atrioventricular.

Table 16.2: Timing of Electrocardiogram Waveforms

Name	How Measured	Significance	Time
PR interval	From start of the P wave to start of QRS complex	Time for wave of excitation to traverse atria and AV node	120–200 ms
QRS duration	From start to finish of QRS complex	Time taken for both ventricles to depolarize fully	60–100 ms
QT interval	From start of QRS complex to end of T wave	Duration of ventricular excitation and recovery	300–430 ms

AV = atrioventricular.

D. Rhythms

The heart of a healthy individual at rest beats with a **normal sinus rhythm** of 60 to 100 beats/min. Some individuals have normal rates that fall below this range (**sinus bradycardia**), whereas strenuous physical activity typically causes the normal rate to exceed 100 beats/min (**sinus tachycardia**). The "sinus" prefix to both rhythms indicates that the rate is established by the SA node. Abnormal rhythms (**dysrhythmias** and **arrhythmias**) can originate from virtually any part of the myocardium (Fig. 16.18).

1. **Atrial arrhythmias:** Atrial fibrillation (**AF**) is an arrhythmia caused by one or more extranodal atrial pacemakers that typically cycle at several hundred times per minute. AF is relatively common, especially in older adults and in patients with heart failure. Loss of atrial pump function reduces CO, and patients typically present with fatigue, dyspnea, and lightheadedness as a result. The AV node acts as a filter that usually protects the ventricles from arrhythmias of atrial origin. The node is re-excited by the chaotic electrical activity running through the atria whenever it emerges from its refractory period, so HR may be irregular and tachycardic, but the QRS is normal because the wave of excitation is still coordinated by the His–Purkinje system.

2. **Atrioventricular block:** Functional and anatomic defects in the AV node can delay or interrupt transmission of signals to the ventricles, a condition known as **AV nodal block**. Block occurs during the PR interval because this is the time when the wave of excitation propagates from the atria to the ventricles. AV block is generally described as being first, second, or third degree, according to severity.

   a. **First degree:** First-degree block is characterized by a lengthening of the PR interval (>0.2 second). It is usually benign and asymptomatic.

   b. **Second degree:** Two types of **second-degree block** are recognized. **Möbitz type I** (also known as **Wenckebach block**) describes a rhythm in which the PR interval lengthens gradually until a complete block occurs, at which point the ventricles fail to excite, and the ECG recording drops a QRS complex. **Möbitz type II** block is characterized by ECG

Clinical Application 16.2: Long QT Syndrome

Long QT syndrome (LQTS) refers to a set of related inherited and acquired disorders that delay phase 3 membrane repolarization and manifest as a prolongation of the QT interval on the ECG. LQTS patients are at risk of developing **torsades de pointes** (French for "twisting around the points"), a characteristic ventricular tachycardia in which the QRS complex rotates about the isoelectric line. Torsades is of concern because it often precipitates sudden cardiac death (SCD). Phase 3 can be delayed by reducing K^+ efflux (I_K) or by prolonging Na^+ influx (I_{Na}) or Ca^{2+} influx (I_{Ca}). The most common form of LQTS (LQT type 1) is caused by *KCNQ1* gene mutation, which reduces the slower K^+ current (I_{KS}). LQT type 3 is caused by *SCN5A* gene mutations that prevent I_{Na} from inactivating fully during depolarization and is a particularly lethal form of LQTS. Arrhythmic events may be precipitated by any of a number of factors, including exercise and abrupt sounds. Patients typically suffer palpitations, syncope, seizures, or SCD as a result.

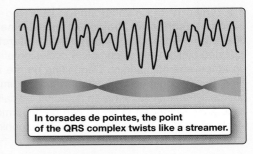

In torsades de pointes, the point of the QRS complex twists like a streamer.

Torsades de pointes.

recordings in which the QRS complex is dropped with no prior warning. Type I is usually benign. Type II may progress rapidly to third-degree block.

   c. **Third degree: Third-degree block** is caused by a defect in the AV node or conduction system that completely prevents electrical signals from reaching the ventricles. In the absence of guidance from the SA node, pacemakers located in the bundle of His or Purkinje network often take over the responsibility for driving ventricular contraction. The ECG typically shows a normal, regular P wave, and a QRS complex that may also be regular but temporally disconnected from the sinus rhythm.

3. **Ventricular dysrhythmias:** Dysrhythmias can also have ventricular origins. **Ectopic ventricular rhythms** originating in the contractile portions of the myocardium propagate via gap junctions until the entire heart is involved. Because the wave of excitation spreads via the myocardial equivalent of slow back streets rather than the Purkinje highway system, the resulting QRS complexes are broad. The excitation sequence is abnormal also, so the QRS complex is highly atypical. Occasional (<6/min) **premature**

Biologic Sex and Aging 16.2: Cardiac Dysrhythmias

Aging is accompanied by declining cardiac myocyte numbers, lost cells being replaced with collagen and other fibrous tissue. Remaining myocytes enlarge. These changes distort and impede conduction pathways and increase susceptibility to ectopic pacemakers and dysrhythmias. More than 80% of individuals older than age 80 years experience premature atrial beats, and atrial fibrillation is common. Aging also increases vulnerability to drugs that prolong the QT interval and trigger torsades de pointes (see Clinical Application 16.2), including various antiarrhythmics, vasodilators, antimicrobials, antihistamines, and psychiatric drugs. Females are at significantly increased risk of drug-induced torsades de pointes because their QT interval is normally prolonged compared with males.

Clinical Application 16.3: Myocardial Infarction

If a region of the myocardium is deprived of adequate blood flow, it becomes **ischemic**. Prolonged or extreme ischemia causes muscle death, an event known as **myocardial infarction** (**MI**). MI is usually precipitated by stenosis or complete occlusion of a coronary supply artery by an atherosclerotic plaque. Depending on the severity of flow impairment, MI may prove immediately fatal or may be limited to focal ventricular wall necrosis. Patients suffering an acute MI typically present with intense ischemic pain. Diagnosis of an MI can be confirmed by measuring circulating levels of cardiac biomarkers, such as troponins, and may often manifest on an ECG as **ST-segment elevation**.

ST-segment elevation occurs because injured and dying cells leak K^+ into the extracellular space. All cells maintain high intracellular K^+ concentrations using the ubiquitous Na^+/K^+-ATPase. When cells die, their membranes disrupt, and K^+ is released. All cells also rely on a steep transmembrane K^+ gradient to maintain membrane potential (V_m) at normal resting levels, so the appearance of K^+ extracellularly causes healthy myocytes peripheral to the ischemic event to depolarize. The ischemic area thereby creates an electrical dipole within the resting myocardium that generates an **injury current**. The current flows in the period between beats and causes a baseline offset on an ECG recording. An observer only becomes aware of the offset during the ST segment, a time during which the entire myocardium is depolarized, and the dipole and its dependent current disappear. In practice, the injury current *tricks* the eye into believing that the ST segment is elevated. Damaged areas eventually necrose and are replaced with scar tissue, at which point the dipole and the injury current disappear.

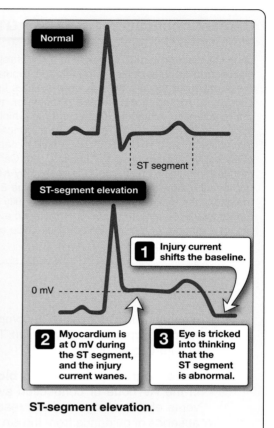

ST-segment elevation.

---

ventricular contractions (**PVCs**) of ectopic origin are common and usually benign (see Fig. 16.18). Frequent (>6/min) PVCs or those that occur in trains can be a sign of underlying myocardial pathology or ischemia. Ectopic pacemakers have the potential to pace the myocardium at high rates (~300 beats/min), a rhythm known as **ventricular tachycardia** (**V-tach**). The onset of V-tach is a grave event because pacing the heart at such high rates disrupts pump function to the point where CO drops to zero. A myocardium deprived of O_2 supply quickly degenerates into **ventricular fibrillation** (**V-fib**) and sudden cardiac death.

E. Mean electrical axis

The **mean electrical axis** (**MEA**) averages the many electrical vectors generated by the wave of excitation as it moves through the heart. It provides a single value that indicates which region of the heart dominates the electrical events (Fig. 16.19). In a healthy individual, the LV dominates because it contains the largest mass of tissue. By convention, a circle is drawn around the heart in the plane of the limb leads, and the left side taken to be at 0°, the right side to be at +180°, the feet to be at +90°, and the head to be at −90°. In a normal, healthy individual, the MEA lies between −30° and +100°. If the MEA is less than −30°, the left side of the heart must be contributing to the MEA to a greater extent than normal. This is known as **left-axis deviation** and is usually an indication of left-ventricular hypertrophy. An MEA of greater than +100° (**right-axis deviation**) indicates right-ventricular hypertrophy.

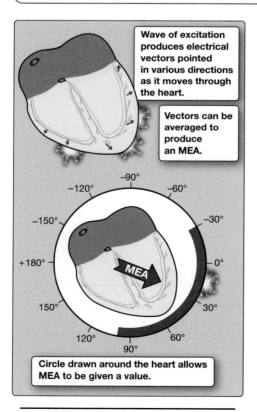

Figure 16.19.
Mean electrical axis (MEA).

Chapter Summary

- The cardiovascular system contains two vascular circuits that are connected in series to form a closed loop. Blood flows around the loop in response to pressure generated by two pumps located within the heart.
- The **left heart** pumps blood at high pressure through the **systemic circulation**. The **right heart** drives blood at a relatively low pressure through the **pulmonary circulation**.
- The two pumps each contain two chambers: one **atrium** and one **ventricle**. One-way valves at the ventricular inlets and outlets help maintain unidirectional flow through the circuits.
- **Cardiac muscle** shares many features in common with skeletal muscle. Both have highly organized and aligned sarcomeres that give them a **striated appearance** under polarized light.
- Skeletal muscle is under voluntary control, but cardiac muscle functions **autonomously**. A specialized **pacemaker** periodically generates APs that spread from myocyte to myocyte to involve the entire heart. The **autonomic nervous system (ANS)** regulates the timing and force of contraction but does not initiate it.
- Contraction of the different regions and chambers within the heart is carefully coordinated by a wave of depolarization. The wave spreads from myocyte to myocyte throughout the myocardium via **gap junctions**. Cardiac myocytes branch extensively to maximize interactions with adjacent cells.
- A heartbeat is initiated by the heart's pacemaker, the **sinoatrial node**, located in the wall of the right atrium. Atria contract first, forcing their contents through the **atrioventricular (AV) valves** toward the ventricles. Excitation of the ventricles is delayed by the **AV node** and then coordinated by high-speed **Purkinje fibers**.
- Conduction velocity through the different regions of the heart is related to the shape of the AP. Contractile myocytes and Purkinje fibers express **fast APs** comprising five phases: **phase 0** (upstroke) is due to a rapidly activating Na^+ current, which then **inactivates (phase 1)**; plateau (**phase 2**) is due to a slow-activating Ca^{2+} **current**, whereas membrane repolarization (**phase 3**) is produced by a K^+ **current.** The interval between beats is termed **phase 4**.
- Nodal cells are slow conducting because they lack the phase 0 Na^+ current, leaving the Ca^{2+} current to drive the AP upstroke.
- Inactivation of the Na^+ current and Ca^{2+} current during depolarization causes myocytes to become **refractory** to further stimulation, which insures against the possibility of tetanus.
- Nodal cells express a pacemaker current (**funny current**) activated during phase 4 that causes slow depolarization toward the threshold for spike formation.
- The funny current is regulated by the **ANS**. The **sympathetic** branch enhances the current, increases the rate of phase 4 depolarization, and speeds heart rate. **Parasympathetic** nerve stimulation has the opposite effect.
- The waves of excitation moving through the heart generate currents that can be recorded at the body surface to produce an **ECG**. A typical ECG recording comprises several distinct waveforms corresponding to excitation of the atria (**P wave**), ventricular septum, apex, and free walls (**Q, R,** and **S**) and then repolarization of the ventricles (**T wave**).
- The timing, magnitude, and form of these waves can be used to diagnose defects in heart function. With a break in the normal conduction pathways, the time between P wave and QRS complex is prolonged ("**heart block**"). An increase in the height of a P wave or QRS complex is indicative of **hypertrophy**. Widening of the QRS complex may be indicative of an ectopic pacemaker. Displacement of a segment may be indicative of **ischemia** and **infarction**.

Study Questions

Choose the ONE best answer.

16.1. A 65-year-old male with a history of hypertension is prescribed a Ca^{2+}-channel blocker to help reduce his blood pressure. What is the most likely effect of this drug on the ventricular myocardium?

    A. It would have no effect.
    B. It would increase contractility.
    C. It would increase heart rate.
    D. Phase 1 would be prolonged.
    E. Phase 2 would be reduced.

Best answer = E. Ca^{2+}-channel blockers reduce Ca^{2+} influx via L-type Ca^{2+} channels during phase 2 of the ventricular action potential, thereby reducing phase 2 and decreasing, not increasing contractility (A, B; see Section V·A). Heart rate is determined by the rate of phase 4 depolarization in sinoatrial nodal cells, which is governed in part by L-type Ca^{2+} channels (see Section V·C·4). Ca^{2+}-channel blockers would be expected to decrease, not increase (C) heart rate. Phase 1 is mediated by Na^+ and K^+ channels (see Section V·B) and would not be affected by a Ca^{2+}-channel blocker (D).

16.2. A mean electrical axis value of −60° would most likely be associated with which of the following conditions?

    A. Aortic stenosis
    B. Left-ventricular infarction
    C. Premature ventricular contractions
    D. Pulmonary edema
    E. Pulmonary hypertension

Best answer = A. Aortic stenosis forces the left ventricle (LV) to work harder and generate higher peak systolic pressures to sustain cardiac output (see Chapter 40·V·A). Over time, this causes LV hypertrophy, which manifests on an ECG as left-axis deviation (normal range = +100° to −30°; see Section VI·E). A left-ventricular myocardial infarction (B) would likely shift the axis to the right, not left, because viable LV muscle mass is reduced. Premature ventricular contractions (C) do not directly alter the mean electrical axis. Pulmonary edema (D) and hypertension (E) promote right-ventricular hypertrophy and right-axis deviation.

16.3. A 50-year-old female reports "thumping" sensations in her chest. An ECG records occasional wide, premature QRS complexes. Which of the following most likely explains the origin of these complexes?

    A. Atrial fibrillation
    B. First-degree heart block
    C. Irritable ectopic focus
    D. Myocardial ischemia
    E. Ventricular fibrillation

Best answer = C. Premature ventricular contractions (PVCs) are characterized by wide and abnormally shaped QRS complexes. They reflect waves of excitation that travel through the myocardium via the slow myocyte-to-myocyte route rather than the fast conduction His–Purkinje system (see Section VI·D·3). PVCs are typically triggered by irritable foci located in the ventricular myocardium (i.e., ectopic) rather than the sinoatrial node. Atrial fibrillation (A) manifests as loss of a P wave, whereas first-degree heart block (B) prolongs the PR interval. Ischemia (D) may affect the ST segment, but QRS complexes still occur in normal position. A fibrillating ventricle (E) shows no organized ECG waveforms.

16.4. Which of the following ECG events most likely coincides with the "reduced ventricular ejection" phase of the cardiac cycle?

    A. P wave
    B. PR interval
    C. QRS complex
    D. ST segment
    E. T wave

Best answer = E. The T wave corresponds to ventricular repolarization (see Section VI·C), which occurs during reduced ejection (see Chapter 17·II). The P wave (A) coincides with atrial systole, which continues during the PR interval (B). The QRS complex (C) is caused by ventricular excitation, which is followed by isovolumic contraction and rapid ejection. The ST segment (D) encompasses isovolumic contraction and persists through rapid ejection.

Cardiac Mechanics 17

I. OVERVIEW

The heart's function is to generate pressure within the arterial compartment. Pressure is required to drive blood flow through the vasculature, delivering O_2 and nutrients to all cells in the body. The amount of blood that the heart ejects is precisely matched to bodily needs. At rest, the needs of the various tissues are modest, and cardiac output (CO) approximates 5 to 6 L/min in an average person. Any increase in tissue activity (digesting a meal, walking, climbing stairs) requires that CO increase to support the active tissue's increased needs. Raising CO is achieved in part by increasing cycle frequency (heart rate [HR]), but the heart, unlike a conventional pump, has the unique ability to increase the amount of blood it ejects on each stroke and to do so with increased force and efficiency. These features allow a fit athlete's heart to ramp up output by as much as five- to sixfold during strenuous exercise. Matching CO to tissue needs is the responsibility of the autonomic nervous system (ANS), which regulates beat frequency, the extent of ventricular filling prior to contraction, and contractile force.

II. CARDIAC CYCLE

The cardiac cycle consists of alternating periods of contraction (**systole**) and relaxation (**diastole**). When describing the cycle, it is useful to correlate four measures of activity: the electrical events that initiate and coordinate contraction (recorded as an **electrocardiogram [ECG]**), pressure within various parts of the system, volume changes, and sounds associated with blood flow and valve function. These four indices are collated in Figure 17.1, which focuses on the left ventricle (LV), but the right ventricle (RV) functions similarly, albeit at lower ejection and filling pressures.

A. Phases

The cardiac cycle can be subdivided into seven discrete phases (numbers corresponding with phases indicated at the top of Fig. 17.1).

1. **Atrial systole:** The cardiac cycle begins with atrial systole, which is initiated by atrial excitation and follows the crest of the P wave on the ECG.

2. **Isovolumic ventricular contraction:** Ventricular systole begins with mitral valve closure, which occurs during the QRS complex. It takes around 50 ms for the ventricle to develop sufficient pressure to force the aortic valve open, during which time

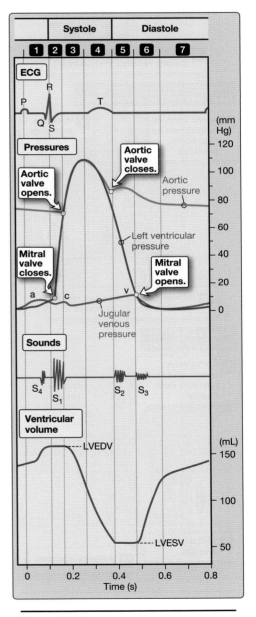

Figure 17.1.
Cardiac cycle. LVEDV and LVESV = left-ventricular end-diastolic and end-systolic volume, respectively.

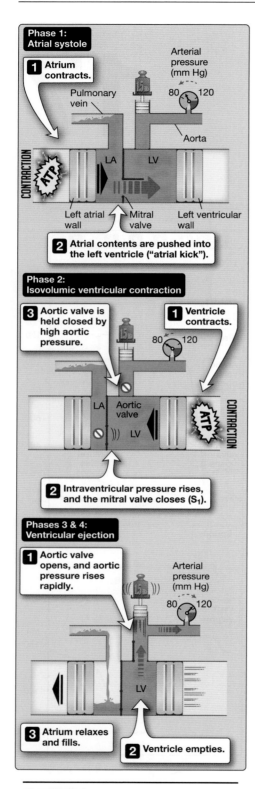

Figure 17.2.
Cardiac cycle in a model heart. LA and LV = left atrium and ventricle, respectively; LVP = left-ventricular pressure.
(Continued)

the myocytes are contracting around a fixed volume of blood. This phase is therefore known as **isovolumic (isovolumetric) contraction**.

3. **Rapid ventricular ejection:** The aortic valve finally opens, and blood exits the ventricle and enters the arterial system at high velocity (**rapid ejection**).

4. **Reduced ventricular ejection:** Ejection velocity decreases as ventricular systole nears completion (**reduced ejection**). Aortic valve closure marks the end of this phase.

5. **Isovolumic ventricular relaxation:** With the ventricle once again a sealed vessel, a period of **isovolumic relaxation** follows.

6. **Rapid ventricular filling:** When the mitral valve opens, blood that had been damming in the atrium during systole surges forward into the ventricle. The **rapid passive filling** phase signals the beginning of diastole.

7. **Reduced ventricular filling:** The cardiac cycle ends with reduced filling. This phase, which is also known as **diastasis**, typically disappears when HR increases because cycle length is shortened largely at the expense of diastole.

B. Ventricular pressure and volume

Figure 17.2 illustrates the following descriptions of blood flow and pressure development in the left heart during a single cardiac cycle. The numbered phases correlate with phases indicated in Figure 17.1.

1. **Diastole:** The LV refills with pulmonary venous blood by way of the left atrium during diastole. By the end of diastole, the ventricle has neared capacity, but atrial contraction forces a small additional blood bolus into the chamber lumen ("**atrial kick**"; see Figs. 17.1 and phase 1 in 17.2), and left-ventricular pressure (LVP) rises to around 10 to 12 mm Hg. LV **end-diastolic volume** (**EDV**) then stands at ~120 mL in an individual at rest, although the range of values considered normal is wide (70–240 mL).

> In a healthy person at rest, atrial systole boosts LV volume by around 10%, but this amount can climb to 30% to 40% when HR is high and the time available for filling is reduced. Patients in heart failure can become so dependent on atrial contribution to resting output that loss of atrial systole due to fibrillation can compromise CO and arterial pressure.

2. **Systole:** LVP climbs at maximal rates during isovolumic contraction (see Fig. 17.2, phase 2). Once LVP meets and exceeds aortic pressure, the aortic valve opens, and blood is ejected into the arterial system (see Fig. 17.2, phases 3 & 4). Pressures continue to climb even though blood is being ejected because the LV myocytes are still actively contracting (see Fig. 17.1). The rapid

ejection phase accounts for ~70% of total ventricular output and drives aortic pressure toward a peak of ~120 mm Hg. The ventricular myocytes now begin to repolarize, contraction wanes, and LVP falls rapidly. The kinetic energy imparted to blood by LV contraction continues to drive ventricular outflow for a brief period, but the rapid fall in LVP soon causes the pressure gradient across the aortic valve to reverse, and the aortic valve slams shut (see Fig. 17.2, phase 5).

3. **Diastole:** Once intraventricular pressure dips below atrial pressure, the mitral valve opens, and filling begins (see Fig. 17.2, phase 6). The LV's helical muscle bands cause it to shorten, twist, and wring blood out through the valves during systole. With relaxation, the myocardium rebounds through natural elasticity, and pressure continues to decrease rapidly even as blood surges in from the atrium (rapid passive filling; phase 6). After a period of reduced filling (see Fig. 17.2, phase 7), the cycle repeats itself.

4. **Ventricular efficiency:** The LV does not empty completely during systole, and **end-systolic volume** (ESV) is usually around 50 mL. Subtracting ESV from EDV gives **stroke volume** (SV), which defines the amount of blood transferred to the arterial system during systole. SV should be >60 mL in a healthy person. Dividing SV by EDV yields **ejection fraction** (EF), normally ~55% to 75%. EF is an important measure of cardiac efficiency and health and is used clinically to assess cardiac status in patients with heart failure, for example.

> EF is typically estimated noninvasively using two- or three-dimensional echocardiography ("echo"). An echo can measure ESV and EDV, wall thickness, muscle shortening velocity, and blood flow patterns during contraction and relaxation.

C. Aortic pressure

The arterial system is composed of small-bore vessels that do not stretch easily (see Chapter 18·II·C). In practice, this means that arterial pressure rises sharply when the LV forces blood into the system, but this pressure readily dissipates when blood flows out of the system and enters capillary beds.

1. **Systole:** Aortic pressure is insensitive to intraventricular events so long as the aortic valve is closed. Once LVP climbs above aortic pressure and the aortic valve opens, aortic pressure rises and falls in near synchrony with LVP (see Fig. 17.1).

2. **Diastole:** Aortic pressure dips briefly immediately following aortic valve closure, creating a characteristic **dicrotic notch** or **incisura** in the aortic pressure curve. The notch is caused by the aortic valve bulging backward into the LV when it closes, creating a transient local pressure drop. Aortic pressure declines slowly throughout diastole, reflecting blood draining out of the arterial system and into the capillary beds (**diastolic runoff**; see Figs. 17.1 and 17.2, phases 6 and 7).

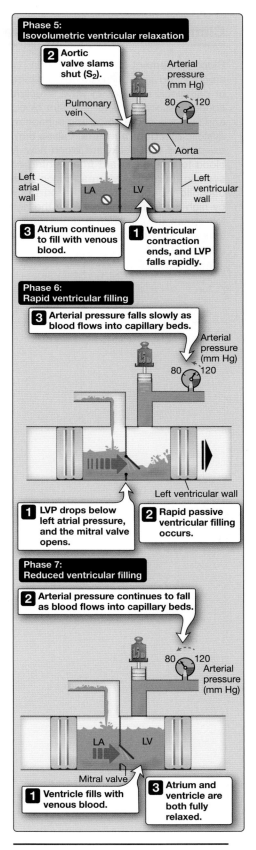

Figure 17.2.
(Continued)

Clinical Application 17.1: Heart Murmurs

Blood movement between the atria, ventricles, and arterial system is usually smooth and silent. Physiologic changes in blood composition and congenital or pathologic changes in valve structure can create **heart murmurs**, which are sounds associated with turbulent blood flow (see Chapter 18·V·A·4). Stiffening and stenosis of the aortic and pulmonary valves obstructs blood flow and causes systolic **ejection murmurs**. Valves that close incompletely also cause murmurs associated with backward blood flow. Mitral and tricuspid **insufficiency** allows blood to flow backward from ventricle to atrium during systole, manifesting as systolic murmurs, whereas incompetent aortic and pulmonary valves cause diastolic murmurs. The figure on the right shows a blood jet (*arrow*) caused by high-pressure backflow from the aorta (Ao) to the left ventricle (LV) into the left atrium (LA) through a regurgitant aortic valve during diastole. The jet was imaged using echocardiography. The trace below the image depicts the sound caused by aortic regurgitation.

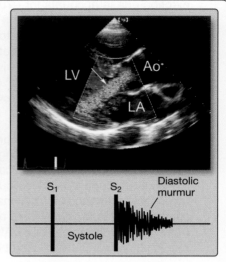

Aortic regurgitation. S_1 and S_2 = first and second heart sounds.

D. Venous pressure

The lack of valves between atria and the venous system allows intra-atrial pressure changes to be transmitted backward into the veins lying close to the heart. Internal jugular venous pressure ([JVP] evidenced by a prominent neck vein) displays three distinct pressure waves during the cardiac cycle (see Fig. 17.1).

1. **a Wave:** Right atrial contraction generates a pressure wave that forces blood forward into the RV and also creates an **a wave** in a jugular venous recording.

2. **c Wave:** Ventricular contraction causes intraventricular pressure to rise sharply, and the atrioventricular (AV) valves bulge backward into the atria. Backward deflection of the tricuspid valve generates a JVP pulse known as the **c wave** (downslope of the c wave is known as the "x descent").

3. **v Wave:** During ventricular systole, blood continues to flow from the venous system into the right atrium and dam against the closed tricuspid valve. Pressure builds as the atrium fills, registered as the upslope of the **v wave** in a jugular venous recording. The downslope ("y descent") correlates with rapid atrial emptying when the tricuspid valve opens and blood surges forward into the RV (a similar surge occurs on the left side of the heart; see Fig. 17.2, phase 6).

E. Heart sounds

Four **heart sounds** (S_1–S_4) are associated with the cardiac cycle (Fig. 17.3). The first two are valve-closing events, and the second two are associated with blood entering the LV.

1. **First: S₁** occurs at the beginning of ventricular systole. LVP develops rapidly during this time, causing blood to start moving backward toward the atria. The movement immediately catches the leaflets of the AV valves and causes them to snap shut (see Fig. 17.2, phase 2). Valve closure and reverberations within the LV wall register as a low rumbling "**lub**" sound, which lasts ~150 ms.

2. **Second:** Aortic and pulmonary valve closure is associated with (**S₂**) that can be heard as a brief "**dup**." S₂ often splits into distinct aortic (A₂) and pulmonary (P₂) components that reflect slight asynchronous closure of the two valves. Splitting is most acute during inspiration, when a drop in intrathoracic pressure enhances the pressure gradient driving blood flow from the systemic veins into the RV. The RV receives more blood than the LV as a result. Because a large blood volume takes longer to eject than a small volume, RV systole is prolonged compared with LV systole. Pulmonary valve closure is delayed to the extent that it can be heard as a separate sound (P₂).

3. **Third:** Auscultation of children and thin adults may reveal a low-intensity rumbling heart sound (**S₃**) during early diastole. S₃ is caused by blood rushing into the LV (i.e., rapid passive filling) and causing turbulence that makes the LV walls reverberate and rumble. S₃ is intensified when transvalvular flow is increased and when the LV is dilated.

4. **Fourth: S₄** is associated with atrial contraction. The force of atrial systole is too weak to be detectable by ear in young, healthy individuals. A ventricle that requires filling assistance can stimulate atrial hypertrophy, however, and then S₄ may become apparent as a brief, low-frequency sound. It reflects blood being forced into a less compliant ventricle at high pressure, causing reverberations within the ventricular wall.

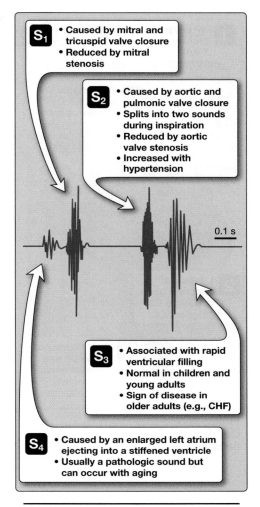

Figure 17.3.
Heart sounds. CHF = congestive heart failure.

Biologic Sex and Aging 17.1: Cardiac Size and Function

A number of significant differences in cardiac size and function between males and females manifest at puberty and persist through old age. In males, heart mass is 15%–30% greater, the left-ventricular (LV) wall is thicker, and end-diastolic volumes are larger. This translates into higher initial ejection rates and improved diastolic filling. Aging hearts stiffen in both males and females, the loss of compliance reflecting a twofold increase in collagen deposition, collagen cross-linking by advanced glycation end products (see Biologic Sex and Aging 4.1), and Ca²⁺ leakage from the sarcoplasmic reticulum that creates a state of chronic partial contraction. In practice, this means that diastolic relaxation and filling is impaired, and the LV becomes increasingly dependent on left atrial kick. Males show additional deleterious changes not seen in females. Myocyte numbers decrease by >60 million per year, and the surviving myocytes enlarge. This leads to systolic dysfunction and increased LV chamber size, changes that have a significant impact on the way male and female hearts fare during cardiovascular stress and cardiac failure.

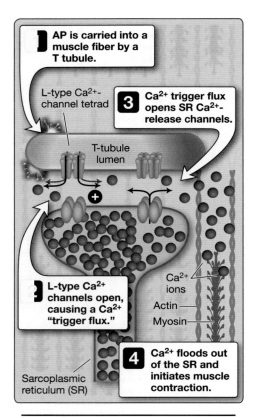

Figure 17.4.
Ca^{2+}-induced Ca^{2+} release. T tubule = transverse tubule.

III. EXCITATION–CONTRACTION COUPLING

The principles underlying contraction and force generation in cardiac muscle are similar to those of skeletal muscle (see Chapter 12·III), with a few differences in membrane system organization, Ca^{2+} dependence of excitation–contraction (EC) coupling, and in Ca^{2+} sensing and handling.

A. Membrane system organization

Action potentials (APs) initiate contraction by triggering Ca^{2+} release from the sarcoplasmic reticulum (SR). APs racing across the surface membrane are funneled into the core of each muscle fiber by transverse (T) tubules (Fig. 17.4). The T tubules communicate with and trigger Ca^{2+} release from the SR. There are two notable differences in cardiac T tubule and SR design compared with skeletal muscle.

1. **Location:** In skeletal muscle, the T tubules align with the ends of the thick filaments, two per sarcomere. Cardiac muscle has fewer tubules, and they run along the Z lines. T tubules tend to be wider in cardiac muscle and branch less extensively.

2. **Diads:** In skeletal muscle, T tubules interface with two SR cisternae at junctional complexes known as **triads**. In cardiac muscle, tubules associate with a single extension of the SR at an analogous structure called a **diad**.

B. Calcium dependence

When an AP arrives at a diad, it encounters and activates L-type Ca^{2+} channels (dihydropyridine receptors) in the T tubule membrane (see Fig. 17.4). In skeletal muscle, this opening event mechanically forces SR Ca^{2+}-release channels open, allowing Ca^{2+} to flood into the sarcoplasm and initiating contraction. In cardiac muscle, L-type Ca^{2+}-channel opening creates a Ca^{2+} **trigger flux** that is alone responsible for activating **Ca^{2+}-induced Ca^{2+} release** (**CICR**) from the SR. The size of the trigger flux and CICR is regulated by the ANS. An approximate 1:1 relationship between the number of L-type Ca^{2+} channels and dependent release channels permits fine adjustments in sarcoplasmic Ca^{2+} concentration and contractility.

C. Calcium sensing and handling

Troponin is a Ca^{2+}-sensitive heterotrimer associated with the actin thin filament. Increases in sarcoplasmic Ca^{2+} concentration cause a conformational change in troponin that rolls tropomyosin away from myosin-binding sites on the thin filament, thereby facilitating actin–myosin interaction. In skeletal muscle, troponin must bind two Ca^{2+} ions before contraction can begin. In cardiac muscle, only one Ca^{2+} ion is required for contraction.

IV. CARDIAC OUTPUT

The body's need for O_2 and nutrients changes constantly with changing activity level. Ensuring that these needs are met requires that CO be adjusted in parallel. To appreciate how and why CO is regulated *in vivo*, it is important to understand the various factors that influence LV output.

Clinical Application 17.2: Cardiac Markers

Cardiac troponin I (cTnI) and troponin T (cTnT) are troponin isoforms unique to heart muscle. cTnI and cTnT both appear in the circulation within 3 hours of acute myocardial infarction (MI), reaching a peak after 10–24 hours. Their appearance in blood indicates ischemia-induced necrosis and loss of cardiac myocyte integrity, so they can be used clinically to detect an acute MI. Blood tests also measure creatine kinase (CK) levels, but CK is not specific to cardiac muscle. Elevated plasma CK levels can be caused by cocaine use or exercise-induced skeletal muscle trauma in the absence of any cardiac damage.

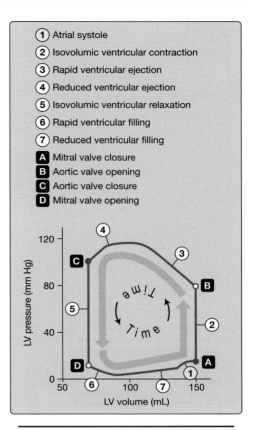

Figure 17.5.
Pressure–volume loop. LV = left ventricle.

A. Determinants

CO (L/min) is calculated from the product of HR (beats/min) and SV (mL):

$$CO = HR \times SV$$

1. **Heart rate:** HR is established by the sinoatrial (SA) node, the cardiac pacemaker. HR is dependent on the ANS, which controls the rate at which the pacemaker generates a wave of excitation. Sympathetic nervous system (SNS) activation raises HR, whereas parasympathetic stimulation decreases it.

2. **Stroke volume:** SV is dependent on LV **preload**, **afterload**, and **contractility**.

    a. **Preload:** Preload refers to a load applied to a myocyte that establishes muscle length before contraction begins. In the LV, preload equates with the volume of blood entering the chamber during diastole (EDV), which is dependent on end-diastolic pressure (EDP).

    b. **Afterload:** Afterload is the load against which a myocyte must shorten. Afterload determines maximal shortening velocity: shortening velocity decreases with increasing afterload. In a healthy individual, the principal component of LV afterload is aortic pressure.

    c. **Contractility:** Contractility is a measure of a muscle's ability to shorten against an afterload. In practice, contractility equates with sarcoplasmic free Ca^{2+} concentration.

B. Pressure–volume loop

The **pressure–volume (PV)** loop (Figs. 17.5 and 17.6) examines the relationship between blood volume and pressure within the LV during a single cardiac cycle. It is particularly useful in demonstrating how preload, afterload, and contractility affect cardiac performance. The loop replots the LVP trace from Figure 17.1 and folds it back on itself in time. All seven phases are reproduced in Figure 17.5, and the four points labeled A through D represent valve events.

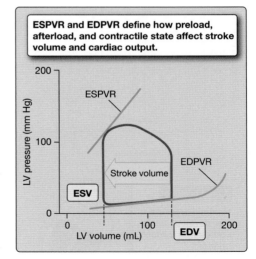

Figure 17.6.
The end-systolic pressure–volume relationship (ESPVR) and the end-diastolic pressure–volume relationship (EDPVR). EDV and ESV = end-diastolic and end-systolic volume, respectively; LV = left ventricle.

Clinical Application 17.3: β-Blockers and Calcium-Channel Blockers

β-Blockers and **Ca²⁺-channel blockers** are two important classes of drug used to control cardiac function and blood pressure.[1] β-Blockers (e.g., propranolol) are β-adrenergic–receptor antagonists that prevent norepinephrine- and epinephrine-mediated increases in myocardial Ca^{2+} concentration, thereby reducing heart rate and contractility and decreasing blood pressure. Ca^{2+}-channel blockers prevent Ca^{2+} influx through L-type Ca^{2+} channels. Verapamil and diltiazem are both relatively myocardium specific and reduce contractility. Nifedipine is a dihydropyridine Ca^{2+}-channel blocker that reduces blood pressure by relaxing vascular smooth muscle.

C. Pressure–volume relationships

The PV loop is constrained by two curves, the **end-diastolic** and **end-systolic pressure–volume relationships** (**EDPVR** and **ESPVR**, respectively), that define how a ventricle performs when presented with any given preload, afterload, or contractile state (see Fig. 17.6).

1. **End-diastolic pressure–volume relationship:** The EDPVR describes passive pressure development during ventricular filling. During early diastole, the ventricle swells with blood relatively easily. The sharp upturn in the EDPVR between 150 and 200 mL reflects the ventricle reaching capacity. Any further volume increases require that significant pressure be applied to stretch the myocardium.

2. **End-systolic pressure–volume relationship:** The ESPVR defines the maximum pressure the LV develops at any given filling volume (LVP$_{max}$). LVP$_{max}$ is an experimental value, determined by initiating a contraction after the aorta has been clamped to prevent outflow during pressure development (Fig. 17.7). The ESPVR demonstrates one of the key properties of the myocardium, namely, that increasing cardiac filling volume increases contractile force and pressure development.

D. Preload

LV preload is determined by EDP, but commonly used surrogates include **right atrial pressure** and **central venous pressure**.

1. **Effect on the sarcomere:** In the absence of preload (i.e., an empty heart), sarcomeric length is minimal (~1.8 μm), and the possibility for further shortening and force development is very limited (Fig. 17.8). Preloading (filling) the LV stretches its walls, pulling the actin and myosin filaments farther apart. In so doing, stretching optimizes the potential for crossbridge formation and sharply increases the amount of tension that can be developed

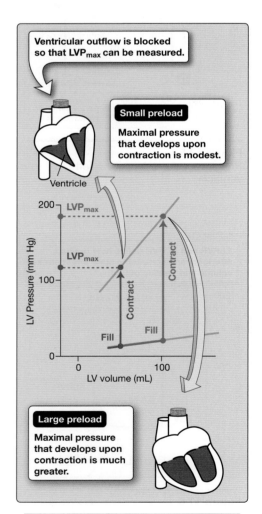

Figure 17.7.
Preload effects on maximal left-ventricular pressure (LVP$_{max}$) development. LV = left ventricle.

[1]For a discussion of β-blockers and Ca^{2+}-channel blockers, see *LIR Pharmacology*, 8e, Chapter 12·IV and V.

during contraction. Strong elastic elements within the myocardium (elastin and collagen) resist lengthening beyond an optimum of ~2.2 μm and account for the abrupt upturn in the whole-heart EDPVR discussed earlier.

> The relationship between preload and force has traditionally been explained in terms of optimizing overlap between thick and thin filaments. However, preloading also causes **length-dependent activation** of the contractile apparatus. Activation sensitizes the crossbridge cycle to Ca^{2+}, allowing for increased force generation. The velocity of contraction increases also. The molecular mechanisms of length-dependent activation remain uncertain.

2. **Effect of preloading on a pressure–volume loop:** Increased ventricular preloading increases LVEDV (A moves to B in Fig. 17.9). Preloading has no direct effect on arterial pressure, so the point at which the aortic valve opens is unchanged (see Fig. 17.9, point C). Myocytes contract down to the same absolute length regardless of preload, so ESV remains unchanged (see Fig. 17.9, point D). Both SV and EF have increased, however, because preloading causes length-dependent sarcomeric activation.

3. **Preload in practice:** The relationship between sarcomeric length and tension shown in Figure 17.8 provides the heart with a near-perfect way of matching contractile force to the volume of blood contained within its chambers. Thus, if the amount of blood returning from the vasculature between beats is high, the chamber walls and sarcomeres are stretched to a greater extent. Stretching increases the amount of force that the muscle is able to generate on the next beat, but this force is actually *needed* to expel the additional blood volume on the next beat. This phenomenon is known as the **Starling law of the heart** or the **Frank-Starling relationship**. In practice, changes in preload are used for both short- and long-term adjustments in CO. Short-term changes involve the SNS, which constricts veins to force their contents forward toward the ventricle. Long-term management of CO is accomplished through sustained increases in circulating blood volume and preload via renal fluid retention (see Chapter 19·IV).

E. Afterload

Afterload is the force against which the ventricle must work in order to eject blood into the arterial system. Under normal circumstances, afterload equates with **mean arterial pressure (MAP)**.

1. **Effects on the sarcomere:** The effects of increasing afterload on sarcomeric function are easiest to demonstrate using a single muscle fiber and a series of weights or loads (Fig. 17.10). The lightest weight will be lifted with little difficulty and at maximum velocity. As the load increases, the rate of contraction slows. At the heavier end of the scale, shortening is both very slow, and the height to which the weight is lifted is reduced.

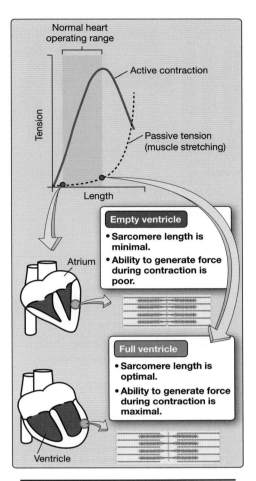

Figure 17.8.
Preloading effects on the ventricular myocardium.

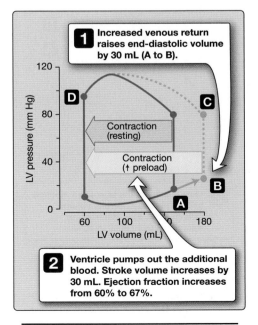

Figure 17.9.
Effects of increasing preload on the pressure–volume loop. LV = left ventricle.

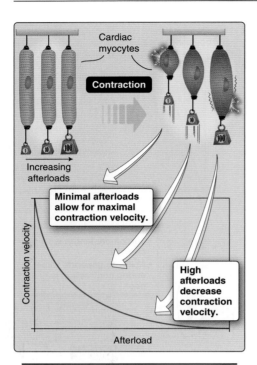

Figure 17.10.
Afterload effects on contraction velocity in cardiac myocytes.

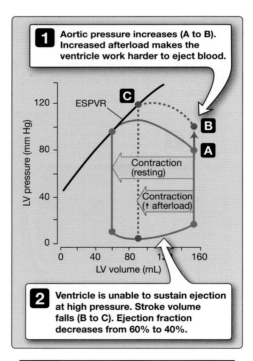

Figure 17.11.
Effects of increasing afterload on the pressure–volume loop. ESPVR = end-systolic pressure–volume relationship; LV = left ventricle.

2. **Effect of afterload on a pressure–volume loop:** Increasing arterial pressure has no direct effect on preload, so EDV remains unchanged (Fig. 17.11). Increasing MAP does increase the amount of pressure that the LV must develop in order to force the aortic valve open, however, as indicated by the upward movement of point A to point B in Figure 17.11. It takes time to develop additional pressure, so isovolumic contraction is prolonged (not apparent from the loop because there is no time index). Ejection is truncated (see Fig. 17.11, points B to C) because the extent to which the myocytes can shorten at such high pressures is limited, as defined by the ESPVR. In simpler terms, myocytes have a limited amount of Ca^{2+} and crossbridge cycles available for developing force during each contraction. If they use more cycles in developing the pressures necessary to force the aortic valve open against a higher afterload, less are subsequently available to sustain ejection. The net result is premature aortic valve closure (see Fig. 17.11, point C). Therefore, even though EDV is unchanged, SV and EF both decrease.

3. **Afterload in practice:** Arterial pressure is changing constantly, but the cardiovascular system can compensate by parallel changes in myocardial contractility. In the short term, this involves the SNS and modulation of Ca^{2+} release, which increases crossbridge cycling (discussed in the next section). Chronic changes in afterload promote myocardial remodeling and hypertrophy.

F. Inotropy

Inotropy refers to the ability of a muscle cell to develop force and is synonymous with contractility. The ventricular myocardium, unlike skeletal muscle, is unusual in that its contractile state can be varied as a way of altering cardiac performance. Agents with a positive inotropic effect (e.g., epinephrine, norepinephrine, digoxin) cause the myocardium to contract faster, develop higher peak systolic pressures, and then relax faster. Negative inotropes (e.g., β-blockers and Ca^{2+}-channel blockers) have the opposite effect.

> A failing heart can be aided using cardiac glycosides such as digitalis.[1] These drugs inhibit the Na^+/K^+-ATPase and raise intracellular Na^+ concentration. This, in turn, reduces the driving force for Ca^{2+} efflux by the Na^+/Ca^{2+} exchanger, making more Ca^{2+} available for the next contraction. Contractility is increased as a result.

1. **Effects on the sarcomere:** Contractility equates with intracellular free calcium concentration. Ca^{2+} binds to and activates troponin to expose myosin-binding sites on actin (see Section III·C). Myosin then binds to and pulls on the actin filament at the expense of one

[1]For a discussion of the actions and side effects of digitalis glycosides, see *LIR Pharmacology*, 8e, Chapter 10·XI·A.

ATP molecule. Thus, a single free calcium ion equates with a single unit of contractile force. Making more Ca^{2+} available increases inotropy, whereas decreasing Ca^{2+} availability decreases cardiac performance and output. Ca^{2+} availability and contractility are regulated by the SNS.

2. **Regulation:** The SNS typically activates when changes in tissue demand for O_2 and nutrients increases the need for CO. CO is increased by raising HR and by making the heart a more effective pump (i.e., increased contractility). SNS postganglionic terminals release norepinephrine at the cardiac neuromuscular junction, whereas preganglionic terminals stimulate epinephrine release from the adrenal medulla. Epinephrine travels via the circulation to the heart. Neurotransmitter and hormone both bind to β_1-adrenergic receptors on the cardiac sarcolemma, which are G protein–coupled receptors. Receptor occupancy activates the cAMP signaling pathway and protein kinase A (PKA). PKA has three principal targets: **L-type Ca^{2+} channels**, **CICR channels** in the SR, and a **Ca^{2+} pump** that refills the SR Ca^{2+} stores (Fig. 17.12).

   a. **Calcium channels:** L-type Ca^{2+} channel phosphorylation by PKA increases channel open probability and thereby increases the size of the Ca^{2+} trigger flux.

   b. **Calcium-release channels:** The Ca^{2+} trigger flux opens CICRs in the SR, allowing Ca^{2+} to flood out of the stores and into the sarcoplasm. Increasing the size of the trigger flux increases Ca^{2+} release from the stores. PKA further magnifies this effect by sensitizing the release channels to a rise in sarcoplasmic Ca^{2+} concentration.

   c. **Calcium pump:** During diastole, Ca^{2+} is returned to the SR by SERCA (sarco[endo]plasmic reticulum Ca^{2+}-ATPase),

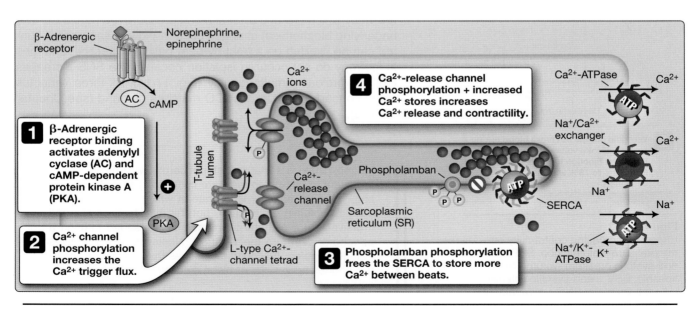

Figure 17.12.
Sympathetic nervous system modulation of contractility.

whereas a Na⁺/Ca²⁺ exchanger in the sarcolemma trans-ports Ca^{2+} out of the cell. SERCA is associated with a small integral SR membrane protein called **phospholamban**, which limits pump speed. Phosphorylating phospholamban inactivates it, which then allows the pump to cycle faster. Two important consequences include faster relaxation times and increased amounts of Ca^{2+} stored for release on the next heartbeat.

  i. **Relaxation:** When the pump runs faster, intracellular Ca^{2+} concentration drops to resting levels (0.1 μmol/L) more quickly. A rapid decrease in sarcoplasmic Ca^{2+} concentra-tion allows the contractile machinery to relax faster, which prolongs the time available for ventricular filling during HR increases (see Chapter 39·V·B).

  ii. **Calcium release:** When the SR pump runs faster, more Ca^{2+} is stored in the SR, and less is returned to the extracel-lular space compared with previously. Stocking the stores with additional Ca^{2+} makes more available for release on the next contraction, and contractile force increases as a result.

3. **Effect of inotropy on a pressure–volume loop:** Increasing Ca^{2+} availability allows the myocardium to contract with greater speed and force at any given filling pressure. The ESPVR shifts upward and to the left (Fig. 17.13). Changes in inotropy have no immediate effect on preload or MAP, so the filling and isovolumic contraction phases of the PV loop are largely unchanged. However, because the myocardium is now working more efficiently, a greater volume of blood is squeezed out of the ventricle compared with previ-ously, and ESV falls (see Fig. 17.13, points A to B). SV and EF are both increased.

4. **Inotropy in practice:** Changes in inotropy are an important mech-anism by which CO and arterial blood pressure are regulated. The underlying biochemical changes occur rapidly, meaning that con-trol can be exerted on a beat-by-beat basis. In practice, changes in inotropy do not occur in isolation because the SNS increases HR and preload simultaneously (Fig. 17.14). All three variables act in concert to ensure that CO matches demand (see Chapter 19·III·D).

Inotropic state is a vital indicator of cardiac well-being, so it is important to be able to measure contractility in a clinical set-ting. The best indicator is the rate at which LVP rises during early isovolumic contraction, but this has to be measured invasively by a catheter-tip manometer threaded through a peripheral vein into the ventricle. Noninvasive alternatives include Doppler ultrasound techniques that estimate myocar-dial-shortening velocity or blood ejection velocity through the aortic valve.

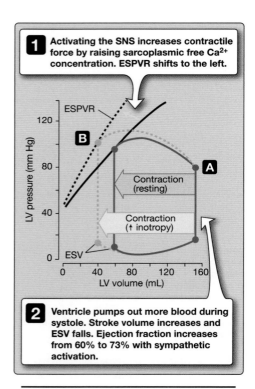

Figure 17.13.
Effects of increasing contractility on the pressure–volume loop. ESPVR = end-systolic pressure–volume rela-tionship; ESV = end-systolic volume; LV = left ventricle; SNS = sympathetic nervous system.

V. ENERGY SOURCE

Cardiac muscle contracts about once a second throughout life, but the contractions are always brief (<1 second). This contrasts with skeletal muscle in which isometric contractions can last for minutes and can be sustained to the point of fatigue, or with smooth muscle in which sphincters may remain tonically constricted for hours. A heart capable of even one prolonged (tetanic) contraction would quickly kill its owner! Thus, cardiac muscle has lost the ability to sustain ATP production and contraction for more than a few seconds. It maintains modest ATP stores that support short contractions and then regenerates these stores using aerobic pathways when relaxed. The limited anaerobic capability creates a high dependence on O_2. If O_2 supply is limited due to reduced arterial blood supply, a creatine phosphate pool can sustain ATP levels for several tens of seconds, and then lactate acid begins to be produced. Prolonged O_2 deprivation (minutes) causes irreversible hypoxic muscle damage and **myocardial infarction**.

VI. CARDIAC WORK

The heart performs work when it moves blood from veins to arteries, so any changes in cardiovascular performance that affect output (i.e., changes in preload, afterload, inotropy, or HR) necessarily affect cardiac workload also. The various determinants of CO are unequal in the way they place demands on cardiac workload.

> A simple way of estimating workload on a heart is using the rate–pressure product, in which HR is multiplied by systolic blood pressure. Although imprecise and contraindicated when there is evidence of aortic stenosis, it suffices in a clinical situation.

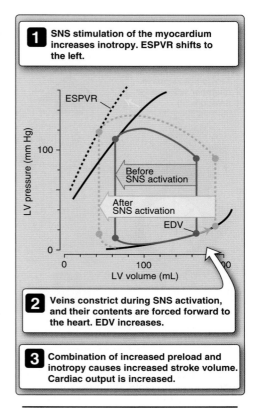

1 SNS stimulation of the myocardium increases inotropy. ESPVR shifts to the left.

2 Veins constrict during SNS activation, and their contents are forced forward to the heart. EDV increases.

3 Combination of increased preload and inotropy causes increased stroke volume. Cardiac output is increased.

Figure 17.14.
Combined effects of sympathetic nervous system activation on the pressure–volume loop. EDV = end-diastolic volume, ESPVR = end-systolic pressure–volume relationship, LV = left ventricle, SNS = sympathetic nervous system.

A. Components

The heart performs two kinds of work: **internal** and **external**.

1. **Internal:** Internal work accounts for >90% of total cardiac workload. Internal work is expended in isovolumic contraction, which generates the force necessary to open the aortic and pulmonary valves. The amount of energy consumed in internal work can be quantified by multiplying the amount of time spent in isovolumic contraction by ventricular wall tension (see following section).

2. **External:** External work, or **pressure–volume work**, is work expended in transferring blood to the arterial system against a resistance. External work accounts for <10% of total cardiac workload, even at maximal levels of output. External work (or **minute work**) can be determined from:

$$\text{Minute work} = \text{MAP} \times \text{CO}$$

External work is represented graphically by the area contained within a PV loop. At rest, ~1% of this work is expended in imparting

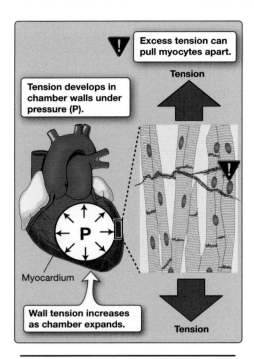

Figure 17.15.
Left-ventricular wall tension caused by increasing intraventricular pressure.

kinetic energy to blood, but the kinetic component can increase to as much as 50% of total at the high levels of output (e.g., during strenuous exercise).

B. Ventricular wall tension

Wall tension is a significant determinant of cardiac workload. Tension is a force that develops within the walls of pressurized chambers (Fig. 17.15) and is counterproductive because it tugs on the ends and sides of the myocytes and contributes to afterload. Wall tension can be quantified using the **law of Laplace**:

$$\sigma = LVP \times \frac{r}{2h}$$

where σ is wall stress, r is ventricular radius, and h is myocardial wall thickness. The Laplace law helps illustrate how differently changes in preload, afterload, and HR impact myocardial performance.

1. **Preload:** The volume of a sphere is proportional to radius cubed $V = 4/3 \times \pi \times r^3$. Therefore, if EDV (preload) were to double, intraventricular radius would rise by ~26%, and wall tension would rise by an equivalent amount. In practice, preloading a heart is a relatively efficient way of increasing CO without substantially increasing cardiac work.

2. **Afterload:** If aortic pressure doubled, LVP would have to rise by a similar amount in order to eject blood. Wall tension would rise by 100%. Changes in afterload stress the myocardium and increase cardiac work to a much greater extent than changes in preload.

3. **Heart rate:** If HR doubles, the amount of time spent in systole and isovolumic contraction doubles also. In effect, doubling HR increases wall tension and cardiac workload by ~100%.

Clinical Application 17.4: Hypertensive Myocardial Hypertrophy

Hypertension (HTN) is a leading risk factor for numerous disorders such as myocardial infarction, heart failure, intracerebral hemorrhage, and chronic kidney disease. Studies suggest that >90% of the population will develop HTN in their fifth decade, the rise in blood pressure (BP) being a physiologic response to an increase in left-ventricular afterload caused by age-related arterial stiffening (see Biologic Sex and Aging 18.1). HTN is defined as a systolic BP of ≥130 mm Hg and a diastolic BP of ≥80 mm Hg. HTN represents increased afterload to the left ventricle (LV), forcing it to generate higher pressures to eject blood into the arterial system. Chronic increases in afterload initiate compensatory pathways that remodel the LV myocardium to increase its contractile strength. New myofibrils are laid down alongside existing ones, causing the LV wall to widen. Increasing ventricular wall thickness decreases lumen capacity and makes the myocardium less compliant and difficult to fill, which increases the likelihood of diastolic failure (see Chapter 40·V·A).

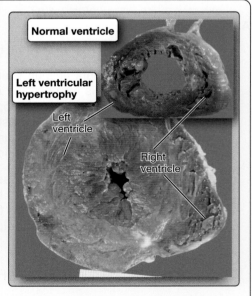

Hypertensive ventricular hypertrophy.

Chapter Summary

- The heart repeatedly contracts (**systole**) to eject blood into the arterial system, and then relaxes and fills (**diastole**).

- The cardiac cycle can be divided into several discrete phases. It begins with **atrial systole**, which pushes blood forward into the ventricle and completes preloading.

- **Ventricular systole** follows. Intraventricular pressure rises rapidly and forces the mitral valve closed (focusing on the left side of the heart, but the right functions similarly). Contraction is initially **isovolumic**, but once luminal pressure exceeds aortic pressure, the aortic valve opens, and **rapid ejection** begins. Aortic pressure, which had been declining during diastole, climbs when blood is forced into the arterial system.

- Rapid ejection gives way to **reduced ejection**. The ventricle then begins to relax, and luminal pressure falls rapidly. **Isovolumic relaxation** begins when the aortic valve is forced closed by high aortic pressures.

- When ventricular pressure drops below atrial pressure, the mitral valve reopens, and **rapid passive ventricular filling** is aided by the influx of blood that has dammed up against the mitral valve during diastole.

- Venous pressure records exhibit an **a wave** during atrial contraction, a **c wave** during ventricular contraction, and a **v wave** caused by venous blood damming within the atria during diastole.

- There are four **heart sounds**: S_1 correlates with atrioventricular valve closure, S_2 is associated with semilunar valve closure, S_3 is a sound of ventricular filling commonly heard in children and in adults with a dilated ventricle, and S_4 is a pathologic sound associated with contraction of a hypertrophied atrium against a stiffened ventricle.

- **Cardiac output** is the product of **stroke volume** and **heart rate**. Stroke volume is determined by ventricular **preload**, **afterload**, and **contractility (inotropy)**. All of these parameters are influenced by the autonomic nervous system (ANS).

- **Preload** is determined by end-diastolic pressure and volume. Preload increases stroke volume through length-dependent activation of the sarcomere (**Frank-Starling mechanism**). **Afterload** is the force that must be overcome for a ventricle to eject blood, and this usually equates with **arterial pressure**. Increases in afterload decrease stroke volume.

- **Inotropy** (contractility) is directly correlated with **sarcoplasmic Ca^{2+} concentration**. Positive inotropes, such as epinephrine and norepinephrine, increase cardiac contractility by increasing both the rate of pressure development and peak systolic pressure.

- Force generation is regulated by the ANS via **β-adrenergic receptors** and cAMP-dependent phosphorylation by **protein kinase A**. Kinase targets include the **L-type Ca^{2+} channel** in the sarcolemma and **SERCA**.

- Cardiac muscle relies on aerobic pathways to supply the ATP needed for contraction. The limited anaerobic capability makes heart muscle vulnerable to decreasing O_2 availability during interruptions in blood supply.

- A heart must perform both internal and external work. Moving blood from the ventricle to the arterial system is termed **external** or **pressure–volume work**.

- Most of the energy used by a heart during the cardiac cycle is consumed during isovolumic contraction (**internal work**). A significant portion of this work is used in overcoming wall tension.

Study Questions

Choose the ONE best answer.

17.1. A 7-year-old male of normal height and build for his age is undergoing a routine physical. The family physician notes an S_3 during auscultation. Which of the following statements most likely describes the cause of the S_3 in this patient?

   A. An ECG would show right-axis deviation.
   B. It coincides with rapid ventricular ejection.
   C. It indicates atrial hypertrophy.
   D. It is caused by aortic valve regurgitation.
   E. It is the sound of ventricular filling.

Best answer = E. The third heart sound (S_3) occurs during ventricular filling and is caused by sudden tensing and reverberation of the ventricular walls (see Section II·E). Although usually a sign of underlying pathology in adults, it is a normal finding in children. Right-axis deviation (A) does not necessarily correlate with a heart sound. Ventricular filling and S_3 occur during diastole, not rapid ejection (B). Atrial hypertrophy (C) would yield an S_4. Regurgitant valves (D) produce murmurs, not heart sounds (see Clinical Application 17.1).

17.2. A 44-year-old female is diagnosed with dilated cardio-myopathy, a condition caused by impaired ventricular contractility and compensatory fluid retention. What is the most likely advantage to fluid retention and preloading?

   A. It decreases ventricular afterload.
   B. It increases ventricular stroke volume.
   C. It increases ventricular wall tension.
   D. It reduces cardiac workload.
   E. It reduces the need for resting cardiac output.

Best answer = B. Increased preloading stretches the myocardium, thereby increasing the amount of force developed upon contraction through length-dependent activation of the sarcomere (see Section IV·D). Preloading increases stroke volume and ejection fraction, which helps compensate for reduced ventricular contractility in this patient. The disadvantage to enhanced preloading is that it increases rather than decreases ventricular afterload (A) by increasing ventricular lumen radius and wall tension (C). An increase in wall tension increases overall workload, according to the law of Laplace (D; see Section VI·B). Resting cardiac output (E) is determined by the metabolic needs of the tissues, not by preload.

17.3. A patient is given an α_1-adrenergic agonist during surgery. The patient has catheters connected in series with pressure transducers in the radial artery and vena cava. Echocardiography is conducted during bolus administration of the drug. Which of the following would most likely be increased during the first heartbeat following drug administration?

   A. Ejection fraction
   B. End-diastolic pressure–volume relationship
   C. Left-ventricular end-systolic volume
   D. Left-ventricular preload
   E. Ventricular shortening velocity

Best answer = C. An α_1-adrenergic agonist would increase systemic vascular resistance (SVR) and left-ventricular (LV) afterload, increasing the pressure needed to open the aortic valve during systole (see Section IV·E). Stroke volume (SV) and ejection fraction (A) both decrease, thereby raising LV end-systolic volume. The end-diastolic pressure–volume relationship (B) is not altered by acute SVR changes. An SVR increase lowers central venous pressure and LV preload (D) at any given cardiac output (see Section V·D). Cardiac inotropy and LV shortening velocity (E) will likely rise soon after the SVR change to maintain SV, but this will take several beats to establish (see Section V·E).

17.4. Cardiac muscle contraction is dependent on a rise in sarcoplasmic Ca^{2+} concentration. The bulk of the Ca^{2+} required for full force generation most likely flows through which of the following Ca^{2+}-channel types?

   A. Dihydropyridine receptors
   B. IP_3-gated channels
   C. Ryanodine receptors
   D. Stretch-activated channels
   E. Transient receptor-potential channels

Best answer = C. Full force development by a cardiac myo-cyte relies on Ca^{2+} release from stores in the sarcoplasmic reticulum ([SR] see Section III·A). Release is mediated by Ca^{2+}-induced Ca^{2+} release (CICR) channels, also known as ryanodine receptors. Dihydropyridine receptors (A) are L-type Ca^{2+} channels that mediate voltage-gated Ca^{2+} fluxes across the T-tubule membrane. Ca^{2+} influx via this pathway acts as a trigger for CICR. IP_3 (B) mediates Ca^{2+} release from the SR in smooth muscle. Stretch-activated channels (D) are also widespread, but ryanodine receptors are the principal pathway for Ca^{2+} fluxes during contraction. Transient receptor-potential channels (E) are found in many tissues, often mediating cellular sensory stimulus transduction (see Chapter 2·VI·D).

Blood and the Vasculature

18

I. OVERVIEW

The functions and design of the cardiovascular system are, in many ways, similar to that of a water utility in a modern city. A water utility is tasked with distributing clean water to its many consumers. The distribution network is vast, and pumping stations are required to ensure that water arrives at sufficiently high pressure for adequate flow from faucets and showerheads (Fig. 18.1). Waste water is collected and returned to treatment plants under low pressure by an elaborate system of drains. The cardiovascular system similarly distributes blood at high pressure to ensure adequate flow to many consumers (cells). Waste (venous) blood travels back to the heart at low pressure for "treatment" by the lungs. Water utilities distribute water, a Newtonian fluid whose flow characteristics behave predictably under pressure. The cardiovascular system circulates blood, a viscous non-Newtonian fluid comprising water, solutes, proteins, and cells. Considerable pressure must be applied to blood to make it flow through the vasculature at rates sufficient to meet the needs of the tissues. The capillaries used to deliver blood to individual cells are extremely leaky, unlike the copper pipe used in household water-distribution systems. Leakiness means that the pressure used to drive flow through the system also drives fluid out of the vasculature and into the intercellular spaces. Lastly, the pipework used to distribute and collect blood from cells is composed of biologic tissue that stretches and causes the vessels to distend when pressure is applied. Distensibility poses a threat to system function because there is always the potential that the entire vascular contents might become trapped in the pipes, thereby allowing the vascular faucets to run dry.

II. VASCULATURE

The systemic vasculature comprises a vast network of blood vessels that channel O_2-rich blood to within a few microns of every cell in the body. Here, O_2 and nutrients are exchanged for CO_2 and other metabolic waste products, and then blood is returned to the heart for reoxygenation by the lungs and redistribution to the tissues. Figure 18.2 provides an overview of the systemic vasculature and its various components.

A. Organization

The human body contains ~100,000,000,000,000 cells, every one of which must be supplied with blood. Creating a vascular distribution

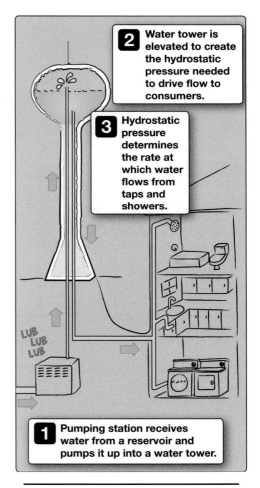

2 Water tower is elevated to create the hydrostatic pressure needed to drive flow to consumers.

3 Hydrostatic pressure determines the rate at which water flows from taps and showers.

1 Pumping station receives water from a reservoir and pumps it up into a water tower.

Figure 18.1.
Hydrostatic pressure drives flow through plumbing systems.

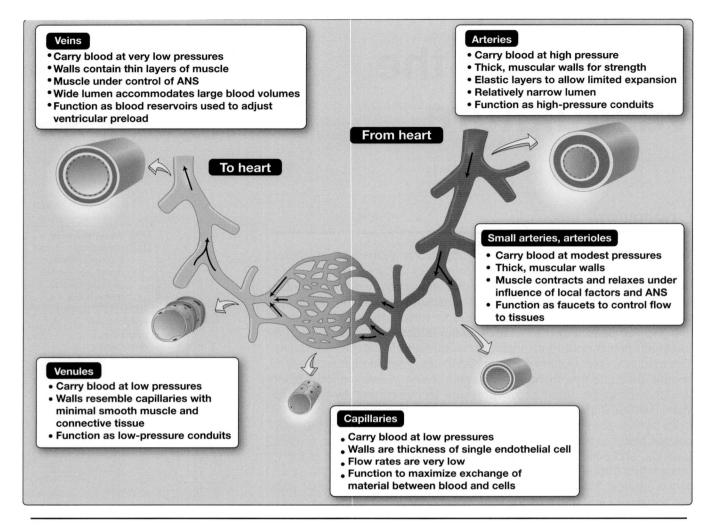

Figure 18.2.
Properties and functions of the vessels comprising the systemic vasculature.

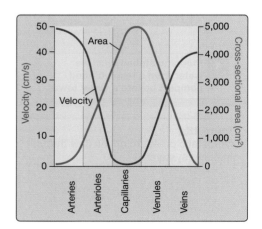

Figure 18.3.
Blood velocity and vessel cross-sectional area across the systemic vasculature.

network capable of such a task requires extensive branching of the vascular tree. Thus, blood leaves the left ventricle (LV) via a single large-diameter vessel (the aorta), which then branches repeatedly to yield ~10,000,000,000 tiny capillaries. The branching pattern greatly increases vascular cross-sectional area, from ~4 cm^2 (aorta) to ~4,000 cm^2 total at the level of the capillaries (Fig. 18.3). Blood flow velocity drops proportionally. Blood exits the LV at up to ~50 cm/s, but velocity has dropped to <1 mm/s by the time blood reaches the capillaries. A low flow rate greatly increases the time available for exchanging materials between blood and tissues during blood's passage through capillary beds. Note that whereas arteries and veins are arranged in series with each other, capillaries are organized in parallel circuits (see Fig. 18.2). This vascular arrangement has important physiologic consequences, as discussed in Section IV. Blood travels back to the heart via venules, which join and fuse to form veins. Smaller veins merge to form larger veins, each fusion decreasing the total cross-sectional area of the system. Blood velocity increases proportionally.

B. Anatomy

Blood vessels all have a common structure, although vessel wall thickness and composition vary with vessel function and location within the vasculature. The lining of all blood vessels consists of a single layer of **endothelial cells** (the **tunica intima**) as shown in Figure 18.4. Arteries and veins also contain layers of **vascular smooth muscle cells** ([**VSMCs**] the **tunica media**) that modify vessel diameter when they contract or relax. Extensive networks of cross-linked **elastic fibers** give all vessels except capillaries and venules the ability to stretch like a rubber hose when blood pressure is raised. Elastic fibers have a central core of coiled **elastin** and an outer covering of **microfibrils** composed of **glycoproteins**. Blood vessels also contain **collagen fibers** that resist stretching and limit vessel expansion when internal pressures rise. A thin outer layer of connective tissue (the **tunica externa** or **tunica adventitia**) maintains vascular integrity and shape.

C. Vessels

A blood vessel's primary function is to provide a conduit for blood flow to and from cells. The different vessel classes (i.e., arteries, capillaries, veins) have additional important functions reflecting their location within the vasculature. Arteries are high-pressure conduits, arterioles are flow regulators, capillaries facilitate exchange of materials between blood and tissues, and venules and veins have a reservoir function.

1. **Large arteries:** The arterial system comprises a network of narrow-bore distribution vessels (see Fig. 18.2). Arteries must carry blood at high pressure (Fig. 18.5), so their walls are thick and their lumens narrow, which limits arterial system capacity. The walls of the larger arteries (also called **elastic arteries**) contain smooth muscle layers and are rich in elastin fibers. The muscle layers have a resting tone, which limits arterial distensibility and helps maintain the pressure of the blood within.

2. **Small arteries and arterioles:** The walls of the smallest arteries and arterioles are dominated by their smooth muscle layers. Collectively known as **resistance vessels**, they act as faucets or stopcocks to control blood flow to capillaries (Fig. 18.6). When a tissue's demand for O_2 and nutrients is high, the VSMCs relax, and flow to the tissues increases. Decreased demand for blood or intervention by the central nervous system constricts the muscular "faucets," and flow to the tissues is reduced (see Chapter 19·II).

3. **Capillaries:** Capillaries bring blood to within 30 μm of virtually every cell in the body. They are designed to keep the blood contained within the vasculature while simultaneously maximizing the opportunity for exchange of materials between blood, interstitium, and tissues. Their walls are the thickness of a single endothelial cell plus the **basal lamina**. Depending on tissue type, capillaries permit direct communication between blood and cells via junctional **clefts** between adjacent cells, **fenestrations** (50–80-nm transmural pores common to the intestine, renal glomerulus, and exocrine organs; Fig. 18.7), or even large gaps in the capillary wall (e.g., liver sinusoids).

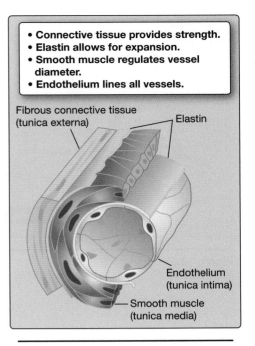

- Connective tissue provides strength.
- Elastin allows for expansion.
- Smooth muscle regulates vessel diameter.
- Endothelium lines all vessels.

Fibrous connective tissue (tunica externa)
Elastin
Endothelium (tunica intima)
Smooth muscle (tunica media)

Figure 18.4.
Blood vessel structure.

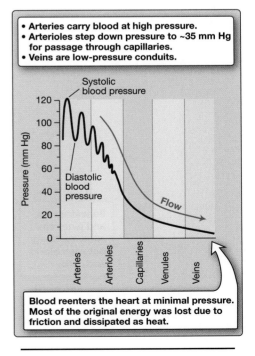

- Arteries carry blood at high pressure.
- Arterioles step down pressure to ~35 mm Hg for passage through capillaries.
- Veins are low-pressure conduits.

Systolic blood pressure
Pressure (mm Hg)
120
100
80
60
40
20
0
Diastolic blood pressure
Flow
Arteries
Arterioles
Capillaries
Venules
Veins

Blood reenters the heart at minimal pressure. Most of the original energy was lost due to friction and dissipated as heat.

Figure 18.5.
Perfusion pressures across the systemic vasculature.

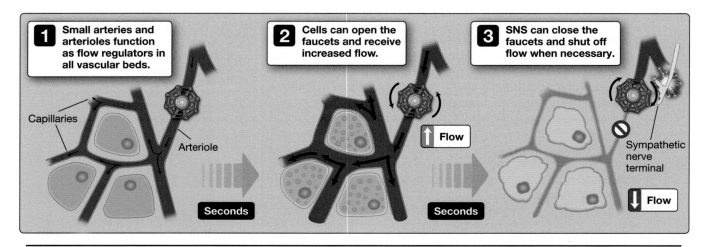

Figure 18.6.
Vascular faucets regulate blood flow. SNS = sympathetic nervous system.

> Although capillaries average only 1 mm in length and 8–10 μm in diameter, they provide a total surface area for exchange of 500–700 m² in an average adult.

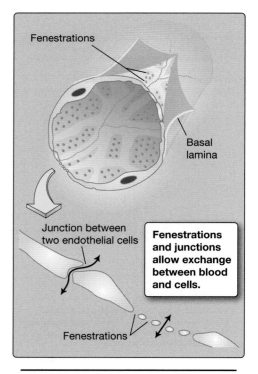

Figure 18.7.
Capillary wall structure.

4. **Venules and veins:** Venules and veins are low-pressure conduits that direct blood back to the heart. Smaller venules are almost indistinguishable from capillaries, which allows them to participate in fluid and metabolite exchange. Venules widen and fuse with each other as they progress toward the heart. Larger venules contain VSMCs within their walls but far fewer than those seen in vessels of equivalent size in the arterial system. The paucity of muscle means that vein walls are thin (see Fig. 18.2), making them highly distensible and able to accommodate large volumes of blood. Under resting conditions, ~65% of total blood volume resides in the venous compartment, creating a reservoir used to boost ventricular preload and cardiac output (CO) when the need arises (see Chapter 19·V). The larger veins contain valves that help maintain unidirectional flow through the system and counter the tendency of blood to be pulled downward under the influence of gravity.

III. BLOOD FLOW DETERMINANTS

Blood will not flow through the vasculature unless it is forced to do so by application of pressure, which is required to overcome **resistance** to flow. Understanding the origins of this resistance is important clinically because it is a primary indicator of cardiovascular status and health. For example, whereas resistance usually goes up during circulatory shock, septic shock is associated with a profound decrease in vascular resistance (see Chapter 40·III·C). The source of resistance is considered in the **Poiseuille law**:

$$Q = \frac{\Delta P \times \pi r^4}{8L\eta}$$

where Q denotes flow, ΔP is the pressure gradient across the ends of the vessel, r is internal radius, L is vessel length, and η is the viscosity of blood. Whereas pressure drives flow, vessel radius, length, and blood viscosity all contribute to flow resistance.

A. Vessel radius

Vessel radius is the primary determinant of vascular resistance. Radius is also a variable because the VSMCs that make up the walls of the small arteries and arterioles contract and relax as a way of controlling flow. Because flow is proportional to r^4, a twofold change in radius causes a 16-fold change in flow. The potency of radius's effect on flow relates to a layer of plasma that clings to the inner surface of all vessels. The layer forms through interactions between blood and the vascular endothelium and, in so doing, impedes flow. Although the depth of the coating layer is essentially the same in all vessels irrespective of bore size, its contribution to total cross-sectional area is much greater in a small-diameter vessel than in a large one; therefore, resistance to flow through a small vessel is correspondingly larger (Fig. 18.8).

B. Vessel length

Blood flow through a vessel is inversely related to vessel length, again reflecting blood's tendency to interact with the vascular endothelium. Changes in vessel length during development and aging have only theoretical considerations and are not addressed further here.

C. Blood viscosity

Blood is a complex fluid whose viscosity varies with flow. A fluid's viscosity is measured relative to water. Adding electrolytes and organic molecules (including proteins) to water raises viscosity from 1.0 to ~1.4 cp. Cells, principally red blood cells (RBCs), have the greatest impact, with viscosity rising at a greater-than-exponential rate with hematocrit ([Hct] Fig. 18.9).

1. **Hematocrit:** Hct measures the percentage of whole blood volume occupied by RBCs. Hct is determined clinically by centrifuging a tube containing a small blood sample to separate cells from plasma. Hct can then be estimated from the height of the packed RBC layer within the tube, which is dependent on both RBC number and volume. Normal values for Hct range between 41% to 53% for males and 36% to 46% for females.

> Hct that falls below a normal range indicates **anemia**. Anemia is more usually defined in terms of hemoglobin (Hb) levels, however. Normal Hb values for males range from 13.5 to 17.5 g/dL, and from 12.0 to 16.0 g/dL in females.

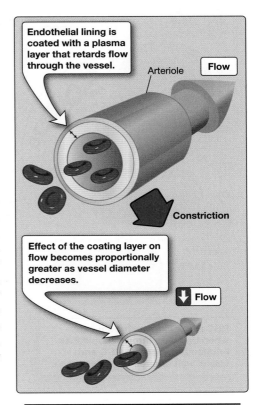

Figure 18.8.
Effects of constriction on flow through a resistance vessel.

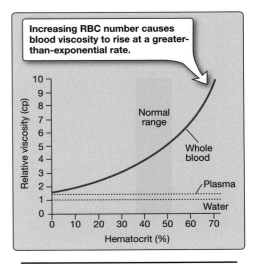

Figure 18.9.
Relationship between hematocrit and blood viscosity. cp = centipoise.

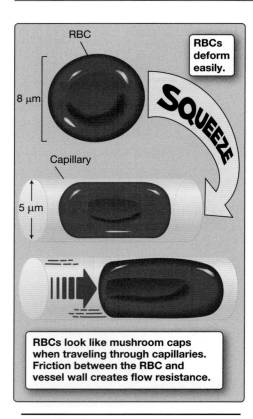

RBC

8 μm

RBCs deform easily.

SQUEEZE

Capillary

5 μm

RBCs look like mushroom caps when traveling through capillaries. Friction between the RBC and vessel wall creates flow resistance.

Figure 18.10.
Red blood cell flexibility.

2. **Flow resistance:** RBCs increase flow resistance by rubbing up against the vessel wall. RBCs travel through capillaries measuring only 2.8 μm in diameter, which is surprising given that these cells are typically pictured as 8-μm diameter disks (Fig. 18.10). RBCs readily distort, however, which allows them to slip through narrow vessels. They also travel in the center of the vessel, which minimizes interactions with the endothelium. Even so, interactions between RBCs and the vessel wall create a resistance to flow that the heart perceives as afterload and requires work to overcome.

Cells can be likened to plastic bags filled with water. Like bags, most cells rupture when deformed mechanically. RBCs more closely resemble plastic bags that are only partially filled with water. They deform easily, which allows them to squeeze through narrow vessels and pores. RBCs swell and deform less readily as they age, which allows the body to target them for destruction. Disposal occurs in the spleen, where RBCs are forced through a filter made of connective tissue fibrils (**splenic cords**). Young, flexible RBCs pass through the filter with relative ease, but older, less distensible RBCs become trapped and are then phagocytosed.

3. **Anemia:** Anemia is associated with fewer RBCs. Blood viscosity and flow resistance decreases also. **Physiologic anemias** occur when blood volume is expanded faster than RBC production, such as during pregnancy (see Chapter 37·IV·C) or exercise training (see Chapter 39·IX·B).

4. **Polycythemia:** Increasing RBC numbers increases blood viscosity and resistance to flow. People living at high altitude demonstrate a physiologic **polycythemia** stimulated by reduced atmospheric O_2 levels (see Chapter 24·V·A). Although increased RBC production helps compensate for reduced O_2 availability, the tradeoff is increased cardiac work, limiting the altitude at which humans can comfortably exist to around 5,000 m.

IV. HEMODYNAMIC OHM LAW

Vascular resistance represents afterload to the LV and determines how hard it must work to generate output. If resistance increases, the heart is forced to work harder to compensate. A healthy heart is well equipped to meet the demands placed on it by changes in vascular resistance under normal conditions, but increased resistance can seriously stress a diseased heart. For this and other reasons, it is important to be able to quantify vascular resistance in a clinical setting. Identifying and summing all the individual components that make up vascular resistance in a typical human is not feasible. As an alternative, vascular resistance can be estimated with relative ease

Clinical Application 18.1: Polycythemia Vera

Polycythemia vera is a neoplastic disorder affecting myeloid RBC precursors. The disease causes uncontrolled RBC production, and hematocrit (Hct) climbs accordingly. Polycythemia is defined by a Hct of >48% in females and >52% in males.

Once Hct reaches around 60%, the RBCs are so closely packed that they collide with each other and begin forming aggregates and clots. Cohesion is dependent on fibrinogen and other large plasma proteins that coat the RBC surface. Viscosity and vascular resistance increase to such a degree that the left ventricle is unable to generate sufficient pressure to maintain even basal flow rates. Patients typically present with headaches, weakness, and dizziness associated with decreased cerebral perfusion. Relentless pruritus (skin itching) is common after taking a warm bath.

Left untreated, the median survival time is 6–18 months. This improves to >10 years if treated with serial phlebotomy, optimally reducing Hct to <42% in females and <45% in males. The main risk factors are thrombotic events (i.e., stroke, deep vein thromboses, myocardial infarction, and occlusion of peripheral arteries).

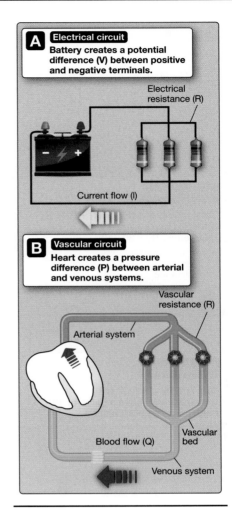

Figure 18.11.
Vascular application of the Ohm law.

using a modified version of the **Ohm law** (Fig. 18.11A). The Ohm law describes the effects of electrical resistance (R) on current flow (I) in a direct current circuit:

$$I = \frac{V}{R}$$

where V is the voltage drop across the resistance. The hemodynamic form of the Ohm law is thus:

$$Q = \frac{P}{R}$$

where Q is blood flow, P is the pressure gradient across a vascular circuit, and R is vascular resistance (see Fig. 18.11B). As discussed earlier, R is defined by vessel radius, length, and blood viscosity ($R = 8L\eta \div \pi r^4$). The hemodynamic form of the Ohm law makes it possible to calculate R for any vessel or vascular circuit, regardless of its size, from measurements of pressure and flow.

A. Systemic vascular resistance

The largest circulation in the body and the one with the greatest resistance is the systemic circulation. The value of **systemic vascular resistance** ([**SVR**] also known as **peripheral vascular resistance** or **total peripheral resistance**) is calculated as:

$$SVR = \frac{MAP - CVP}{CO}$$

where MAP − CVP represents the pressure difference between the aorta (**mean arterial pressure [MAP]**) and vena cavae (**central venous pressure [CVP]**). MAP is a time-averaged value that

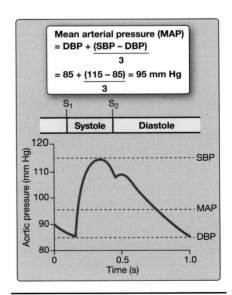

Figure 18.12.
Derivation of mean arterial pressure. DBP = diastolic blood pressure; SBP = systolic blood pressure.

recognizes that arterial pressure rises and falls in step with the cardiac cycle (Fig. 18.12). MAP is calculated as:

$$MAP = DBP + \frac{(SBP - DBP)}{3}$$

where **SBP** = **systolic blood pressure** and **DBP** = **diastolic blood pressure** (SBP − DBP is also known as **pulse pressure**). Using typical normal values for MAP (95 mm Hg), CVP (5 mm Hg), and CO (6 L/min), SVR is calculated to be 15 mm Hg·min·L^{-1}.

> SVR typically varies between 11 and 15 mm Hg·min·L^{-1} in an average person. SVR may also be expressed clinically in units of dyn·s·cm^{-5}, calculated by multiplying the values above by 80. Thus, a normal SVR range is between ~900 and 1,200 dyn·s·cm^{-5}.

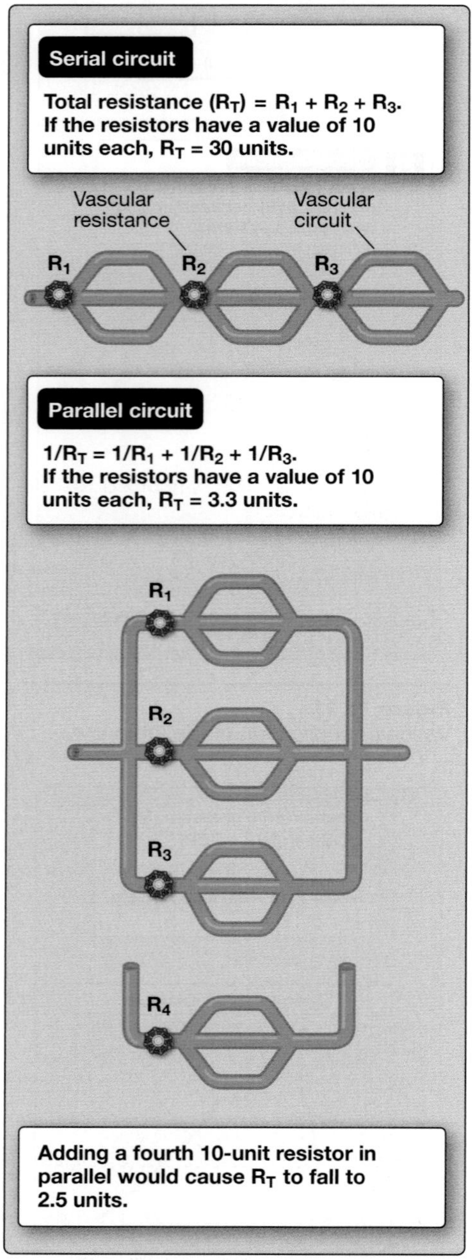

Pulmonary vascular resistance (**PVR**) can be calculated in a similar manner (by using mean pulmonary artery and left atrial pressures), amounting to ~2 to 3 mm Hg·min·L^{-1} (150–250 dyn·s·cm^{-5}) in an average person.

B. Serial and parallel circuits

Hemodynamic circuits are treated in the same way as electrical circuits when calculating the combined resistance of multiple individual components (Fig. 18.13). The total resistance (R$_T$) of a circuit containing three resistors (R$_1$ – R$_3$) arranged in series is equal to the sum of the individual components. If each of the resistors below has a resistance of 10 units, R$_T$ = 30 units:

$$R_T = R_1 + R_2 + R_3 = 10 + 10 + 10 = 30$$

Calculating the total resistance of the same three resistors arranged in parallel requires that the reciprocal of each component be summed (see Fig. 18.13):

$$\frac{1}{R_T} = \frac{1}{R_1} + \frac{1}{R_2} + \frac{1}{R_3} = \frac{1}{10} + \frac{1}{10} + \frac{1}{10} = 0.3$$

Note that at 3.3 units, R$_T$ of the parallel circuit is significantly less than that of any individual component. Thus, even though the systemic circulation contains approximately 10^{10} capillaries that individually have a very high resistance to flow, their parallel arrangement means that their *combined* resistance is relatively low.

> Adding capillaries to a vascular circuit causes SVR to decrease, not increase, because they provide additional pathways for blood flow (see Fig. 18.13).

Serial circuit

Total resistance (R$_T$) = R$_1$ + R$_2$ + R$_3$. If the resistors have a value of 10 units each, R$_T$ = 30 units.

Vascular resistance   Vascular circuit

R$_1$ R$_2$ R$_3$

Parallel circuit

1/R$_T$ = 1/R$_1$ + 1/R$_2$ + 1/R$_3$. If the resistors have a value of 10 units each, R$_T$ = 3.3 units.

R$_1$

R$_2$

R$_3$

R$_4$

Adding a fourth 10-unit resistor in parallel would cause R$_T$ to fall to 2.5 units.

Figure 18.13.
Calculating the resistance of vascular circuits.

V. LIMITS OF THE POISEUILLE LAW

The Poiseuille law helps identify the sources of flow resistance in the cardiovascular system, but system complexity limits its application to smaller arterial vessels and to capillaries. Confounding cardiovascular design features include the predilection for turbulent flow, the fact that blood viscosity is velocity dependent, and the compliance of blood vessels.

A. Turbulence

When blood flows through the vasculature, it experiences drag caused by its various components interacting with the vessel wall. As discussed earlier, the vascular endothelium is coated with a layer of immobilized plasma. This coating exerts drag on blood that is flowing closer to the center of the vessel, creating another slowed layer that exerts its own drag, and so on toward the center of the vessel. Thus, flow through vessels occurs in concentric layers that slip over each other, with the fastest flow at the center and the slowest up against the walls of the vessel. This flow pattern is referred to as **laminar** or **streamline flow** (Fig. 18.14). Laminar flow is observed in most regions of the cardiovascular system, and the Poiseuille law is valid for as long as it is maintained. When streamline flow is disrupted, kinetic energy is squandered on chaotic motion, a pattern known as **turbulence** (Fig. 18.15).

1. **Reynolds equation:** The likelihood of turbulence can be predicted using the **Reynolds equation**:

$$Re = \frac{v \times d \times \rho}{\eta}$$

where Re is Reynolds number, v is mean blood velocity, d is vessel diameter, ρ (rho) is blood density, and η is blood viscosity. Blood density does not change within the parameters of normal human physiology. Many blood vessels constrict and relax, so their internal diameter changes constantly but not to the extent that they cause turbulence *in vivo*. Velocity and viscosity are both physiologically relevant variables, however.

2. **Blood velocity effects:** Turbulence is most likely to be observed within the heart chambers or within the vessels that enter and leave the heart. These are regions where large volumes of blood are moving at high flow velocities. Turbulence occurs once a certain critical velocity is achieved, causing the orderly streamline flow to become chaotic and inefficient.

3. **Equation of continuity:** Congenital and pathologic heart valve defects are common causes of turbulence. The aortic valve is located in a high-pressure, high-velocity region of the cardiovascular system where it is subject to constant wear and tear. It is not uncommon for the valve leaflets to calcify and stiffen with age or perhaps fuse along their commissures as a result of repeated inflammation. Such changes reduce the cross-sectional area of the valve orifice and obstruct outflow. Because CO must be

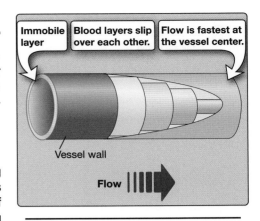

Figure 18.14.
Laminar blood flow.

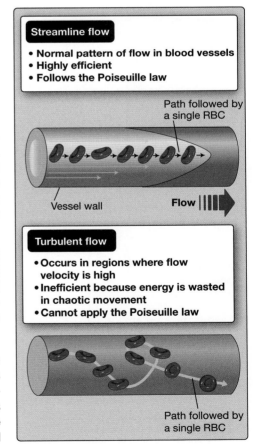

Figure 18.15.
Streamline and turbulent flow.

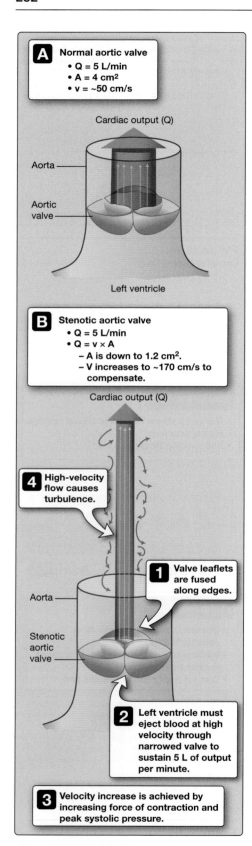

A Normal aortic valve
- Q = 5 L/min
- A = 4 cm²
- v = ~50 cm/s

Cardiac output (Q)

Aorta

Aortic valve

Left ventricle

B Stenotic aortic valve
- Q = 5 L/min
- Q = v × A
  - A is down to 1.2 cm².
  - V increases to ~170 cm/s to compensate.

Cardiac output (Q)

4 High-velocity flow causes turbulence.

1 Valve leaflets are fused along edges.

Aorta

Stenotic aortic valve

2 Left ventricle must eject blood at high velocity through narrowed valve to sustain 5 L of output per minute.

3 Velocity increase is achieved by increasing force of contraction and peak systolic pressure.

Figure 18.16.
Effect of aortic stenosis on ventricular ejection velocity. A = area; Q = flow; v = ejection velocity.

maintained at a basal 5 to 6 L/min regardless of circumstance, LV pressure increases and drives flow at higher velocity through the narrowed outlet (Fig. 18.16). The extent to which velocity is increased by stenosis is defined by the **equation of continuity**:

$$Q = v_{nl} \times A_{nl} = v_s \times A_s$$

where Q is flow, v_{nl} and v_s are flow velocities through normal and stenotic valves, respectively, and A_{nl} and A_s are valve cross-sectional areas. If Q is constant and A_s is reduced, velocity must increase to compensate.

4. **Sounds:** Turbulent flow creates crosscurrents and eddies and causes kinetic energy to be expended when blood impacts the vessel wall. Impacts cause vibrations that travel to the body surface where they can be heard as sounds. Common examples encountered clinically include **murmurs**, **bruits**, and **Korotkoff sounds**.

   a. **Murmurs:** Blood being forced at high velocity through a stenotic aortic or pulmonary valve yields a **systolic murmur**. Valves that fail to close completely also produce murmurs. Such murmurs are caused by blood being forced backward through an incompetent valve and impacting blood contained within the atria or ventricles (see Clinical Application 17.1).

   b. **Bruits:** The term "bruit" usually refers to a pathologic sound associated with turbulent arterial blood flow. For example, a carotid bruit typically indicates vessel narrowing due to atherosclerosis.

   c. **Korotkoff sounds:** Turbulence can be induced artificially for diagnostic purposes. Partial occlusion of the brachial artery with a pressure cuff causes murmurs that reflect blood being ejected at high velocity through the compressed area and impacting the column of blood beyond. These murmurs (Korotkoff sounds) can be heard with a stethoscope placed downstream of the cuff. Sounds are first heard when cuff pressure drops just below SBP, allowing small amounts of blood to jet through the occluded artery. The murmurs typically disappear when cuff pressure drops below DBP, and the artery is fully patent. These sounds thereby provide a convenient way of approximating SBP and DBP.

5. **Hematocrit:** Because blood velocity is inversely related to viscosity and Hct, anemia can also increase the likelihood of turbulence. For example, the physiologic anemia that accompanies pregnancy causes **functional murmurs**, sounds associated with ejection of blood at high velocity through a normal valve (see Chapter 37·IV·C).

6. **Occurrence of turbulence *in vivo*:** In an ideal system, turbulence can be expected when Re exceeds 2,000. When Re is below 1,200, laminar flow prevails. The cardiovascular system is less than ideal. Many factors, especially the extensive branching inherent to the vascular tree, lower the threshold for turbulence to around 1,600. Vessel branches disrupt laminar flow and create foci for local eddy currents to form.

Clinical Application 18.2: Thrombi and Anticoagulation Therapy

The combination of blood stasis, hypercoagulability, and endothelial damage (Virchow triad) predisposes patients to **venous thromboembolism**. **Thrombi** are fibrin and platelet aggregates adhered to a vessel wall. Thrombi may be reabsorbed, enlarge, or break loose to form **emboli**. Emboli are carried through the vasculature until they encounter a vessel too small to traverse, where they become lodged, occluding the vessel. In a healthy person, thrombi can form during prolonged periods of immobility such as during long-haul airplane flights. Cramped cabins and hard seats restrict mobility and compress the vasculature that returns blood from the lower extremities. Blood trapped within the deep veins of the legs may form **deep vein thromboses**. Disembarkation restores flow, and emboli may then be carried through the veins to the right side of the heart and become lodged in the pulmonary vasculature (**pulmonary embolism**). Thrombi can be treated with fibrin-specific thrombolytics, such as alteplase, which activate plasminogen and thereby dissolve the clot.

Atrial fibrillation (AF) and heart valve replacement also put patients at risk of thromboembolism. AF prevents orderly contraction and flow through the atria, creating regions of blood stasis within the affected chamber. Prosthetic ("mechanical") heart valves may also allow pockets of stagnation to form behind their leaflets, increasing the incidence of thrombus formation. If this valve is located on the left side of the heart, a freed embolus potentially can enter the cerebral vasculature and cause a stroke. The risk of thromboembolism can be reduced in AF and mechanical valve replacement patients by oral anticoagulants such as warfarin.[1] Warfarin is a vitamin K antagonist that prevents formation of several coagulation factors required for clot formation.

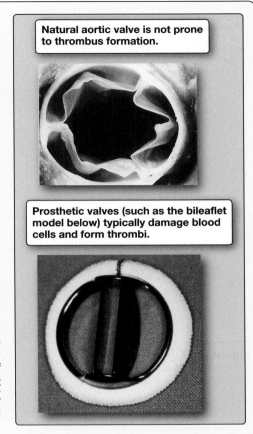

Natural aortic valve is not prone to thrombus formation.

Prosthetic valves (such as the bileaflet model below) typically damage blood cells and form thrombi.

Natural and prosthetic aortic valves.

B. Velocity effects

When blood is stationary or barely moving, RBCs have time to adhere to each other and form aggregates that resemble stacks of coins (**rouleaux**). Aggregates require more effort to move through the circulation than individual cells, thereby increasing resistance to flow. Rouleaux begin to break apart as flow velocity increases, and viscosity falls in parallel (Fig. 18.17).

C. Vessel compliance

The Poiseuille law assumes that blood vessels are rigid tubes. Although smaller vessels (capillaries, arterioles, and small arteries) are relatively nondistensible, most veins and larger arteries expand when internal pressure rises (Fig. 18.18). Distensibility invalidates application of the Poiseuille law, but it does provide cardiovascular benefits.

1. **Venous reservoir:** Distensibility defines the ease with which a vessel swells when filling pressure rises. Recall that veins have thinner walls than arteries. Therefore, whereas a pressure increase of 1 mm Hg might cause a 1-mL increase in arterial

[1]For more information on anticoagulants and thrombolytics, see *LIR Pharmacology*, 8e, Chapter 13·VII·A.

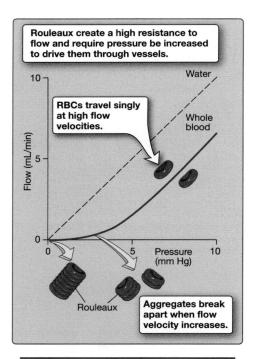

Figure 18.17.
Effects of rouleaux on the pressure required to induce blood flow.

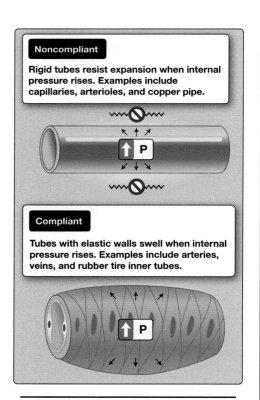

Figure 18.18.
Relative compliance of blood vessels.
P = pressure.

volume, it would cause a similarly sized vein to swell by 6 to 10 mL. The distensibility of veins means that the venous system as a whole has a much higher **compliance** or **capacitance** than the arterial system, which allows it to function as a blood reservoir. Compliance is a measure of a vessel's ability to accommodate volume (V) when filling pressure (P) increases:

$$\text{Compliance} = \frac{\Delta V}{\Delta P}$$

2. **Arterial pump:** During systole, the ventricle ejects blood into the arterial tree faster than it can be passed on to capillaries. The distensibility of large arteries allows them to expand to accommodate a full ventricular stroke volume (Fig. 18.19) and then transmit it to the capillary beds during diastole when the ventricle is relaxing, and the aortic valve is closed (**diastolic runoff**). The energy driving flow during diastole was stored in the elastic elements of the arterial wall by the ventricle during systole. This storage and runoff effect (known as a **windkessel**) is advantageous in that it evens out pressure and flow through the vasculature over time even though ventricular ejection is overtly phasic.

Windkessel is German for "air chamber" and refers to a feature of early fire engine design. Prior to invention of the internal combustion engine, fire engines were horse drawn and hand operated. In the event of fire, water was pumped by hand into the air chamber, causing pressure to develop within. Pressurized water from the windkessel was then directed from a hose at the fire. The inclusion of the windkessel in the design ensured a continuous, steady stream of water from the hose, even when fire personnel were unable to pump.

Biologic Sex and Aging 18.1: Arterial Compliance

Aging is accompanied by changes in the extracellular matrix (ECM) that result in an approximate fourfold decrease in arterial distensibility by the eighth decade (see Biologic Sex and Aging 4.1). Briefly, elastin breaks down and is replaced by collagen, which is a nonelastic fiber. Calcification of elastin, collagen, and the ECM, along with formation of crossbridges between adjacent collagen fibers, all contribute to age-related increases in arterial stiffness (**arteriosclerosis**). Arterial aging is also characterized by a change in vascular smooth muscle cell phenotype, from a quiescent contractile cell to a migratory cell that invades the intima and secretes proinflammatory factors that accelerate the ECM changes described above. Stiffening of the arterial tree reduces the amount of blood that can be stored within arteries during systole for subsequent diastolic runoff. The left ventricle is forced to make up for this deficit by generating higher pressures to drive increased flow during systole. This pressure manifests as an essential hypertension that is common in older adults. In males, systolic blood pressure (SBP) increases 5 mm Hg per decade up until age 60 years and then rises 10 mm Hg per decade thereafter. SBP in females is typically lower initially but increases at an accelerated rate after menopause.

VI. EXCHANGE BETWEEN BLOOD AND TISSUES

The primary function of the cardiovascular system is to deliver O_2 and nutrients to all cells in the body. Blood is distributed by capillaries, vessels with exceptionally thin walls (~0.5 μm, or the width of an endothelial cell) designed to facilitate diffusional exchange between blood and cells. Blood moves through capillaries very slowly (~1 mm/s), which maximizes the opportunity for exchange during passage. In discussions of how exchange occurs at the cellular level, it is helpful to recognize four general mechanisms: transcytosis, bulk flow, diffusion via pores, and diffusion via cells (Fig. 18.20).

A. Transcytosis

Transcytosis refers to the exchange of materials between blood and interstitium using membrane-bound vesicles. Endothelial cells take up material on one side of the cell by endocytosis, transport the endocytic vesicle across the cell interior, and then release the contents on the opposite side by exocytosis (see Fig. 18.20[1]). Transcytosis is not a major pathway for exchange, but it does provide for transit of large, charged molecules such as antibodies.

B. Bulk flow

Capillaries are typically very leaky. Depending on tissue type, transmural gaps, **fenestrations**, and **clefts** between adjacent cells can provide ready pathways for exchange of ions and solutes (see Fig. 18.20[2]). Blood entering capillaries is pressurized, so these same pathways allow water and anything dissolved within it to be driven out of the vasculature and into the interstitium. This **bulk flow** of fluid is not completely unregulated, however. Intercellular junctions

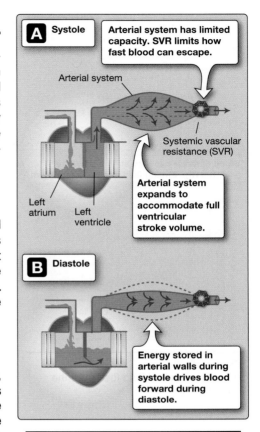

Figure 18.19.
Arterial walls expand during systole and then drive flow during diastole.

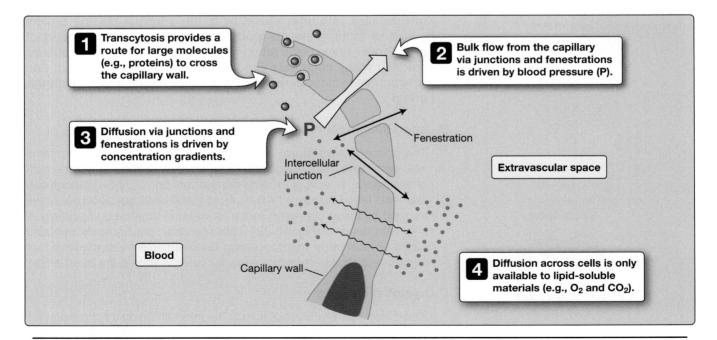

Figure 18.20.
Four general mechanisms by which materials are exchanged across the capillary wall.

typically contain proteinaceous barriers (adherens junctions and tight junctions) that both cement cells together and filter fluid as it leaves the blood. Proteins are too large to escape via junctions or pores and remain trapped within the vasculature.

C. Diffusion via fenestrations and pores

The same pathways that allow for bulk flow also provide pathways for simple **diffusion** of water and other small molecules (see Fig. 18.20[3]). Movement is driven by chemical concentration gradients between blood, interstitium, and cells.

D. Diffusion across endothelial cells

Lipid-soluble materials cross between blood and interstitium by simple diffusion across endothelial cells and their plasma membranes (see Fig. 18.20[4]). This is the primary means by which O_2 and CO_2 are exchanged.

VII. FLUID MOVEMENT

The leakiness of capillaries is problematic. Blood must enter capillaries under pressure to ensure that it has sufficient energy to traverse the capillaries and veins and return to the heart, yet this same pressure also drives fluid out of the vasculature (Fig. 18.21). The severity of the problem is such that, in the absence of any counter measure, we would lose our entire plasma volume to the interstitium within an hour or two!

A. Vascular water retention

The main force holding water in the bloodstream is an osmotic potential generated by proteins that are trapped in blood by virtue of their size. **Albumin** (60 kDa) is the principal plasma protein (~80% of total), although **globulins** (54–725 kDa) are important also. Proteins help blood retain water through direct osmotic effects, but their many negatively charged groups secondarily attract and concentrate osmotically active cations such as Na^+ and K^+ (**Donnan effect**).

B. Interstitial water retention

The space (**interstitium**) between blood vessels and cells contains collagen fibers that provide structural support to tissues, but the bulk of the space is occupied by a dense network of fine proteoglycan filaments (see Chapter 4·IV·B·3). Fluid that filters from blood becomes trapped by these filaments, much as water is trapped by filaments in a gelatin dessert (Fig. 18.22). The interstitial gel normally contains ~25% of total body water, creating an invaluable fluid reservoir that can be recruited to reinforce vascular volume should the need arise.

C. Lymphatic system

Blood loses several liters of fluid to the interstitium on a typical day, far more than the proteoglycan gel can absorb. It is the lymphatic system's responsibility to retrieve excess fluid and return it to the circulation, along with any proteins that may have escaped the vasculature.

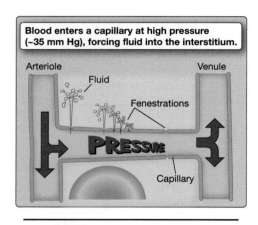

Figure 18.21.
Pressure-induced fluid loss from the vasculature.

Figure 18.22.
Contents of a gelatin dessert.

1. **Structure:** Lymphatic capillaries are simple, blind-ended endothelial cell tubes that arise in the interstitium (Fig. 18.23). Adjacent endothelial cells overlap to create flap valves that allow fluid influx but discourage retrograde flow, and protein filaments tethered to the cell margins maintain vessel patency.

2. **Flow:** The larger lymphatic vessels are structurally similar to veins. They contain valves that help maintain unidirectional flow, and their walls contain smooth muscle layers that contract spontaneously in response to rising fluid pressure within. Contraction propels lymph onward and simultaneously creates a slight negative pressure within the lymphatic capillaries that allows them to suction fluid and protein from the interstitium. The lymphatics ultimately drain into the left and right subclavian veins.

D. Starling forces

Maintaining a balance between the forces governing fluid filtration and reabsorption from the vasculature is vital for continued health. Excess filtration causes edema, whereas an inability to recover filtered fluid can compromise LV preload and MAP. There are four principal **Starling forces** governing fluid movement, which are related in the **Starling law of the capillary:**

$$Q = K_f \left[\left(P_c - P_{if} \right) - \left(\pi_c - \pi_{if} \right) \right]$$

where Q is net fluid flow across the capillary wall, K_f is a filtration coefficient that recognizes that total surface area and permeability of capillary beds varies from tissue to tissue, P_c is capillary hydrostatic pressure, P_{if} is interstitial fluid pressure, π_c is plasma colloid osmotic pressure, and π_{if} is interstitial colloid osmotic pressure.

1. **Capillary hydrostatic pressure:** Blood enters capillaries at a pressure of ~35 mm Hg. Blood exits capillaries and enters veins at a pressure of ~15 mm Hg (Fig. 18.24). Mean capillary hydrostatic pressure (P_c) is typically closer to venous pressure than to arteriolar pressure but still is usually a positive pressure that drives fluid out of the capillary and into the interstitium.

2. **Plasma colloid osmotic pressure:** The main force opposing P_c is the osmotic pressure created by plasma proteins (colloid osmotic pressure, also known as oncotic pressure). Values for π_c typically average ~25 mm Hg.

3. **Interstitial fluid pressure:** P_{if} is typically between 0 and −3 mm Hg normally, due largely to lymphatic suctioning. The lymphatic system does have a finite capacity for fluid removal, however, and if fluid filters from the vasculature faster than it can be removed, the tissue swells. Tissues that are enclosed within skin, bone, or another physical boundary have limited opportunity for expansion, so P_{if} climbs and can become a significant force driving fluid back into the capillary.

4. **Interstitial colloid osmotic pressure:** The interstitium always contains a small amount of protein that creates an osmotic pressure of <5 mm Hg favoring fluid movement out of the capillary. The lymphatic system removes proteins along with fluid, but capillaries continually leak proteins via larger fenestrations and intercellular clefts.

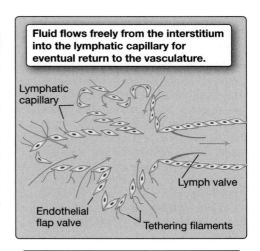

Figure 18.23.
Lymph vessels.

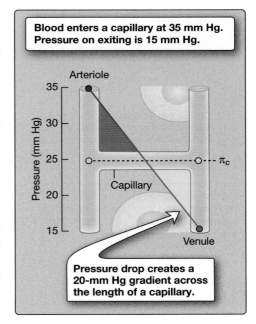

Figure 18.24.
Hydrostatic pressure gradient across the length of a capillary. π_c = plasma colloid osmotic pressure.

Clinical Application 18.3: Lymphatic Filariasis

Lymphatic filariasis results from infection by one of three parasitic nematodes, most commonly *Wuchereria bancrofti* (>90% of total). Also known familiarly as **elephantiasis**, infection can cause gross disfiguration of the legs, arms, and genitalia. The infection is suggested to affect as many as 60 million individuals worldwide and is endemic in the developing regions of Asia, Africa, and South America. Infection occurs by way of a mosquito bite, which injects its host with larval nematodes. The larvae migrate to and establish themselves in lymphatic vessels, where they mature, mate, and breed to produce microfilariae (larvae). The presence of larvae within the lymphatic vessels interferes with drainage and causes pitting edema. Patients typically are infected in childhood but do not become symptomatic until adulthood after they have accumulated large numbers of parasites during repeated infections. Treatment involves prolonged (>1 year) dosing with antihelminthic drugs such as ivermectin.[1]

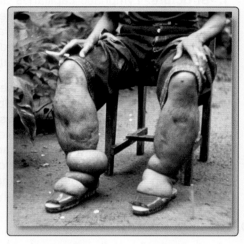

Elephantiasis.

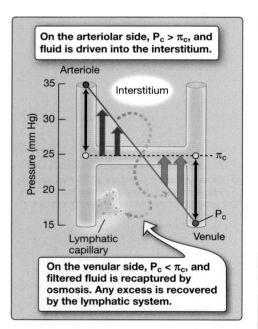

On the arteriolar side, $P_c > \pi_c$, and fluid is driven into the interstitium.

On the venular side, $P_c < \pi_c$, and filtered fluid is recaptured by osmosis. Any excess is recovered by the lymphatic system.

Figure 18.25.
The Starling equilibrium. P_c = capillary hydrostatic pressure; π_c = plasma colloid osmotic pressure.

E. Starling equilibrium

The balance of forces governing fluid movement across the capillary wall is so perfect that net flow is close to zero in most tissues. Any excess filtrate is returned to the circulation by the lymphatics, which collect <4 L daily. This number disguises the fact that an additional 16 to 18 L leaves and is then reabsorbed by capillaries daily. This fluid turnover occurs because of local imbalances between P_c and π_c. At the arteriolar end of the capillary, P_c exceeds π_c by ~10 mm Hg, causing fluid to filter from the capillary and enter the interstitium. By the time blood has traversed the capillary, P_c has dropped below π_c. Absorption is now favored, and most of the filtered fluid is recovered. The near-precise balancing between filtration and reabsorption across the capillary wall has been termed the **Starling equilibrium** (Fig. 18.25).

Capillary endothelial cell membranes contain an aquaporin (AQP1) that allows water to diffuse back and forth between the blood and the interstitium. A single water molecule has been estimated to exit and return to a capillary ~80 times before blood finally drains into a venule, creating a net diffusional flow of 80,000 L/d!

VIII. STARLING EQUILIBRIUM DISTURBANCES

Because water traverses the capillary wall so easily, disturbances in the Starling equilibrium can rapidly cause large amounts of fluid to leave the bloodstream and enter the interstitium, or vice versa. This feature is put to good use in several aspects of cardiovascular design.

 [1]For more information on ivermectin, see *LIR Pharmacology*, 8e, Chapter 36·II·C.

A. Renal circulation

Kidneys cleanse blood of surplus water, electrolytes, and various waste products. As shown in Figure 18.26, blood enters and leaves the glomerulus, a specialized renal capillary network, at a pressure that exceeds π_c across the entire length of the capillary (P_c = ~50 mm Hg; see Chapter 25·III·A). The pressure excess causes 180 L/d of protein- and cell-free fluid to filter into the Bowman space. Most of the water and essential ions and other solutes are later recovered, leaving waste products concentrated in urine.

B. Pulmonary circulation

Mean pulmonary vascular pressures are far lower than in the systemic circulation. P_c for pulmonary capillaries averages 7 mm Hg (compare with ~25 mm Hg in the systemic circulation). π_{if} tends to be higher (~14 mm Hg), but the net driving force for fluid movement is still directed inward (Fig. 18.27). This is advantageous because it ensures that the lungs remain relatively fluid free. Fluid accumulating in the pulmonary interstitium and alveolar sacs interferes with O_2 and CO_2 exchange and makes lungs stiff and difficult to expand.

C. Decreased blood volume

The interstitium contains 10 L of fluid on average. This represents a readily accessible fluid reservoir that can be recruited by the vasculature to support CO when circulating blood volume decreases. Causes include dehydration and hemorrhage. The heart maintains MAP by drawing down the venous blood reservoir. Hydrostatic pressure on the venular side of the capillary falls as a result, causing the pressure gradient across the capillary to steepen. With π_c now dominating along much of the capillary's length, the Starling forces favor fluid recovery from the interstitium. Circulating blood volume rises as a result (Fig. 18.28).

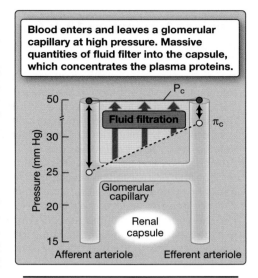

Figure 18.26.
Fluid filtration from renal glomerular capillaries. P_c = capillary hydrostatic pressure; π_c = plasma colloid osmotic pressure.

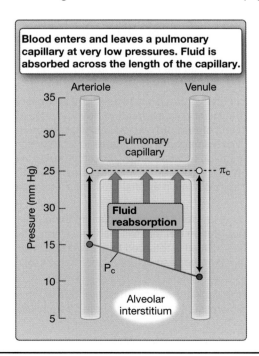

Figure 18.27.
Fluid absorption by pulmonary capillaries. See text for details.

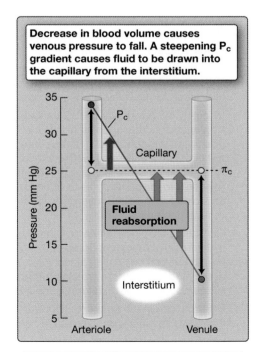

Figure 18.28.
Use of Starling forces to recruit fluid from the interstitium. See text for details.

Clinical Application 18.4: Congestive Heart Failure

Edema is encountered frequently in a clinical setting, and there are many causes. One of the most common is **congestive heart failure**. Left-ventricular failure presents as an inability to maintain arterial pressure at levels that ensure adequate tissue perfusion. The body compensates by retaining fluid (see Chapter 19·IV) to increase blood volume and raise central venous pressure (CVP). The enhanced preload helps compensate for the failure-induced decrease in output through the Frank-Starling mechanism, but raising CVP raises mean capillary hydrostatic pressure. The lymphatic system helps compensate for large quantities of fluid that now filter from the vasculature into the interstitium, but the tendency for edema (tissue **congestion**) is increased greatly. Initially, this may manifest as swelling of the feet and ankles, but, in the later stages of failure, pulmonary edema may occur also.

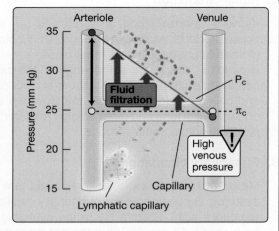

Excess fluid filtration causes edema during congestive heart failure. P_c = capillary hydrostatic pressure; π_c = plasma colloid osmotic pressure.

Chapter Summary

- **Arteries** and **arterioles** are thick-walled, narrow-bore vessels designed to carry blood under high pressure. The smaller arteries and arterioles contract and relax to modulate flow to the capillaries.
- **Capillaries** are simple tubes of endothelial cells designed to facilitate exchange of materials between blood and tissues.
- **Veins** are low-pressure drainage vessels with thin walls and a high capacity that allows them to function as a blood reservoir.
- The heart pumps blood at high pressure to overcome several factors that resist flow. These are summarized in the **Poiseuille law**, which states that flow is proportional to the pressure gradient driving flow, vessel radius to the fourth power, and the inverse of vessel length and blood viscosity.
- Vessel radius is the main determinant of vascular resistance, which is why the smaller arteries and arterioles (**resistance vessels**) control flow so effectively through contraction and relaxation. Blood viscosity is largely a reflection of **hematocrit**.
- Estimates of flow resistance recognize that several factors can invalidate the Poiseuille law, including blood **viscosity changes** with flow velocity, the occurrence of **turbulence**, and vessel **capacitance**.
- Viscosity increases when RBC density is high, or when low flow rates give red blood cells an opportunity to aggregate. Aggregates increase resistance to flow. **Turbulent flow** is less efficient than **streamline flow** because kinetic energy is dissipated through chaotic motion. It usually only occurs in regions of the cardiovascular system where flow velocities are high, such as when blood is forced through a heart valve. **Vessel capacitance** relates to a tendency to distend when filling pressure is increased. **Distensibility** allows arteries to store blood under pressure during systole and then release it to the capillary beds during diastole (**diastolic runoff**). The high capacitance of veins allows them to function as a blood reservoir for use when cardiac output increases or to help sustain arterial pressures when circulating blood volume decreases.
- Capillaries leak fluid continually through pores, fenestrations, and junctions between adjacent endothelial cells, driven by hydrostatic pressure. Some of this fluid is trapped by a **proteoglycan gel** that fills the interstitium. The gel gives up fluid to support circulating blood volume when needed. Most filtered fluid is returned to the circulation by oncotic forces associated with plasma proteins (**albumin** and **globulins**) trapped in the vasculature by their large size. Excess filtrate is returned to the vasculature by the **lymphatic system**.
- Disturbances in the forces that control fluid movement across the capillary wall can have serious repercussions. Increases in venous pressure can raise net capillary hydrostatic pressure to the point where filtered fluid overwhelms the lymphatics. The result is **edema**.
- Decreases in capillary hydrostatic pressure allow fluid to be pulled from the interstitium. This provides a means of supporting cardiac output during a circulatory emergency.

Study Questions

Choose the ONE best answer.

18.1. An 11-year-old male's dentist gives him nitrous oxide gas (N_2O) via a face mask to anesthetize him. The main route by which N_2O reaches the brain is most likely which of the following?

A. Bulk flow through fenestrations
B. Diffusion through intercellular junctions
C. Diffusion through endothelial cells
D. Specialized endothelial transporters
E. Transcytosis across the capillary wall

Best answer = C. N_2O is a small, highly soluble molecule that, like O_2 and CO_2, easily diffuses through endothelial cell membranes (see Section VI). Fenestrations (A) and intercellular junctions (B) provide routes for passage of water and any dissolved ions. Transporters (D) are typically used to transport charged molecules against a concentration gradient, whereas transcytosis (E) is used primarily as a means of moving large proteins between the bloodstream and tissues.

18.2. A 24-year-old female in the 16th week of gestation comes to the office for prenatal care. She is noted to have developed a soft mid-systolic heart murmur. The most likely cause of this murmur is which of the following?

A. Decreased blood density
B. Decreased hematocrit
C. Increased blood volume
D. Increased systemic vascular resistance
E. Increased vascular compliance

Best answer = B. RBC production lags behind volume retention during pregnancy, leading to a functional anemia. Anemia decreases blood viscosity and increases the likelihood of turbulence in regions where flow velocity is high, such as during passage across heart valves (see Section III·C). Blood density (A) does not change physiologically. Blood volume increases (C) during pregnancy to support increased cardiac output but such increases do not necessarily cause murmurs. Systemic vascular resistance decreases, not increases (D), during pregnancy. Changes in vascular compliance (E) do not cause heart murmurs.

18.3. A 67-year-old male comes to the office for a physical examination. He has no previous medical history and does not currently take any medications. His blood pressure is 138/88 mm Hg. The most likely cause of his elevated blood pressure is which of the following?

A. Decreased arterial compliance
B. Decreased heart rate
C. Increased diastolic runoff
D. Increased venous distensibility
E. Left-ventricular failure

Best answer = A. Arterial compliance decreases with age due to stiffening of the extracellular matrix. Systolic blood pressure increases to compensate for the reduced ability to store blood in the arterial tree during systole (see Biologic Sex and Aging 18.1). Decreased heart rate (B) would decrease blood pressure. An increase in diastolic runoff (C) would tend to decrease diastolic blood pressure. Increased venous distensibility (D) would decrease venous return, cardiac output, and blood pressure. Left-ventricular failure (E) would also tend to decrease cardiac output and blood pressure.

18.4. Kwashiorkor is a severe form of childhood malnutrition seen mostly in developing countries. Symptoms include hepatomegaly and pitting edema of the lower extremities. The pitting edema is most likely due to which of the following?

A. Decreased hematocrit
B. Excessive fluid retention
C. Inadequate cardiac output
D. Plasma protein deficit
E. Reduced interstitial pressure

Best answer = D. Plasma proteins create an osmotic potential (plasma colloid osmotic pressure) that holds fluid in the vasculature (see Section VII·D). Kwashiorkor results from inadequate protein intake, which impairs the liver's ability to synthesize proteins. Fluid filters into the interstitium and causes edema as a result. Hematocrit (A) does not affect the tendency for fluid to filter from the vasculature. Excessive fluid retention (B) would exacerbate edema caused by the plasma protein deficit. Reduced cardiac output (C) and arterial pressure would reduce fluid filtration. Interstitial pressure (E) is increased by and counters fluid filtration.

19 Cardiovascular Regulation

I. OVERVIEW

The volume of blood contained within the cardiovascular system represents only around 20% of its total capacity. Carrying a limited blood volume has clear energetic advantages, but such frugality requires that flow to the different organs be carefully metered with an eye to the needs of the system as a whole if cardiovascular catastrophe is to be avoided. The threat stems from the cardiovascular system's absolute dependency on pressure to drive flow. Just as a broken main in a city's infrastructure can deprive consumers of fresh water (Fig. 19.1), uncontrolled flow through a low-resistance circuit (e.g., an exercising muscle) can cause perfusion pressure and flow through the vasculature to drop precipitously. Because some tissues (e.g., the brain and heart) are highly dependent on sustained arterial blood flow for normal function, loss of arterial pressure is a potentially fatal event. Thus, although the cardiovascular system includes flow regulators (**resistance vessels**) that can be operated by the tissues themselves if their nutrient needs increase, it also incorporates mechanisms by which the central nervous system (CNS) can monitor and maintain arterial pressure by reapportioning flow for the benefit of the system as a whole.

Figure 19.1.
Pressure is required to drive flow through vessels.

II. VASCULAR CONTROL

The cardiovascular system functions very much like a public water utility. Pumping stations ensure that there is always sufficient volume and pressure in the mains to supply consumer needs. Consumers typically do not leave taps running but rather turn them on and then off as their need to bathe or fill kettles has been satisfied. Consumers recognize that fresh water is a valuable commodity and that supplies are limited. Similarly, resistance vessels allow tissues (the consumers) to draw arterial blood from the vascular system on the basis of metabolic need. Because resistance vessels are located at key positions within the vasculature, they are subject to a multitude of controls. Four general control mechanisms can be recognized: **local**, **central** (neural), **hormonal**, and **endothelial**.

A. Local

All tissues are able to regulate their blood supply through local control of resistance vessels. Flow is tied to tissue need. Increased activity causes the resistance vessels to dilate, and blood flow increases proportionately. If supply exceeds prevailing needs, the resistance vessels constrict reflexively. There are two broad classes of local control mechanisms: **metabolic** and **myogenic**.

1. **Metabolic:** Cells continually release various metabolic byproducts, including adenosine, lactate, K^+, H^+, and CO_2. When tissue activity increases, metabolites are produced in greater quantities, and interstitial concentrations rise (Fig. 19.2). Local O_2 levels simultaneously fall. Resistance vessels lie close to the cells they serve and are sensitive to the appearance of these metabolites in extracellular fluid (ECF) and to PO_2. Some metabolites act directly on **vascular smooth muscle cells** (**VSMCs**), whereas others act through endothelial cells, but all cause the VSMCs to relax and the vessel to dilate. Blood flow increases as a result, simultaneously providing the tissues with the nutrients they need and also carrying away metabolites (Fig. 19.3). When activity ceases, metabolite concentrations fall, and a reflex vasoconstriction again matches flow with need.

2. **Myogenic:** Resistance vessels in many circulations constrict reflexively when intraluminal pressures rise. Contraction is mediated by stretch-activated Ca^{2+} channels in the VSMC membranes and may protect capillaries from surges in arterial pressure. Postural changes can cause sudden gravity-induced pressure spikes of >200 mm Hg in the pedal vasculature, for example.

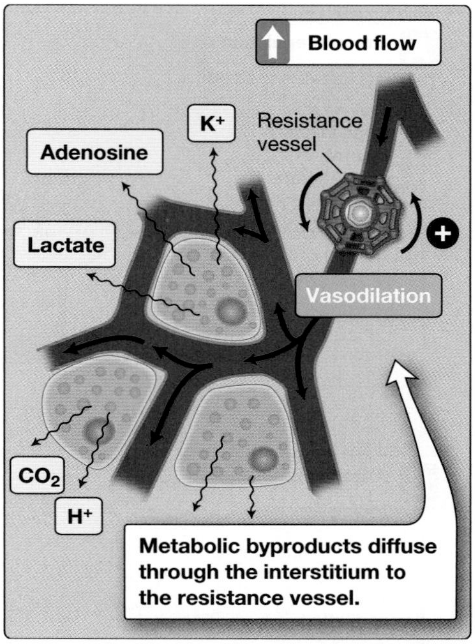

B. Physiologic consequences

Local and myogenic control mechanisms operate independently of external influence, which frees the CNS from having to micromanage circulatory control. This autonomy manifests in several ways, including flow **autoregulation** and **hyperemia**.

1. **Autoregulation:** Autoregulation is the intrinsic ability of an organ to maintain stable blood flow in the face of changing perfusion pressures (Fig. 19.4). If arterial pressure increases suddenly (e.g., during a pressure spike), flow increases also. Metabolites are washed away faster than they are produced, and the resistance vessels constrict reflexively. The myogenic response potentiates this effect, so that, over the course of several seconds, flow rates are restored to levels approximating those observed prior to the pressure change. Conversely, a sudden drop in arterial pressure leads to reflex vasodilation, and flow is restored within a few seconds. When presented graphically (see Fig. 19.4B), pressure extremes are seen to overwhelm the resistance vessels' autoregulatory powers, but flow remains relatively stable over a wide range of pressures.

2. **Hyperemia: Active hyperemia** is a normal vasodilatory response to increased tissue activity and rising metabolite levels (Fig. 19.5). Muscles also demonstrate **postexercise hyperemia**, a period of increased blood flow that persists even after activity has ceased. This reflects a time during which metabolite levels are still high, and the muscles are repaying oxygen debts accumulated during exercise (see Chapter 39·VI·C). **Reactive hyperemia** is a period of increased blood flow that follows transient ischemia, caused by arterial occlusion, for example.

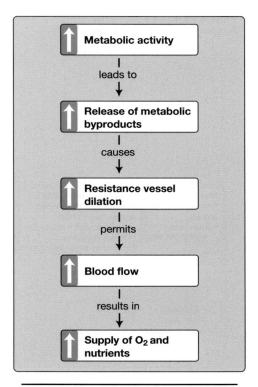

Figure 19.2.
Metabolic control of resistance vessels.

Figure 19.3.
Metabolic control concept map.

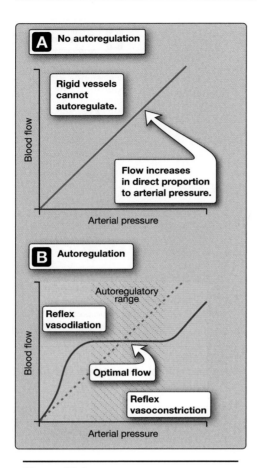

Figure 19.4.
Autoregulation of blood flow.

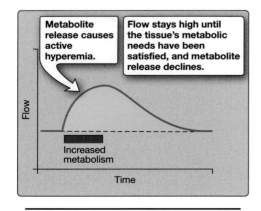

Figure 19.5.
Active hyperemia.

C. Central control

All resistance vessels are innervated by the sympathetic nervous system (SNS). When arterial pressure falls, SNS nerve terminals release norepinephrine onto the VSMCs, causing them to contract. Contraction is mediated by α_1-adrenergic receptors acting via the IP_3 signaling pathway, causing Ca^{2+} release from the sarcoplasmic reticulum (see Chapter 1·VII·B·3).

D. Hormonal control

Many circulating hormones modulate resistance vessels, including **antidiuretic hormone (ADH)**, **angiotensin II (Ang-II)**, **epinephrine**, and **atrial natriuretic peptide (ANP)**.

1. **Antidiuretic hormone:** ADH, also known as **arginine vasopressin**, is released from the posterior pituitary when tissue osmolality rises or blood volume decreases (see Chapter 28·II·C). Its principal role is in ECF volume regulation through control of renal water retention, but if circulating levels are sufficiently high (e.g., during hemorrhage), it can vasoconstrict also. ADH binds to vasopressin V_{1A} receptors on VSMCs. V_{1A} receptors couple to the IP_3 signaling pathway.

2. **Angiotensin II:** Ang-II is a potent vasoconstrictor. It appears in the bloodstream when renal artery pressure falls, although sympathetic activity can also trigger Ang-II release (see Section IV·C). Ang-II binds to angiotensin AT_1 receptors on VSMCs, which couple to the IP_3 signaling pathway.

3. **Epinephrine:** Epinephrine is produced and released by the adrenal medulla during SNS activation. Its primary effect is to increase myocardial contractility and heart rate (HR), but it also binds to α_1-adrenergic receptors on VSMCs to potentiate the direct SNS-mediated vasoconstriction. Resistance vessels in some circulations (e.g., skeletal) express a β_2-adrenergic receptor that mediates epinephrine-mediated vasodilation. In the skeletal vasculature, this pathway may facilitate increased blood flow to muscles during "fight-or-flight" responses.

4. **Atrial natriuretic peptide:** ANP is released from atrial myocytes when stretched by high blood volumes (see Chapter 28·III·C). ANP acts as a safety valve to limit blood pressure increases resulting from ECF volume retention. It is a vasodilator that binds to natriuretic peptide receptors (NPR1) on VSMCs. NPR1 is a guanylyl cyclase (GC) that relaxes VSMCs by activating protein kinase G (PKG).

E. Endothelial control

The endothelial lining of resistance vessels acts as an intermediary for a number of vasoactive compounds, including **nitric oxide (NO)**, **prostaglandins (PGs)**, **endothelium-derived hyperpolarizing factor (EDHF)**, and **endothelins ([ETs]** Fig. 19.6).

1. **Nitric oxide:** NO is a potent vasodilator that acts on both arteries and veins. Also known as **endothelium-derived relaxing factor (EDRF)**, it is synthesized by a constitutive endothelial NO

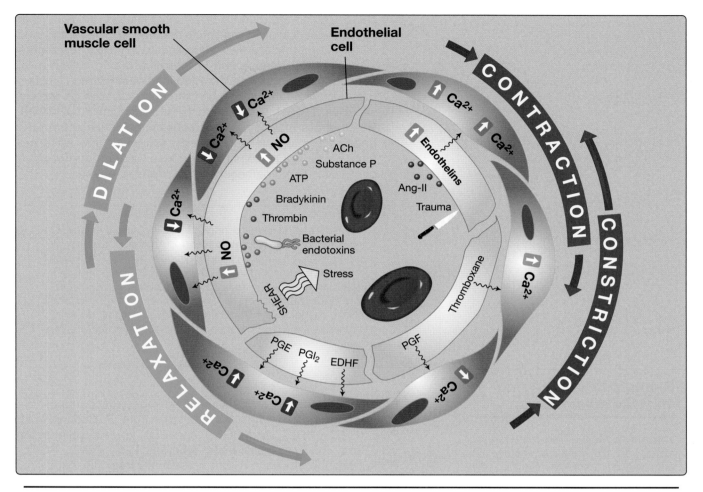

Figure 19.6.
Role of endothelium in controlling resistance vessels. Ang-II = angiotensin II; EDHF = endothelium-derived hyperpolarizing factor; NO = nitric oxide; PGE, PGF, and PGI$_2$ = prostaglandin E, F, and I$_2$.

synthase (eNOS, or NOS3), following a rise in intracellular Ca^{2+} concentrations. NO is a gas with a half-life of less than 10 seconds *in vivo*, meaning that its actions remain highly localized. It diffuses through the endothelial cell membrane to adjacent VSMCs and then binds to and activates a soluble GC. cGMP levels rise as a result, which, in turn, activate PKG. PKG then phosphorylates and thereby inhibits **myosin light-chain kinase**. It also phosphorylates and increases the activity of a **SERCA** (sarco[endo]plasmic reticulum Ca^{2+}-ATPase) **Ca^{2+} pump** and causes intracellular Ca^{2+} concentration to fall. The net result is vasodilation and increased blood flow. NO helps mediate the actions of many vasodilators, including **local modulators** and **stressors**.

a. **Local modulators:** NO mediates vasodilation caused by **acetylcholine** (**ACh**) binding to endothelial M$_3$ muscarinic receptors, **ATP** stimulation of P2Y purinergic receptors, and by **adenosine** A$_1$ receptor activation.

b. **Stressors: Substance P** initiates NO-mediated vasodilation in many circulations by activating an endothelial neurokinin

NK$_1$ receptor. **Bradykinin** is an inflammatory mediator that causes vasodilation by binding to endothelial B$_2$ receptors and stimulating release of NO and vasodilatory PGs. **Thrombin** stimulates NO release and vasodilation during activation of the coagulation cascade. Flow-induced **shear stress** initiates a protective reflex vasodilation through activation of stretch-sensitive TRPV4 Ca^{2+} channels and stimulation of eNOS. Shear stress may also initiate substance P release from the endothelium to potentiate vasodilation when flow rates are high. Finally, NO release in response bacterial **endotoxins** contributes to the pathogenesis of **septic shock**.

> Nitroglycerin and related nitrates are commonly used to relieve the pain of **angina pectoris**. Angina is caused by inadequate myocardial O$_2$ supply, typically because coronary arteries have become narrowed by plaque (**atherosclerosis**). Nitrates break down to release NO *in vivo*, causing arterial and venous vasodilation to lower ventricular afterload and preload, respectively. Reducing cardiac workload restores the balance between O$_2$ demand and supply and relieves the angina.

Biologic Sex and Aging 19.1: Vascular Control

The sex hormones have wide-ranging effects on the vascular endothelium and vascular smooth muscle cells (VSMCs), the scope of which is not yet fully appreciated. Estrogen (E$_2$) has long been recognized to stimulate nitric oxide (NO) release from the endothelium, but it also has direct effects on VSMCs, decreasing resting tone and blunting responses to pressors such as norepinephrine and endothelin-1 (ET-1). The net result is that resting blood pressure (BP) is lower in young females than in males, and females are less likely to develop hypertension (HTN). The protective effect of E$_2$ is lost during menopause, whereupon both resting BP and HTN incidence rises above that of age-matched males. Note that although the HTN associated with aging can be explained in terms of decreasing arterial compliance (see Biologic Sex and Aging 18.1), changes in local control mechanisms contribute. For example, aging increases vascular ET-1 receptor numbers and circulating ET-1 levels, resulting in a chronic increase in resting arterial tone.

2. **Prostaglandins:** The endothelium is an important source of several vasoactive PGs, which it synthesizes from arachidonic acid. PGE (PGE$_1$, PGE$_2$, and PGE$_3$) and PGI$_2$ (prostacyclin) relax VSMCs in many vascular beds, whereas PGF (PGF$_1$, PGF$_{2\alpha}$, PGF$_{3\alpha}$) and thromboxane A$_2$ are vasoconstrictors.

3. **Endothelium-derived hyperpolarizing factor:** EDHF opens K$^+$ channels in the plasma membranes of VSMCs. The ensuing membrane hyperpolarization reduces membrane Ca^{2+} permeability, causing intracellular Ca^{2+} levels to fall and therefore vasodilation. A number of EDHF candidate molecules (e.g., epoxyeicosatrienoic acids and C-type natriuretic peptide) and pathways may all contribute to the phenomenon to varying degrees.

4. **Endothelins:** ETs are a group of related peptides synthesized and released by endothelial cells in response to many factors, including Ang-II, mechanical trauma, and hypoxia. ET-1 is a potent vasoconstrictor that binds to ET$_A$ receptors on VSMC membranes and triggers intracellular Ca^{2+} release via the IP$_3$ pathway (see Fig. 19.6).

F. Circulatory hierarchy

The discussions earlier delineate several mechanisms by which flow to individual vascular beds is regulated. In practice, most of the moment-by-moment control involves a simple weighing of the amount of flow that a tissue needs to support prevailing activity levels versus the amount that the CNS is willing to make available based on the needs of the organism as a whole. Thus, if there is a threat to arterial pressure, the CNS has the ability to deprive certain vascular beds of cardiac output (CO) in an effort to preserve flow to the more important organs. In reviewing the relative ability of the different organs to demand and receive flow, a circulatory hierarchy emerges.

1. **Hierarchy:** At the head of the list are the circulations that supply the brain, myocardium, and skeletal musculature (during exercise). Here, local control mechanisms dominate, and central controls have little or no effect. At the bottom of the hierarchy are organs such as the gut, kidneys, and skin that receive blood flow under optimal conditions, but flow is sacrificed if there is a need to preserve arterial pressure.

2. **Rationale:** The rationale for such a hierarchy can best be understood in evolutionary terms. One of the greatest challenges that the cardiovascular system faces involves intense physical activity, of the kind that might be required for chasing after prey or running away from predators (Fig. 19.7). Maintaining optimal flow to the three organs at the top of the hierarchy is critical for sustaining such activities. Meeting the challenge requires that the CNS temporarily divert flow away from organs at the bottom of the hierarchy to support the needs of the skeletal musculature (see Chapter 39·V). Fortunately, these same organs also have relatively low metabolisms so their temporary deprivations do not threaten survival.

III. ARTERIAL PRESSURE CONTROL

Survival of the individual requires that pressure be maintained in the arterial system at all times. Because all organs of the body have the ability to demand increased flow, they could easily cause arterial pressure to collapse if their arteriolar supply vessels were not strictly regulated. The cardiovascular system includes two distinct pathways for monitoring and maintaining arterial pressure. The first is fast activating and helps compensate for short-term pressure changes. Known as a **baroreceptor reflex** (**baroreflex**), it employs simple feedback loops that include **sensors** to monitor pressure and flow, an **integrator** to compare current with preset pressure values, and **effector mechanisms** that make any necessary adjustments. A second, slow-activating system manipulates **mean arterial pressure** ([**MAP**] see Chapter 18·IV·A) through changes in circulating blood volume by modifying renal function (discussed in Section IV).

A. Sensors

Two main groups of sensors provide the integrator (located in the brainstem medulla) with information about pressure and flow in the cardiovascular system: **high-pressure arterial baroreceptors** located in the **carotid sinus** and **aortic arch** and low-pressure **cardiopulmonary receptors**.

1. **Arterial baroreceptors:** The carotid and aortic baroreceptors are the primary means of detecting changes in MAP. They monitor pressure indirectly by responding to arterial wall stretch.

   a. **Anatomy:** Baroreceptors are clusters of bare sensory nerve endings buried within the elastic layers of the carotid sinus and aorta (Fig. 19.8A). Information from the former is relayed

During exercise, blood is directed away from less essential organs:
- Gastrointestinal
- Renal
- Reproductive

Redirected flow is used to support essential organs:
- Brain
- Heart
- Skeletal muscle

Figure 19.7.
Blood flow redistribution during exercise.

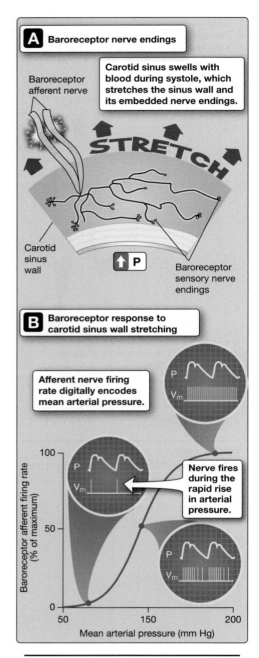

Figure 19.8.
Arterial baroreceptors. P = pressure,
V_m = membrane potential.

to the brain via sensory afferents traveling in the **sinus nerve**, which joins with the **glossopharyngeal nerve** (**cranial nerve** [**CN**] **IX**) en route to the brainstem. Afferents from the **aorta travel** in the aortic nerve and the **vagus nerve** (**CN X**).

b. **Function:** In the absence of stretch, the baroreceptors are inactive. When MAP increases, the walls of the aorta and carotid sinus expand, and the embedded nerve endings are stretched. The nerves respond with graded receptor potentials. If the degree of deformation is sufficiently high, the receptor potentials trigger spikes in the sensory nerve (see Fig. 19.8B). Baroreceptors are especially sensitive to *changes* in pressure, responding to the sharp rise in pressure that occurs during rapid ejection with strong depolarization and a train of high-frequency spikes. During reduced ejection and diastole, the depolarization abates, and spike frequency drops to a new steady-state level that reflects diastolic pressure.

c. **Sensitivity:** Stretch sensitivity varies from one nerve ending to the next, thereby allowing for responsiveness over a wide pressure range (see Fig. 19.8B). The carotid baroreceptors have a response threshold of around 50 mm Hg and saturate at 180 mm Hg. The aortic baroreceptors operate over a range of 100 to 300 mm Hg, meaning that they are usually inactive under resting conditions. In practice, this means that the carotid baroreceptors bear the brunt of providing responses to moment-by-moment changes in blood pressure.

2. **Cardiopulmonary receptors:** A second set of baroreceptors is found in low-pressure regions of the cardiovascular system. They provide the CNS with information about the "fullness" of the vascular system, and their principal role is in modulating renal function. However, because fullness correlates with ventricular preload, they also have a role in maintaining MAP.

a. **Anatomy:** The receptors are similar to those found in the arterial system: bare sensory nerve endings embedded in walls of the vena cavae, the pulmonary artery and veins, and the atria. They relay information back to the CNS via the vagal nerve trunk.

b. **Function:** Atria contain two functionally distinct populations of baroreceptors. **A receptors** respond to tension that develops in the atrial wall during contraction. **B receptors** are sensitive to atrial wall stretching during filling. B receptors are also involved in raising HR when central venous pressure (CVP) is high, a response known as the **Bainbridge reflex**.

B. **Central integrator**

The sensory afferents converge on the **medulla oblongata**. Here, arterial pressures are compared with preset values and decisions then made about the nature and intensity of a compensatory response.

1. **Control centers:** The medulla contains a collection of nuclei that together comprise a **cardiovascular control center**. Some cells in this area cause vasoconstriction when active and are known as the **vasomotor center**. Another group comprises a **cardioacceleratory center**, which increases HR and myocardial inotropy when activated. A third group (**cardioinhibitory center**) slows HR when active. The

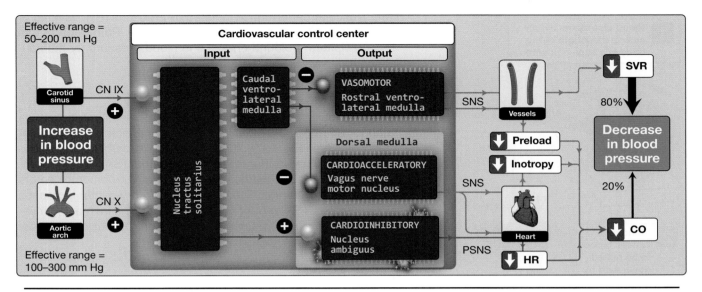

Figure 19.9.
Organization and output of the medullary cardiovascular control center. CN = cranial nerve, CO = cardiac output, HR = heart rate, PSNS and SNS = parasympathetic and sympathetic nervous systems, SVR = systemic vascular resistance.

three control centers are interlinked extensively so as to generate a unified response to changes in arterial pressure (Fig. 19.9).

2. **Feedback loops:** Arterial pressure is a product of CO and **systemic vascular resistance** (**SVR**) (MAP = CO × SVR), and the control centers adjust both parameters simultaneously. Control is exerted using simple feedback loops (Fig. 19.10). Sensory afferents project to the **nucleus tractus solitarius** within the medulla and synapse with interneurons that, in turn, project to the three control centers (see Fig. 19.9). The sensory afferents are all excitatory, but the interneurons may be either excitatory (glutamatergic) or inhibitory (γ-aminobutyric acid [GABA]ergic). The cardioinhibitory center, which includes the dorsal motor nucleus of the vagus and the nucleus ambiguus, receives inputs from excitatory interneurons, so this area is excited when MAP is high (a **positive feedback loop**). The cardioacceleratory center (dorsal medulla) and vasomotor center (rostral ventrolateral medulla) are innervated by inhibitory interneurons. When MAP is high, they suppress the activity of the nerves they innervate. Inhibition is required because the cardioacceleratory and vasomotor centers control sympathetic nerves that are tonically active in the absence of external input. This arrangement creates a **negative feedback loop** between MAP and SNS output.

3. **Integration with other central and peripheral pathways:** There are numerous inputs to the cardiovascular center from other regions of the brain and periphery. The brainstem also contains a **respiratory center** that controls breathing. The respiratory center monitors arterial O_2 and CO_2 using peripheral and central chemoreceptors (see Chapter 24·III). The cardiovascular and respiratory centers work in close cooperation to optimize tissue oxygenation and blood flow. **Hypothalamic control centers** help coordinate vascular responses to changes in external and internal body temperatures. **Cortical control centers** account for changes in

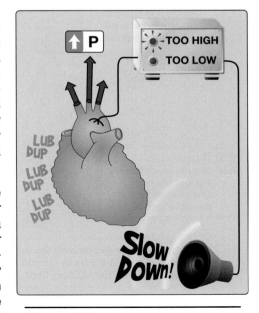

Figure 19.10.
Cardiovascular feedback loop.

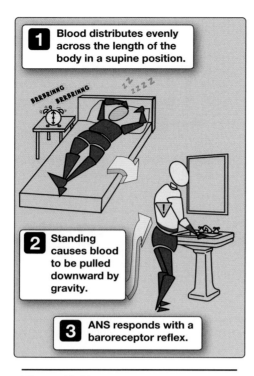

Figure 19.11.
Orthostatic reflex. ANS = autonomic nervous system.

cardiovascular performance induced by emotions (e.g., fainting or anticipatory changes associated with exercise). **Pain centers** can also precipitate profound changes in blood pressure by manipulating cardiovascular center output.

C. Effector pathways

The cardiovascular centers adjust cardiac and vascular function via the autonomic nervous system (ANS). The cardioinhibitory center depresses HR (see Fig. 19.9). It acts via parasympathetic fibers traveling in the vagus nerve that target the sinoatrial (SA) and atrioventricular (AV) nodes. The cardioacceleratory and vasomotor centers act via sympathetic nerves. The cardioacceleratory center increases HR by manipulating SA and AV nodal excitability and increasing myocardial contractility. The vasomotor center controls resistance vessels, veins, and adrenal glands.

D. Response

To understand how the various effector pathways work toward a common goal, it is helpful to dissect the ANS response to a sudden drop in arterial pressure. Such events are triggered routinely when climbing out of bed and assuming an erect position (an **orthostatic reflex**). When a person stands, 600 to 800 mL of blood is forced downward under the influence of gravity and begins to pool in the legs and feet (Fig. 19.11). Venous return (VR) and CVP fall as a consequence. The drop in preload reduces left-ventricular stroke volume (SV), causing CO to fall. The net result is that MAP drops by ~40 mm Hg (Fig. 19.12). The baroreceptor nerve endings are stretched to a lesser

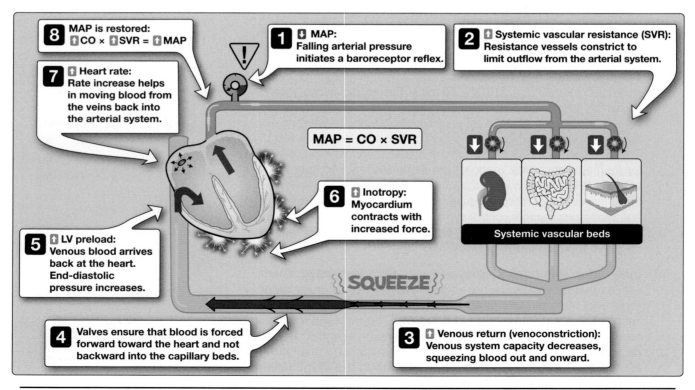

Figure 19.12.
Baroreceptor reflex. CO = cardiac output, LV = left ventricle, MAP = mean arterial pressure.

degree, and spike frequency in the afferent arm of the reflex pathway falls. Within the cardioinhibitory center, the loss of excitatory input causes withdrawal of parasympathetic output to the SA node and myocardium. Within the two pressor regions of the cardiovascular center, decreased baroreceptor output weakens inhibitory interneuron influence on SNS effector pathways. With the brakes removed, sympathetic nerves now drive up arterial pressure by constricting resistance vessels and veins, increasing myocardial contractility, and increasing HR.

1. **Resistance vessels:** All resistance vessels are innervated by SNS nerve terminals that cause vasoconstriction when active. SVR increases, and outflow from the arterial tree is reduced as a result. The SVR increase accounts for ~80% of the compensatory response to a fall in MAP under resting conditions (i.e., the cardiac component contributes ~20%).

2. **Veins:** Veins and larger venules contract when the SNS is active, reducing the capacity of the venous reservoir and causing intravenous pressures to rise. Valves ensure that this pressure increase forces blood forward toward the heart. Here, it increases left-ventricular preload (end-diastolic pressure) and increases SV on the next beat.

3. **Myocardium:** SNS activation increases myocardial contractility by increasing intracellular Ca^{2+} release. The myocardium now works with increased efficiency and contributes to the increased SV caused by preloading. The SNS also speeds myocardial relaxation rate by increasing the rate at which Ca^{2+} is released from the contractile machinery and then removed from the sarcoplasm. Faster relaxation times make more time available for preloading during diastole and thereby facilitate a concurrent increase in HR.

4. **Nodes:** SA and AV nodes are innervated by both parasympathetic and sympathetic nerve terminals, both of which are active at rest. A drop in MAP simultaneously causes parasympathetic activity to be withdrawn and sympathetic activity to increase, accelerating the rate at which SA nodal cell membrane potential slides toward the threshold for action potential formation (phase 4 depolarization; see Chapter 16·V·C). HR and CO both increase as a result.

5. **Adrenal glands:** SNS activation causes the adrenal glands to secrete epinephrine into the circulation. Epinephrine binds to the same receptors on blood vessels and the myocardium as does neurally derived norepinephrine.

E. Baroreceptor reflex limitations

The baroreceptor reflex is an extremely effective short-term control mechanism, but pressure changes that are sustained for more than a few minutes cause a parallel shift in system sensitivity (Fig. 19.13). The advantage to the shift is that it allows the system to retain responsiveness over a wide range of pressures, even if MAP is increased to a level that would previously have been saturating. The disadvantage is that, although the reflex is ideal for moment-by-moment adjustments in arterial pressure, its contribution to long-term pressure control (>1–2 days) is limited.

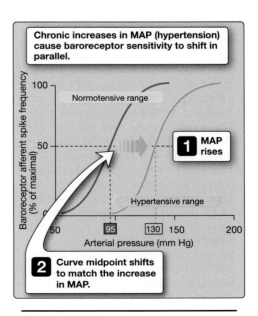

Figure 19.13.
Shifts in baroreceptor sensitivity. MAP = mean arterial pressure.

Clinical Application 19.1: Orthostatic Hypotension

Orthostatic or postural hypotension is common in older adults. The condition describes the drop in arterial pressure that accompanies standing from a sitting position. The resulting decrease in cerebral perfusion pressure can lead to momentary lightheadedness, dizziness, weakness, and darkening of the vision. In extreme cases, patients may not be able to rise from a supine position without fainting (**syncope**). Orthostatic hypotension can result from low circulating blood volume, but it is more frequently associated with decreases in baroreceptor sensitivity caused by hardening of the arteries (**arteriosclerosis**). Arteriosclerosis is an inevitable consequence of aging (see Biologic Sex and Aging 18.1) and is caused by deposition of collagen and other fibrous materials in the arterial wall, which lowers wall compliance. Atherosclerosis is a form of arteriosclerosis associated with lipid deposition and plaque formation, which thickens the arterial walls and encroaches on the arterial lumen. The baroreceptor reflex relies on aortic and carotid artery distensibility in order to transduce changes in arterial pressure. Atherosclerosis prevents the sensory nerve endings from detecting the pressure drop that accompanies translocation of blood to the lower extremities upon standing, and compensatory responses are thus delayed for several seconds.

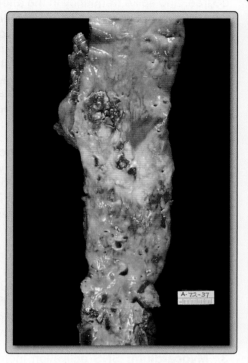

Intima of a sclerotic aorta scarred by lesions.

Biologic Sex and Aging 19.2: Central Control of Blood Pressure

Younger females show an increased incidence of orthostatic intolerance and syncope (an inability to adequately control blood pressure during any change in posture) compared with males, reflecting sex hormone effects on both local and central cardiovascular regulatory mechanisms (see also Biologic Sex and Aging 19.1). The pathways involved are not well delineated, but, in animal models, estrogen decreases sympathetic output from the cardiovascular control centers and decreases blood pressure. This agrees well with observations that resting sympathetic tone is decreased in premenopausal females compared with males. The net effect of central and local control differences is that baroreflex sensitivity is shifted right, which decreases system sensitivity to sudden changes in arterial pressure.

 Hypertension is not the only condition that causes baroreceptor resetting. Shifts in sensitivity also occur during exercise, thermoregulation, sleep, pregnancy, and with aging.

IV. LONG-TERM CONTROL PATHWAYS

A drop in arterial pressure activates the baroreceptor reflex outlined earlier, but it also initiates pathways that require 24 to 48 hours to become fully effective. These pathways converge on the kidney, which is primarily responsible for long-term control of blood pressure through regulation of vascular fullness (**circulating blood volume**). Because blood is principally water, this necessarily involves regulation of water output and intake, but it also requires regulation of Na⁺ levels because this is the ion that governs how water partitions between the intracellular and extracellular compartments. These concepts are discussed in more detail in Chapter 3·III·B and Chapter 28·II and III.

A. Water output

Water output is controlled by ADH, a peptide that is synthesized by the hypothalamus and then transported to the posterior pituitary for release. It stimulates water reabsorption by the renal collecting tubule and collecting ducts. At high concentrations, ADH also increases

SVR by constricting resistance vessels (Fig. 19.14). Several sensors and pathways regulate ADH release including **osmoreceptors**, baroreceptors, and Ang-II.

1. **Osmoreceptors:** The brain contains several regions that have the potential to monitor plasma osmolality, including areas surrounding the third ventricle in close proximity to the hypothalamus (see Chapter 7·VII·C). Tissue osmolality is a reflection of total body water and salt concentration. When osmolality exceeds 280 mOsmol/kg H_2O, the receptors cause ADH to be released into the circulation.

2. **Baroreceptors:** A decrease in circulating blood volume causes CVP to fall, which is sensed by the cardiopulmonary receptors. Loss of preload also causes arterial pressure to fall and triggers a baroreceptor reflex. The CNS cardiovascular control centers respond by increasing sympathetic activity and promoting ADH release.

3. **Angiotensin II:** Activating the **renin–angiotensin–aldosterone system (RAAS)** causes circulating Ang-II levels to rise. The list of target organs for Ang-II includes the hypothalamus (see Table 28.2), where it stimulates ADH release.

B. Water intake

Water enters the body along with food, but the bulk of liquid intake occurs through drinking, driven by thirst. The sensation is triggered by decreased blood volume and arterial pressure and by increased plasma osmolality (see Chapter 28·II). Thirst sensation is increased by Ang-II and also by relaxin at certain times during the menstrual cycle and pregnancy.

C. Sodium output

Osmoreceptors control water retention and excretion, but they sense the "saltiness" of body fluids rather than water per se. Thus, if tissue osmolality remains high, they will urge retention of water regardless of total accumulated volume. The primary determinant of circulating blood volume is Na^+ concentration, which is regulated through the RAAS, as follows (Fig. 19.15).

1. **Renin–angiotensin–aldosterone system:** Renin is a proteolytic enzyme synthesized by granular cells (also known as juxtaglomerular cells) in the wall of glomerular afferent arterioles (see Chapter 25·IV·C). The cells form a part of the juxtaglomerular apparatus (JGA), which monitors flow of Na^+ and Cl^- through the renal tubule. When the JGA is stimulated appropriately, it releases renin into the bloodstream. Here, renin breaks down **angiotensinogen** (a circulating plasma protein formed in the liver) to release **angiotensin I**. The latter serves as a substrate for **angiotensin-converting enzyme (ACE)**. ACE is expressed by many tissues, including the kidney, but conversion largely occurs during transit through the lungs. The product is Ang-II, which constricts resistance vessels, stimulates ADH release from the posterior pituitary, stimulates thirst, and promotes aldosterone release from the adrenal cortex.

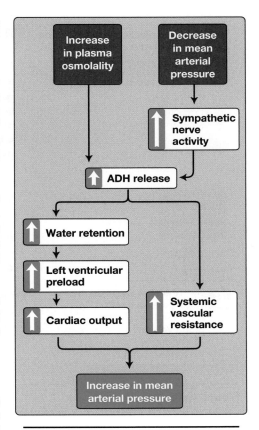

Figure 19.14.
Antidiuretic hormone (ADH) effects on blood pressure.

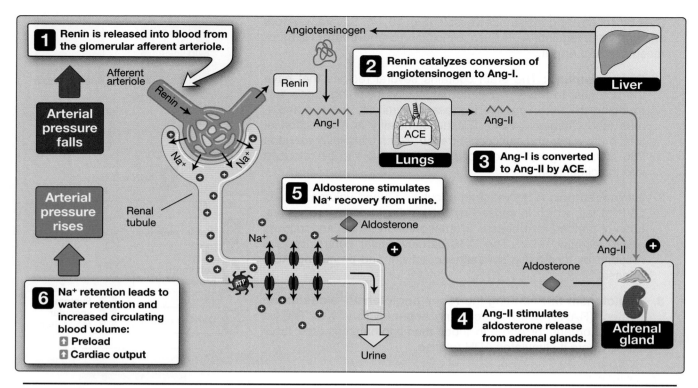

Figure 19.15.
Renin–angiotensin–aldosterone system. ACE = angiotensin-converting enzyme, Ang-I and Ang-II = angiotensin I and II.

2. **Aldosterone:** Aldosterone targets **principal cells** in the renal tubule distal segments (see Chapter 27·IV·B). It has multiple actions, all of which promote recovery of Na$^+$ and osmotically obligated water from the tubule. Aldosterone upregulates expression of genes encoding Na$^+$ channels and pumps, which is why it takes up to 48 hours for this pressure control pathway to become maximally effective.

3. **Renin:** The afferent arteriole of the renal glomerulus has baroreceptive properties. When renal perfusion pressure decreases, the afferent arteriole releases renin, initiating RAAS. Release is potentiated by the SNS, which activates following a drop in MAP.

4. **Atrial natriuretic peptide:** Atrial myocytes synthesize and store ANP, releasing it when stretched by high filling volumes. ANP has multiple sites of action along the length of the kidney tubule, all of which are geared toward excretion of Na$^+$ and water. The ventricles release a related compound, **brain natriuretic peptide** (BNP). Originally isolated from porcine brain, BNP has similar release characteristics and actions as ANP.

D. Sodium intake

Just as thirst stimulates water intake, salt craving triggers a need to ingest NaCl. Salt appetite is controlled by CNS areas that include the nucleus accumbens in the forebrain (an area associated with reward-seeking behaviors) and is stimulated by aldosterone and Ang-II.

V. VENOUS RETURN

The long-term arterial pressure–control pathways are all geared toward increasing circulating blood volume. The added volume finds its way to the venous compartment, where it raises CVP and increases left-ventricular preload. Preloading produces handsome payoffs in terms of ability to generate and sustain MAP. Blood localizes to the venous compartment because veins are thin-walled vessels that readily distend to accommodate volume. By contrast, the arterial system comprises a series of high-pressure, narrow-bore tubes that have a very limited capacity (~11% of total blood volume) as shown in Figure 19.16. Capillaries are numerous but hold even less blood than arteries (~4% of total). The cardiopulmonary system also has a very limited capacity. The venous system at rest typically contains >65% of total blood volume, creating an invaluable reservoir that can be mobilized by venoconstriction for use elsewhere should the need arise. However, the features that make veins a good reservoir also allow them to trap blood and limit VR under certain circumstances. When VR is reduced, CO is reduced also. Thus, any consideration of how the cardiovascular system functions as a unit must include an understanding of the role and limitations of the venous system.

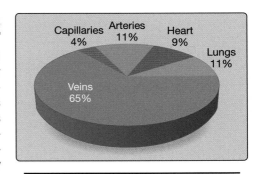

Figure 19.16.
Blood distribution within the cardiovascular system.

A. Venous reservoir

Veins' thin walls allow them to collapse easily when intraluminal pressures fall (Fig. 19.17). Conversely, increasing venous pressures by a few mm Hg causes veins to distend with minimal resistance. Once the system reaches capacity, veins must be stretched to accommodate additional volume.

Clinical Application 19.2: Coronary Artery Bypass Grafts

Coronary heart disease (CHD) is a leading cause of death in the western hemisphere. Patients usually present with angina or myocardial infarction caused by coronary artery occlusion, typically as a result of atherosclerosis. Treatment options may include revascularization with coronary artery bypass grafting ([CABG] or, familiarly, "cabbage") surgery. Although the walls of veins are much thinner and less muscular than those of arteries, they do have considerable strength. This makes it possible to use them as surrogate coronary supply vessels during CABG. Surgery involves removing a donor vein (usually the greater saphenous leg vein) and grafting a segment between the aorta and a site distal to the occlusion. A grafted vein is reversed in orientation to allow blood to flow freely through the valves.

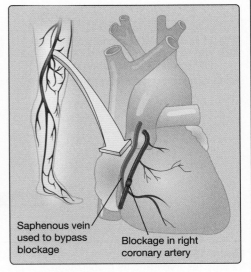

Saphenous vein used to bypass blockage

Blockage in right coronary artery

Use of veins for coronary artery bypass grafting.

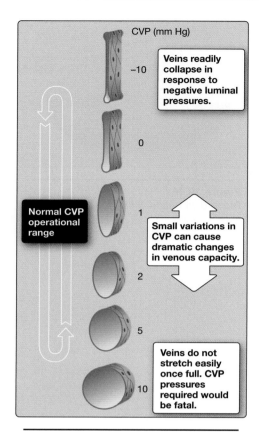

CVP (mm Hg)

−10 — **Veins readily collapse in response to negative luminal pressures.**

0

Normal CVP operational range

1 — **Small variations in CVP can cause dramatic changes in venous capacity.**

2

5

10 — **Veins do not stretch easily once full. CVP pressures required would be fatal.**

Figure 19.17.
Effects of filling pressure on venous capacity. CVP = central venous pressure.

B. Venoconstriction

Vein walls contain layers of VSMCs that are innervated by and contract during sympathetic activation. However, whereas arteries are capable of contracting down to the point of occlusion, venoconstriction is limited by a vein's unique microanatomy.

1. **Vein anatomy:** The VSMCs contained within the walls of veins are attached in series with collagen filaments. The filaments are folded and coiled in a relaxed vein, slowly unfolding and tensing as the vessel fills (Fig. 19.18). The filaments effectively limit the extent to which internal diameter can be reduced by venoconstriction. Note that veins also contain valves that prevent retrograde flow during venoconstriction.

2. **Effects of venoconstriction:** Venoconstriction has three principal effects: It mobilizes the blood reservoir, reduces overall capacity, and decreases transit time. It has minimal effect on flow resistance.

   a. **Mobilization:** Venoconstriction raises venous pressure by a few mm Hg and drives blood out of the reservoir. Valves ensure that blood is forced forward toward the heart, where it preloads the left ventricle and increases CO through the Frank-Starling mechanism (see Chapter 17·IV·D).

   b. **Capacity:** Venoconstriction decreases the internal diameter of veins and thereby decreases system capacity. Blood that had previously resided in veins is ultimately transferred to the capillary beds that supply active tissues.

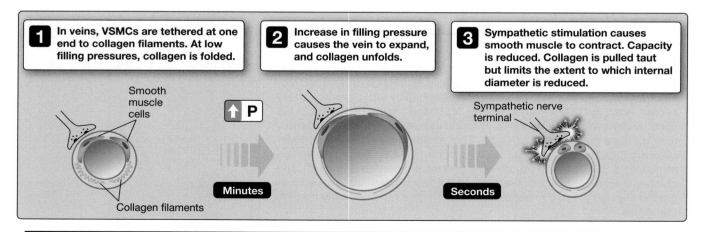

1 In veins, VSMCs are tethered at one end to collagen filaments. At low filling pressures, collagen is folded.

2 Increase in filling pressure causes the vein to expand, and collagen unfolds.

3 Sympathetic stimulation causes smooth muscle to contract. Capacity is reduced. Collagen is pulled taut but limits the extent to which internal diameter is reduced.

Smooth muscle cells

↑ P

Minutes

Collagen filaments

Sympathetic nerve terminal

Seconds

Figure 19.18.
Venous reservoir and its mobilization. P = pressure, VSMC = vascular smooth muscle cell.

At rest, vein walls do not exert appreciable force on the contents within. This is referred to as **unstressed blood volume**. Venoconstriction increases intravenous pressure (thus creating wall stress) and forces the stressed blood volume onward to the right atrium.

c. **Transit time:** Reducing system capacity reduces the amount of time that blood takes to traverse the system and thereby increases the rate at which it can be reoxygenated and forwarded to active tissues.

d. **Resistance:** The venous system remains a low-resistance pathway even after sympathetic activation, and there is no significant effect on flow resistance.

C. Venous pump

The high capacity of veins and their tendency to swell in response to even mild filling pressures means that large volumes of blood can easily become trapped in the lower extremities under the influence of gravity. Venoconstriction can reduce system capacity, but, in the absence of any additional motive force, the resultant loss of VR can eventually threaten CO and the ability to maintain arterial pressure. Blood pooling in veins is normally prevented by a **venous pump** (or skeletal **muscle pump**). Whenever skeletal muscles contract, they compress the blood vessels running between the fibers. Intravenous pressures are low, so compression readily collapses the veins and expels their contents. Valves ensure that resultant flow is in the direction of the heart (Fig. 19.19). Rhythmic contraction and relaxation of leg muscles effectively pumps blood upward against gravity, simultaneously "milking" blood from the pedal vasculature and ensuring continued VR.

D. Cardiac output and venous return

The example earlier demonstrates that CO is limited by the rate at which blood traverses the vasculature. Movement through the vasculature is, in turn, dependent on CO. Thus, a true understanding of how the cardiovascular system functions *in vivo* requires that the interdependence between CO and VR be appreciated.

1. **Return supports output:** The preload dependence of CO is defined by the **cardiac function curve** (Fig. 19.20). Increases in left-ventricular filling pressure increase CO through length-dependent activation (see Chapter 17·IV·D), and filling pressure is dependent on CVP. Changes in ventricular inotropy modify this relationship: Positive inotropes shift the curve upward and to the left, whereas negative inotropes shift the curve downward and to the right (see Chapter 17·IV·F).

2. **Output creates return:** Quantifying how CO affects CVP requires that the heart and lungs be replaced with an artificial pump whose output can be controlled (Fig. 19.21A). Prior

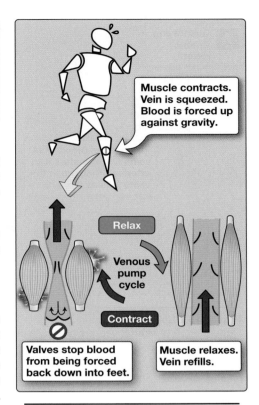

Figure 19.19.
Venous pump.

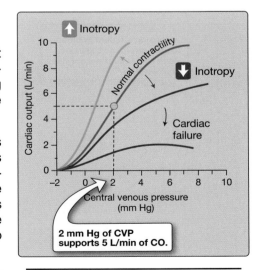

Figure 19.20.
Cardiac function curves. CO = cardiac output, CVP = central venous pressure.

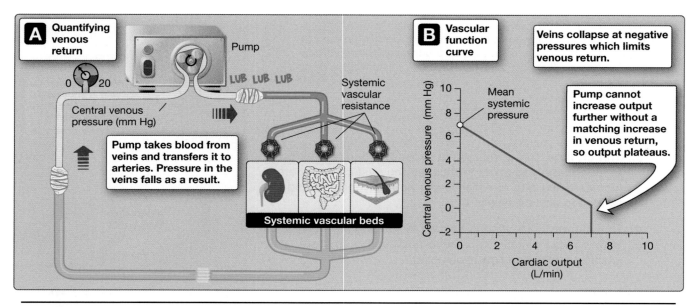

Figure 19.21.
Dependence of central venous pressure on cardiac output.

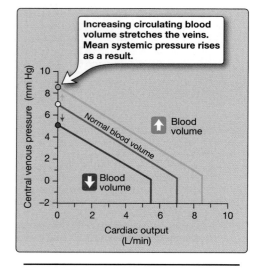

Figure 19.22.
Vascular function curves.

to turning the pump on, normal circulating blood volume (5 L) must be restored. The vasculature stretches when accommodating this much blood, creating a pressure of ~7 mm Hg (see Fig. 19.21B) known as **mean systemic pressure (MSP)**. MSP is the pressure that exists in the vasculature when the heart is arrested, and all parts of the system have come into equilibrium. When the pump is turned on, it translocates blood from the veins to the arteries. Because the arterial compartment has a relatively small volume, and outflow is limited by resistance vessels, translocation generates significant pressure within the arterial system. It simultaneously causes CVP to fall because blood is being withdrawn. Driving the pump faster causes CVP to fall further until it finally becomes negative (see Fig. 19.21B). At this point, the great veins collapse and limit any additional increases in CO.

MSP is also known as mean circulatory filling pressure (MCFP). Mean systemic filling pressure (MSFP) is the pressure that exists within the systemic circulation in human subjects a few seconds after cardiac arrest or after inhibiting venous return using positive airway pressures during mechanical ventilation. MSFP is commonly used as a surrogate for MSP, although MSFP is higher (~20 mm Hg).

3. **Circulating blood volume:** The vascular function curve is dependent on circulating blood volume (Fig. 19.22). If blood volume increases, MSP necessarily increases also because the vasculature stretches to a greater degree to accommodate the extra

volume. When the pump is turned on, CVP falls as before, but, because overall pressure in the system is higher, collapse of the large veins is delayed. Conversely, if circulating blood volume is decreased, then MSP decreases, and collapse of the great veins occurs at lower output levels.

4. **Venous capacity:** Venoconstriction and venodilation caused by changes in SNS activity produce similar effects to changes in circulating blood volume. Initiating a baroreflex reduces venous system capacity, and MSP rises. Venodilation increases system capacity, and MSP falls.

5. **Systemic vascular resistance:** Constriction and relaxation of resistance vessels has little or no effect on MSP because the contribution of the small arteries and arterioles to overall vascular capacity is small. Changes in SVR do impact CVP, however. When resistance vessels constrict, they reduce flow through the capillary beds. This translates into less VR, and CVP falls (Fig. 19.23). Conversely, vasodilation allows blood to surge through capillary beds and into the venous system, which raises CVP.

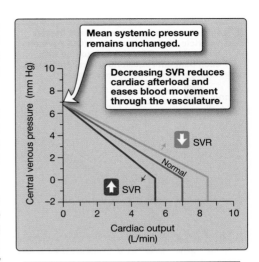

Figure 19.23.
Systemic vascular resistance (SVR) effects on vascular function curves.

E. Heart and vein interdependence

The cardiac and vascular function curves can be combined to create a single **cardiovascular function curve** (Fig. 19.24). The two plots overlap at an equilibrium point that defines how much CO can be supported by the vasculature for any given contractility and blood volume. In the example shown, the equilibrium point resides at a CVP of 2 mm Hg and a CO of 5 L/min. In the absence of any changes, the system cannot stray permanently from this equilibrium point because 2 mm Hg of pressure is required to support 5 L/min of output, and any increase in CO would drop CVP below 2 mm Hg (Fig. 19.25). If HR were suddenly slowed to reduce CO, the reduced amount of blood being translocated from veins to arteries would cause it to dam up in the right atrium, and CVP would rise. CVP equates with preload, so SV and CO would increase on the next beat. Re-equilibration might require several beats to accomplish, but eventually, CO and CVP would settle back to 5 L/min and 2 mm Hg.

F. Moving the equilibrium point

Sustained increases and decreases in CO require that the cardiac and vascular function curves be modified to establish a new equilibrium point. The former is accomplished through changes in inotropy, the latter through changes in circulating blood volume.

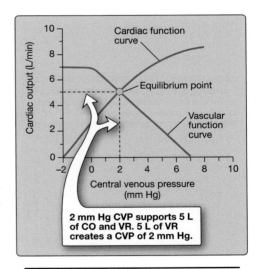

Figure 19.24.
Cardiovascular function curve. CO = cardiac output, CVP = central venous pressure, VR = venous return.

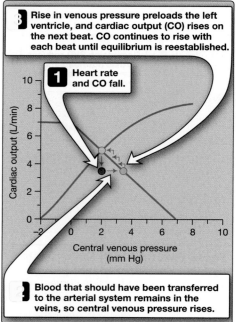

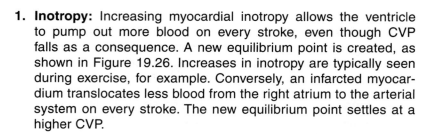

Figure 19.25.
Changes in output cannot be sustained away from equilibrium.

1. **Inotropy:** Increasing myocardial inotropy allows the ventricle to pump out more blood on every stroke, even though CVP falls as a consequence. A new equilibrium point is created, as shown in Figure 19.26. Increases in inotropy are typically seen during exercise, for example. Conversely, an infarcted myocardium translocates less blood from the right atrium to the arterial system on every stroke. The new equilibrium point settles at a higher CVP.

2. **Circulating blood volume:** Transfusing a patient with blood, for example, raises CVP and increases preload, permitting a higher CO in the absence of any change in inotropy. Conversely, hemorrhage reduces preload, and the equilibrium point shifts to a lower value for CO (see Fig. 19.26). Similar effects can be achieved acutely with venoconstriction and venodilation, respectively.

3. **Sympathetic activation:** Figure 19.26 shows the effects of isolated changes in cardiac inotropy and CVP. However, in practice, when the MAP falls and the SNS activates, both parameters change simultaneously, along with SVR. Figure 19.27 shows how these changes might affect the cardiovascular function curve. Note that the effects on venoconstriction or venodilation on MSP are partly offset by simultaneous changes in SVR.

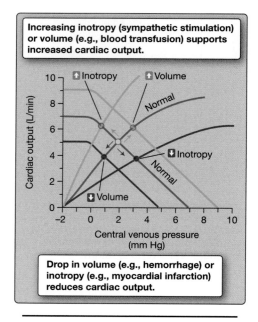

Figure 19.26.
Changes to the cardiovascular function curves.

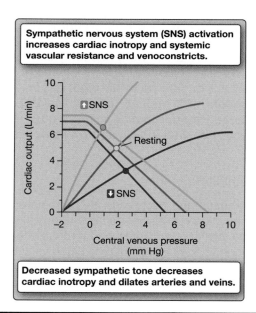

Figure 19.27.
Effects of changes in sympathetic tone on cardiovascular function curves.

Chapter Summary

- Cardiovascular system capacity greatly exceeds its contents. Blood flow to organs must be carefully metered to maintain pressure and flow through the entire system. Flow to tissues is regulated by **resistance vessels** (small arteries and arterioles).

- All tissues can demand additional flow to support increased activity. Control is accomplished locally through release of **metabolites** and **paracrines**, which either constrict or dilate resistance vessels. Resistance vessels are also **stretch sensitive**, allowing for reflex vasoconstriction during pressure surges. The ability of tissues to match their own blood supply to prevailing needs is called **autoregulation**.

- Resistance vessels are innervated by the **sympathetic nervous system** (**SNS**) which vasoconstricts when active. Sympathetic innervation allows the SNS to override local controls when arterial pressure is low.

- Arterial pressure is maintained via a **baroreceptor reflex**, a simple neural feedback pathway that adjusts cardiac output (CO) and vascular resistance in response to arterial pressure changes.

- Arterial pressure is sensed primarily by **baroreceptors** located in the carotid sinus and aortic wall. Secondary sensors include **cardiopulmonary baroreceptors**, located within the atria and pulmonary circulation.

- The **cardiovascular control center** is located in the brainstem. The **cardioinhibitory** and **cardioacceleratory centers** control heart rate and inotropy. The **vasomotor center** controls blood vessels and adrenal glands. A decrease in blood pressure initiates a reflex increase in heart rate, myocardial contractility, vascular resistance, and ventricular preload (through **venoconstriction**). Blood pressure rises as a consequence.

- The baroreceptor reflex is used for short-term pressure adjustments. Long-term pressure control involves the kidneys and modulation of total body water (TBW) and Na^+.

- TBW is controlled by **antidiuretic hormone** (**ADH**) release from the **posterior pituitary**.

- Na^+ retention is controlled by the **renin–angiotensin–aldosterone system** (**RAAS**). Renal glomerular afferent arterioles release **renin** when blood pressure is low. Renin proteolyzes **angiotensinogen** to **angiotensin I**, which is then converted to **angiotensin II** (**Ang-II**) by **angiotensin-converting enzyme**. Ang-II promotes Na^+ recovery from the kidney tubule, stimulates ADH release, and constricts resistance vessels.

- Na^+ and water retention increases **circulating blood volume** and raises **central venous pressure** (**CVP**). Increased CVP increases left-ventricular preload and CO.

- Veins contain valves that ensure unidirectional blood flow and help move blood upward against gravity. Rhythmic compression of leg veins by active muscles milks blood out of the lower extremities (**venous pump**) when walking or running.

- The dependence of CO on **venous return** means that CO can only be increased if CVP is sufficient to support the new level of output.

Study Questions

Choose the ONE best answer.

19.1. Skeletal muscle metabolism increases dramatically during physical activity, sustained by equally dramatic perfusion increases. Which of the following mechanisms most likely facilitates activity-induced increases in muscle blood flow?

  A. Antidiuretic hormone release
  B. Flow-induced nitric oxide release
  C. Histamine release
  D. Norepinephrine-induced vasodilation
  E. Rising metabolite levels

Best answer = E. Increased blood flow to active tissues ("active hyperemia") is mediated by local accumulation of metabolic byproducts, including CO_2, H^+, and adenosine, which cause reflexive dilation of resistance vessels (see Section II·A). Antidiuretic hormone (A) from the posterior pituitary and norepinephrine (D) from sympathetic nerve terminals both constrict resistance vessels. Flow-induced nitric oxide release (B) may contribute to increased flow at high levels of cardiac output (see Section II·E), but this effect is secondary to the influence of metabolic byproducts. Histamine (C) can cause vasodilation but usually only as a part of an allergic reaction.

19.2. An 11-year-old female wrestling with her younger brother causes him to faint when inadvertently applying pressure on his left carotid sinus. Syncope most likely occurred as a result of which of the following?

  A. Carotid baroreceptor stimulation
  B. Carotid chemoreceptor stimulation
  C. Cerebral vasculature vasoconstriction
  D. Occlusion of his carotid artery
  E. Occlusion of his jugular vein

Best answer = A. Applying pressure to the carotid sinus stimulates the baroreceptors within the vessel walls, mimicking the effects of increased blood pressure (see Section III·A). This promotes a reflex decrease in cardiac output and systemic vascular resistance (SVR). Arterial pressure falls as a result, causing cerebral hypotension and syncope. Carotid body chemoreceptors (B) do not directly sense blood pressure, but, when stimulated, they increase SVR and blood pressure. Occlusion of only one of the cerebral arteries (C, D) or jugular veins (E) is unlikely to decrease cerebral perfusion or change cerebral CO_2 levels sufficiently to cause syncope.

19.3. A 45-year-old female faints when standing up following a 90-minute Medical Physiology lecture. Which of the following variables most likely increases in a healthy compensating person after rising to an upright position?

  A. Aortic baroreceptor firing rate
  B. Cerebral venous pressure
  C. Left-ventricular preload
  D. Right-ventricular preload
  E. Systemic vascular resistance

Best answer = E. When a person stands upright, blood pools in the lower extremities (see Section III·D). Decreased venous return causes right- (D) and then left-ventricular preload (C) to fall, which reduces ventricular stroke volume and cardiac output. Arterial pressure begins to fall as a result, which is sensed via a decrease in arterial (carotid sinus and aortic) baroreceptor firing rate (A). If the pressure drop is severe, cerebral blood flow (B) can be compromised. Normally compensating individuals tolerate the upright position by initiating a baroreceptor reflex, which includes an increase in systemic vascular resistance.

19.4. A 60-year-old female is semi-recumbent in a hospital bed. One of her functional goals is to retain muscle mass during her hospital stay. Physical therapy prescribes and oversees quad sets (isometric contractions of her quadriceps) and ankle pumps (rhythmic contractions of her soleus and gastrocnemius muscles) as part of her rehabilitation plan. Her central venous pressure increases during these exercises. Which of the following is the most likely mechanism?

  A. Cardiac autoregulation
  B. Decrease in cardiac stroke volume
  C. Frank-Starling effect
  D. Increased heart rate
  E. Venous pump

Best answer = E. Muscle contractions engage the venous pump, which pushes blood from the lower extremities back toward the right heart (see Section V·C). Retrograde flow between contractions is prevented by venous valves. Cardiac autoregulation (A) helps to match coronary blood flow to myocardial needs, not venous blood flow (see Section III·B). Blood volume moving into the vena cava increases central venous pressure (CVP). Increasing CVP increases, not decreases, cardiac stroke volume (B) via the Frank-Starling relationship (C; see Section IV·D). Heart rate increases (D) during exercise, for example, and tends to decrease, not increase, CVP.

Special Circulations

20

I. OVERVIEW

The peripheral vasculature serves a variety of organs whose functional diversity has required specialization of circulatory design and control (Fig. 20.1). The heart and brain have minimal anaerobic capabilities, which makes them highly dependent on their blood supply for normal activity. Coronary and cerebral vascular control is, therefore, dominated by local regulatory mechanisms, which facilitate accurate matching of O_2 supply with tissue demand. By contrast, splanchnic circulatory control is dominated by central regulatory mechanisms. The organs that comprise the gastrointestinal (GI) system require significant amounts of blood to support their digestive and absorptive functions when active (see Fig. 20.1B). Digestion involves secreting copious amounts of fluid into the GI tract lumen. This fluid is derived from blood that must be supplied by the splanchnic vasculature. The digestive organs are able to communicate their need for increased blood flow through local control mechanisms, but the central nervous system (CNS) retains the ability to shut off splanchnic blood flow completely if an urgent need for blood arises elsewhere in the cardiovascular system. Regulation of blood supply to skeletal muscles is unusual in that the relative dominance of central and local controls changes according to the needs of the muscle relative to those of body as a whole.

II. CEREBRAL CIRCULATION

The cerebral circulation supplies the brain, an organ that represents only 2% of body weight yet commands 15% of resting cardiac output (CO; see Fig. 20.1A). This demand for blood flow reflects the brain's high rate of metabolism. Brain tissue has few metabolic stores and is highly dependent on oxidative pathways for energy production. In practice, this dependence means that the brain is highly reliant on the cerebral circulation for normal function.

A. Anatomy

Four major arterial branches supply the brain: the left and right vertebral arteries and the left and right internal carotid arteries (Fig. 20.2). The vertebral arteries join to form the basilar artery, which travels via the brainstem to the base of the brain and then splits again and links via communicating arteries to the internal carotids, creating the **circle of Willis**. Extensive interconnections between adjacent arteries allow for continued flow around potential sites of blockage. Because gravity normally assists rather than hinders return of cerebral venous blood to the heart (unlike venous return from regions below the heart), cerebral veins do not contain valves.

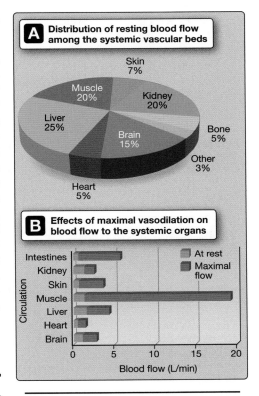

Figure 20.1.
Blood flow to organs of the systemic circulation.

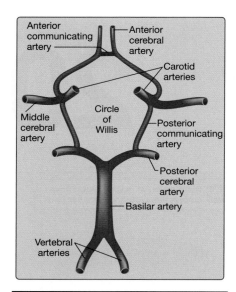

Figure 20.2.
Major arteries supplying the cerebral circulation.

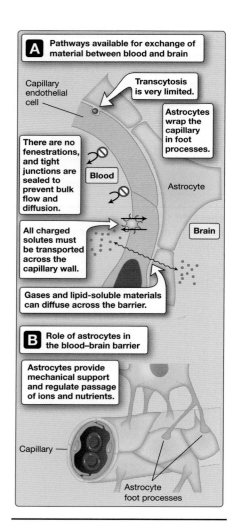

Figure 20.3.
Blood–brain barrier.

B. Blood–brain barrier

Blood contains a variety of chemicals that could adversely affect brain function. These include many hormones and neurotransmitters, such as epinephrine, glycine, glutamate, and ATP. The brain is protected from these and other chemicals by the **blood–brain barrier** (**BBB**) that provides three layers of defense: physical, chemical, and cellular.

1. **Physical barrier:** Capillaries in most circulations are relatively leaky, allowing materials to exchange across their walls by four general mechanisms (see Chapter 18·VI). Cerebral capillaries are uniquely modified to deny passage by most of these routes.

    a. **Barrier structure:** Cerebral capillary endothelial cells rarely exhibit transcytotic vesicles, which are a major route for transport in skeletal muscle vasculature, for example. Cerebral capillary walls contain no fenestrations, and adjacent endothelial cells are fused together by impermeable tight junctions (Fig. 20.3A). This effectively blocks bulk flow and diffusion of ions and water.

    b. **Barrier permeability:** Lipid-soluble molecules such as O_2 and CO_2 can readily diffuse across the capillary wall, but all other materials must be carried across. Cerebral endothelial membranes are dense with transporters that deliver glucose, amino acids, choline, monocarboxylic acids, nucleotides, and fatty acids to the brain. Movement of ions and protons across the BBB is regulated by channels, exchangers, and pumps. Aquaporins (AQPs) allow water to migrate freely between the blood and the brain in response to changes in extracellular fluid osmolality, however.

    c. **Barrier breaks:** Notable breaks occur in the BBB. Circumventricular organs (CVOs) are CNS specializations tasked with sampling blood composition or secreting hormones (see Chapter 7·VII·C). Both functions require easy neuronal access to blood, so cerebral capillaries in these regions are fenestrated and leaky to facilitate exchange of ions and small proteins.

2. **Chemical barrier:** Cerebral capillary endothelial cells contain monoamine oxidase, peptidase, acid hydrolase, and a variety of other enzymes that are capable of degrading hormones, transmitters, and other biologically active molecules. They provide a chemical barrier to bloodborne factors. The apical (luminal) membrane of cerebral capillary endothelial cells contains P-glycoprotein, also known as "multidrug resistance transporter," first discovered in cancer cells. It protects the brain against potentially toxic lipophilic drugs by returning them to the blood.

3. **Supporting structures:** Cerebral capillaries have a thick basal lamina and are supported mechanically by astrocytes (see Fig. 20.3B). Astrocytes also maintain tight-junction integrity and regulate exchange of material between the blood and the brain (see Chapter 5·V).

> The impenetrability of the BBB presents a logistical problem when trying to treat intracranial infections or tumors. In such cases, it may be necessary to shock the endothelium osmotically in order to create a temporary breach that allows passage of antibiotics and chemotherapy agents.

C. Regulation

The cerebral circulation is adept at responding to changes in mean arterial pressure. Autoregulation of cerebral blood flow is accomplished primarily through local control mechanisms.

1. **Local controls:** Cerebral resistance vessels dilate in response to the same metabolic factors that allow for local control in other circulations, but they are especially sensitive to changes in the partial pressure of CO_2 in blood ([P_{CO_2}] Fig. 20.4). Small increases in P_{CO_2} cause profound vasodilation, whereas decreases in P_{CO_2} cause vasoconstriction. P_{CO_2} sensitivity explains why hyperventilating can cause loss of consciousness. When a person breathes at rates beyond physiologic needs, blood P_{CO_2} falls. Loss of vasodilator influence causes the cerebral resistance vessels to constrict, and the individual becomes lightheaded. Normal cerebral flow can be restored by rebreathing expired air to increase the CO_2 content of alveolar air. Arterial P_{CO_2} increases as a result, the cerebral vessels dilate, and flow renormalizes.

2. **Central controls:** Cerebral resistance vessels are innervated by both branches of the autonomic nervous system (ANS). The nerves are active, but their effects on the vascular smooth muscle cells are typically minor compared with responses to metabolites.

D. Regional flow patterns

The cranium affords mechanical protection to the brain, but it also physically constrains cerebral blood flow and volume. All tissues of the body swell when metabolic activity rises, reflecting an increased volume of blood flowing through the vasculature. In the brain, demand for blood flow similarly increases and decreases as mental focus shifts, and the neurons become more or less active. The brain resides within the skull, however, which prevents any changes in intracranial volume. Brain tissues also resist volume changes because they are largely composed of fluid, which is incompressible. Thus, regional increases in cerebral flow that accompany increased activity are typically matched by opposing changes in a different brain area (Fig. 20.5).

E. Flow interruption

The brain has a very low tolerance for ischemia. A 20% to 30% decrease in cerebral flow causes lightheadedness. A 40% to 50% decrease causes fainting (**syncope**). Complete interruption of flow for >4 to 5 minutes can cause organ failure and death. Cerebral vessels that have narrowed with age or disease may cause a **transient ischemic attack (TIA)**, a localized reduction in flow and loss of cerebral function lasting minutes or hours. Interruptions in cerebral flow (**cerebrovascular accidents**, or **strokes**) occur when a cerebral vessel is occluded. Such events cause infarction and more permanent neurologic defects.

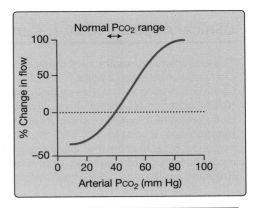

Figure 20.4.
Cerebral blood flow dependence on P_{CO_2}.

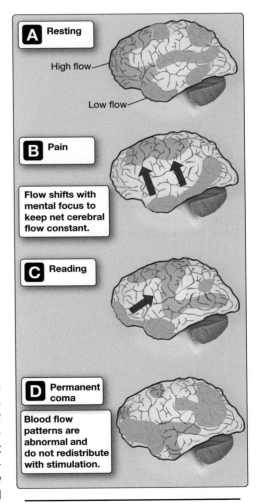

Figure 20.5.
Shifting regional cerebral flow patterns.

Clinical Application 20.1: Central Nervous System Ischemic Response

The brainstem houses the cardiovascular control center, a region that has autonomic connections to the heart and peripheral vasculature (see Chapter 7·VI). Loss of blood flow to the cardiovascular center causes the neurons to depolarize and become spontaneously active as their ion pumps fail and ion gradients dissipate. The result is a **central nervous system (CNS) ischemic response**, a massive sympathetic outpouring that shuts down flow to the peripheral organs and causes blood pressure to rise to maximal levels. It represents the brain's last-ditch effort to preserve its own supply. The **Cushing reaction** is a special kind of CNS ischemic response caused by an increase in intracranial pressure. The hypothalamus responds by activating the sympathetic nervous system and raising arterial pressure, but this initiates a baroreflex-mediated slowing of heart rate. Bradycardia is usually accompanied by systolic hypertension and respiratory depression (**Cushing triad**) and is an indication of extreme intracranial pressure and impending herniation and death.

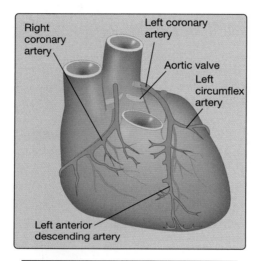

Figure 20.6.
Major coronary blood vessels.

III. CORONARY CIRCULATION

The coronary circulation supplies the myocardium, a tissue that rivals the brain in terms of its nutritional demands and the body's critical dependence on continued flow for normal function.

A. Anatomy

The cardiac vasculature is notable for its extensive cross-connections, the density of coronary capillaries, and venous drainage.

1. **Arteries:** The myocardium is supplied by left and right coronary arteries that originate from the root of the ascending aorta immediately above the aortic valve (Fig. 20.6). The right coronary artery generally supplies the right heart, whereas the left coronary artery supplies the left. The arteries course over the heart's surface and then dive down through the muscle layers, an arrangement that has functional consequences (see Section C). Adjacent arteries are extensively interconnected by numerous ~100-μm diameter collateral vessels.

2. **Capillaries:** Cardiac muscle has a capillary density of ~3,300/mm^2, more than eight times greater than that in skeletal muscle (~400/mm^2). The capillary-to-muscle fiber ratio is the same in both types of muscle (1:1), the density difference reflecting the thinness of cardiac muscle fibers as compared with their skeletal muscle counterparts (<20 μm vs. 50 μm, respectively). The high density maximizes the efficiency with which O_2 and nutrients can be delivered to cardiac muscle.

3. **Veins:** The great cardiac vein and anterior cardiac veins drain into the right atrium via the coronary sinus. Thebesian veins drain directly into the four cardiac chambers via small ostia, but their contribution to overall drainage is small. Cardiac veins are also extensively interconnected by collaterals.

B. Regulation

At rest, the coronary circulation receives ~5% of CO. Cardiac muscle extracts >70% of available O_2 from blood, and it has a very low capacity for anaerobic metabolism, much like the brain. This O_2 dependence means that any increase in work must be matched by an increase in coronary flow, achieved entirely through local control mechanisms.

1. **Local controls:** Coronary resistance vessels are exceptionally sensitive to adenosine. Local control mechanisms allow for a four- to fivefold increase in coronary flow when CO increases, a phenomenon called **coronary flow reserve** (Fig. 20.7).

2. **Central controls:** Coronary resistance vessels are innervated by both branches of the ANS, but their influence is overridden by local controls. Some individuals have resistance vessels that are abnormally susceptible to sympathetic vasoconstrictor influence, causing their coronary arteries to spasm during high sympathetic drive. The ischemia that results from these temporary flow interruptions is experienced as pain known as **vasospastic angina** (previously known as **Prinzmetal**, or **variant**, **angina**).

C. Extravascular compression

Blood flow through most systemic vascular beds follows the aortic pressure curve, rising during systole and falling during diastole. Flow through the left coronary artery drops sharply during systole and then rises sharply with the onset of diastole (Fig. 20.8). This unique flow pattern occurs because ventricular myocytes collapse their arterial supply vessels when they contract (**extravascular compression**) as shown in Figure 20.9. The effect is felt strongest during early systole because aortic pressure, the main force maintaining vascular patency, is at a low point. During diastole, the compressive forces are

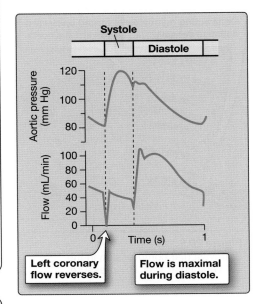

Figure 20.7.
Coronary flow reserve.

Clinical Application 20.2: Angina

Angina is a specific form of chest discomfort or pain associated with myocardial ischemia that usually occurs during activities or events that increase myocardial workload. **Typical (stable) angina** is not usually life threatening, and its symptoms can be reversed with drugs that reduce cardiac workload.[1] **Unstable angina** indicates a danger of the vessel becoming blocked completely. Typical interventions might include

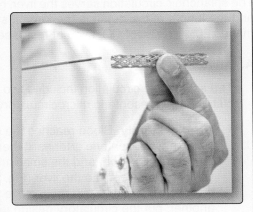

A metal stent.

balloon angioplasty or implanting of a metal sleeve (**stent**) to open the stenotic vessel, or coronary artery bypass graft surgery (see Clinical Application 19.2).

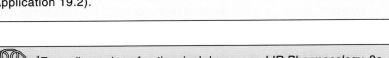
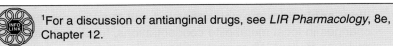

Figure 20.8.
Left coronary blood flow.

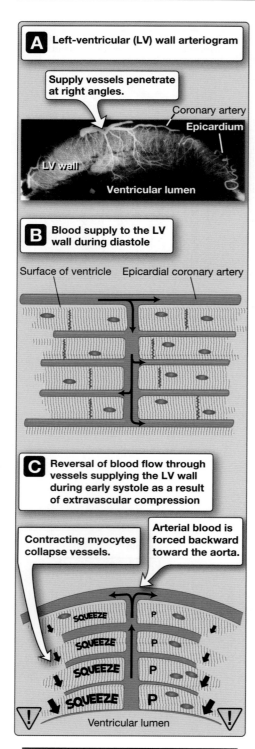

A Left-ventricular (LV) wall arteriogram

Supply vessels penetrate at right angles.

Coronary artery

Epicardium

LV wall

Ventricular lumen

B Blood supply to the LV wall during diastole

Surface of ventricle　　Epicardial coronary artery

C Reversal of blood flow through vessels supplying the LV wall during early systole as a result of extravascular compression

Contracting myocytes collapse vessels.

Arterial blood is forced backward toward the aorta.

SQUEEZE　P
SQUEEZE　P
SQUEEZE　P
SQUEEZE　P

Ventricular lumen

Figure 20.9.
Extravascular compression in the left-ventricular wall. P = pressure.

removed, and blood surges through the musculature at peak rates (see Fig. 20.8).

D. Flow interruption

Because the ventricular myocytes extract such high levels of O_2 from blood, a delicate balance exists between myocardial workload and coronary supply. If the balance is disturbed, then myocytes become ischemic and potentially infarcted. Most commonly, this occurs due to **atherosclerosis** and **coronary artery disease**.

1. **Atherosclerosis:** Atherosclerotic lesions appear at an early age in the populations of most Western countries. They evolve to become complex plaques of lipids, hypertrophied vascular myocytes, and fibrous material. Plaques enlarge at the expense of the vascular lumen and impair blood flow. This causes an imbalance between coronary supply and myocardial demand, resulting in ischemia. Ischemic myocytes release large quantities of vasoactive compounds, such as adenosine, but vasodilators have no effect on plaque. As the O_2 deficit continues, the myocytes release lactic acid, which stimulates pain fibers within the myocardium and causes angina pectoris.

2. **Collaterals:** Collaterals are usually constricted in a healthy heart, but, if a supply vessel becomes occluded, they dilate in response to rising metabolite levels. Flow through collaterals may prevent **infarction** if the occluded vessel is small. In time, these channels enlarge and may provide near-normal flow to the ischemic area.

IV. SPLANCHNIC CIRCULATION

The splanchnic circulation serves the liver, gallbladder, spleen, pancreas, and the length of the gut. It is the largest systemic circulation and commands 20% to 30% of CO, even at rest (see Fig. 20.1).

A. Anatomy

The splanchnic circulation includes two notable features: a serial circuit and two mesenteric microcirculatory specializations.

1. **Arterial supply:** Blood reaches the splanchnic circulation via the celiac and mesenteric arteries. It then passes through the spleen, pancreas, stomach, and the small and large intestines and drains via the portal vein into the liver (Fig. 20.10). This is one of the few body regions where two organs are arranged in series with each another, but the circulatory design is functional because it allows the liver to filter out bacteria and begin processing absorbed nutrients before the blood is returned to the general circulation.

2. **Specializations:** The intestines are served by the mesenteric circulation, which is a subdivision of the splanchnic circulation. Flow through individual mesenteric capillaries is regulated by single smooth muscle cells wrapped around their inlet. These **precapillary sphincters** are not innervated, but they are responsive to changes in local metabolite concentrations, functioning as on/off switches to flow. When metabolite levels are low, the sphincters

are contracted ("off"), and flow is inhibited. They relax intermittently ("on") as local metabolite levels rise, allowing blood to flow through the capillary. At rest, only a small proportion of sphincters is relaxed, but the pattern of capillary flow shifts continually (**vasomotion**). When intestinal workload increases following a meal, the sphincters spend a much greater percentage of time in the "on" position. When demand for intestinal flow is low, blood is able to bypass nonperfused capillaries via **metarterioles**, which provide cross-connections between mesenteric arterioles and venules.

B. Regulation

Blood flow in the splanchnic circulation is controlled locally and by the ANS, but the enteric nervous system is heavily involved also.

1. **Local controls:** The mechanisms by which the gut autoregulates are poorly delineated. Increases in flow during a meal may be triggered by metabolites, GI hormones, vasodilator kinins released from the intestinal epithelia, bile acids, osmolality increases, and by the products of digestion.

2. **Central controls:** The splanchnic vasculature is regulated by both ANS divisions. The parasympathetic nervous system increases blood flow both in anticipation of and while digesting a meal (classic "rest-and-digest" response). The sympathetic nervous system (SNS) constricts all splanchnic vascular beds during "fight-or-flight" responses, thereby shunting blood away from the GI tract for use elsewhere in the circulation.

C. Splanchnic reservoir

At rest, the splanchnic circulation holds ~15% of total circulating blood volume, representing a significant reservoir that can be tapped by the SNS should a more urgent need for blood arise elsewhere. The consequences of sympathetic activation on splanchnic flow depend on stimulation intensity.

1. **Mild sympathetic activation:** Mild SNS activation curtails splanchnic flow, but circulation renormalizes within minutes by reflexive resistance vessel dilation caused by rising metabolite levels, a phenomenon known as **autoregulatory escape** (Fig. 20.11). Moderate SNS activation (e.g., as seen during moderate exercise) produces a more persistent curtailment of splanchnic flow.

2. **Maximal sympathetic activation:** Vigorous exercise places intense demands on the cardiovascular system. Strong SNS stimulation of splanchnic resistance vessels reduces total flow to <25% of basal values, whereas venoconstriction forces 200 to 300 mL of blood from the splanchnic vasculature. Dependent tissues compensate for reduced flow by increasing O_2 extraction from the residual supply.

D. Circulatory shock

Severe hemorrhage and other forms of circulatory shock precipitate extreme levels of sympathetic activity that reduce splanchnic flow to minimal levels for prolonged periods (see Fig. 20.11). If flow is not

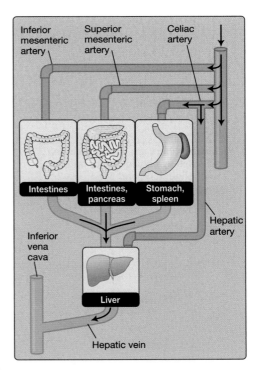

Figure 20.10.
Splanchnic circulation.

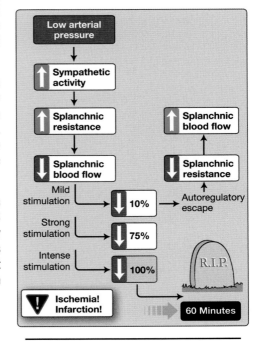

Figure 20.11.
Sympathetic dominance over splanchnic blood flow.

restored within an hour or so, the epithelial lining of the small intestine infarcts and the blood–gut barrier disintegrates. Breakdown allows toxic materials from the gut (bacterial enterotoxins and endotoxins) to enter the bloodstream, resulting in toxemia and septic shock (see Chapter 40·IV·C).

V. SKELETAL MUSCLE CIRCULATION

Resting flow to skeletal muscle is modest considering the mass of tissue it serves (~20% of CO). Such modesty belies the profound intrinsic vasodilatory capabilities of muscle during exercise, however.

A. Anatomy

When skeletal muscles are active, they are highly dependent on the vasculature to deliver O_2 and nutrients and to carry away heat, CO_2, and other metabolic waste products. These functions are facilitated by a dense capillary network. The skeletal vasculature is supplied by superficial feed arteries that branch multiple times within the muscle groups until they become terminal arterioles. Each arteriole gives rise to numerous capillaries that travel parallel to individual muscle fibers within a fascicle. Each fiber is typically associated with three or four capillaries, which reduces the range over which O_2 must diffuse to reach the innermost myofibrils to ~25 μm (Fig. 20.12).

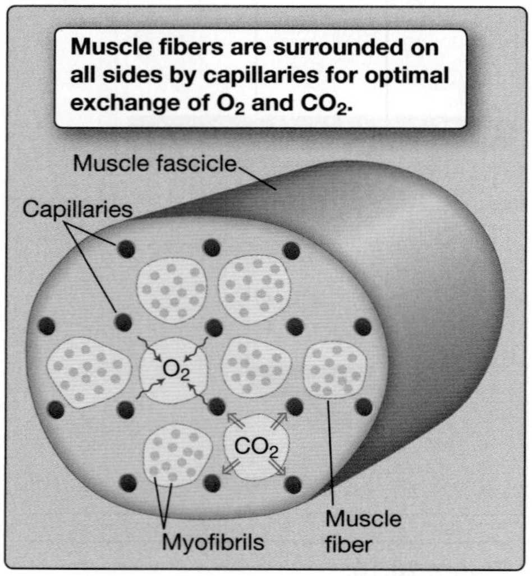

Figure 20.12.
Relationship between a skeletal muscle fiber and its blood supply.

B. Regulation

Local and central control mechanisms affect the skeletal vasculature with equal potency. These mechanisms can yield dramatic flow extremes depending on circumstance.

1. **Local controls:** In resting muscle, only a small percentage of capillaries are actively perfused because the terminal arterioles (resistance vessels) that feed them are constricted. When the muscle becomes active, metabolite concentrations rise, the arterioles dilate, and previously inactive capillaries now course with blood (**capillary recruitment**).

2. **Central controls:** Skeletal resistance vessels are richly innervated by SNS fibers whose resting tone maintains flow at minimal levels governed largely by a muscle's metabolic needs. When arterial pressure falls, SNS activity increases as a normal part of a baroreceptor reflex (see Fig 19.12). The consequences depend on whether the muscle is being exercised at the time.

   a. **At rest:** SNS activation decreases flow to a resting muscle. The increased resistance contributes to the rise in systemic vascular resistance that accompanies a baroreceptor reflex.

   b. **During exercise:** During exercise, local metabolite concentrations dominate vascular control. The potency of these mechanisms is such that flow through the skeletal vasculature may rise to >21 L/min during intense exercise, which amounts to >400% of resting CO (see Chapter 39·VI)!

C. Extravascular compression

Contracting muscles compress the blood vessels that run between the fibers and cause temporary interruptions in flow. Isometric contractions (e.g., induced by lifting a weight) can inhibit flow for several tens of seconds and are followed by a vigorous **reactive hyperemia** upon relaxation. Isotonic exercises, such as jogging and swimming, involve rhythmic cycles of contraction and relaxation and produce a phasic flow pattern (Fig. 20.13). Note that flow oscillates between two extremes with each contraction, but flow is increased overall (**active hyperemia**).

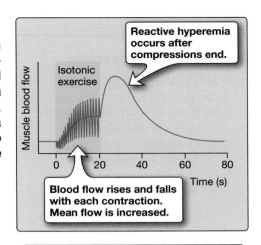

Figure 20.13.
Phasic blood flow pattern in skeletal vasculature during aerobic exercise.

Chapter Summary

- The **cerebral circulation** supplies the brain, a tissue notable for a high O_2 demand and its dependence on continued blood flow for function. Loss of flow initially causes fainting, with irreversible cellular changes following within minutes.

- The brain is protected from bloodborne agents by the **blood–brain barrier**. **Tight junctions** between adjacent endothelial cells create a physical barrier to ions and other water-soluble chemicals. Endothelial enzymes that degrade potentially threatening molecules provide a chemical barrier, and **astrocytes** provide mechanical support. All substrates needed for cerebral metabolism must be transported across the barrier.

- Flow through the cerebral vasculature is regulated by local metabolite concentrations. Cerebral resistance vessels are exceptionally sensitive to P_{CO_2}. Flow volume is limited by the cranium, so flow increases in one area of the brain are balanced by decreases in another area to keep overall volume constant.

- The **coronary circulation** similarly supplies tissues that are highly dependent on O_2 for continued function, and any increase in myocardial workload must be matched by increased flow. Coronary resistance vessels are regulated by local factors, primarily **adenosine**. Sympathetic nerve terminals produce no significant physiologic effect.

- Flow through the left coronary arteries decreases during early systole, reflecting the effects of **extravascular compression** during ventricular contraction. The left ventricle receives the bulk of its supply during diastole. If flow through the coronary vasculature becomes limiting, lactic acid concentrations rise and cause pain (**angina**). Interruptions in coronary flow cause ischemia and may result in **myocardial infarction**.

- The splanchnic circulation supplies all the organs involved in digestion and absorption of nutrients, including the liver, which is **arranged in series** with the other splanchnic organs.

- The **splanchnic vasculature** holds 25% of total circulating blood volume at rest. GI organs have low O_2 requirements, so flow can be diverted elsewhere during a hypotensive crisis without short-term risk of ischemia.

- The sympathetic nervous system dominates splanchnic vascular control. Prolonged splanchnic vasoconstriction (>1 hour) can destroy the intestinal epithelium, creating breaches in the barrier separating blood and intestinal contents.

- The **skeletal vasculature** is controlled by both central and local factors. When muscle activity increases, rising metabolite levels can increase flow >20-fold. When muscle is inactive, sympathetic constrictor influence can divert flow for use elsewhere.

Study Questions

Choose the ONE best answer.

20.1. A 14-year-old female feels faint while hyperventilating during a panic attack. Her light-headedness is most likely due to which of the following?

    A. Cerebral vascular spasms
    B. Decreased cardiac output
    C. Decreased cerebral P_{CO_2}
    D. Hypoxia
    E. Sympathetic activation

Best answer = C. Hyperventilation results in CO_2 being blown off to the atmosphere at inappropriately high rates, causing arterial P_{CO_2} to fall. Cerebral resistance vessel patency is closely related to cerebral arterial P_{CO_2}, such that hypoventilation can cause cerebral vasoconstriction, reduced cerebral blood flow, and syncope (see Section II·C). Coronary vasospasm (A) can cause angina. Hyperventilation can cause a reflex decrease in cardiac output (B), but the syncope is due to local effects of a P_{CO_2} decrease. Hypoxia (D) results from hypoventilation. Sympathetic activation (E) increases cerebral perfusion pressure.

20.2. A 55-year-old male with severe angina is scheduled for quadruple bypass surgery. The coronary arterioles downstream of the stenotic regions during the anginal episodes are dilated fully. What is the most likely cause of this vasodilation?

    A. Adenosine
    B. High-velocity flow
    C. Lactic acid
    D. Norepinephrine
    E. Parasympathetic activity

Best answer = A. Coronary resistance vessels are controlled by the needs of the myocardium via changes in local metabolite concentrations, especially adenosine (see Section III·B). The shear stress caused by high-velocity flow (B) can cause vasodilation via nitric oxide release but is unlikely in the setting of decreased perfusion and angina. Lactic acid (C) also causes vasodilation but to a lesser degree than adenosine. Norepinephrine (D) constricts blood vessels. The parasympathetic nervous system (E) does not have any significant role in coronary vessel regulation.

20.3. Coronary catheterization studies have indicated that flow can increase four- to fivefold to support an increased demand for cardiac output. Such an increase is most likely possible because of which of the following changes?

    A. Increased flow during diastole
    B. Increased O_2 extraction from blood
    C. Recruitment of collaterals
    D. Resistance vessel dilation
    E. Sympathetic stimulation of supply arteries

Best answer = D. Increases in cardiac output and coronary flow during exercise, for example, are supported by reflex resistance vessel dilation in response to rising local metabolite levels (see Section III·B). Flow during diastole does increase (A) with activity, but this is not the underlying cause of the flow increase. O_2 extraction from blood (B) is maximal at rest and cannot increase further. Collaterals (C) provide pathways for flow around sites of coronary occlusion. Sympathetic activation (E) has minimal effects on coronary supply vessels.

20.4. A 22-year-old male is participating in a 1,500-m freestyle swimming event. While swimming, blood flow to the active muscles increases by several orders of magnitude. This increase most likely results from which of the following?

    A. Decreased sympathetic activity
    B. Epinephrine release
    C. Increased venous pump activity
    D. Reactive hyperemia
    E. Rising metabolite levels

Best answer = E. The increase in skeletal muscle blood flow during exercise results from a rise in the concentration of various metabolites released by the active muscle fibers (see Section V·B). Sympathetic activity increases, not decreases (A), during exercise to support increased cardiac output and flow to active muscles. Epinephrine release (B) during exercise similarly increases cardiac output to support increased muscle blood flow. Venous pump activity increases (C) venous return to support cardiac output. Reactive hyperemia (D) refers to the increase in blood flow that follows cessation of exercise.

Lung Mechanics

21

I. OVERVIEW

All cells generate ATP to fuel their many activities. The preferred pathway for ATP formation is aerobic glycolysis, which requires a constant supply of carbohydrates and molecular oxygen (O_2), which are consumed in the process of **internal (cellular) respiration**. Carbohydrates must be hunted and gathered, but O_2 is freely available from the atmosphere and will readily enter the blood by diffusion if blood and atmosphere can be brought into close proximity. The main challenge to diffusional uptake is that meeting the O_2 demands of all dependent tissues requires a **blood–gas interface** with a surface area roughly half the size of a singles' tennis court! This engineering challenge has been met by developing a system of branching tubes (**airways**) that terminate in ~300,000,000 thin-walled sacs called **alveoli** (Figs. 21.1 and 21.2). Alveolar walls contain the blood–gas interface, and their combined surface area (~80 m^2) is more than adequate to meet the body's O_2 needs, even during strenuous exercise. The second main challenge to ensuring adequate O_2 uptake is in constantly refreshing alveolar gas contents. Air, like blood or water, requires that pressure be applied to it to make it move. The solution has been to fashion an air pump comprising the chest wall and diaphragm. The pump alternately creates negative and positive pressures within the thorax that cause air to flow in and out of the airways (**ventilation**). O_2 rapidly dissolves in blood and is then sequestered within red blood cells for transport to the various tissues. Aerobic respiration consumes O_2 but generates CO_2, which must be excreted. Fortunately, CO_2 also readily enters blood by diffusion, which allows it to be whisked away in the veins along with other waste products. On reaching the lungs, blood releases CO_2 to alveolar air, which is then pumped out of the body during expiration (**external respiration**). This chapter considers the features of the blood–gas interface and air pump. Later chapters in this Respiratory System unit discuss factors affecting gas exchange across the blood–gas interface (see Chapter 22), blood and gas transport mechanisms (see Chapter 23), and respiratory regulation (see Chapter 24).

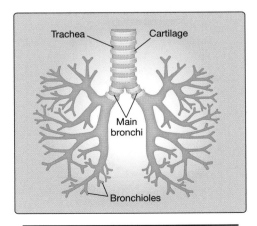

Figure 21.1.
Branching structure of the lung airways.

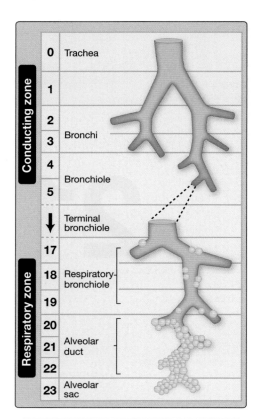

Figure 21.2.
Airway generations.

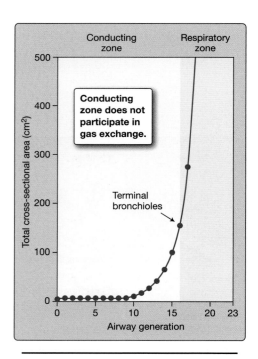

Figure 21.3.
Amplification of lung surface area.

II. AIRWAY ANATOMY

Each lung houses an elaborate system of **airways** and **alveolar sacs** (see Fig. 21.1). The airways channel air from the external atmosphere to the blood–gas interface. The airways begin with the trachea (generation 0) and then branch repeatedly to yield a bronchial tree. The tree contains ~23 branch **generations** (Fig. 21.2) and comprises two functionally distinct zones, a **conducting zone** and a **respiratory zone**.

A. Conducting zone

Airways in the conducting zone do not participate in gas exchange, they simply channel airflow. The larger airways (generations 0 through ~10) are supported structurally with **cartilage** to help maintain patency. Generations 10 through 16 are called **bronchioles**, with the terminal bronchioles (~generation 16) demarcating the end of the conducting zone. The conducting zone is lined with a mucus-secreting, ciliated epithelium. The cilia beat constantly, sweeping mucus and trapped particulates up and out of the lungs (**mucociliary escalator**).

> Tobacco smoke impairs respiratory cilia function. Ciliary arrest allows bacteria and other inhaled particulates to accumulate in the lungs, causing local irritation and epithelial inflammation. Smokers are prone to coughing episodes and bronchitis as a result. Ciliary function is typically restored with tobacco cessation.

B. Respiratory zone

The **respiratory zone** (generations 17–23) is characterized by a tremendous amplification of cross-sectional area even as the passages narrow (Fig. 21.3). The respiratory zone is the location of the blood–gas interface. The transition begins with **respiratory bronchioles**, which are conducting airways with walls so thin that they actively participate in gas exchange. Alveolar ducts are elongated airways distinguished by the presence of an occasional smooth muscle cell. Their walls are shared with a succession of alveolar sacs, much like a long hotel hallway in which doors to individual rooms stand wide open.

C. Alveolar sacs

Alveoli are thin-walled, polyhedral sacs with internal diameters of 75 to 300 µm (Fig. 21.4) interconnected via 2- to 3-µm **pores of Kohn**. The alveolar lining separates atmospheric air from the vasculature. It comprises two types of respiratory epithelial cell, or **pneumocyte**.

1. **Type I pneumocytes: Type I pneumocytes** are thin and flat. They make up the bulk of alveolar surface area (~90%).

2. **Type II pneumocytes: Type II**, or **granular**, **pneumocytes** are present in equal numbers but are more compact and therefore occupy less area. They are filled with numerous **lamellar inclusion bodies** that contain **pulmonary surfactant**. Type II cells are capable of rapid division, which allows them to repair alveolar wall damage. They subsequently transform into type I cells, which divide rarely.

3. **Blood–gas interface:** Pulmonary capillaries meander between adjacent alveolar sacs. Their density is so great that they create a near-continuous sheet of blood covering alveolar surfaces. The distance separating red blood cells from atmospheric air approximates the width of a capillary endothelial cell plus a pneumocyte (~300 nm total).

III. BLOOD SUPPLY

The lung receives blood from two different sources: **pulmonary** and **bronchial** circulations.

A. Pulmonary circulation

The **pulmonary circulation** brings O_2-poor venous blood from the right ventricle via pulmonary arteries to the blood–gas interface for gas exchange. Pulmonary veins then carry O_2-rich blood to the left side of the heart for delivery to the systemic circulation. The pulmonary circulation has a low vascular resistance, and, thus, mean pulmonary arterial pressures are low (~16 mm Hg; see Fig. 16.2C). The pulmonary circulation receives the entire output of the heart (~5 L/min at rest, ~25 L/min during strenuous exercise).

B. Bronchial circulation

The **bronchial circulation** is a systemic vascular bed that supplies the conducting airways with O_2 and nutrients. Bronchial arteries arise from the aorta and feed capillaries that drain either via bronchial veins or via anastomoses with pulmonary capillaries into veins of the pulmonary circulation. These connections allow small amounts of deoxygenated blood to bypass the blood–gas interface and reenter the systemic circulation without being oxygenated. This **venous admixture** represents a **physiologic shunt** that decreases pulmonary vein O_2 saturation by 1% to 2%.

IV. SURFACE TENSION AND SURFACTANT

Subdividing the lung into 300 million alveoli creates a large surface area for gas exchange, but there is a significant tradeoff. Each alveolus is moistened with a thin film of **alveolar lining fluid**. The fluid generates **surface tension**, which has consequences for lung performance.

A. Surface tension

Water molecules are much more strongly attracted to each other than to air. This attraction creates surface tension as the individual molecules within alveolar lining fluid draw close to each other and away from the air–water interface. Surface tension always minimizes the area of an exposed surface, which is why soap bubbles or falling raindrops become roughly spherical (Fig. 21.5). The moisture film within an alveolus means that it behaves much like a bubble, even though it maintains a connection with the pulmonary lumen during normal breathing. Surface tension is such a powerful force that alveoli and airways would collapse unless provided with a means of diminishing its effects (Fig. 21.6A).

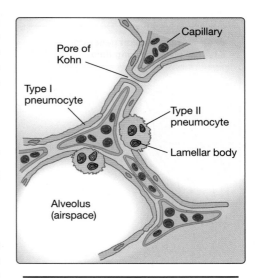

Figure 21.4.
Alveolar wall structure.

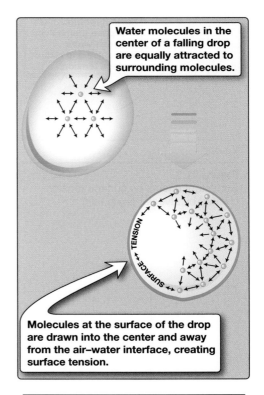

Water molecules in the center of a falling drop are equally attracted to surrounding molecules.

Molecules at the surface of the drop are drawn into the center and away from the air–water interface, creating surface tension.

Figure 21.5.
Origins of surface tension.

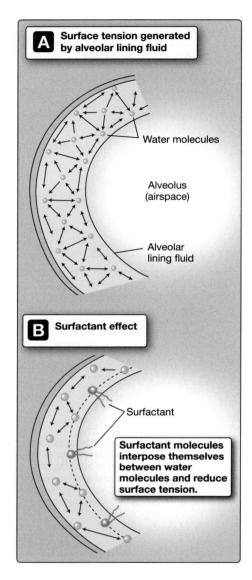

Figure 21.6.
Surfactant effects on surface tension created by alveolar lining fluid.

B. Surfactant

Type II pneumocytes synthesize and release pulmonary surfactant specifically to counter the effects of surface tension. Surfactant is a complex mix of lipids and proteins (see Fig. 21.6B).

1. **Composition:** Surfactant's principal component is dipalmitoyl phosphatidylcholine (DPPC), a phospholipid. Other components alter secretion rate, aid its distribution within the surface film, or defend the lung against pathogens. Surfactant is stored in lamellar bodies and exocytosed onto the alveolar surface as needed.

2. **Effects on surface tension:** DPPC and other surfactant phospholipids have hydrophilic head groups and hydrophobic tails. When secreted onto the alveolar surface, the molecules localize to the air–water interface, where they spread to form a monolayer (see Fig. 21.6B). Their tail groups orient toward the air-filled alveolar lumen, whereas the head groups remain immersed in the superficial aqueous layer. The polar nature of the head groups allows them to interact with and interpose themselves between adjacent water molecules, thereby weakening surface tension. The intensity of surfactant's effects increases in direct proportion to the density of molecules in the surface film.

3. **Functions:** Surfactant's importance in pulmonary function cannot be overstated. There are three main functions: stabilizing alveolar size, increasing lung compliance, and keeping lungs dry.

   a. **Stabilizing alveolar size:** When two bubbles of unequal size are connected, the smaller bubble collapses, and the larger one inflates (Fig. 21.7A). This phenomenon is explained by the **Laplace law**:

$$P = \frac{2T}{r}$$

   where P is pressure, T is surface tension, and r is bubble radius.

   The Laplace law predicts that pressure within a sealed bubble rises when its radius is reduced. If the collapsing bubble communicates with a larger bubble, rising pressure within the small bubble drives air into the larger bubble. Alveoli approximate bubbles (although their exact shape is more polyhedral), and all alveoli are interconnected via the pulmonary lumen. The Laplace law predicts sequential collapse of all but one alveolus! Although alveolar collapse (**atelectasis**) occurs with regularity *in vivo*, surfactant greatly reduces its extent. A decrease in alveolar volume decreases surface area, which concentrates surfactant molecules within the surface film (see Fig. 21.7B). Concentrating the molecules further weakens the forces that create surface tension, thereby preventing collapse. Conversely, alveolar expansion decreases surfactant molecule density and allows surface tension to dominate control of alveolar volume. Surfactant keeps alveolar diameter relatively stable throughout the lung.

Clinical Application 21.1: Infant Respiratory Distress Syndrome

Infants born prematurely have underdeveloped lungs that are incapable of producing levels of surfactant necessary to stabilize alveolar volume. Atelectasis is common, as are hyperexpanded regions of the lung. Collapsed alveoli cannot participate in gas exchange, and the infant becomes cyanotic as a result. The infant's lungs also have low compliance, which increases the work of breathing. **Infant respiratory distress syndrome (IRDS)** is characterized by hypoxia, tachypnea, tachycardia, and exaggerated breathing movements. Ventilatory failure is a likely outcome in the absence of medical intervention. IRDS infants are supported with mechanical ventilation and by delivering surfactant to the lungs until the lungs are sufficiently developed to produce adequate surfactant.

b. **Increasing compliance:** Lung compliance is a measure of the amount of pressure needed to inflate lungs to a given volume ($\Delta V/\Delta P$; see also Chapter 18·V·C). Surface tension decreases lung compliance and thus increases the effort required for inflation. Surfactant reduces surface tension's adverse effect on compliance, making the lungs easier to inflate.

c. **Keeping lungs dry:** The collapsing fluid bubble within an alveolus exerts negative pressure on the alveolar lining. This pressure creates a driving force for fluid movement from the interstitium onto the alveolar surface. The presence of fluid within an alveolar sac interferes with gas exchange and negatively impacts lung performance. Surfactant reduces the pressure gradient and thereby helps keep lungs fluid free.

V. MECHANICS OF BREATHING

The blood–gas interface is separated from the external atmosphere by a distance of ~30 cm (i.e., the length of the trachea and other intervening airways). O_2 cannot diffuse over such a distance fast enough to meet the demands of internal respiration, so air must be drawn into the lungs by an air pump.

A. Pump structure

The pump functions much like an accordion, a musical instrument comprising a bellows operated using two handholds ("manuals") as shown in Figure 21.8. When the manuals are drawn apart, the bellows expand, and pressure inside them drops. This creates a pressure gradient that drives airflow over a set of reeds that give the instrument its familiar sound. Lung tissue (the pulmonary equivalent of bellows) is too fragile to be attached to the muscles and tendons that might function as manuals. Instead, they are hermetically sealed to the lining of the thoracic cavity. This allows the chest wall and diaphragm to expand the bellows while keeping the force per unit area applied to the lungs minimal (Fig. 21.9). The seal relies on **pleurae** and pleural fluid.

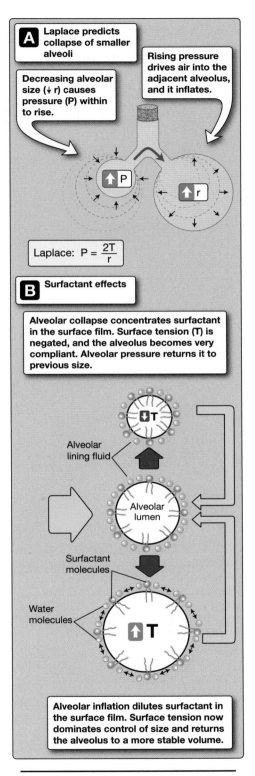

A Laplace predicts collapse of smaller alveoli

Decreasing alveolar size (↓ r) causes pressure (P) within to rise.

Rising pressure drives air into the adjacent alveolus, and it inflates.

Laplace: $P = \dfrac{2T}{r}$

B Surfactant effects

Alveolar collapse concentrates surfactant in the surface film. Surface tension (T) is negated, and the alveolus becomes very compliant. Alveolar pressure returns it to previous size.

Alveolar lining fluid

Alveolar lumen

Surfactant molecules

Water molecules

Alveolar inflation dilutes surfactant in the surface film. Surface tension now dominates control of size and returns the alveolus to a more stable volume.

Figure 21.7.
Surfactant stabilizes alveolar size.

Figure 21.8.
Expanding and contracting an accordion's bellows creates airflow and musical notes.

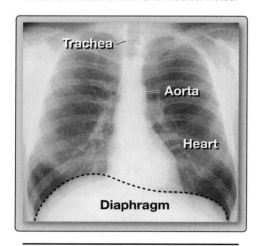

Figure 21.9.
Normal posteroanterior chest x-ray.

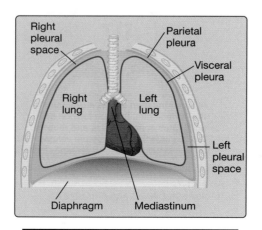

Figure 21.10.
Pleurae.

B. Pleurae

Pleurae are thin, serous membranes that cover the lungs. Similar membranes cover the heart (**pericardium**) and viscera (**peritoneum**). Pulmonary pleurae have two essential air-pump functions: creating the hermetic seal and secreting pleural fluid.

1. **Hermetic seal:** The lungs are enveloped in **visceral pleura** (Fig. 21.10). Each lung is individually shrink-wrapped within its own pleura, and there is no connection between the two pleural cavities. The chest wall, diaphragm, and mediastinum (heart, larger blood vessels, airways, and associated structures) are covered with **parietal pleura**. The visceral and parietal pleurae are physically attached to their respective underlying structures but not to each other, and the two membranes are separated by the **intrapleural space**. The pleurae effectively exclude air from the intrapleural space to hermetically seal the lungs to the diaphragm and rib cage.

2. **Pleural fluid:** The parietal pleura is innervated and vascular. It is believed to be the source of a viscous **pleural fluid** that is secreted into the intrapleural space. Pleural fluid has two important functions: lubrication and cohesion.

    a. **Lubrication:** Pleural fluid lubricates the pleural surfaces and allows the lungs to slide freely over the chest wall and diaphragm during normal breathing movements.

    b. **Aiding inspiration:** Pleural fluid is secreted and reabsorbed constantly. The volume contained within the intrapleural space at any one time is ~10 mL total, but it spreads to create a thin film that covers all surfaces, making the two pleurae almost inseparable under physiologic circumstances. The same cohesive force makes two glass microscope slides difficult to separate when a drop of water is caught between them. Cohesion allows forces generated by movement of the chest wall and diaphragm to be transferred directly to the lung surface.

C. Pump cycling

Respiration involves repeated cycles of inspiration and expiration. Inspiration draws air into the lungs and increases O_2 availability at the blood–gas interface.

1. **Inspiration:** The air pump is operated by skeletal muscles (Table 21.1). Most important of these is the diaphragm (see Fig. 21.10), a dome-shaped muscle that separates the thoracic and abdominal cavities and is innervated by the phrenic nerve. When the muscle contracts, intrathoracic volume increases.

    a. **Vertical dimensions:** Diaphragm contraction pushes downward on the abdominal contents and increases the vertical dimensions of the thoracic cavity by between 1 and 10 cm, depending on contractile force (Fig. 21.11A).

    b. **Cross-sectional area:** Diaphragm contraction also increases thoracic cross-sectional area by pulling upward on the lower

Table 21.1: Muscles Used in Breathing

Inspiration	
Diaphragm	**Costal fibers** are attached to the ribs; **crural fibers** course around the esophagus and are tethered via ligaments to the vertebrae. Contraction pushes downward on the abdominal contents and raises the chest wall.
External intercostals	Connect adjacent ribs and are angled forward so that contraction raises the chest wall.
Accessory muscles	Accessory muscles are used during forced inspiration and exercise. **Scalenes** elevate the first two ribs, **sternomastoids** lift the sternum, and muscles in the upper part of the respiratory tract dilate the upper airways.

Expiration	
Abdominal muscles	Muscles of the abdominal wall (**rectus abdominis, transversus abdominis**, and **internal** and **external oblique muscles**) contract during forced expiration to compress the abdominal cavity and push the diaphragm upward. These muscles are also activated during vomiting, coughing, and defecation.
Internal intercostals	Connect adjacent ribs. They pull the ribs downward and inward when they contract.

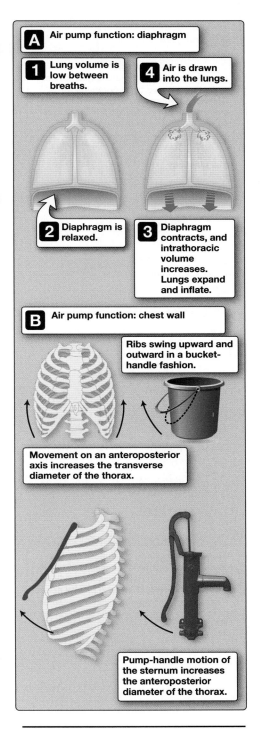

Figure 21.11.
Thoracic volume changes during inspiration.

ribs (see Fig. 21.11B). The ribs rise in a bucket-handle fashion to increase transverse dimensions, a motion aided by contraction of the external intercostal muscles. The upper ribs are attached to the sternum, which rises to increase the anterior–posterior dimensions of the thorax (pump-handle action).

2. **Expiration:** Expiration is generally passive and is driven both by the effects of surface tension on alveolar volume and energy stored in a lung's elastic elements during inspiration (**elastic recoil**). Elasticity reflects an abundance of elastin and collagen fibers in both airway and alveolar walls. Expiratory muscles are typically used during exercise or when airway resistance is increased by disease, for example (see Table 21.1).

VI. STATIC LUNG MECHANICS

Surface tension's influence on lung volume is tempered by surfactant, but it remains a significant force that impacts lung behavior during normal breathing.

A. Forces acting on a static lung

A healthy lung at rest is subject to two equal and opposing forces, one directed inward and the other outward.

1. **Inward:** As discussed previously, a lung's elasticity and surface tension effects generate an inwardly directed force that favors smaller lung volumes (Fig. 21.12A).

2. **Outward:** The muscles and various connective tissues associated with the rib cage also have elasticity. At rest, the elastic elements favor outward movement of the chest wall.

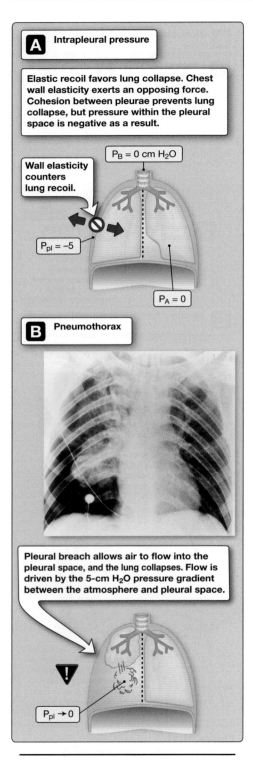

A Intrapleural pressure

Elastic recoil favors lung collapse. Chest wall elasticity exerts an opposing force. Cohesion between pleurae prevents lung collapse, but pressure within the pleural space is negative as a result.

$P_B = 0$ cm H_2O

Wall elasticity counters lung recoil.

$P_{pl} = -5$

$P_A = 0$

B Pneumothorax

Pleural breach allows air to flow into the pleural space, and the lung collapses. Flow is driven by the 5-cm H_2O pressure gradient between the atmosphere and pleural space.

$P_{pl} \rightarrow 0$

Figure 21.12.
Pneumothorax. P_A = alveolar pressure, P_B = barometric pressure, P_{pl} = intrapleural pressure.

3. **Net effect:** The two opposing forces create negative pressure within the intrapleural space (**intrapleural pressure**, or **P_{pl}**). P_{pl} is measured relative to the atmosphere and averages several centimeters of water, depending on vertical position within the lung (discussed later). If either pleura is breached, air rushes into the pleural space, driven by the pressure difference between atmosphere and pleural space (**pneumothorax**) as shown in Figure 21.12B. P_{pl} falls to zero, the lung collapses, and elastic elements in the chest wall cause it to spring outward.

> The human thorax contains two pleural cavities. In practice, this means that injury resulting in pneumothorax usually affects only one lung at a time. Therefore, although pneumothorax is still a serious condition, it is not immediately fatal. By contrast, the American buffalo (bison) contains a single pleural cavity that is relatively easy to breach with a gunshot or arrowhead. This immense animal once roamed the plains of North America in large numbers, but its vulnerability to pneumothorax allowed hunters and farmers to decimate herds during the late 1800s.

B. Pressure–volume curves

A collapsed lung can help us understand how much effort is required to inflate it during normal breathing.

1. **Inflation:** A collapsed lung can be inflated in one of two ways, both of which modify **transpulmonary pressure** (**P_L**). P_L is the difference between **intraalveolar pressure** (**P_A**) and P_{pl}:

$$P_L = P_A - P_{pl}$$

A mechanical positive-pressure ventilator can be used to raise pressure inside the lung ($P_A > P_{pl}$), inflating it much as an individual inflates a balloon. Alternatively, air can be withdrawn from the pleural space to create negative pressure outside the lung ($P_{pl} < P_A$). Both maneuvers raise P_L. Lung collapse allows the airways to close off and seal with fluid. Restoring patency requires that the surface tension seal be broken, which requires considerable effort. P_L must be increased by several cm H_2O before any significant increase in volume occurs (Fig. 21.13, phase 1). Once P_L exceeds ~7 to 10 cm H_2O, airways pop open, and volume increases linearly with inflation pressure (see Fig. 21.13, phase 2). At ~20 cm H_2O, the lung reaches its maximal volume, known as **total lung capacity** (**TLC**), as shown in Figure 21.13, phase 3.

2. **Deflation:** A lung allowed to deflate from TLC yields a different pressure–volume curve from that seen during inflation (a phenomenon known as **hysteresis**). This is because surfactant is recruited from pneumocytes to the alveolar surface film during

lung inflation. The added surfactant decreases elastic recoil and thereby resists lung deflation. Note that the **hysteresis loop** begins and ends at a positive volume (normally ~500 mL; see Fig. 21.13). This is because the larger airways collapse at zero pressure and trap air within the more distal regions.

3. **Normal breathing:** Normal breathing involves changes in lung volume that are only a fraction of total, but hysteresis is still evident (see Fig. 21.13). When the lung is resting between breaths, the chest wall prevents it from collapsing, and holds it at ~50% of TLC. At 50% TLC, all alveoli are patent, and the lung is positioned on the steepest portion of the pressure–volume curve. In practice, this means the increase in P_L during inspiration is maximally effective in increasing alveolar volume.

4. **Surface tension effects:** Surface tension's influence on the pressure–volume loop can be estimated by filling lungs with fluid to eliminate air–liquid interfaces (Fig. 21.14). Fluid-filled lungs are more compliant, and the hysteresis associated with surface tension disappears.

C. Gravitational effects

Lungs and blood have mass and are therefore subject to the influence of gravity. Gravity causes significant regional differences in P_L and alveolar volume.

1. **Apex:** When the thorax is positioned vertically, such as when an individual is seated or standing erect, a lung within can be imagined as hanging suspended by its apical pleura. Suspension creates a strongly negative P_{pl} (and strongly positive P_L) locally and causes apical alveoli to inflate to ~60% of their maximal volume (Fig. 21.15). Gravity similarly stretches the coils at the top of a Slinky (the famous toy) farther apart than those at the base (Fig. 21.16). In practice, gravitational influences force the lung apex to function near the top of the pressure–volume curve, where the opportunity for further expansion during inspiration is very limited.

2. **Base:** The lung base supports the mass of pulmonary tissue above it. Alveoli in this region are compressed, much like coils at the base of the Slinky. The weight of tissue above also pushes outward against the chest. P_{pl} and P_L both approach zero (see Fig. 21.15). In practice, this means that alveoli at the base of the lung respond to increases in P_L with large changes in volume because they occur over the lower, steepest part of the pressure–volume curve.

D. Lung compliance

The amount that lung volume increases in response to changes in P_L is a measure of its **compliance**. Lungs are highly compliant organs, increasing volume by ~200 mL for every cm H_2O in P_L. Compliance is governed both by surface tension and the elastic properties of the lungs and chest wall.

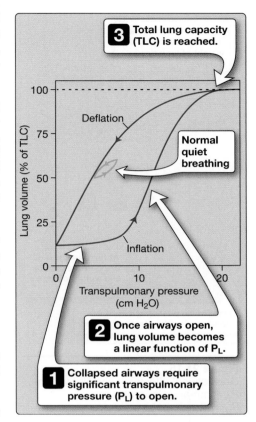

Figure 21.13.
Pulmonary pressure–volume loop.

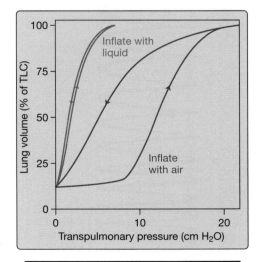

Figure 21.14.
Pressure–volume loop for a fluid-filled lung. TLC = total lung capacity.

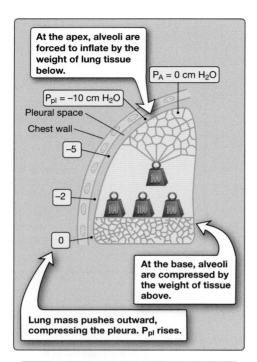

Figure 21.15.
Gravitational effects on alveolar volume.
P_A = intraalveolar pressure,
P_{pl} = intrapleural pressure.

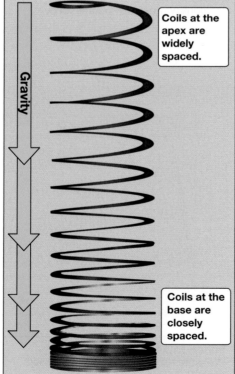

Figure 21.16.
Gravitational effects on a Slinky.

Biologic Sex and Aging 21.1: Lung Mechanics

The lungs' ability to function as an air pump relies on respiratory muscles and the elastic properties of the lung and chest wall. Lung performance increases and peaks at around age 20 years in females and 25 years in males and then decreases in an approximately linear manner thereafter. The intercostal muscles lose mass and strength, which limits chest expansion. The diaphragm flattens and weakens by ~25% by age 70 years. Extracellular matrix changes discussed previously (see Biologic Sex and Aging 4.1) are particularly detrimental for air-pump performance. Elastin degradation and collagen cross-linking stiffens the lung and causes elastic recoil to drop by 1–2 cm H_2O per decade. Kyphotic changes in the spine and remodeling of the chest wall change the shape of the thorax to the detriment of pump function. Calcification and collagen cross-linking also stiffens the chest wall and increases the burden on the respiratory muscles. Together, these age-related changes in air-pump structure increase the work of breathing by ~30% by age 75 years.

VII. LUNG DISEASES

Obstructive and **restrictive pulmonary diseases** are two broad disease groups that cause significant changes in static lung properties. The two groups are typified, respectively, by **emphysema** and **pulmonary fibrosis**. We will revisit these diseases frequently to help illustrate the mechanical principles involved in normal breathing.

A. Obstructive pulmonary disease

Emphysema, chronic bronchitis, and asthma all represent **obstructive pulmonary diseases**, which increase airway resistance to airflow. Because the former two examples often coexist and may be difficult to distinguish clinically, they are commonly grouped and discussed as **chronic obstructive pulmonary disease (COPD)**. COPD is extremely common and has become the fourth leading cause of death in the United States (see Table 40.1). There are three general obstructive mechanisms: airway occlusion, wall thickening, and loss of mechanical tethering.

1. **Airway occlusion:** Secretions that are excessive or difficult to expel may occlude airways (Fig. 21.17B). Occlusive diseases include **chronic bronchitis**, **asthma**, **cystic fibrosis**, and **bronchiectasis** (airway dilation associated with inflammation and copious sputum formation).

2. **Wall thickening:** When the airway wall hypertrophies or becomes edematous, it encroaches on the lumen and reduces its cross-sectional area (see Fig. 21.17B).

3. **Loss of mechanical tethering:** All structures in the lung are linked mechanically. Together, they form a dependent network, much like the fabric of a nylon stocking (**interdependence**; see Fig. 21.17A). Interdependence maintains airway patency when external forces may favor collapse. **Emphysema** develops when alveolar walls (the fabric of the lung) erode, allowing surrounding airways to collapse and obstruct airflow during normal breathing (see Fig. 21.17C). Emphysema is commonly caused by heavy smoking.

> **Emphysema** denotes anatomic tissue loss, although the term is commonly used to describe smoking-related lung disease. It is a finding evident on computed tomography imaging of the lung. It is also a pathologic finding seen at autopsy or lung tissue biopsy. Postmortem examination of an emphysematous lung shows enlarged cystic air spaces replacing normal lung. Alveolar loss reduces elastic recoil, increases pulmonary compliance, and reduces the surface area available for O_2 uptake.

B. Restrictive pulmonary disease

Pulmonary fibrosis is a **restrictive pulmonary disease**. Others include pleural diseases and problems affecting breathing muscles, all of which limit lung expansion. Pulmonary fibrosis (scarring) results from any one of several **interstitial lung diseases** (**diffuse parenchymal lung diseases**). Scarring typically begins with an injury to the alveolar epithelium. Causes include certain compounds inhaled in the workplace (e.g., asbestos, coal dust, sawdust), circulating drugs (e.g., antibiotics and chemotherapeutic agents), systemic diseases (e.g., rheumatoid arthritis, systemic lupus erythematosus, scleroderma, and sarcoidosis), or may be idiopathic. The initial insult causes the alveolar wall to thicken and fill with an exudate containing lymphocytes, platelets, and other immune effector cells (see Fig. 21.17D). The wall is then infiltrated by fibroblasts, which lay down bundles of collagen and other fibers between the alveolar sacs. Scar tissue is relatively noncompliant, so the lung becomes stiff and expands with difficulty during inspiration. Scarring also decreases O_2 uptake, and hypoxemia may develop as the disease progresses.

VIII. DYNAMIC LUNG MECHANICS

During inspiration, air moves from the external environment through a set of branching tubes of ever-decreasing diameter. Flow is driven by the pressure difference between the external atmosphere (barometric pressure; P_B) and alveolus ($\Delta P = P_B - P_A$). Flow is inversely proportional to airway resistance (R):

$$\dot{V} = \frac{\Delta P}{R}$$

where $\dot{V}$ is airflow (volume ÷ unit time).

A. Pressures driving airflow

Airflow occurs in response to pressure gradients set up between the alveoli and the external atmosphere. The body has no way of controlling P_A directly. Instead, the diaphragm and other respiratory muscles manipulate P_{pl}. When P_{pl} falls, P_L rises, and the alveoli expand. P_A becomes negative because the product of pressure and volume of a fixed number of air molecules remains constant (as per the **Boyle law**).

$$P_{A1}V_{A1} = \downarrow P_{A2} \uparrow V_{A2}$$

where P_{A1} and P_{A2} denote alveolar pressure before and after alveolar expansion (V_{A1} and V_{A2}). Alveolar expansion thus creates a $P_B > P_A$

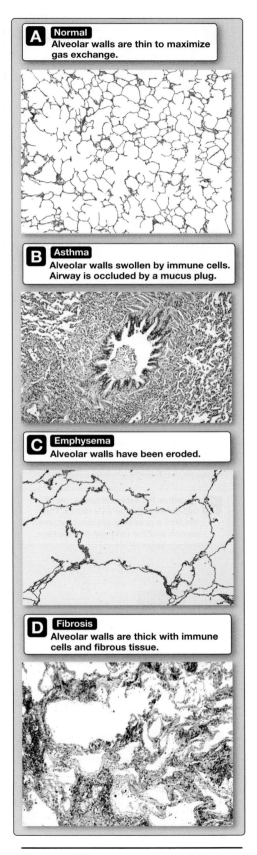

A **Normal**
Alveolar walls are thin to maximize gas exchange.

B **Asthma**
Alveolar walls swollen by immune cells. Airway is occluded by a mucus plug.

C **Emphysema**
Alveolar walls have been eroded.

D **Fibrosis**
Alveolar walls are thick with immune cells and fibrous tissue.

Figure 21.17.
Normal and diseased lungs.

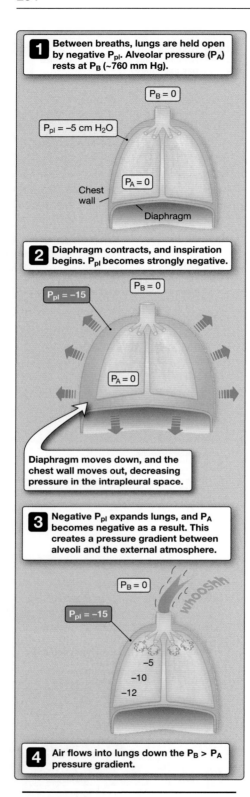

Figure 21.18.
Pressure gradients driving airflow during inspiration. All values are given in cm H_2O. P_B = barometric pressure, P_{pl} = intrapleural pressure.

pressure gradient that drives airflow into the lungs (Fig. 21.18). Because flow occurs against airway resistance, it takes time for air to move in or out of the lungs and for the pressure gradient to dissipate, particularly at the points farthest removed from the site of highest resistance.

B. Resistance to airflow

In a healthy individual, breathing is usually an effortless and unconscious act, so it seems surprising that airways offer resistance to flow. The resistance has several origins. Resistance is proportional to airway length (L) and the viscosity (η) of the gas moving through it and inversely proportional to the fourth power of airway radius (r), as stated in the **Poiseuille law**:

$$R = \frac{8L\eta}{\pi r^4}$$

The Poiseuille law was derived using rigid glass tubes. Airways are compliant, so the law cannot be strictly applied, but it does help us understand the variables influencing airflow in the lungs (see also Chapter 18·III). Airway radius is the primary determinant of lung resistance, although air viscosity and turbulence are considerations also.

1. **Airway radius:** Airway radius decreases with each successive generation within the bronchiolar tree. Decreasing radius increases resistance, but the negative impact on net airflow through the lung is more than offset by the gain in airway numbers with each successive generation. In other words, although individual bronchioles have very high resistance, their combined resistance is almost negligible (calculated from the sum of reciprocals; see also Chapter 18·IV·B). The site of greatest resistance in the lung is in the pharynx and larger airways (generations 0 through ~7) as shown in Figure 21.19.

2. **Air viscosity:** Air viscosity is dependent on air density. Air density increases when compressed, as during a deep-sea dive, for example. Increasing density increases flow resistance and the **work of breathing (WOB)**. Breathing an O_2/helium mixture partly offsets this density increase. Helium has a fraction of atmospheric air's density (~14%) and therefore reduces WOB.

3. **Turbulence:** The Poiseuille law assumes that airflow through the lungs is streamline, but this is generally not the case. Each branch point in the bronchial tree creates a local eddy current that disrupts streamline flow and increases airway resistance. In practice, the eddy currents cause flow through the airways to be proportional to ($\Delta P + \sqrt{\Delta P}$) rather than ΔP alone.

C. Changes in airway resistance

Changes in airway radius can significantly impact airflow. Airway radius is governed by airway musculature and by lung volume.

1. **Smooth muscle:** Bronchioles are lined with smooth muscle cells. When the muscles contract, they decrease airway radius and increase resistance to airflow. Smooth muscle relaxation and

bronchiolar dilation reduces resistance and increases airflow. Airway muscles are regulated by the autonomic nervous system (ANS) and by local factors.

   a. **Autonomic control:** Airways are controlled by both parasympathetic (PSNS) and sympathetic (SNS) branches of the ANS.

      i. **Parasympathetic:** PSNS nerve fibers from the vagus nerve release acetylcholine (ACh) from their terminals when active. ACh binds to M_3 muscarinic ACh receptors and causes bronchoconstriction, which reduces airflow.

      ii. **Sympathetic:** SNS activation causes bronchodilation, mainly by inhibiting ACh release rather than through direct effects on the musculature. SNS terminals release norepinephrine, which binds to a presynaptic β_2-adrenergic receptor. This receptor is also activated by epinephrine released from the adrenal medulla during SNS activation. SNS-mediated bronchodilation facilitates increased airflow during exercise, for example.

   b. **Local factors:** Local irritants and allergens constrict bronchioles and obstruct airways. Airway muscle contraction is a response to histamine and other inflammatory mediators.

2. **Lung volume:** Airway resistance is highly dependent on lung volume. The decrease in P_{pl} that inflates alveoli during inspiration is transmitted to airways also. Airways dilate, and resistance falls as a consequence. Airways are also dilated by **radial traction**. Traction results from mechanical tethering between the alveoli and all surrounding structures. Thus, when lungs expand, radial traction increases airway patency and lowers their resistance (Fig. 21.20). At low lung volumes, radial traction is reduced, and airway resistance increases.

D. Airway collapse during expiration

Airways tend to collapse and limit flow during expiration, an effect known as **dynamic compression of the airways**. The reasons and consequences of collapse are easiest to appreciate during a forced expiration after a deep inspiration (Fig. 21.21). Forced expiration begins with contraction of the abdominal muscles and internal intercostals, which forces the chest wall downward and inward, and causes P_{pl} to become positive. The positive pressure is transferred to and compresses the alveoli, decreasing their volume and causing P_A to rise above P_B. Compression thus establishes the pressure gradient that drives expiratory outflow. The larger airways have a relatively high resistance to flow that limits lung-emptying rates, so alveoli remain filled with pressurized air for a time. High P_A maintains patency, even though P_{pl} may be positive and favoring alveolar collapse. Airway pressure falls with distance from the alveoli and proximity to the main site of resistance (bronchi and trachea). Therefore, whereas P_A may be strongly positive (relative to P_B), pressure within the larger airways may be much closer to zero (i.e., P_B) and thus more susceptible to collapse by P_{pl} (see Fig. 21.21[3]). The larger airways are equipped with cartilage that helps maintain patency during forced expiration, but it may be inadequate to prevent collapse. As air leaves the lungs

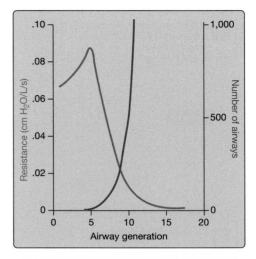

Figure 21.19.
Resistance to airflow within the bronchial tree.

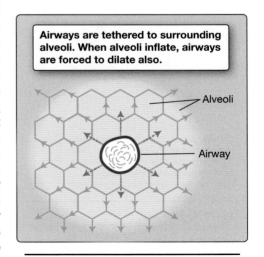

Figure 21.20.
Radial traction on airways during lung inflation.

and local P_A drops, the collapse zone moves distally and involves increasingly smaller airways. Compression and collapse of conducting airways is the self-regulating, limiting factor that determines how fast air escapes the lungs during expiration. If a subject attempts to speed outflow with a more forceful muscular contraction, the pressure gradient driving outflow is raised, but so are the forces favoring airway collapse, with a net zero sum gain (Fig. 21.22).

E. Work of breathing

Breathing requires that the respiratory muscles contract to expand the lungs against resistance (discussed in the following section). WOB normally accounts for ~5% of total energy usage at rest, but it can rise to >20% of total during exercise. Such workloads are normally insignificant in a healthy individual, but some patients with pulmonary disease have difficulty expanding their lungs, and even resting breathing movements can fatigue their respiratory muscles and precipitate respiratory failure (see Chapter 40·VI).

1. **Work components:** Many factors contribute to the WOB. The two principal factors are elastic work and resistive work. **Elastic work** includes the work required to counter a lung's elastic recoil during inspiration, which is proportional to its compliance. Work is also required to displace the chest wall outward and the abdominal organs downward. **Resistive work** involves moving air back and forth through the airways against airway resistance.

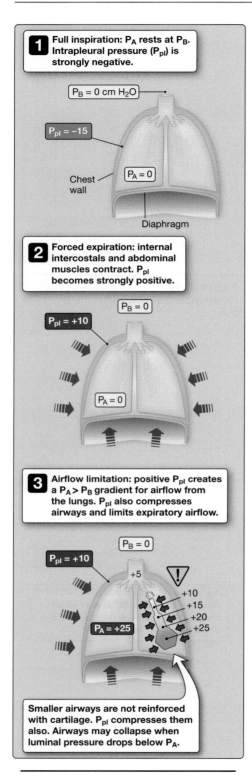

Figure 21.21.
Airway collapse during forced expiration. All values are given in cm H_2O. P_A = intraalveolar pressure, P_B = barometric pressure.

Clinical Application 21.2: Pursed-Lipped Breathing

Chronic obstructive pulmonary disease (COPD) is characterized by airflow limitation (obstruction). Spirometry testing reveals reduced vital capacity with a flow–volume loop whose expiratory limb appears "scooped-out" (concave upward). There may also be a long tail on the expiratory limb because patients with COPD have a hard time exhaling due to loss of elastic recoil and airway collapse. Patients can partly compensate for the loss of mechanical support by

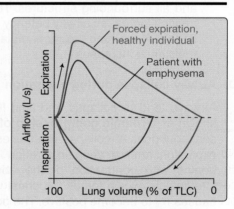

Effects of emphysema on airflow. TLC = total lung capacity.

pursing their lips (as if whistling) during expiration, a behavior known as **pursed-lipped breathing**, or puffing. This behavior is effective because it moves the site of main airway resistance closer to the mouth and extends the time during which airway pressure remains high and the airways patent. Patients with anatomic tissue loss (emphysema) in addition to airflow obstruction tend to hyperventilate and use accessory muscles to help with expiration, giving them a characteristic pink complexion (formerly known as "pink puffers"). This contrasts with patients with COPD whose disease is characterized by chronic bronchitis and excessive mucus production that interferes with oxygen uptake (these patients have previously been described as "blue bloaters").

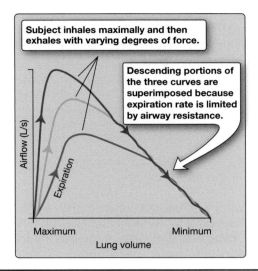

Figure 21.22.
Airway resistance limits flow during forced expiration.

2. **Measuring work:** Work is calculated as the amount of force required to move an object a given distance. In pulmonary terms, the WOB is calculated from the product of the force needed to change the P_L gradient and the air volume moved per unit time. Work can be represented graphically as the area to the left of the inspiratory phase of the pressure–volume loop (Fig. 21.23).

3. **Pulmonary diseases:** COPD and pulmonary fibrosis both increase the WOB (see Fig. 21.23). Patients with COPD work harder to exhale against high airway resistance (increased resistive work). Pulmonary fibrosis stiffens the lung and requires that a patient generate a higher P_L than normal to expand the lungs during inspiration (increased elastic work).

IX. LUNG VOLUMES AND CAPACITIES

Normal quiet breathing uses <10% of TLC. Exercise increases this amount significantly, but even at maximal levels of exercise, there is always a small percentage of lung volume that communicates with the ventilated space but that does not itself participate in ventilation. Clinically, it is important to determine the contribution of this volume to the mix of gases in the lungs and to assess how lung volume(s) may be impacted by the progression of various pulmonary diseases. In addition to airflow measurement with **spirometry**, **pulmonary function tests** (**PFTs**) typically measure four primary **lung volumes**, which are then combined to derive several **lung capacities** (Fig. 21.24). PFTs also assess the efficiency of the blood–gas interface ("diffusing capacity" is discussed in Chapter 22·V).

A. Volumes

The volume of air inspired or expired with each breath, typically ~500 mL in an average adult, is called the **tidal volume** (**TV**). **Inspiratory reserve volume** (**IRV**) and **expiratory reserve volume** (**ERV**) are the volumes that can be inspired or expired, respectively,

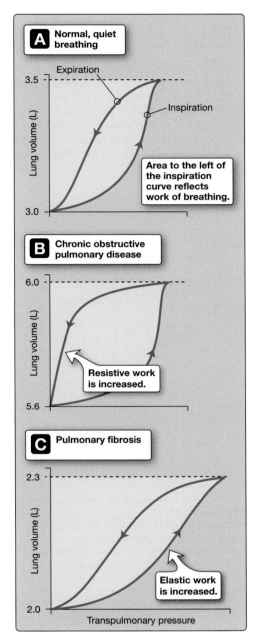

Figure 21.23.
Effects of pulmonary disease on the work of breathing. Values used for lung volume are illustrative.

Biologic Sex and Aging 21.2: Pulmonary Function Tests

The alveoli and alveolar ducts are rich in elastin (see Fig. 4.16), which allows for repeated cycles of inflation and deflation. Aging-associated elastin fragmentation causes alveolar distension and ductal dilation, leading to a 30% decrease in blood–gas interface surface area by age 90 years. The consequences for lung volumes and capacities are not unlike those seen with smoking. Loss of radial traction allows airway collapse and air-trapping during expiration. Residual volume increases by ~50% at the expense of vital capacity by age 70 years. FEV_1 declines steadily by 300 mL per decade after age 30 years, and then accelerates at 65 years. The expiratory limb of the flow–volume loop takes on a scooped-out appearance, as seen with emphysema (see Clinical Application 21.2).

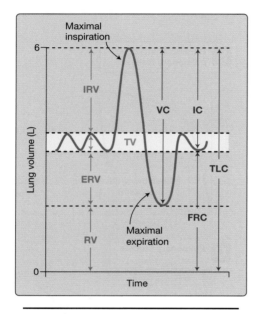

Figure 21.24.
Spirometry. ERV and IRV = expiratory and inspiratory reserve volume, respectively; FRC = functional residual capacity; IC = inspiratory capacity; RV = residual volume; TLC = total lung capacity; TV = tidal volume; VC = vital capacity.

over and above TV. **Residual volume (RV)** is the volume of air remaining in the lung after a maximal expiration (~1.2 L in a normal individual). A spirometer is unable to provide information about RV, so PFTs often include more specialized techniques such as body plethysmography or helium-dilution and nitrogen-washout assays that monitor intrapulmonary concentrations of gases over time.

B. Capacities

The sum of all four lung volumes (TLC) amounts to ~6 L in a normal individual (see Fig. 21.24). **Functional residual capacity (FRC)** is the volume remaining in the lungs after expelling a tidal breath. **Inspiratory capacity (IC)** is the sum of the TV and the IRV. **Vital capacity (VC)** is the sum of the TV, IRV, and ERV and is the maximal TV achievable (i.e., the biggest breath an individual can take). **Forced vital capacity (FVC)** is the volume of air that can be **forcibly** expired after a maximal inspiration.

C. Forced expiratory volume

Forced expiratory volume in 1 second (FEV_1) is the volume of air that can be forcibly expired in 1 second following a maximal inspiration and is an important clinical measure of lung function (see Clinical Application 21.3). FEV_1 is usually expressed as a function of FVC (FEV_1/FVC) to correct for differences in lung size and capacity.

X. DEAD SPACE AND VENTILATION

Gas exchange largely occurs at the alveolar surface. By the time inspired air contacts the gas exchange interface, its O_2 and CO_2 concentration has been modified through mixing with gases lingering in the RV, which itself is influenced by how often the contents of the lung are refreshed (i.e., ventilation). Alveolar gas concentration is also influenced by the amount of inhaled air that does not participate in gas exchange because it fills **dead space**.

A. Dead space

The lung contains two types of dead space: anatomic and physiologic.

1. **Anatomic:** The pharynx, trachea, bronchi, and other conducting airways contain ~150 mL of air that is inspired but never reaches the blood–gas interface. This represents **anatomic dead space**.

Clinical Application 21.3: Pulmonary Function Tests

Pulmonary function tests are useful in detecting the presence of obstructive and restrictive pulmonary pathophysiology. Chronic obstructive pulmonary disease (COPD) is best identified by measuring airflow with spirometry and documenting obstruction (reduced forced expiratory volume in 1 second [FEV_1] in a setting of an FEV_1/FVC [forced vital capacity] ratio of <70% or <5th percentile). Patients with COPD typically also operate at very high lung volumes because exhalation is impaired by airway obstruction. Patients with pulmonary fibrosis work at low volumes because the lung is noncompliant and difficult to expand; therefore, total lung capacity is reduced. They typically take shallow breaths and breathe rapidly.

	FEV_1	FVC	FEV_1: FVC (%)
Normal	~4.0	~5.0	>70
Obstructive	~1.3	~3.1	<70
Restrictive	~2.8	~3.1	>70

FEV_1 = volume of air forcibly expelled in 1 second; FVC = forced vital capacity. Values for FEV_1 and FVC are in liters. FVC is generally slightly less than vital capacity, hence the use of a distinguishing term.

2. **Physiologic:** In a diseased lung, some alveoli may be ventilated but unable to participate in gas exchange because the blood–gas interface is damaged, or pulmonary blood flow to these regions has been interrupted. These regions represent dead space. The term physiologic dead space includes anatomic dead space and contributions from these nonfunctional alveoli. In a healthy lung, physiologic and anatomic dead spaces are approximately equal. In a diseased lung, physiologic dead space may amount to >1,500 mL.

3. **Calculating dead space:** Dead-space volume (V_D) can be calculated by measuring the amount of CO_2 contained in expired air ($P_{E}CO_2$). Because dead space does not participate in gas exchange, it contains negligible CO_2. The amount of CO_2 in air originating from regions of the lung involved in gas exchange equals that of arterial blood (P_aCO_2) because blood gases equilibrate with alveolar gases during transit through the lungs (i.e., $P_ACO_2 = P_aCO_2$). Thus, V_D can be determined from the extent to which the amount of CO_2 in expired air has been decreased by CO_2-free air originating from dead space:

$$V_D = TV \times \frac{P_aCO_2 - P_ECO_2}{P_aCO_2}$$

B. Ventilation

Ventilation can be expressed as **minute ventilation (V_E)** or **alveolar ventilation (V_A)**. V_E is the total volume of air inhaled and exhaled per minute:

$$\text{Minute ventilation} = TV \times \text{breaths/min}$$

V_A is the volume of air per minute that enters the areas participating in gas exchange:

$$V_A = (TV - V_D) \times \text{breaths/min}$$

where V_A represents alveolar ventilation and V_D is dead space.

Chapter Summary

- Lungs facilitate exchange of O_2 and CO_2 between blood and air. The **blood–gas interface** is located within **alveoli**, thin-walled sacs that amplify interface surface area and bring the pulmonary circulation into close proximity to inhaled air.

- Alveoli are moistened with a thin fluid film that generates surface tension. Surface tension favors lung collapse. The alveolar epithelium produces **surfactant** to counter this surface tension. Surfactant is a **phospholipid** complex that helps stabilize alveolar size and increases lung compliance.

- Breathing involves repeated cycles of **inspiration** and **expiration**. Air is drawn into the lungs by contracting the **diaphragm** and other **respiratory muscles**. Contraction increases the volume of the thoracic cavity and lungs.

- The diaphragm, chest wall, and lungs move as one unit. They are linked by a thin film of **pleural fluid**, which lubricates the **visceral** and **parietal pleurae** and provides the cohesive force required to expand the lungs.

- At rest, a lung is subject to two opposing forces. Surface tension and elastic elements in lung tissue favor collapse (**elastic recoil**). Elastic elements in the chest wall favor expansion and thus prevent collapse. Introducing air between the two pleurae (**pneumothorax**) breaks the connection between lungs and chest wall and allows a lung to collapse.

- Gravity causes significant regional differences in alveolar size in an upright lung. The base of the lung is compressed by its own mass, whereas alveoli at the apex may be expanded to 60% of their maximal volume.

- Airflow between the alveoli and the external atmosphere is driven by **pressure gradients**. Flow occurs against a **resistance** that depends largely on an airway's internal radius.

- Airway resistance is modulated by the **autonomic nervous system** but also changes passively with lung volume. During lung expansion, the airways are forced to dilate by surrounding structures acting via **mechanical tethers**, and dilation causes airway resistance to fall. When lung volumes are low, the airways are compressed by the mass of surrounding tissue, and their resistance is high.

- Airways are also sensitive to transmural pressures developed during expiration, such that their resistance becomes a pressure-dependent limiting factor on outflow.

- Air movement between lungs and atmosphere is measured using **spirometry**, one of several **pulmonary function tests** (**PFTs**) used to assess lung health. PFTs derive four lung volumes (i.e., **tidal volume, inspiratory reserve volume, expiratory reserve volume**, and **residual volume**) and capacities (i.e., **total lung capacity, functional residual capacity, inspiratory capacity**, and **vital capacity**).

- Air that is enclosed within regions of the lung that do not participate in gas exchange is known as **dead space**.

Study Questions

Choose the ONE best answer.

21.1. A 26-year-old female is participating in astronaut training that allows trainees to experience 20–25 seconds of weightlessness during the descent phase of a parabolic flight. Which of the following parameters is most likely to decrease during this period of weightlessness compared with standing erect at ground level?

   A. Alveolar volume at the lung base
   B. Intrapleural pressure at the lung apex
   C. Intrapleural pressure at the lung base
   D. Tidal volume
   E. Transpulmonary pressure at the lung base

Best answer = C. Gravity causes the mass of the lung to be pulled downward when standing erect. The lung base is compressed and pushed against the chest wall by the mass of tissue above (see Section VI·C). Intrapleural pressure (P_{pl}) and transpulmonary pressure (P_L) are close to zero as a result. When freed of gravitational influences, P_{pl} decreases and P_L (E) increases. Alveoli at the lung base are compressed by the mass of tissue above in an erect lung, so weightlessness allows them to expand and their volume increases rather than decreases (A). P_{pl} at the lung apex is strongly negative in the erect lung and increases, rather than decreases (B) when gravitational effects are removed. Tidal volume (D) is determined by metabolic needs and would be unlikely to be affected by a brief period of weightlessness.

21.2. A 55-year-old male with a history of interstitial pulmonary fibrosis undergoes pulmonary function testing. What parameter would most likely be decreased in this patient?

   A. FEV_1
   B. FEV_1/FVC
   C. FVC
   D. Fraction of expired O_2
   E. Peak expiratory flow rate

Best answer = C. Restrictive lung disease is associated with lung stiffening, which limits lung expansion (see Section VII·B). This manifests as a decrease in forced vital capacity (FVC) in pulmonary function tests. In practice, such patients can voluntarily inhale and exhale less air volume than a healthy person of comparable age, sex, and height. Peak expiratory flow rate (E) and forced expiratory volume in 1 second, or FEV_1 (A), may or may not be normal. The FEV_1/FVC ratio (B) is increased because FVC is typically reduced significantly. The fraction of expired O_2 (D) would not be changed by restrictive lung disease.

21.3. A 63-year-old female has a history of chronic obstructive pulmonary disease. Her medications currently include a β-adrenergic receptor agonist. The most likely effect of this medication on pulmonary function includes which of the following?

   A. Bronchiolar constriction
   B. Decreased airway resistance
   C. Decreased forced vital capacity
   D. Decreased total lung capacity
   E. Increased diffusing capacity

Best answer = B. β-Adrenergic receptor agonists relax airway smooth muscle, promoting bronchiolar dilation, not constriction (A; see Section VIII·C) and decreased airway resistance. Relaxation occurs due to inhibition of acetylcholine release from parasympathetic nerve terminals. The increased airway luminal diameter improves flow, as measured by the forced expiratory volume in 1 second (FEV_1). Forced vital capacity (C), the maximal air volume that can be forcibly expired; total lung capacity (D); and diffusing capacity (E), a measure of blood–gas interface exchange capacity, are not affected by β-adrenergic drugs.

21.4. A 16-year-old male presents to the office with shortness of breath. History reveals that his family has adopted a new pet. His pulmonologist suspects an underlying allergy-induced asthma and orders a pulmonary function test. Which of following is most likely to have decreased in this boy?

   A. Expiratory reserve volume
   B. FEV_1
   C. Forced vital capacity
   D. Inspiratory capacity
   E. Tidal volume

Best answer = B. Allergy-induced asthma is associated with airway narrowing and obstruction (see Section VII·A), which impairs the volume of air that can be forcibly expired per unit time (i.e., forced expiratory volume in 1 second). Static lung volumes and capacities, including expiratory reserve volume (A), inspiratory capacity (D), and tidal volume (E), do not change appreciably with airflow obstruction, although residual volume may be increased by obstructive physiology when air-trapping occurs. In most cases, forced vital capacity (C) would be unaffected because this parameter is not time dependent.

22 Gas Exchange

I. OVERVIEW

Lungs facilitate O_2 and CO_2 exchange between blood and air. O_2 helps fuel ATP synthesis by aerobic respiration, whereas CO_2 is produced as a metabolic byproduct of aerobic respiration. Lungs make this exchange possible by bringing blood into close proximity to atmospheric air at a blood–gas interface. When the diaphragm and other inspiratory muscles contract, the lungs inflate. O_2-rich, CO_2-poor air flows into the lungs, flushing the blood–gas interface and sustaining steep O_2 and CO_2 pressure gradients that optimize gas exchange. Gas exchange occurs rapidly, enhanced by the thin divide between blood and air (<1 μm) and by the large interface surface area. The efficiency of exchange is critically dependent on the pulmonary circulation, which brings CO_2 to the lungs for disposal and carries away O_2 (Fig. 22.1). Physiologic and pathologic changes in either ventilation or perfusion of the blood–gas interface can negatively impact lung performance.

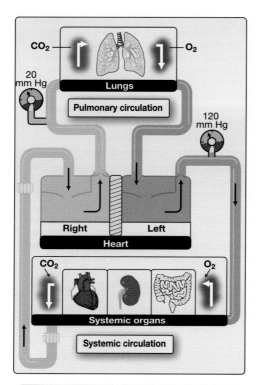

Figure 22.1.
Pulmonary and systemic circulations.

II. PARTIAL PRESSURES

Gases move between air and blood by passive diffusion. The basic principles involved are similar to those described for solutes (see Chapter 1·IV): gases tend to move from areas of high gas concentration to low gas concentration. When discussing gas transport by blood, however, the issue is complicated by the need to factor in how soluble a gas might be in water (Fig. 22.2). If a gas is water insoluble, it cannot enter the circulation except under extreme (nonphysiologic) pressures. Also, gas concentrations are discussed in terms of **partial pressures** instead of moles/liter.

A. Gas pressures

The random motion of gas molecules exerts pressure on the walls of the vessel that contains it. The amount of pressure is directly proportional to the number of molecules within the vessel (i.e., the concentration of gas molecules), as described by the **ideal gas law**:

$$P = \frac{nRT}{V}$$

where P = pressure, n = number of molecules, R = universal gas constant, T is temperature, and V is container volume.

B. Partial pressure

The term "partial pressure" recognizes that atmospheric air is a mix of several different gases. The total pressure exerted by gas mixtures is equal to the sum of partial pressures of each of the individual components (**Dalton law**).

1. **Atmospheric air composition:** Atmospheric air is composed of 78.09% N_2, 20.95% O_2, 0.93% argon (Ar), 0.03% CO_2, and trace amounts of various other inert gases and pollutants. The fractional composition does not change with height above sea level or with temperature.

2. **Inspired air composition:** Air composition does change during inspiration because mucous membranes lining the nose and mouth add water vapor. By the time air reaches alveoli, it contains 6.18% water by volume (100% relative humidity). The fractional composition of the other gases is reduced correspondingly: 73.26% N_2, 19.65% O_2, 0.87% Ar, and 0.03% CO_2.

3. **Partial pressure of inspired air:** Atmospheric pressure at sea level is 760 mm Hg, reflecting the mass of air molecules stacked above. The partial pressure of the individual gases that make up inspired air reflects their fractional composition. Thus, the partial pressure of O_2 in inspired air arriving at the alveolar membrane ($P_{I_{O_2}}$) is the product of atmospheric pressure (760 mm Hg) and fractional composition ($F_{I}O_2 = 19.65\%$, or 0.197):

$$P_{I}o_2 = 760 \times 0.197 = 150 \text{ mm Hg}$$

The partial pressure of CO_2 in inspired air ($P_{I}CO_2$) is 0.21 mm Hg. The latter is negligible in physiologic terms and therefore is usually rounded down to 0 mm Hg (Table 22.1).

C. Blood gases

Alveolar ventilation brings atmospheric air to the blood–gas interface. The amount of O_2 and other air constituents that dissolve in blood is proportional to their partial pressures and their solubility in blood (**Henry law**). O_2 and CO_2 are both soluble gases that rapidly equilibrate across the blood–gas interface during inspiration. P_{O_2} in alveolar gas ($P_{A}O_2$) necessarily falls as the fractional composition of CO_2 increases and as O_2 molecules cross the interface and dissolve in blood: when the two compartments have equilibrated, $P_{A}O_2$ has dropped from 150 to 100 mm Hg. At equilibrium, the concentration of O_2 dissolved in blood can be calculated from

$$[O_2] = P_{A}o_2 \times s = 100 \text{ mm Hg} \times 0.0013 \text{ mmol/L/mm Hg} = 0.13 \text{ mmol/L}$$

where $[O_2]$ is dissolved O_2 concentration, and s is solubility of O_2 in blood.

The Henry law thus predicts that if blood O_2 concentration is 0.13 mmol/L and in equilibrium with a gas compartment, P_{O_2} in that compartment must be 100 mm Hg. Therefore, we consider the partial pressure of O_2 in blood to be 100 mm Hg, which allows us to discuss the pressure gradients driving gas movement between gas and liquid phases.

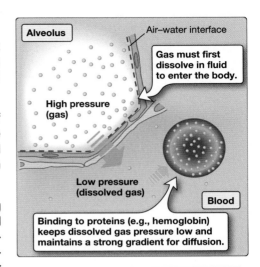

Figure 22.2.
Gas diffusion between air and blood.

Table 22.1: Partial Pressures of Oxygen and Carbon Dioxide

Location	O_2 (mm Hg)	CO_2 (mm Hg)
External air	160	0
Conducting airways (during inhalation)	150	0
Alveoli	100	40
Pulmonary capillary	100	40
Systemic artery	95*	40
Pulmonary artery	40	45

*Value is slightly less than that of pulmonary capillary because of physiologic shunts.

Partial pressures reflect the amount of free gas dissolved in fluid and do not provide information about how much additional gas might be bound to hemoglobin (Hb), for example.

III. PULMONARY CIRCULATION

The pulmonary circulation, like the systemic circulation, receives 100% of cardiac output, but there the similarities end. Several features make the pulmonary vasculature unique, reflecting its location within the general circulation and adaptations designed to facilitate gas exchange.

A. Overview

The pulmonary circulation has a resistance of 2 to 3 mm Hg·min·L^{-1}, or about fivefold less than the systemic circulation (see Chapter 18·IV). Mean pulmonary arterial pressures are reduced correspondingly (10–17 mm Hg), as is supply-artery wall thickness. Pulmonary arterioles contain a fraction of the smooth muscle that characterizes systemic resistance vessels and makes them difficult to distinguish from veins. The paucity of muscle means that pulmonary vessels readily distend in response to minor changes in filling pressure. The pulmonary vasculature can accommodate up to 20% of circulating blood volume, and changes in posture routinely cause gravity-induced shifts of ~400 mL between the pulmonary and systemic circulations.

B. Blood–gas interface

Red blood cells (RBCs) are separated from atmospheric gas by the width of a capillary endothelial cell plus an alveolar epithelial cell (~0.15–0.30 μm; see Fig. 21.4). The density of pulmonary capillaries is so great that the alveolar surface is bathed in a near-continuous sheet of blood, which optimizes gas exchange. Pulmonary capillaries have an average length of 0.75 mm, providing ample opportunity for gas equilibration between blood and air, even at high flow rates. At rest, a single RBC traverses the length of a capillary and flows past five to seven alveoli in ~0.75 s.

C. Lung volume

The high compliance of pulmonary blood vessels means that they readily collapse when compressed by surrounding tissues. In practice, this means that changes in airway pressure during the respiratory cycle have a major impact on alveolar perfusion rates. The nature and timing of the change depends on vessel location within the bronchial tree.

1. **Supply vessels:** Flow through pulmonary arteries and arterioles is very sensitive to changes in intrapleural pressure (P_{pl}). P_{pl} becomes strongly negative during inspiration to inflate the alveoli and draw in air from the atmosphere (Fig. 22.3; see also Chapter 21·VIII for details). The negative pressure simultaneously dilates blood vessels that are embedded in the lung parenchyma. Because

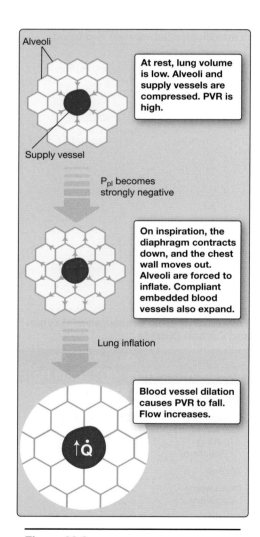

At rest, lung volume is low. Alveoli and supply vessels are compressed. PVR is high.

P_{pl} becomes strongly negative

On inspiration, the diaphragm contracts down, and the chest wall moves out. Alveoli are forced to inflate. Compliant embedded blood vessels also expand.

Lung inflation

Blood vessel dilation causes PVR to fall. Flow increases.

Figure 22.3.
Effects of inspiration on pulmonary supply vessels. P_{pl} = pleural pressure, PVR = pulmonary vascular resistance, $\dot{Q}$ = pulmonary blood flow.

vascular resistance is inversely proportional to vessel radius ($R \propto 1/r^4$), supply-vessel dilation during inspiration decreases pulmonary vascular resistance ([PVR] see also Chapter 18·IV).

2. **Capillaries:** Pulmonary capillaries course between adjacent alveoli. When alveoli expand during inspiration, their walls stretch. The embedded capillaries are stretched longitudinally, and their internal diameter decreases (Fig. 22.4). The same effect causes skin to blanch when stretched. Stretching capillaries increases their resistance to flow and increases PVR.

3. **Pulmonary vascular resistance dependence on volume:** The differential effects of inspiration on supply-vessel and capillary resistance summate to produce a U-shaped plot of PVR against lung volume (Fig. 22.5). PVR is very high at low lung volumes (supply vessels are compressed) and at total lung capacity (capillaries are stretched), but resistance is lowest during normal quiet breathing.

D. Gravity

Because the pulmonary vasculature has low resistance overall, pulmonary arterial pressures are also very low. This makes flow through the pulmonary vasculature extremely susceptible to gravitational influences.

1. **Pulmonary blood pressures:** The heart is located within the mediastinum, nestled between the right and left lungs (see Fig. 21.10). The pulmonary valve (where the pressure available to drive flow through the pulmonary circulation is measured) is located approximately 20 cm below the lung apex. The right ventricle generates a mean pulmonary arterial blood pressure (P_{pa}) of ~15 mm Hg, or ~20 cm H_2O. When an individual is in a prone position, arterial pressures at the lung apex and base should both approximate 20 cm H_2O. When erect, gravity exerts a downward force that decreases arterial pressure above the heart by ~1 cm H_2O for each cm of vertical distance. Gravity increases pressures below the heart by the same amount.

2. **Regional differences:** The effects of gravity on P_{pa} mean that when a person is upright, pulmonary flow is lowest at the apex and increases progressively with decreasing height (Fig. 22.6). Although there is a flow continuum from apex to base, we can distinguish three distinct zones (1–3) based on flow characteristics.

   a. **Zone 1—minimal flow:** At the apex, **alveolar pressure > arterial pressure > venous pressure.** Because P_{pa} falls with height above the heart, pressure within an arteriole located ~20 cm above the ventricle is zero. Pulmonary venular pressure (P_{pv}) is less than zero at the same height (–9 cm H_2O). This creates a pressure gradient of 9 cm H_2O available to drive flow through apical capillaries, but, in practice, they are collapsed. Collapse occurs because pressure within an alveolus (P_A) at rest is 0 cm H_2O (i.e., barometric pressure), which is greater than the perfusion pressure maintaining capillary patency (see Fig. 22.6, upper panel). Zone 1 only exists at the

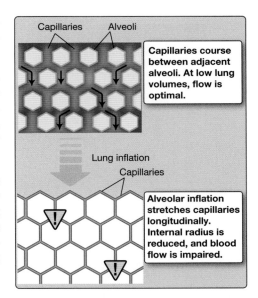

Figure 22.4.
Pulmonary capillary patency during inspiration.

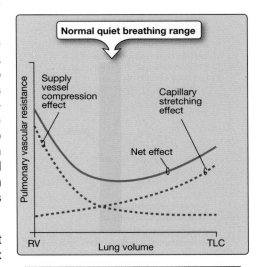

Figure 22.5.
Effects of lung volume on pulmonary vascular resistance. RV = residual volume, TLC = total lung capacity.

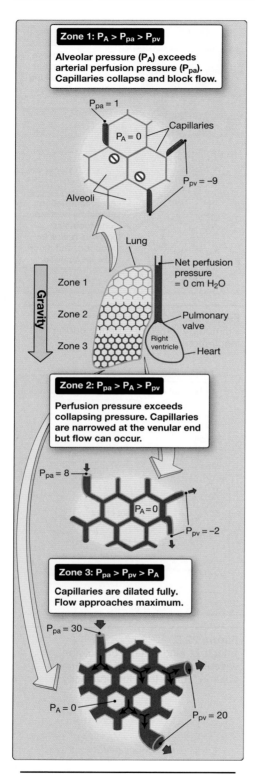

Figure 22.6.
Regional perfusion and flow patterns in a static, upright lung. Values are given in cm H_2O. P_{pv} = pulmonary venular pressure.

very pinnacle of the lung when pulmonary vascular pressures are critically low (e.g., during hemorrhage or other form of circulatory shock) or when alveolar pressure is raised artificially by **positive pressure ventilation**.

b. **Zone 2—moderate flow:** In zone 2, **arterial pressure > alveolar pressure > venous pressure** (see Fig. 22.6, middle panel). Zone 2 includes the apex and the middle of the lung, regions in which P_{pa} and mean capillary pressure (P_c) are greater than P_A. P_{pv} in zone 2 is still lower than P_A, so a capillary tends to be compressed at the venular end, but flow continues nonetheless. The resistance created by extravascular compression gradually decreases with lung height, reflecting the concomitant rise in both P_{pa} and P_{pv}. (Note: P_A is responsive only to barometric pressure and changes in intrathoracic volume and is insensitive to position.)

c. **Zone 3—maximal flow:** The lung base is located below the pulmonary valve. Gravity enhances perfusion pressures in this region, so **arterial pressure > venous pressure > alveolar pressure** (see Fig. 22.6, lower panel). Vascular collapse is no longer an issue here. Instead, capillaries at the lung base are typically distended by high, gravity-enhanced perfusion pressures. In the systemic circulation, resistance vessels tightly control P_c through reflex constriction and dilation of smooth muscle layers within the vessel walls. Pulmonary arterioles are endowed with relatively little smooth muscle, which makes them ineffective pressure regulators. Thus, P_c rises in concert with P_{pa} and P_{pv}, and the capillary swells beyond normal capacity. Flow through blood vessels is proportional to the fourth power of internal radius; therefore, flow is disproportionately high also.

E. Flow regulation

Blood flow through systemic resistance vessels is controlled by the sympathetic nervous system, bloodborne agents, rising metabolite levels, and other factors. By contrast, pulmonary resistance vessels are relatively insensitive to sympathetic activity or humoral factors. The vasculature is mildly sensitive to rising interstitial CO_2 and H^+ levels, but, whereas systemic resistance vessels would dilate reflexively, pulmonary vessels **constrict** when CO_2 and H^+ levels rise. A predominant force controlling pulmonary resistance vessels is P_{O_2}. Low O_2 levels promote **hypoxic vasoconstriction**. This reflex is again the opposite of how systemic resistance vessels respond to hypoxia, but it has clear advantages for optimizing pulmonary function. Hypoxic vasoconstriction steers blood away from poorly ventilated areas, redirecting it to well-ventilated regions where gas exchange can occur.

F. Venous admixture

Ideally, blood would leave the pulmonary circulation and enter the systemic circulation at 100% O_2 saturation. In practice, this never occurs because there is always some degree of **venous admixture**, or the mixing of deoxygenated (venous) and oxygenated blood prior to blood entering the systemic arterial system. The two main causes are **shunts** and **low ventilation/perfusion $\left(\dot{V}_A/\dot{Q}\right)$ ratios**.

1. **Shunts: Shunts** allow venous blood to bypass the normal process of gas exchange (Fig. 22.7). There are two types: **anatomic shunts** and **physiologic shunts**.

    a. **Anatomic:** Anatomic shunts have a structural basis, such as a fistula or a blood vessel. Examples include an atrial septal defect that allows blood from the right atrium to enter the left atrium, or an anastomosis between a pulmonary artery and a pulmonary vein. These are also known as **right-to-left shunts**.

    b. **Physiologic: Physiologic shunting** occurs when atelectasis, pneumonia, or some other problem affecting ventilation of the blood–gas interface prevents gas exchange. Hypoxic vasoconstriction redirects flow, but there is always some residual perfusion of a nonfunctional interface. Blood from these regions escapes oxygenation and reduces arterial O_2 saturation levels when it enters the systemic circulation.

2. **Low ventilation/perfusion ratios:** $\dot{V}_A/\dot{Q}$ ratios are discussed in more detail in the next section, but if the blood–gas interface is perfused at rates that exceed its diffusional limits, O_2 saturation cannot occur. Venous admixture is the result.

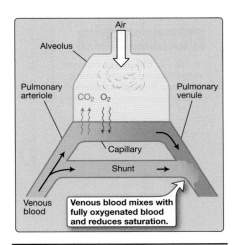

Figure 22.7.
Shunts allow venous admixture.

IV. VENTILATION/PERFUSION RATIO

At rest, the pulmonary circulation is perfused with ~5 L/min of blood $\left(\dot{Q}\right)$, representing the entire output of the right heart. Lung inflation maximally draws ~4 L of air into the alveoli sacs during this time (alveolar ventilation is abbreviated as $\dot{V}_A$, so the pulmonary $\dot{V}_A/\dot{Q}$, = 0.8. In an ideal lung, all alveoli would be ventilated and perfused optimally, but there are many physiologic causes of mismatch.

A. Model lung mechanics

The function of alveolar ventilation is to bring outside air into close proximity to blood so that O_2 may be loaded and CO_2 off-loaded. At sea level, the humidified atmospheric air that enters the respiratory zone contains 150 mm Hg O_2 and negligible CO_2 (Fig. 22.8). Blood arriving at the alveolus from pulmonary arterioles (**mixed venous blood**) is rich in CO_2 (Pco_2 = 45 mm Hg) but O_2 poor (Po_2 = 40 mm Hg). During normal quiet breathing, equilibration of both gases between air and blood completes before blood has progressed even a third of the way through the capillary, raising P_{Aco_2} to 40 mm Hg and lowering P_{Ao_2} to 100 mm Hg. Alveoli have no means of modifying these values further, so blood exiting a pulmonary capillary also has a Pco_2 of 40 mm Hg and Po_2 of 100 mm Hg O_2. Changes in either ventilation or perfusion affect these values, however.

1. **Normal changes in ventilation:** Physiologic changes in respiratory rate (normal range = 12–16 breaths/min) modify alveolar gas proportions (Fig. 22.9). Hypoventilation allows CO_2 to accumulate because it is being expelled at a lower rate. Alveolar O_2 content simultaneously falls. Conversely, hyperventilation refreshes

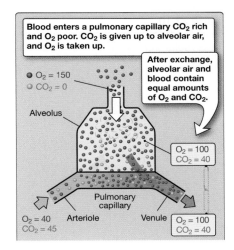

Figure 22.8.
CO_2 and O_2 exchange between pulmonary blood and alveolar air. Partial pressures are given in mm Hg.

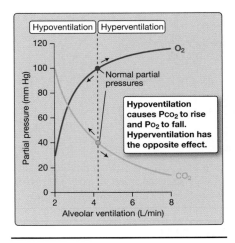

Figure 22.9.
Effect of ventilation on alveolar and arterial blood gas composition.

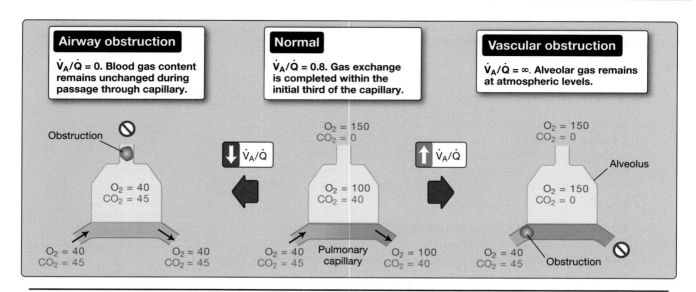

Figure 22.10.
Effect of obstructing either ventilation or perfusion on P_{O_2} and P_{CO_2} in the lung. All partial pressures are given in mm Hg. $\dot{Q}$ = alveolar perfusion, $\dot{V}_A$ = alveolar ventilation.

alveolar gas contents at an increased rate, so CO_2 levels fall and O_2 levels rise.

2. **Airway obstruction:** If an airway is obstructed by a mucus plug, for example, the $\dot{V}_A/\dot{Q}$ ratio in the affected area drops to zero. In the absence of ventilation, alveolar gas equilibrates with mixed venous blood at a $P_{A_{CO_2}}$ of 45 mm Hg and a $P_{A_{O_2}}$ of 40 mm Hg. Blood leaving the area of obstruction has no opportunity to exchange O_2 or CO_2, and levels remain unchanged during transit (Fig. 22.10, left). This creates a physiologic shunt, as discussed earlier.

3. **Blood flow obstruction:** If blood flow is prevented by an embolus, for example, the $\dot{V}_A/\dot{Q}$ ratio approaches infinity. Alveolar gas composition remains unchanged following inspiration because there is no blood contact (see Fig. 22.10, right).

B. Ventilation/perfusion ratios in an upright lung

Gravity significantly affects alveolar ventilation and perfusion (see Figs. 21.15 and 22.6). This creates a broad spectrum of $\dot{V}_A/\dot{Q}$ ratios in an upright lung (Fig. 22.11).

1. **Zone 1—highest ratio:** Alveoli at the lung apex ventilate poorly because they are inflated to 60% of maximal volume even at rest (see Fig. 21.15). Perfusion in this region is minimal, however, because the vasculature is compressed by alveolar pressures that exceed perfusion pressures (see Fig. 22.6, upper panel). Thus, P_{O_2} and P_{CO_2} in the small volumes of blood exiting this region approach that of inspired air $\left(\dot{V}_A/\dot{Q} \sim \infty \right)$.

2. **Zone 2—moderate ratio:** Ventilation improves slowly with decreasing lung height. Perfusion increases more steeply, however, causing the $\dot{V}_A/\dot{Q}$ ratio to fall rapidly toward the base.

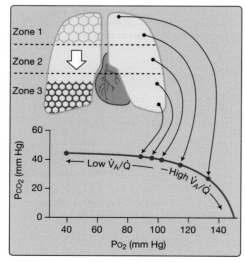

Figure 22.11.
Distribution of $\dot{V}_A/\dot{Q}$ ratios in an upright lung. $\dot{Q}$ = alveolar perfusion, $\dot{V}_A$ = alveolar ventilation.

3. **Zone 3—lowest ratio:** Alveoli at the lung base are compressed at rest and ventilate very well upon inspiration (see Fig. 21.15). Pulmonary perfusion pressures are also very high in this region, so flow rates are maximal (see Fig. 22.6, lower panel).

4. **Net effect:** The extent to which the different regions contribute to the composition of the blood leaving the lung is determined by their perfusion rates. Thus, the $\dot{V}_A/\dot{Q}$ extremes seen at the apex have minimal effect on overall saturation levels. The O_2 and CO_2 content of systemic arterial blood is determined largely by the heavily perfused regions at the base.

C. Ventilation/perfusion mismatches

Blood flow through the lung base is so high that it exceeds the ventilatory capacity of the blood–gas interface and causes a local $\dot{V}_A/\dot{Q}$ mismatch. Blood leaving the area has a P_{O_2} of ~88 mm Hg, or 12 mm Hg below optimum, whereas P_{CO_2} is higher by ~2 mm Hg. Some degree of physiologic shunting caused by $\dot{V}_A/\dot{Q}$ mismatch occurs normally even in a healthy individual but can become severe when an airway is obstructed by, for example, aspiration of a foreign body, tumor growth, or during an asthma exacerbation. The $\dot{V}_A/\dot{Q}$ ratio is an important measure of pulmonary function and health. Both parameters can be visualized clinically using radioactive tracers, but imaging techniques are generally used only if gross deficiencies in either ventilation or perfusion are suspected such as those caused by pulmonary embolism (Fig. 22.12).

D. Alveolar–arterial oxygen difference

Potential problems with either ventilation or perfusion can also be assessed using the alveolar–arterial difference for O_2 (A–aDO_2), which

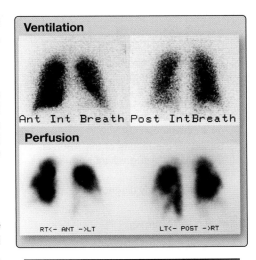

Figure 22.12.
This ventilation scan (visualized radiographically using radioactive xenon gas) is normal, but the perfusion scan (visualized radiographically using radiolabeled albumin) shows many areas devoid of radioisotope, a pattern characteristic of pulmonary embolism.

Clinical Application 22.1: Pulmonary Tuberculosis

The microorganism that causes the majority of pulmonary tuberculosis (TB) cases, *Mycobacterium tuberculosis*, favors lung regions where O_2 levels are high and typically establishes itself at the apices, where alveolar gas composition most closely resembles that of atmospheric air. Most individuals with intact immune systems prevent further bacterial replication following initial infection, and the disease enters a latent phase. About 10% of infected individuals show active disease, characterized by chronic cough, weight loss (which explains why TB was once called "consumption"), and fatigue. In advanced cases, lung tissue is destroyed, and large cavities develop. The cavities are avascular, which can make infection difficult to treat. Multiple drugs must be given together for a long period to fully eradicate tubercular organisms from the tissue. The incidence of TB is higher in elderly populations compared with young adults, and the risk of associated mortality increases.

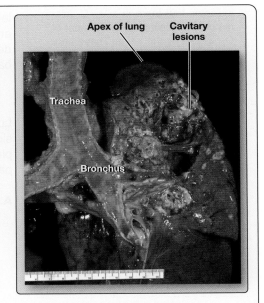

Postmortem specimen showing apical lung lesions caused by tuberculosis.

Example 22.1

A 50-year-old female with a history of prior deep vein thrombosis presents in the emergency department with shortness of breath. A room air arterial blood gas (ABG) sample is obtained, and the patient is placed on supplemental O_2.

Based on the results of the ABG, is the patient's A-a gradient normal or abnormal?

What is her likely diagnosis?

ABG results:

$P_aO_2 = 70$ mm Hg

$P_aCO_2 = 32$ mm Hg

pH = 7.47

P_IO_2 (room air, sea level) = 150 mm Hg

Based on age, the patient's A-a gradient would be predicted to be $2.5 + (0.21 \times 50) = 13$ mm Hg

Using ABG values above and Equation 22.1

$$P_AO_2 = P_IO_2 - \frac{P_aCO_2}{RER}$$

$$= 150 - 32/0.8$$

$$= 150 - 40$$

$$= 110 \text{ mm Hg}$$

Her observed A-a difference (P_AO_2–P_aO_2) is $110 - 70 = 40$ mm Hg, or 27 mm Hg higher than predicted.

The abnormally wide A-a gradient indicates that there is a $\dot{V}_A/\dot{Q}$ mismatch, suggesting an impairment of O_2 uptake by the lungs. These findings are consistent with a diagnosis of pulmonary embolism.

---

compares P_{O_2} in alveoli with that of systemic arterial blood. Ideally, the two values should be the same. In practice, there is always a difference between alveolar gas and blood of 5 to 15 mm Hg P_{O_2}, depending on age. P_{AO_2} is assessed using a simplified form of the **alveolar gas equation**:

Equation 22.1 $\qquad P_IO_2 = P_IO_2 - \frac{P_ACO_2}{RER}$

where P_IO_2 is the partial pressure of O_2 in inspired air, P_ACO_2 is alveolar P_{CO_2}, and RER is the respiratory exchange ratio. P_ACO_2 can be determined by analyzing gas captured at the very end of expiration but, in practice, is assumed to be equal to P_aCO_2. "RER" (normally 0.8) represents the ratio of CO_2 produced to O_2 consumed by internal respiration. P_aO_2 (and P_aCO_2) can be measured by arterial blood gas analysis. The difference between P_AO_2 and P_aO_2 for a healthy individual can be predicted as:

A-a gradient $= 2.5 + 0.21 \times$ Age in years

An A-a difference wider than predicted indicates that O_2 uptake at the blood–gas interface is impaired (see Example 22.1).

V. GAS EXCHANGE

The rate at which gases diffuse across the blood–gas interface (i.e., gas flow, or $\dot{V}$) is determined by the pressure difference across the interface (ΔP), the surface area available for exchange (A), and barrier thickness (T):

$$\dot{V} = \frac{\Delta P \times A \times D}{T}$$

where D is a diffusion coefficient that takes into account the molecular weight and solubility of a gas. In practice, surface area, thickness, and the diffusion coefficient can be combined to yield a constant that describes the lung's **diffusing capacity** (D_L) for gas. Gas flow across the barrier can then be estimated from:

$$\dot{V} = \Delta P \times D_L$$

Lung design maximizes flow by providing a large surface area for diffusion and by restricting barrier thickness to the width of a pneumocyte plus a capillary endothelial cell. Ventilation and perfusion maintain steep partial pressure gradients across the interface.

A. Diffusion-limited exchange

Blood traverses the length of a pulmonary capillary in ~0.75 seconds at rest. Equilibration of O_2 between alveolar gas and blood occurs within a fraction of this time, so uptake is not normally limited by the rate at which O_2 diffuses across the exchange barrier (Fig. 22.13A). During maximal exercise, however, cardiac output increases and capillary transit time decreases to <0.4 seconds. Blood may exit the capillary before being fully O_2 saturated (see Fig. 22.13B). O_2 uptake is now considered to be **diffusion limited** because exchange has been limited by the rate at which O_2 diffuses across the blood–gas interface. In practice, diffusion limitations rarely occur except in elite athletes capable of achieving cardiac outputs and corresponding pulmonary flow rates in excess of 35 L during maximal exercise (see Chapter 39). The effects of diffusion limitations can best be appreciated by studying the characteristics of carbon monoxide (CO) uptake, which is always diffusion limited (Fig. 22.14).

> The "CO" abbreviation is generally used to denote cardiac output in cardiovascular and other areas of physiology, but CO indicates carbon monoxide in pulmonary physiology.

1. **Carbon monoxide uptake:** Hb binds CO with an affinity that is ~240 times greater than that for O_2 (see Chapter 23·IV·C·2). In practice, this means CO molecules bind to Hb as fast as they can diffuse across the exchange barrier, and alveolar CO never has the chance to equilibrate with plasma CO. A diffusion limitation such as this might be offset by increasing the pressure gradient driving diffusion or by increasing the D_L for CO (D_{LCO}).

2. **Changing perfusion rate:** Intuitively, slowing blood flow through the capillary would seem beneficial in terms of increasing net uptake. A slower flow rate would allow more time for the gas and liquid phases to come into equilibrium before blood exits the capillary. In practice, although decreasing perfusion rate does allow for greater saturation, net uptake actually decreases because the volume of blood exiting the capillary per unit time is reduced also.

> Changing perfusion rate has no net effect on gas transport in a diffusion-limited exchange scenario.

B. Perfusion-limited exchange

Blood becomes fully O_2 saturated shortly after entering a pulmonary capillary (at rest). Because more O_2 could be transferred if flow were increased (even though this transfer may exceed body requirements), exchange is considered to be **perfusion limited**. The characteristics of perfusion-limited gas exchange can be best appreciated by studying N_2O uptake. Hb does not bind N_2O, so blood and alveolar partial pressures for N_2O equilibrate in <100 milliseconds (Fig. 22.15). Modest changes in barrier architecture have little effect on net uptake. Instead, net N_2O uptake is tied to flow.

> In a perfusion-limited system, gas will saturate whatever amount of blood is presented to it over a wide range of values.

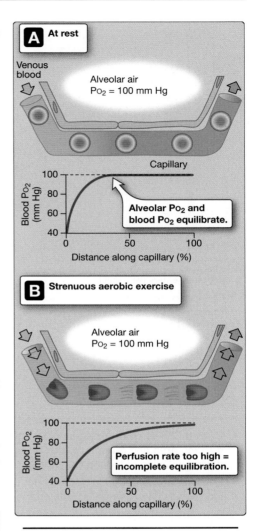

Figure 22.13.
Effect of capillary perfusion rate on oxygenation saturation.

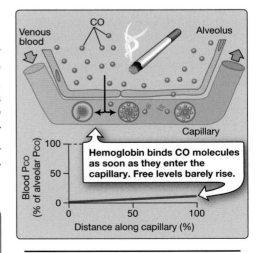

Figure 22.14.
Diffusion-limited gas exchange. P_{CO} = partial pressure of carbon monoxide.

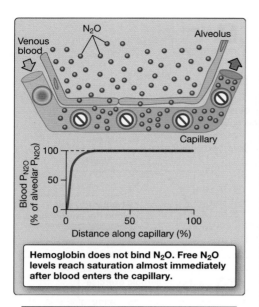

Figure 22.15.
Perfusion-limited gas exchange. P_{N2O} = partial pressure of N_2O.

Biologic Sex and Aging 22.1: Gas Exchange

Aging is accompanied by a progressive decline in P_{aO_2}. Although estimates vary, studies suggest that a P_{aO_2} of 80 to 85 mm Hg should be considered normal after age 65 years. Multiple factors contribute.

- **Ventilation/perfusion imbalances:** Aging-associated elastin fragmentation deprives small airways of mechanical support from surrounding structures (i.e., loss of radial traction), which can cause air trapping. Additionally, stiffening of the pulmonary vasculature can lead to pulmonary hypertension. These changes together cause a net increase in alveolar ventilation/perfusion $\left(\dot{V}_A/\dot{Q}\right)$ mismatching.

- **Decreased lung diffusing capacity:** Elastin fragmentation also causes dilation of the alveolar ducts and alveoli. This results in a 30% decrease in blood–gas interface surface area and reduced lung diffusing capacity for carbon monoxide (D_{LCO}) by age 90 years. In males, D_{LCO} decreases by 2.0–3.2 mL·min^{-1}·mm Hg^{-1} per decade. Estimates are less for females (0.6–1.8 mL·min^{-1}·mm Hg^{-1} per decade).

- **Decreased capillary density:** Reduced blood–gas interface surface area also decreases capillary density and capillary blood volume.

Chapter Summary

- **O_2 and CO_2 exchange** occurs across the **blood–gas interface** within the lung. Exchange is enhanced by the **large surface area** and **minimal width** of the interface. Ventilation and perfusion ensure that the partial pressure gradients driving diffusion of O_2 and CO_2 across the barrier are kept high.

- The interface is perfused by blood from the **pulmonary circulation**. Pulmonary perfusion pressures are very low, and the vessels have thin walls. Pulmonary vessels readily expand and collapse in response to **extravascular forces**.

- **Gravitational effects** on pulmonary blood vessels in an upright lung create three different zones of flow. **Perfusion pressures and flow** are lowest at the lung apex (zone 1). Flow is highest at the lung base (zone 3). Gravity also affects **alveolar ventilation**. Alveoli at the lung apex ventilate poorly, whereas alveoli at the lung base ventilate very well. The combined effects of gravity on perfusion and ventilation mean that most O_2 uptake occurs at the base of an upright lung.

- In an ideal lung, **alveolar ventilation $\left(\dot{V}_A\right)$** and **perfusion $\left(\dot{Q}\right)$** should be perfectly matched $\left(\dot{V}_A/\dot{Q} \text{ ratio} = 1.0\right)$. **Mismatches** occur because of airway obstruction or decreased perfusion.

- O_2 and CO_2 exchange occurs by diffusion, driven by partial pressure gradients for both gases. **Diffusion of gases** across the alveolar wall is influenced by blood–gas barrier thickness and surface area, both of which may become limiting in a diseased lung (**diffusion-limited exchange**). Net uptake may also be limited by inadequacy of perfusion (**perfusion-limited exchange**).

Study Questions

Choose the ONE best answer.

22.1. A pharmaceutical company is interested in developing a drug that lowers pulmonary vascular resistance (PVR) in individuals with pulmonary hypertension. The experimental protocol involves assessing candidate drug effects when the effect of lung volume on PVR and pulmonary flow is minimal. When is this most likely to occur?

A. At functional residual capacity
B. At high alveolar pressures
C. At high intrapleural pressures
D. At residual volume
E. At total lung capacity

Best answer = A. Pulmonary blood vessels are thin walled, which makes them susceptible to extravascular compression (see Section III). PVR is highest and perfusion is lowest when lung volumes are very high or very low. When alveolar pressures are high (B) and at total lung capacity (E), PVR is high because the capillaries are stretched and compressed between adjacent alveoli. When intrapleural pressures are high (C) and at residual volume (D), arterial supply vessels are collapsed by external pressure. The nadir in the PVR-to-lung-volume curve occurs at functional residual capacity, because the combined effect of capillary and supply vessel compression is minimal.

22.2. A 58-year-old female presents with a right-to-left shunt caused by a pulmonary arteriovenous malformation. Which of the following variables would most likely be increased in this individual?

A. Alveolar–arterial O_2 difference
B. Arterial dissolved O_2 content
C. Arterial P_{O_2}
D. Oxyhemoglobin levels
E. Venous P_{O_2}

Best answer = A. Right-to-left shunts allow blood to pass from the right to the left heart without being oxygenated (see Section III·F). The shunted blood lowers the P_{O_2} of arterial blood (C), thereby widening the alveolar–arterial O_2 difference. The amount of O_2 that blood carries in dissolved form (B) is minimal normally but would be decreased further by a shunt. A right-to-left shunt would decrease oxyhemoglobin levels (D) and venous P_{O_2} (E).

22.3. A hypoxemic 50-year-old male with an increased alveolar–arterial O_2 gradient is given 100% O_2 via a facemask, causing arterial P_{O_2} to increase to >500 mm Hg. Results of a lung-diffusing capacity test were normal. What is the most likely cause of the hypoxemia?

A. Alveolar hypoventilation
B. Diffusion limitation
C. Hypobaric conditions
D. Right-to-left shunt
E. Ventilation/perfusion mismatch

Best answer = E. Hypoxemia with an increased alveolar–arterial O_2 gradient (A–aDO_2) could be due to either a ventilation/perfusion mismatch or a diffusion limitation (see Section IV·D). The lung-diffusing capacity test eliminates a diffusion limitation (B). Hypoventilation (A) and hypobaric conditions (C) can result in hypoxemias, but they do not change A–aDO_2. Right-to-left shunts (D) also cause an increased A–aDO_2, but 100% O_2 would not increase arterial P_{O_2} to the levels observed here.

22.4. A 75-year-old male with a history of interstitial pulmonary fibrosis presents with increased dyspnea on exertion. A carbon monoxide (CO) uptake test is ordered. Which of the following most likely describes pulmonary CO uptake during this test?

A. Binding is limited.
B. Diffusion is limited.
C. Perfusion is limited.
D. Solubility is limited.
E. Ventilation is limited.

Best answer = B. Net carbon monoxide (CO) uptake by the lungs is limited by the rate at which it diffuses across the blood–gas interface (see Section V·A). CO uptake binds to hemoglobin (Hb) with high affinity, so uptake is not binding limited (A). Uptake is relatively insensitive to changes in pulmonary perfusion (C), unlike uptake of a perfusion-limited gas such as N_2O, which is why the test is used to assess lung-diffusing capacity. The avidity with which Hb binds CO means that blood rarely carries appreciable amounts of gas in dissolved form (D). Uptake is not limited by ventilation (E) under physiologic conditions.

23 Blood and Gas Transport

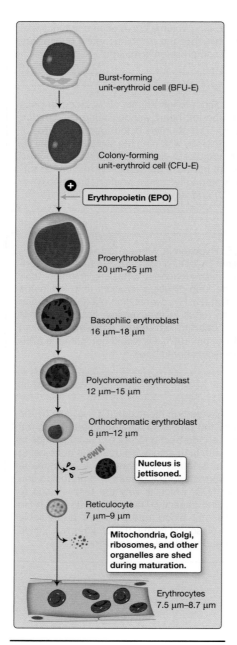

I. OVERVIEW

O_2 diffuses across the thin (~300 nm) barrier separating alveolar gas from blood relatively rapidly. Because O_2 cannot be ferried to dependent tissues in gaseous form (gas bubbles would block smaller blood vessels), it must be carried in solution. Unfortunately, O_2 has a very poor water solubility compared with many other gases, which limits transport volume to ~3 mL of gaseous O_2 per liter of blood. An average adult consumes ~250 mL O_2/min at rest, so resting cardiac output would have to be maintained at 83 L/min and then climb to >1,000 L/min during exercise if transport relied on O_2-solubility properties alone! The solution to this problem has been to cram blood full of **hemoglobin** (**Hb**), a protein uniquely designed to bind O_2 in the lungs and then release it to tissues upon arrival. Hb increases blood's O_2-carrying capacity to ~200 mL/L, which is more than adequate to meet all tissue needs, even during exercise. Because Hb is a protein, and capillaries tend to leak proteins, Hb is confined within membrane-bound packages, otherwise known as red blood cells (RBCs), or **erythrocytes**. RBCs start life as bone marrow stem cells (Fig. 23.1), but then they are then stripped of all cellular organelles to allow for maximal Hb packing. They are also sculpted to maximize surface area for gas exchange and given a flexibility that allows them to squeeze through capillaries a third of their diameter. Their cytoplasm is rich in **carbonic anhydrase** (**CA**), an enzyme that facilitates interconversion between $CO_2 + H_2O$ and $HCO_3^- + H^+$. HCO_3^- formation allows blood to ferry large amounts CO_2 from tissues (CO_2 is a metabolic byproduct) back to the lungs for release to the atmosphere. HCO_3^-, coincidentally, is an important buffer that helps maintain extracellular pH.

II. HEMATOPOIESIS

Blood is a connective tissue. Its cellular components (red cells, white cells, platelets) are formed in red bone marrow. In the fetus and newborns, all bones participate in hematopoiesis.[1] By age 20 years, cellular production is largely restricted to the vertebrae, sternum, ribs, and pelvis.

Figure 23.1.
Erythropoiesis.

 [1]For more information on hematopoiesis, see *LIR Immunology*, 3e, Chapter 4·I and *LIR Integrated Systems*, Chapter 4·V.

A. Erythropoiesis

RBCs have an average lifespan of 120 days, so new RBCs must be generated at a rate of ~2×10^{11} per day to maintain a stable hematocrit. **Erythropoiesis** begins when a pluripotent stem cell commits to the erythroid line to form a burst-forming unit-erythroid (BFU-E; see Fig. 23.1) capable of generating tens of thousands of cells. Maturation of a BFU-E creates a colony-forming unit-erythroid (CFU-E) that forms a small colony of macrophages and erythroid cells. The latter progresses through a series of stages classified according to their staining characteristics (proerythroblast, basophilic erythroblast, polychromatic erythroblast, orthochromatic erythroblast), during which time nuclear diameter decreases, and Hb concentration rises. RBC production takes about a week.

1. **Erythrocyte birth:** Orthochromatic erythroblasts expel their nucleus to form a **reticulocyte**, which then leaves the marrow sinus and enters the circulation (see Fig. 23.1). Over the next 24 to 48 hours, the reticulocyte matures, a process that involves shedding any remaining organelles (ribosomes, Golgi, mitochondria) and structuring a cytoskeleton to hold an RBC in the familiar discoid shape.

Hematocrit, Hb levels, RBC morphology, and reticulocyte counts are important in differentiating causes of anemia. Reticulocytes usually comprise 0.5%–1.5% of RBC numbers, but this rises in anemia during compensation for RBC loss.

B. Erythropoiesis regulation

Erythropoiesis is tightly regulated by a number of hormones, cytokines, and transcription factors. The single most important influence is hypoxia, which stimulates RBC production through hypoxia-inducible factor 1 (HIF-1) activation and release of **erythropoietin** (**EPO**) from the kidney.

1. **Hypoxia-inducible factor 1:** HIF-1 is a dimeric transcription factor comprising HIF-1α and HIF-1β subunits. It is found in all tissues and mediates responses to hypoxia. HIF-1β is expressed constitutively. Under normal conditions, HIF-1α is maintained at very low levels by prolylhydroxylases. Hydroxylation targets the α-subunit for rapid degradation and blocks HIF-1 transcription factor activity. When O_2 levels are low, the prolylhydroxylases are inactive, allowing HIF-1α and HIF-1β to dimerize and interact with hypoxia-response elements on numerous genes. The gene products help cells survive reduced O_2 availability, reduce the rate of reactive O_2 species formation, activate glycolytic pathways, and increase EPO expression.

2. **Erythropoietin:** EPO is 30.4-kDa α-globulin synthesized by specialized interstitial cells in the renal cortex and outer medulla (liver and brain are also capable of limited EPO production). EPO is produced on demand and then released into the circulation, where it has a half-life of 4 to 12 hours. EPO prevents progenitor cell apoptosis and stimulates CFU-E cell division and differentiation.

Biologic Sex and Aging 23.1: Erythropoiesis and Anemia

Adult females have ~12% lower hematocrit (Hct) and hemoglobin (Hb) levels compared with males. The difference is unrelated to menstrual blood loss but instead relates to differences in the set point for erythropoiesis. The physiologic reason for the difference is not understood fully, but the cause is clearly hormonal. Erythropoiesis is stimulated by androgens and inhibited by estrogens. The red blood cell (RBC) complement of females is estimated to be >200% of critical levels, so females are not disadvantaged by the lower Hct. At the level of the microcirculation, females have a superior ability to deliver O_2 to tissues compared with males, an effect related to estrogen-induced vasodilation. Considering that RBC production and maintenance of Hct requires a tremendous allocation of resources, the lower Hct in females may be viewed as an advantage rather than a deficiency.

Basal rates of RBC production are not significantly affected by age, although responses to erythropoietic stress (e.g., hemorrhage) may be blunted in older adults. Anemia is a common finding among older adults (age ≥75 years), with the incidence in males being higher than that in females. Although the causes may be multifactorial, nutritional deficiencies account for most cases. Iron deficiency is common among older adults, often due to chronic gastrointestinal losses.

Clinical Application 23.1: Nutritional Anemias

Erythropoiesis is dependent on a number of factors, the most important being iron, folate (vitamin B_9), and vitamin B_{12}. Deficiencies in any of these factors lead to **anemia**, characterized by decreased or increased red blood cell (RBC) size (**microcytic** or **macrocytic** anemia, respectively). Normal mean corpuscular volume (MCV) is 80–100 fL.

Iron is needed for hemoglobin (Hb) synthesis. If dietary intake is inadequate, or iron loss from the body is excessive (e.g., due to blood loss), microcytic, hypochromic anemia results (MCV <80 fL). Hypochromic RBCs are notable for their pallor. In addition to containing reduced amounts of Hb, iron-deficient RBCs are prone to hemolysis, which lowers hematocrit.

Folate and **vitamin B_{12}** (**cobalamin**) are essential for DNA synthesis and RBC maturation, among many other functions. Folate deficiency is common in individuals not taking nutritional supplements. Vitamin B_{12} deficiency is rare except in individuals adhering to a strict vegan diet. Deficiency in either vitamin results in macrocytic anemia (MCV >100 fL), characterized by megaloblast accumulation in bone marrow and blood. Megaloblasts are enlarged erythroblasts in which DNA synthesis is arrested, even as cytoplasmic growth and Hb accumulation continues unabated. Blood samples also reveal the presence of hypersegmented neutrophils. Macrocytic RBCs are large, contain immature nuclei, and are commonly oval shaped. These cells have reduced deformability and fragile membranes, which predisposes them to rupture in areas such as the spleen, where flexibility is required to squeeze through narrow slits between adjacent endothelial cells (splenic cords) to progress from the arterial to the venous system. Hemolysis leads to anemia. A principal cause of vitamin B_{12} deficiency is autoimmune destruction of gastric parietal cells, which secrete acid (see Chapter 30·IV·D) and intrinsic factor (IF). IF is required for intestinal vitamin B_{12} absorption (see Chapter 31·II·F). This condition is known as pernicious anemia.

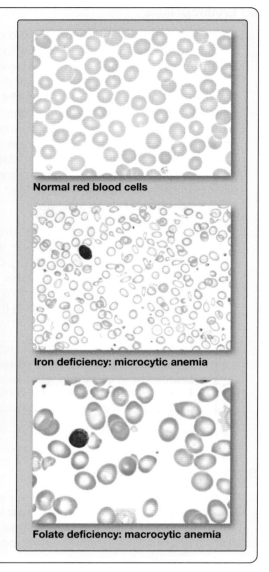

Normal red blood cells

Iron deficiency: microcytic anemia

Folate deficiency: macrocytic anemia

III. ERYTHROCYTES

Erythrocyte shape and composition is uniquely adapted to life in the vasculature, for delivering O_2 to tissues, and for helping maintain acid–base homeostasis.

A. Shape

RBCs are thin, biconcave disks with a diameter of 7.5–8.7 μm (Fig. 23.2). The shape greatly increases the surface area–volume ratio and thereby facilitates rapid gas exchange. Young, healthy RBCs have a surface area of ~135 μm² and a volume of 90 fL. The unique shape gives RBCs great resilience and flexibility, allowing them to slip through capillaries with internal diameters of ≤3 μm (see Fig. 18.10). RBC shape is maintained by a complex cytoskeleton assembled from hexagonal scaffolding subunits comprised largely of **spectrin** and

Figure 23.2.
Electron micrograph of red blood cells.

actin (Fig. 23.3). Ankyrin bolts the resulting scaffolding to the plasma membrane.

B. Composition

RBCs lack a nucleus and other organelles. They have no synthetic or oxidative phosphorylation capabilities, so they must generate ATP by glycolysis. RBC protein content is dominated by Hb (>98.5% of total). Other significant proteins are an anion exchanger (AE1) that facilitates HCO_3^- movement across the plasma membrane and CAs that facilitate interconversion between HCO_3^- and CO_2.

1. **Hemoglobin:** Hb is a metalloprotein comprising two dimeric subunits (Fig. 23.4). Each subunit contains a globin chain linked to a heme group. The heme group binds O_2, so each Hb can carry four O_2 molecules.

   a. **Heme:** "Heme" is a generic name for a metal ion within a porphyrin ring (see Fig. 23.4). The Hb heme group contains a single ferrous iron atom (Fe^{2+}) coordinated by four nitrogens. Fe^{2+} allows heme to bind O_2. (Note: O_2 binding is oxygenation rather than oxidation (to ferric iron, Fe^{3+}), meaning that it is reversible.) The heme group confers color to Hb. In the O_2-unbound state, it has a bluish purple color characteristic of venous blood. The O_2-bound form is bright red (arterial blood).

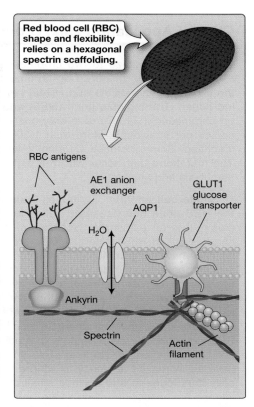

Figure 23.3.
Red blood cell cytoskeleton.

Oxidation of Fe^{2+} to Fe^{3+} creates methemoglobin (metHb). MetHb is unable to bind O_2. MetHb forms continuously *in vivo*, yet it amounts to only ~1.5% of total Hb. This is because RBCs contain metHb reductase, an enzyme that reduces Fe^{3+} back to the ferrous form.

   b. **Globin:** An Hb tetramer contains two α globins (or α-like globins, such as ζ-globin; see Biologic Sex and Aging 23.2) and two β globins (or β-like globins, such as δ-, ε-, and γ-globin; see Fig. 23.4). α Globins are polypeptides comprising 141 amino acid residues. β Globins are slightly larger (146 amino acid residues). Adult Hb, designated HbA, contains two α-globin chains and two β chains ($\alpha_2\beta_2$). Adult blood also contains a small proportion of HbA_2, fetal Hb ([HbF] see Biologic Sex and Aging 23.2), and HbA_{1c}, which forms spontaneously as a result of HbA glycation (Table 23.1). The numerous other named Hb variants (e.g., HbS; see Clinical Application 23.2) are mutational forms.

   c. **Quaternary structure:** A single heme molecule is held by covalent bonds within a cleft in the surface of a folded globin monomer. An α globin and a β globin with their associated heme groups bind tightly to form an αβ dimer. Two such dimers then associate to create the Hb $\alpha_2\beta_2$ tetramer. The two αβ dimers are tethered by relatively weak bonds, however, which allows conformational shifts and affinity changes in response to O_2 binding (Fig. 23.5).

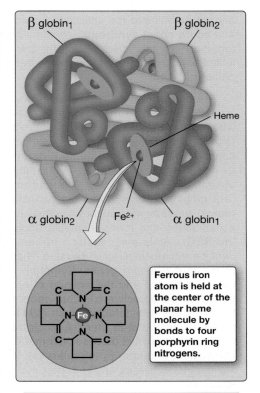

Ferrous iron atom is held at the center of the planar heme molecule by bonds to four porphyrin ring nitrogens.

Figure 23.4.
Hemoglobin structure.

Clinical Application 23.2: Globin Gene Mutations and Hemoglobinopathies

Globin genes localize to two chromosomes. The genes for α globin and α-like globins are clustered on chromosome 16. There are two copies of the α globin gene (*HBA1* and *HBA2*) per chromosome (i.e., four copies per individual). The β-globin gene (*HBB*) and β-like globin genes are clustered on chromosome 11. There is only one copy of the β-globin gene (i.e., two copies per individual). More than 1,300 globin gene mutations have been identified. Some increase O_2 affinity, whereas others decrease O_2 affinity. The mutations that occur with the highest frequency present clinically as hemolysis and anemia.

Sickle cell anemia: Sickle cell anemia results from a point mutation in the *HBB* gene that causes a valine-for-glutamate substitution to yield β^S globin. It is an autosomal recessive disorder. Hb that contains a β^S-globin chain is called HbS. Heterozygotes are carriers of the sickle cell trait and typically have no symptoms. Homozygotes with two copies of the defective β-globin gene ($\alpha_2\beta^S_2$) develop sickle cell anemia, which is a severe condition associated with increased morbidity. The mutation causes Hb tetramers to polymerize when deoxygenated. The polymers then associate to form long, rigid fibers that bundle together and cause the red blood cells (RBCs) to distort and assume a characteristic "sickle" shape. Sickling decreases RBC deformability, a key feature that normally allows RBCs to slip through tiny capillaries with a diameter of $\leq 3\ \mu m$ (see Fig. 18.10). Sickled cells also have an increased tendency to adhere to the vascular endothelium. Sickled cells thus lodge in and occlude small blood vessels, causing an acute and intense pain associated with tissue ischemia. Sickled cells are also prone to lysis, which lowers their lifespan from ~120 to ~20 days and leads to hemolytic anemia. Repeated vasoocclusive episodes cause ischemic damage to the spleen, which increases susceptibility to infection. Damage to joints and bones causes chronic pain. Bone marrow infarction releases necrotic tissue and fat emboli into the vasculature, which then causes pulmonary infarction and **acute chest syndrome** (chest pain and pain in the ribs, arms, and legs, together with shortness of breath, and chest radiographs show pulmonary infiltrates). Acute chest syndrome is the leading cause of death in sickle cell patients.

Thalassemias: Normally, synthesis of α-globin and β-globin chains is tightly matched to ensure that each α chain is partnered by a β chain with no excess of either. Thalassemia, named after the Greek for "sea" (*thalassa*) and a reference to the Mediterranean populations in which the condition was first documented, is an anemic condition that results from α- and β-globin imbalances. Point mutations in the *HBB* gene yield β thalassemias. Heterozygotes (**thalassemia minor**) have reduced amounts of β globin, but symptoms are typically mild. Homozygotes (**thalassemia major**) present with severe anemia. In the absence of β globin, α globin accumulates and forms insoluble aggregates that precipitate and damage the RBC membrane, resulting in hemolysis. Such individuals are dependent on transfusions for survival. α Thalassemias result from defects in one of the four α-globin genes. Symptom severity depends on the number of defective alleles and the amount of normal α globin produced. Individuals with a single defective α-globin gene are asymptomatic carriers, whereas homozygotes with defects in both *HBA1* and *HBA2* develop a severe anemia that is usually fatal *in utero*. A single functional α-globin gene leads to the development of HbH disease with moderate anemia. Excess β-globin chains accumulate and form β_4 tetramers, an Hb form known as HbH. HbH is soluble but ineffective as an O_2 carrier. HbH tends to precipitate when exposed to oxidants, causing RBC damage and hemolysis.

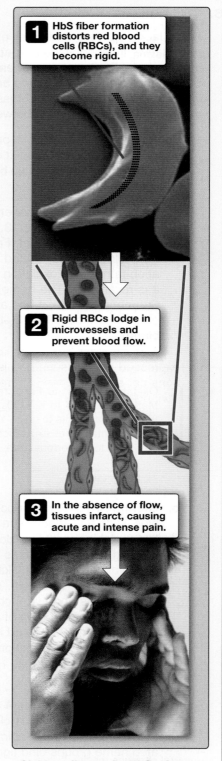

1 HbS fiber formation distorts red blood cells (RBCs), and they become rigid.

2 Rigid RBCs lodge in microvessels and prevent blood flow.

3 In the absence of flow, tissues infarct, causing acute and intense pain.

Sickle cell anemia. HbS = hemoglobin with a β^S-globin chain.

(Continued)

2. **Anion exchanger:** The RBC membrane contains 1 million copies of AE1, an anion exchanger that exchanges HCO_3^- for Cl^- and mediates the "chloride shift" observed during CO_2 loading (see later). Historically known as "band 3," AE1 is a product of the *SLC4A1* gene and accounts for ~25% of total membrane protein, creating a transport capacity of 5×10^{10} ions/sec. It also serves as a membrane anchor point for the cytoskeleton via ankyrin binding (see Fig. 23.3). The second most abundant RBC membrane protein is aquaporin 1 (AQP1), which facilitates water and CO_2 movement across the membrane.

3. **Carbonic anhydrase:** RBCs contain two CAs: CA-I and CA-II. CA-I is expressed at unusually high levels in RBCs. CA-II is expressed in all tissues and is notable for catalyzing $>10^6$ reactions per second. These impressive AE1 transport and CA reaction numbers attest to RBC efficiency in removing CO_2 from tissues and transporting it to the lungs for excretion.

Table 23.1: Normal Adult Hemoglobins

Form	Chain Composition	Fraction of Total Hemoglobin
HbA	$\alpha_2\beta_2$	90%
HbA$_2$	$\alpha_2\delta_2$	2%–3%
HbF	$\alpha_2\gamma_2$	<1%
HbA$_{1c}$	$\alpha_2\beta_2$-glucose	4%–6%

Hb = hemoglobin.

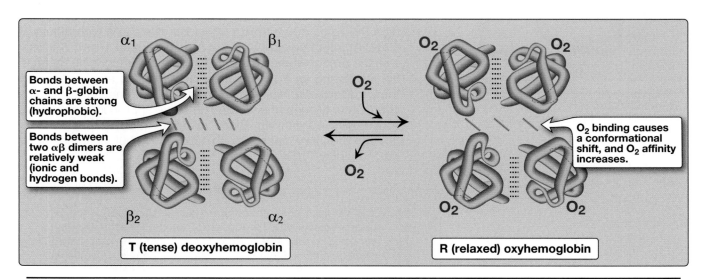

Figure 23.5.
Structural changes resulting from hemoglobin oxygenation and deoxygenation.

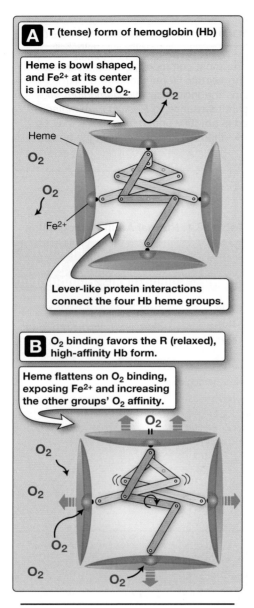

Figure 23.6.
Cooperativity between heme groups.

IV. OXYGEN TRANSPORT

The limited solubility of O_2 in water means that supplying all tissues with the O_2 needed to support internal respiration relies wholly on Hb and its unique O_2-binding properties.

A. Oxygen binding

Oxygenation causes Hb to shift from a low-affinity form to a high-affinity form (Fig. 23.6).

1. **Deoxyhemoglobin:** In the deoxygenated state, movement between the polypeptide chains within the Hb tetramer is constrained, and Hb exists in a tense ([T] formerly known as "taut") form. The porphyrin ring is bowl shaped and the Fe^{2+} at its center relatively inaccessible to O_2 (see Fig. 23.6A). The T form has a low affinity for O_2.

2. **Oxyhemoglobin:** O_2 binding to heme puts stress on the porphyrin ring and favors a shape change that causes it to flip from a bowl to a plate. When a shape change occurs, it is transmitted through a key histidine residue (H8) to the other heme moieties via a series of lever-like interactions within the protein. Hb now assumes a relaxed (R) form (see Fig. 23.5B). When the porphyrin ring assumes a plate shape, O_2 accessibility to heme is greatly increased. In practice, this means that the R form of Hb has an O_2 affinity ~300 times greater than the T form.

3. **Binding cooperativity:** A single O_2-binding event increases the likelihood of Hb assuming the high-affinity form but does not guarantee it. Each successive O_2-binding event (up to a maximum of four) further increases the probability of a shape change, a phenomenon termed "**cooperativity**." Binding cooperativity yields a sigmoidal O_2 dissociation curve. In practice, this means that, at a low Po_2, Hb exists largely in a low-affinity T state. At a high Po_2 (e.g., 100 mm Hg), most Hb exists in a high-affinity R state. At intermediate O_2 levels, Hb is a mix of T and R forms, with the predominance of the R form (and therefore Hb O_2 affinity) increasing as Po_2 increases. Note that the steepest part of the O_2 dissociation curve coincides with the range of Po_2 values common to tissues (Fig. 23.7). The curve approaches saturation at a Po_2 of 60 mm Hg. Blood **arterialization** during passage through the lungs raises Po_2 to 100 mm Hg but increases saturation by only ~10%.

The Hb color change from dark blue to bright red when O_2 binds makes it possible to monitor arterial O_2 saturation levels using noninvasive pulse oximetry (SpO_2). A light-emitting probe is attached to a finger or ear, and the relative amounts of saturated and desaturated Hb are calculated from the amount of light absorbed at 660 nm and 940 nm, respectively.

4. **Hemoglobin concentration:** The amount of O_2 that blood can carry depends on Hb concentration.

   a. **Oxygen capacity:** Blood contains ~150 g of Hb/L, or 15 g/dL (normal range is 12.0–16.0 g/dL for females and 13.0–17.5 g/dL for males). Each Hb molecule is capable of binding four O_2 molecules, which is equivalent to 1.39 mL O_2/g of Hb. Thus, blood's theoretical **O_2 capacity** is 20.8 mL/dL, a value that increases and decreases in direct proportion to blood Hb concentration (Fig. 23.8).

   b. **Oxygen saturation:** O_2 **saturation** is a measure of the number of occupied O_2-binding sites on the Hb molecule. At 100% saturation (arterial blood), all four heme groups are occupied. At 75% saturation (venous blood), three are occupied. Only two sites are occupied at 50% saturation. The degree of O_2 saturation is not dependent on Hb concentration, at least within a physiologic range.

B. Hemoglobin–oxygen dissociation curve

The O_2 dissociation curve's shape explains Hb's ability to bind O_2 in the lungs and then release it on demand to the tissues.

1. **Association:** Mixed-venous blood arrives at an alveolus with a Po_2 of 40 mm Hg but an O_2 saturation of ~75%. The cooperative nature of O_2 binding to Hb means that any unoccupied heme groups have a very high affinity for O_2. This allows Hb to capture O_2 as fast as it can diffuse across the blood–gas interface, simultaneously maintaining a steep pressure gradient for O_2 diffusion across the exchange barrier, even as equilibration with alveolar gas occurs. Notice that the plateau region of the O_2 dissociation curve begins at a Po_2 of around 60 mm Hg (see Fig. 23.7). In practice, this ensures that saturation still occurs if ventilation is impaired and P_{AO_2} is suboptimal (i.e., 60 mm Hg).

2. **Dissociation:** Once blood arrives at a tissue, Hb must release bound O_2 to make it available to mitochondria. Transfer is facilitated by the steepness of the pressure gradient between blood and mitochondria, which maintain a local Po_2 of ~3 mm Hg. Hb begins releasing O_2 at a Po_2 of 60 mm Hg and delivers ~60% of total as Po_2 falls to 20 mm Hg. Each O_2-dissociation event lowers the affinity of the remaining heme groups for bound O_2, so that if a tissue's metabolic rate is very high and its need for O_2 is increased, unloading occurs with increased efficiency.

C. Dissociation-curve shifts

Hb is uniquely sensitive to tissue needs, allowing it to deliver increasing amounts of O_2 when metabolism increases. This is made possible through allosteric changes that decrease the protein's O_2 affinity and promote unloading. These changes manifest as a rightward shift in the Hb–O_2 dissociation curve (Fig. 23.9).

1. **Rightward shifts:** Metabolism generates heat and CO_2 and acidifies the local environment. All three changes reduce Hb's O_2 affinity and cause it to unload O_2. The liberated O_2 keeps free

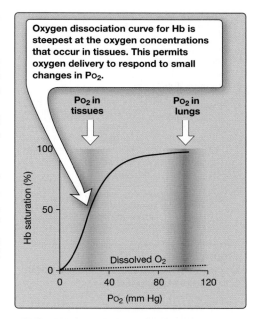

Figure 23.7.
O_2 dissociation curve for hemoglobin (Hb).

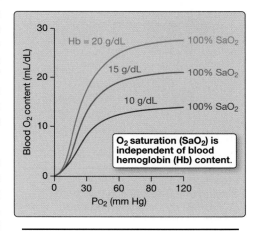

Figure 23.8.
Effects of hemoglobin concentration on blood O_2 concentration.

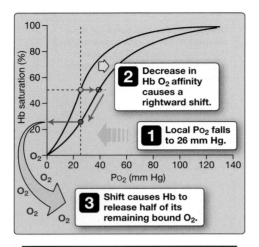

Figure 23.9.
Decreasing the O$_2$ affinity of hemoglobin (Hb) causes O$_2$ unloading.

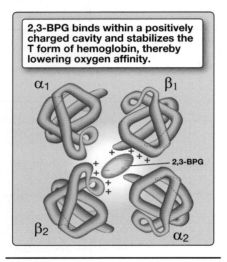

Figure 23.10.
Binding of 2,3-bisphosphoglycerate (2,3-BPG) by deoxyhemoglobin. T form = tense form.

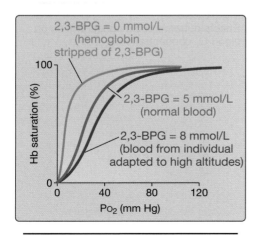

Figure 23.11.
2,3-Bisphosphoglycerate (2,3-BPG) decreases hemoglobin's O$_2$ affinity.

(dissolved) O$_2$ levels high and maintains a steep pressure gradient between blood and mitochondria, even as blood's O$_2$ stores are being emptied.

  a. **Temperature:** During strenuous exercise, muscle temperature rises by 2°C–5°C (see Chapter 39). The Hb–O$_2$ dissociation curve shifts by ~5 mm Hg to the right as a result, causing more O$_2$ to be released to the metabolically active tissue.

  b. **Carbon dioxide:** Aerobic metabolism generates CO$_2$ and causes tissue Pco$_2$ to rise. CO$_2$ binds to terminal globin amino groups and decreases Hb's O$_2$ affinity. The Hb–O$_2$ dissociation curve shifts to the right, and O$_2$ is unloaded. CO$_2$ also dissolves in water to yield free acid, which promotes further O$_2$ unloading via the **Bohr effect**.

  c. **Protonation:** Protonation stabilizes deoxyhemoglobin and thereby decreases its O$_2$ affinity. Metabolism generates several different acids in addition to carbonic acid, and the amount produced is proportional to metabolic activity. The Hb–O$_2$ dissociation curve shifts to the right, and O$_2$ is released (Bohr effect).

  d. **2,3-Bisphosphoglycerate:** 2,3-Bisphosphoglycerate (2,3-BPG) is synthesized from an intermediate in the glycolytic pathway. 2,3-BPG is abundant in RBCs, its concentration rivaling that of Hb. 2,3-BPG binds to the T form of Hb. It nestles within a pocket located at the center of the Hb tetramer and stabilizes it, thereby favoring a low-affinity state (Fig. 23.10). The Hb–O$_2$ dissociation curve shifts to the right (Fig. 23.11).

> Chronic hypoxemia caused by pathologic changes in lung function or living at high altitude stimulates 2,3-BPG production. Increased 2,3-BPG levels shift the Hb–O$_2$ dissociation curve even further to the right, which increases tissue accessibility to available O$_2$ (see Fig. 23.11). Although 2,3-BPG does reduce the efficiency of Hb O$_2$ loading in the lungs, the effects are minor and more than offset by the beneficial effects of assisting O$_2$ delivery to tissues.

2. **Leftward shifts:** Hb's O$_2$ affinity increases, and the Hb–O$_2$ dissociation curve shifts left when body temperature decreases or when CO$_2$ or H$^+$ levels decrease. These changes reflect decreased metabolic activity and a decreased need for O$_2$ delivery to tissues. A leftward shifted Hb–O$_2$ dissociation curve is also observed in the fetus, or when 2,3-BPG levels decrease, or as a result of carbon monoxide (CO) binding to Hb.

  a. **Fetal hemoglobin:** HbF contains γ chains in place of the two β chains ([α$_2$γ$_2$] Fig. 23.12; see also Biologic Sex and Aging 23.2). This causes the HbF–O$_2$ dissociation curve to shift left compared with adult Hb.

b. **2,3-Bisphosphoglycerate:** If HbA is stripped of 2,3-BPG, its O_2 dissociation curve resembles that of HbF. Storing blood causes 2,3-BPG concentrations to decline over the course of a week, causing a leftward shift in the dissociation curve (see Fig. 23.11). Although RBCs replenish lost 2,3-BPG within hours to days of transfusion, giving a critically ill patient large volumes of 2,3-BPG–depleted blood presents some difficulties because such blood does not readily give up its O_2.

c. **Carbon monoxide:** CO is formed by combustion of hydrocarbons. Common sources of exposure include automobile exhaust, poorly ventilated heating systems, and smoke. Hb binds CO with high affinity to produce carboxyhemoglobin (CO-Hb), which is bright red in color. CO-Hb comprises up to ~3% of total Hb in nonsmokers and 10% to 15% of total Hb in smokers. CO occupancy of the O_2-binding sites severely reduces Hb's ability to bind and carry O_2. Inhaling the gas at a concentration of only 0.1% reduces O_2-carrying capacity by 500%. CO simultaneously stabilizes the high-affinity Hb form and shifts the Hb–O_2 dissociation curve to the left (Fig. 23.13). These changes dramatically reduce Hb's ability to release O_2 to tissues and make CO an extremely deadly gas (Fig. 23.14). CO poisoning is a leading cause of poisoning deaths in the United States.

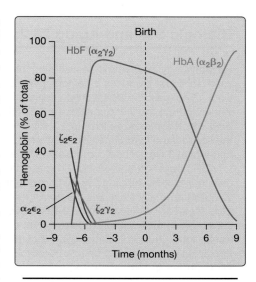

Figure 23.12.
Fetal hemoglobins. See Biologic Sex and Aging 23.2 for more details.

V. CARBON DIOXIDE TRANSPORT

Metabolism generates ~200 mL CO_2/min in an average person at rest. CO_2 is carried away from tissues in venous blood and then exhaled from the lungs. CO_2 handling differs from the way O_2 is transported in two notable respects. First, CO_2 is highly soluble in water and therefore does not require a carrier protein for transport. Second, CO_2 generates substantial amounts of acid when in solution, requiring the presence of a buffering system.

A. Methods of transport

CO_2 is transported through the vasculature in three principal forms: in solution, as HCO_3^-, and in association with Hb.

1. **Dissolved:** CO_2 is >20 times more soluble in blood than O_2, so ~5% of total CO_2 is transported in solution.

2. **Bicarbonate:** Ninety percent of CO_2 is carried as HCO_3^-. HCO_3^- is generated by spontaneous dissociation of H_2CO_3 (see reaction below), the formation of carbonic acid from CO_2 and water having been facilitated by CA:

Equation 23.1

$$H_2O + CO_2 \leftrightharpoons H_2CO_3 \leftrightharpoons HCO_3^- + H^+$$
$$\text{CA}$$

3. **Carbamino compounds:** Five percent of total blood CO_2 is carried as carbamino compounds, which form by reversible reaction of CO_2 with the amine groups of proteins, principally Hb. CO_2 also binds to plasma proteins, but the amount involved is insignificant.

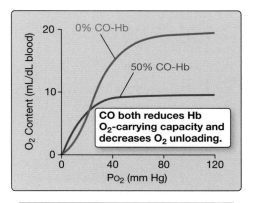

Figure 23.13.
Carbon monoxide (CO) effects on hemoglobin (Hb) O_2 affinity. CO-Hb = carboxyhemoglobin.

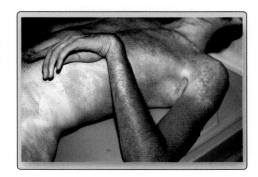

Figure 23.14.
Carbon monoxide poisoning turns skin cherry red, a color that persists after death.

Biologic Sex and Aging 23.2: Fetal Hemoglobins

Responsibility for erythropoiesis during fetal development initially falls on the yolk sac. It first produces a hemoglobin (Hb) $\zeta_2\varepsilon_2$ tetramer, which appears at weeks 5–6 (see Fig. 23.12). Two additional fetal Hb forms ($\alpha_2\varepsilon_2$ and $\zeta_2\gamma_2$) are evident during the first trimester. Red blood cell production now shifts to the liver, which synthesizes HbF ($\alpha_2\gamma_2$). HbF is the predominant Hb form during the second and third trimester. HbA, which is produced by bone marrow, first appears at 9 weeks and comprises up to 30% of total Hb at term.

HbF has a left-shifted Hb–O_2 dissociation curve compared with HbA, due to γ globins binding 2,3-bisphosphoglycerate (2,3-BPG) very weakly. 2,3-BPG normally stabilizes the deoxygenated form of HbA and reduces its affinity. HbF's inability to bind 2,3-BPG favors O_2 loading at low partial pressures, which allows fetal blood to achieve an O_2 saturation of ~75% during transit through the placenta, even though P_{O_2} is only 30–35 mm Hg (see Fig. 37.11). HbF levels drop to <1% of total Hb within 6–12 months of birth, although levels may again rise when erythropoiesis is stressed by acute blood loss, hemolysis, or chemotherapy, for example. Some individuals with hereditary persistence of HbF continue to express HbF at high levels (up to 30% of total Hb) into adulthood with no apparent ill effect.

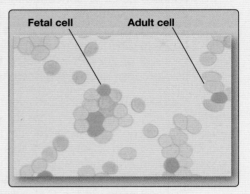

Hereditary persistence of fetal hemoglobin (HbF). Red blood cells containing HbF appear bright pink.

B. Loading

Blood contains more than twice the amount of CO_2 than O_2 (~23 mmol/L CO_2 vs. 9.5 mmol/L O_2). Much of this CO_2 resides in blood stores, and passage through the systemic capillary beds increases its total content by only 8%. CO_2 that has been newly picked up from tissues is transported to the lungs principally as HCO_3^- (~60%) as shown in Figure 23.15. The remainder is carried in dissolved form (~10%) or in association with hemoglobin or another protein (~30%). CO_2 uptake from the tissues occurs by simple diffusion, driven by the partial pressure gradient for CO_2. The subsequent fate of CO_2 can be divided into several discrete steps that correspond to the sections below (Fig. 23.16).

1. **Uptake by red blood cells:** Uptake by RBCs is facilitated by AQP1 and Rh complex proteins, which function as gas channels. RBCs contain high levels of CA-I that convert CO_2 to H_2CO_3 as fast as it enters cells (see Fig. 23.16). This helps maintain a strong partial pressure gradient between tissues and blood that drives continued CO_2 diffusion. H_2CO_3 then rapidly dissociates to form HCO_3^- and H^+ (see Equation 23.1).

2. **Bicarbonate transport:** HCO_3^- is transported out of the RBC in exchange for Cl^- by the AE1 Cl^-/HCO_3^- exchanger (see Fig. 23.16). The resultant **Cl^- shift (Hamburger shift)** causes a slight increase in RBC osmolality and produces mild swelling, but this is reversed in the lungs.

3. **Hydrogen-ion buffering:** The H^+ released during HCO_3^- formation remains trapped in RBCs by the cell membrane, which is relatively impermeable to cations. This might be expected to lower intracellular pH, but H^+ accumulation occurs at the precise moment that Hb is releasing O_2 and undergoing a conformational change that favors H^+ binding. As noted earlier (i.e., the Bohr effect), H^+ binding actually facilitates O_2 unloading by shifting the

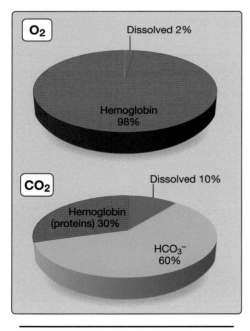

Figure 23.15.
Comparison of the ways O_2 and CO_2 are transported between lungs and tissues.

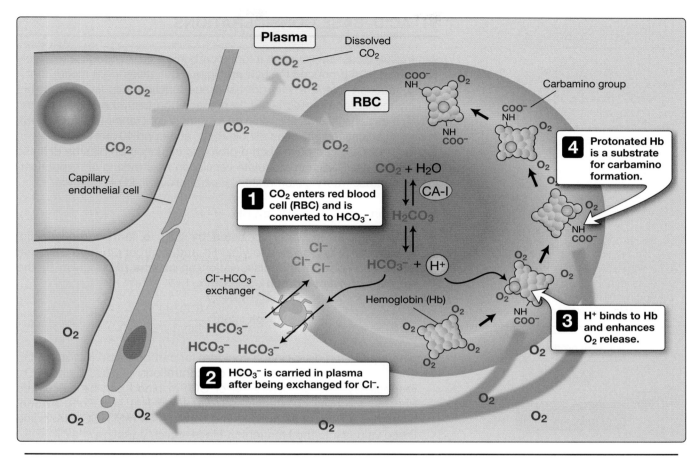

Figure 23.16.
Transport of CO_2 in blood. CA = carbonic anhydrase.

Hb–O_2 dissociation curve to the right and reducing Hb's affinity for O_2 (see Fig. 23.16). Virtually all the acid excess caused by loss of HCO_3^- to plasma is buffered by Hb. With intracellular H^+ kept low by Hb and the Cl^-/HCO_3^- exchanger keeping HCO_3^- low, the reaction catalyzed by CA remains biased in favor of increased H^+ and HCO_3^- formation. The CO_2-carrying capacity of blood increases as a result.

4. **Carbamino formation:** When Hb binds H^+, it becomes a more favorable substrate for carbamino compound formation (**Haldane effect**; Fig. 23.17; see also Fig. 23.16). Hb carries appreciable amounts of CO_2 in carbaminohemoglobin form.

C. Unloading

When blood arrives at the lungs, the partial pressure gradients for both O_2 and CO_2 are reversed compared with tissues. A high P_{O_2} causes H^+ to dissociate from Hb (Haldane effect), and the reaction in Equation 23.1 now favors H^+ and HCO_3^- association to form H_2O and CO_2 (see Fig. 23.17). HCO_3^- reenters RBCs in exchange for Cl^- and combines with H^+ to form H_2CO_3 which dissociates to release CO_2 and H_2O. CO_2 then diffuses out of the blood, driven by the partial pressure gradient for CO_2 between blood and the alveolar lumen.

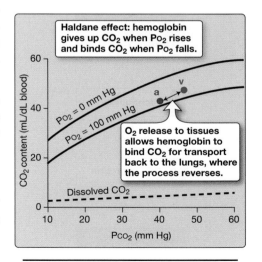

Figure 23.17.
Effect of P_{O_2} on the CO_2 dissociation curve (Haldane effect). a and v = arterial and venous blood, respectively.

VI. ACID–BASE CONSIDERATIONS

When CO_2 dissolves in water, it forms carbonic acid. Although a relatively weak acid, it is produced in such prodigious quantities (>20 mol/d) that it could seriously interfere with normal tissue function if its levels were not closely monitored and regulated. In practice, the central nervous system (CNS) maintains plasma pH within an extremely tight range (pH 7.35–7.45), in part by adjusting ventilation to hold P_{aCO_2} at around 40 mm Hg. However, the fact that CO_2 can have such a profound influence on plasma pH also means that the CNS can modulate ventilation as a means of compensating for nonrespiratory disturbances in extracellular fluid pH balance.

A. Carbon dioxide effects on pH

CO_2 dissolves in water (assisted by CA) to form carbonic acid, which quickly dissociates to yield H^+ and bicarbonate (see Equation 23.1). The effect of this dissociation on plasma pH is given by the **Henderson-Hasselbalch** equation:

$$pH = pK + \log \frac{\left[HCO_3^- \right]}{\left[CO_2 \right]}$$

where pK is the dissociation constant for carbonic acid (6.1 at 37°C), and $[HCO_3^-]$ and $[CO_2]$ denote concentrations of HCO_3^- and CO_2, respectively. The concentration of CO_2 in blood can be calculated from its solubility constant (0.03 mmol/mm Hg) and P_{CO_2}. Arterial blood has a P_{CO_2} of 40 mm Hg and contains 24 mM HCO_3^-. Inserting these values into the Henderson-Hasselbalch equation:

$$pH = 6.1 + \log \frac{24}{0.03 \times 40} = 7.4$$

Note that any increase in P_{CO_2} will cause pH to fall (**acidosis**), whereas decreases will cause pH to rise (**alkalosis**).

B. Causes of changes in extracellular fluid pH

Changes in pH caused by the lungs are referred to as **respiratory acidosis** or **respiratory alkalosis**. Nonrespiratory changes are **metabolic acidosis** or **metabolic alkalosis**.

1. **Respiratory acidosis:** Increased P_{aCO_2} results from hypoventilation, alveolar ventilation $\left(\dot{V}_A \right)$/perfusion $\left(\dot{Q} \right)$ mismatches, or an increase in the diffusional distance between the alveolar sac and pulmonary blood supply (e.g., due to edema or pulmonary fibrosis).

2. **Respiratory alkalosis:** P_{aCO_2} decreases with hyperventilation, typically because of anxiety or other emotional state. It can also result from hypoxemia precipitated by ascent to high altitude.

C. Compensation

Cells are defended against excess acid accumulation in the short term by buffers, most notably the HCO_3^- buffer system and intracellular proteins such as Hb (see Chapter 3·IV·B). Buffers operate on a time scale of seconds or less. Correction of an altered acid–base

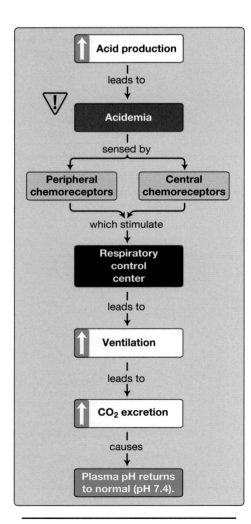

Figure 23.18.
Concept map for ventilatory response to acidemia.

status ultimately requires a change in lung or kidney function. CNS respiratory centers continually monitor plasma pH (Fig. 23.18). If pH falls, they increase pulmonary ventilation to transfer CO_2 to the atmosphere, and pH renormalizes. Conversely, a rise in plasma pH initiates a reflex ventilatory decrease, and CO_2 is retained. Ventilatory responses require several minutes to take effect and, if the underlying cause is a metabolic disturbance, may never be sufficient to compensate fully. Respiratory control pathways are considered in Chapter 24. The role of the kidneys in pH balance is detailed in Unit VI, *Urinary System*.

Chapter Summary

- Gas transport through the vasculature is the responsibility of red blood cells (RBCs), which are a product of bone marrow (**erythropoiesis**). RBCs contain no nucleus or other organelles, but they are enriched with **hemoglobin (Hb)**.

- Hb is a tetrameric O_2-binding metalloprotein containing two α-globin and two β-globin chains ($\alpha_2\beta_2$). O_2 binds to one of four heme groups, which contains a ferrous iron atom (Fe^{2+}).

- O_2 binding causes a conformational change in Hb that increases its affinity for O_2. The cooperative nature of O_2 binding ensures blood O_2 saturation during passage through the lungs and facilitates O_2 release as blood passes through tissues served by the systemic circulation.

- **CO_2** is transported in solution, in association with Hb, and as HCO_3^-. HCO_3^- forms by dissociation of carbonic acid.

- Because CO_2 dissolves in water to form **carbonic acid**, ventilatory changes that cause CO_2 to be excreted at rates that exceed or fail to keep up with CO_2 production can result in **respiratory alkalosis or acidosis**, respectively.

Study Questions

Choose the ONE best answer.

23.1. A 38-year-old male presents to the emergency department with shortness of breath and fatigue. His medical history is significant for iron deficiency anemia. Laboratory studies reveal low hypoxia-inducible factor levels. The patient's symptoms are most likely due to which of the following?

A. Decreased red blood cell count
B. Hereditary persistence of hemoglobin F
C. Increased erythropoietin levels
D. Leftward shift in the hemoglobin–O_2 dissociation curve
E. Vitamin B_{12} deficiency

Best answer = A. Hypoxia-inducible factor (HIF) is a transcription factor activated in response to low O_2 levels (see Section II·B). It increases erythropoietin (EPO) expression, which, in turn stimulates red blood cell (RBC) production. Reduced HIF availability impairs responses to hypoxia and results in a decreased RBC count, dyspnea, and fatigue. Hemoglobin F (HbF) is a fetal Hb whose expression may be increased in adults under certain conditions. Hereditary persistence of HbF (B) is usually asymptomatic. EPO expression levels (C) would be decreased, not increased, in this patient. Leftward shifts in the Hb–O_2 dissociation curve (D) may result from 2,3-bisphosphoglycerate deficiency. Vitamin B_{12} deficiency (E) is a cause of macrocytic anemia.

23.2. A 25-year-old female with normal lung function presents with anemia following childbirth (hemoglobin = 8.6 g/dL). Which of the following parameters is most likely to be reduced?

A. Arterial O_2 content
B. Arterial O_2 saturation
C. Arterial Po_2
D. Minute ventilation
E. Right ventricular output

Best answer = A. Anemia, as defined by blood hemoglobin (Hb) content (normal female Hb = 12.0–16.0 g/dL) reduces the total amount of O_2 that can be carried by blood (see Section IV·A). Arterial O_2 saturation (B) is a measure of Hb's O_2-binding state, which is largely independent of blood Hb concentration under physiologic conditions. Arterial Po_2 (C) is a measure of dissolved O_2 concentration and is not significantly affected by Hb concentration. A decrease in arterial O_2 content would stimulate compensatory increases in minute ventilation (D) and right ventricular output (E).

23.3. A 26-year-old African American male with a history of sickle cell disease presents to the emergency department with severe pain in his lower back, arms, and legs. He reports feeling short of breath and having chest pain. Laboratory studies reveal a hemoglobin level of 7.2 g/dL (normal range 13.5–17.5 g/dL). What additional finding would most likely be expected in this patient?

A. α-Globin aggregates
B. β-Globin tetramers
C. $\alpha_2\delta_2$ Hb tetramers
D. *HBB* gene mutation
E. High HbA_2 levels

Best answer = D. Sickle cell disease is caused by a point mutation in the *HBB* gene, which encodes the β-globin subunit of hemoglobin (Hb). The mutation causes a valine-for-glutamate substitution (see Clinical Application 23.2). The result is an $\alpha_2\beta^S_2$ Hb tetramer that polymerizes and forms rigid fibers when deoxygenated. These fibers distort red blood cells and cause microvascular blockage and the ischemic pain that characterizes a sickle-cell crisis, as seen in the patient. α-Globin aggregates (A) form as a result of α-globin gene mutations and cause α thalassemia. β-Globin tetramers (B) are seen in β thalassemias, which are also caused by *HBB* gene mutation. $\alpha_2\delta_2$ Hb tetramers (C) describe HbA_2, which is a normal Hb variant. HbA_2 normally accounts for 2%–3% of total Hb. Increases in HbA_2 (E) do not cause sickle-cell anemia.

23.4. An unidentified male is brought to the emergency department after having been found unconscious. Laboratory analysis of a blood sample reveals a leftward shift in the hemoglobin (Hb)–O_2 dissociation curve. The shift is most likely due to which of the following conditions?

A. Acidemia
B. Anemia
C. Carbon monoxide exposure
D. Hypercapnia
E. Increased 2,3-bisphosphoglycerate levels

Best answer = C. Carbon monoxide (CO) is a toxic byproduct of hydrocarbon combustion that readily occupies the O_2-binding sites on hemoglobin (Hb) and stabilizes the high affinity form of the protein (see Section IV·C). The Hb–O_2 dissociation curve shifts leftward as a result. Hb's ability to deliver O_2 to tissues is decreased, causing hypoxia. CO poisoning is common and can be fatal. Hydrogen ions (A); increased PCO_2, or hypercapnia (D); and increased 2,3-bisphosphoglycerate levels (E) all reduce Hb's affinity for O_2 and shift the dissociation curve to the right. Anemia (B) would not be expected to affect the Hb–O_2 dissociation curve.

Respiratory Regulation

24

I. OVERVIEW

Normal breathing (**eupnea**) is usually an unconscious act overseen by the autonomic nervous system (ANS). The cyclical pattern is established by a respiratory control center within the brainstem that coordinates contraction of the diaphragm and other muscles involved in inspiration and expiration (Fig. 24.1). Because the needs of internal respiration change with activity level, the pattern generator must also change its output to match prevailing needs. Sensors located within the central nervous system (CNS) and throughout the periphery continuously monitor blood and tissue P_{CO_2}, P_{O_2}, and pH levels and feed this information back to the respiratory control center for processing. The control center also receives information from mechanoreceptors located in the lungs and chest wall. The control center then adjusts ventilation as needed, using motor outputs to the diaphragm, intercostals, and other muscles involved in breathing. The principal goal of the neural network regulating respiration is to maintain arterial P_{CO_2} at a stable level while simultaneously ensuring adequacy of O_2 flow to the tissues. The dominance of CO_2 in respiratory control reflects the body's need to maintain extracellular fluid (ECF) pH within a narrow range.

II. NEURAL CONTROL CENTERS

Several regions of the brain influence breathing. The basic respiratory rhythm is established by a pattern generator located in the medullary respiratory control center.

A. Medullary control center

The medulla contains several discrete groups of neurons involved in respiratory control (Fig. 24.2). Although functioning as a single unit, control-center neurons and their input/output pathways are mirrored on either side of the medulla. Either side can generate independent respiratory rhythms if the brainstem is transected. Within the control center are two concentrations of neurons that fire in phase with the inspiration–expiration cycle and are assumed to have a key role in establishing the respiratory rhythm. These are called the **dorsal respiratory group** (**DRG**) and the **ventral respiratory group** (**VRG**).

1. **Dorsal respiratory group:** The DRG is located in the **nucleus tractus solitarius** ([**NTS**] Fig. 24.3), a key area involved in

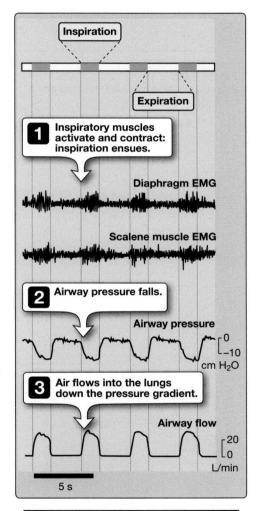

Figure 24.1.
Rhythmic inspiration–expiration cycle.
EMG = electromyogram.

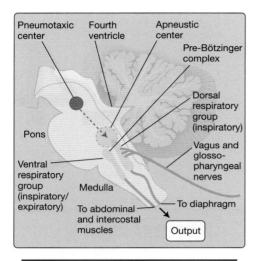

Figure 24.2.
Brainstem areas involved in respiratory control.

processing and coordinating sensory information received via the ANS from the periphery (see Chapter 7·VI·B). It consists largely of inspiratory (I) neurons.

a. **Function:** The DRG receives inputs from various sensors via cranial nerves (CNs) X and IX, including chemoreceptors, lung and chest wall mechanoreceptors, and arterial baroreceptors (see Chapter 19·III). This information is then processed using interneuron networks. If the sensors report suboptimal CO_2 or O_2 levels, the DRG formulates and executes an appropriate response.

b. **Output:** Two populations of inspiratory neurons can be distinguished. During lung inflation, one group (Iβ cells) activates, whereas a second group (Iα cells) ceases firing. When inflation is suppressed, Iα cells activate, and Iβ cells fall silent. DRG premotor neurons project to groups of spinal motor neurons, including the phrenic nerve (diaphragm) and intercostal nerves (external intercostals). Activating these neurons causes inspiration, although the pacemaker (respiratory rhythm generator or central pattern generator [CPG]) appears to reside elsewhere (discussed later). DRG neurons also project to the VRG, an area controlling inspiratory and expiratory muscle contraction during forced ventilatory maneuvers.

2. **Ventral respiratory group:** The VRG can be subdivided into rostral, intermediate, and caudal regions (see Fig. 24.3). These areas correspond anatomically to the nucleus ambiguus (rostral-intermediate), retrofacial nucleus (rostral), nucleus para-ambigualis (intermediate), and nucleus retroambigualis (caudal).

a. **Function:** The VRG primarily coordinates accessory muscles of inspiration and expiration. It is largely quiescent during quiet breathing but becomes highly active during exercise and any forced ventilatory maneuvers, such as during pulmonary function testing.

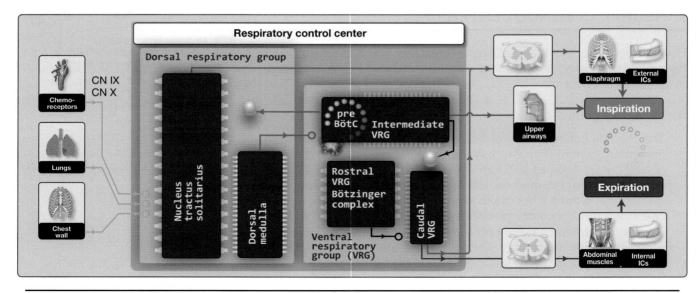

Figure 24.3.
Respiratory control center organization. BötC = Bötzinger complex, ICs = intercostal muscles.

   b. **Output:** The VRG's intermediate region contains motor efferents to accessory muscles in the pharynx and larynx that dilate the upper airways during inspiration. The caudal region contains premotor neurons that synapse within the spinal cord to control the internal intercostals and other accessory muscles of expiration. The rostral region (**Bötzinger complex**) communicates via interneurons with the DRG and the caudal region of the VRG. It may be involved in coordinating VRG output.

3. **Central pattern generator:** The VRG intermediate region contains a small area known as the **pre-Bötzinger complex** composed of cells that exhibit pacemaker-like activity. Ablation of the complex abolishes rhythmic breathing, suggesting that it constitutes a respiratory CPG.

B. Pontine centers

The pons contains areas that influence medullary output, including the **pontine respiratory groups** (formerly known as the pneumotaxic and apneustic centers) located in the Kölliker-Fuse and parabrachial nucleus. Transectioning the brain between the medulla and pons causes prolonged inspiratory gasping (**apneuses**), suggesting that pontine respiratory group neurons normally limit lung expansion. Stimulating these areas shortens inspiration and increases breathing rate. A role for the pontine respiratory groups in normal breathing has not been established, but they are thought to fine-tune the inspiratory–expiratory cycle. The pontine respiratory groups also receive input from chemoreceptors and are involved in responses to hypocapnia and hypoxia.

C. Cerebral cortex and other brain regions

Emotions such as fear, excitement, and rage can alter respiratory rate, reflecting the ability of the hypothalamus and limbic system to modulate the CPG. The cerebral cortex can adjust CPG output to allow for talking, playing a musical wind instrument, and other activities that require fine conscious control of breathing movements. A person can also consciously override the CPG to increase or decrease breathing frequency (tachypnea and bradypnea, respectively), or cease breathing altogether (apnea). The effects of breath holding on blood gas concentrations are generally tolerated for a relatively short time before chemical control feedback pathways override voluntary control.

III. CHEMICAL CONTROL OF VENTILATION

A primary mission of the respiratory system is to optimize ECF pH through manipulation of P_{CO_2}. It must also maintain steep O_2 and CO_2 partial pressure gradients to maximize transfer of these two gases between tissues and the external environment. Fulfilling these roles requires that information about the chemical composition of ECF be sensed and relayed back to the respiratory control center so that ventilation can be modified accordingly. The body employs central and peripheral chemoreceptors for this purpose. Central chemoreceptors are primarily sensitive to arterial P_{CO_2}

Clinical Application 24.1: Sleep Apnea

Sleep is usually accompanied by a pattern of normal quiet breathing, but some individuals may cease breathing for prolonged periods (tens of seconds) several times per hour (**sleep apnea**). Not surprisingly, apnea usually manifests as **excessive daytime sleepiness**. There are multiple causes. **Central sleep apnea** results from a complete loss of respiratory drive and is relatively uncommon. **Obstructive sleep apnea** is more prevalent, especially in obese individuals. Loss of ventilation results from obstruction (by soft tissues, such as the tongue and uvula) and collapse of the upper airways during sleep. Airways normally lose an active dilator influence during sleep, which makes them more prone to collapse, even in a healthy person. Fat deposits around the airways greatly increase the likelihood of collapse, however, and may require use of a nighttime breathing aid (i.e., a continuous positive airway pressure mask) that provides relief by pneumatically splinting the airways to maintain patency.

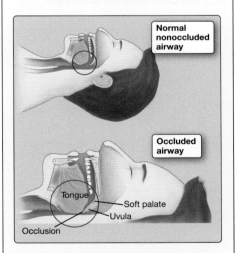

Normal nonoccluded airway

Occluded airway

Tongue
Soft palate
Uvula
Occlusion

Airway obstruction causes sleep apnea.

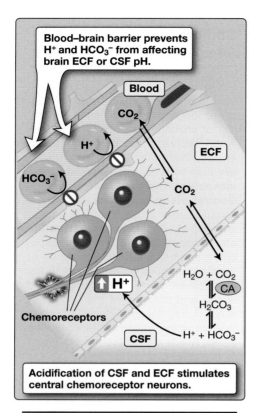

Figure 24.4.
Central chemoreceptors monitor arterial P_{CO_2} through effects of CO_2 on CSF and ECF pH. CA = carbonic anhydrase.

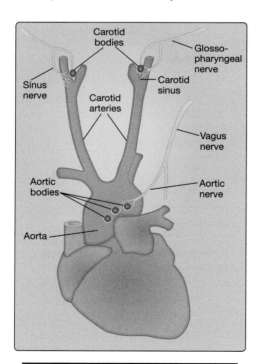

Figure 24.5.
Peripheral chemoreceptors are contained within carotid bodies and aortic bodies.

(P_{aCO_2}) and mediate 60% to 80% of the ventilatory response to changing arterial blood gas levels. Although peripheral chemoreceptors have a lesser role overall, they are the body's principal arterial P_{O_2} (P_{aO_2}) sensors.

A. Central chemoreceptors

Ventilatory control under resting conditions is dominated by the central chemoreceptors, which are responsive primarily to changes in P_{aCO_2}. The chemoreceptors are CNS neurons located behind the blood–brain barrier ([BBB] see Chapter 20·II·B) in the brainstem, cerebellum, hypothalamus, and midbrain. Known chemoreceptive regions include the retrotrapezoid nucleus in the ventral medulla, which receives input from the peripheral chemoreceptors and projects to the DRG and VRG. The BBB is impermeable to virtually all blood constituents except for lipid-soluble molecules, such as O_2 and CO_2. Once inside the barrier, CO_2 dissolves to form carbonic acid, which acidifies brain ECF and cerebrospinal fluid (CSF), as shown in Figure 24.4. CSF contains minimal amounts of protein to buffer pH (see Chapter 6·VII·D). The consequence is that even modest changes in P_{aCO_2} cause significant CSF and ECF acidosis. The chemoreceptor neurons respond to acid with excitatory impulses that impel the respiratory center to increase breathing rate. The BBB acts as an important information filter because, by excluding bloodborne ions such as H^+, it provides a way for the chemoreceptors to distinguish changes in P_{aCO_2} from any background changes in ECF pH.

> Systemic changes in pH ultimately affect all tissues regardless of cause or BBB ion permeability. Central chemoreceptors play a critical role in the integrated response to changing blood pH, although reactions to metabolic acidosis or alkalosis may be slower and less intense than respiratory responses triggered by changes in P_{aCO_2}.

B. Peripheral chemoreceptors

Peripheral chemoreceptors are located within carotid bodies located at the bifurcation of the two common carotid arteries (Fig. 24.5) and within aortic bodies distributed along the underside of the aortic arch. Peripheral chemoreceptors sense changes in P_{aO_2}, P_{aCO_2}, and arterial pH.

1. **Structure:** Carotid and aortic bodies are notable for their small size (3–5 mm), high metabolic rate, high capillary density (~6-fold higher than in the brain), and the highest rates of blood flow relative to their mass of all tissues (up to 25-fold higher than the brain). The high flow rate minimizes the effect of high chemoreceptor metabolic rate on blood gas content, thus allowing for a truer reading of arterial O_2 and CO_2 levels. Carotid bodies contain two types of cells. **Type I**, or **glomus, cells** are chemosensors (Fig. 24.6). They are arranged in clusters and in close apposition to fenestrated capillaries. **Type II**, or **sustentacular, cells** play a supportive role similar to glia. Carotid body glomus cells signal the respiratory center via the carotid sinus and glossopharyngeal (CN IX) nerves. Aortic bodies appear to be organized similarly. They communicate with the CNS via the vagus nerve (CN X).

2. **Sensory mechanism:** Glomus cell membranes contain a large conductance, BK-type K^+ channel whose open probability is Po_2 dependent. When P_aO_2 is high, the channel is open and allows a K^+ efflux that maintains glomus-cell membrane potential (V_m) at strongly negative levels. When P_aO_2 falls, the channel closes, and V_m depolarizes. The change in V_m activates L-type Ca^{2+} channels and elicits a Ca^{2+} influx that stimulates neurotransmitter release onto the sensory afferents (see Fig. 24.6). Glomus cells are also excited by increasing P_aCO_2 and H^+ concentrations independently of changes in P_aO_2. They help fine-tune the information that the respiratory center receives from central chemoreceptors. Changes in P_aCO_2 and pH also appear to influence O_2-dependent K^+ channel open probability through changes in cytoplasmic H^+ concentration.

3. **Output:** Carotid body chemoreceptors increase output when either P_aO_2 falls or P_aCO_2 rises above normal (100 mm Hg and 40 mm Hg, respectively; Fig. 24.7). The aortic bodies respond similarly but to a lesser degree and after a delay. The physiologic role of these chemoreceptors during normal ventilation is uncertain.

C. Integrated ventilatory responses

The respiratory control center processes information about P_aO_2, P_aCO_2, and arterial pH. Changes in these variables rarely occur in isolation, which forces the respiratory center to make choices about an appropriate ventilatory response (Fig. 24.8). Under most circumstances, the output from the respiratory center is designed to optimize P_aCO_2, but concomitant changes in P_aO_2 and pH influence system sensitivity to changes in P_aCO_2.

1. **Changing carbon dioxide levels:** The nature of control-center response to a rise in P_aCO_2 depends on whether the change is acute or chronic.

    a. **Acute:** When tissue metabolism increases, P_aCO_2 rises, and the respiratory center compensates by increasing alveolar ventilation (see Fig. 24.8). CO_2 is an extremely potent ventilatory stimulus (Fig. 24.9). The peripheral receptors are fast acting and trigger an immediate response. The central receptors take several minutes to activate fully, but their effects ultimately dominate the ventilatory response. In practice, ventilation increases linearly with any rise in P_aCO_2 above resting values and decreases when P_aCO_2 falls below 40 mm Hg.

    b. **Chronic:** Patients with chronic pulmonary disease may not be able to ventilate at levels required to hold P_aCO_2 at 40 mm Hg. Ventilation increases initially when P_aCO_2 begins to rise because the pH of CSF falls, but the choroid plexus responds by secreting HCO_3^- into the CSF, which, over a period of 8 to 24 hours, largely offsets the effect of the higher P_aCO_2 on plasma pH. Therefore, although P_aCO_2 may remain high, the medullary chemoreceptors no longer register the change, and the respiratory control system adapts to a new, higher P_aCO_2.

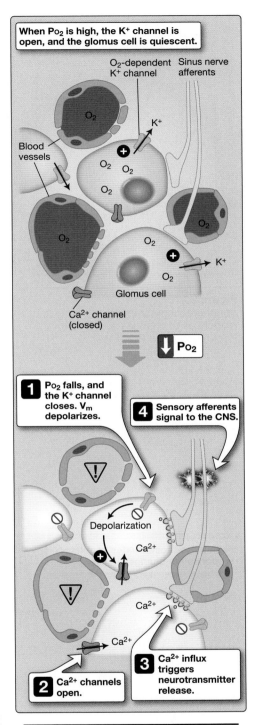

When Po_2 is high, the K^+ channel is open, and the glomus cell is quiescent.

O_2-dependent K^+ channel
Sinus nerve afferents
Blood vessels
O_2
K^+
O_2
Glomus cell
Ca^{2+} channel (closed)
↓ Po_2

1 Po_2 falls, and the K^+ channel closes. V_m depolarizes.

4 Sensory afferents signal to the CNS.

Depolarization
Ca^{2+}

2 Ca^{2+} channels open.

3 Ca^{2+} influx triggers neurotransmitter release.

Figure 24.6.
Mechanism of carotid body glomus-cell response to falling Po_2. V_m = membrane potential.

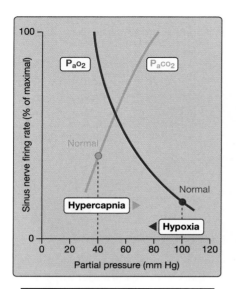

Figure 24.7.
Peripheral chemoreceptor
responses to changes in arterial
P_{CO_2} (P_aCO_2) and P_{O_2} (P_aO_2).

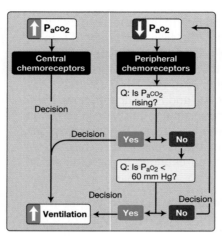

Figure 24.8.
Decision tree governing ventilatory
responses to changes in arterial
P_{CO_2} (P_aCO_2) and P_{O_2} (P_aO_2).

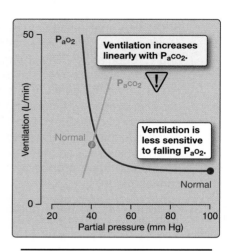

Figure 24.9.
Ventilatory responses to changes in
arterial P_{CO_2} (P_aCO_2) and P_{O_2} (P_aO_2).

> Patients who have adapted to hypercapnia may rely on the
> effects of hypoxia on the peripheral chemoreceptors to sustain
> their ventilatory drive. When significant hypercapnia exists,
> administering O_2 may help normalize P_aO_2, but it can also
> reduce the drive to breathe, potentially increasing P_aCO_2 further
> and inducing acute respiratory acidemia due to hypoventilation.

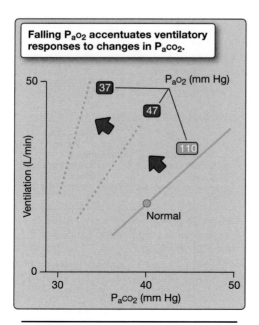

Figure 24.10.
Sensitization of ventilatory responses
to changes in arterial P_{CO_2} (P_aCO_2) by
decreasing arterial P_{O_2} (P_aO_2).

2. **Changing oxygen levels:** The peripheral chemoreceptors promote a rapid compensatory increase in ventilation if P_aO_2 falls to
dangerously low levels (see Fig. 24.9). More modest decreases in
P_aO_2 (between 60 and 100 mm Hg) have little effect on ventilation,
even though spike frequency in the peripheral chemoreceptor nerve
afferents increases in direct proportion to the drop in P_aO_2 (see Fig.
24.7). The reason why is that the peripheral and central chemoreceptors work against each other for respiratory center control, with
the central chemoreceptors retaining the upper hand until P_aO_2 falls
to 60 mm Hg. It is interesting to note that 60 mm Hg marks the point
at which hemoglobin (Hb) begins to desaturate (i.e., the steep portion of the dissociation curve; see Fig. 23.7). Thus, Hb's O_2-binding
properties allow the respiratory center to adjust ventilation over a
wide range to maintain a stable P_aCO_2 with little adverse effects on
O_2 delivery. Control center output in response to changes in P_aO_2 is
influenced both by blood pH and P_{CO_2}.

a. **pH:** When P_aO_2 falls, deoxyhemoglobin concentration rises.
Deoxygenation makes Hb a more favorable substrate for H^+
binding (the Bohr effect; see Chapter 23·IV·C) and plasma H^+
concentrations fall as a result. The rise in pH decreases chemoreceptor sensitivity to a fall in P_aO_2.

b. **Carbon dioxide:** When ventilation increases, P_aO_2 increases,
and P_aCO_2 falls (see Fig. 22.9). Because the respiratory control

centers are designed to optimize P_aCO_2, the response to mild hypoxia is overridden in favor of a stable CO_2 concentration.

3. **Synergism:** Conditions that decrease P_aO_2 typically cause a concomitant rise in P_aCO_2. Because P_aCO_2 equates with H^+ concentration, pH falls also. It is therefore not surprising to find that changes in P_aO_2, P_aCO_2, and arterial pH may act synergistically to elicit a ventilatory response that is greater than the sum of their individual actions. Hypoxia increases chemoreceptor sensitivity to hypercapnia (Fig. 24.10; see previous page), and rising P_aCO_2 and H^+ concentrations sensitize the receptors to hypoxia (Fig. 24.11).

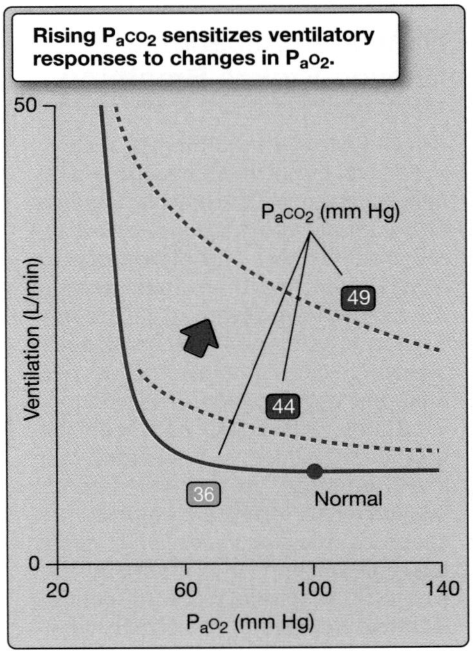

Biologic Sex and Aging 24.1: Respiratory Regulation

Aging is associated with an increase in the respiratory rate and a decrease in tidal volume, resulting, at least in part, from increasing lung and chest wall stiffness. However, aging also brings about dramatic decreases in responsiveness to hypoxia and hypercapnia. Compared with young adults, ventilatory responses to hypoxia drop by 75%, and responses to hypercapnia are halved by age 65–73 years. Mouth occlusion pressures, which are a measure of inspiratory drive, decrease to an equivalent degree (50% decreased response to hypoxia and 60% decrease to hypercapnia). These changes reflect a deteriorating ability to sense, and for the respiratory control centers to respond to, changing blood gas levels. Note, however, that these changes may be more of a reflection to adopting a sedentary lifestyle as opposed to physiologic aging *per se* because they can be offset by exercise.

IV. PULMONARY RECEPTOR ROLE

The lung and airways contain a variety of receptors that help protect the system from foreign bodies and provide the respiratory center with feedback about lung volume (Fig. 24.12). Information flow from these receptors travels primarily in the vagus nerve.

A. Rapidly adapting receptors (irritant receptors)

Rapidly adapting receptors (RARs) are stretch receptors embedded in the airway walls. During inspiration, the lung inflates, and the airways stretch, exciting the RARs and initiating trains of action potentials in the sensory afferent nerve. Bursting intensity is proportional to the rate and degree of stretch, which conveys information about the rate and extent of lung expansion. The RARs adapt rapidly, however, and afferent nerve activity trails off within a second of holding the lung at the new volume. A subset of RARs is also sensitive to lung deflation. RARs also mediate responses to irritants and noxious stimuli, such as ammonia, smoke, pollen, dust, and cold air (see Fig. 24.12). The receptors trigger bronchoconstriction, mucus secretion, and coughing, presumably to prevent foreign materials from reaching the respiratory zone. Irritant responses have also been implicated in the bronchoconstriction that results from histamine release during an allergic asthma attack.

B. Slowly adapting receptors

Embedded within the smooth muscle layers of conducting airways are myelinated sensory fibers. They also respond to stretch, with output intensity reflecting the extent of inflation (see Fig. 24.12). In contrast to

Rising P_aCO_2 sensitizes ventilatory responses to changes in P_aO_2.

Figure 24.11.
Sensitization of ventilatory responses to changes in arterial Po_2 (P_aO_2) by decreasing arterial Pco_2 (P_aCO_2).

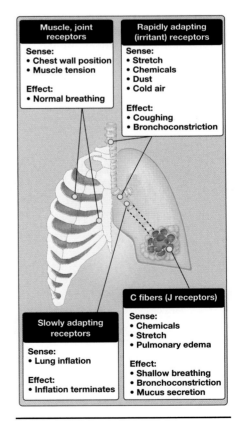

Figure 24.12.
Chest wall and lung sensory receptors. J receptors = juxtacapillary receptors.

Clinical Application 24.2: Cheyne-Stokes Breathing

Cheyne-Stokes breathing is a cyclic breathing pattern characterized by periods of apnea followed by a series of breaths of progressively increasing respiratory effort and airflow toward a maximum and then again waning toward apnea. Although Cheyne-Stokes breathing is observed occasionally in normal individuals at high altitudes during sleep, it is common in patients with stroke and heart failure. Heart failure is associated with low perfusion rates, which impairs the brainstem's ability to monitor the effects of changing ventilation on blood gas composition. It is theorized that when the respiratory control centers increase ventilation to correct hypercapnia, there is a time delay before the centers are able to see the results of this change, during which they impel further increases in respiratory effort and cause a hypocapnic overshoot. The centers compensate by decreasing ventilation, with the same hypoperfusion delay causing respiratory effort to decline inappropriately and resulting in apnea. The cyclic breathing pattern is repeated with a variable periodicity of ~30–100 seconds.

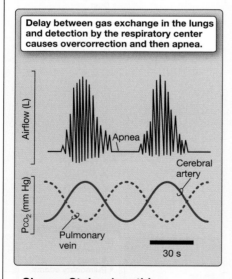

Delay between gas exchange in the lungs and detection by the respiratory center causes overcorrection and then apnea.

Cheyne-Stokes breathing.

RARs, they adapt very slowly when the stimulus is sustained. As with RARs, a subset of these **slowly adapting receptors (SARs)** is sensitive to deflation. SARs are largely insensitive to chemical stimulation. If tidal volume is sufficiently high, SARs can terminate inspiration and prolong the exhalation that follows (**Hering-Breuer reflex**).

> The stretch receptors that mediate the Hering-Breuer reflex relay sensory signals to the respiratory control centers via the vagus nerve. Lung-transplant patients breathe normally despite having lost this pathway, showing that feedback from the receptors is not required to sustain the breathing cycle. The Hering-Breuer reflex is prominent in newborns, however, suggesting that it may have a physiologic role in preventing lung overinflation during infancy.

C. C fibers

C fibers are unmyelinated nerve fibers with similar functions and response characteristics to RARs. They are sensitive to lung inflation, injury, pulmonary vascular congestion, and certain chemicals (see Fig 24.12). When stimulated, C fibers cause bronchoconstriction, mucus secretion, and rapid, shallow breathing. Two types of C fibers serve the bronchial tree. Bronchial C fibers are associated with the conducting airways and are perfused by the systemic (bronchial) circulation. Pulmonary C fibers, also known as **juxtacapillary receptors (J receptors)**, are located more distally and exposed to a different microenvironment influenced by the pulmonary circulation.

D. Joint and muscle receptors

Stretch and tension receptors located in the chest wall sense wall movement and the amount of effort involved in breathing (see Fig. 24.12). Output from these receptors allows for increased force of inspiration and expiration when wall movement is impeded. Limb joints contain similar receptors that contribute to increased ventilation during exercise.

V. RESPIRATORY ADAPTATION TO ENVIRONMENT

O_2 diffusion from the atmosphere to mitochondria is driven by a partial pressure gradient. Because atmospheric pressure decreases with height above sea level, moving to higher altitude forces the respiratory system to adapt to a reduced partial pressure gradient. By contrast, diving dramatically increases the pressure gradient driving uptake of O_2 and other atmospheric gases, which can have severe physiologic consequences.

A. Altitude

The fractional composition of air does not change with increasing altitude, but the partial pressures of its various constituents all fall with decreasing barometric pressure during ascent. The P_{O_2} of air at the peak of Mt. Everest (8,848 m), for example, is 43 mm Hg (Fig. 24.13). Alveolar P_{O_2} is lower than in inspired air because airways add water, which reduces the fractional composition of the other components. P_{O_2} is reduced by altitude to a greater extent than might be predicted based on barometric pressure alone, because the rate at which airways add water and the blood–gas interface adds CO_2 does not change with altitude. Falling $P_{A_{O_2}}$ reduces the partial pressure gradient driving O_2 uptake and causes hypoxia. Sensory and cognitive functions deteriorate rapidly with altitude as a result, reflecting the acute dependence of CNS neurons on O_2 availability (Fig. 24.14). The physiologic response to hypoxia can be divided into three phases: acute responses, adaptive responses, and long-term acclimation (Fig. 24.15).

1. **Acute (minutes):** Hypoxia is sensed by the peripheral chemoreceptors. The respiratory center responds by increasing ventilatory drive, but this causes $P_{a_{CO_2}}$ to fall (see Fig. 22.9), which activates the central chemoreceptors. Ventilatory drive is blunted as a consequence. The respiratory center also suppresses the cardioinhibitory center (see Fig. 19.9) and allows heart rate to rise (see Fig. 24.15A). Resting cardiac output increases, facilitating increased O_2 uptake by increasing pulmonary perfusion. Hypoxia causes concomitant pulmonary vasoconstriction, which increases pulmonary vascular resistance and forces the right heart to generate higher pressures to maintain output.

2. **Adaptive (days to weeks):** The central chemoreceptors adapt slowly over a period of 8 to 24 hours, allowing ventilation rates to climb in order to address the altitude-induced hypoxia (see Fig. 24.15B). The resulting drop in $P_{a_{CO_2}}$ causes respiratory alkalosis, but the kidneys compensate by decreasing acid excretion, and blood pH renormalizes (see Fig. 24.15C). Alkalosis also stimulates 2,3-bisphosphoglycerate (2,3-BPG) production. 2,3-BPG decreases Hb's O_2 affinity, causing the Hb–O_2 dissociation curve to shift rightward (see Fig. 23.11). This change enhances O_2 unloading to the tissues.

3. **Acclimation (months to years):** Long-term acclimation to living at high altitude involves changes in the properties of blood, the vasculature, and the cardiopulmonary system.

   a. **Blood:** Hypoxia stimulates erythropoietin release from kidneys and promotes red blood cell production (see Chapter 23·II·B). Hb concentration increases proportionately, from ~15 g/dL to ~20 g/dL (see Fig. 24.15D). Concurrent increases in circulating blood volume can yield an overall increase in blood's O_2-carrying capacity by >50%.

   b. **Vasculature:** Hypoxia stimulates angiogenesis. Capillary density increases throughout the body, allowing for improved tissue perfusion.

   c. **Cardiopulmonary system:** The increase in pulmonary arterial pressures required to perfuse the lungs in the face of hypoxic vasoconstriction promotes vascular and ventricular remodeling. Smooth-muscle proliferation increases vascular wall thickness, and the right ventricle hypertrophies to counter the increased afterload. Although the pressure increase stresses the pulmonary circulation, it is also beneficial in that it increases perfusion of the lung apex and allows apical alveoli to participate in O_2 uptake.

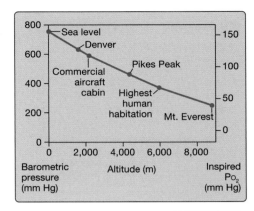

Figure 24.13.
Effects of altitude on barometric pressure and P_{O_2}.

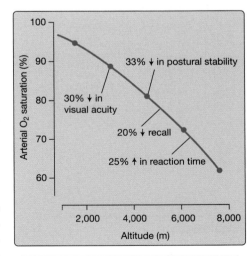

Figure 24.14.
Decrease in sensory and cognitive functions caused by the decrease in arterial O_2 saturation that occurs with ascent to altitude.

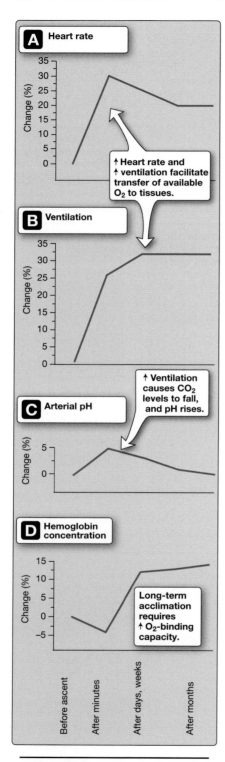

Figure 24.15.
Changes in various physiologic parameters after ascent and adaptation to altitude (3,000 m above sea level) shown as percentages relative to levels recorded prior to ascent.

4. **Adverse effects:** Many individuals develop **acute mountain sickness** when they ascend to high altitude, a temporary condition characterized by headache, irritability, insomnia, dyspnea, dizziness, nausea, and vomiting. Symptoms usually dissipate over a period of several days. **Chronic mountain sickness** develops after prolonged residence at high altitude and reflects the adverse cardiovascular consequences of the adjustments noted earlier. Polycythemia increases blood viscosity and resistance to blood flow, forcing both ventricles to operate at higher pressures (see Clinical Application 18.1). Decreased $P_{A O_2}$ causes bronchoconstriction, which stresses the right side of the heart. If the hypoxia is sufficiently severe or prolonged, pulmonary veins also constrict, and the arteries become narrowed by vascular remodeling. Ultimately, this may cause pulmonary edema, right heart failure, and death.

B. Diving

Diving presents several challenges to the respiratory system, most of which are associated with external hydrostatic pressure at increased depth. Water is denser than air, so pressure rises quickly with depth beneath the surface. It takes a water column of only ~10 m to exert a pressure equivalent to that of the atmosphere (760 mm Hg), so a diver at ~30 m is subject to pressures approximating four atmospheres.

1. **Effects of depth:** Water squeezes and compresses a diver from all sides. It also compresses gases within the alveoli, which increases the partial pressures driving uptake of all gases and decreases alveolar volume, causing two significant challenges.

   a. **Partial pressures:** At sea level, O_2 and CO_2 are the only components of atmospheric air to dissolve in blood to any significant extent. Diving can increase the partial pressure on all constituents to such a degree that all are forced to dissolve in potentially lethal excess.

   b. **Volume:** Pressurizing a gas decreases its volume (Fig. 24.16). At 30 m, 1 L of gas (sea level volume) occupies ~250 mL. Conversely, 1 L of gas expands to fill 4 L when the diver surfaces from 30 m, which potentially can cause severe damage to any tissue that contains it.

2. **Gas toxicity:** Air consists principally of N_2 (78%) and O_2 (21%), both of which become toxic when inhaled under pressure. The CO_2 composition of inspired air is insignificant and is not of concern unless the diver's breathing apparatus traps exhaled gas, allowing CO_2 to rise.

   a. **Nitrogen narcosis:** N_2 has no significant effect on bodily function at sea level because it does not dissolve in tissues. However, at depths of ~40 m and below, P_{N_2} rises to the point where it dissolves in cell membranes in amounts sufficient to disrupt ion-channel function. Its effects are narcotic and similar to those of ethanol (**nitrogen narcosis**). The severity of its actions is related to depth and pressure, initially causing a feeling of well-being but ultimately causing loss of function at ~80 m and below.

b. **Oxygen poisoning:** O_2 is an inherently toxic molecule because of its tendency to form free radicals. At sea level, the amount of O_2 being delivered to tissues is closely regulated by Hb, which acts both as a vehicle for transporting O_2 through the circulation and also as an O_2 buffer. The delivery system is essentially saturated under normal circumstances. Breathing O_2 at high pressure causes it to dissolve in blood in amounts that exceed the buffering capacity of Hb. The tissues subsequently are exposed to a P_{O_2} that exceeds a normal range, causing a variety of neurologic effects, including visual disturbances, seizures, and coma.

c. **Deep-sea diving:** Divers who work at depth breathe a helium/oxygen mix (**heliox**), with the percentage of O_2 carefully tailored to yield a partial pressure that is supportive rather than harmful. Helium replaces N_2 because it dissolves in body tissues less readily, is less narcotic, and has a density that is considerably reduced compared with N_2 (14%). Inhaling the helium mix reduces airway resistance and decreases the work of breathing (see Chapter 21·VIII·E).

> Heliox can also be used clinically to support patients with anatomic or physiologic airway obstruction. The gas mix's decreased density allows it to slip past the obstruction site more easily than does atmospheric air and thus helps improve patient oxygenation.

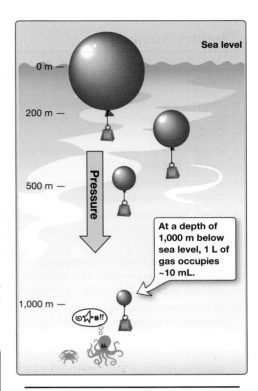

Figure 24.16.
Changes in gas volume caused by water pressure at various depths below sea level (0 m).

3. **Decompression sickness:** A diver breathing air at pressure for prolonged periods can accumulate significant amounts of N_2 within their tissues. The average amount of N_2 contained within the body at sea level is ~1 L. A prolonged dive at 30 m raises this amount to as much as 4 L. N_2 is taken up by diffusion across the blood–gas interface and then distributed by the circulation to all tissues, but it preferentially partitions in body fat. When a diver ascends back to the surface, N_2 is no longer subject to the pressure that forced it to dissolve at depth, so it comes out of solution and forms pure N_2 bubbles. The presence of bubbles in the blood stream causes decompression sickness, or "**the bends.**" Bubbles block blood vessels, and, as small bubbles coalesce to form large bubbles, progressively larger vessels are affected. Dependent tissues become ischemic, typically manifesting as pain in joints and in limb musculature. More severe symptoms may include neurologic deficits, dyspnea, and death. Slowing the rate of ascent gives the ~3 L of excess gas dissolved in the aqueous phase more time to diffuse out of the tissues and into the circulation for transport to the lungs for exhalation (Fig. 24.17). Fat is relatively avascular, however. This increases the distance over which N_2 must diffuse before it can be carried away by the circulation, thereby slowing the rate at which it can be removed. Complete renormalization of tissue N_2 levels may take several hours after ascent.

Figure 24.17.
Deep-sea divers carefully time their ascent to avoid "the bends."

Chapter Summary

- A cyclical pattern of inspiration and expiration is established and controlled by a **central pattern generator (CPG)**. The CPG resides in a **brainstem respiratory control center** (medulla).

- The medulla contains two groups of cells involved in respiratory control. The **dorsal respiratory group** drives inspiration during quiet breathing. The **ventral respiratory group** coordinates accessory muscles and is believed to house the CPG.

- Higher control centers can override unconscious breathing to allow for talking, coughing, and other voluntary acts that use the same musculature.

- Central and peripheral chemoreceptors relay information about the chemical composition of blood to the control centers. **Central chemoreceptors** monitor P_aCO_2. **Peripheral chemoreceptors** are sensitive to P_aO_2, P_aCO_2, and arterial pH.

- When P_aCO_2 rises, both the peripheral and central chemoreceptors are excited and cause the control center to respond with an immediate increase in ventilation. Decreases in P_aO_2 are a less effective stimulus for ventilation unless CO_2 rises simultaneously.

- Other sensors feeding information to the medulla include **rapidly adapting** and **slowly adapting receptors** that are sensitive to lung inflation and deflation. **C fibers** are sensitive to chemical stimuli that could damage the lungs. **Joint** and **muscle receptors** provide information about muscle contraction and movement.

- Atmospheric pressure decreases with increasing altitude. P_O2 decreases also, causing hypoxia. The respiratory and cardiovascular control centers compensate by increasing ventilation and perfusion. Full acclimation to high altitude requires months and involves increases in lung perfusion, hematocrit, and systemic capillary density.

- Descent to depth in water increases the partial pressures of inspired gases. N_2 and O_2, which have limited solubility at sea level, become toxic when forced to dissolve in tissues at depth. N_2 preferentially partitions in fat and requires many hours to extract before returning to the surface. Premature ascent allows pure N_2 gas bubbles to form in blood, occluding vessels and causing severe ischemia-associated pain known as "**the bends**."

Study Questions

Choose the ONE best answer.

24.1. A 78-year-old male is brought to the emergency department after sudden loss of focal brain function. A brain CT confirms a cerebrovascular accident (CVA). If the CVA primarily affects forced expirations during rest and exercise, which of the following neural areas is most likely to be affected?

A. Apneustic center
B. Dorsal respiratory group
C. Phrenic nerve center
D. Pneumotaxic center
E. Ventral respiratory group

Best answer = E. Ventral respiratory group neurons are involved in forced expiration and coordination of labored inspiration and expiration (see Section II·A). The pontine centers limit lung expansion (apneustic center [A]) and cause rapid shallow breathing (pneumotaxic center [D]), although the role of either center during normal breathing is uncertain. Dorsal respiratory group neurons (B) regulate inspiration and implement the resting respiratory rhythm. The phrenic nerve (C) contains motor neurons that control the diaphragm, which is a principal inspiratory muscle (see Chapter 21·V·C).

24.2. A researcher is investigating mechanisms of respiratory control. Carotid body glomus cells are noted to respond to low arterial P_{O_2} with Ca^{2+} influx, causing release of neurotransmitters that stimulate sensory nerve afferents. An increase in which of the following most likely triggers the Ca^{2+} influx in the glomus cells?

A. Brain interstitial H^+
B. K^+ conductance
C. Membrane depolarization
D. Na^+ conductance
E. Na^+ equilibrium potential

Best answer = C. Arterial P_{O_2} is sensed by an O_2-dependent K^+ conductance in glomus cells (see Section III·B). A fall in arterial P_{O_2} allows the K^+ channel to close, not open (B), decreasing K^+ efflux and causing membrane depolarization. Depolarization activates voltage-gated Ca^{2+} channels and Ca^{2+} influx. Changes in brain interstitial H^+ (A) initiate ventilatory responses mediated by central chemoreceptors (see Section III·A), not peripheral chemoreceptors. Na^+ conductances (D) and changes in the Na^+ equilibrium potential (E) do not have a role in this response.

24.3. A 23-year-old female presents to the office because of coughing paroxysms when air temperature is below freezing. She has no significant medical history. Which of the following types of sensory receptors are most likely to have triggered this response?

A. Central chemoreceptors
B. Irritant receptors
C. Juxtacapillary receptors
D. Peripheral chemoreceptors
E. Pulmonary stretch receptors

Best answer = B. Irritant receptors protect the lung from noxious stimuli, such as dust, chemicals, and cold air (see Section IV·A). These receptors may trigger coughing, bronchoconstriction, and mucus production when stimulated. Central (A) and peripheral (D) chemoreceptors respond to changes in arterial blood gas composition (see Section III). Juxtacapillary receptors (C) respond to capillary engorgement and interstitial edema (see Section IV·C). Pulmonary stretch receptors (E) activate during lung inflation (see Section IV·B).

24.4. A 29-year-old male living at sea level experiences headache and nausea after traveling to a ski resort (resort elevation = 2,500 m). Within a day, his symptoms improve, and he feels well enough to ski. Which of the following most likely accounts for his physiologic accommodation?

A. Angiogenesis
B. Central chemoreceptor adaptation
C. Hemoglobin isoform alteration
D. Pulmonary stretch receptor stimulation
E. Red blood cell synthesis

Best answer = B. The initial response to hypobaric hypoxia at high altitudes is hyperventilation, which causes arterial P_{CO_2} to fall, suppressing the normal drive to breathe (see Section V·A). The resulting hypoxemia causes the symptoms associated with acute mountain sickness. Central chemoreceptors adapt slowly to lowered arterial P_{CO_2} over 8 to 24 hours, allowing ventilation rate to rise and the symptoms improve. Red blood cell production (E) and angiogenesis (A) require weeks to months to compensate for the effects of hypoxia. There is no evidence that altitude causes a change in hemoglobin isoform (C) or pulmonary stretch responses (D).

24.5. A 58-year-old female comes to the physician because of shortness of breath on exertion and a dry cough for the past several months. She denies ever having been a smoker. Her past medical history is unremarkable. Pulmonary function tests reveal a restrictive pattern consistent with idiopathic pulmonary fibrosis (IPF). Which of the following changes is most likely in this patient?

A. Decreased work of breathing
B. Hypercapnia-induced bronchiolar dilation
C. Hypoxia-induced bronchiolar dilation
D. Increased radial traction on the airways
E. Increased total lung capacity

Best answer = D. Airways and surrounding tissues are linked mechanically (see Chapter 21·VIII·C). When lungs expand, increased radial traction leads to an increase in airway patency. Pulmonary fibrosis is associated with scar tissue deposition within the lung parenchyma, which strengthens these mechanical connections and increases radial traction on inspiration. IPF decreases pulmonary compliance, which makes the lungs difficult to expand and increases rather than decreases the work of breathing (A). Bronchiolar patency (B, C) is largely under the control of the sympathetic nervous system and circulating hormones, not P_{O_2} or P_{CO_2}. IPF makes lungs difficult to expand and reduces total lung capacity (E).

24.6. A 69-year-old female has participated in a longitudinal study on the effects of physiologic aging on the respiratory system since she was 22 years old. Respiratory function has been assessed at 7-year intervals. Which of the following measures of respiratory system health is most likely to have increased with aging?

A. Diffusing capacity
B. Elastic recoil
C. Inspiratory drive
D. Residual volume
E. Vital capacity

Best answer = D. Aging-associated elastin degradation causes distension and dilation of alveoli and alveolar ducts, which reduces radial traction on the airways (see Biologic Sex and Aging 21.2). This allows airways to collapse and trap air on expiration, thereby increasing residual volume and decreasing vital capacity (E). Elastin degradation and resultant erosion of alveolar walls reduces blood–gas interface surface area, which reduces lung diffusing capacity (A). Tissue loss also reduces elastic recoil (B). Mouth occlusion pressure, which is a measure of inspiratory drive (C) and inspiratory muscle strength, decreases with age.

24.7. A 70-year-old male presents to the office because of a racing heart and fatigue for the past 2 weeks. He is a competitive cyclist, and he admits to taking erythropoietin (EPO) to enhance endurance. Vital signs are notable for tachycardia and elevated blood pressure. Imaging studies reveal mild cardiac hypertrophy. Which of the following values is most likely to be increased in this patient?

A. Arterial O_2 content
B. Arterial O_2 saturation
C. Arterial pH
D. Arterial P_{O_2}
E. Venous P_{CO_2}

Best answer = A. EPO is a hormone produced by the kidney in response to hypoxia (see Chapter 23·II·B). It stimulates red blood cell (RBC) production to increase blood's O_2-carrying capacity and arterial O_2 content. Arterial O_2 saturation (B) reflects the number of occupied O_2-binding sites on hemoglobin (Hb) and should not be affected by EPO. Arterial pH (C) increases when P_{CO_2} decreases and would not be affected by EPO. Arterial P_{O_2} (D) is determined by the amount of O_2 dissolved in blood, not RBC numbers. Venous P_{CO_2} (E) would be expected to increase with reduced O_2 availability, but not by EPO.

24.8. A 35-year-old male with a history of chronic pain is brought to the emergency department in a coma. His wife reports finding him unconscious with an empty bottle of prescription opioids nearby. His vital signs are notable for a pulse of 56 beats/min, respirations of 7 breaths/min and shallow, and a blood pressure of pressure of 84/52 mm Hg. Which of the following sets of arterial blood values are most likely to be recorded in this patient?

	pH	P_{O_2}	P_{CO_2}
A.	7.10	35	30
B.	7.25	50	65
C.	7.40	35	40
D.	7.40	85	65
E.	7.50	85	30

Best answer = B. The patient has overdosed on opioids, causing respiratory depression (normal respiration rate = 12–20 beats/min) through binding to μ-opioid receptors within the pre-Bötzinger complex, a respiratory rhythm–generating area of the brainstem (see Chapter 24·II). Hypoventilation allows arterial P_{CO_2} to rise above a normal value of ~40 mm Hg and P_{O_2} to fall below normal (~98 mm Hg). The increase in P_{CO_2} results in respiratory acidosis (normal arterial pH = 7.35–7.45; see Chapter 23·VI). Decreased arterial P_{CO_2} values (A, E) are not consistent with hypoventilation. The patient's arterial pH would not be expected to be within a normal range (C, D) or alkalemic (E).

Unit VI: Urinary System

Filtration and Micturition

25

I. OVERVIEW

Metabolism generates many acids, toxins, and other waste products that can severely impair cell function if allowed to accumulate. Metabolic wastes are passed from cells to the circulation and then on to the kidneys, where they are removed by filtration and excreted in urine. **Excretion** is only one of three essential kidney functions, however. The kidney is also an **endocrine organ** that controls red blood cell production by bone marrow. The kidney also has a vital **homeostatic** role in controlling blood pressure, tissue osmolality, electrolyte and water balance, and plasma pH. Excretory and homeostatic functions both begin when blood is forced at high pressure through a filtration membrane to separate plasma from cells and proteins (Fig. 25.1). The ultrafiltrate is then funneled into a tubule lined with a specialized transport epithelium. Channels and transporters in the epithelium's luminal (apical) surface then recover any *useful* components from the ultrafiltrate as it progresses toward the bladder and, ultimately, the external environment. If we were to take the same approach to domestic duties, we would carry the entire contents of our household, including laptop, flat-screen TV, plants, clothing, and other accoutrements of life out to the street and then carry back all items that we wished to retain. The empty pizza boxes, cans, napkins, and stray socks would be left on the curb for the city sanitation department. We would repeat this process 48 times a day! This remarkable approach to housekeeping has two major advantages. The first is speed because toxins (metabolic or ingested) can be effectively cleared from the circulation in as little as 30 minutes. The second is that the kidney need only be selective about what it recovers from the filtrate, not what it excretes, because anything not reabsorbed is automatically excreted. The kidney also efficiently employs osmotic gradients to recover filtered water, so that, despite the massive volume of fluid handled every day (~180 L), kidney energy use is only slightly greater than that of the heart (10% of total body energy consumption, compared with 7% for the heart).

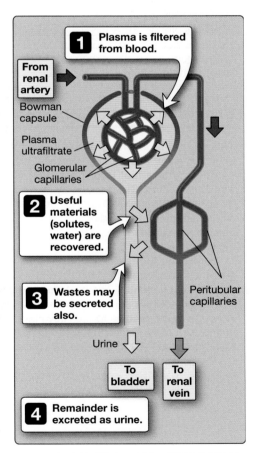

Figure 25.1.
Overview of kidney function.

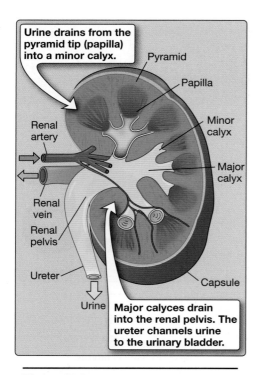

Urine drains from the pyramid tip (papilla) into a minor calyx.

Pyramid

Papilla

Renal artery

Minor calyx

Major calyx

Renal vein

Renal pelvis

Ureter

Capsule

Urine

Major calyces drain into the renal pelvis. The ureter channels urine to the urinary bladder.

Figure 25.2.
Gross anatomy of the kidney.

II. ANATOMY

The functional unit of the kidney is the **nephron**, comprising a blood filtration component (**renal corpuscle**) and a filtrate recovery component (**renal tubule**). Each kidney contains ~1,000,000 nephrons.

A. Gross

Kidneys are paired, bean-shaped organs that lie one to either side of the vertebral column close up against the abdominal wall behind the peritoneum. Each is ~11 cm long, 6 cm wide, and 4 cm deep; weighs ~115 to 170 g, depending on sex; and is enclosed within a resilient connective tissue **capsule** that protects its inner structures (Fig. 25.2). The capsule is penetrated at the hilum by a **ureter**, a **renal artery** and **renal vein**, lymphatic vessels, and nerves. Viewed in cross section, the kidney is seen to be composed of several distinct bands. An outer band (**cortex**) lies beneath the capsule and is the site of blood filtration. The middle band (**medulla**) is divided into 8 to 18 conical **renal pyramids**. The pyramids contain thousands of tiny ducts that each collect urine from multiple nephrons and guide it toward the ureter. The pyramid tip (**papilla**) inserts into a collecting vessel known as a **minor calyx**. Minor calyces join to form **major calyces**, which drain into a common **renal pelvis**. The pelvis forms the head of a ureter, which propels urine to the urinary bladder for storage and voluntary release.

Clinical Application 25.1: Tubulointerstitial Disorders

The normal functioning of the renal tubule and interstitium can be negatively impacted either acutely or chronically.

Acute interstitial nephritis (**AIN**) is an inflammatory condition affecting the renal interstitium. Symptoms may include an acute rise in plasma creatinine levels and proteinuria (protein in urine), both reflecting general renal dysfunction. AIN usually results from drug exposure, β-lactam antibiotics (e.g., penicillin and methicillin) being the most common.[1] Kidneys typically recover normal function after discontinuing drug use.

Polycystic kidney disease (**PKD**) is an inherited disorder characterized by the presence of innumerable fluid-filled cysts within the kidneys and, to a lesser degree, the liver and pancreas. The cysts form within the nephron and progressively enlarge and compress the surrounding tissues, preventing fluid flow through the tubules. Although many patients remain asymptomatic, others may begin to show symptoms of impaired renal function (such as hypertension) in their fourth decade. There is no treatment, and PKD may ultimately cause renal failure.

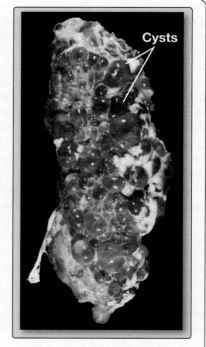

Cysts

Hereditary polycystic kidney disease.

[1]For more information on adverse reactions to β-lactam antibiotics, see *LIR Pharmacology*, 8e, Chapter 29·II·E.

B. Functional

Kidneys are composed largely of fluid, as are most tissues. Although the fluid within a kidney is compartmentalized (i.e., vascular, luminal, interstitial), and flow between the compartments is limited by cellular barriers, water moves relatively freely between the three compartments, driven by osmotic pressure gradients. A survey of tissue osmolality in different regions of the kidney shows gross differences between the cortex and medulla (Fig. 25.3). The cortex has an osmolality that approximates that of plasma, but the osmolality of the inner medulla is increased severalfold. This osmotic gradient is essential to normal kidney function because it is used to recover virtually all the water filtered from the vasculature each day (average urinary water excretion is 1–2 L/d).

C. Vasculature

The renal nephron's *modus operandi* is to channel blood at relatively high pressure through a network of leaky blood vessels (**glomerular capillary network**). Pressure forces plasma out of the vasculature through a filtration barrier during passage. The plasma filtrate is channeled into the renal tubule, whose function is to recover essential solutes and >99% of the fluid and return it to the vasculature. Reabsorption is the primary responsibility of the **peritubular capillary network**. The peritubular network is plumbed in series with and receives blood from the **glomerulus** (Fig. 25.4).

1. **Glomerular network:** Blood enters the glomerulus from an **interlobular artery** at relatively high pressure (~50 mm Hg) via an **afferent arteriole**. Blood courses through a tuft of specialized **glomerular capillaries**. The capillaries branch and interconnect extensively via anastomoses to maximize the surface area available for filtration. Spaces between capillaries are filled with **mesangial cells**, an epithelial cell type (**myoepithelial cell**) that contracts and relaxes as a way of modulating glomerular capillary surface area and fluid filtration rate. Blood leaves the glomerulus, not by a venule, but by an **efferent arteriole**, still at high pressure. The afferent and efferent arterioles are both resistance vessels that regulate glomerular blood flow and fluid filtration rates through vasoconstriction and vasodilation (see Section IV).

2. **Peritubular network:** The peritubular capillary network surrounds and closely follows the renal tubule as it tracks through the kidney, sustaining the tubule with O_2 and nutrients. The network also carries away fluids and dissolved electrolytes that have been reabsorbed from the tubule lumen. Prompt removal of these materials helps maintain the concentration gradients for chemical and osmotic diffusion across the tubule epithelium that are required for normal renal function.

D. Tubule

A renal tubule comprises a long, thin tube of renal epithelial cells. It can be divided into several distinct segments based on morphology and function (Fig. 25.5). At its head is the **glomerular (Bowman) capsule**, which completely envelops and isolates the glomerulus from its surrounds. The capsule captures and contains fluid filtering from the glomerular capillaries. Plasma filtrate flows from glomerular

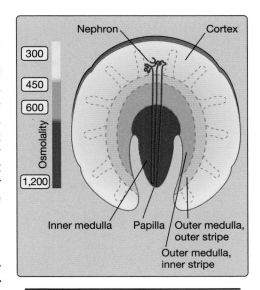

Figure 25.3.
Corticopapillary osmolality gradient. Osmolality values are in mOsmol/kg H_2O.

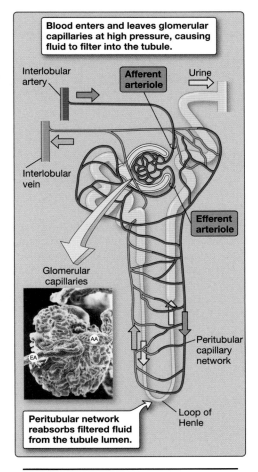

Figure 25.4.
Glomerular and peritubular capillary networks. AA and EA = afferent and efferent arteriole, respectively.

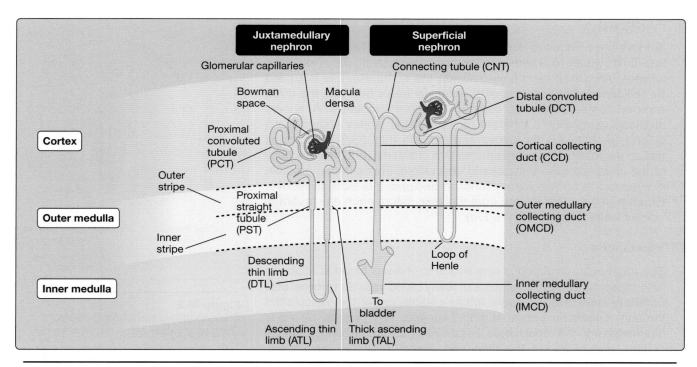

Figure 25.5.
Nephron types and the collecting duct system.

capillaries into the **Bowman space** and then enters the **proximal tubule** (**PT**). The PT contains both a **convoluted** and a **straight** section (**PCT** and **PST**, respectively). After exiting the PT, filtrate begins a long descent into the medulla. The tubule then executes a hairpin turn and returns to the cortex. This hairpin structure is known as the **nephron loop**, or **loop of Henle**. The descending portion of the loop is known as the **descending thin limb**. The ascending portion of the loop can be divided into the **ascending thin limb** and **thick ascending limb** (**TAL**). Fluid then passes through the **distal convoluted tubule** (**DCT**), and the **connecting tubule**, before emptying into a common collecting duct. The collecting duct conveys urine to the calyx and comprises three sections: a **cortical collecting duct**, an **outer medullary collecting duct**, and an **inner medullary collecting duct**.

E. Nephron types

The kidney contains two different types of nephron: **superficial** (**cortical**) **nephrons** and **juxtamedullary nephrons**.

1. **Superficial:** Superficial nephrons receive ~90% of renal blood supply, and they reabsorb a similarly large percentage of the fluid that filters from the vasculature. Their glomeruli are located in the outer cortical regions, and their nephron loops are short. The loops dip into the outer medulla but do not enter the inner medulla (see Fig. 25.5).

2. **Juxtamedullary:** Juxtamedullary nephrons receive ~10% of total renal blood supply. Their glomeruli are located within the inner cortex, and they have very long nephron loops that reach deep

into the inner medulla. The peritubular network that serves juxtamedullary nephrons is a specialized capillary network that follows the tubule down into the medulla to create a long looping vascular structure called the **vasa recta**. Juxtamedullary nephrons are designed to concentrate urine (see Chapter 27·II).

III. FILTRATION

In Chapter 18 (see Section 18·VII·D), we discussed the delicate balance that exists between the forces that favor fluid filtration from blood (mean **capillary hydrostatic pressure [P_c]**) and the forces that favor fluid retention (**capillary colloid osmotic pressure [π_c]**). In most tissues, increasing capillary hydrostatic pressure causes edema. In the kidney, however, increasing **glomerular P_c (P_{GC})** is the first step in urine formation (Fig. 25.6).

A. Starling forces

The forces that control fluid movement across the glomerular capillary wall are the same as for any vascular bed. These microcirculatory forces are considered in the **Starling equation**:

$$ GFR = K_f \left[(P_{GC} - P_{BS}) - (\pi_{GC} - \pi_{BS}) \right] $$

where GFR is glomerular filtration rate (i.e., net fluid flow across the capillary wall), which is measured in mL/min. K_f is a glomerular filtration coefficient, π_{GC} is **glomerular capillary oncotic pressure**, and P_{BS} and π_{BS} are the hydrostatic pressure and colloid osmotic pressure, respectively, of fluid contained within the Bowman space and entering the tubule. Changes in any of these variables can have dramatic effects on GFR and urine formation.

1. **Filtration barrier:** K_f is a measure of glomerular permeability and surface area. The barrier comprises three distinct layers that, together, create a three-step molecular filter that produces a cell- and protein-free plasma ultrafiltrate. The barrier is composed of a **capillary endothelial cell**, a **thick glomerular basement membrane**, and a **filtration slit diaphragm** (see Fig. 25.6, lower).

    a. **Layer 1:** Glomerular capillary endothelial cells are dense with fenestrations, resembling a sieve. The pores are ~70 nm in diameter, which allows free passage to water, solutes, and proteins. Cells are too large to fit through the pores, so they remain trapped in the vasculature.

    b. **Layer 2:** The glomerular basement membrane comprises three layers. An inner **lamina rara interna** is fused to the capillary endothelial cell layer. A middle layer, the **lamina densa**, is the thickest of the three. An outer **lamina rara externa** is fused to podocytes (see next section). The basement membrane is rich in negatively charged **heparan sulfate glycoproteins** that repel proteins (which also carry a negative charge) and reflect them back into the vasculature.

    c. **Layer 3:** Glomerular capillaries are completely ensheathed in tentacle-like foot processes that project from **podocytes**.

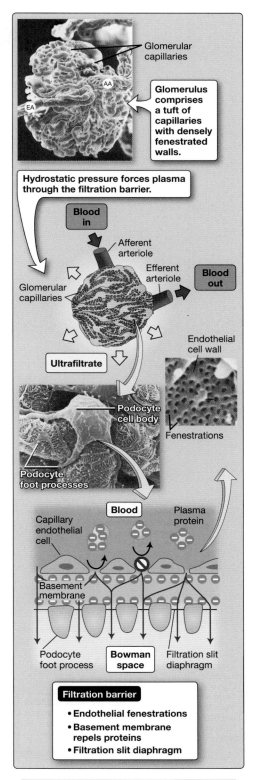

Figure 25.6.
Glomerulus and its filtration barrier. AA and EA = afferent and efferent arteriole, respectively.

Table 25.1: Glomerular Filtration Barrier Selectivity

Substance	Molecular Weight (Da)	Molecular Radius (nm)	Permeability*
Na⁺	23	0.10	1.0
Water	18	0.15	1.0
Glucose	180	0.33	1.0
Inulin	5,000	1.48	0.98
Myoglobin	17,000	1.88	0.75
Hemoglobin	68,000	3.25	0.03
Serum albumin	69,000	3.55	<0.01

*Permeability compares the plasma concentration of a substance with that in urine.

Podocytes are specialized epithelial cells. The covering is not continuous, however. The foot processes end in "toes" that interdigitate, leaving narrow slits between them. The slits are bridged by a proteinaceous **filtration slit diaphragm** that prevents proteins and other large molecules from entering the Bowman space. The fluid that finally enters the tubule is a plasma ultrafiltrate containing electrolytes, glucose, and other small organics, but anything larger than ~5,000 Da is excluded (Table 25.1).

2. **Hydrostatic pressure:** The glomerular capillary network is a physiologic oddity in that it is located within the renal arterial system rather than at its terminus. Blood enters the glomerulus via the afferent arteriole at a pressure of ~50 mm Hg, or ~15 mm Hg higher than in most capillary beds (Fig. 25.7; also see Chapter 18·VIII). Pressure does not fall significantly as blood travels through the glomerulus.

3. **Colloid osmotic pressure:** Capillary colloid osmotic pressure is proportional to plasma protein concentration. Blood enters the glomerulus at a π_{GC} of ~25 mm Hg, as in any other circulation. Blood loses ~15% to 20% of its total volume to filtrate during passage through the capillary network. The plasma proteins are prevented from leaving the vasculature by the filtration barrier, so π_{GC} rises with distance along a capillary. Blood entering the efferent arteriole has a π_{GC} of ~33 mm Hg.

4. **Bowman space:** The plasma filtrate entering the Bowman space is essentially protein free, so π_{BS} is 0. The sheer volume of fluid being expressed into the Bowman space causes a significant pressure to build, however. This pressure (~15 mm Hg) opposes filtration but is beneficial in that it creates a positive pressure gradient between the Bowman space and the renal sinus that propels fluid through the tubule.

5. **Net force:** Using values cited earlier, we note a net positive pressure favoring **ultrafiltration** (P_{UF}) that gradually decreases from ~15 mm Hg to ~2 mm Hg across the length of the glomerular capillary (see Fig. 25.7; also see Fig. 18.26).

B. Glomerular filtration rate

In a healthy person, P_{BS}, π_{BS}, and π_{GC} are all relatively invariant. The main factor affecting GFR is P_{GC}, which is determined by aortic

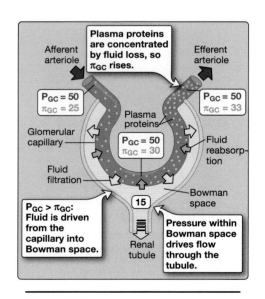

Figure 25.7.
Forces favoring glomerular ultrafiltration. All values are given in mm Hg. π_{GC} = glomerular capillary colloid oncotic pressure, P_{GC} = glomerular capillary hydrostatic pressure.

pressure, renal arterial pressure, and by changes in afferent and efferent arteriolar resistance (Fig. 25.8). (Note: Because the afferent and efferent arterioles are resistance vessels controlling flow through the glomerulus and all points south of the glomerulus [see next section], their contraction state significantly impacts net renal blood flow [RBF].)

> Mesangial cells may also regulate GFR through changes in glomerular capillary surface area, which affects K_f. The role of mesangial cells is minor compared with that of glomerular arterioles, however.

1. **Afferent arteriole:** Constricting the afferent arteriole decreases glomerular blood flow, just as it would in any other circulation. P_{UF} and GFR fall as a result. Afferent arteriolar dilation increases P_{UF} and GFR.

2. **Efferent arteriole:** Efferent arteriolar patency determines how easily blood exits the glomerular vasculature. Arteriolar constriction increases resistance to outflow from the glomerulus, causing P_{UF} and GFR rise. Arteriolar dilation allows pressure to bleed out of the glomerular network, and P_{UF} and GFR both fall as a result.

3. **Regulation:** In practice, changes in afferent and efferent arteriolar resistance rarely occur in isolation. The two resistance vessels are regulated by a multitude of factors, as are resistance vessels in other vascular beds (see Chapter 19·II). How they are regulated is considered in Section IV.

C. Values

GFR increases with body size and decreases with age (see Biologic Sex and Aging 25.1). A normal GFR range (adjusted to reflect body surface area) averages ~100 to 130 mL/min/1.73 m². This represents a ~1,000-fold greater filtration rate than observed in, for example, skeletal musculature, all due to the high P_{UF} and leakiness of the filtration barrier.

IV. REGULATION

RBF and GFR are governed by two overriding needs that are sometimes at odds with one another. **Local** vascular autoregulatory pathways maintain RBF at rates that optimize GFR. However, **central** control pathways may assume control and divert blood away from the kidney when it is needed elsewhere. Central control is exerted hormonally and through the autonomic nervous system (ANS).

A. Autoregulation

Autoregulation stabilizes RBF during changes in mean arterial pressure (MAP). All circulations in the body autoregulate to some degree, but the kidney's autoregulatory prowess is particularly well developed. RBF remains relatively stable over a MAP range of

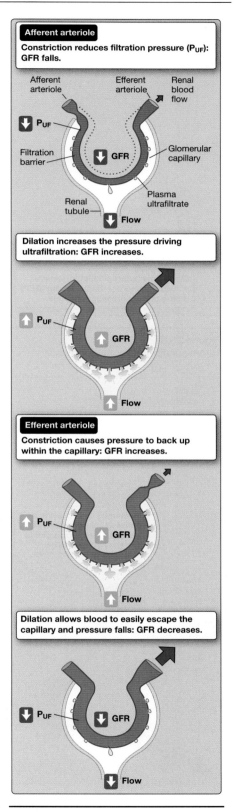

Figure 25.8.
Effects of changing afferent and efferent arteriolar resistance on glomerular filtration rate (GFR).

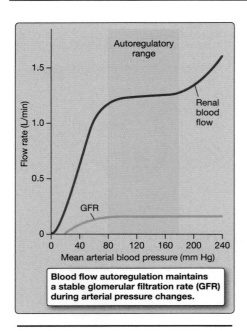

Figure 25.9.
Autoregulation of renal blood flow.

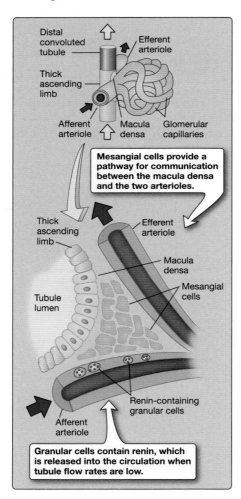

Mesangial cells provide a pathway for communication between the macula densa and the two arterioles.

Granular cells contain renin, which is released into the circulation when tubule flow rates are low.

Figure 25.10.
Juxtaglomerular apparatus.

~80 to 180 mm Hg (Fig. 25.9). Autoregulation helps maintain a constant GFR over a similarly wide MAP pressure range.

B. Myogenic response

The **myogenic response** is an autoregulatory mechanism. The smooth muscle cells lining the afferent arteriole contain Ca^{2+}-permeant mechanosensory channels that activate when the vessel wall is stretched (e.g., by an increase in luminal pressure). Ca^{2+} influx initiates contraction, and the arteriole constricts reflexively. The myogenic response stabilizes RBF and GFR during changes in posture, for example.

> The myogenic response is common to all vascular beds. Rising from a prone position can elicit arterial pressure pulses of >100 mm Hg in the lower extremities, reflecting gravitational effects on the blood columns contained within the vasculature (see Chapter 19·II·A·2).

C. Tubuloglomerular feedback

Tubuloglomerular feedback (**TGF**) is an autoregulatory mechanism mediated by the **juxtaglomerular apparatus** (**JGA**) that adjusts RBF and GFR to optimize fluid flow through the renal tubule. The JGA is a functional complex that includes the renal tubule, mesangial cells, and the afferent and efferent arterioles (Fig. 25.10).

1. **Tubule:** The loop of Henle comes into direct contact with the glomerulus after returning from the medulla and a short distance before transitioning to the distal tubule. The TAL wall is modified at the contact site to form a specialized sensory region called the **macula densa** (see Fig. 25.10). The macula densa monitors Na^+ and Cl^- flowing through the tubule, which, in turn, reflect RBF and GFR. Na^+ and Cl^- permeate macula densa cells via a $Na^+/K^+/2Cl^-$ cotransporter located in the apical membrane. Cl^- immediately exits via a basolateral Cl^- channel, causing a membrane depolarization whose magnitude is a direct reflection of tubule fluid NaCl concentration.

2. **Mesangial cells:** Mesangial cells provide a physical pathway for communication between the sensory (macula densa) and effector (arterioles) arms of the TGF system (see Fig. 25.10). All cells in the JGA are interconnected via gap junctions (see Chapter 4·II·F), which allow for direct chemical communication between the system components.

3. **Afferent arteriole:** The afferent arteriole is notable for its adenosine receptor and for renin-producing **granular cells** (also known as **juxtaglomerular cells**) within its walls (see Fig. 25.10).

   a. **Adenosine receptor:** Adenosine receptors are members of the G protein–coupled receptor superfamily that act through the cAMP signaling pathway (see Chapter 1·VII·B·2). Afferent arteriolar smooth muscle cells express a type A_1 receptor, which couples to an inhibitory G protein and decreases cAMP

Clinical Application 25.2: Glomerular Disease

Normal renal function can be severely impacted by pathologic changes in the glomerular filtration coefficient (K_f), hydrostatic pressure, and colloid osmotic pressure. Glomerular disease damages the filtration barrier and increases K_f, thereby allowing cells and proteins to pass into the tubule. It is the leading cause of renal failure in the United States. Glomerular disease can be divided into two broad and overlapping syndromes based on the characteristics of proteins and cellular debris contained within urine (**urine sediments**) and the associated symptoms: **nephritic syndrome** and **nephrotic syndrome**.

Nephritic syndrome is associated with diseases that cause inflammation of the glomerular capillaries, mesangial cells, or podocytes (**glomerulonephritis**). Inflammation creates localized breaches in the filtration barrier and allows cells and modest amounts of protein to escape into the tubule and appear in urine (**proteinuria**). Red cells typically collect and aggregate in the distal convoluted tubule and then appear in urine as tubular red cell **casts**.

Nephrotic syndrome refers to a set of clinical findings that include heavy proteinuria (>3.5 g/d), lipiduria, edema, and hyperlipidemia. Cell casts, which are characteristic of an inflammatory process, are absent. Nephrotic syndrome reflects a general deterioration of the renal tubule (**nephrosis**) that includes degradation of glomerular barrier function and is a frequent cause of mortality in patients with diabetes mellitus. Loss of albumin and other plasma proteins in urine causes plasma oncotic pressure to fall and accounts for the generalized edema associated with nephrotic syndrome. Decreased oncotic pressure also causes hyperlipidemia, but its benefits, if any, are unknown.

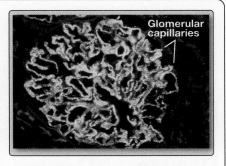

Immunoglobulin G–deposit imaging during glomerulonephritis.

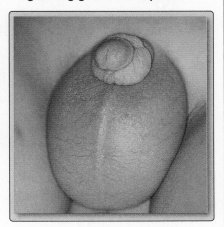

Scrotal edema in a 7-year-old male with nephrotic syndrome.

levels when occupied. cAMP normally inhibits smooth-muscle contractility via a protein kinase A–dependent pathway (see Chapter 13·III·C). Thus, when the afferent arteriole binds adenosine, it constricts.

    b. Granular cells: Granular cells are specialized secretory cells that produce renin, a proteolytic enzyme that initiates the renin–angiotensin–aldosterone system ([RAAS] see Chapter 19·IV·C) by converting angiotensinogen to angiotensin I. Angiotensin I is subsequently converted by angiotensin-converting enzyme (ACE) to angiotensin II (Ang-II), which is vasoactive.

4. Efferent arteriole: The efferent arteriole expresses a type A_2 adenosine receptor, which increases intracellular cAMP levels when occupied. Adenosine binding causes efferent arteriolar dilation.

5. Regulation: Macula densa cell output and arteriolar response depends on whether tubule fluid flow rates are high or low (Fig. 25.11).

    a. High flow rates: When GFR is high, increased amounts of NaCl are delivered to the macula densa. The resulting membrane depolarization activates nonspecific cation channels in the cell membrane, causing Ca^{2+} influx. Because all cells in the JGA are coupled via gap junctions, when intracellular Ca^{2+} concentrations

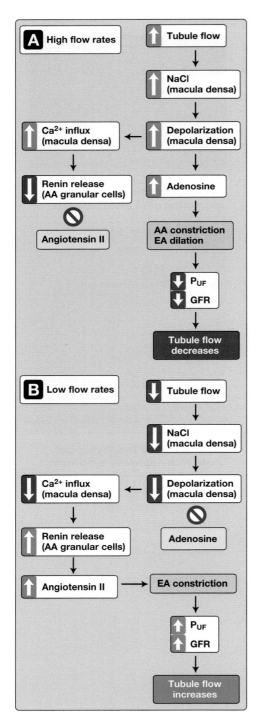

Figure 25.11.
Tubuloglomerular feedback. AA and EA =
afferent and efferent arteriole, respectively; GFR = glomerular filtration rate;
P_{UF} = capillary hydrostatic ultrafiltration
pressure.

rise in the macula densa, they rise in the renin-secreting granular cells of the afferent arteriole also. Ca^{2+} strongly inhibits renin
release. Na^+ and Ca^{2+} influx also cause macula densa adenosine
levels to rise. Adenosine acts as a paracrine that signals to the
afferent and efferent arterioles. The afferent arteriole constricts,
and the efferent arteriole dilates, reducing blood flow into the glomerular capillaries and facilitating outflow. Both effects depress
P_{GC}, and GFR falls (see Fig. 25.11A).

   b. Low flow rates: When GFR decreases, the amount of NaCl
   arriving at the macula densa falls. Macula densa cells hyperpolarize as a result, reducing Ca^{2+} influx via the nonspecific cation channel. Ca^{2+} concentrations in the granular cells fall also,
   thereby removing the brakes on renin release and allowing it to
   be deposited into the circulation (see Fig. 25.11B). Renin then
   activates RAAS, and circulating Ang-II levels rise. Ang-II is a
   potent vasoconstrictor in all vascular beds, but its effects on
   the glomerular arterioles are not equal. The efferent arteriole is
   more sensitive to Ang-II. The net effect is to limit outflow from
   the glomerular capillaries, causing P_{GC} and GFR to rise.

D. Paracrines

Adenosine is only one of several autoregulatory paracrines produced
by the kidney (Table 25.2). **Prostaglandins (PGs)** and **nitric oxide**
both dilate glomerular arterioles and increase RBF and GFR. They
may help offset intense Ang-II–mediated vasoconstriction during
circulatory shock, for example (see Chapter 40·IV·B·3). PG-induced
vasodilation becomes increasingly important for maintaining RBF
during development of chronic kidney disease, such that inhibiting
PG synthesis pharmacologically can precipitate acute renal failure.
Endothelins are local vasoconstrictors released in response to
Ang-II or when glomerular flow rates are damagingly high.

E. Angiotensin II

All of the components necessary for Ang-II formation and response
(including ACE) are inherent to the kidney, suggesting that the RAAS
constitutes a primary renal autoregulatory system that optimizes tubule
fluid flow through manipulation of renal plasma flow (RPF) and GFR.
However, one of the kidney's primary homeostatic functions is helping
control blood pressure, and Ang-II creates an important hormonal link
between blood pressure and renal function (see Chapters 19 and 28).

F. Central controls

The kidney governs total body water and Na^+ content, which, in turn,
determines blood volume and MAP. The kidney also receives ~20%
of cardiac output at rest, a significant blood volume that might be
used to sustain more critical circulations (e.g., cerebral and coronary
circulations) in the event of circulatory shock (see Chapter 40·IV·B).
Therefore, RBF is subject to control by the ANS, acting through neural
and endocrine pathways.

   1. **Neural:** Glomerular arterioles are innervated by noradrenergic
      sympathetic terminals that activate when MAP falls. Sympathetic

activation raises systemic vascular resistance by restricting blood flow to all vascular beds, including the kidneys. Mild sympathetic stimulation preferentially constricts the efferent arteriole, which reduces RBF while simultaneously maintaining GFR at sufficiently high levels to ensure continued kidney function. Intense sympathetic stimulation severely curtails blood flow through both glomerular arterioles, and urine formation ceases. In cases of severe hemorrhage, prolonged occlusion of arteriolar supply vessels can cause renal ischemia, infarction, and failure (see Chapter 40·IV·C).

2. **Endocrine:** Hormonal regulation of RBF is mediated principally by **epinephrine, atrial natriuretic peptide** (**ANP**), and Ang-II (see earlier discussion). Epinephrine is released into the circulation following sympathetic activation and stimulates the same pathways as does norepinephrine released from sympathetic nerve terminals. ANP is released from cardiac atria when they are stressed by high blood volumes. The ANP receptor has intrinsic guanylyl cyclase activity that dilates the afferent arteriole and increases RBF. It also relaxes mesangial cells to increase filtration barrier surface area. The net result is an increase in RBF and GFR and salt and water excretion (see Chapter 28).

Table 25.2: Glomerular Regulators

Stimulus	Mediator	Effects	
		RBF	GFR
Tubule flow rates			
↓ NaCl	Renin (Ang-II)	↑	↑
↑ NaCl	Adenosine	↓	↓
Sympathetic activation			
Mild	Norepinephrine	↓	None
Intense	Norepinephrine, epinephrine, renin (Ang-II)	↓	↓
↑ Blood volume	Atrial natriuretic peptide	↑	↑
Uncertain	Dopamine	↑	↑
Vascular endothelium			
↓ Flow?	Prostaglandins	↑	↑
Shear stress	Nitric oxide	↑	↑
Stress, trauma, vasoconstrictors	Endothelins	↓	↓
Inflammation	Leukotrienes	↓	↓

Ang-II = angiotensin II, GFR = glomerular filtration rate, RBF = renal blood flow.

V. ASSESSING RENAL FUNCTION

Patients with kidney disease can present with an array of symptoms, including hypertension, edema, and bloody urine (**hematuria**), or they may be asymptomatic. It is important to be able to assess GFR in these situations in order to narrow the range of possible pathologies. GFR cannot be measured directly, but filtration barrier health can be assessed from studies of plasma (renal) clearance.

A. Clearance

Clearance measures the kidney's ability to "clear" plasma of any given substance and then excrete it in urine. Plasma flows through the kidneys at ~625 mL/min. If the kidney could remove every last molecule of substance X (for example) from plasma during its encounter with a nephron, substance X clearance would be ~625 mL/min. In practice, such a high degree of clearance is not possible because the glomerulus filters only a fraction of the total amount of plasma that passes through the capillary network (~20%, or ~125 mL/min). However, it is still useful to know how much of the 125 mL/min does get cleared in a healthy individual because this parameter can be used to help diagnose problems with kidney function. Clearance is calculated as:

$$C_X = \frac{U_X \times V}{P_X}$$

where C_X is clearance of substance X (mL/min); U_X and P_X are urinary and plasma concentrations of X, respectively (mmol/L or mg/mL);

Example 25.1

A 35-year-old female is being evaluated for renal surgery. Her plasma creatinine concentration (P_{Cr}) is 0.8 mg/dL. A 24-hour urine collection has a creatinine concentration (U_{Cr}) of 90 mg/dL and a total volume (V) of 1,425 mL. What is her glomerular filtration rate (GFR)?

GFR can be estimated from creatinine clearance (C_{Cr}):

$$GFR = C_{Cr} = \frac{U_{Cr} \times V}{P_{Cr}}$$

Using values provided above:

$U_{Cr} = 90$ mg/dL $= 0.9$ mg/mL

$P_{Cr} = 0.8$ mg/dL $= 0.008$ mg/mL

$V = 1{,}425$ mL/24 hr $= 0.99$ mL/min

$$GFR = \frac{0.9 \times 0.99}{0.008} = 111.4 \text{ mL/min}$$

Example 25.2

A healthy 22-year-old male volunteers for a research study evaluating the effects of a new drug on renal blood flow (RBF). The protocol required a urinary catheter to measure kidney output while infusing *para*-aminohippuric acid (PAH) intravenously. PAH concentration (P_{PAH}) stabilized at 0.025 mg/mL. Urine flow rate (V) was then measured at 1.2 mL/min, and urine PAH concentration (U_{PAH}) was 18 mg/mL. Hematocrit (Hct) was 48%. What was the subject's RBF?

$U_{PAH} = 18$ mg/mL

$V = 1.2$ mL/min

$P_{PAH} = 0.025$ mg/mL

RBF is calculated from renal plasma flow (RPF) and Hct. RPF is calculated from PAH clearance (C_{PAH}):

$$C_{PAH} = \frac{U_{PAH} \times V}{P_{PAH}} = \frac{18 \times 1.2}{0.025}$$
$$= 864 \text{ mL/min} = RPF$$

RBF can now be calculated as

$$RBF = \frac{RPF}{1-Hct} = \frac{864}{0.52}$$
$$= 1{,}661.5 \text{ mL/min}$$

and V is urine flow rate (mL/min). In practice, clearance is usually measured over a 24-hour period to reduce urinary sampling errors (see following text).

> Clearance is defined as the plasma volume that is *completely* cleared of any given substance per unit time.

B. Glomerular filtration rate

If there were a substance that crossed the filtration barrier unhindered and then traversed the tubule without interference (i.e., no secretion, no reabsorption), we could use the rate of its appearance in urine to calculate GFR. One such substance is inulin, an inert fructose polymer synthesized by many plants. Inulin is infused intravenously to establish a known plasma concentration, and then its rate of appearance in urine is measured. GFR can then be calculated from:

$$GFR = C_{in} = \frac{U_{in} \times V}{P_{in}}$$

where C_{in} is inulin clearance; U_{in} and P_{in} are urinary and plasma concentrations of inulin, respectively; and V is urine flow.

> Inulin is the gold standard for filtration markers. Alternatives include radioactive iothalamate, iohexol, diethylenetriamine pentaacetic acid (DTPA), and the related ethylenediaminetetraacetic acid (EDTA).

C. Creatinine clearance

Inulin clearance is an expensive and cumbersome test to perform, so a preferred (albeit less accurate) alternative involves measuring creatinine clearance (Example 25.1). Creatinine is derived from creatine breakdown in skeletal muscle and is produced and excreted constantly. Creatinine filters freely from the glomerulus and is not reabsorbed by the tubule, but it is secreted. The PCT contains organic acid transporters ([OATs] see Chapter 26·IV·B) that secrete creatinine into the tubule, causing a 10% to 40% overestimate of GFR. An increase in plasma creatinine is generally an indication that GFR has decreased.

D. Renal plasma flow

Theoretically, if a substance could be identified that *was* completely cleared from plasma during a single pass (i.e., none leaves the kidney via the renal vein), similar techniques could be used to quantify RPF. There is no known substance, but ***para*-aminohippurate (PAH)** comes close. PAH is avidly removed from plasma and secreted into the renal tubule by OATs in the PCT epithelium. PAH clearance underestimates RPF by ~10% (i.e., 10% of the PAH that passes through the kidney escapes excretion). Although not used clinically,

PAF is used in research and in preclinical drug trials to estimate RBF before and after interventions:

$$RPF = C_{PAH} = \frac{U_{PAH} \times V}{P_{PAH}}$$

where C_{PAH} is PAH clearance (mL/min); U_{PAH} and P_{PAH} (mmol/L) are urinary and plasma concentrations of PAH, respectively; and V is urine flow (mL/min) as shown in Example 25.2.

Biologic Sex and Aging 25.1: Renal Function

Glomerular filtration rate (GFR) is estimated clinically (eGFR) using correction factors that recognize that GFR is lower in females compared with males (120 and 130 mL/min/1.73 m², respectively) and that renal function declines with age:

$$eGFR\ (mL/min/1.73\ m^2) = \frac{(140 - age\,[years]) \times weight\,[kg] \times 0.85\,(if\ female)}{72 \times P_{Cr}\,[mg/dL]}$$

A more recent iteration also considers the effects of race on eGFR (not shown).

The impact of normal age-related changes in renal function is so severe that normal eGFR values for healthy octogenarians overlap with those indicative of chronic kidney disease in a younger adult (60 mL/min/1.73 m²). In brief, renal blood flow (RBF) drops by 10% per decade after age 30 years, eGFR drops by 7.5–10 mL/min/ 1.73 m² per decade, and kidney weight falls by 25%–30% by age 80 years. Although aging does not significantly impact normal fluid and electrolyte balance, it does erode renal reserve and thereby limits the kidneys' ability to respond to fluid and electrolyte challenges.

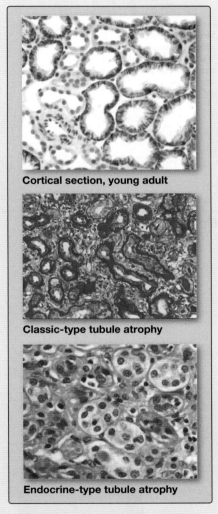

Cortical section, young adult

Classic-type tubule atrophy

Endocrine-type tubule atrophy

Age-related renal tubule atrophy.

- **Renal blood flow:** The decline in RBF is due to increased glomerular arteriolar resistance. RBF and GFR are maintained through a chronic rise in prostaglandin (PG) levels, meaning that cyclooxygenase inhibitors (e.g., nonsteroidal anti-inflammatory drugs) potentially can cause acute kidney injury in older adults. PGs are vasodilators (see Section IV·D).

- **Glomerular filtration rate:** Declining eGFR is due to a combination of declining RBF, focal segmental glomerulosclerosis (FSGS), and renal tubule atrophy.

  - FSGS first becomes apparent at around age 30 years. Initially, a few glomeruli contain segments in which cells have been replaced with collagen-rich extracellular matrix (ECM), which exposes the underlying basement membrane. The membrane in these regions then becomes inflamed, and it thickens. These changes progress until they involve the entire glomerulus, at which point, the occluded glomerular capillaries shrivel. Deprived of its glomerulus, the nephron degenerates and is replaced with ECM. Remaining glomeruli hypertrophy in response to the increased workload, but the podocytes cannot keep pace with the expanding filtration barrier. This leaves some regions denuded and exposed, triggering a chain of events that results in sclerosis. By age 80 years, ~75% of glomeruli are sclerotic, compared with ~3% at age 30 years.

  - GFR is estimated based on creatinine clearance, which is dependent both on RBF and tubule function. Aging is associated with renal tubule atrophy, in which the basement membrane thickens and may be duplicated, causing the lumen to narrow (classic atrophy; middle panel of figure). The tubule lumen fills with casts composed of hyaline, a mucoprotein secreted by the epithelium (Tamm-Horsfall mucoprotein). In another form of atrophy (endocrine atrophy; bottom panel), the tubule loses any segment-specific features, and the lumen is minimal.

- **Kidney weight:** At birth, a male kidney contains as many as 1,100,000 nephrons (females have ~15% fewer), and they are not replaced if damaged. FSGS gradually reduces this number after age 30 years, and kidney weight declines in parallel. Nephron loss occurs primarily within the renal cortex, which may then fill with cysts.

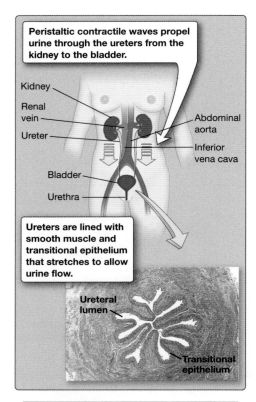

Peristaltic contractile waves propel urine through the ureters from the kidney to the bladder.

Kidney
Renal vein
Ureter
Abdominal aorta
Inferior vena cava
Bladder
Urethra

Ureters are lined with smooth muscle and transitional epithelium that stretches to allow urine flow.

Ureteral lumen
Transitional epithelium

Figure 25.12.
Ureters.

E. Renal blood flow and filtration fraction

Knowing RPF makes it possible to calculate RBF and filtration fraction (see Example 25.2). RBF is calculated from:

$$RBF = \frac{RPF}{1 - Hematocrit}$$

Hematocrit is the volume of blood occupied by red blood cells:

$$Filtration\ fraction = \frac{GFR}{RPF}$$

Filtration fraction measures the amount of plasma that filters into the Bowman space (usually ~0.2, or 20% of RPF).

VI. MICTURITION

The fluid flowing from the collecting ducts and entering the calyces is urine in its final form. No further modification occurs en route to or in the bladder. Urine is produced constantly and is stored in the urinary bladder until voided (**micturition**).

A. Ureters

Ureters convey urine from the kidneys to the bladder (Fig. 25.12). Ureters are lined with a transitional epithelium designed to stretch without tearing to accommodate intraluminal volume increases. The ureter wall contains circular and longitudinal smooth muscle layers. The muscles are stimulated to contract by waves of depolarization that originate in pacemaker regions of the calyces and renal sinus. The waves sweep down the ureters, triggering a peristaltic contraction that increases intraluminal pressure locally and drives urine toward the bladder. Wave propagation through the musculature is facilitated by gap junctions that electrically couple adjacent smooth muscle cells. Waves propagate at ~2 to 6 cm/s and typically recur several times per minute.

B. Bladder

The urinary bladder is a hollow, muscular organ comprising a large urine storage area (**body**) and a **neck** (or **posterior urethra**) that funnels urine to the **urethra** (Fig. 25.13).

1. **Body:** The bladder is lined on its interior surface by transitional epithelium. When the bladder is empty, the wall is thrown up into a series of ridges called **rugae** (see Fig. 25.13). The bladder wall is composed of three indistinct layers of bundled smooth muscle fibers known as **detrusor muscle**. The fibers within the layers are arranged in circular, spiral, or longitudinal fashion, so that they decrease bladder size and raise intraluminal pressure when stimulated to contract by the ANS.

2. **Valves and sphincters:** A full bladder develops considerable internal (**intravesical**) pressure, which potentially could force urine backward through the ureters. So, the ureters enter the bladder at an oblique angle, creating a valve that prevents ureteral reflux (Fig. 25.14). The neck of the bladder comprises a mix of detrusor muscle

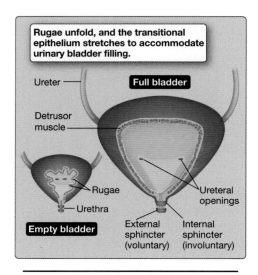

Rugae unfold, and the transitional epithelium stretches to accommodate urinary bladder filling.

Ureter
Full bladder
Detrusor muscle
Rugae
Urethra
Empty bladder
External sphincter (voluntary)
Internal sphincter (involuntary)
Ureteral openings

Figure 25.13.
Urinary bladder.

and elastic tissue that together form an **internal sphincter** controlled by the ANS. The sphincter remains contracted to prevent urine from entering the urethra until micturition. A second **external sphincter** surrounds the urethra below the bladder neck. The external sphincter is composed of skeletal muscle and is under voluntary control.

C. Innervation

The only part of the urinary system under voluntary control is the external sphincter (see Fig. 25.13). The sphincter is innervated by the **pudendal** nerve, which originates in the sacral spinal cord (S2–S4). The ureters and lower urinary tract (bladder, urethra, and internal sphincter) are all under ANS control. Efferents originate in spinal T11–L2 segments and travel to the urinary tract via the hypogastric nerve or descend in the paravertebral chain and then travel in the pelvic nerve. Sympathetic nervous system (SNS) activity relaxes detrusor muscle and constricts the bladder neck and urethra. Additionally, preganglionic fibers originating in the sacral spinal cord (S2–S4) travel in the pelvic nerve to the pelvic plexus and bladder wall. These fibers stimulate voiding by detrusor contraction and relaxation of the urethra and internal sphincter.

D. Spinal micturition reflex

Bladder capacity is ~500 mL. It fills passively, and the rugae unfold to accommodate the volume increase during this initial "**guarding**" phase (smooth muscle length adaptation; see also Chapter 13·IV). Filling occurs with minimal increase in intravesical pressure. Once bladder capacity reaches ~300 mL, the bladder wall begins to stretch, activating mechanoreceptors in the detrusor layers and urothelium (Fig. 25.15). Sensory afferents relay this information to the spinal cord

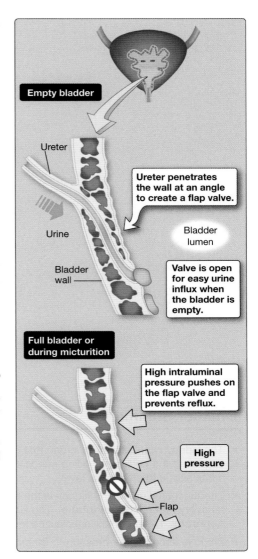

Figure 25.14.
Intravesicular ureteral valve preventing urine reflux.

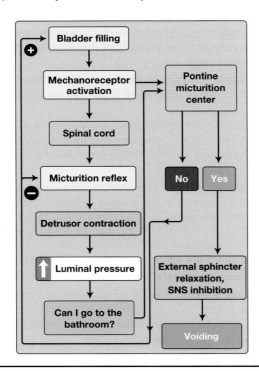

Figure 25.15.
Micturition reflex initiated by bladder filling. SNS = sympathetic nervous system.

via the pelvic and hypogastric nerves and initiate a reflex increase in parasympathetic nervous system (PSNS) efferent activity. The detrusor muscle contracts as a result, causing intravesical pressure to rise sharply and creating a sense of "urgency," with contraction frequency and discomfort level increasing with bladder volume. Sensory signals also travel rostrally to the brain. Central nervous system (CNS) control of micturition is complex and involves many different loci, including a **pontine micturition center (PMC)**, which is believed to coordinate brain output to the lower urinary system. If voiding is inconvenient, the pontine center suppresses the presynaptic PSNS nerves that stimulate bladder contraction. Meanwhile, tonic contraction of the external urinary sphincter prevents urine flow until relaxed voluntarily.

E. Voiding

Voiding begins with voluntary relaxation of the external sphincter. The PMC relaxes the internal sphincter and allows urine to enter the bladder neck and urethra. The PMC then activates PSNS outputs to the detrusor muscle and suppresses SNS outflow, and a sustained detrusor contraction begins. Mechanosensors in the urethra are stimulated by the presence of urine within the lumen, and their output to the PMC reinforces voiding. Detrusor contraction continues until the bladder is empty, although a small volume of urine (6–12 mL) typically remains after voiding is complete.

Chapter Summary

- The kidney is an **excretory** organ that cleanses blood of metabolic end products, toxins, water, and ions that may be surplus to the body's prevailing needs. It is also an **endocrine** and **homeostatic** organ controlling blood pressure, tissue osmolality, and electrolyte levels.

- The functional unit of the kidney is the **nephron**, which comprises a blood filtration module (**glomerulus**) and a filtrate recovery module (**renal tubule**). The kidney contains two types of nephrons: **superficial** (**cortical**) and **juxtamedullary nephrons**. The latter are specialized for formation of concentrated urine.

- Blood is forced under pressure (~50 mm Hg) through a **glomerular filtration barrier** to separate plasma from cells and proteins. The kidney receives ~20% of cardiac output and filters ~20% of the plasma it receives for a total of ~180 L/day (**glomerular filtration rate [GFR]**).

- GFR is a function of **glomerular capillary ultrafiltration pressure**. GFR is controlled by constriction and dilation of **afferent** and **efferent** glomerular arterioles. Afferent arteriolar dilation and efferent arteriolar constriction increase GFR. Afferent arteriolar constriction and efferent arteriolar dilation decrease GFR. Both arterioles are resistance vessels subject to multiple controls of local and central origin.

- The **juxtaglomerular apparatus (JGA)** is a functional complex comprising a section of the thick ascending limb (TAL) and the glomerular arterioles. The JGA is a sensory system that allows renal perfusion and filtration pressures to be modulated to stabilize tubule fluid flow. Flow is sensed by the **macula densa**, a specialized region of the tubule wall, via changes in luminal Na^+ and Cl^- concentrations. If flow is too high, signals from the macula densa cause afferent arteriolar constriction and a decrease in GFR. If flow is too low, the afferent arteriole releases **renin**, which activates the **renin–angiotensin–aldosterone system**. **Angiotensin II** constricts the efferent arteriole and raises GFR.

- Formed urine is channeled by **renal calyces** and the **renal sinus** into **ureters** and conveyed to the **urinary bladder**.

- The bladder stores urine until emptying (**micturition**) is convenient. Valves prevent urinary reflux into the ureters, whereas **inner** and **outer sphincters** control outflow via the **urethra**. The outer sphincter is under voluntary control, but inner sphincter and bladder contraction is controlled by spinal reflexes and the central nervous system (CNS).

- Bladder filling stretches its muscular wall and initiates a **spinal micturition reflex**. The reflex causes autonomic efferents to stimulate bladder contraction and emptying. Emptying is suppressed by tonic contraction of an external urethral sphincter until convenient. Voluntary relaxation of the external sphincter and reflex relaxation of the internal sphincter allows voiding to proceed.

- The efficiency of renal function can be assessed from plasma clearance of **inulin** and **creatinine**. **Clearance** refers to the amount of plasma completely cleared of a substance per unit time. Clearance of **para-aminohippurate** provides an estimate of renal plasma flow.

Study Questions

Choose the ONE best answer.

25.1. A patient taking penicillin for a bacterial infection presents with nausea and vomiting. Urinalysis reveals mild proteinuria and cell casts, suggestive of acute interstitial nephritis. Which of the following glomerular structures most likely prevents cells from entering the tubule normally?

A. Capillary endothelial cells
B. Glomerular basement membrane
C. Mesangial cells
D. Podocytes
E. Smooth muscle cells

Best answer = A. The glomerular filtration barrier comprises capillary endothelial cells, a basement membrane, and a filtration slit diaphragm located between podocyte foot processes (see Section III·A). Capillary endothelial cells are fenestrated, but the pores are small (~70 nm), effectively trapping the cells in the vasculature. The other barrier layers help prevent proteins from entering the tubule. Negative changes on the glomerular basement membrane (B) reflect proteins back into blood. Mesangial cells (C) are not directly involved in fluid filtration. A filtration slit diaphragm between adjacent podocyte (D) foot processes prevents proteins from entering the tubule. Smooth muscle cells (E) are located in glomerular arteriolar walls.

25.2. Studies on animal models have identified a gene that prevents a nonspecific cation channel that mediates NaCl sensing by macula densa cells from closing completely upon membrane hyperpolarization. What would be the most likely effect of expressing of such a gene in human subject?

A. Glomerular filtration rate would be insensitive to tubule flow.
B. Glomerular filtration rate would increase.
C. It would cause acute kidney injury.
D. It would cause loss of glomerular blood flow autoregulation.
E. The renin–angiotensin–aldosterone system would fail to activate.

Best answer = A. The juxtaglomerular apparatus (macula densa, glomerular arterioles, and mesangial cells) provides a mechanism by which glomerular perfusion pressure can be adjusted to optimize glomerular filtration rate (GFR) and tubule flow. The nonspecific cation channel mediates a Ca^{2+} influx that suppresses renin release and decreases GFR (B). When GFR is suboptimal, the channel closes and renin is released from afferent arteriolar granular cells to activate the renin–angiotensin–aldosterone system (RAAS) and increase GFR. A channel that permits Ca^{2+} influx even when GFR is low will uncouple GFR from tubule flow. Glomerular autoregulatory reflexes and the RAAS should be affected minimally (D, E) because they act independently of tubule flow. Acute kidney injury (C) is unlikely.

25.3. A 65-year-old male with a family history of nephrolithiasis presents with flank pain. A creatinine clearance assessment is performed. Creatinine clearance most likely equates with which of the following?

A. Amount of creatinine entering the urinary bladder per minute
B. Amount of creatinine traversing the glomerulus per minute
C. Plasma volume completely cleared of creatinine per minute
D. Renal blood flow
E. Renal plasma flow

Best answer = C. "Clearance" defines the kidneys' ability to remove a substance from blood over time (see Section V·A). Creatine clearance is used clinically to estimate glomerular filtration rate (see Section V·C). Clearance of other substances (e.g., para-aminohippuric acid) can be used to estimate renal plasma flow (E) and, if hematocrit is known, renal blood flow (D; see Section V·D). A change in clearance might affect how much creatinine enters the bladder, but excretion rate does not equate with clearance (A). Clearance is unrelated to the amount of a substance traversing the glomerular network per unit time (B).

25.4. A 17-year-old male presents with urethral burning following urination. He is asked to provide a urine sample and swabbed to test for a possible bacterial infection. Which of the following is most likely responsible for initiating micturition when providing a urine sample?

A. Internal urethral sphincter relaxation
B. Pontine micturition center
C. Rising intravesical pressure
D. Spontaneous detrusor contractions
E. Uroepithelial mechanoreceptors

Best answer = B. Voiding is initiated and coordinated by the pontine micturition center (PMC), which relaxes the internal (involuntary) urethral sphincter and facilitates detrusor muscle contraction once voluntary relaxation of the external urethral sphincter has occurred (see Section VI·D). Although internal sphincter relaxation (A) is required for urine flow, it does not initiate voiding. Rising intravesical pressure (C) during bladder filling triggers spontaneous detrusor contractions (D) in response to uroepithelial mechanoreceptor (E) stimulation, but bladder emptying is suppressed by the PMC until voiding is convenient.

26 Reabsorption and Secretion

I. OVERVIEW

The ultrafiltrate entering the **proximal tubule** (**PT**) from the Bowman space has a composition almost identical to that of plasma. It contains over 150 different components, but the major constituents are inorganic ions (Na^+, K^+, Mg^{2+}, Ca^{2+}, Cl^-, HCO_3^-, H^+, and phosphates), sugars, amino acids and peptides, creatinine, and urea. It also contains large amounts of water. The renal tubule's function is to recover >99% of the water and the majority of the solutes before they reach the bladder. Most is recovered within the first few millimeters of the PT, including virtually all organic compounds (sugars, amino acids, peptides, and organic acids) and two thirds of the filtered ions and water. Much of this material is recovered paracellularly by osmosis, made possible by the tubule wall's inherently leaky nature. The PT also actively secretes several organic compounds into the tubule lumen for subsequent urinary excretion. Principal sites for reabsorption, secretion, and regulation of various solutes along the nephron are summarized in Figure 27.19.

II. PRINCIPLES

The PT is a high-capacity, "leaky" transport epithelium drawn into a ~50-μm tube (Fig. 26.1A). The initial portion of the tube is coiled (proximal convoluted tubule [PCT]), and it then straightens to form the proximal straight tubule (PST). The PT's primary function is isosmotic fluid reabsorption.

> An epithelium's "leakiness" is a reflection of the ease with which solutes and water permeate the tight junctions between adjacent epithelial cells. Leaky epithelia are highly permeable, whereas intercellular junctions in tight epithelia are relatively impermeable (see Chapter 4·II·E·2).

A. Cellular structure

The PT reabsorbs ~120 L of fluid and solutes per day. The enormity of this load is reflected in the ultrastructure of the epithelial cells that make up its walls, which are packed with mitochondria, and their surface membranes are specialized to amplify surface area (see Fig. 26.1B).

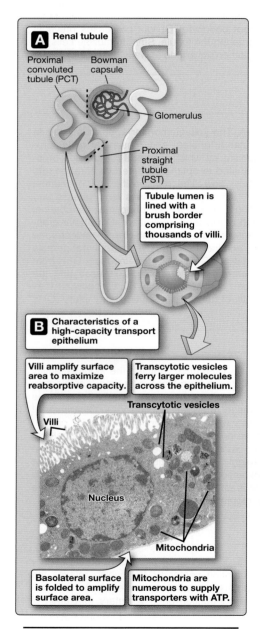

A Renal tubule

Proximal convoluted tubule (PCT)

Bowman capsule

Glomerulus

Proximal straight tubule (PST)

Tubule lumen is lined with a brush border comprising thousands of villi.

B Characteristics of a high-capacity transport epithelium

Villi amplify surface area to maximize reabsorptive capacity.

Transcytotic vesicles ferry larger molecules across the epithelium.

Transcytotic vesicles

Villi

Nucleus

Mitochondria

Basolateral surface is folded to amplify surface area.

Mitochondria are numerous to supply transporters with ATP.

Figure 26.1.
Proximal tubule structure.

1. **Metabolism:** The apical and basolateral membranes of PT epithelial cells are packed with channels and transporters for retrieval and secretion of inorganic ions and other solutes. Reabsorption is driven by ion gradients generated by basolateral ATP-dependent pumps, so the cytoplasm is dense with mitochondria to supply the PT's high metabolic needs.

2. **Surface area:** PT apical and basolateral membranes are extensively modified to increase their surface area. The membrane expanse is required to accommodate high numbers of channels and transporters and also to maximize area for contact between the epithelial cell and tubule contents. The numerous, densely packed microvilli that sprout from the apical surface create a **brush border** that is structurally and functionally similar to that found in the small intestine (see Chapter 31·II).

3. **Junctions:** Adjacent epithelial cells are connected apically by tight junctions to form a continuous sheet. PT tight junctions are highly permeable to solutes and water, so the epithelium has a very low electrical resistance.

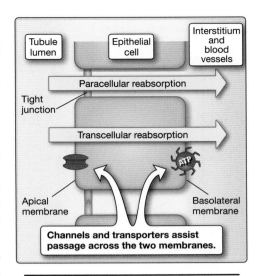

Figure 26.2.
Pathways for reabsorption from the tubule lumen.

B. Reabsorption

Reabsorption involves transferring water and solutes from the tubule lumen to the interstitium. Once in the interstitium, these materials are free to enter the peritubular capillary network by simple diffusion. There are two paths by which materials can cross epithelia (Fig. 26.2). The paracellular route lies between two adjacent epithelial cells. The permeability of the paracellular route is determined by tight junction structure. The transcellular route takes a solute through the inside of an epithelial cell and usually requires the assistance of channels or transporters to traverse the apical and basolateral membranes. The principal mechanisms involved in reabsorption are summarized in Unit I (see Chapter 4·III).

C. Peritubular network

Fluid and solute reabsorption by the PT are highly dependent on support from the peritubular capillary network, which closely follows the tubule through the kidney (see Fig. 25.4). The network sustains the tubule with O_2 and nutrients, but, just as importantly, it also clears recovered fluid from the interstitium before it has a chance to accumulate and reduce the gradients favoring reabsorption. The Starling forces governing fluid movement across the peritubular capillary wall are configured so as to promote reabsorption from the renal interstitium (Fig. 26.3; see also Chapter 18·VII·D). The main force favoring fluid reabsorption is plasma colloid osmotic pressure (π_{PC}). Capillary hydrostatic pressure (P_{PC}) is the principal force opposing reabsorption.

1. **Plasma colloid osmotic pressure:** π_{PC} averages 25 mm Hg in virtually all other regions of the body, but blood entering the peritubular network has just traversed the glomerulus where ~20% of plasma was removed by filtration. The plasma proteins are concentrated as a result, which raises π_{PC} to ~33 mm Hg.

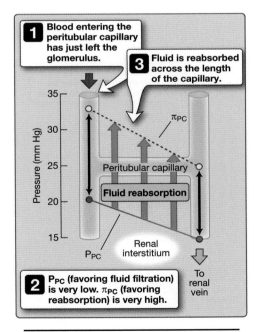

Figure 26.3.
Forces controlling fluid reabsorption by peritubular capillaries. P_{PC} = capillary hydrostatic pressure, π_{PC} = plasma colloid osmotic pressure.

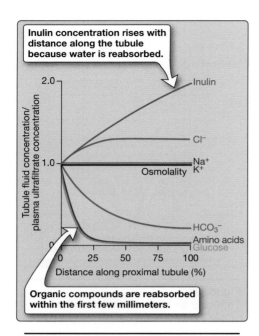

Figure 26.4.
Changes in proximal tubule fluid composition with distance from the Bowman capsule.

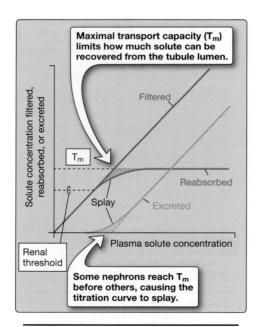

Figure 26.5.
Limits to transporter-mediated solute reabsorption.

2. **Capillary hydrostatic pressure:** Blood has to pass through an efferent arteriole before reaching peritubular capillaries. Efferent arterioles have a relatively high resistance, which decreases the pressure of blood entering the peritubular network to ~20 mm Hg. This is much lower than in other systemic capillary beds (~35 mm Hg). P_{PC} then declines over the length of the capillary. The combination of a high π_{PC} and low P_{PC} means that the driving force for fluid absorption is strongly positive across the entire length of the capillary (Flow $\propto \pi_{PC} - P_{PC}$, or ~15 mm Hg; see Fig. 26.3).

III. ORGANIC SOLUTES: REABSORPTION

Plasma is laden with glucose (~5 mmol/L), amino acids (~2.5 mmol/L), small peptides, and organic acids (e.g., lactate, pyruvate), all of which filter freely into Bowman space. These compounds represent a significant resource that must be recovered from the filtrate before it reaches the bladder. In practice, >98% of organic compounds are recovered in the early PCT and the remaining ~1% to 2% are reabsorbed in the PST (Fig. 26.4). Most organics are recovered by apical transporters, traverse cells by diffusion, and then are transported across the basolateral membrane to the interstitium and vasculature. Transporter involvement means that reabsorption shows saturation kinetics (Fig. 26.5).

A. Kinetics

The renal epithelium expresses a finite number of transporters, which limits solute reabsorption capacity. If the glomerulus filters solutes in excess of maximal transporter capacity (T_m), the excess continues through the tubule and appears in urine. Plasma solute concentrations vary with intake and tissue use, but a healthy nephron is usually well equipped to recover filtered loads within a normal physiologic range. Solutes start appearing in urine in small amounts even before T_m is reached (**renal threshold**; see Fig. 26.5). This region of the titration curve is said to show **splay**, reflecting transporter and nephron heterogeneity.

> "**Filtered load**" is the amount of any substance that filters from the glomerulus and enters the Bowman space per unit time (mmol/min or mg/min). Filtered loads are the products of glomerular filtration rate (GFR) and plasma concentration of the substance in question.

1. **Transporters:** Nephrons typically contain multiple transporter classes capable of transferring the same organic solute across the surface membrane. The combined activity of pathways with different T_m values contributes to splay (see Fig. 26.5).

2. **Nephrons:** Nephrons show anatomic diversity, which causes differences in single-nephron GFR, transporter capacity, and transporter location along the tubule. These differences also contribute to splay.

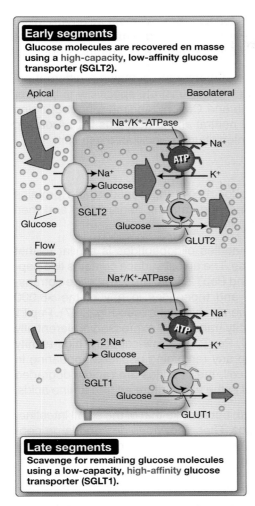

Early segments
Glucose molecules are recovered en masse using a high-capacity, low-affinity glucose transporter (SGLT2).

Late segments
Scavenge for remaining glucose molecules using a low-capacity, high-affinity glucose transporter (SGLT1).

Figure 26.6.
Glucose reabsorption strategies. SGLT1 and SGLT2 = sodium-dependent glucose cotransporter family members 1 and 2.

B. Glucose

Glucose plasma concentrations vary between ~3.8 and 6.1 mmol/L in a healthy person. Glucose filters freely into the tubule, and ~98% is reabsorbed in the early PT. Uptake occurs transcellularly and is transporter mediated (Fig. 26.6).

1. **Apical:** Glucose is recovered using two different Na$^+$ cotransporters. They both harness the transmembrane Na$^+$ gradient to simultaneously reabsorb Na$^+$ and a glucose molecule. One of these cotransporter classes localizes primarily to the early part of the PCT, the other to the PST.

    a. **Convoluted tubule:** The PCT expresses a high-capacity, low-affinity Na$^+$-glucose cotransporter (SGLT family; SGLT2) that recovers the bulk of filtered glucose immediately after it enters the tubule.

    b. **Straight tubule:** The PCT reabsorbs all but a small fraction of the glucose-filtered load. The PST expresses a high-affinity, low-capacity 2 Na$^+$-glucose cotransporter (SGLT1), designed to recover the remaining glucose before it enters the nephron loop.

2. **Basolateral:** Glucose uptake by the epithelial cells generates a concentration gradient that drives facilitated diffusion (see Chapter 1·V·C·2) via GLUT family glucose transporters (GLUT2 and GLUT1 in the PCT and PST, respectively) across the basolateral membrane to the interstitium.

C. Amino acids

Plasma contains all of the common amino acids, and all are filtered into the renal tubule. The early PT recovers >98% of the filtered amino acid load (see Fig. 26.4). The amount filtered approaches T_m even under resting conditions, so urine always contains trace amounts of most amino acids. Physiologic increases in plasma amino acid levels easily overwhelm the nephron reabsorptive capacity, and significant amounts are then excreted. There are multiple pathways for amino acids to cross the apical and basolateral membranes.

1. **Apical:** There are several classes of amino acid transporter in the apical membrane. They generally have broad substrate specificity, so a single species of amino acid may have several recovery options. Anionic (acidic) amino acids are recovered by an excitatory amino acid transporter (EAAT3) that exchanges H$^+$, two Na$^+$,

---

Clinical Application 26.1: Diabetes Mellitus

Plasma glucose concentration can rise to ~10 mmol/L before renal reabsorptive capacity is exceeded in normal, healthy individuals. Once transporter capacity is exceeded, significant quantities of glucose begin to spill over into the urine. The presence of unrecovered glucose within the renal tubule lumen causes an osmotic diuresis, manifesting as polyuria (urine output of >3 L/d). The frequent need to urinate gives rise to the term "diabetes," which is derived from a Greek verb (*diabainein*) having a similar meaning. The presence of glucose in urine gives it a sweet taste, providing a ready (albeit somewhat distasteful) means of diagnosing **diabetes mellitus** in the early days of medicine.

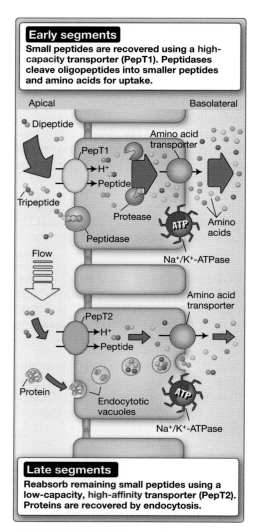

Early segments

Small peptides are recovered using a high-capacity transporter (PepT1). Peptidases cleave oligopeptides into smaller peptides and amino acids for uptake.

Late segments

Reabsorb remaining small peptides using a low-capacity, high-affinity transporter (PepT2). Proteins are recovered by endocytosis.

Figure 26.7.
Oligopeptide and protein reabsorption.

and an amino acid for K^+. Cationic (basic) amino acids are taken up in exchange for a neutral amino acid. Neutral amino acids are taken up either by a Na^+ cotransporter or a H^+ cotransporter.

2. **Basolateral:** The basolateral membrane contains a different set of amino acid transporters whose substrate specificity is broader than that of the apical membrane. Cationic and many neutral amino acids are exchanged for a neutral amino acid plus Na^+. Aromatic amino acids cross to the interstitium by facilitated diffusion.

D. Peptides and proteins

The PT has three strategies for recovering peptides and proteins (Fig. 26.7): uptake via small-peptide carriers, degradation and then uptake via carriers, and endocytosis.

1. **Transport:** How the PT handles oligopeptides is similar to that described earlier for glucose. The apical surface contains two peptide transporters: PepT1 and PepT2. Both are H^+-peptide cotransporters that transport di- and tripeptides in any of the >8,000 possible amino acid residue combinations (see Fig. 26.7). PepT1 is a low-affinity, high-capacity transporter expressed preferentially in the early part of the PT. PepT2 is a low-capacity, high-affinity transporter that scoops up any remaining peptides appearing in the PST. Once inside the cell, the peptides are rapidly degraded by proteases and returned to the vasculature as free amino acids.

2. **Degradation:** The PT brush border, like that of the small intestine, expresses many peptidases. These enzymes degrade large peptides (including hormones) into small peptides or their constituent amino acids to be reabsorbed.

3. **Endocytosis:** PT epithelial cells also internalize numerous proteins by receptor-mediated endocytosis (see Fig. 26.7). This pathway is largely responsible for recovering filtered albumin, such that defects in this pathway can lead to significant proteinuria. Albumin binds to **megalin** (a member of the low-density lipoprotein receptor family) and **cubilin** (a vitamin B_{12}–intrinsic factor receptor also found in the terminal ileum). These two proteins form a complex with **amnionless**, a trafficking protein. Once internalized, most recovered proteins are targeted to lysosomes, where they are digested and then released on the basolateral side as free amino acids or small peptides. The PT also expresses receptors that recognize and internalize specific hormones such as somatostatin. The pharmaceutical industry has been exploring the possibility of using these receptors as vehicles for drug delivery.

E. Organic acids

Plasma contains significant quantities of lactate; pyruvate; and other mono-, di-, and tricarboxylates that are freely filtered by the glomerulus and then reabsorbed by the PT using two different Na^+ cotransporters. One is specific for monocarboxylates (e.g., lactate, pyruvate), and the other for di- and tricarboxylates (e.g., citrate, succinate). Monocarboxylates then exit the cell via a basolateral H^+/carboxylate cotransporter. Di- and tricarboxylates are exchanged for an organic anion by a member of the **organic anion transporter (OAT)** family.

IV. ORGANIC SOLUTES: SECRETION

Blood that has traversed the glomerular capillary network still contains metabolic end products that are undesirable and, possibly, toxic. Although these waste products would eventually be excreted during subsequent passes, the kidney supplements its passive filtration and cleansing functions with an active secretory process. Secretion occurs in the late PT and is almost 100% effective in ridding the body of several organic anions and cations in a single pass. Uric acid, for example, is a relatively insoluble end product of nucleotide metabolism actively secreted by the PT. Other secreted waste products include creatinine, oxalate, and bile salts. Secretion also helps clear exogenous toxins from the body. The secretory transporters have a very broad substrate specificity, which allows them to handle a wide array of potential chemical threats. These pathways also clear a wide range of pharmaceutical drugs from the vasculature (Table 26.1).

> The tendency for the PT to take up pharmaceuticals from the circulation via organic anion and other transporters puts it at grave risk because intracellular concentrations can quickly rise to toxic levels. Therefore, the transporters responsible for uptake have themselves become high-priority targets for pharmaceutical intervention. Inhibiting the uptake systems not only reduces drug toxicity but also decreases the rate of drug elimination from the body, thus reducing dosing frequency.

A. Kinetics

Secretion is transporter mediated and, therefore, shows saturation kinetics, as demonstrated using *para*-aminohippurate (PAH) in Figure 26.8. PAH is a hippuric acid derivative used in studies of renal plasma flow (see Chapter 25·V·D) that is both filtered from the glomerulus and secreted from the PT via the OAT pathways mentioned earlier.

1. **Filtration:** PAH is freely filtered by the glomerulus in amounts directly proportional to GFR (i.e., ~20% of total plasma PAH).

2. **Secretion:** Blood entering the peritubular network still contains 80% of the original arterial PAH load. All but 10% is taken up by transporters in the late PT basolateral membrane and then secreted into the tubule lumen. PAH excretion rises accordingly. Transporter capacity is finite, however, so the secretion curve flattens and plateaus as plasma PAH concentration approaches T_m (see Fig. 26.8). The secretion curve exhibits splay due to transporter and nephron heterogeneity, as discussed earlier in reference to glucose reabsorption.

B. Transporters

PT epithelia express a number of broad-specificity transporters for organic anions and cations. Organic anions are taken up from the blood by several members of the OAT family. OAT1 exchanges an organic ion

Table 26.1: Drugs Secreted by the Proximal Tubule

Drug Name	Drug Class
Anions	
Acetazolamide	Diuretic, various
Chlorothiazide	Diuretic
Furosemide	Diuretic
Probenecid	Uricosuric
Penicillin	Antimicrobial
Methotrexate	Anticancer
Indomethacin	Anti-inflammatory
Salicylate	Anti-inflammatory
Saccharin	Sweetener
Cations	
Amiloride	Diuretic
Quaternary ammonium compounds	Antimicrobial
Quinine	Antimalarial
Morphine	Analgesic
Chlorpromazine	Antipsychotic
Atropine	Cholinergic antagonist
Procainamide	Antiarrhythmic
Dopamine	Pressor
Epinephrine	Pressor
Cimetidine	Antacid (H_2 blocker)
Paraquat	Herbicide

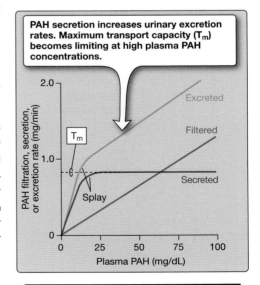

PAH secretion increases urinary excretion rates. Maximum transport capacity (T_m) becomes limiting at high plasma PAH concentrations.

Figure 26.8.
Effects of plasma *para*-aminohippurate (PAH) concentration on secretion and excretion rates.

Clinical Application 26.2: Gout

Organic anion transporters are one of several transporter families involved in reabsorption and excretion of uric acid. Most mammals metabolize urate to allantoin, but primates lost the necessary enzyme (uricase) during evolution. Unlike allantoin, urate is relatively insoluble, and, when blood concentrations rise, it forms spiked, rod-shaped crystals that are commonly deposited in joints. The result is a painful inflammatory arthritis known as "gout," which typically affects the great toe (podagra). Uric acid crystals demonstrate birefringence when viewed under polarized light, which can prove diagnostic. Gout treatment options include drugs that inhibit the transporters that normally reabsorb urate as it passes down the tubule, thereby increasing excretion rates.

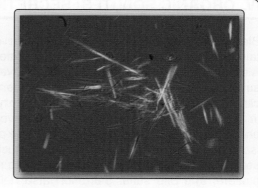

Birefringent uric acid crystals in synovial fluid from a patient with gout.

for a dicarboxylate such as α-ketoglutarate. A related family of **organic cation transporters** takes up amine and ammonia compounds from blood. Anions and cations are then both extruded into the tubule lumen by one of several **multidrug-resistant proteins** (**MRPs**). MRPs are members of the ATP-binding cassette pump superfamily that includes the cystic fibrosis transmembrane conductance regulator. Organic anions can also cross the apical membrane by one of several OATs.

V. UREA

Urea is a small organic molecule comprising two amide groups joined by a carbonyl group. It is formed in the liver[1] and excreted in urine as a way of disposing of unwanted amino acids and nitrogen (Fig. 26.9). Normal plasma concentrations average 2.5 to 6.0 mmol/L. The PT reabsorbs ~50% of the filtered load, largely via the paracellular route. Two forces drive movement. The first is solvent drag, created by the large volumes of water being reabsorbed in the PT. Loss of water from the tubule lumen secondarily concentrates solutes in the tubule lumen, which enhances the driving force for urea diffusion across the epithelium. The kidney ultimately excretes ~30% to 50% of the filtered urea load, but it first serves an important role in helping concentrate urine. The pathways involved are discussed in Chapter 27 (see Section V·D).

> Urea is the principal means by which nitrogenous wastes are excreted from the body, so plasma urea levels are a useful indicator of renal health and function. Clinical laboratories cite urea levels in the form of blood urea nitrogen (BUN). Normal BUN values are in the range of 1.2–3.0 mmol/L (7–18 mg/dL).

[1]The role of urea in nitrogen excretion and details of the urea cycle are discussed at length in *LIR Biochemistry*, 8e, Chapter 19·V.

Figure 26.9.
Urea formation.

Figure 26.9 diagram

Dietary protein
↓ Proteases
Amino acid surplus
↓ — α-Keto acids
Aminotransferases
↓
Aspartate + Glutamate
↓ Glutamate dehydrogenase
NH₃ + α-Ketoglutarate
↓
Urea cycle
↓
Urea

VI. PHOSPHATE AND CALCIUM

Plasma contains a total of ~1.0 to 1.5 mmol/L inorganic phosphorus (P_i) and ~2.1 to 2.8 mmol/L Ca^{2+}. Both are critically important for normal cell function. P_i is a component of RNA and DNA, powers metabolism in the form of ATP, and associates with numerous lipids and proteins. Ca^{2+} is a vital second messenger that activates enzymes, initiates muscle contraction, and triggers neurotransmitter secretion. About half of total plasma phosphorus and calcium exists in ionized form (as HPO_4^{2-}, $H_2PO_4^-$, and Ca^{2+}), the remainder being complexed with proteins and other molecules. Plasma contains only a tiny fraction of total body phosphorus and calcium, however. The vast majority of phosphorus (>80%) and calcium (>99%) is locked in hydroxyapatite crystals in a mineral vault called bone. Plasma P_i and Ca^{2+} concentrations are regulated by parallel mechanisms. Total body concentrations represent a precise balance between bone deposition and resorption, intestinal secretion and absorption, and renal filtration and reabsorption. All three processes are regulated by parathyroid hormone ([PTH] discussed in more detail in Chapter 35).

A. Phosphate

The kidney tubule reabsorbs ~90% of the filtered P_i load, of which ~80% is reclaimed in the PT and the remaining 10% in the distal convoluted tubule (DCT). The PT is the principal site of P_i regulation, effected through PTH and plasma P_i concentrations (see Fig. 27.19).

1. **Reabsorption:** P_i is reabsorbed using three apical Na^+/P_i cotransporters. The most important of these is Na^+/P_i IIc, which exchanges two Na^+ ions for HPO_4^{2-}. Na^+/P_i IIa exchanges three Na^+ ions for HPO_4^{2-}, whereas PiT2 exchanges two Na^+ ions for $H_2PO_4^-$. The route by which P_i crosses the basolateral membrane is uncertain.

2. **Regulation:** PTH blocks P_i recovery from the tubule lumen by promoting endocytosis and subsequent degradation of Na^+/P_i IIa and Na^+/P_i IIc (Fig. 26.10). In the absence of a recovery pathway, P_i passes through the tubule and is excreted. If dietary P_i intake is restricted, Na^+/P_i cotransporters are inserted into the apical membrane, thereby facilitating P_i reabsorption.

B. Calcium

Plasma-free Ca^{2+} concentrations are tightly regulated in the range of 1.0 to 1.3 mmol/L, and virtually all filtered Ca^{2+} is reabsorbed by the nephron (see Fig. 27.19). The PT recovers ~65%, largely via the paracellular route. The motive force is partly solvent drag and, in the later stages of the PT where the lumen is positively charged with respect to blood, the transepithelial voltage difference. Most of the remaining 35% of filtered load is reabsorbed in the thick ascending limb ([TAL] ~25%) and the DCT (~8%). The DCT is the main site of Ca^{2+} regulation (see Chapter 27·III·C).

> In the early regions of the PT, the lumen is negatively charged with respect to the renal interstitium. In the later regions of the PT, the transepithelial voltage gradient reverses polarity to become lumen-positive. Reversal occurs because Cl^- is reabsorbed preferentially in the later regions, leaving behind a net positive charge.

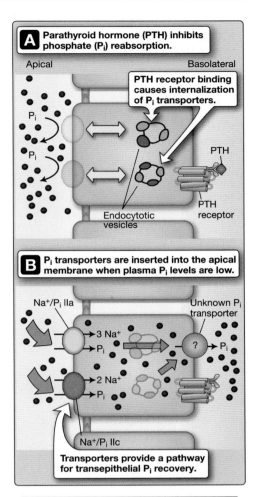

Figure 26.10.
Regulation of phosphate reabsorption.

VII. MAGNESIUM

Mg^{2+} is a vital cofactor required for the normal function of hundreds of enzymes, its positive charge helping stabilize protein structural integrity. It also regulates ion flow through ion channels, so physiologic decreases in plasma-free concentrations cause membrane hyperexcitability, arrhythmias, and muscle tetany. The majority of total body Mg^{2+} is complexed in bone or associated with proteins and other small molecules. Plasma concentrations are normally maintained in a range of ~0.75 to 1.00 mmol/L, of which ~60% is in the free form. Mg^{2+} is a common ingredient in most foods, so ~2% to 5% of the filtered load typically is excreted in urine to balance daily intake. The PT recovers ~15% of the filtered load. Reabsorption occurs paracellularly by solvent drag and diffusion. Reabsorption is favored by the small lumen-positive potential difference that exists across the more distal regions of the PT epithelium. The bulk of filtered Mg^{2+} (~70%) is recovered in the TAL. The principal site of Mg^{2+} homeostatic regulation is the early distal tubule (see Chapter 27·III·B and Fig. 27.19).

VIII. POTASSIUM

K^+ is unique among electrolytes in that even modest changes in plasma K^+ concentrations can be life threatening, causing potentially fatal cardiac dysrhythmias and arrhythmias (see Clinical Application 2.1). Plasma concentrations are tightly regulated within the range of 3.5 to 5.0 mmol/L. K^+ is filtered freely across the glomerulus, so the nephron handles a daily load of ~0.9 mol (35 g). The PT reabsorbs ~80% of the filtered load, primarily via the paracellular route (Fig. 26.11). As is the case for Ca^{2+} and Mg^{2+}, absorption occurs through solvent drag and by diffusion that is enhanced by a transepithelial voltage gradient. Another 10% is recovered in the TAL (see Chapter 27·II·B), but regulation of K^+ reabsorption (and excretion) occurs primarily in the distal segments (see Chapter 27·IV·C and Fig. 27.19).

IX. BICARBONATE AND HYDROGEN IONS

One of the kidney's most important functions is to help maintain extracellular fluid (ECF) pH at ~7.40. Metabolism generates immense quantities of volatile acid (H_2CO_3) that is expelled via the lungs and another ~50 to 100 mmol/d of nonvolatile acid (sulfuric, phosphoric, nitric, and other minor acids; see Chapter 3·IV·A) that must be excreted by the kidneys. Although all portions of the nephron are involved in acid–base homeostasis to some degree (see Fig. 27.19), the PT is a principal site for HCO_3^- recovery and H^+ secretion.

A. Bicarbonate

Excreting HCO_3^- causes the ECF to become acidic, so the first goal of pH homeostasis is to recover 100% of the filtered HCO_3^- load. The PT recovers ~80% of total. Because HCO_3^- is anionic, it cannot diffuse freely across membranes, so the PT secretes molar amounts of H^+ into the tubule lumen to titrate the HCO_3^- and then uses carbonic anhydrase (CA) to convert the H_2CO_3 to CO_2 and H_2O. Both molecules

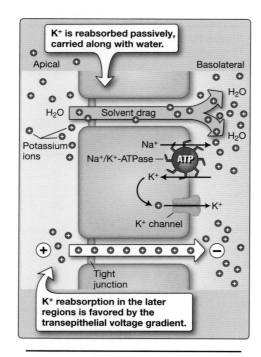

Figure 26.11.
Potassium reabsorption pathways in the proximal tubule.

are then recovered by simple diffusion. Reclamation is a four-step process (numbers below correspond to steps shown in Fig. 26.12):

Step 1: H$^+$ is transported into the tubule lumen by an apical Na$^+$/H$^+$ exchanger (NHE3). The exchange is powered by the transmembrane Na$^+$ gradient.

Step 2: H$^+$ combines with luminal HCO$_3^-$ to form H$_2$CO$_3$, which dissociates to form H$_2$O and CO$_2$. The reaction is catalyzed by CA-IV, which is expressed on the epithelium's apical surface:

$$HCO_3^- + H^+ \leftrightharpoons \underset{CA}{H_2CO_3} \leftrightharpoons CO_2 + H_2O$$

Step 3: CO$_2$ diffuses into the cell and combines with H$_2$O to reform HCO$_3^-$ and H$^+$. The reaction is catalyzed by intracellular CA-II.

Step 4: HCO$_3^-$ is reabsorbed across the basolateral membrane to the interstitium and then into the vasculature, largely via an electrogenic Na$^+$/HCO$_3^-$ cotransporter (NBCe1). A Cl$^-$/HCO$_3^-$ exchanger (AE1) assists HCO$_3^-$ reabsorption in the later part of the PT. H$^+$ is pumped back into the tubule lumen to repeat the reabsorption cycle.

HCO$_3^-$ reabsorption causes a slight acidification of the tubule contents, from pH 7.4 at the glomerulus to ~pH 6.8 in the late PT.

> Acetazolamide is a CA inhibitor that blocks HCO$_3^-$ and Na$^+$ reabsorption by the PT, causing diuresis. The drug acts on both the apical (CA-IV) and intracellular form (CA-II) of the enzyme. As a class, the CA inhibitors are relatively ineffective as diuretics because distal regions of the tubule compensate for their effects on PT function.[1] The main indication for CA inhibitor use is in patients with metabolic alkalosis because the drugs impair the tubule's ability to reabsorb HCO$_3^-$, thereby causing excess base to be excreted in urine.

B. Hydrogen ions

The PT is a principal site for H$^+$ secretion, although final determination of urine pH and regulation of ECF pH occurs in the distal segments (see Chapter 27·V·E). H$^+$ is secreted by the NHE3 Na$^+$/H$^+$ exchanger mentioned earlier and by a H$^+$ pump (Fig. 26.13).

1. **Sodium–hydrogen ion exchange:** The NHE3 Na$^+$/H$^+$ exchanger uses the Na$^+$ gradient created by the basolateral Na$^+$/K$^+$-ATPase to power H$^+$ secretion. The dependence on the Na$^+$ gradient means that its ability to *concentrate* H$^+$ in the lumen is limited, but it has a very high *capacity* that accounts for ~60% of net H$^+$ secretion in the PT.

[1]For more information on acetazolamide use, see *LIR Pharmacology*, 8e, Chapter 9·VI.

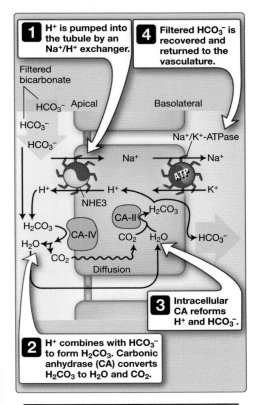

Figure 26.12.
Bicarbonate reabsorption pathway in the proximal tubule.

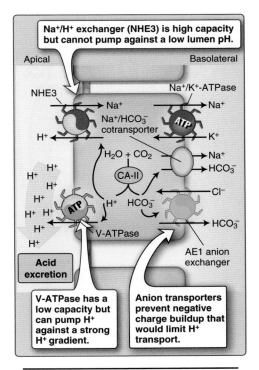

Figure 26.13.
Acid secretion by the proximal tubule.
CA-II = carbonic anhydrase II.

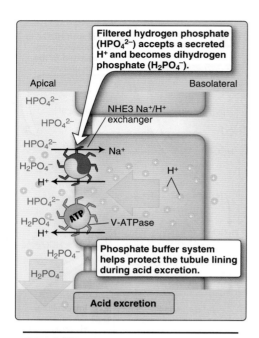

Figure 26.14.
Phosphate buffer system.

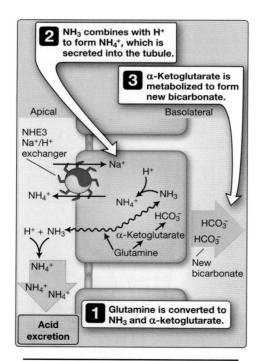

Figure 26.15.
Excreting acid in the form of the ammonium ion.

The NHE3 exchanger is also a principal pathway by which the PT recovers Na^+ from the tubule lumen (see following text).

2. **Proton pump:** The PT also actively secretes H^+ into the tubule using a vacuolar-type H^+ pump (V-ATPase). The V-ATPase accounts for ~40% of net secretion in the PT and can establish a strong H^+-concentration gradient across the apical membrane. The pump is **electrogenic**, meaning that it causes a negative charge to build within the cell. This charge can become limiting to further transport, so H^+ secretion is balanced by HCO_3^- movement across the basolateral membrane via a Na^+/HCO_3^- cotransporter and an anion exchanger (see Fig. 26.13).

C. Nonvolatile acid

Ideally, the H^+ excess created by nonvolatile acid formation would be transported to the kidney and then dumped into the tubule and excreted without further ado. In practice, the amount of nonvolatile acid generated is large, and the ability of available H^+ transporters to pump H^+ against a concentration gradient is limited. The V-ATPase mentioned earlier can create a lumen pH of ~4.0 at best (i.e., 0.1 mmol/L H^+), which is insufficient to handle the daily acid excess. Two different workarounds have evolved to allow H^+ to be excreted in the quantities required to maintain pH balance. The first is to simultaneously excrete urinary buffers (**titratable acids**) that limit a rise in free H^+ concentration even as acid is being pumped into the tubule lumen. The second is to attach H^+ to ammonia (NH_3) and excrete it as an ammonium ion (NH_4^+).

1. **Titratable acids:** The plasma filtrate contains several buffers, and the PT secretes several more. These include hydrogen phosphate (pK = 6.8), urate (pK = 5.8), creatinine (pK = 5.0), lactate (pK = 3.9), and pyruvate (pK = 2.5). Collectively, these buffers are known as "titratable acids" that complex with and, thereby, limit rises in tubule H^+ concentration. Hydrogen phosphate's pK makes it a more effective urinary buffer than the other titratable acids. Hydrogen phosphate accepts H^+ to become dihydrogen phosphate (Fig. 26.14):

$$H^+ + HPO_4^{2-} \leftrightarrows H_2PO_4^-$$

The PT reabsorbs ~80% of filtered phosphate, with the remaining 20% helping buffer lumen pH during nonvolatile H^+ excretion.

2. **Ammonia:** Plasma does not normally contain NH_3, but PT cells are able to synthesize it from glutamine, converting it to NH_3 and α-ketoglutarate. NH_3 is lipid soluble, so it readily diffuses out of the epithelial cell into the tubule lumen where it combines with H^+ to form NH_4^+. Some NH_4^+ is formed inside the PT cells and moved into the tubule lumen by the Na^+/H^+ exchanger, which is able to bind NH_4^+ in place of H^+ (Fig. 26.15). Unlike NH_3, NH_4^+ is charged and not lipid soluble, which traps it within the tubule

lumen ("**diffusion trapping**"). Some NH_4^+ is reabsorbed in more distal regions of the tubule (see Chapter 27·V·E), but the remainder remains trapped until excreted in urine.

3. **New bicarbonate:** Excreting ~50 to 100 mmol of nonvolatile acid generated every day creates a sizable deficit in the body's buffer systems. This must be matched precisely by the formation of new buffer, or ECF would rapidly become acidotic. Excreted buffer is replaced by generation of "new" HCO_3^-. Some is formed *de novo*, and some is created from α-ketoglutarate after NH_3 is formed from glutamine. α-Ketoglutarate is metabolized to glucose and then to CO_2 and H_2O. CA then catalyzes H_2CO_3 formation, which dissociates to yield HCO_3^- and H^+. The newly formed HCO_3^- diffuses into blood and is ultimately used to buffer nonvolatile acid at its formation site within tissues.

X. SODIUM, CHLORIDE, AND WATER

Plasma Na^+ concentration is maintained at between ~136 and 145 mmol/L, primarily as a way of controlling how water distributes between the three body compartments (intracellular, interstitial, and plasma; see Chapter 3·III·B). Na^+ moves freely across the glomerular filtration barrier, so the daily filtered load exceeds 25 mol (575 g). Approximately 99.6% of the filtered load is reabsorbed during passage through the renal tubule, the bulk (~67%) being recovered by the PT (see Fig. 27.19). Cl^- follows Na^+ across the epithelium, driven inward by sodium's positive charge. Reabsorption of Na^+, Cl^-, and organic solutes creates a strong osmotic potential that also drives water from the tubule lumen toward the interstitium. The net effect of these and all the other reabsorptive and secretory processes described in the previous sections is that the fluid reabsorbed by the PT is isosmotic and has a composition that resembles plasma. There are regional differences in the way that Na^+ and Cl^- are reabsorbed between the early and late regions of the PT, however.

A. Early proximal convoluted tubule

Early PT epithelial cells are specialized to recover virtually all useful organic solutes and HCO_3^- in association with Na^+, which leads to significant transcellular Na^+ reabsorption. Some of this Na^+ then leaks backward paracellularly (Fig. 26.16).

1. **Transcellular:** The primary force driving reabsorption is the basolateral Na^+/K^+-ATPase, which establishes a Na^+ gradient that drives Na^+-coupled glucose, amino acid, organic acid, and phosphate reabsorption from the tubule. Large quantities of Na^+ also enter cells via the NHE3 Na^+/H^+ exchanger. Na^+ is then moved to the interstitium by the Na^+/K^+-ATPase and, to a lesser degree, by a basolateral Na^+/HCO_3^- cotransporter. Cotransport is driven by high intracellular HCO_3^- concentrations following reabsorption and *de novo* synthesis.

2. **Paracellular:** The cotransporters that recover organic solutes from the plasma filtrate are electrogenic, leaving an excess of negative charges in the tubule lumen. These charges create a ~3-mV difference between tubule and interstitium, which creates

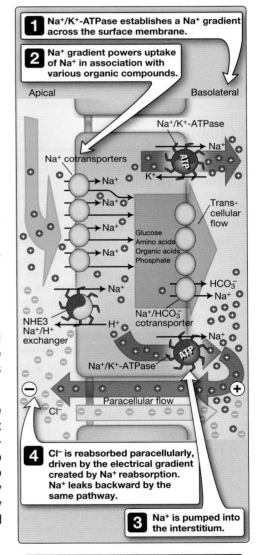

1 Na^+/K^+-ATPase establishes a Na^+ gradient across the surface membrane.

2 Na^+ gradient powers uptake of Na^+ in association with various organic compounds.

Apical   Basolateral

Na^+/K^+-ATPase

Na^+ cotransporters

Na^+
Na^+
Na^+
Glucose
Amino acids
Organic acids
Phosphate
Na^+

K^+

Trans-cellular flow

HCO_3^-
Na^+

NHE3
Na^+/H^+
exchanger
H^+
Na^+/HCO_3^-
cotransporter

Na^+/K^+-ATPase

Na^+

Paracellular flow
Cl^-

4 Cl^- is reabsorbed paracellularly, driven by the electrical gradient created by Na^+ reabsorption. Na^+ leaks backward by the same pathway.

3 Na^+ is pumped into the interstitium.

Figure 26.16.
Pathways for Na^+ reabsorption and backflow in the early proximal convoluted tubule.

a significant force that drives paracellular Cl^- reabsorption. The paracellular route also permits significant amounts of reabsorbed Na^+ (~30%) to leak backward from the interstitium to tubule lumen. Movement is driven by the voltage gradient.

B. Proximal straight tubule

The fluid entering the PST has been stripped of most useful organic solutes and HCO_3^- but contains relatively high concentrations of Cl^-. Na^+ and Cl^- are reabsorbed via both transcellular and paracellular routes.

1. **Transcellular:** The late PT takes up Na^+ in exchange for H^+, which creates a transcellular Na^+ flux. This region of the PT also contains a Cl^--base exchanger (CFEX) that allows for significant transcellular Cl^- uptake. CFEX exchanges Cl^- for formate, oxalate, OH^-, or HCO_3^-.

2. **Paracellular:** High luminal Cl^- concentrations drive diffusion of Cl^- out of the lumen via the paracellular route. This leaves an excess of positive charge in the lumen that favors Na^+ reabsorption, so Na^+ follows Cl^- across the tight junctions and into the interstitium.

Chapter Summary

- The proximal tubule (PT) recovers ~67% of the fluid and up to 100% of some solutes that are filtered into the renal tubule by the glomerulus. PT epithelial cells possess apical **microvilli** that increase surface area, and the junctions between cells are leaky to maximize free flow of water and dissolved solutes.

- The PT reabsorbs fluid **isosmotically. Transcellular absorption** is powered mainly by the transmembrane Na^+ gradient established by a basolateral Na^+/K^+-ATPase. Reabsorption also occurs by diffusion across tight junctions (**paracellular absorption**), aided by bulk fluid movement (**solvent drag**).

- Reabsorbed fluid is returned to the vasculature via the **peritubular network**. Blood reaches the peritubular capillaries by way of the glomerulus. Glomerular filtration concentrates the plasma proteins, so **plasma colloid osmotic pressure** is high. By the time blood enters the peritubular network, **capillary hydrostatic pressure** is low. These features combine to create a strong motive force for fluid uptake from the interstitium, which facilitates reabsorption.

- The PT recovers almost 100% of filtered **glucose** and **amino acids**, principally via Na^+ cotransport. The PT also recovers **small peptides** by H^+ cotransport. Larger peptides and proteins are degraded to small peptides and are then reabsorbed or taken up by **endocytosis**.

- The PT actively secretes **organic acids**, **toxins**, and **drugs** using **organic anion** or **cation transporters** or **multi-drug-resistant proteins**.

- Phosphate is recovered from the PT by **Na^+/phosphate cotransporters**. Reabsorption is regulated by **parathyroid hormone**. Ca^{2+} reabsorption by the PT occurs paracellularly.

- Mg^{2+} reabsorption by the PT is minimal (~15% of filtered load) and occurs paracellularly.

- Approximately 80% of the filtered K^+ load is recovered in the PT.

- The lungs and kidneys together are responsible for maintaining the pH of ECF within a narrow range (pH 7.35–7.45). Lungs excrete the daily load of **volatile acids** (CO_2) generated during metabolism. Kidneys excrete **nonvolatile acids** (sulfuric, phosphoric, nitric, and other minor acids).

- **pH homeostasis** begins in the PT with recovery of 80% of filtered HCO_3^-, the body's primary pH buffer. Excretion of nonvolatile acid requires that buffers be excreted also to control luminal free H^+ concentration. The primary **urinary buffers** are **phosphate** and **ammonium**, the latter newly synthesized from glutamine in the PT.

- Na^+ reabsorption by the PT is driven by the basolateral Na^+/K^+-ATPase through cotransport with organic solutes and in exchange for H^+. Cl^- absorption occurs principally in the late PT by the paracellular route or by a Cl^--base exchanger. Water reabsorption occurs by osmosis, driven by influx of Na^+, Cl^-, and solutes.

Study Questions

Choose the ONE best answer.

26.1. A 31-year-old male with a body mass index of 35 kg/m² is found to have glycosuria during a routine physical. Elevated urinary glucose levels correlate with unmanaged type 2 diabetes mellitus. Which of the following most likely explains why glucose appears in the urine of patients with untreated diabetes?

A. Glucose causes an osmotic diuresis.
B. High plasma insulin inhibits Na⁺/K⁺-ATPases.
C. High plasma insulin levels are nephrotoxic.
D. Hyperglycemia downregulates glucose transporters.
E. Tubule glucose levels exceed transport capacity.

> Best answer = E. Transporters exhibit saturation kinetics, which limits the tubule's ability to reabsorb solutes (see Section III·A). Although glucose transport maximum (T_m) is seldom reached in a healthy individual, the plasma ultrafiltrate of patients with untreated diabetes may contain glucose levels that exceed the tubule's reabsorptive capability, causing it to appear in urine. Glucose can cause an osmotic diuresis (A), but such an event would be a consequence of exceeding T_m, not the cause. Insulin does modulate the Na⁺/K⁺-ATPase, but it increases pump activity rather than inhibiting it (B). Possible effects of hyperglycemia on insulin-induced nephrotoxicity (C) and on transporter numbers (D) are not of significant physiologic concern.

26.2. As a part of a research study, proximal tubule (PT) function in patients with Fanconi syndrome is compared with a control group of healthy volunteers. Fanconi syndrome is associated with PT dysfunction. Symptoms include polyuria, glycosuria, hypocalcemia, hypomagnesemia, and hypophosphatemia. Which of the following filtered solutes is most likely to be fully recovered by the PT in the control group?

A. Ca^{2+}
B. Na^+
C. Peptides
D. PO_4^{3-}
E. Uric acid

> Best answer = C. The PT reabsorbs a large percentage of most materials filtering from blood, including Ca^{2+}, Na^+, and PO_4^{3-}, but it reabsorbs close to 100% of proteins, peptides, amino acids, and glucose (see Section III). The PT recovers 65% of Ca^{2+} (A), the remainder being recovered in the thick ascending limb and distal segments. The PT recovers 67% of Na^+ (B), although this amount can increase in the presence of angiotensin II. The PT recovers 80% of the PO_4^{3-} filtered load (D), with the remainder being recovered distally. The PT secretes uric acid (E), oxalate, and other wastes (see Section IV).

26.3. A 22-year-old male develops acute mountain sickness during a Breckenridge skiing vacation. He is prescribed a carbonic anhydrase (CA) inhibitor to stimulate a metabolic acidosis to help alleviate the symptoms. The drug most likely blocks which of the following?

A. Conversion of HPO_4^{2-} to $H_2PO_4^-$
B. H⁺-ATPase activity
C. H⁺/Na⁺ exchange
D. HCO_3^- reabsorption
E. Volatile acid excretion

> Best answer = D. CA is an enzyme that catalyzes H⁺ and HCO_3^- formation from H_2O and CO_2. In the kidney, CA facilitates HCO_3^- reabsorption from the tubule lumen. CA is not involved in the phosphate buffer system (A) and CA inhibitors have no direct effect on the H⁺-ATPase (B) that secretes H⁺ into the tubule lumen. CA has no physiologic effect on H⁺/Na⁺ exchange (C). Ca^{2+} facilitates excretion of volatile acid (H_2CO_3) in the lungs, not the kidneys (E).

26.4. A pharmaceutical company is investigating the influence of transepithelial voltage on drug clearance by the proximal tubule. Electrical recordings suggest that the tubule transitions from being lumen-negative in early sections of the tubule to lumen-positive in the proximal straight tubule. This transition is most likely due to which of the following?

A. HCO_3^- reabsorption
B. Na⁺/glucose exchange
C. Organic anion secretion
D. Paracellular Cl⁻ reabsorption
E. Solvent drag

> Best answer = D. Reabsorption of Na^+ in the early proximal tubule (PT) leaves the lumen negative. Cl⁻ has become relatively concentrated by the late PT, creating a strong electrochemical gradient for paracellular Cl⁻ reabsorption (see Section X·B). Cl⁻ reabsorption leaves the lumen positively charged with respect to the interstitium. HCO_3^- is reabsorbed by titrating with secreted H⁺ to yield CO_2 and H_2O (A; see Section IX·A). Glucose is reabsorbed by Na⁺/glucose cotransport, not exchange (B; see Section III·B). Organic anion secretion (C) would increase lumen negativity. Solvent drag (E) refers to isosmotic reabsorption of water and any dissolved solutes via the paracellular pathway (see Chapter 4·III·C). Solvent drag does not establish voltage gradients.

27 Urine Formation

I. OVERVIEW

The fluid leaving the **proximal tubule** (**PT**) and entering the **loop of Henle** (**nephron loop**) has been stripped of almost all useful organic molecules, such as glucose, amino acids, and organic acids. The residual fluid (~60 L/d) comprises water, inorganic ions, and excretory products. The function of the loop and distal nephron segments is to recover remaining useful components (principally water and electrolytes) before the fluid reaches the bladder and is excreted as urine. The amount of fluid and electrolytes recovered is determined by homeostatic needs and is heavily regulated (see Chapter 28; the main sites of water and solute recovery and regulation are summarized in Fig. 27.19). The first step is to begin extracting water. One way to achieve this would be to pump water out of the tubule, much as bailing a waterlogged boat. Nature has yet to devise the cellular equivalent of a bilge pump, however, so, as an alternative, the tubule contents are forced to run an osmotic gauntlet (two- to fourfold > plasma) created within the renal medulla expressly for the purpose of extracting water from the tubule lumen. The tubule contents are exposed to the osmotic torments of the medulla twice before finally being deposited in the bladder. The first trip involves passage around the loop of Henle.

Note that the corticopapillary osmotic gradient is established by juxtamedullary nephrons (see Fig. 25.5), which are the focus of this chapter. Superficial nephrons are important contributors to the transport processes discussed previously (see Chapters 25 and 26) but contribute little to the osmotic gradient.

II. LOOP OF HENLE

The loop of Henle comprises three sections: a **descending thin limb** (**DTL**), an **ascending thin limb** (**ATL**), and a **thick ascending limb** (**TAL**), as shown in Figure 27.1. Thin limb function is very simple: it conveys fluid down through the inner reaches of the medulla and exposes it to the corticopapillary osmotic gradient (see Chapter 25·II·B). Water and solutes exit and reenter passively during the fluid's passage. The ATL transitions gradually at the inner–outer medullary junction to become the TAL. The increasing wall thickness reflects an abundance of mitochondria and other cellular machinery required to support the activity of numerous ion pumps. The TAL establishes the corticopapillary gradient.

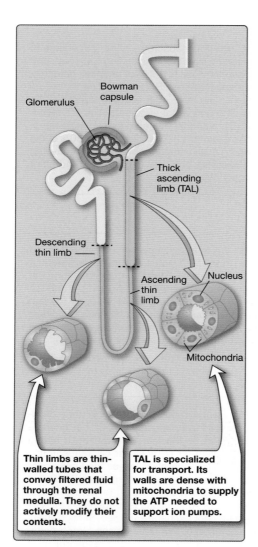

Thin limbs are thin-walled tubes that convey filtered fluid through the renal medulla. They do not actively modify their contents.

TAL is specialized for transport. Its walls are dense with mitochondria to supply the ATP needed to support ion pumps.

Figure 27.1.
Loop of Henle tubule structure.

A. Thin limbs

The PT ends abruptly at the border between outer and inner stripes of the outer medulla (see Fig. 25.5). The DTL and ATL are both composed of thin epithelial cells with a few stubby microvilli. Adjacent cells are extensively coupled by wide tight junctions. Water and solutes move across the tubule wall (transcellularly and paracellularly) passively, driven by a pronounced interstitial corticopapillary osmotic gradient, although selectivity of passage is regulated and changes from one section to the next (see Fig. 26.2).

1. **Corticopapillary gradient:** The osmotic gradient is established within the medullary interstitium by a **countercurrent multiplication** mechanism (described in Section C). Cortical osmolality approximates that of plasma (~290–300 mOsmol/kg H_2O) but increases progressively with distance toward the papillary tips (Fig. 27.2). The gradient's magnitude varies according to the body's need to conserve or excrete water (**diuresis**). When water conservation is necessary, papillary tip osmolality may increase to ~1,200 mOsmol/kg H_2O, whereas, during hypervolemic conditions, tip osmolality may be closer to 600 mOsmol/kg H_2O.

2. **Descending thin limb:** The DTL is relatively impermeable to solutes, but the epithelial cell membranes contain aquaporins (AQPs) that allow free passage of water (see Fig. 27.2A). Water exits the tubule by osmosis as the fluid is carried deeper into the medulla, causing luminal Na^+ and Cl^- to become progressively more concentrated. About 27 L of water is recovered from the DTL per day, or 15% of glomerular filtrate.

3. **Ascending thin limb:** The tubule epithelium transitions at the turn of the loop from being water permeable to water impermeable (the ATL does not express AQPs), which prevents further water movement until the tubule contents reach the collecting ducts ([CDs] see Fig. 27.2B). ATL epithelial cells *are* permeable to Cl^-, however. Cl^- leaves the tubule lumen during fluid's passage back up to the cortex, driven by the transepithelial electrochemical gradient. Na^+ follows Cl^- paracellularly.

> Forcing fluid around the loop of Henle extracts water but does not increase its osmolality because solutes are extracted also. Urine only becomes concentrated when exposed to the corticopapillary gradient a second time, which occurs during passage through the CDs.

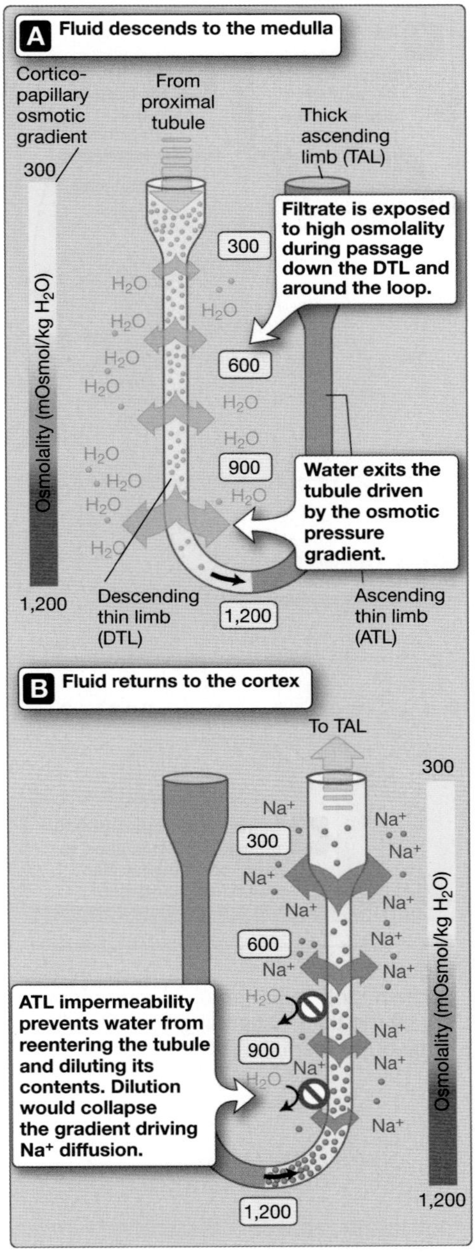

Figure 27.2.
Water and Na^+ reabsorption in the loop of Henle.

B. Thick ascending limb

The TAL actively recovers significant amounts of Na^+, Cl^-, K^+, Ca^{2+}, and Mg^{2+} from the tubule lumen (summarized in Fig. 27.19).

1. **Sodium, chloride, and potassium:** The TAL reabsorbs ~25% of the filtered load of Na^+ and Cl^- and 10% of the K^+ load. Reabsorption occurs via both transcellular and paracellular routes and is so efficient

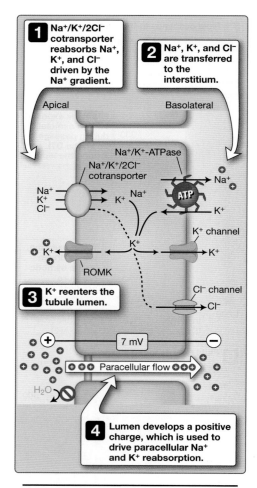

Figure 27.3.
Sodium, potassium, and chloride recovery by the thick ascending limb. ROMK = renal outer medullary K+ channel.

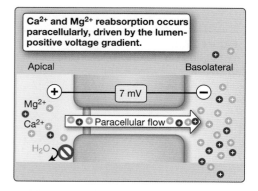

Figure 27.4.
Calcium and magnesium recovery by the thick ascending limb.

Clinical Application 27.1: Loop Diuretics

Physiologic regulation and fine-tuning of urine composition occurs in segments distal to the thick ascending limb (TAL), but drugs that inhibit the Na+/K+/2Cl− cotransporter have proved to be very powerful clinical tools for treating edema. As a class, these drugs are known as **loop diuretics** and include furosemide, bumetanide, ethacrynic acid, and torsemide.[1] Inhibiting the cotransporter prevents reabsorption of Na+, Cl−, and K+ directly and indirectly prevents reabsorption of water. Inhibition also prevents a positive charge developing within the tubule lumen, thereby reducing Ca^{2+} and Mg^{2+} reabsorption. Segments distal to the TAL do not have the ability to compensate for loss of cotransporter function, so the loop diuretics all cause copious urine formation. Although reduced salt and water retention effectively reduces circulating blood volume and helps prevent edema, the concomitant loss of K+ and Mg^{2+} to urine can cause hypokalemia and hypomagnesemia.

that it leaves the tubule contents hyposmotic relative to plasma, even though no net water movement has occurred. The TAL is sometimes referred to as the **diluting segment** for this reason.

a. **Transcellular:** Reabsorption is facilitated by the transmembrane Na+ gradient generated by the basolateral Na+/K+-ATPase. Ion reabsorption can be broken down into several steps (Fig. 27.3).

   i. **Sodium, potassium, and chloride reabsorption:** An apical Na+/K+/2Cl− cotransporter (NKCC2, an *SLC12A1* gene product) facilitates Na+, K+, and Cl− reabsorption (see Fig. 27.3[1]). Na+ is pumped out of the cell by the Na+/K+-ATPase, whereas K+ and Cl− flow into the interstitium down their respective electrochemical gradients via basolateral Cl− and K+ channels (see Fig. 27.3[2]).

   ii. **Potassium secretion:** The apical membrane also contains a renal outer medullary K+ channel (ROMK), which allows K+ to cross back to the tubule lumen (see Fig. 27.3[3]). This pathway is necessary to prevent luminal K+ depletion, which would bring NKCC2 activity to a halt.

b. **Paracellular:** K+ secretion creates an ~7-mV electrical gradient between tubule lumen and interstitium that drives paracellular Na+ and K+ reabsorption (see Fig. 27.3[4]).

2. **Calcium and magnesium:** The TAL reabsorbs ~25% of the filtered Ca^{2+} load and ~65% to 70% of filtered Mg^{2+}. Most of this reabsorption occurs paracellularly (see Clinical Application 4.2) and is driven by the voltage difference between the tubule lumen and the interstitium (Fig. 27.4). TAL epithelial cells express a basolateral Ca^{2+}-sensing receptor (CaSR) that modulates this voltage difference. Hypercalcemia (or hypermagnesemia) activates CaSR, and the transepithelial voltage gradient weakens. Reabsorption of Ca^{2+} and Mg^{2+} is reduced as a result.

 [1]For more information on the mechanism of action and use of loop diuretics, see *LIR Pharmacology*, 8e, Chapter 9·IV.

3. **Bicarbonate and acid:** Fluid leaving the PT still contains ~20% of the filtered HCO_3^- load. Virtually all of this is recovered, either in the TAL or in the distal segments.

   a. **Bicarbonate:** HCO_3^- is reabsorbed using the same strategies seen in the PT (see Fig. 26.12). Carbonic anhydrase (CA) facilitates H^+ and HCO_3^- formation from H_2O and CO_2. H^+ is pumped across the apical membrane by a H^+-ATPase and a Na^+/H^+ exchanger, where it combines with filtered HCO_3^- to form CO_2 and H_2O, the reaction again catalyzed by CA. HCO_3^- is reabsorbed across the basolateral membrane in exchange for Cl^- and via a Na^+/HCO_3^- cotransporter.

   b. **Acid:** The PT generates NH_3 as a way of excreting H^+ in the form of NH_4^+ (see Fig. 26.15). The TAL reabsorbs a portion of the NH_4^+ via NKCC2 (NH_4^+ substitutes for K^+) and then transfers it to the interstitium, where, like Na^+, it helps form the corticopapillary osmotic gradient through countercurrent multiplication.

C. Corticopapillary osmotic gradient

Loop diuretics are effective because they collapse the corticopapillary osmotic gradient used to draw water from the DTL and to later concentrate urine during its passage through the CDs. The gradient is established by the TAL, but it affects all vessels traveling through the medulla.

1. **Tubule arrangement:** Textbook figures (e.g., see Fig. 25.5) traditionally separate the various nephron segments across the width of a page to make labeling easier, but, in real life, the DTL, ATL, CDs, and vasa recta are all bundled together like a handful of drinking straws (Fig. 27.5). The interstitial space between them is minimal, so the interstitium and contents of most tubules are in osmotic equilibrium. Changes in one compartment affect the others almost instantaneously. The fact that some tubules (e.g., DTLs) carry fluid down toward the papilla at the same time as other tubules (e.g., ATLs) in the bundle carry fluid back up to the cortex allows for amplification of an osmotic difference between tubule lumen and interstitium generated by TAL epithelial cells.

2. **Countercurrent multiplication:** The corticopapillary gradient is easiest to understand when broken down into a series of theoretical steps. Prior to multiplication, tubule contents and the interstitium are all assumed to be in equilibrium at 300 mOsmol/kg H_2O (Fig. 27.6[1]).

   a. **Single effect:** The TAL reabsorbs Na^+ from the tubule via the NKCC2 cotransporter and transfers it to the interstitium using the basolateral Na^+/K^+-ATPase. This transfer generates a maximal 200-mOsmol/kg H_2O osmolality difference between the tubule lumen and interstitium (see Fig. 27.6[2]). Thus, if interstitial and tubule osmolality are both initially at 300 mOsmol/kg H_2O, Na^+ reabsorption causes lumen osmolality to fall to 200 mOsmol/kg H_2O and interstitial osmolality

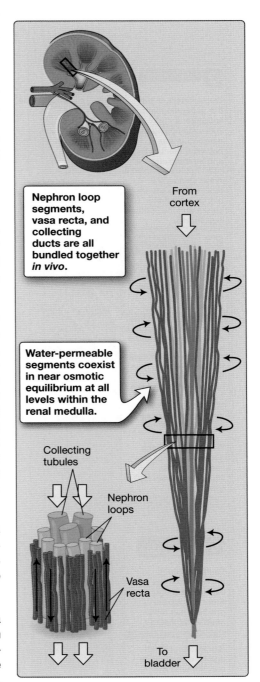

Nephron loop segments, vasa recta, and collecting ducts are all bundled together *in vivo*.

From cortex

Water-permeable segments coexist in near osmotic equilibrium at all levels within the renal medulla.

Collecting tubules

Nephron loops

Vasa recta

To bladder

Figure 27.5.
Arrangement of tubule segments and vasa recta in renal medulla.

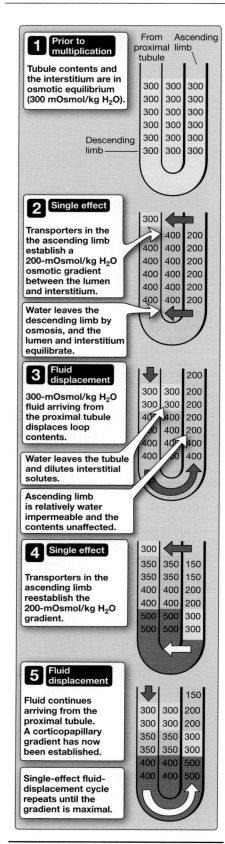

Figure 27.6.
Countercurrent multiplication in the loop of Henle.

to rise to 400 mOsmol/kg H_2O. The DTL, which lies next to the TAL, is filled with fluid arriving from the PT, which has an osmolality of 300 mOsmol/kg H_2O. Because the DTL is highly water permeable, water is drawn from its lumen by the osmotic pressure gradient until it equilibrates with the interstitium at 400 mOsmol/kg H_2O (the reabsorbed water is subsequently carried away by the peritubular vasculature). This same phenomenon occurs simultaneously down the length of the TAL and DTL and is known as a "**single effect.**"

   b. **Fluid displacement:** Fluid continues arriving at the DTL from the PT, displacing the 400-mOsmol/kg H_2O fluid downward and around the loop tip (see Fig. 27.6[3]). The interstitium at the corticomedullary junction reequilibrates at 300 mOsmol/kg H_2O. The TAL is water impermeable, so the fluid within remains at 200 mOsmol/kg H_2O. At the loop tip, the two limbs and interstitium remain equilibrated at 400 mOsmol/kg H_2O.

   c. **Single effect:** TAL cells continue transferring Na^+ from tubule lumen to interstitium, but lumen osmolality begins this cycle at 200 mOsmol/kg H_2O (see Fig. 27.6[4]). Na^+ reabsorption reestablishes the 200-mOsmol/kg H_2O gradient across the tubule wall, causing lumen osmolality to fall to 150 mOsmol/kg H_2O and interstitial osmolality to rise to 350 mOsmol/kg H_2O. Further down the TAL toward the medulla, lumen osmolality drops from 400 mOsmol/kg H_2O to 300 mOsmol/kg H_2O, and interstitial osmolality rises to 500 mOsmol/kg H_2O. Even after only two conceptual cycles, a corticopapillary gradient has begun to form. Each cycle multiplies the gradient further.

   d. **Fluid displacement:** Fluid with an osmolality of 300 mOsmol/kg H_2O continues to arrive at the DTL from the PT, decreasing the local interstitial osmolality and pushing high-osmolality fluid around the tip of the loop (see Fig. 27.6[5]). The next transport cycle reduces the tubule osmolality at the top of the TAL and further increases osmolality at the tip.

3. **Urea:** Countercurrent multiplication eventually generates a papillary osmolality of 600 mOsmol/kg H_2O, but this can rise to 1,200 mOsmol/kg H_2O when water must be conserved. The gradient's magnitude determines how much water can be reclaimed from the filtrate and is regulated according to prevailing needs. Achieving a 1,200-mOsmol/kg H_2O gradient is only possible with assistance from urea. When water conservation is necessary, the CDs allow urea to pass from the duct lumen to the medulla, further enhancing its osmolality and water reabsorptive capabilities. The pathways involved are described in Section V.

D. Vasa recta

The nephron loops require extensive vascular support not only to supply O_2 and nutrients but also to carry away reabsorbed water and electrolytes. Because plasma has an osmolality of ~300 mOsmol/kg H_2O and capillaries are inherently leaky vessels, there is a risk that blood entering the medulla could wash out the corticopapillary osmotic gradient and prevent urine concentration. Washout is largely prevented by two important features of the vasa recta (Latin for "straight

The corticopapillary osmotic gradient is created by juxtamedullary nephrons, which represent a relatively small proportion of total nephron number (~10%). The remaining ~90% are superficial and have short loops, which limits the maximal degree to which urine can be concentrated. Desert rodents such as the Australian hopping mouse (Genus *Notomys*; Fig. 27.7) can produce ~10,000-mOsmol/kg H_2O urine. Their kidneys contain a much higher proportion of juxtamedullary nephrons compared with superficial nephrons, and concentrating ability is increased accordingly. This remarkable ability to conserve fluid means that hopping mice are able to subsist on water extracted from their food (e.g., roots, leaves, and berries) and never need to drink, which has clear survival advantages in an arid environment.

Figure 27.7.
Australian hopping mouse.

vessels"). Flow rate is slow, and the vessels form a hairpin loop that creates a **countercurrent exchange system** (Fig. 27.8).

1. **Flow rate:** The medulla receives <10% of total renal blood flow. The vasa recta has a high intrinsic resistance due to its length, which keeps flow to a nutritional minimum. The slow flow rate allows for near-complete equilibration of water and solutes as the blood is carried through the medulla.

2. **Countercurrent exchange:** The vasa recta is intimately associated with the nephron loop and the CDs, closely paralleling the DTL down to the papilla and then back up to the cortex alongside the ATL and TAL (see Figs 25.4 and 27.5). The blood vessels are leaky, so water leaves and solutes enter passively, which maintains an osmotic equilibrium between blood and interstitium (see Fig. 27.8). On the way back up to the cortex, water reenters the blood vessels, and solutes leave passively. Therefore, blood flow through the vasa recta has minimal net effect on the corticopapillary gradient when perfusion rates are low.

III. EARLY DISTAL TUBULE

The transition from the TAL to the distal convoluted tubule (DCT) is marked by a fivefold increase in tubule wall thickness. The epithelial cells are filled with platelike structures packed with mitochondria. The apical surface bears slender microvilli, and the basolateral membrane is folded, both modifications designed to increase surface area. These anatomic features all point to the early DCT as being the site of active solute reabsorption. The DCT has very low water permeability and is the principal site for homeostatic regulation of Mg^{2+} and Ca^{2+}.

A. Sodium and chloride

The early DCT reabsorbs only a small fraction of the Na^+ and Cl^- filtered load, primarily via an apical Na^+/Cl^- cotransporter ([NCC] encoded by *SLC12A3*). Na^+ is then pumped out of the cell to the interstitium by the basolateral Na^+/K^+-ATPase, whereas Cl^- exits via a Cl^- channel. The DCT is impermeable to water, so extracting NaCl from the tubule lumen further dilutes its contents.

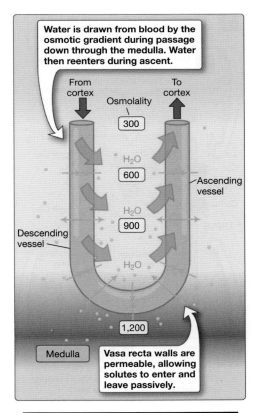

Water is drawn from blood by the osmotic gradient during passage down through the medulla. Water then reenters during ascent.

From cortex

To cortex

Osmolality

300

H_2O

600

Ascending vessel

Descending vessel

H_2O

900

H_2O

1,200

Medulla

Vasa recta walls are permeable, allowing solutes to enter and leave passively.

Figure 27.8.
Countercurrent exchange in the vasa recta. Osmolality values are in mOsmol/kg H_2O.

Clinical Application 27.2: Thiazide Diuretics

Thiazide diuretics (e.g., hydrochlorothiazide) inhibit Na^+ reabsorption by the distal convoluted tubule (DCT) Na^+/Cl^- cotransporter. The DCT reabsorbs relatively small amounts of NaCl, so thiazide diuresis is of limited help in reducing edema, although the DCT does help determine the final plasma Na^+ content, which, in turn, helps determine blood pressure. Thiazides are, therefore, useful for treating hypertension. Inhibiting Na^+ influx causes DCT epithelial cell hyperpolarization, which increases the electrochemical gradient driving Ca^{2+} reabsorption. Thiazide diuresis sometimes causes hypercalcemia for this reason.[1]

B. Magnesium

By the time tubule fluid reaches the DCT, 85% of filtered Mg^{2+} load has been reabsorbed, principally in the TAL via the paracellular pathway (see Clinical Application 4.2). The DCT is the only segment that recovers Mg^{2+} in a regulated fashion, the amount recovered reflecting homeostatic needs. There are no further opportunities for recovery once Mg^{2+} leaves the DCT. Mg^{2+} is reabsorbed from the tubule lumen via TRPM6 (a member of the transient receptor potential [TRP] superfamily; see Chapter 2·VI·D), which is expressed in the early part of the DCT. Mg^{2+} recovery is regulated by epidermal growth factor through increased TRPM6 activity. Influx is passive, driven by the electrochemical gradient across the apical membrane. The means by which Mg^{2+} crosses the basolateral membrane remains uncertain.

C. Calcium

The late DCT reabsorbs ~8% of the filtered Ca^{2+} load. Reabsorption occurs passively via an apical membrane channel, but net uptake is regulated by parathyroid hormone (PTH) as shown in Figure 27.9.

1. **Apical:** Ca^{2+} crosses the apical membrane via TRPV5 and TRPV6, two TRP family channel members, reabsorption being powered by the electrochemical gradient for Ca^{2+} (see Fig. 27.9). All cells must maintain a very low intracellular Ca^{2+} concentration (see Chapter 1·II) and could easily be overwhelmed by the amount of Ca^{2+} crossing the apical membrane. Therefore, DCT epithelial cells contain large amounts of a high-affinity Ca^{2+}-binding protein (**calbindin**) that buffers Ca^{2+} influx until it can be pumped across the basolateral membrane. Intracellular buffering also maintains a steep electrochemical gradient favoring Ca^{2+} reabsorption from the tubule lumen.

2. **Basolateral:** Interstitial Ca^{2+} concentrations are ~10,000 times higher than intracellular concentrations, so Ca^{2+} must be actively pumped out of the epithelial cell by a basolateral Ca^{2+}-ATPase (see Fig. 27.9). The Ca^{2+}-ATPase functions much like a sump pump. When intracellular Ca^{2+} levels are rising, its activity increases, and the excess is deposited into the interstitium. The basolateral

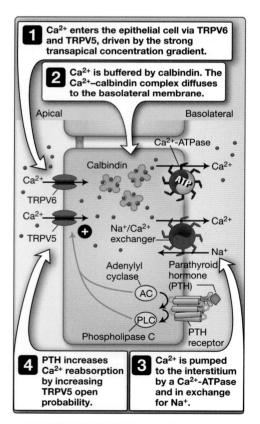

1 Ca^{2+} enters the epithelial cell via TRPV6 and TRPV5, driven by the strong transapical concentration gradient.

2 Ca^{2+} is buffered by calbindin. The Ca^{2+}–calbindin complex diffuses to the basolateral membrane.

3 Ca^{2+} is pumped to the interstitium by a Ca^{2+}-ATPase and in exchange for Na^+.

4 PTH increases Ca^{2+} reabsorption by increasing TRPV5 open probability.

Figure 27.9.
Calcium reabsorption by the distal convoluted tubule. TRPV5 and TRPV6 = transient receptor-potential channels.

[1]For more information on the mechanism of action and use of thiazide diuretics, see *LIR Pharmacology*, 8e, Chapter 9·III.

membrane also contains a Na^+/Ca^{2+} exchanger that helps support Ca^{2+}-pump activity when intracellular Ca^{2+} concentrations are high.

3. **Regulation:** Ca^{2+} reabsorption is regulated by PTH. When plasma Ca^{2+} concentrations are suboptimal, PTH is released into the circulation from the parathyroid glands (see Chapter 35·IV). PTH binds to a G protein–coupled receptor (GPCR) on the basolateral membrane of DCT cells, which activates the cAMP and the IP_3 signaling pathways (see Fig. 27.9). Both increase TRP open probability and Ca^{2+} reabsorption. PTH effects on Ca^{2+} reabsorption are potentiated by vitamin D, which increases the expression of most (perhaps all) of the proteins involved in Ca^{2+} transport, including calbindin.

IV. DISTAL SEGMENTS

The late DCT, connecting tubule (CNT), and cortical collecting duct (CCD) have similar structures and functions and are referred to collectively as the **distal segments** (Fig. 27.10). These segments are notable for their **intercalated cells**, which comprise ~20% to 30% of the tubule epithelium. There are two types of intercalated cells: α-intercalated cells secrete acid, whereas β-intercalated cells secrete HCO_3^-. The other 70% to 80% of the epithelium comprises Na^+-reabsorbing cells. In the CCD, these cells are known as **principal cells**.

A. Epithelial structure

The late DCT is the most distal portion of the renal nephron. The CNT connects the DCT to the CD system and, ultimately, the renal pelvis. Several CNTs fuse before joining a CCD, each CCD draining ~11 nephrons. The epithelium in these regions is characterized by cells filled with mitochondria and a basolateral membrane that is amplified by extensive infoldings.

B. Sodium and chloride

The fluid arriving at the late DCT is relatively dilute and contains low concentrations of Na^+ and Cl^-. The distal segments together reabsorb only ~5% of the NaCl filtered load, but this is the principal site for regulation by hormones concerned with extracellular fluid (ECF) Na^+ homeostasis and is therefore one of the more critical stages of recovery (see Chapter 28·III·C).

1. **Pathways:** Na^+ and Cl^- reabsorption occurs transcellularly, driven by the transmembrane Na^+ gradient established by the basolateral Na^+/K^+-ATPase. The late DCT expresses the NCC noted in Section III·A, but the predominant pathway for Na^+ reabsorption in the distal segments is via **ENaC** (an epithelial sodium channel), which appears in the late DCT (Fig. 27.11). Na^+ crossing the apical membrane via ENaC leaves the tubule lumen very negative. K^+ reentry via the apical ROMK partly offsets the charge, but, even so, the tubule lumen rests at around −40 mV compared with blood. This creates a strong driving force for paracellular Cl^- reabsorption. Cl^- is also recovered transcellularly via α-intercalated cells. An apical Cl^- channel allows influx from the tubule lumen, and the ion then crosses to the interstitium via a Cl^-/HCO_3^- exchanger.

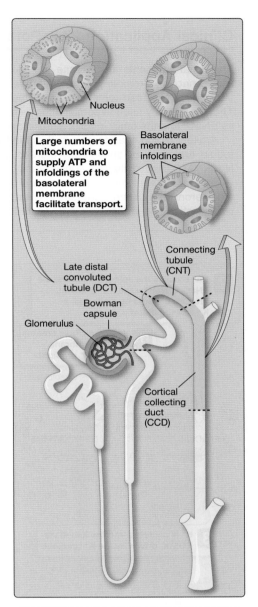

Figure 27.10.
Renal tubule distal segments.

Large numbers of mitochondria to supply ATP and infoldings of the basolateral membrane facilitate transport.

Nucleus

Mitochondria

Basolateral membrane infoldings

Late distal convoluted tubule (DCT)

Connecting tubule (CNT)

Bowman capsule

Glomerulus

Cortical collecting duct (CCD)

Clinical Application 27.3: Potassium-Sparing Diuretics

The distal segments are the site of action of two general classes of drugs that promote Na$^+$ and water excretion while simultaneously promoting K$^+$ retention, which is why they are referred to as "K$^+$-sparing" diuretics.[1] Because very little of the original Na$^+$ load remains by the time filtrate reaches the distal segments, these drugs have limited natriuretic effect. They are usually used in combination with loop or thiazide diuretics to limit K$^+$ loss. One class of K$^+$-sparing diuretics inhibits ENaC, and the other inhibits aldosterone binding to the mineralocorticoid receptor (MR).

Amiloride and triamterene both inhibit ENaC and prevent Na$^+$ reabsorption by principal cells. Na$^+$ remains in the tubule and acts as an osmotic diuretic. Reducing the amount of Na$^+$ entering principal cells secondarily decreases Na$^+$/K$^+$-ATPase activity and, consequently, reduces K$^+$ uptake and subsequent secretion.

Spironolactone and eplerenone are competitive inhibitors of aldosterone binding to the MR. They act by reducing aldosterone-stimulated increases in ENaC, Na$^+$/K$^+$-ATPase, and K$^+$-channel expression. The net result is a decrease in Na$^+$ reabsorption and a decrease in K$^+$ secretion.

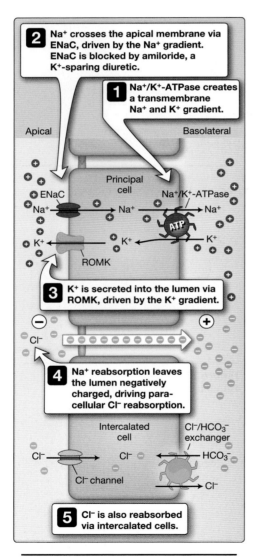

Figure 27.11.
Sodium and chloride reabsorption by the distal segments. ROMK = renal outer medullary K$^+$ channel.

2. Regulation: Na$^+$ recovery by principal cells is regulated by aldosterone (Fig. 27.12). Aldosterone is released from the adrenal cortex in response to angiotensin II (Ang-II) or an increase in plasma K$^+$ concentration (hyperkalemia). Ang-II is formed during activation of the **renin–angiotensin–aldosterone system (RAAS)** when blood pressure and renal blood flow is low (see Chapter 19·IV·C). Aldosterone binds to a basolateral mineralocorticoid receptor (MR) and then is internalized and translocated to the nucleus, where it upregulates transcription and expression of numerous proteins involved in Na$^+$ reabsorption and K$^+$ secretion (see Fig. 27.12). These include ENaC, ROMK, and the Na$^+$/K$^+$-ATPase. Aldosterone also stimulates basolateral membrane elaboration to increase its surface area and facilitate increased Na$^+$/K$^+$ transport capacity. Synthesis of new channel and transporter subunits is relatively slow, requiring ~6 hours to implement, but aldosterone also has short-term effects mediated by a serum- and glucocorticoid-activated kinase (SGK). SGK increases apical Na$^+$ permeability by reducing ENaC turnover rates and increasing basolateral Na$^+$/K$^+$-ATPase activity.

C. Potassium

Hyper- and hypokalemia both adversely affect cardiac excitability and function (see Clinical Application 2.1), so the kidneys must excrete K$^+$ when dietary intake exceeds homeostatic needs and conserve K$^+$ when dietary intake is limited. Plasma K$^+$ concentration is determined in the distal segments and in the outer medullary collecting duct (OMCD).

1. Secretion: K$^+$ is secreted by principal cells using the same pathway used to reabsorb Na$^+$ (see Fig. 27.11). K$^+$ is taken up from blood by the basolateral Na$^+$/K$^+$-ATPase and transferred to the tubule lumen via the apical ROMK. Secretion is favored both by a

[1] For more information on the mechanism of action and use of K$^+$-sparing diuretics, see *LIR Pharmacology*, 8e, Chapter 9·V.

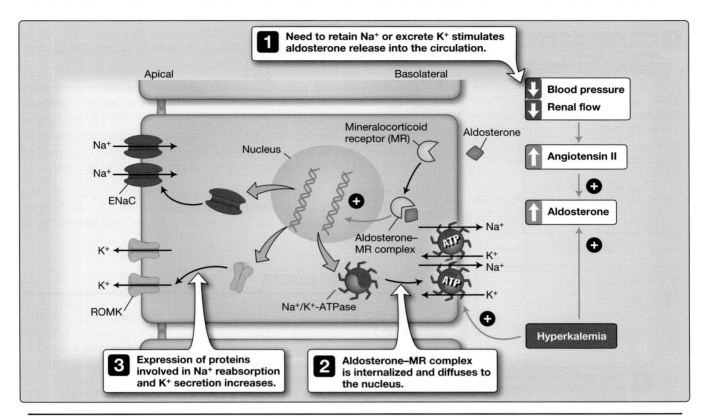

Figure 27.12.
Aldosterone regulation of sodium reabsorption and potassium secretion by principal cells in the distal segments. ROMK = renal outer medullary K^+ channel.

high intracellular K^+ concentration and by the net negative charge within the tubule lumen (see Fig. 27.11). Hyperkalemia promotes K^+ secretion directly by increasing basolateral Na^+/K^+-ATPase activity. Hyperkalemia is also a potent stimulus for aldosterone release from the adrenal cortex. Aldosterone increases expression of proteins involved in Na^+ reabsorption and K^+ secretion, as detailed in Figure 27.12.

2. **Reabsorption:** K^+ reabsorption relies on α-intercalated cells, which express a H^+/K^+-ATPase on their apical membrane (Fig. 27.13). The ATPase pumps H^+ into the tubule lumen in exchange for K^+, which subsequently exits the cell via basolateral K^+ channels. K^+ reabsorption increases during hypokalemia and involves regulation of both principal cells and α-intercalated cells.

   a. **Principal cells:** Hypokalemia decreases circulating aldosterone levels, thereby reducing expression of proteins involved in K^+ secretion. Hypokalemia also reduces K^+ uptake by principal cells through direct effects on basolateral Na^+/K^+-ATPase activity (see Fig. 27.12).

   b. **α-Intercalated cells:** Hypokalemia upregulates H^+/K^+-ATPase numbers in the apical membrane, which increases the α-intercalated cell reabsorptive capability. Because the pump links K^+ absorption with H^+ secretion and excretion, increased K^+ reabsorption may be accompanied by metabolic alkalosis.

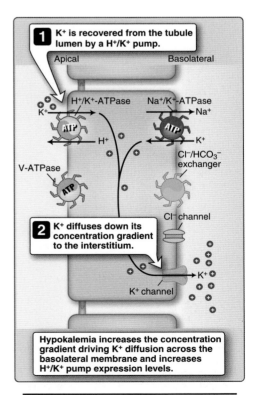

Figure 27.13.
Potassium reabsorption by α-intercalated cells in the distal segments.

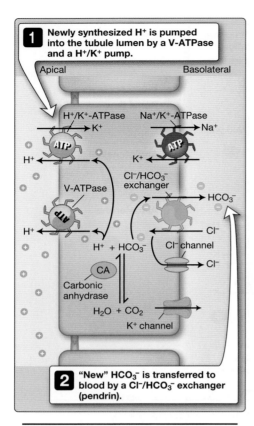

Figure 27.14.
Acid secretion by α-intercalated cells in the distal segments.

D. Acid and bicarbonate

H^+ and HCO_3^- handling by the distal segments is largely the responsibility of α- and β-intercalated cells.

1. **α-Intercalated cells:** α-Intercalated cells are the predominant form. They secrete H^+ into the tubule lumen via a H^+/K^+-ATPase that is also found in the gastric lining (Fig. 27.14) and via a V-ATPase. Newly synthesized HCO_3^- is secreted across the interstitium via a Cl^-/HCO_3^- anion exchanger (AE1, which is a member of the SLC4 transporter family and a product of the *AE1* gene).

2. **β-Intercalated cells:** β-Intercalated cells secrete HCO_3^- into the tubule lumen using **pendrin**, an apical Cl^-/HCO_3^- exchanger. Newly synthesized acid is pumped into the interstitium by a H^+/K^+-ATPase.

V. COLLECTING DUCTS

The fluid that enters the CD system has been denuded of all valuable solutes and is very dilute (~100 mOsmol/kg H_2O) compared with the surrounding cortex (~300 mOsmol/kg H_2O). The fluid is now poised to once again run the corticopapillary osmotic gauntlet to extract water. If body water intake exceeds homeostatic needs, the tubule fluid will flow through the CDs to the renal sinus and bladder without further water recovery, potentially at a rate of up to 20 L/d. If water intake is limited (as is usually the case), AQPs are inserted into the CD epithelium to allow water to flow out of the ducts and back to the vasculature. The driving force for movement is the osmotic potential created by the corticopapillary gradient, which becomes ever more powerful as urine flows toward the renal sinus.

A. Epithelial structure

The OMCD is a straight, unbranching tube that passes through the outer medulla (Fig. 27.15). Inner medullary collecting ducts (IMCDs) fuse successively toward the papillary tip, gaining diameter and increasing wall thickness with each fusion. IMCD epithelial cells bear stubby microvilli on both their apical and basolateral surfaces, and their basolateral membrane is folded extensively, consistent with their high potential reabsorptive capacity.

B. Urine volume determinants

A normal, healthy individual excretes ~1 to 2 L of ~300- to 500-mOsmol/kg H_2O urine every day. This range can change considerably depending on the amount of water ingested and the amount lost to the environment through evaporation (skin, mucous membranes, lungs) and nonurinary excretion (i.e., feces; see Chapter 28·II·A), but there are physiologic limits to output.

1. **Maximal output:** Maximal urinary output is around 20 L/d. Although excretion rates can go higher, flow volumes in excess of 20 L/d exceed the kidneys' ability to recover Na^+ and K^+ from the tubule lumen, resulting in hyponatremia and hypokalemia. Hyponatremia causes nausea, headaches, confusion, and seizures (all symptoms of cerebral edema) and, like hypokalemia (see earlier discussion), can be fatal.

2. **Minimal output:** The human body generates ~600 mOsmol of solutes every day that must be excreted in urine. The kidney's ability to concentrate urine is limited by the corticopapillary gradient to ~1,200 mOsmol/kg H_2O, so the 600 mOsmol of excreted solutes are accompanied by at least 0.5 L of water per day. If additional solutes must be excreted (e.g., as the result of eating too many salty chips), then urine volume increases accordingly.

3. **Free water clearance:** Free water clearance (C_{H2O}) is a measure of the kidney's water-handling ability. For the purposes of this discussion, dilute urine (i.e., with an osmolality less than that of plasma, or ~300 mOsmol/kg H_2O) can be considered to comprise two components. The first is the volume needed to dissolve excretory solutes to a final osmolality of 300 mOsmol/kg H_2O. The second is **free water**, or the amount of water in urine in excess of that required to dissolve the excreted solutes. C_{H2O} cannot be calculated directly and must be determined by measuring total urine volume and then subtracting out the amount of water needed to create an isosmotic solution from the amount of excreted osmolytes contained in urine. This latter component is measured from osmolal clearance (C_{Osm}):

$$C_{Osm} = \frac{U_{Osm} \times V}{P_{Osm}}$$

where U_{Osm} is urine osmolality, V is urine flow rate, and P_{Osm} is plasma osmolality. C_{H2O} is then calculated as:

$$C_{H_2O} = V - C_{Osm} = V \times \frac{(1 - U_{Osm})}{P_{Osm}}$$

A negative C_{H_2O} indicates that urine is concentrated (hyperosmotic). A positive value indicates that urine is dilute (hyposmotic).

C. Water reabsorption

When water intake exceeds homeostatic needs, dilute urine passes through the CDs to the bladder largely unchanged, as if flowing through a cast-iron downspout. If there is a need to conserve water, virtually all of the fluid (aside from the ~0.5 L/d of obligatory loss) can be recovered. Recovery of water and final urine concentration is governed by the presence of AQPs in the ductal epithelium and is regulated by **antidiuretic hormone** ([**ADH**] also known as **arginine vasopressin**).

1. **Aquaporins:** AQPs form pores that allow water to cross the lipid bilayer (see Chapter 1·V·A). AQPs are expressed constitutively in the apical and basolateral membranes of the PT and DTL, which gives these segments high water permeability. The ATL and TAL do not express AQPs, so they are water impermeable. Principal cells in the CNT, CCD, OMCD, and IMCD all express AQP2, but the channels are not inserted into the apical membrane until ADH binds to a basolateral vasopressin (V_2) receptor (Fig. 27.16A).

2. **Antidiuretic hormone:** ADH is released from the posterior pituitary in response to an increase in plasma osmolality or a

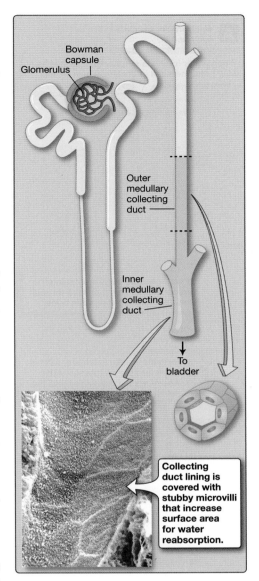

Figure 27.15.
Renal medullary collecting ducts.

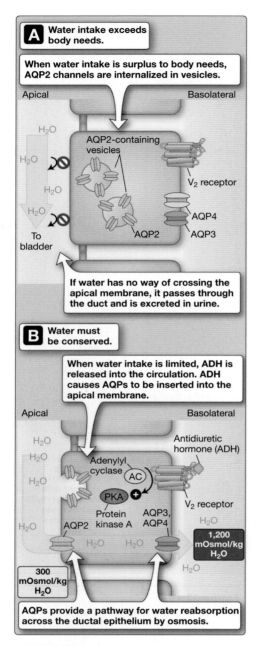

Figure 27.16.
Water reabsorption by the collecting ducts. V_2 = vasopressin type 2.

decrease in mean arterial blood pressure. The hormone is carried via the peritubular capillary network to the CD system, where it binds to ADH V_2 receptors. V_2 receptors are GPCRs, which, when occupied, activate protein kinase A (PKA) via the cAMP signaling pathway (see Chapter 1·VII·B·2). PKA phosphorylates intracellular trafficking proteins, causing AQP2-containing vesicles to shuttle to the cell surface and fuse with the apical membrane (see Fig. 27.16A).

3. **Water reabsorption:** The basolateral membrane of CD principal cells also contains AQP isoforms (AQP3 and AQP4) that are not ADH dependent. AQP2, AQP3, and AQP4 together provide a pathway for transcellular reabsorption of water, driven by the osmotic gradient between tubule and interstitium (see Fig. 27.16B). Note that the fluid entering the CD system from the DCT has a lower osmolality than that of the cortex (~100 mOsmol/kg H_2O, compared with ~300 mOsmol/kg H_2O). This difference causes substantial amounts of water to be reabsorbed even before the fluid runs the corticopapillary osmotic gauntlet. As the tubule contents progress toward the papilla, more water is reabsorbed, and urine osmolality reaches its maximal value.

4. **Aquaporin recycling:** When water intake increases and circulating ADH levels fall, AQPs are removed from the membrane by endocytosis and returned to subapical vesicles. Principal cells then remain water impermeable until ADH release resumes, and AQPs are returned to the apical surface.

D. Urea recycling

The fluid that enters the IMCD is now close to urine in its final form. The principal excretory components (in relative order based on molar amounts) are urea, creatinine, ammonium salts, and organic acids. The final step in urine formation is reabsorption of urea, which is regulated by ADH.

1. **Reabsorption:** Urea is reabsorbed by facilitated diffusion, driven by high ductal concentrations and facilitated by urea transporters (UTs) in the IMCD (Fig. 27.17). The basolateral membrane contains a UT (UT-A3) that is constitutively active. The apical membrane contains a UT (UT-A1) that is minimally active unless ADH is circulating. ADH causes PKA-dependent phosphorylation of UT-A1, activating it and thereby creating a pathway for urea to leave the duct and reenter the medullary interstitium. Allowing urea to equilibrate across the tubule wall also helps prevent an osmotic diuresis that might otherwise result from the presence of highly concentrated fluid within the ductal lumen. It also contributes to the corticopapillary osmotic gradient used to concentrate urine (Fig. 27.18).

2. **Recycling:** Remembering the close anatomic arrangement between the CDs, blood vessels, and limbs of the nephron loop (see Fig. 27.5), urea that reenters the interstitium from the IMCD could potentially be reabsorbed by earlier tubule segments or be carried away by the circulation. In practice, it does both.

a. **Loop of Henle:** Urea reenters the tubule by facilitated diffusion via UT-A2, which is present in the epithelium lining both DTL and ATL (see Fig. 27.18). Urea then recycles back through the distal segments and the CDs. From here, it either may be excreted in urine or take another trip through the medulla.

b. **Vasa recta:** The descending vasa recta expresses UT-B transporters (UT-B1 and UT-B2) that allow urea to enter the vasculature by facilitated diffusion. Uptake by the vasa recta is beneficial because it increases the osmolality of blood during its passage through the medulla, thereby preventing washout of the osmotic gradient. Urea exits the vasa recta and reenters the interstitium during the return trip to the cortex (see Fig. 27.18), so the amount returned to the systemic circulation is minimal (~5% of original filtered load).

3. **Excretion:** Ultimately, the amount of urea excreted in urine depends on the need to conserve water. When water intake is limited, urea is recycled through the medulla, and excretion rates are minimal. When water intake is unlimited, ADH release is suppressed, and urea has no significant pathway to escape the IMCD. It is excreted in urine as a result.

E. Acid handling

α-Intercalated cells continue secreting H⁺ during urine's passage through the CD and can cause significant urine acidification (pH 4.4, the minimal value attainable). Creatinine (pK = 5.0) becomes a viable buffer at such low pH values, allowing it to assist in H⁺ excretion, but the bulk of acid is excreted in the form of NH_4^+. NH_4^+ is excreted as a result of "**diffusion trapping**" or through direct secretion.

1. **Diffusion trapping:** NH_4^+ is formed from NH_3 as a result of glutamine metabolism in the PT (see Chapter 26·IX·C). NH_3 is lipid soluble, allowing it to diffuse out of PT epithelial cells and enter the interstitium, where it accumulates in relatively high concentrations. Some NH_3 may then diffuse into the tubule or CD lumen, where it immediately combines with H⁺ to form NH_4^+. NH_4^+ is not lipid soluble and is therefore now trapped in the tubule or duct unless provided with a carrier that facilitates reabsorption (diffusion trapping). NH_4^+ trapped in the proximal segments is actively reabsorbed by the NKCC2 cotransporter in the TAL and then transferred to the interstitium to help generate the corticopapillary osmotic gradient through countercurrent multiplication (see Section II·B·3).

2. **Transport:** Some of the NH_4^+ transferred to the interstitium by the TAL enters the vasa recta and is carried away by the circulation (washout). This portion eventually reaches the liver, where it is converted to urea. A significant proportion is also transferred to the CD lumen for excretion in urine, although the pathways involved are not well defined.

The principal sites of solute reabsorption, secretion, and regulation in the renal tubule are summarized in Fig. 27.19.

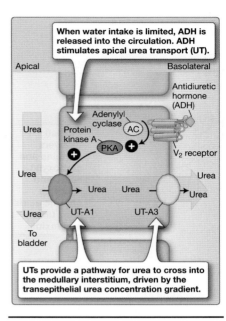

Figure 27.17.
Urea reabsorption by the inner medullary collecting duct.

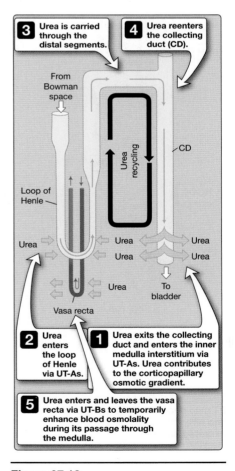

Figure 27.18.
Urea recycling. UT = urea transporter.

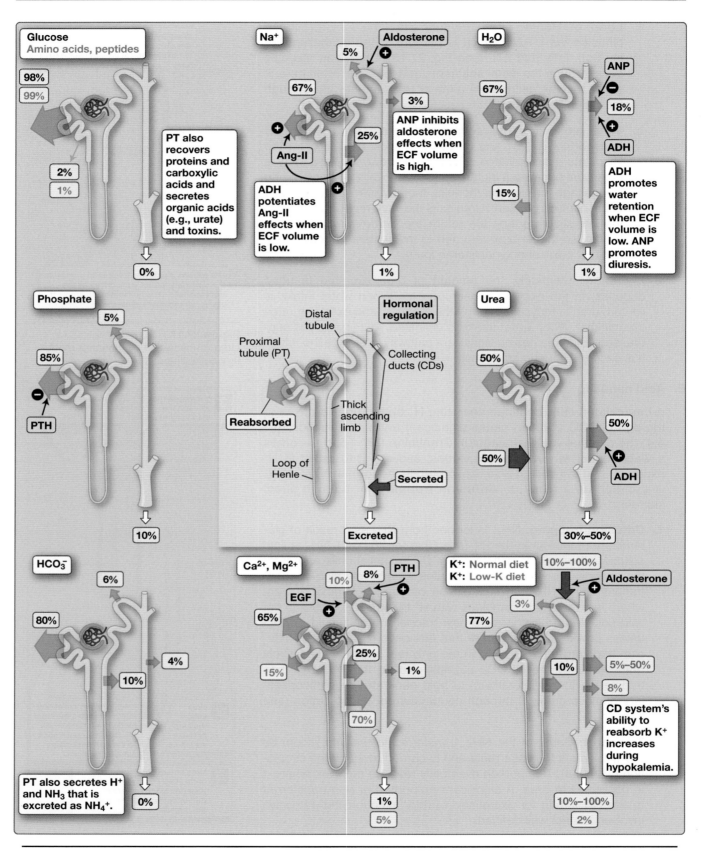

Figure 27.19.
Principal sites of solute and water recovery and secretion by the renal nephron. ADH = antidiuretic hormone;
Ang-II = angiotensin II; ANP = atrial natriuretic peptide; EGF = epidermal growth factor; PTH = parathyroid hormone.

Chapter Summary

- The function of the nephron segments distal to the proximal tubule is to recover inorganic ions and to concentrate urine. These segments are the primary sites of **homeostatic regulation** of Na^+, K^+, Ca^{2+}, Mg^{2+}, Cl^-, and water.

- The **loop of Henle** comprises three segments that convey the tubule contents through a **corticopapillary osmotic gradient** designed to extract water from the filtrate.

- The corticopapillary osmotic gradient forms by **countercurrent multiplication** of a transepithelial osmotic gradient created by the **thick ascending limb (TAL)** epithelium. The TAL pumps Na^+ and other ions (e.g., NH_4^+) into the cortical interstitium. These ions then diffuse into the **descending thin limb**, increasing the osmolality of the fluid within. The ions are carried toward the papilla, around the tip of the loop, and back up toward the TAL via the **ascending thin limb**. At the TAL, they are pumped back out into the interstitium to return to the medulla. Therefore, the loop traps ions in the medulla and causes this region to develop a high osmolality.

- The corticopapillary osmotic gradient extracts water and ions from the tubule lumen. Reabsorbed fluid is carried away by the **vasa recta**. Flow through the descending and ascending limbs of the vasa recta occurs in opposite directions, thus creating a **countercurrent exchange system** that prevents incoming arterial blood from washing out the osmotic gradient.

- The **early distal convoluted tubule** is the primary site of regulated Ca^{2+} and Mg^{2+} reabsorption. Reabsorption is regulated by **parathyroid hormone**.

- The **late distal convoluted tubule**, **connecting tubule**, and **collecting duct** (**distal segments**) are the primary site of homeostatic regulation of Na^+ and K^+. Na^+ reabsorption is regulated by **aldosterone**. Aldosterone increases epithelial Na^+ permeability by upregulating expression of Na^+ channels and pumps in the apical and basolateral membranes.

- K^+ may be secreted or reabsorbed depending on plasma K^+ concentrations. Secretion is stimulated by hyperkalemia acting through aldosterone. Aldosterone increases epithelial K^+ permeability and Na^+/K^+-ATPase activity.

- The **collecting ducts** (cortical and inner and outer medullary collecting ducts) determine final urine osmolality by reabsorbing water. Reabsorption is regulated by **antidiuretic hormone (ADH)**.

- ADH release is stimulated by increased plasma osmolality or decreased blood pressure. The hormone exerts its antidiuretic effects by stimulating insertion of an aquaporin (AQP2) into the apical membranes of ductal epithelial cells. Water is reabsorbed by osmosis, driven by the corticopapillary osmotic gradient.

- The outer medullary collecting duct is also the site of urea reabsorption. Urea is recovered by facilitated diffusion. When water intake is limited, the tubule fluid becomes highly concentrated, and there is a strong driving force for urea movement into the medullary interstitium. The presence of urea in the medulla contributes to the corticopapillary osmotic gradient.

Study Questions

Choose the ONE best answer.

27.1. Increasing which of the following variables would most likely decrease the magnitude of the renal corticopapillary osmotic gradient that allows for urine concentration?

 A. Blood flow through the vasa recta
 B. Renin release from the afferent arteriole
 C. Sympathetic nervous system activation
 D. Thick ascending limb $Na^+/K^+/2Cl^-$ cotransport
 E. Urea reabsorption by the collecting ducts

Best answer = A. The corticopapillary osmotic gradient is established by countercurrent multiplication in the loop of Henle (see Section II·C). The countercurrent multiplier relies on $Na^+/K^+/2Cl^-$ cotransport by the thick ascending limb (D), so the gradient collapses when the cotransporter is inhibited by loop diuretics. Increasing flow through the vasa recta washes ions out of the medulla, thereby diminishing the osmotic gradient. Renin is released (B) when arterial pressure falls or when the sympathetic nervous system activates (C), conditions that signal a likely need to conserve water. Gradient magnitude increases as a result, in part through increased urea reabsorption from the collecting ducts (E).

27.2. Genetic evaluation of a 6-year-old boy with impaired growth and intellectual disability identified alleles associated with Bartter syndrome. Bartter syndrome type 1 mimics loop diuretics by causing thick ascending limb (TAL) dysfunction. Which of the following most likely describes the TAL in healthy individuals?

 A. Fluid leaves the TAL at ~600 mOsmol/kg H_2O.
 B. It extracts Na^+, K^+, and Cl^- from the lumen.
 C. It has a high water permeability.
 D. It is known as the "concentrating segment."
 E. It is the primary site of Ca^{2+} reabsorption.

Best answer = B. The TAL reabsorbs Na^+, K^+, and Cl^- from the tubule lumen via $Na^+/K^+/2Cl^-$ cotransport and transfers these ions to the interstitium, where they help form the corticopapillary osmotic gradient (see Section II·B). The TAL has a low, not high, water permeability (C) that prevents H_2O from following ions into the interstitium, so the tubule fluid becomes relatively dilute at <300 mOsmol/kg H_2O, not ~600 mOsmol/kg H_2O (A). The TAL may be referred to as the "diluting (not "concentrating") segment" (D) for this reason. Ca^{2+} is reabsorbed primarily in the proximal tubule (E), with regulated reabsorption occurring in the distal tubule (see Section III·C).

27.3. A genetic mutation affecting the $Na^+/K^+/2Cl^-$ cotransporter (NKCC2) in the thick ascending limb (TAL) of the loop of Henle prevents the development of a transepithelial voltage gradient. Which of the following blood conditions would most likely result?

 A. Hypercalcemia
 B. Hyperkalemia
 C. Hypernatremia
 D. Hypomagnesemia
 E. Hypophosphatemia

Best answer = D. Approximately 70% of the Mg^{2+} filtered load is reabsorbed paracellularly in the TAL. NKCC2 and other TAL transporters establish a significant transepithelial voltage gradient (lumen positive) to facilitate reabsorption of cations such as Mg^{2+}. The lack of a transepithelial gradient would decrease reabsorption of not only Mg^{2+}, but also Ca^{2+}, K^+, and Na^+. Blood levels of these cations would therefore most likely decrease, not increase (A, B, and C). Phosphate is reabsorbed in the proximal and distal convoluted tubules and so is less affected by transport issues in the TAL (E).

27.4. A 77-year-old female is taking a thiazide diuretic to treat hypertension but has become hypercalcemic. Thiazides inhibit Na^+/Cl^- reabsorption by the distal convoluted tubule. Which of the following most likely describes why thiazide diuretics also cause hypercalcemia?

 A. Apical Na^+/Ca^{2+} exchange increases.
 B. Paracellular Ca^{2+} uptake increases.
 C. The gradient driving Ca^{2+} uptake steepens.
 D. The Na^+/Cl^- cotransporter also carries Ca^{2+}.
 E. Thiazide diuretics also inhibit Ca^{2+}-ATPases.

Best answer = C. Ca^{2+} reabsorption by the distal convoluted tubule (DCT) is mediated by a Ca^{2+} channel (TRPV5, a TRP channel) and driven by the electrochemical gradient across the tubule epithelium's apical membrane (see Section III·C). Inhibiting the Na^+/Cl^- cotransporter reduces Na^+ influx into the epithelial cell, so the interior becomes more negative. This negativity increases the driving force for Ca^{2+} reabsorption and causes hypercalcemia. The DCT does not reabsorb significant amounts of Ca^{2+} via apical Na^+/Ca^{2+} exchange (A), paracellularly (B), or via Na^+/Cl^- cotransport (D). Thiazides have no significant effect on Ca^{2+}-ATPases (E).

Water and Electrolyte Balance

28

I. OVERVIEW

When our evolutionary ancestors emerged from the oceans and laid claim to the land, they carried within them a small sea in which to bathe their cells (Fig. 28.1). This sea, which we know as extracellular fluid (ECF), is composed largely of Na^+, Cl^-, and water. It also contains smaller amounts of HCO_3^-, K^+, Ca^{2+}, Mg^{2+}, and phosphates. All these constituents have specific roles to play in human physiology, and the concentrations of each must be maintained within a limited range if we are to survive and thrive (see Table 1.1). We continuously lose water and electrolytes to the environment as a result of secretion, excretion, and evaporation. Maintaining water and electrolyte balance requires that these losses be replaced through drinking fluids and ingesting food, but ingestion and subsequent absorption of salts by the gastrointestinal (GI) system is largely unregulated, being tied to the absorption of nutrients (e.g., glucose and peptides). The central nervous system (CNS) can modify behavior to increase intake if ECF salt and water levels fall below optimal (through salt cravings and thirst), but the main regulated step in salt and water balance is excretion, which is mediated by the kidneys. Although the body includes pathways that maintain stable plasma concentrations of all the common electrolytes, discussions of ECF homeostasis are dominated by Na^+ and water. Na^+ and water together determine ECF volume, which, in turn, determines plasma volume, cardiac output (CO), and mean arterial pressure (MAP).

II. WATER BALANCE

Maintaining water balance is one of the body's most fundamental and important homeostatic functions. Because water is the universal solvent, when total body water (TBW) levels fall, solute concentrations rise, to the detriment of bodily function. TBW's role in supporting CO and MAP (discussed in more detail in Section III) means that TBW regulatory pathways are layered and influence both intake and output.

A. Tally sheet

Individuals ingest and lose ~2.5 L of water per day on average. Actual water requirements are less (1.6 L/d), as shown in Table 28.1, dictated by the amount of **insensible** water loss (evaporation) and obligatory water loss (water needed for urine formation; see Chapter 27·V·B).

1. **Intake:** The tally sheet for intake includes water formed through metabolism ($C_6H_{12}O_6 + 6\ O_2 \rightarrow 6\ CO_2 + 6\ H_2O$), ingested with

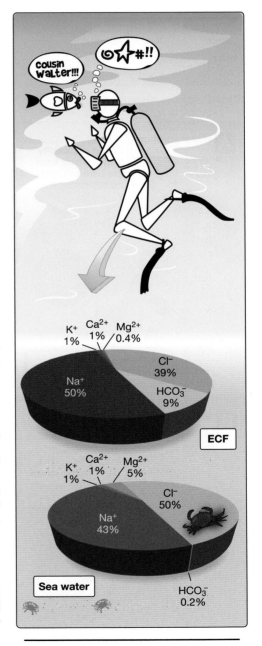

Figure 28.1.
Electrolyte composition of the internal and external seas.

Table 28.1: Water Intake and Output Routes

Pathway	mL/d
Intake	
Metabolism	300
Food	800
Beverages	500*
Total	1,600
Output	
Feces	200*
Skin	500
Lungs	400
Urine	500*
Total	1,600

*Primary regulated steps.

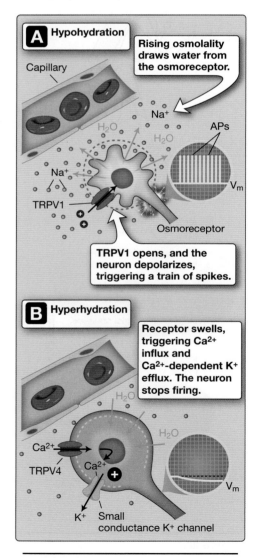

Figure 28.2.
Effect of increasing osmolality on osmoreceptor output. V_m = membrane potential.

food, and imbibed by drinking fluids. Drinking is the primary regulated intake step.

2. **Output:** Output includes water evaporation from respiratory and cutaneous epithelia (i.e., insensible losses), sweat, fecal water content, and urine. Water loss from the respiratory epithelia is dependent on respiratory rate and air humidity, but such losses stress water balance only under extreme conditions (e.g., climbing at high altitude). Cutaneous evaporation remains relatively constant under normal conditions. Sweat formation reaches 1.5 to 2.0 L/h during heat stress. Vomiting and diarrhea (see Clinical Application 4.5) can greatly accelerate water loss from the GI tract, but normal fecal water loss is modest. Urine formation is the primary regulated output mechanism.

B. Sensory mechanism

TBW is sensed indirectly through changes in ECF osmolality and MAP. ECF osmolality is normally maintained at 275 to 295 mOsmol/kg H_2O. Osmolality is monitored by the **lamina terminalis**, an area bordering the third ventricle composed of three nuclei: the **subfornical organ** (**SFO**), the **organum vasculosum of the lamina terminalis** (**OVLT**), and the **median preoptic nucleus** (**MnPO**). Together, they comprise a **thirst center**. The SFO and OVLT are **circumventricular organs** (see Fig. 7.10) located outside of the blood–brain barrier so that they might sample ECF composition. The SFO and OVLT both contain osmoreceptors. Osmoreceptors are neurons responsive to changes in cell volume, which is dependent on ECF osmolality (Fig. 28.2; also see Chapter 3·II·E). Osmosensing uses two different mechanosensitive transient receptor-potential channels ([TRPs] see Chapter 2·VI·D). The MnPO helps integrate sensory signals from the SFO and OVLT before they are forwarded to other regions of the hypothalamus and CNS.

1. **Decreased cell volume:** Increasing ECF osmolality causes cell shrinkage, which deforms the cytoskeleton and activates an N-terminal variant of TRPV1. The resulting cation influx depolarizes the neuron and initiates signaling (see Fig. 28.2A).

2. **Increased cell volume:** When TBW rises, the osmoreceptors swell, and the resulting surface membrane deformation activates TRPV4. TRPV4 mediates a Ca^{2+} influx that activates a small-conductance Ca^{2+}-activated K^+ channel (SK). The resulting K^+ efflux hyperpolarizes the neuron, and signaling is suppressed (see Fig. 28.2B).

C. Regulation

OVLT neurons project to the **supraoptic nucleus** (**SON**) and **paraventricular nucleus** (**PVN**), which function as an **osmostat** (osmolality regulation center) located within the anterior hypothalamus. Responses to changes in osmolality are accomplished by neurosecretory cells that are themselves osmosensitive, which creates an additional layer of osmoregulatory control.

Water balance is achieved by modulating water intake and urinary output.

> **Reset osmostat** is a disorder in which the osmosensory threshold is shifted toward lower ECF osmolalities than normal. Reset osmostat is an important cause of the syndrome of inappropriate antidiuretic hormone release ([SIADH] see Clinical Application 28.1).

1. **Intake:** A need to drink water is perceived as thirst, which causes an individual to seek a thirst-quenching beverage. The emotion originates in cortical areas that include the **anterior** and **posterior cingulate cortex** and **insular cortex**. The threshold and linear rise in intensity for thirst sensation is essentially identical to that for secretory responses to changes in ECF osmolality (~280 mOsmol/kg H_2O), reflecting the influence of the thirst center osmoreceptors. Hormones such as **angiotensin II (Ang-II)** and relaxin increase thirst by activating the SFO, whereas **atrial natriuretic peptide (ANP)** inhibits it. Thirst is also increased by input from the area postrema and nucleus tractus solitarius in response to low blood pressure (signaled by arterial baroreceptors and cardiopulmonary receptors) and visceral osmoreceptors. Thirst is normally sated before any change in ECF osmolality due to input from oropharyngeal sensors. These sensors are cold sensitive, which explains why cooled beverages are more thirst quenching than hot beverages.

2. **Output:** The glomerulus filters water into the renal tubule at a rate of ~125 mL/min. Approximately 67% of the filtrate is immediately reabsorbed by the proximal tubule (PT), another 15% is recovered in the nephron loop's descending thin limb (DTL), and most of the remaining 18% is reabsorbed in the outer and inner medullary collecting ducts (OMCD and IMCD, respectively). Output regulation occurs via **antidiuretic hormone (ADH)**. ADH is a small peptide hormone produced by SON and PVN neurosecretory neurons and moved by fast axonal transport for release from terminals in the posterior pituitary, as shown in Figure 28.3 (also see Chapter 7·VII·D). Release is triggered by a rise in ECF osmolality (Fig. 28.4). The release threshold is 280 mOsmol/kg H_2O, so small amounts of ADH circulate even when plasma osmolality is within a normal range. ADH has a half-life of 15 to 20 minutes before being metabolized by kidney and liver proteases.

3. **Integration with other indicators of hypovolemia:** CO and MAP are dependent on TBW, so lamina terminalis output is heavily influenced by sensory input from baroreceptors, which monitor MAP, and by Ang-II, whose levels rise rapidly following a drop in renal artery pressure (see Section III). Both MAP and Ang-II can sustain ADH release by resetting thresholds and increasing sensitivity to changing MAP, even though ECF osmolality may have renormalized (see Fig. 28.4).

   a. **Actions:** ADH provides pathways for water to flow out of the renal collecting ducts and rejoin the ECF (see Chapter 27·V·C). In the absence of such pathways, water is channeled to the bladder and excreted.

   b. **Negative feedback:** ADH-mediated ECF volume expansion is limited by ANP. ANP is released from atrial myocytes when ECF and blood volumes are high. ANP has many actions

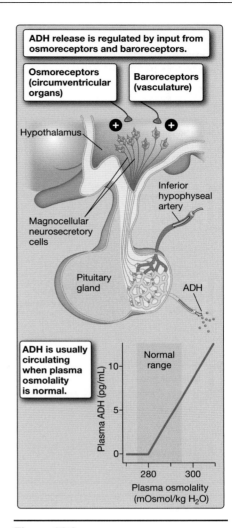

Figure 28.3.
Regulation of antidiuretic hormone (ADH) release.

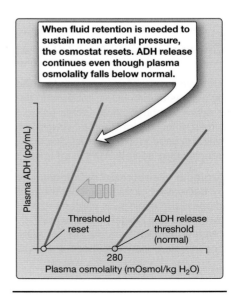

Figure 28.4.
Effect of arterial pressure on antidiuretic hormone (ADH) release.

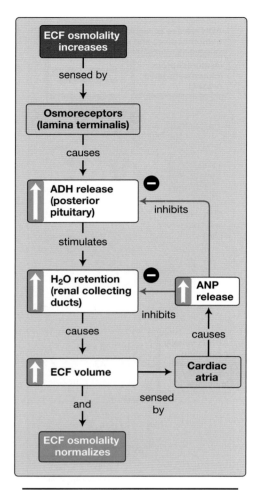

Figure 28.5.
Regulation of antidiuretic hormone
(ADH) release. ANP = atrial natriuretic
peptide.

Female thirst thresholds are influenced by hormonal changes associated with the ovarian cycle and pregnancy. The thirst threshold decreases by ~3 mOsmol/kg H_2O during the luteal phase of the ovarian cycle, a shift dependent on rising estrogen and progesterone levels. During pregnancy, the thirst threshold decreases by ~10 mOsmol/kg H_2O by the end of the first trimester and remains low throughout gestation. This decrease is attributable to relaxin, a hormone produced at high levels by the corpus luteum. Relaxin has numerous physiologic effects, including vasodilation, increased arterial compliance, and increased cardiac stroke volume. Relaxin increases thirst by affecting osmoreceptor neuron output from the two lamina terminalis circumventricular organs. Increased thirst and water retention leads to increased blood volume which, together with vasodilation, facilitates a sustained increased in cardiac output needed to support pregnancy (see Chapter 37·IV·B).

Dehydration is common in older adults (>50% patients in elderly care facilities are dehydrated), which is a significant concern due to increased morbidity and mortality. Water intake decreases with age. Although aging impairs the kidneys' ability to concentrate urine, the principal cause of dehydration is decreased water intake. By age 70 years, water intake drops by ~65% compared with that of 25-year-olds, with concurrent increases in ECF osmolality to >300 mOsmol/kg H_2O. The reason appears to be related to earlier satiety and decreased water-seeking behaviors because plasma osmolality thresholds for thirst and antidiuretic hormone release are unaffected by age.

(detailed in the next section), including decreasing thirst, antagonizing ADH release, and ADH-mediated water retention (Fig. 28.5).

III. SODIUM BALANCE

Although it is possible to identify mechanisms and sites involved in Na^+ balance, the pathways involved are so closely interwoven with those controlling water balance and MAP that Na^+ balance cannot be discussed in isolation.

Clinical Application 28.1: Antidiuretic Hormone–Release Disorders

The **syndrome of inappropriate antidiuretic hormone** (**SIADH**) secretion is a relatively common disorder characterized by increased circulating antidiuretic hormone (ADH) levels and water retention. Patients typically develop hyponatremia as a result. Although some cases are idiopathic, common causes of SIADH include central nervous system disorders (e.g., stroke, infection, or trauma), drugs (anticonvulsants, such as carbamazepine and oxcarbazepine, and cyclophosphamide, which is used to treat certain cancers), and some pulmonary diseases and carcinomas. For example, small-cell lung malignancies may secrete ADH in an unregulated manner, thereby causing SIADH.

Arginine vasopressin deficiency (**AVP-D**), previously known as central diabetes insipidus, describes polyuria caused by decreased circulating ADH levels. Patients may also present with nocturia and polydipsia. AVP-D, most commonly idiopathic in etiology, is characterized by degeneration of hypothalamic ADH-secreting cells. AVP-D may also be caused by trauma and surgery. Familial AVP-D is a dominant hereditary form of AVP-D caused by ADH gene mutation. The most common familial form causes an ADH-processing defect and accumulation of misfolded hormone. The secretory cells degenerate because of these accumulations, although the cause is still under investigation.

A. Tally sheet

The body contains ~75 g of Na^+ on average, almost half of which is associated with bone. Na^+ is obtained from dietary sources and leaves the body via feces, sweat, and urine.

1. **Intake:** The average diet contains far more Na^+ than is required to offset losses. The U.S. recommended dietary allowance (RDA) is 1.5 g/d, but per capita consumption worldwide is typically much greater (up to 7 g/d). A low-Na^+ diet triggers salt craving, which manifests as a need to seek out and ingest salty foods. The pathways involved are not well defined.

2. **Output:** A small amount of ingested Na^+ is lost in feces. Sweat is also a minor pathway for Na^+ loss unless sweating is prolonged and profuse (see Chapter 15·VI·D). Urine is the primary route for Na^+ output. Because there is no obligatory Na^+ loss as there is for water, urinary Na^+ excretion typically balances the amount ingested. When Na^+ intake is limited, however, the renal tubule can recover 100% of the filtered load and generate Na^+-free urine for several weeks. The principal sites for recovery are the PT (67% of filtered load) and the thick ascending limb ([TAL] ~25%). The distal segments and collecting ducts recover the remaining 8% and are the principal sites of output regulation (see Chapter 27·IV·B).

B. Sodium and blood pressure relationship

When Na^+ is ingested, most of it ends up in the ECF (~85%) because, although Na^+ exchanges freely between ECF and intracellular fluid (ICF), all cells actively eliminate Na^+ from the ICF via the Na^+/K^+-ATPase (see Chapter 3·III·B). Ingesting Na^+ thus increases ECF osmolality, creating an urge to drink water and stimulating water retention (see Fig. 28.5). Because plasma is an ECF component, Na^+ ingestion also increases circulating blood volume, which raises central venous pressure ([CVP] see Fig. 19.22). A rise in CVP increases left-ventricular (LV) preload. The LV responds with an increase in stroke volume (SV) and CO via the Frank-Starling mechanism, which raises MAP (see Chapter 17·IV·D). Increasing MAP has immediate and wide-ranging repercussions for both the cardiovascular system and the kidneys, all aimed at excreting the Na^+ and water excess to reduce ECF volume and renormalize MAP.

C. Regulation

ECF volume is determined by four different but interdependent pathways that regulate Na^+ balance, water balance, and MAP (Fig. 28.6). They include the **renin–angiotensin–aldosterone system** ([**RAAS**] see Fig. 19.15), the sympathetic nervous system (SNS), ADH, and ANP. Three of the four pathways activate when ECF volume and MAP are low, such as might occur when a marathon runner has become hypohydrated due to inadequate replenishment of salt and water loss. When ECF volume and MAP are high, these pathways are inhibited.

1. **Renin–angiotensin–aldosterone system:** The RAAS activates following a decrease in glomerular afferent arteriolar perfusion

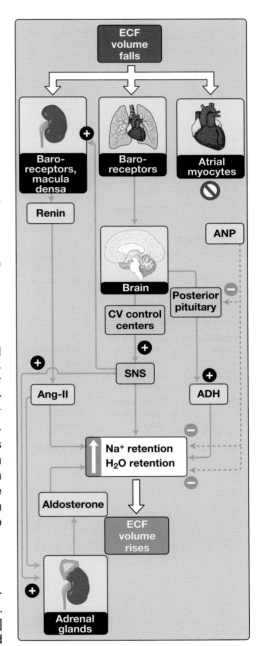

Figure 28.6.
Pathways regulating ECF volume under normal conditions. Note: Hormones may have additional effects when the system is stressed. ADH = antidiuretic hormone, Ang-II = angiotensin II, ANP = atrial natriuretic peptide, CV = cardiovascular.

Table 28.2: Angiotensin II Effects

Target	Action	Effects
Adrenal cortex		
	Aldosterone release	↑ Na⁺ reabsorption, renal distal segments
		↑ ENaC
		↑ ROMK
		↑ Na⁺/K⁺-ATPase
Kidney		
	Proximal tubule, thick ascending limb	↑ Na⁺ reabsorption
		↑ Na⁺/H⁺ exchanger
	Distal segments	↑ ENaC
Vasculature		
	Vasoconstriction (resistance vessels)	↑ Systemic vascular resistance
Central nervous system		
Hypothalamus	Antidiuretic hormone release (posterior pituitary)	↑ H₂O reabsorption, renal distal segments
Cortex	Thirst and salt craving	↑ H₂O and NaCl intake

ENaC = epithelial Na⁺ channel, ROMK = renal outer medullary K⁺ channel.

pressure (sensed by renal baroreceptors), a decrease in fluid flow past the macula densa (see Chapter 25·IV·C), and an increase in sympathetic activity triggered by a decrease in MAP (see Chapter 19·IV). RAAS effects are mediated by Ang-II, whose actions are all geared toward retaining Na⁺ and water and raising MAP (Table 28.2).

2. **Sympathetic nervous system:** The SNS activates when the brainstem cardiovascular control centers detect a need to raise MAP, sensed by arterial baroreceptors (see Chapter 19·III·A). The SNS stimulates Na⁺ reabsorption both through direct effects on Na⁺ transport by the proximal tubule and through glomerular vasoconstriction, which reduces glomerular filtration rate (GFR) and urinary Na⁺ excretion.

3. **Antidiuretic hormone:** ADH's primary role is in maintaining water balance, but severe cardiovascular stress (e.g., hemorrhage) can cause circulating ADH levels to rise to the point that they constrict resistance vessels. ADH's effects on glomerular blood flow are similar to those of Ang-II. ADH can also stimulate Na⁺ reabsorption from the TAL and cortical collecting duct (CCD), which further enhances fluid retention.

4. **Atrial natriuretic peptide:** ANP is released when cardiac atria are stretched by high blood volumes and provides a negative feedback pathway that limits ECF volume expansion. Its principal effects are to antagonize Ang-II and ADH action and to stimulate natriuresis through direct effects on the glomerulus.

   a. **Angiotensin II:** ANP inhibits Na⁺/H⁺ exchange in the PT, Na⁺/Cl⁻ cotransport in the early distal tubule, and Na⁺ channels in the distal segments, all of which promote natriuresis.

   b. **Antidiuretic hormone:** ANP suppresses ADH release and prevents ADH-stimulated insertion of aquaporins into the apical membranes of the collecting ducts. These actions prevent water reabsorption and promote diuresis.

   c. **Vasculature:** ANP vasodilates to increase glomerular capillary hydrostatic pressure. GFR increases markedly as a result, causing diuresis.

> Brain natriuretic peptide (BNP) is a related peptide released from the ventricles when stressed by high preloads. Although ANP is rapidly metabolized by the liver, circulating BNP is more stable and provides an early, sensitive indicator of heart failure. BNP measurements are rapid and inexpensive. They are used clinically to determine the presence and severity of failure and to help exclude congestive heart failure as a possible cause of dyspnea.

D. Glomerulotubular balance

Na⁺ balance is maintained, in part, by a phenomenon known as **glomerulotubular (GT) balance**. GT balance refers to the PT's tendency to reabsorb a constant fraction of the filtered Na⁺ load regardless of GFR. Normally, fractional reabsorption is ~67%, although this value may change during ECF volume contraction and expansion. Therefore, when GFR increases (e.g., due to a rise in glomerular filtration pressure), the PT increases net Na^+ reabsorption to compensate for the increased amounts of Na^+ appearing in the tubule lumen. GT balance helps ensure that Na^+ is not excreted inappropriately when GFR rises. GT balance relies on changes in peritubular and tubule function.

1. **Peritubular:** When GFR increases due to an increase in filtration fraction (filtration fraction = GFR ÷ renal plasma flow), blood leaving the glomerulus via the efferent arteriole has a higher colloid osmotic pressure (π_c) compared with previously because the plasma proteins have been concentrated by glomerular filtration to a greater degree. A higher π_c enhances fluid reabsorption by the peritubular network serving the PT. Conversely, when GFR falls due to a decrease in filtration fraction, the osmotic potential favoring fluid reabsorption by peritubular blood is reduced, thereby facilitating GT balance.

2. **Tubule:** The PT's reabsorptive capacity typically exceeds the normal filtered load for most organic and inorganic solutes, including Na^+. When GFR and filtered load increases, the reabsorptive reserve allows the PT to compensate by increasing net uptake, which helps maintain GT balance.

E. Pressure-induced natriuresis

Hypertension produces a **pressure natriuresis** that occurs independently of the pathways delineated above. Pressure natriuresis acts as a safety valve to reduce ECF volume through Na^+ and water excretion, thereby bringing MAP back down to normotensive levels. Natriuresis results primarily from a hypertension-induced removal of Na^+/H^+ exchangers from PT villi, which reduces the segment's Na^+ reabsorptive capacity.

> Under resting conditions, most adjustments to blood volume are accomplished through the kidney's autoregulatory mechanisms in concert with osmoreceptor-mediated release of ADH. Other pathways are only called into action when the cardiovascular system is stressed.

IV. POTASSIUM BALANCE

The body contains ~3.6 mol (~140 g) of K^+, ~98% of which is concentrated within cells by the plasma membrane Na^+/K^+-ATPase. However, all cells express K^+ channels and K^+ transporters on their surface membranes that allow K^+ to move relatively freely between the ICF and ECF. These pathways make significant shifts in K^+ localization possible (e.g., during changes in pH balance; see Section D), causing an **internal K^+ balance** disturbance (Table 28.3). Left unchecked, such disturbances can have serious physiologic consequences, up to and including death

Clinical Application 28.2: Sodium Reabsorption and Blood Pressure

The relationship between Na^+, ECF volume, and blood pressure means that mutations in any of several key proteins involved in renal Na^+ reabsorption may potentially cause hypo- or hypertension.

Liddle syndrome is a very rare congenital disorder that increases epithelial sodium channel (ENaC) expression by the distal segments, resulting in increased Na^+ reabsorption. The syndrome is characterized by hypertension and may be associated with hypokalemia and metabolic alkalosis.

Pseudohypoaldosteronism type 1 mutations cause hyponatremia, hypotension, and hyperkalemia. The dominant form prevents mineralocorticoid receptor expression, causing aldosterone resistance. Recessive forms inhibit ENaC activity, thereby preventing Na^+ reabsorption in the distal segments.

Table 28.3: Conditions Affecting Internal Potassium Balance

Causal Event	Mechanism
Shifts from ECF to ICF (hypokalemia)	
↑ Insulin	↑ Na^+/K^+-ATPase
↑ Epinephrine	↑ Na^+/K^+-ATPase
Alkalosis	↑ Na^+/K^+-ATPase to compensate for cation (H^+) efflux
Shifts from ICF to ECF (hyperkalemia)	
Exercise	↑ Excitation and K^+ channel opening
Acidosis	H^+ displaces K^+ and inhibits uptake pathways.
↑ ECF osmolality	Cells lose water and K^+ follows by diffusion.
Cell trauma, necrosis	Loss of cellular K^+ containment

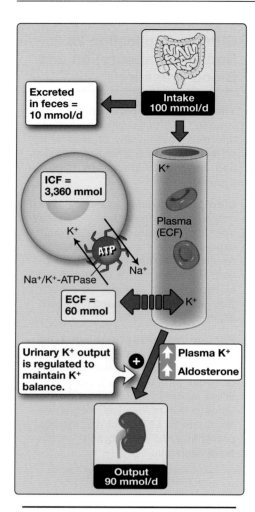

Figure 28.7.
Potassium balance.

(see Clinical Application 2.1). Despite the challenges, kidneys maintain plasma K^+ concentrations within a narrow range (3.5–5.0 mmol/L).

A. Tally sheet

Maintaining stable plasma K^+ concentration involves a simple balance between ingestion and urinary excretion (Fig. 28.7).

1. **Intake:** The U.S. RDA for potassium is 4.7 g (~120 mmol/d). Net intake varies widely with diet. Fruits and vegetables are particularly rich in K^+ and are more than adequate to meet daily K^+ needs. Most K^+ ingested is subsequently absorbed during transit through the GI tract.

2. **Output:** Kidneys are the only significant route for K^+ output. 80% of filtered K^+ is reabsorbed isosmotically in the PT. Another 10% is reabsorbed in the TAL. Regulation of K^+ balance occurs in the distal segments (see Chapter 27·IV·C). When dietary K^+ intake exceeds need (which is generally the rule), the distal segments secrete K^+ for urinary excretion. When the body is severely K^+ depleted, the tubule can reabsorb >99% of the filtered load.

B. Regulation

K^+ balance is achieved primarily by the distal segments. The responsibility for maintaining K^+ balance shifts from principal cells to α-intercalated cells depending on whether K^+ intake is high, and the excess must be secreted, or K^+ intake is restricted, requiring reabsorption.

1. **Potassium secretion:** The GI tract can transfer several tens of millimoles of K^+ to the vasculature during a typical meal. Processing such a significant K^+ load while maintaining plasma K^+ levels within a safe range requires that the excess be stored temporarily to give the kidneys time to excrete the surplus.

   a. **Interim storage:** Ingesting a meal causes circulating insulin levels to rise. Insulin has many effects on cell metabolism (see Chapter 33·III), including stimulation of Na^+/K^+-ATPase activity. Ingested K^+ moves temporarily from ECF to ICF as a result.

   b. **Decreased reabsorption:** When plasma K^+ rises, so does the concentration of K^+ entering the PT, which reduces reabsorption of all cations, including K^+ and Na^+.

   c. **Increased secretion:** Hyperkalemia stimulates aldosterone release from the adrenal cortex. Aldosterone targets principal cells in the distal segments, promoting an increase in basolateral Na^+/K^+-ATPase and apical renal outer medullary K^+ channel (ROMK) expression. The Na^+/K^+ pump creates the driving force, and ROMK provides a pathway for increased K^+ secretion into the tubule lumen.

2. **Potassium reabsorption:** Hypokalemia suppresses aldosterone release, inhibiting K^+ secretion. Hypokalemia simultaneously stimulates H^+/K^+-ATPase activity in collecting duct α-intercalated cells, promoting K^+ reabsorption from the ductal lumen.

C. Sodium and potassium balance relationship

The pathways regulating Na^+ balance and K^+ balance converge on basolateral Na^+/K^+-ATPase activity in the distal segments (Fig. 28.8).

The Na^+/K^+-ATPase exchanges Na^+ for K^+, simultaneously enhancing Na^+ reabsorption and K^+ secretion. Because there are situations in which secreting K^+ during Na^+ reabsorption (or vice versa) would be deleterious, the two processes must be functionally uncoupled. Uncoupling is achieved through the potent effects of tubule flow rate on K^+ excretion.

1. **Flow:** K^+ secretion by principal cells is powered by the transepithelial K^+ electrochemical gradient. When tubule flow rates are low, K^+ diffusing from principal cells causes luminal K^+ concentrations to rise significantly, which blunts the driving force for further diffusion and secretion (Fig. 28.9A). When tubule flow rates are high, K^+ is flushed out of the distal segments at an accelerated rate, and the gradient favoring K^+ secretion remains high (see Fig. 28.9B). Note that high flow rates also open large-conductance Ca^{2+}-activated K^+ channels (maxi-K^+ channels, or BK channels) located in the apical membrane of both principal cells and intercalated cells. The role of BK channels in K^+ balance is uncertain.

2. **Diuresis:** When ECF volume is too high, the kidney excretes Na^+ and water at increased rates. The first step in excretion is to increase GFR and reduce reabsorption from the PT, which raises flow rates through the distal tubule. High flow rates would be expected to cause excessive K^+ secretion, but when ECF volume and MAP are high, the RAAS is suppressed. In the absence of aldosterone, K^+ secretion by the distal segments is attenuated, thereby preventing excessive flow-induced K^+ loss (see Fig. 28.9C).

3. **Volume expansion:** When ECF volume and MAP are low, the RAAS is activated, and the tubule's Na^+ reabsorptive capacity increases. This allows for Na^+ and water retention, but it simultaneously upregulates the pathway that mediates K^+ secretion by principal cells. Although this might be expected to increase K^+ excretion, it occurs in the context of low flow rates through the tubule, which blunts the driving force for K^+ secretion.

D. pH and potassium balance relationship

K^+ balance is very sensitive to changes in pH balance. Acidosis causes hyperkalemia, whereas alkalosis causes hypokalemia. These disturbances reflect combined effects of H^+ on ICF and renal function.

1. **Intracellular fluid:** H^+ has a variety of ways of crossing cell membranes, so, when plasma H^+ concentrations rise, ICF concentration rises also. Because H^+ carries a positive charge, H^+ influx would be expected to depolarize membrane potential (V_m), but the cell responds with a counterbalancing K^+ efflux to maintain V_m at resting levels. ECF K^+ concentration rises as a result (hyperkalemia). Alkalosis has the opposite effect. When plasma H^+ concentrations fall, H^+ diffuses out of the cell, and K^+ is taken up from the ECF to redress the charge imbalance. The result is hypokalemia.

2. **Renal function:** Although the kidneys might be expected to correct such K^+ imbalances, in practice, H^+ has simultaneous effects on tubule function that cause the imbalance to worsen. Acidosis inhibits K^+ secretion by the distal segments by inhibiting principal cell Na^+/K^+-ATPase activity. Inhibition reduces K^+ uptake from blood and reduces the concentration gradient driving K^+ efflux

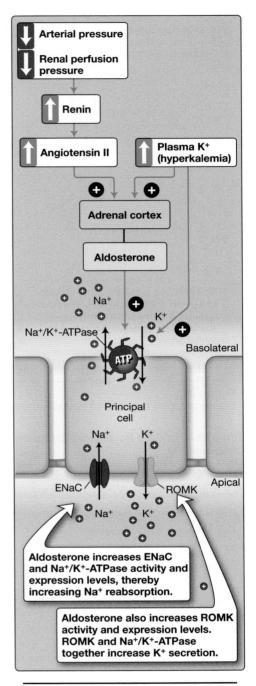

Figure 28.8.
Convergence of pathways regulating Na^+ reabsorption and K^+ secretion in the distal segments. ENaC = epithelial Na^+ channel, ROMK = renal outer medullary K^+ channel.

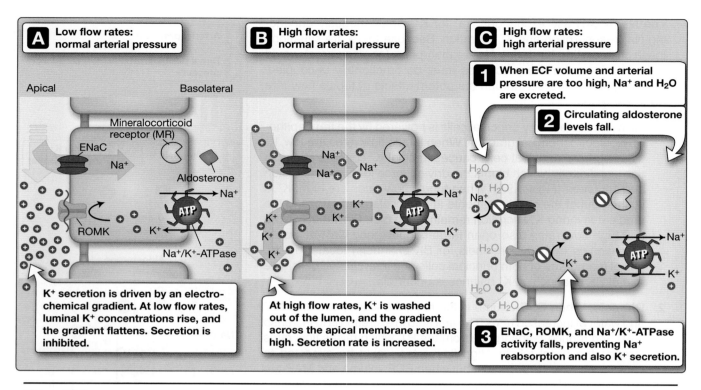

Figure 28.9.
Effects of tubule flow rate on potassium excretion. ENaC = epithelial Na$^+$ channel, ROMK = renal outer medullary K$^+$ channel.

Table 28.4: Urinary Potassium Excretion Determinants

Variable	Net Effect on K$^+$ Excretion
Hypokalemia	Inhibit
Hyperkalemia	Increase
Alkalosis	Increase
Acidosis	Decrease
↓ Tubule flow	Decrease
↑ Tubule flow	Increase
Aldosterone	Increase

across the apical membrane into the tubule lumen. H$^+$ also inhibits apical K$^+$ channels in principal cells, reducing K$^+$ secretion directly and further potentiating hyperkalemia. Alkalosis has the opposite effect, promoting K$^+$ secretion and hypokalemia. Factors affecting K$^+$ excretion are summarized in Table 28.4.

E. Sodium, potassium, and pH balance relationship

Na$^+$ reabsorption by the PT occurs, in part, via an apical Na$^+$/H$^+$ exchanger (NHE3), which uses the transmembrane Na$^+$ gradient to power H$^+$ secretion. This Na$^+$/H$^+$ coupling means that pathways modulating Na$^+$ reabsorption can also affect pH balance. When MAP or ECF volume falls, Na$^+$/H$^+$ exchanger activity increases due to Ang-II release, and the ensuing H$^+$ secretion increase results in **contraction alkalosis**. Aldosterone further potentiates the alkalosis by increasing NHE3 expression levels in the PT and by stimulating H$^+$/K$^+$-ATPase activity in the distal segments. Conversely, when MAP or ECF volume rises, Na$^+$ reabsorption and H$^+$ secretion are attenuated, causing acidosis. Alterations in K$^+$ balance also affect pH balance. Hypokalemia causes alkalosis by stimulating NHE3 activity and NH$_3$ production in the PT and by stimulating H$^+$/K$^+$ pump activity in the distal segments. Conversely, hyperkalemia causes acidosis.

V. pH BALANCE

pH balance is achieved through the combined actions of the lungs and kidneys. The lungs excrete volatile acid (H$_2$CO$_3$, which is expired as CO$_2$). The kidneys excrete nonvolatile acid (Fig. 28.10).

A. Tally sheet

The tally sheet for acid is unusual insofar as most acid is generated internally by metabolism rather than being ingested.

1. **Intake:** The majority of daily acid "intake" (~15–22 mol) is formed from carbohydrate metabolism. An additional 70 to 100 mmol/d of nonvolatile acid (nitric, sulfuric, and phosphoric acid) is generated through breakdown of amino acids and phosphate compounds.

2. **Output:** Most of the CO_2 generated during metabolism and converted to H^+ and HCO_3^- for blood transport is subsequently excreted by the lungs. A small amount of volatile acid remains trapped in the body when HCO_3^- is lost in feces and must be excreted by the kidneys as nonvolatile acid. Nonvolatile acid is excreted primarily as titratable acid and ammonium in the PT (see Chapter 26·IX·C).

B. Regulation

Volatile acid is sensed by chemoreceptors (see Chapter 24·III·A) and regulated by adjusting respiratory rate. All nephron segments are involved in excreting nonvolatile acid, but the PT and intercalated cells in the distal segments play prominent roles.

1. **Volatile acid:** Brainstem respiratory control centers monitor arterial Pco_2 (P_aco_2) via changes in CSF pH and input from other chemoreceptor populations. If either parameter is higher than optimal, the control centers increase alveolar ventilation to transfer additional acid to the atmosphere. If P_aco_2 or H^+ levels are lower than normal, alveolar ventilation rates and CO_2 transfer decrease.

2. **Nonvolatile acid:** The PT secretes the bulk of the daily nonvolatile acid load. Acidosis increases H^+ secretion and upregulates NH_3 synthesis by the PT, whereas alkalosis has the opposite effect. The primary effectors of pH balance are intercalated cells in the distal segments (see Chapter 27·V·E). Chronic metabolic acidosis increases the proportion of acid-secreting α-intercalated cells, whereas metabolic alkalosis increases the proportion of HCO_3^--secreting β-intercalated cells.

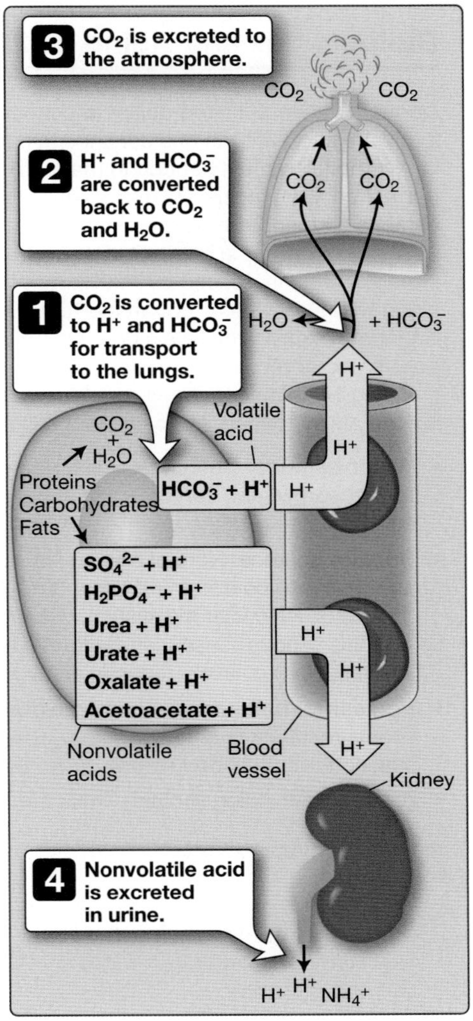

Figure 28.10.
Volatile and nonvolatile acid excretion.

VI. ACID–BASE DISORDERS

pH balance can be upset by changes in pulmonary, GI, and renal function or can be triggered by changes in acid or base production. In practice, this means that acid–base disorders are encountered frequently in clinical medicine.

A. Types and compensation

There are four basic types of "simple" acid–base disorders. Respiratory acidosis and alkalosis are primary disorders of CO_2 handling by the lungs. Metabolic acidosis and alkalosis manifest as a primary disorder in plasma HCO_3^- levels, although there may be many underlying causes (discussed in the following text).

Cells are protected from acid–base changes by three primary defense mechanisms with varying time courses and efficacy: buffers (immediate), the lungs (minutes), and the kidneys (days; see also Chapter 3·IV).

Biologic Sex and Aging 28.2: Renal Homeostatic Functions

Neonates have a limited ability to concentrate urine. Nephrogenesis is completed during gestation, but neonatal nephron loops are short at birth and continue to develop postnatally, thereby increasing the ability to establish the strong corticopapillary osmotic gradient needed to reabsorb water. Nephron length is not the only consideration, however. Although antidiuretic hormone (ADH) levels are normal at birth, neonates have limited urea- and Na^+-reabsorptive capabilities, meaning that the osmotic gradient with its urine-concentrating prowess does not achieve adult levels until about age 1 year.

Aging witnesses a reversal of these trends. By the eighth decade, urine-concentrating ability has decreased by 20%, and urine flow has increased by 100%. Again, ADH levels are normal, but urea transporter numbers have decreased by 50%, along with $Na^+/K^+/2Cl^-$ cotransporter levels in the thick ascending limb, which limits formation of a strong osmotic gradient. Aquaporin expression levels drop by 80%, so, not only is the motive force for water reabsorption decreased, pathways available for water recovery are reduced also.

Aging-related changes in renal function echo those seen in other organs: Basal homeostatic functionality is preserved, but the ability to respond to challenge is greatly reduced.

- **Water balance:** Decreased thirst and water intake (see Biologic Sex and Aging 28.1), combined with decreased ability to reabsorb water, can cause fatal hypernatremia when water intake is restricted. Conversely, older adults also have a limited ability to dilute urine, so excessive water intake can result in a severe hyponatremia.
- **Sodium balance:** Normal plasma Na^+ levels fall by 1 mEq/L per decade, reflecting a decreased ability to reabsorb Na^+ with age. This may be due, in part, to a 30%–60% decrease in basal circulating levels of the various components of the renin–angiotensin–aldosterone system, together with a blunted response to stimuli that would normally increase renin secretion. This has serious consequences for the ability to maintain mean arterial pressure when Na^+ intake is limited. Conversely, aging kidneys also have a reduced ability to excrete an excessive Na^+ load, which may lead to hypervolemia and hypertension.
- **Potassium balance:** Decreased aldosterone secretion in response to stressors similarly impairs responsiveness to events resulting in K^+ excess (e.g., medications), leading to hyperkalemia. Disturbances in pH balance can exacerbate susceptibility to hyperkalemia, given the linkage between K^+ balance and pH balance.
- **pH balance:** Aging similarly limits the ability of kidneys to respond to increased or decreased acid loads.

When more than one type of simple acid–base disturbance is present, a "mixed" acid–base disorder is said to exist. The number of identifiable disorders never exceeds three because a body cannot simultaneously over- and under-excrete CO_2. A "triple" disorder is, therefore, metabolic alkalosis, metabolic acidosis, plus a respiratory disorder.

---

1. **Buffers:** Buffers limit the effects of acid–base changes until compensation can occur. The principal intracellular buffers include proteins (including hemoglobin in red blood cells) and phosphates. The principal buffer in ECF is HCO_3^-, which combines with H^+ to form H_2O and CO_2 via a reaction catalyzed by carbonic anhydrase ([CA] Fig. 28.11).

   Equation 28.1 $$H^+ + HCO_3^- \leftrightarrows H_2CO_3 \overset{CA}{\leftrightarrows} CO_2 + H_2O$$

   An increase in nonvolatile acid production is buffered by HCO_3^-, causing ECF (including plasma) HCO_3^- levels to fall. Conversely, increased CO_2 production increases plasma HCO_3^- levels even as plasma pH falls (see Fig. 28.11).

2. **Lungs:** Respiratory control centers located in the brainstem adjust alveolar ventilation to increase or decrease CO_2 transfer

to the atmosphere. Because respiratory rate is normally 12 to 15 breaths/min, compensation occurs rapidly.

3. **Kidneys:** Kidneys are the third and final line of acid defense, adjusting the amount of H⁺ they secrete to maintain strict control over pH balance. Upregulation of the necessary enzymatic pathways takes hours to implement, making renal compensation much slower than respiratory compensation (up to 3 days).

B. Clinical assessment

Assessing a patient's acid–base status requires data from an arterial blood gas (ABG) sample and a basic metabolic panel. The ABG analysis provides data on pH, P_aCO_2, arterial Po_2, and HCO_3^-, which provides a starting point from which the nature of the acid–base disturbance (e.g., simple vs. mixed, respiratory vs. metabolic) may be ascertained. The metabolic panel provides contiguous data that help interpret the metabolic origin of an acid–base disturbance.

1. **Davenport diagram:** Davenport diagrams are typically not used clinically, but they are helpful to illustrate how acid–base disorders manifest as changes in arterial pH, Pco_2, and HCO_3^-. The diagram is a pictorial representation of the Henderson-Hasselbalch equation (Fig. 28.12):

$$pH = pK + \log\frac{\left[HCO_3^-\right]}{\left[CO_2\right]}$$

where $[HCO_3^-]$ and $[CO_2]$ represent plasma HCO_3^- and CO_2 concentrations, respectively (the latter calculated from the product of Pco_2 and CO_2 solubility). At a plasma pH of 7.4 and P_aCO_2 of 40 mm Hg, $[HCO_3^-]$ is ~26 mmol/L. Changes in plasma H⁺ and Pco_2 cause HCO_3^- concentrations to shift in a predictable manner (see Equation 28.1 and Fig. 28.12).

2. **Anion gap:** The anion gap is an important clinical determination that helps identify and distinguish among types of metabolic acidosis (see Section E in the following discussion). An anion gap is calculated by comparing measured serum cation and anion concentrations, which, according to the principle of bulk electroneutrality, must always be equal. The principal measured cation is Na⁺ (see Fig. 28.1). The principal measured anions are Cl⁻ and HCO_3^-. Typical serum values for these ions are 140 mmol/L Na⁺, 100 mmol/L Cl⁻, and 25 mmol/L HCO_3^-. The difference between serum Na⁺ and the sum of serum Cl⁻ and HCO_3^- is the anion gap (Fig. 28.13). The anion gap represents all the unmeasured anions, including proteins and organic ions such as phosphate, citrate, and lactate:

$$\text{Normal anion gap} = \left[Na^+\right] - \left(\left[Cl^-\right] + \left[HCO_3^-\right]\right)$$
$$= 140 - (100 + 25) = 15 \text{ mmol/L}$$

The anion gap is normally 8–16 mmol/L. Some forms of metabolic acidosis are caused by accumulation of lactate, keto acids, or other such unmeasured anions, which causes the gap to widen.

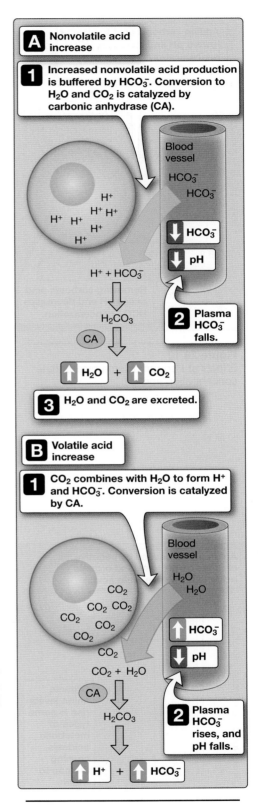

Figure 28.11.
Effects of nonvolatile and volatile acids on plasma bicarbonate and pH.

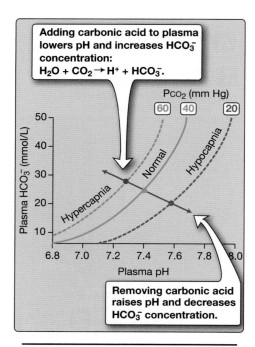

Figure 28.12.
Effects of P_{CO_2} on plasma bicarbonate concentration and pH.

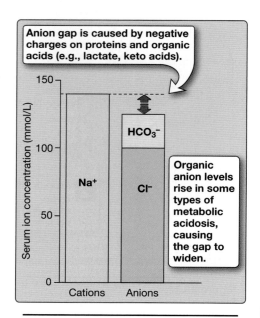

Figure 28.13.
Serum anion gap.

C. Respiratory acidosis

Respiratory acidosis is usually caused by hypoventilation but can result from any condition that allows P_{aCO_2} to rise. Respiratory acidosis is characterized by an elevated P_{aCO_2} and a low arterial pH.

1. **Causes:** Causes of respiratory acidosis include decreased ventilatory drive, an air pump disorder, and processes that interfere with gas exchange.

    a. **Ventilatory drive:** Because ventilatory drive and the respiratory rhythm originate in the brainstem, congenital CNS disorders or tumors affecting brainstem function can potentially cause respiratory acidosis. For example, **Ondine curse** is a rare form of congenital central hypoventilation syndrome in which ventilatory drive and respiratory reflexes are absent. Drugs that suppress CNS function (e.g., opiates, barbiturates, and benzodiazepines[1]) can also cause respiratory depression and increase P_{aCO_2}.

> In medicine, the term "Ondine curse" is synonymous with hypoventilation associated with loss of autonomic respiratory drive. The term has its origins in European mythology, referring to a water nymph (Undine), who became mortal in order to marry a man she had fallen in love with. When she aged, her husband fell into the arms of a younger woman. Ondine punished him with a curse that forced him to have to remember to breathe. Once he finally fell asleep, he died. In reality, there is no record of any such curse: the myth was misquoted in the medical literature.

    b. **Air pump:** Inspiration is accomplished through contraction of inspiratory muscles (diaphragm and external intercostals) that expand the lungs and create the pressure gradient that drives air into the alveoli. Any disease process that affects these muscles or their motor command pathways potentially may cause respiratory acidosis. Common examples include amyotrophic lateral sclerosis, myasthenia gravis (see Clinical Application 12.2), muscular dystrophy (see Clinical Application 12.1), and infectious diseases such as poliovirus (see Clinical Application 5.1).

    c. **Gas exchange:** Airway obstruction can prevent normal alveolar ventilation and cause P_{aCO_2} to rise. Causes include aspiration of a foreign body, bronchospasm, chronic obstructive lung diseases, and obstructive sleep apnea (see Clinical Application 24.1). Conditions that cause the alveoli to fill with fluid (pulmonary edema), pus (pneumonia), or other infiltrates (e.g., acute respiratory distress syndrome; see Chapter 40·VI·D) may also increase P_{aCO_2}.

2. **Compensation:** The effects of acidosis caused by an acute rise in P_{aCO_2} are limited by the HCO_3^- buffer system. Hypercapnia biases Equation 28.1 in favor of HCO_3^- formation, so plasma HCO_3^- rises

[1]For more information on drugs that cause central nervous system and respiratory depression, see *LIR Pharmacology*, 8e.

even as pH falls (Fig. 28.14). Compensation occurs over a period of several days and involves an increase in renal H^+ secretion and NH_3 production. The "new" HCO_3^- generated during H^+ secretion and NH_3 synthesis is transferred to ECF, so plasma HCO_3^- rises further during compensation.

D. Respiratory alkalosis

Respiratory alkalosis is *always* caused by hyperventilation and is characterized by a low P_aCO_2 and an elevated arterial pH.

1. **Causes:** There are fewer primary causes of respiratory alkalosis compared with respiratory acidosis. They include increased ventilatory drive and hypoxemia.

   a. **Ventilatory drive:** Hyperventilation is a common response to anxiety, such as might be induced by fear or pain, panic attacks, and hysteria. A mild respiratory alkalosis may also occur during pregnancy (see Chapter 37·IV·E). Aspirin poisoning causes respiratory alkalosis by stimulating the respiratory control centers directly.[1]

   b. **Hypoxemia:** Hypoxemia raises respiratory rate and can cause respiratory alkalosis in some circumstances. Ascent to high altitude stimulates hyperventilation to compensate for reduced O_2 availability and can precipitate respiratory alkalosis (see Chapter 24·V·A). Respiratory alkalosis may also occur when O_2 uptake is impaired due to pulmonary embolism or severe anemia.

2. **Compensation:** An acute fall in P_aCO_2 is accompanied by a decrease in plasma HCO_3^- levels (Fig. 28.15). Compensation involves reduced H^+ secretion and decreased NH_3 synthesis by the kidneys. Because less "new" HCO_3^- is formed, plasma HCO_3^- level falls further during compensation.

E. Metabolic acidosis

Metabolic acidosis is caused by increased nonvolatile acid accumulation. It may also result from excessive HCO_3^- loss from the body. Metabolic acidosis is characterized by a low plasma HCO_3^- and a low arterial pH.

1. **Causes:** Metabolic acidosis can result from a number of endogenous and exogenous mechanisms, including excess nonvolatile acid production, poisoning, HCO_3^- loss, and an impaired ability to excrete H^+.

   a. **Acid production:** The body normally generates ~1.5 mol of lactic acid per day, almost all of which is metabolized, mostly by the liver. Strenuous muscle activity can increase lactate production temporarily, but the liver has a high metabolic capacity, and lactate levels typically renormalize within 30 minutes. When the liver is damaged, lactate accumulates and causes a lactic acidosis. Ketoacidosis is a metabolic acidosis resulting from ketone body production (i.e., acetone, acetoacetic acid, β-hydroxybutyrate) and metabolism. Ketoacidosis usually is associated with an insulin deficiency (**diabetic ketoacidosis**;

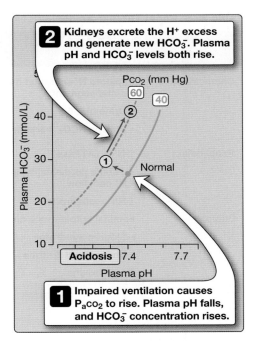

Figure 28.14.
Respiratory acidosis and renal compensation.

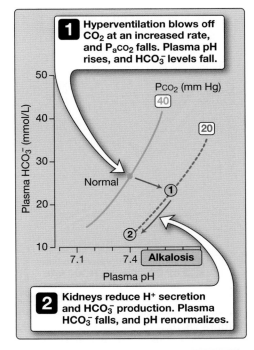

Figure 28.15.
Respiratory alkalosis and renal compensation.

[1]For more information on the side effects of salicylates, see *LIR Pharmacology*, 8e, Chapter 40·III·A.

Table 28.5: Renal Tubular Acidosis (RTA)

	Type 1 RTA (Distal RTA)	Type 2 RTA (Proximal RTA)	Type 4 RTA (Hypoaldosteronism)
Characteristics	Impaired H$^+$ secretion by the distal segments • Urine pH >5.3 • Plasma HCO$_3^-$ variable	Impaired proximal HCO$_3^-$ reabsorption • Urine pH variable • Plasma HCO$_3^-$ 12–20 mmol/L	Impaired aldosterone release or response • Urine pH <5.3 • Plasma HCO$_3^-$ >17 mmol/L • Hyperkalemia
Renal defect	↓ H$^+$/K$^+$-ATPase ↑ Tubule permeability, allowing H$^+$ backflow ↓ Na$^+$ reabsorption	Nonspecific tubule dysfunction or mutations in genes involved in HCO$_3^-$ reabsorption	Impaired Na$^+$ reabsorption via epithelial Na$^+$ channel (ENaC)
Etiology	Familial autoimmune disorders • Sjögren syndrome • Rheumatoid arthritis Drugs, toxins	Familial Fanconi syndrome Drugs, toxins Carbonic anhydrase inhibitors	Congenital hypoaldosteronism (Addison disease) Aldosterone resistance Diabetic nephropathy Drugs (including K$^+$-sparing diuretics)

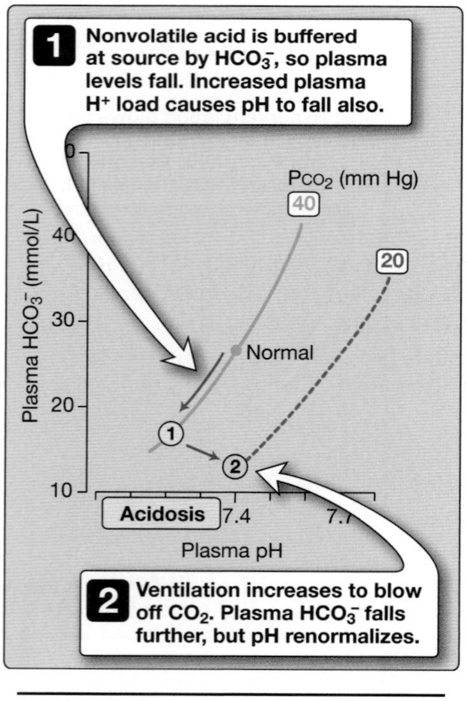

Figure 28.16.
Metabolic acidosis and respiratory compensation.

see Clinical Application 33.1). Lactic acidosis and ketoacidosis both cause a high anion gap metabolic acidosis.

b. **Drugs and poisons:** Aspirin is an acid that can produce a mixed disorder with a high anionic gap when ingested at toxic levels. Other common causes of **toxic acidosis** include methanol and ethylene glycol ingestion. Methanol is often consumed as a cheap ethanol substitute, whereas ethylene glycol is an antifreeze typically ingested accidently. Neither is toxic until metabolized. Methanol is converted to formaldehyde and formic acid, whereas ethylene glycol is metabolized to glycoaldehyde and glycolic and oxalic acids. Both toxins cause a high anionic gap metabolic acidosis.

c. **Bicarbonate loss:** The small and large intestines secrete HCO$_3^-$, which may be excreted inappropriately during episodes of severe diarrhea, causing metabolic acidosis. HCO$_3^-$ loss may also result from congenital or acquired disorders that impair HCO$_3^-$ reabsorption by the PT. The result is a normal anion gap acidosis known as **type 2 renal tubular acidosis** ([**RTA**] Table 28.5). Diuretics, especially CA inhibitors (see Chapter 26·IX·A), can also cause HCO$_3^-$ loss via urine.

d. **Impaired acid excretion:** Type 1 and type 4 RTA are both characterized by a decreased ability to excrete H$^+$, which results in a normal anion gap acidosis. Type 1 RTA is usually due to a congenital inability to acidify urine in the distal segments. Type 4 RTA results from hypoaldosteronism or an impairment of the renal tubule's ability to respond to aldosterone (see Table 28.5 and Chapter 28·IV·E).

2. **Compensation:** Nonvolatile acid excesses are buffered by plasma HCO$_3^-$, causing plasma concentrations to fall (Fig. 28.16). Acidosis initiates a reflex increase in ventilation to transfer volatile acid to the atmosphere, and plasma HCO$_3^-$ falls further.

F. Metabolic alkalosis

Metabolic alkalosis results when the body takes in HCO_3^- or loses H^+ and is characterized by elevated plasma HCO_3^- and arterial pH.

1. **Causes:** Although metabolic alkalosis can be caused by excessive $NaHCO_3$ intake ($NaHCO_3$ is used as an antacid), the most common causes include diuretics, vomiting, and nasogastric (NG) suctioning. Vomiting and NG suctioning both transfer stomach acid to the body's exterior, leaving an HCO_3^- excess and alkalosis (Fig. 28.17).

2. **Compensation:** Metabolic alkalosis acutely increases plasma HCO_3^- levels and raises pH, but the respiratory system soon compensates by decreasing ventilation and allowing P_aCO_2 to rise. The kidneys may also assist compensation by reducing H^+ secretion and allowing filtered HCO_3^- to pass through the tubule to the bladder. Fluid loss during prolonged vomiting may also result in ECF volume contraction, which favors HCO_3^- reabsorption and manifests as a contraction alkalosis.

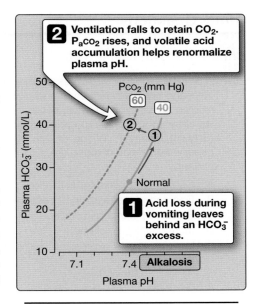

Figure 28.17.
Metabolic alkalosis and respiratory compensation.

Chapter Summary

- **Total body water (TBW)** is sensed through changes in the **osmolality** of ECF. **Osmoreceptors** are located within **circumventricular organs** within the brain, near the **hypothalamus**.

- A decrease in TBW increases ECF osmolality. The osmoreceptors respond by stimulating **antidiuretic hormone (ADH)** release from the **posterior pituitary**. ADH causes **aquaporins** to be inserted into the epithelial lining of the **collecting ducts**, which permits water reabsorption. Osmoreceptor activation also increases **thirst**.

- ECF osmolality also depends on plasma Na^+ levels. Na^+ balance is controlled principally by **aldosterone**-induced increases in renal Na^+ retention. Aldosterone is released during **renin–angiotensin–aldosterone system (RAAS)** activation.

- Both water and Na^+ balance are dominated by the need to optimize **mean arterial pressure (MAP)**. MAP is determined, in part, by ECF volume. When ECF volume is low, MAP falls, and the RAAS activates. **Angiotensin II (Ang-II)** is the primary RAAS effector hormone. Ang-II stimulates aldosterone release, modulates glomerular filtration rate, stimulates Na^+ reabsorption from the renal tubule, promotes water retention via ADH, and causes thirst.

- **Atrial natriuretic peptide (ANP)** provides a negative feedback pathway that limits ECF volume expansion. ANP is released from cardiac atria when ECF volume is high. ANP antagonizes the actions of Ang-II and promotes **natriuresis** and **diuresis**.

- K^+ balance is controlled by aldosterone. Aldosterone is released as a direct response to **hyperkalemia** and stimulates K^+ secretion by the distal tubule. **Hypokalemia** stimulates K^+ reabsorption, primarily in the distal tubule.

- The kidneys and lungs together maintain pH balance. The lungs excrete **volatile acid** (H_2CO_3). The kidneys excrete **nonvolatile acid** and can help compensate for changes in pH balance caused by respiratory disorders.

- Plasma pH is normally maintained within a narrow range (7.35–7.45). A rise in P_aCO_2 causes **respiratory acidosis** and **acidemia** (pH < 7.35). The kidneys compensate by excreting additional H^+. **Hyperventilation** causes **respiratory alkalosis** and **alkalemia** (pH > 7.45). The kidneys compensate by reducing H^+ secretion.

- Accumulation of nonvolatile acids (e.g., lactic acid and keto acids), toxins, and renal disturbances in H^+ secretion or HCO_3- reabsorption can cause **metabolic acidosis**. The lungs compensate by increasing ventilation and transferring CO_2 to the atmosphere. Loss of stomach H^+ from prolonged vomiting causes **metabolic alkalosis**. The lungs compensate by retaining CO_2, and the kidneys decrease H^+ secretion.

Study Questions

Choose the ONE best answer.

28.1. A 66-year-old male recovering from a stroke is nauseated and reports headaches and malaise. Laboratory studies reveal hyponatremia, consistent with a diagnosis of syndrome of inappropriate antidiuretic hormone (SIADH) secretion. Which of the following additional observations would most likely be expected in this patient?

A. Central osmoreceptor activation
B. Decreased mean arterial pressure
C. Elevated plasma osmolality
D. Increased ECF volume
E. Low aquaporin expression levels in the collecting ducts

Best answer = D. Antidiuretic hormone (ADH) secretion at rates that exceed the need to conserve water causes water retention. ECF volume rises as a result. Retained water dilutes out ECF Na^+ to cause hyponatremia. Na^+ is the primary determinant of ECF osmolality which would therefore be decreased, not increased (C) in this patient, and central osmoreceptors would be inactive (A). ECF volume is a determinant of left ventricular preload and cardiac output, so mean arterial pressure would increase, not decrease (B). ADH increases aquaporin 2 expression in the renal collecting ducts, so expression levels would not be low (E) in this patient with SIADH.

28.2. A researcher observes a consistent 75% decrease in renal blood flow in subjects performing a cardiopulmonary exercise test. The test involves pedaling a stationary bike for 8–15 min. Pedal resistance is slowly increased until the test subjects are no longer able to continue. Which of the following most likely accounts for the decreased flow?

A. Antidiuretic hormone release
B. Decrease in mean arterial pressure
C. Decrease in renal arterial pressure
D. Increased renal sympathetic nerve activity
E. Sweat-induced hypovolemia

Best answer = D. The sympathetic nervous system (SNS) increases cardiac output and decreases flow to less essential organs (such as the kidney) to maintain a normal mean arterial pressure (MAP) during skeletal muscle vasodilation (see Section III·C and Chapter 39·V), not decrease it (B). The SNS constricts resistance vessels, including glomerular arterioles, which reduces renal blood flow. Although antidiuretic hormone (A) can vasoconstrict under some circumstances, these effects are secondary to SNS activation. Renal arterial pressure is closely tied to MAP and should not be affected to a significant degree (C); the SNS-induced decrease in renal blood flow is due to vasoconstriction of arterioles that are downstream of the renal artery. Sweat-induced hypovolemia (E) may potentiate SNS-induced attenuation of renal flow during prolonged exercise, but not during brief exercise tests with a low thermal load.

28.3. A physician notes that an underweight teenage girl's tooth enamel is eroded. A basic metabolic panel reveals hypokalemia and metabolic alkalosis, suggestive of an eating disorder and repeated purging. Which of the following additional findings would most likely be consistent with such a diagnosis?

A. Decreased β-intercalated cell activity
B. High plasma aldosterone levels
C. Hypoventilation
D. Increased NH_4^+ excretion
E. Renal tubular acidosis

Best answer = C. Loss of stomach acid during repeated vomiting leaves an HCO_3^- excess that manifests as metabolic alkalosis (see Section VI·F). The respiratory centers help compensate by decreasing volatile acid (H_2CO_3) transfer to the environment by decreasing alveolar ventilation (hypoventilation). β-Intercalated cells secrete HCO_3^- into the tubule lumen, so their activity would be increased, not decreased (A), during alkalosis. Aldosterone (B) is involved in Na^+ balance, not pH balance. NH_4^+ excretion helps dispose of nonvolatile acid, so excretion rates would fall, not increase (D), during alkalosis. Renal tubular acidosis (E) is a metabolic acidosis that may have several underlying causes.

Principles and Signaling

29

I. OVERVIEW

We are what we eat. All animals, including *Homo sapiens*, derive the energy substrates needed to fuel metabolism and growth by ingesting biologic tissue and breaking it down into simple chemical structures (e.g., carbohydrates, amino acids, and lipids). Disassembling tissues is the job of the gastrointestinal (GI) tract and its support organs (pancreas, gallbladder, and liver). The GI tract is a 30-foot-long hollow tube divided into a series of compartments separated by sphincters (Fig. 29.1). Food enters the GI system via the mouth, which is where disassembly begins. It travels through the esophagus to the stomach, which sterilizes food and grinds it into smaller pieces. The small intestine reduces food to simple biologic building blocks and then absorbs them. The large intestine is a fermentation compartment where bacteria break down plant fiber to release energy substrates. The large intestine also stores wastes until they can be eliminated. Food disassembly and absorption is a complex process: food must be physically moved between compartments, a series of fluids must be added at the appropriate moment, and nutrient absorption may be regulated according to specific bodily needs. Coordinating these various activities is the **enteric nervous system** (**ENS**). The ENS functions largely independently of the central nervous system (CNS), made possible by a sophisticated network of ~200 million neurons and many times that number of support cells. Given the system's complexity and how powerfully the need to ingest food influences human behavior and emotions, the ENS is commonly referred to as the "second brain."

II. GASTROINTESTINAL LAYERS

Each of the compartments within the GI tract has a different and distinct function, yet they share a common tubular structure that is modified according to the task the compartment must perform. The tube is

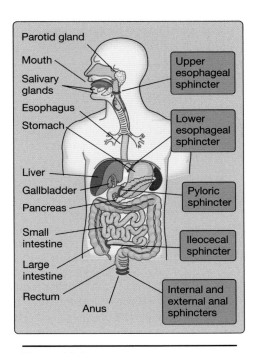

Figure 29.1.
GI system anatomy.

Clinical Application 29.1: Microbiome

Every external body surface, including the integumentary, respiratory, and GI systems, is colonized by microorganisms. The GI tract contains a complex community of microbes (gut **microbiota**) that includes archaea, viruses, protists, yeast, fungi, and more than 1,000 species of bacteria. These microbes and their unique gene complement are referred to as the **microbiome**. The number of bacteria in the gut microbiome is estimated to be more than 100 trillion. These bacteria produce several beneficial chemicals, including secondary bile acids and vitamins K and B_{12}. They also produce neurotransmitters such as serotonin and γ-aminobutyric acid (GABA) that are absorbed by the intestinal epithelium and circulate freely in the body, which has led to the suggestion that the microbiome constitutes a "third brain."

In health, we have a symbiotic relationship with our microbiomes. Although the microbiome includes many pathogens (e.g., *Clostridioides difficile*), their numbers are held in check by commensals and symbionts through cell-to-cell communication and release of antimicrobials that limit pathogen growth. Prolonged antibiotic use can cause an imbalance in microbial populations, leading to chronic or recurrent *C. difficile* infection. Symptoms include diarrhea, abdominal pain, and nausea and can result in severe complications such as sepsis. Chronic *C. difficile* infections can be effectively treated by transplanting colonic microbiota from healthy individuals to restore normal microbial balance.

Disturbances in the microbiome are associated with several diseases, including inflammatory bowel disease ([IBD] e.g., Crohn disease and ulcerative colitis), cardiovascular disease, type 2 diabetes mellitus, obesity, and neuropsychiatric disease. For example, *Bacteroidetes* spp., which make up >50% of gut microbiota, produce short-chain fatty acids such as butyrate (see Chapter 31·III·B). Butyrate binds to an epithelial G protein–coupled receptor, activation of which decreases inflammation and increases expression of proteins that increase epithelial barrier tightness. A decrease in the number of *Bacteroidetes* spp. results in epithelial leakiness and symptoms associated with IBD. Proving that changes in the microbiome affect human health and disease has been challenging, but this emerging research area could provide interesting insights into many disease etiologies.

composed of multiple anatomically distinct layers (Fig. 29.2) comprising the **mucosa**, **submucosa**, **muscularis externa**, and the **serosa**.

A. Mucosa

The epithelium, lamina propria, and lamina muscularis mucosae together form the mucosa, which is specialized for **protection**, **secretion**, and **absorption**.

1. **Epithelium:** The **epithelium** forms a continuous barrier that separates GI tract contents from the rest of the body and prevents entry of pathogens. In most sections of the GI tract, the epithelium comprises a single cell layer. The esophagus is an exception, being lined with a stratified epithelium that protects it from mechanical abrasion associated with swallowing food. GI epithelial cells are shed and replaced every 2 to 3 days. In segments of the GI tract tasked with nutrient digestion and absorption, the apical surface is enhanced with **villi** and **microvilli** to increase epithelial surface area and maximize contact between epithelium and intestinal contents (see Fig. 4.4). **Crypts** (epithelial invaginations) increase the surface area available for secretion (Fig. 29.3).

2. **Lamina propria:** The **lamina propria** is a loose connective tissue composed of elastin and collagen fibers that contains sensory nerves, blood and lymphatic vessels, and mucus-secreting glands (see Fig. 29.2).

3. **Lamina muscularis mucosae:** The **lamina muscularis mucosae** comprise two thin layers of smooth muscle (see Fig. 29.2). The myocytes are arranged in an inner circular layer and an outer

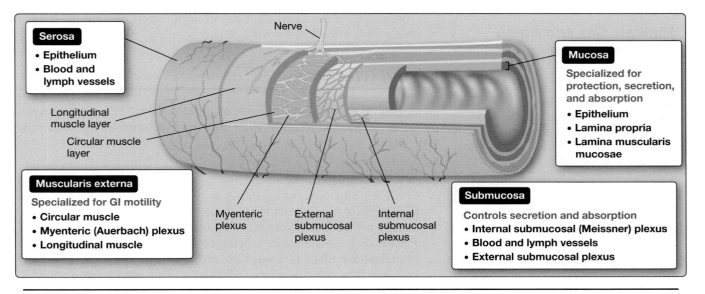

Figure 29.2.
GI tract layers.

longitudinal layer. The muscle layers contract spontaneously to push the mucosa up into a series of ridges and folds to increase surface area, agitate the intestinal contents, and expel secretions from crypts.

B. Submucosa

The **submucosa** contains blood and lymphatic vessels. The submucosa of the small and large intestine also houses two interconnected **submucosal nerve plexuses** containing sensory afferents, ganglia, and autonomic motor nerves controlling blood vessel, intestinal smooth muscle, glandular, and immune cell activity. The outer plexus comprises a sparse monolayer. The inner plexus (**Meissner plexus**) is dense and multilayered.

C. Muscularis externa

The **muscularis externa** contains two thick layers of smooth muscle separated by the **myenteric plexus (Auerbach plexus)**. An inner **circular muscle** layer is arranged in rings and pinches the tube when it contracts. An outer longitudinal muscle layer shortens the tube when it contracts. Circular muscle also forms sphincters, which are barriers regulating movement of solids and liquids between compartments. Cells within the muscle layers are connected via gap junctions, which allow electrical signals to propagate the length and width of the GI tract.

D. Serosa

The **serosa** comprises an outermost layer of squamous epithelial cells overlying a layer of connective tissue (see Fig. 29.2). Some portions of the GI tract (e.g., the lower esophagus) do not have a serosal layer but, instead, connect directly to the adventitia, which is connective tissue that anchors them to the abdominal or pelvic wall.

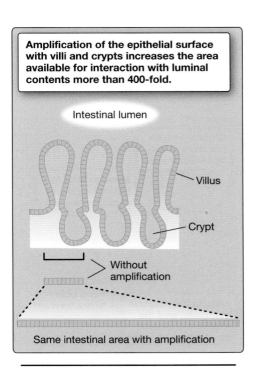

Figure 29.3.
Surface area amplification.

Biologic Sex and Aging 29.1: Enteric Nervous System

GI motility disorders such as dysphasia, gastroesophageal reflux, constipation, and fecal incontinence are common in older adults. Although these disorders may be partly attributable to smooth muscle atrophy and sphincteric thinning, alterations in the timing and coordination of peristalsis and sphincteric reflexes also occur, suggesting that enteric nervous system (ENS) function is impaired. In the esophagus, myenteric ganglion cell numbers decrease by up to 60% by age 70 years. The myenteric plexus coordinates peristalsis. Muscle contraction amplitude is also reduced, which impairs clearance of refluxed acid. In the colon, myenteric plexus ganglion cell numbers decrease by ~40% between ages 20 and 65 years. Due to the presence of a functional reserve of ENS neurons, a direct loss of neurons is not necessarily causative, but reduced ENS function is a likely contributing factor in the etiology of constipation.

The serosa is also associated with large blood and lymph vessels that supply the intestinal wall.

III. INNERVATION

GI function is regulated by three ANS divisions: the ENS, the **parasympathetic nervous system** (**PSNS**), and the **sympathetic nervous system** (**SNS**).

A. Enteric nervous system

The ENS begins in the esophagus and extends uninterrupted down the length of the GI tract to the anus. It is capable of coordinating GI activity without input from the CNS using **intrinsic reflex pathways**. PSNS and SNS neurons synapse with ENS neurons and can modulate the reflex pathways but are not required for GI function. In practice, however, the ENS, PSNS, and SNS work closely together to optimize GI activity based on the individual's nutritional needs. The ENS reflex pathways contain readily identifiable sensory afferents, interneurons, and motor efferents, but there is extensive cross-connectivity and crosstalk between the different pathway components. ENS neurons are supported by **enteric glial cells**, which structurally and functionally resemble CNS astrocytes.

1. **Sensory afferents:** ENS sensory afferents monitor various GI parameters. Mechanosensors monitor changes in gut wall tension, stretch, and pressure associated with movement of luminal contents. Chemosensors monitor the chemical composition of the lumen, including pH, osmolality, and nutrient levels. Thermoreceptors sense the temperature of ingested food and liquids. Sensory afferents project circumferentially and longitudinally in both directions (i.e., toward the mouth and toward the anus). Sensory signals are relayed to interneurons located within small myenteric and submucosal ganglia, but they may also be relayed to motor efferents directly.

2. **Interneurons:** Interneurons project orally and anally (**ascending** and **descending** interneurons, respectively). Ascending interneurons project to excitatory motor neurons, whereas descending interneurons project to inhibitory motor neurons and submucosal ganglia.

3. **Motor neurons:** ENS motor neurons regulate all aspects of GI function. Circular and longitudinal muscle layers are regulated by motor neurons originating in the myenteric plexus. Motor neurons that project orally are usually excitatory, whereas those that project anally are usually inhibitory. This organization facilitates peristaltic movements (Fig. 29.4). Submucosal motor neurons regulate splanchnic blood flow (vasodilator neurons) and the activity of secretory glands (secretomotor neurons), endocrine cells, and the absorptive epithelium.

B. Parasympathetic nervous system

Parasympathetic innervation is derived from the vagus nerve (cranial nerve X) and pelvic nerves (S2–S4) and includes both sensory and

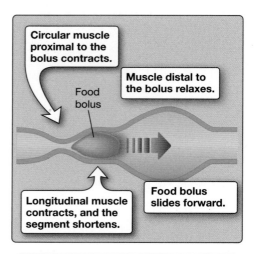

Figure 29.4.
Peristalsis.

motor components (Fig. 29.5). Sensory afferents respond to wall stretch and pressure and the temperature and osmolality of luminal contents. PSNS motor efferents synapse with ENS neurons to modulate glandular secretory activity and intestinal motility. These pathways facilitate **vagovagal reflexes** (a reflex in which both afferent and efferent signals travel in the vagus nerve), such as the reflex relaxation that occurs when food enters and distends the stomach. As a general rule, the PSNS stimulates secretion and increases GI motility.

C. Sympathetic nervous system

Sympathetic nerves originate in the thoracic (T5–T12) and lumbar (L1–L3) regions and synapse in the celiac, superior mesenteric, or inferior mesenteric ganglia (Fig. 29.6). SNS innervation does not include a direct sensory arm. As a general rule, SNS activation inhibits glandular secretions and decreases GI motility.

> The relationship between the ENS and ANS allows for bidirectional neural communication between the GI system and brain, a relationship referred to as the gut–brain axis. Communication may also involve humoral and immunologic communication. More recently, the term has been expanded to include the role of the microbiota in regulating CNS function (gut–brain–microbiome axis).

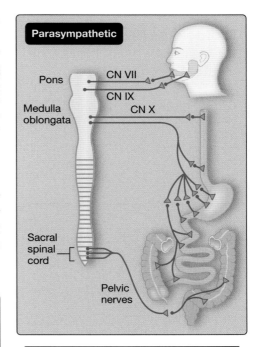

Figure 29.5.
Parasympathetic innervation.

Clinical Application 29.2: Chagas Disease

Chagas disease is caused by infestation with the protozoan parasite *Trypanosoma cruzi*, which is often transmitted by a triatomine vector (e.g., "kissing bug") during feeding. Chagas disease manifests clinically as cardiomyopathy and GI disease, symptoms resulting from parasite-induced inflammation and tissue damage. GI symptoms largely reflect destruction of submucosal and myenteric plexus neurons. Loss of sphincter control results in tonic contraction of both the lower esophageal

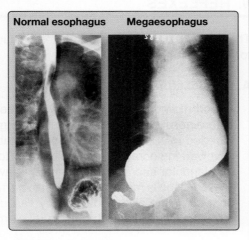

Normal esophagus   Megaesophagus

Megaesophagus.

sphincter (LES) and the internal anal sphincter. In the esophagus, an inability to relax the LES during swallowing causes ingested food to accumulate and pressure to build within the distal esophagus. In time, the esophagus becomes grossly distended (megaesophagus). A similar process occurs in the colon, resulting in megacolon.

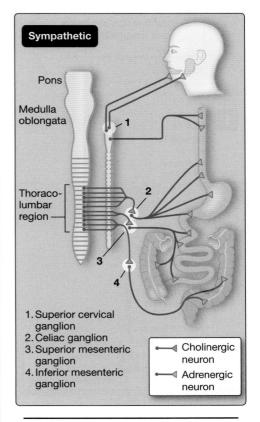

1. Superior cervical ganglion
2. Celiac ganglion
3. Superior mesenteric ganglion
4. Inferior mesenteric ganglion

Cholinergic neuron
Adrenergic neuron

Figure 29.6.
Sympathetic innervation.

IV. MOTILITY

The GI tract exhibits several different preprogrammed motility patterns that assist in food staging or disassembly. For example, the stomach and colon both have reservoir functions that facilitate food and fecal storage, respectively, until the luminal contents can be moved forward. The stomach also exhibits churning motions that break down food mechanically. Two motility patterns are common to most GI segments: peristalsis and segmentation contractions.

A. Peristalsis

Peristalsis is a propagating contraction that moves intestinal contents down the GI tract from one compartment to the next (see Fig. 29.4). Peristaltic contractions are coordinated by ascending and descending interneurons. Ascending interneurons synapse with excitatory cholinergic motor neurons and stimulate contractions proximal to the site of propulsion (Fig. 29.7). Descending neurons excite inhibitory motor neurons distal to the site of propulsion, causing muscle relaxation, and facilitating forward movement. Inhibitory neurons release nitric oxide (NO) and vasoactive intestinal paptide.

B. Segmentation contractions

Segmentation contractions move intestinal contents back and forth within the GI lumen. In the stomach and small intestine, their function is to mix and churn luminal contents. Mixing movements maximize contact between ingested food and intestinal secretions. In the large intestine, segmentation contractions maximize contact between wastes and the intestinal wall to facilitate extraction of residual water.

V. REFLEXES

The ENS facilitates several reflexes, including vasodilatory, secretomotor, and motor reflexes.

A. Vasodilator

Blood flow to the intestines increases from ~0.3 L/min in the interdigestive period to ~6 L/min during a meal. This flow increase is needed to support the increased metabolic demands of enterocytes involved in secretion and absorption, to provide the fluid component of glandular secretions, and to carry away the products of digestion. Vasodilation is mediated, at least in part, by local ENS reflex pathways that stimulate NO production by vascular endothelial cells. NO is a potent vascular smooth muscle relaxant (see Chapter 19·II·E).

B. Secretomotor

Food ingestion stimulates reflex increases in glandular secretions. Sensory afferents include mechanoreceptors, chemoreceptors, and osmoreceptors that notify the submucosal plexus that food is present within the intestinal lumen. The submucosal plexus coordinates secretomotor reflexes.

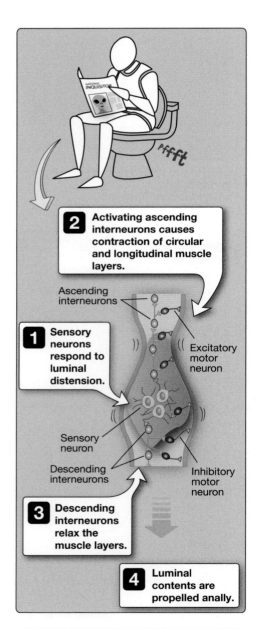

2 Activating ascending interneurons causes contraction of circular and longitudinal muscle layers.

Ascending interneurons

1 Sensory neurons respond to luminal distension.

Excitatory motor neuron

Sensory neuron

Descending interneurons

Inhibitory motor neuron

3 Descending interneurons relax the muscle layers.

4 Luminal contents are propelled anally.

Figure 29.7.
Neuronal coordination of peristalsis.

C. Extrinsic motor reflexes

The GI tract is a long tube with no bypass routes. Sometimes, movement of luminal contents out of a more proximal compartment must be delayed to allow time for a later compartment to complete its digestion cycle. At other times, more distal compartments must be evacuated in order to make room for food arriving in a more proximal compartment. Most of us are familiar with an urge to defecate after ingesting food, for example (gastrocolic reflex). Such staging and clearance movements and coordinated by extrinsic motor reflexes. Four such reflexes are summarized in Table 29.1.

VI. SIGNALING MOLECULES

GI regulation also involves dozens of GI **hormones**, **paracrines**, and **neurotransmitters**. The more important of these and their actions are detailed in Table 29.2. GI peptide hormones are produced by enteroendocrine cells located in the intestinal mucosa. Enteroendocrine cell distribution and density varies along the length of the GI tract (Fig. 29.8). For example, gastrin-secreting G cells are primarily located in the stomach antrum, but they spill over into the proximal small intestine, with G-cell numbers decreasing in more distal regions.

Table 29.1: Gastrointestinal Reflexes

Reflex	Trigger	Response
Gastroileal	Gastric distension	Ileal peristalsis and opening of the ileocecal sphincter
Gastrocolic	Gastric distension	Peristaltic mass movements and urge to defecate
Enterogastric	Duodenal distension	Inhibition of gastric motility and acid secretion
Intestino-intestinal	Excessive distension of the small intestine	Inhibition of intestinal motility

Table 29.2: Gastrointestinal Signaling Molecules

Signal	Source	Target	Function
Hormones			
Amylin	Pancreatic β cells	Stomach, pancreas	Slows gastric emptying and acid secretion; inhibits glucagon secretion
Cholecystokinin (CCK)	I cells in duodenum and jejunum	Pancreas, gallbladder, stomach	Increases enzyme secretion; contracts gallbladder; slows gastric emptying
Gastrin	G cells in stomach antrum	Stomach	Increases gastric acid secretion; stimulates mucosal hypertrophy
Glucagon-like peptide 1 (GLP-1)	L cells in small intestine	Stomach, pancreas	Decreases gastric acid secretion, slows gastric emptying, increases insulin secretion
Glucose-dependent insulinotropic peptide (GIP)	K cells in duodenum and jejunum	Pancreas, stomach	Releases insulin; inhibits acid secretion
Motilin	M cells in upper GI tract	GI smooth muscle	Increases contractions and migrating motor complexes
Peptide YY	L cells in small intestine	Stomach, pancreas	Decreases gastric acid secretion, slows gastric emptying, decreases pancreatic fluid and enzyme secretion
Secretin	S cells in small intestine	Pancreas, stomach	Releases HCO_3^- and pepsin
Somatostatin	D cells in stomach and duodenum, δ in pancreas	Stomach, pancreas	Inhibits peptide hormone and gastric acid secretion
Paracrines			
Histamine	Enterochromaffin-like cells, mast cells	Stomach	Increases gastric acid secretion
Prostaglandins (PGs)	GI epithelium	Mucosa	Increase mucus and HCO_3^- secretion and blood flow

(Continued)

Table 29.2: Continued

Signal	Source	Target	Function
Neurotransmitters			
Acetylcholine (ACh)	PSNS (cholinergic), ENS	Smooth muscle, glands	Contracts wall muscle; relaxes sphincters; increases salivary, gastric, and pancreatic secretions
Enkephalins	ENS	Smooth muscle, glands	Constrict sphincters; decrease intestinal secretions
Gastrin-releasing peptide (GRP)	PSNS (cholinergic), ENS	Glands	Increases gastrin secretion
Neuropeptide Y	SNS (adrenergic), ENS	Smooth muscle, glands	Relaxes wall muscle; decreases intestinal secretions
Nitric oxide (NO)	ENS	Smooth muscle	Relaxes wall and sphincter muscle; increases blood flow
Norepinephrine (NE)	SNS (adrenergic)	Smooth muscle, glands	Relaxes wall muscle; contracts sphincters; decreases salivary secretions
Substance P	PSNS (cholinergic), ENS	Smooth muscle, glands	Contracts wall muscle; increases salivary secretions
Vasoactive intestinal peptide (VIP)	PSNS (cholinergic), ENS	Smooth muscle, glands	Relaxes sphincters; increases pancreatic and intestinal secretions

ENS = enteric nervous system, PSNS = parasympathetic nervous system, SNS = sympathetic nervous system.

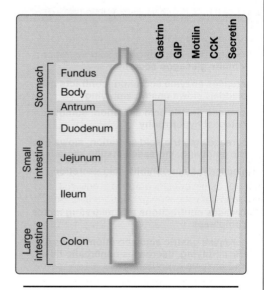

Figure 29.8.
Principal sites of GI hormone release. CCK = cholecystokinin, GIP = glucose-dependent insulinotropic peptide.

Chapter Summary

- The GI system is responsible for ingesting, **digesting,** and **absorbing** nutrients and for then expelling wastes.

- GI activities are coordinated by the **enteric nervous system** (**ENS**) using intrinsic reflex pathways. The parasympathetic nervous system and sympathetic nervous system modulate ENS function.

- Movement through the GI tract involves sequenced contractions of circular and longitudinal muscle layers coordinated by a **myenteric** nerve plexus. Principal motility patterns involve **peristaltic** (propulsive) contractions and **segmentation** contractions (mixing and churning).

- Secretion and absorption are controlled by a **submucosal** nerve plexus responding to input from sensory afferents that monitor intestinal wall stretch and tension, chemical composition and temperature of the luminal contents, and luminal osmolality.

- Digestion involves breakdown of food into smaller, simpler particles, aided by **secretion** of fluids and enzymes. Absorption is aided by epithelial specializations that increase surface area.

- GI hormones include **cholecystokinin, gastrin, glucose-dependent insulinotropic peptide, secretin**, and **somatostatin**.

- GI paracrines include **histamine** and **prostaglandins**.

- GI transmitters include **ACh, nitric oxide, vasoactive intestinal peptide, norepinephrine, substance P**, and **enkephalins**.

Study Questions

Choose the ONE best answer.

29.1. A neurobiologist is using electrophysiologic techniques to investigate the circuits involved in enteric nervous system motor reflexes. Glass capillary microelectrodes are used to stimulate or inhibit ascending and descending interneurons, or the motor neurons controlling circular and longitudinal muscle layers within a target site measuring <5 mm². Which of the following would most likely initiate anally directed peristaltic contractions?

   A. Inhibition of ascending and descending interneurons
   B. Inhibition of ascending and stimulation of descending interneurons
   C. Inhibition of circular and stimulation of longitudinal muscle layers
   D. Stimulation of ascending and inhibition of descending interneurons
   E. Stimulation of ascending and descending interneurons

Best answer = E. Peristaltic contractions are coordinated by ascending and descending interneurons (see Section IV·A). Ascending interneurons project to excitatory motor neurons that contract both circular and longitudinal muscle, whereas descending motor neurons project to motor neurons that inhibit muscle contractions. Stimulating these interneurons simultaneously results in an anally directed peristaltic contraction. Inhibition of ascending and descending interneurons (A) would cause intestinal relaxation. Inhibition of ascending and stimulation of descending interneurons (B) would result in muscle relaxation distal to the site of stimulation but no propulsive force development, as would inhibition of circular and stimulation of longitudinal muscle layers (C). Stimulation of ascending and inhibition of descending interneurons (D) would generate propulsive force but there would be no directed movement in the absence of relaxation distal to the site of stimulation.

29.2. A researcher is conducting a survey of medical literature prior to writing a grant proposal on the potential role of disturbances in the gut microbiome in development of psychiatric disease. The survey reveals that gut microbes release several transmitters, vitamins, and energy substrates. Known beneficial products of gut microbiota most likely include which of the following?

   A. Butyric acid
   B. Lactic acid
   C. Vitamin C
   D. Vitamin D
   E. Zinc

Best answer = A. Gut microbiota include *Bacteroidetes* spp., which synthesize short-chain fatty acids (SCFAs) such as acetate, propionate, and butyrate (see Clinical Application 29.1 and Chapter 31·III·B). SCFAs are absorbed by the colonocytes and used to fuel metabolism. Lactic acid (B) is not a SCFA; it is a byproduct of carbohydrate fermentation. Gut microbes produce vitamins K and B_{12}, not vitamins C (C) and D (D). Zinc (E) is an inorganic elemental metal.

29.3. A 40-year-old male with uncontrolled Crohn disease undergoes an ileal resection to remove damaged tissue. Synthesis and release of which GI hormone would most likely be affected by this surgery?

   A. Cholecystokinin
   B. Gastrin
   C. Glucose-dependent insulinotropic peptide
   D. Motilin
   E. Prostaglandins

Best answer = A. Cholecystokinin (CCK) is secreted by I cells throughout the small intestine, including the ileum (see Section IV·A). CCK acts on the stomach, pancreas, and gallbladder to delay gastric emptying and promote release of pancreatic enzymes and gallbladder bile. Gastrin (B) is secreted by G cells in the stomach and proximal small intestine. Glucose-dependent insulinotropic peptide (C) and motilin (D) are secreted by K and M cells, respectively, in the duodenum and jejunum, but not the ileum. Prostaglandins (E) are GI paracrines, not hormones.

Mouth, Esophagus, and Stomach

30

I. OVERVIEW

We ingest food in bite-size pieces that are too large to be absorbed by the gastrointestinal (GI) tract in a timely fashion. Therefore, the role of the initial GI sections (i.e., **mouth** and **stomach**) is to break down food into smaller, more readily digested pieces. The **esophagus** creates a passage between mouth and stomach. As everyone knows from experience, swallowing dry food can be a challenge, so the mouth also adds a watery lubricant (saliva) to ease transit. The esophagus drops ingested food into an acid bath (gastric juice), in part to sterilize it, but also to begin the job of disassembling food into its essential component parts.

II. MOUTH

The mouth mechanically chops and grinds food into smaller fragments by **mastication** (chewing) to make it easier to swallow and increase its surface area to facilitate chemical digestion. The mouth also adds saliva to hydrate food and lubricate the mouth.

A. Teeth

Teeth cut (incisors), tear, pierce (canines), grind, and crush (premolars and molars) food (Fig. 30.1A). The crown portion of teeth is coated with enamel, which is >95% calcium hydroxyapatite (see Chapter 14·II·A). Enamel creates an extremely hard shell that facilitates chewing and, along with dentin (a hard but less mineralized connective tissue), protects the pulp cavity (containing nerves and blood vessels) and root canal (see Fig. 30.1B). Both enamel and dentin are harder than compact bone. Jaw muscles provide the mechanical force and motions needed for biting and chewing.

B. Tongue

The tongue grips and repositions food during mastication. It contains intrinsic skeletal muscle fibers oriented longitudinally, vertically, and in a transverse plane, which allows for shape changes. Extrinsic skeletal muscles are used to change tongue position (e.g., tongue protrusion and side-to-side movements). The tongue also contains taste buds (see Chapter 10·II·A) and serous and mucous glands, but glandular output does not make a significant contribution to food hydration.

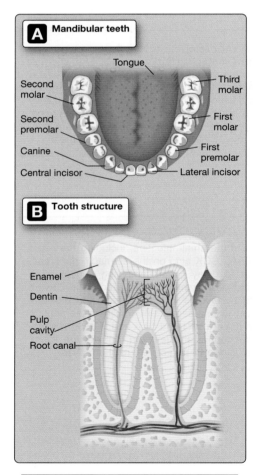

Figure 30.1.
Teeth classification and anatomy.

C. Salivary glands

Salivary glands produce a **serous fluid** (i.e., a watery fluid containing proteins that resembles serum) that lubricates the mouth, facilitates taste, begins food digestion, and is protective. Daily saliva production approximates 1.0 to 1.5 L, the majority of which is secreted by the **sublingual**, **submandibular**, and **parotid glands** (see Fig. 29.1).

1. **Anatomy:** A salivary gland is composed of **lobules**. Each lobule contains numerous grape-shaped secretory **acini**, which are the functional units of the salivary gland. An acinus is lined with epithelial (**acinar**) cells that are specialized for secreting protein and fluid. The **primary secretion** flows from the acinus via an **intercalated duct** into a larger **striated duct** (**interlobular duct**). Acinar flow is assisted by **myoepithelial cells**, which surround the acinus and proximal portion of the ductal system. When stimulated appropriately, myoepithelial cells contract to squeeze fluid out of the acinus and through the ducts. Acinar and ductal epithelia also contain mucous cells that secrete mucin, a glycoprotein that gives mucus its lubricant properties. Sublingual and submandibular glands secrete a mixed serous and mucous solution, whereas the parotid gland mainly secretes serous fluid.

2. **Primary secretion:** The primary secretion is an isotonic NaCl solution resembling plasma. Secretion begins with the basolateral Na^+/K^+-ATPase, which establishes the ion gradients used to drive movement of other ions (Fig. 30.2). The Na^+ gradient powers basolateral K^+ and Cl^- (and Na^+) uptake via a $Na^+/K^+/2Cl^-$ cotransporter, with K^+ then diffusing back across the basolateral membrane via K^+ channels. Cl^- diffuses across the apical membrane into the lumen via TMEM16A (anoctamin 1), a Ca^{2+}-dependent anion channel. Na^+ follows Cl^- paracellularly. NaCl secretion creates an osmotic gradient that drives H_2O into the lumen via AQP5, a Ca^{2+}-activated aquaporin. These pathways account for the bulk of serous volume, but basolateral Cl^- uptake by a Cl^-/HCO_3^- exchanger (AE2) may contribute (not shown in the figure). HCO_3^- forms from CO_2 and H_2O in a reaction catalyzed by carbonic anhydrase. H^+ leaves the cell via a basolateral Na^+/H^+ exchanger (NHE1).

3. **Ductal modification:** Ductal cells modify the primary secretion by reabsorbing Na^+ and Cl^- and secreting K^+ and HCO_3^-. Na^+ is reabsorbed from the ductal lumen via an epithelial sodium channel (ENaC) and a NHE1 located in the apical membrane and then is pumped across the basolateral membrane by the Na^+/K^+-ATPase (Fig. 30.3). Cl^- reabsorption and HCO_3^- secretion involves an apical Cl^-/HCO_3^- exchanger and the cystic fibrosis transmembrane conductance regulator (CFTR) Cl^- channel. Cl^- is then transferred to the interstitium via a Cl^- channel. Secreted HCO_3^- enters ductal cells via a basolateral Na^+/HCO_3^- cotransporter (NBCe1-B). Ductal epithelia are relatively water impermeable, so NaCl reabsorption causes saliva to become hypotonic. The effects of ductal modification are most obvious at low salivary secretion rates (Fig. 30.4). Ductal cell transporters follow saturation kinetics, and their capacity is limited, so salivary composition increasingly resembles the primary secretion as secretion rates rise.

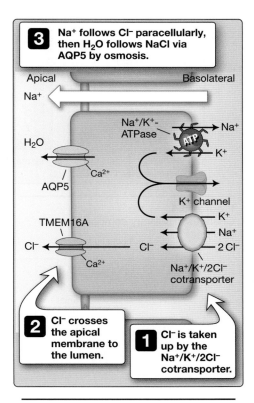

Figure 30.2.
Acinar cell ion secretion pathways.

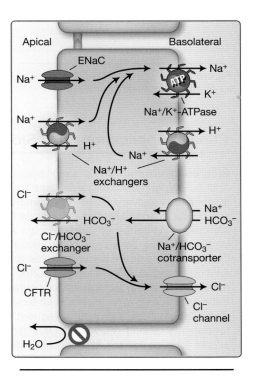

Figure 30.3.
Ductal cell modification of the primary secretion.

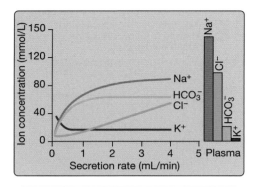

Figure 30.4.
Changes in salivary composition with flow rate.

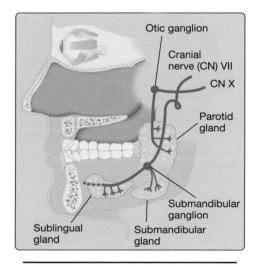

Figure 30.5.
Innervation of salivary glands.

4. **Proteins:** Saliva contains low concentrations of protective proteins and enzymes that originate from acinar, mucous, and ductal cells.

   a. **Protective proteins:** Salivary proteins with antimicrobial activity include lysozyme to disrupt the cell wall of susceptible bacterial species and **lactoferrin** to inhibit microbial growth. Lactoferrin is an iron-binding protein. Salivary **immunoglobulin A** is active against both bacteria and viruses. **Proline-rich proteins** have antimicrobial properties and help tooth enamel formation.

   b. **Enzymes:** Salivary enzymes include **salivary amylase**, which digests carbohydrates. Salivary amylase is denatured by stomach acid, but the pancreas reintroduces amylase in the duodenum. **Lingual lipase** hydrolyzes lipids, but the oral phase of lipid digestion is not significant physiologically.

5. **Regulation:** Salivary flow is controlled by both the parasympathetic nervous system (PSNS) and sympathetic nervous system (SNS). Although activation of either autonomic arm increases salivary gland output, the sympathetic component is transient and yields a lower salivary volume compared with the PSNS. PSNS effects are mediated through salivary nuclei located in the brainstem medulla. Salivary flow is increased by smell, taste, mechanical pressure in the mouth, and various reflexes (e.g., classical conditioning), whereas it is decreased by stress, dehydration, and during sleep. In addition to salivary flow, neural stimulation increases blood flow and myoepithelial cell contraction and causes gland hypertrophy. Salivary gland innervation pathways are summarized in Figure 30.5. Parasympathetic nerves release acetylcholine (ACh), which binds to muscarinic M_3 receptors acting via the IP_3 signaling pathway (see Chapter 1·VII·B·3). IP_3 triggers Ca^{2+} release from intracellular stores and activates TMEM16A and AQP5. The sympathetic nerves release norepinephrine (NE), which binds to α- and β-adrenergic receptors coupled to IP_3 and cAMP signaling pathways, respectively (see Chapter 1·VII·B·2).

Clinical Application 30.1: Sjögren Syndrome

Sjögren syndrome (SS) is a multisystem autoimmune disorder that presents as dry, gritty, irritated eyes and a dry mouth (xerostomia), often with inflammatory salivary gland enlargement. Xerostomia reflects inadequate saliva production. Affected individuals have difficulties swallowing food, increased incidence of dental caries and oral candidiasis, decreased taste, a burning sensation in the mouth and throat, and speech issues. The pathogenesis of SS is complex and not well delineated, but autoantibodies to the ACh receptor (AChR) have been detected in some SS patients. The AChR is required for neural stimulation of salivary secretion. Patients with mild symptoms can be treated with salivary stimulants (AChR agonists). Artificial saliva and frequent sips of water are recommended for patients lacking residual salivary gland function. SS most commonly affects females in their fifth and sixth decades. Aging is typically accompanied by decreases in lacrimal and salivary gland function that produce similar symptoms to SS (dry eyes and mouth), but these changes are physiologic rather than pathologic.

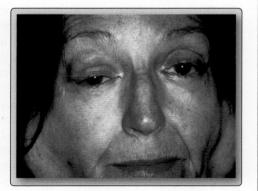

Irritated eyes and parotid gland enlargement in a patient with Sjögren syndrome.

III. ESOPHAGUS

The oropharynx and esophagus convey ingested food, fluids, and oral secretions to the stomach through the act of swallowing (**deglutition**). Swallowing is a PSNS reflex response involving 30 or more muscles. Swallowing can be divided into three phases: oral, pharyngeal, and esophageal.

A. Oral and pharyngeal phases

Once chewing has reduced food to pieces small enough to enter the pharynx and sufficient lubricant has been added to facilitate passage, the tongue voluntarily forces a food bolus backward into the oropharynx. Receptors lining the pharynx detect the presence of solids or liquids and send afferent signals via cranial nerves (CNs) IX and X to a **swallowing center** in the medulla. Swallowing now becomes involuntary. Effector pathways include the **nucleus ambiguus** and the **dorsal motor nucleus** and CNs V, IX, X, and XII. The soft palate elevates to seal the nasopharynx to prevent nasal regurgitation. The larynx and hyoid bones elevate, the glottis closes, and the epiglottis seals off access to the larynx. Respiration is inhibited centrally to prevent food aspiration. Sequenced contractions of superior, middle, and inferior pharyngeal constrictor muscles force the bolus distally, the **upper esophageal sphincter** (**UES**) relaxes, and the esophageal phase begins. The oral and pharyngeal phases together last less than a second.

B. Esophageal phase

The esophageal phase is relatively slow (10–15 seconds). Peristaltic contractions generate a positive pressure wave that propels the bolus from the UES toward the stomach, much as an ocean wave drives a surfer to the shoreline (Fig. 30.6). The esophagus is unusual in that the proximal third is lined with **skeletal muscle** under somatic motor control, whereas the distal two thirds is lined with **smooth muscle** controlled by the myenteric plexus (enteric nervous system [ENS] see Chapter 29·III·A). The **lower esophageal sphincter** (**LES**), which is also composed of smooth muscle, opens reflexively ahead of the bolus to permit entry to the stomach. Changes in sphincter tone are mediated by ACh, **nitric oxide** (**NO**), and **vasoactive intestinal peptide** (**VIP**). If food is not cleared by the first pressure wave (**primary peristalsis**), repetitive waves may be initiated (**secondary peristalsis**). The latter are restricted to the smooth muscle layers and are mediated by local ENS reflexes.

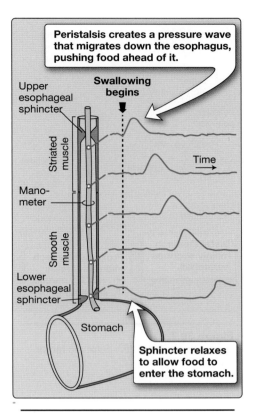

Figure 30.6.
Esophageal peristalsis.

Clinical Application 30.2: Gastroesophageal Reflux Disease

Reflux of gastric contents into the esophagus is normal during belching, for example, but is usually countered by secondary peristalsis, and esophageal exposure is brief. In **gastroesophageal reflux disease** (**GERD**), lower esophageal sphincter (LES) dysfunction (e.g., decreased LES tone or hiatal hernia) causes prolonged or repeated reflux events and may lead to esophagitis and inflammatory damage to the mucosa. If exposure is chronic, esophageal remodeling can occur (Barrett esophagus), which predisposes patients to developing esophageal adenocarcinoma.

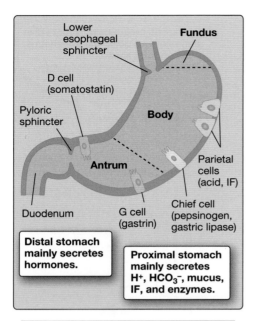

Figure 30.7.
Stomach anatomy. IF = intrinsic factor.

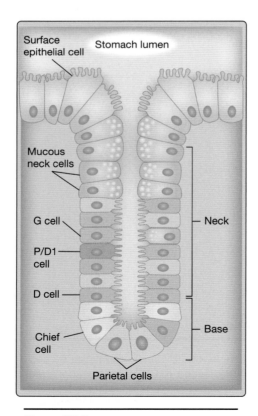

Figure 30.8.
Gastric gland structure.

IV. STOMACH

The amount of material ingested during a meal typically exceeds the intestine's ability to process solids, so food must be stored temporarily until it can be passed on to the small intestine at a rate optimized for digestion and absorption. Food storage is one of the stomach's primary functions. The stomach also sterilizes food, adds fluids and digestive enzymes, and then grinds and churns the mixture to produce **chyme** (a semisolid fluid containing partially digested food) for metered delivery to the duodenum.

A. Functional anatomy

The stomach can be divided into three gross anatomic areas: the **fundus**, **body** (or **corpus**), and **antrum** (Fig. 30.7). The general structure of the stomach wall is similar to that seen in the rest of the GI tract (see Fig. 29.2), but the mucosa contains numerous invaginations called **gastric pits** (Fig. 30.8). The base of a pit opens onto a tubular **gastric (oxyntic) gland**, which contains a long neck region and short base (fundus). The neck is lined with **mucous neck cells**, which secrete mucus. The base contains **chief cells**, which produce pepsinogen and gastric lipase. Scattered throughout the gastric gland are **parietal cells (oxyntic cells)**, which secrete acid and intrinsic factor, and enteroendocrine cells producing gastrin (**G cells**) or somatostatin (**D cells**; see Table 29.2). There are regional differences in gastric secretory output. The proximal stomach primarily secretes H^+, HCO_3^- (a product of the surface epithelial cells), mucus, enzymes, and intrinsic factor, whereas the distal stomach primarily secretes hormones.

> Oxyntic is a term often applied to gastric anatomy. It is derived from the Greek word *oxos*, meaning soured wine (i.e., the sharp taste of vinegar, or acetic acid).

B. Storage

Between meals, the proximal stomach has a relatively high resting tone, and the wall is thrown into a series of folds called **rugae**. Distension of the distal esophagus by incoming food triggers **receptive relaxation**, a vagovagal reflex that causes a passive decrease in muscle tone. Rugae unfold, allowing ingested food to collect without increasing gastric pressure. Relaxation is mediated by NO and VIP release from vagal nerve terminals. Food accumulation is also made possible by gastric **accommodation**, a reflex relaxation mediated locally by the ENS. The proximal stomach has a relatively sparse musculature, so food collects in layers and remains undisturbed, much like layers of refuse in a landfill. The average stomach at rest has a capacity of ~50 mL but can accommodate ~1.5 L of food during a meal.

C. Motility

The distal stomach is muscular and specialized for mixing, churning, and grinding actions. Contractions are initiated by **interstitial cells of**

Cajal (**ICCs**), which generate pacemaker currents (**slow waves**) that excite the smooth muscle layers.

1. **Pacemaker:** ICCs are found in the stomach corpus and antrum and throughout the small intestine. Their cell bodies are located between the longitudinal and circular muscle layers, and they extend processes that contact adjacent ICCs and smooth muscle fibers to form a loose network. Gap junctions allow electrical and chemical signals to propagate between ICCs and the muscle fibers. ICCs located mid-stomach function as **pacemakers** that generate a slow wave once every ~10 seconds. Depolarization is mediated by Ca^{2+} influx via T-type Ca^{2+} channels and sustained by Cl^- efflux via TMEM16A (a Ca^{2+}-dependent anion channel). Slow waves initiate a contraction once every ~20 seconds. Slow-wave amplitude and duration is decreased by NE release from SNS neurons and increased by mechanical stretch and ACh released by PSNS and enteric neurons.

2. **Gastric mill:** Waves of contraction propagate at ~2.5 mm/s from the corpus toward the antrum, pushing stomach contents toward the pylorus. The pyloric sphincter contracts and closes ahead of the wave, however. The wave of contraction speeds up to ~7 mm/s in the antrum, races up ahead of the stomach contents, and then stops short of the sphincter by ~6 mm. Antral contraction now pushes stomach contents backward toward the corpus (**retropulsion**). These motions will be immediately familiar to anyone who has watched ocean waves roll into a cove, race up a beach, and then recede. Repetitive gastric waves grind food into smaller pieces (the gastric mill), in much the same way that ocean waves reduce rocks to pebbles and pebbles to sand. Gastric motions also help mix and churn food with gastric secretions. Gastric emptying normally takes about 3 hours, which is the time required to reduce food to pieces small enough to pass through the pyloric sphincter into the duodenum (<2 mm). Liquids pass through the stomach relatively quickly. The rate of gastric emptying is heavily influenced by the composition of chyme entering the duodenum, mediated by duodenal chemoreceptors, mechanoreceptors, and local reflex pathways. Emptying is slowest following a fatty meal (see Section V).

Clinical Application 30.3: Vomiting

The physical act of vomiting (**emesis**) begins in the ileum rather than the stomach. A wave of retrograde peristalsis sweeps through the small intestine and the stomach toward the esophagus, pushing contents of both compartments before it. Deep inhalation is then followed by contraction of the diaphragm and abdominal muscle against a closed glottis, causing intraabdominal pressure to rise. Gastric contents are forced into the esophagus and out through the mouth, passage assisted by reflex relaxation of the lower and then the upper esophageal sphincter. Vomiting is a reflex coordinated by a **vomiting center** and **chemoreceptor trigger zone** (**CTZ**) in the brainstem medulla. The CTZ is located within the area postrema, which is a circumventricular organ that monitors the composition of extracellular fluid. The vomiting reflex is commonly triggered by the presence of toxins or other irritants within the stomach or small intestine and therefore has an obvious protective function. Other common triggers include GI pathology, inner ear dysfunction, CNS disturbances, and many drugs that stimulate the CTZ (e.g., cisplatin).

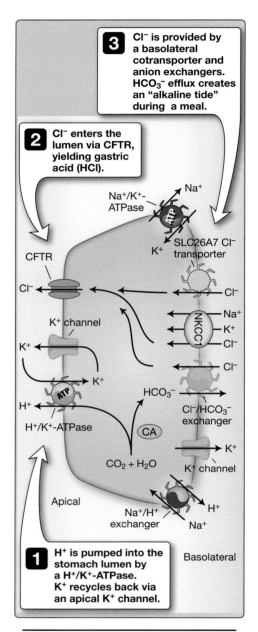

Figure 30.9.
Parietal cell H^+ secretion.
CA = carbonic anhydrase,
NKCC1 = $Na^+/K^+/2Cl^-$cotransporter.

3. **Housekeeping function:** When empty, the stomach becomes quiescent. Once every 90 to 120 minutes, a series of contractions (**migrating motility complexes [MMCs]**) migrate through the stomach, the pylorus, and the length of the small intestine. The pyloric sphincter dilates fully during passage of an MMC, so particles larger than 2 mm (e.g., corn kernels, accidently ingested coins, and Lego blocks) are swept out of the stomach and into the large intestine, along with intestinal bacteria, secretions, and sloughed epithelial cells. MMCs are initiated by motilin release from duodenal M cells.

D. **Acid secretion**

Stomach pH between meals is maintained at ~1.0 to 2.5. Gastric acid has two principal functions. The first is to kill most of the bacteria ingested along with food, and the second is to denature proteins to aid their digestion. Acid is produced by parietal cells.

1. **Secretory pathway:** Acid is generated from H_2O and CO_2, facilitated by carbonic anhydrase, and then pumped into the gastric lumen by an apical H^+/K^+-ATPase (Fig. 30.9). A K^+ channel in the apical membrane ensures that the gastric lumen contains the K^+ needed for H^+/K^+ exchange. HCO_3^- exits the parietal cell via a basolateral anion exchanger (AE2), which simultaneously provides Cl^- for secretion into the stomach via the CFTR to yield hydrochloric acid (HCl). Cl^- entry may also be facilitated by a basolateral $Na^+/K^+/2Cl^-$cotransporter (NKCC1) and an SLC26A7 anion exchanger. Gastric acid secretion increases during a meal, causing a surge of HCO_3^- secretion across the basolateral membrane (**"postprandial alkaline tide"**) that causes the interstitial space to become slightly basic.

2. **Regulatory pathways:** Acid secretion is highly regulated, involving multiple feedforward and feedback control pathways (Fig. 30.10). Parietal cells contain numerous subapical vesicular organelles (**tubulovesicles**) that contain the H^+/K^+-ATPase and the various ion channels needed for acid secretion. When the parietal cell is stimulated by a secretagogue, the vesicles fuse with apical canaliculi and insert proton pumps into the apical membrane.

   a. **Feedforward:** Food entering the stomach typically has a neutral pH and buffers the small amount of acid present between meals. Therefore, eating stimulates gastric acid secretion to restore the acid environment. ACh, gastrin, and histamine all promote acid secretion through second-messenger pathways that converge on H^+/K^+-ATPase expression. ACh is released from vagal nerve endings and binds to an M_3 muscarinic ACh receptor that couples via G_q and phospholipase C to the IP_3 signaling pathway. Diacylglycerol activates protein kinase C, which increases H^+/K^+-ATPase activity. Gastrin binds to a CCK_2 gastrin receptor, which also couples to G_q. Histamine acts through an H_2 receptor that couples to G_s, which activates the cAMP signaling pathway and protein kinase A. (Note: The histamine pathway is dominant. ACh and gastrin augment histamine actions to yield a net increase in acid production that exceeds the sum of their individual actions [**potentiation**].)

   b. **Feedback:** Somatostatin and prostaglandins (PGs) inhibit acid production. D cells are located in the stomach mucosa and release somatostatin in response to low stomach pH. The somatostatin receptor couples to G_i, which inhibits the cAMP signaling pathway. PGs also inhibit acid production via G_i.

3. **Phases of acid secretion:** We can distinguish three phases of gastric acid secretion: cephalic, gastric, and intestinal. Regulatory pathway prominence shifts with each successive phase.

   a. **Cephalic phase:** The **cephalic phase** of acid secretion is triggered by the thought, sight, smell, and taste of food and accounts for ~20%–30% of total acid secretion. The cephalic phase is mediated primarily by the vagus nerve and ACh. ACh stimulates H^+ production through direct effects on parietal cells, but it also stimulates histamine release from **enterochromaffin-like** (ECL) cells and gastrin release from G cells (vagal terminals synapsing with G cells release gastrin-releasing peptide rather than ACh). Gastrin not only stimulates parietal cell activity directly (see Fig. 30.10), but it also does so indirectly by binding to ECL cells and stimulating histamine secretion. The vagus simultaneously facilitates gastrin and histamine secretion by inhibiting somatostatin release from D cells. Somatostatin normally inhibits G cell and ECL secretory activity. The vagus also increases mucus secretion to help protect the stomach lining from H^+.

   b. **Gastric phase:** The gastric phase accounts for ~60%–70% of gastric secretions and is triggered by distension of the gastric wall. Distension is sensed by vagal afferents and increases vagal efferent activity to increase acid secretion (a **vagovagal reflex**). Distension also activates local enteric reflexes. Food arrival in the stomach is also sensed by G cells, which release gastrin in response to amino acids and partially digested proteins.

   c. **Intestinal phase:** The intestinal phase accounts for ~10%-20% of gastric secretions. Amino acids and partially digested peptides entering the small intestine stimulate duodenal G cells. Negative feedback during the intestinal phase is provided by glucose-dependent insulinotropic peptide release from duodenal K cells, which inhibits acid secretion.

E. Gastric juice

Between meals, the stomach produces small amounts of gastric juice containing NaCl, HCl, and K^+. NaCl is secreted by nonparietal cells whose activity is not influenced by food intake. Ingesting food causes secretion rate to increase from ~0.5–1.5 mL/min to ~5–10 mL/min, with the relative contribution of parietal cells to fluid composition increasing and that of nonparietal cells decreasing (Fig. 30.11).

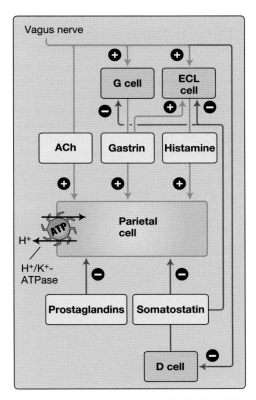

Figure 30.10.
Control of acid secretion. ECL = enterochromaffin-like cell.

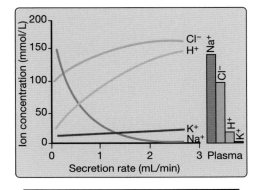

Figure 30.11.
Changes in gastric juice composition with flow rate.

Clinical Application 30.4: Gastric Ulcers

The gastric mucosal barrier protects the stomach lining against acid and enzymes so effectively that gastric ulcers are uncommon under physiologic conditions. Ulcers are associated with nonsteroidal anti-inflammatory drug (NSAID) use and *Helicobacter pylori* infection. NSAIDs inhibit synthesis of prostaglandins that help maintain the mucous layer. *H. pylori* is well adapted to inhabit the HCO_3^--rich layer protecting the mucosal surface. Here, the bacterium disrupts tight junctions and invades the epithelium, causing inflammatory damage and ulceration. Left untreated, ulcers may perforate the gastric lining, as shown. Treatment involves eradicating *H. pylori* and an H^+/K^+-ATPase inhibitor to decrease acid secretion during healing.

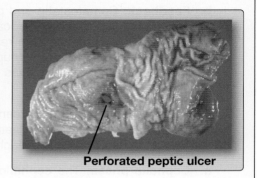

Perforated peptic ulcer

Gastric ulcer.

F. Gastric enzymes and intrinsic factor

Chief cells secrete gastric lipase and pepsinogen. Gastric lipase has a low pH optimum (3–6). It targets triacylglycerols, cleaving fatty acids from the glycerol backbone to leave diacylglycerols (see Chapter 31·II·E·2). It accounts for up to 30% of total GI lipid digestion. Pepsinogen is an inactive proenzyme activated by low pH to yield pepsin, an endopeptidase. Pepsin cleaves aromatic amino acids and has an optimal pH between 1 and 3. Pepsin is deactivated in the duodenum once pH increases to >3.5. Intrinsic factor is the only essential gastric secretory product. It facilitates vitamin B_{12} absorption in the ileum (see Chapter 31·II·F·1).

G. Gastric mucosal barrier

The stomach produces several extremely corrosive substances (i.e., acid and enzymes) that require the mucosa protect itself, both during secretion and when digesting a meal. Protection involves a physical barrier and a mucous coat.

1. **Physical barrier:** The gastric epithelium's apical membrane is impermeable to H^+, which prevents back-diffusion during acid secretion. Tight junctions between adjacent cells are also highly resistant to H^+ back flow.

2. **Mucous coat:** The gastric epithelium constructs a general-purpose protective mucous coat. The coat is <1 mm thick, but surface epithelial cells secrete HCO_3^- beneath it that maintains a neutral environment at the epithelial surface, even though stomach luminal pH may be <1.0. The neutral pH also inactivates any pepsin that may back diffuse from the lumen. Coat integrity is dependent on PGs secreted by the gastric mucosa in response gastrin and other local factors. PGs stimulate synthesis and secretion of mucus and bicarbonate, and they promote vasodilation to increase blood flow to the epithelium. PGs such as PGE_2 can also inhibit acid secretion by parietal cells (see Fig. 30.10).

V. APPETITE

"My stomach is rumbling" and "I'm stuffed" are common expressions conveying the sensations of hunger and satiety, respectively. Since these sensations describe either an empty or overly distended stomach, it might seem that appetite would be controlled by gastric mechanoreceptors and the ENS. Although the stomach does have an important role in appetite regulation, hunger and satiety are controlled by the hypothalamus via the gut–brain axis.

A. Hypothalamic control centers

Although many areas of the central nervous system (CNS) are involved in appetite control, they all converge on three key hypothalamic areas. The lateral hypothalamic area (LHA) constitutes a **hunger center** (Fig. 30.12). The ventromedial nucleus of the hypothalamus (VMH) constitutes a **satiety center**. The LHA and VMH collate incoming hormonal and sensory signals and relay this information to the **arcuate nucleus (ARC)** which is a comparator that formulates an appropriate feeding-related response via **orexigenic** and **anorexigenic** neurons. Orexigenic neurons stimulate appetite and the need to ingest food. Anorexigenic neurons suppress food intake.

B. Hunger

When the stomach is empty, P/D1 cells within oxyntic glands secrete **ghrelin**. Ghrelin is a 28–amino acid **hunger hormone** that stimulates vagal afferents and is also released into the blood stream for transport to the ARC. The ARC responds by increasing appetite.

C. Satiety

Distension of the stomach and the small intestine releases several hormones that cause the ARC to inhibit orexigenic neurons and stimulate anorexigenic neurons, thereby inhibiting further food intake. The more important satiety signals include **cholecystokinin** from I cells, and **glucagon-like peptide 1** and **peptide YY** from L cells. These hormones are also secreted in response to nutrient-rich chyme entering the small intestine. **Amylin**, which is released along with insulin by pancreatic β cells, also decreases appetite. All these hormones delay gastric emptying and decrease intestinal motility to allow time for ingested food to be digested and absorbed (see Table 29.2).

D. Long-term control of appetite

The hunger and satiety pathways described above are all short-term responses. In the long-term, appetite control is achieved through leptin. Leptin is a 146–amino acid peptide hormone produced by adipose tissue, and circulating levels reflect total body fat content. Leptin's targets include the ARC, where it inhibits orexigenic and stimulates anorexigenic pathways, thereby reducing food intake.

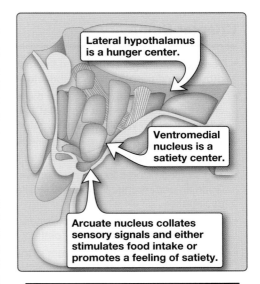

Lateral hypothalamus is a hunger center.

Ventromedial nucleus is a satiety center.

Arcuate nucleus collates sensory signals and either stimulates food intake or promotes a feeling of satiety.

Figure 30.12.
Hypothalamic areas involved in regulation of appetite.

Chapter Summary

- Teeth chop and grind food, and the mouth lubricates it with secretions from three salivary glands: **sublingual**, **submandibular**, and **parotid**. Saliva lubricates food, begins digestion, and is antimicrobial.

- Salivary secretion is controlled by the sympathetic and parasympathetic nervous systems and involves a two-step process. First, acinar cells secrete NaCl and water into the duct lumen. Ductal cells then reabsorb Na^+ and Cl^- and secrete K^+ and HCO_3^- to yield a watery, hypotonic fluid.

- The esophagus transports food from the mouth to the stomach. **Swallowing** moves food from the mouth, through the upper esophageal sphincter, and then to the stomach by **esophageal peristalsis**.

- The stomach stores food (via receptive relaxation and **accommodation**), grinds food, and mixes in gastric secretions. Grinding and mixing motions are facilitated by slow waves initiated by **interstitial cells of Cajal**, which act as a gastric pacemaker.

- Gastric secretions include electrolytes, water, mucus from **mucous neck cells**, pepsinogen from **chief cells**, and intrinsic factor and H^+ from **parietal cells**.

- Regulation of H^+ secretion occurs at the level of parietal cell H^+/K^+-ATPase expression. **ACh** from nerves, **gastrin** from G cells, and **histamine** from enterochromaffin-like cells increase secretion, whereas **somatostatin** from D cells and **prostaglandins** decrease secretion.

- Appetite is controlled by **hunger** and **satiety centers** in the hypothalamus. The arcuate nucleus coordinates feeding behaviors with feedback from gastric and intestinal hormones.

Study Questions

Choose the ONE best answer.

30.1. A 52-year-old female taking scopolamine (a cholinergic antagonist) for motion sickness during airplane travel also develops symptoms consistent with xerostomia as a side effect. Which of the following changes is most likely to support this diagnosis?

A. Decreased salivary Cl^- concentration
B. Decreased salivary K^+ concentration
C. Increased mucus production
D. Increased parotid cell IP_3 levels
E. Stimulated parotid cell adenylyl cyclase

Best answer = A. Salivary secretion is controlled primarily by the parasympathetic nervous system (see Section II·C). When active, ACh release increases salivary secretion via the IP_3-signaling pathway. Blocking cholinergic signaling reduces, not increases, IP_3 levels (D) and decreases salivary flow. Salivary ionic composition is dependent on flow rates. When flow rate decreases, Cl^- content decreases, whereas K^+ concentration (B) increases. Anticholinergics decrease mucus production by salivary glands (C). Scopolamine does not stimulate adrenergic receptors; therefore, no change in adenylyl cyclase would be expected (E).

30.2. A 58-year-old male presents with persistent abdominal pain, bloating, and fatigue for several weeks. He has a history of pernicious anemia. Upper endoscopy reveals diffuse gastric atrophy, and biopsy results confirmed the diagnosis of autoimmune atrophic gastritis. A proton pump inhibitor is prescribed. Loss of gastric D-cell function in this patient has most likely caused gastric acid secretion to increase through which of the following mechanisms?

A. Decreased G-cell activity
B. Increased ACh release
C. Increased prostaglandin E_2 synthesis
D. Loss of parietal cell inhibition
E. Reduced potentiation

Best answer = D. Gastric D cells secrete somatostatin, which normally inhibits H^+ secretion from parietal cells (see Section IV·D). Reducing somatostatin levels would set up the potential for increased H^+ secretion. G cells (A) secrete gastrin, which stimulates H^+ secretion from parietal cells. ACh (B) also increases H^+ secretion by several direct and indirect routes. Prostaglandins (C) decrease H^+ secretion normally but through pathways that do not require D cells. Potentiation (E) refers to the observation that H^+ secretion increases to a greater extent when two stimulatory factors bind simultaneously (e.g., gastrin plus ACh) than might be expected from the sum of their individual actions.

30.3. A 35-year-old female reports heartburn and stomach pains, which frequently wake her at night. Upper gastrointestinal endoscopy and biopsy reveal a peptic ulcer. Which of the following most likely explains how the duodenum normally protects itself against ulcer formation?

A. Histamine release from enterochromaffin-like cells
B. Its thick apical membrane
C. Its thick layer of viscous mucus
D. Release of peptidases in inactive form
E. Secretin release by S cells

Best answer = E. Peptic ulcers occur in the stomach and duodenum (see Clinical Application 30.3). They are often precipitated by *Helicobacter pylori*, but intestinal wall erosion is due to acid and enzymes. The duodenum's main defense against acid is secretin released from S cells when stimulated by acid. Secretin triggers HCO_3^- release from the pancreas (see Section II·B). Histamine (A) is a paracrine released by the stomach to increase acid secretion. The duodenum does not possess a thick apical membrane (B), which would impair nutrient absorption. Unlike the stomach, the duodenum does not have a thick protective mucus layer (C), which increases vulnerability to acid. Pancreatic peptidases are released in inactive form but activate immediately in the intestinal lumen (D).

30.4. An 8-year-old male with Prader-Willi syndrome presents with excessive daytime sleepiness and behavioral problems. He has class II obesity (BMI = 38 kg/m^2) as a result of an insatiable appetite and hyperphagia. The hyperphagia is most likely caused by high circulating levels of which of the following hormones?

A. Amylin
B. Cholecystokinin
C. Gastrin
D. Ghrelin
E. Leptin

Best answer = D. Ghrelin is known as the hunger hormone because it stimulates appetite (see Section V·B). It is secreted by P/D1 cells located in gastric oxyntic glands when the stomach is empty. Individuals with Prader-Willi syndrome (a genetic disorder localizing to chromosome 15) have high circulating ghrelin levels, which causes hyperphagia. Amylin (A) is secreted by pancreatic β cells. It decreases appetite. Cholecystokinin (B) is secreted by I cells in the small intestine in response to fatty meals. It slows gastric emptying and suppresses appetite. Gastrin (C) is secreted by gastric G cells. It stimulates gastric acid production. Leptin (E) is a satiety hormone secreted by adipocytes.

31 Small and Large Intestines

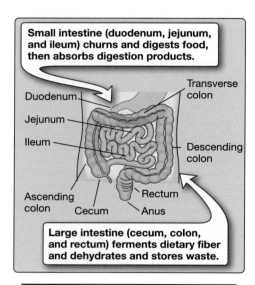

Small intestine (duodenum, jejunum, and ileum) churns and digests food, then absorbs digestion products.

Duodenum
Jejunum
Ileum
Transverse colon
Descending colon
Ascending colon
Cecum
Rectum
Anus

Large intestine (cecum, colon, and rectum) ferments dietary fiber and dehydrates and stores waste.

Figure 31.1.
Small and large intestines.

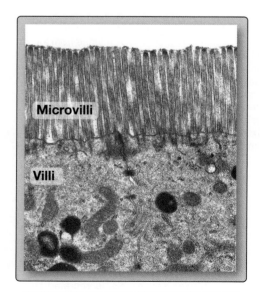

Microvilli

Villi

Figure 31.2.
Villi and microvilli.

I. OVERVIEW

The stomach liquefies and churns food to yield a semisolid slurry called **chyme**. The task of the next intestinal compartment (the **small intestine**; Fig. 31.1) is to enzymatically reduce the small particles, emulsified lipid droplets, and macromolecules within chyme to simpler forms (e.g., **amino acids**, **glucose**, and **short-chain fatty acids** [**SCFAs**]) that the intestinal epithelium can absorb. Absorption involves diversion of considerable resources. Consider, for example, that absorbing the 55 g of sugar contained within a piece of chocolate cake involves transporting 96,763,261,670,420,700,000,000 (nearly 97 sextillion) glucose molecules and a similar number of fructose molecules (the other product of sucrose digestion)! After digestion and absorption has been completed, food wastes are transferred to the final compartment (**large intestine**). The cecum and colon together form a fermentation vessel in which bacteria break down indigestible dietary fiber to release SCFAs. The colon absorbs these energy substrates, along with water. The distal colon also stores and desiccates waste (feces) until it can be eliminated from the body.

II. SMALL INTESTINE

The small intestine is the longest gastrointestinal (GI) tract segment at ~6 m. It comprises three functional segments: **duodenum** (~0.3 m), **jejunum** (~2.3 m), and **ileum** (~3.4 m). Digestion and absorption are facilitated by circular folds (folds of Kerckring), villi (finger-like epithelial projections), and microvilli (apical membrane specializations) that amplify epithelial surface area from ~2.5 m² to ~400 m². The microvilli resemble bristles of a paintbrush, which is why the intestinal epithelium is said to possess a **brush border** (Fig. 31.2).

A. Motility and mixing

During food digestion, the small intestine performs **segmentation** contractions that churn food back and forth to further reduce particle size and to mix in secretions from the intestinal epithelium, pancreas, and gallbladder (see Chapter 32). As in the stomach, contractions are initiated by slow waves generated by **interstitial cells of Cajal**. Unlike in the stomach, however, slow waves do not trigger contractions unless they cross a threshold for action potential formation. Intestinal slow waves are also twice as frequent compared with those in the stomach (~12 cycles per minute). Slow wave frequency is increased by parasympathetic activation and reduced by sympathetic activation. During fasting, contractions occur infrequently and involve short intestinal segments. Once every 90 to 120 minutes, **migrating motor complexes (MMCs)** create a peristaltic wave that sweeps through the length of the small intestine to clear intestinal secretions and residual food particles. MMCs begin in the stomach or duodenum and are initiated by the peptide hormone **motilin**. Feeding disrupts these complexes in favor of segmentation contractions.

B. Secretions

The small intestine secretes ~1 L of fluid and mucus per day. Mucus helps lubricate chyme, facilitates mixing, and protects the epithelial surface. The small intestine also contains enteroendocrine cells that secrete **cholecystokinin (CCK)**, **secretin**, and **glucose-dependent insulinotropic peptide** (see Table 29.2).

C. Carbohydrate digestion and absorption

Carbohydrates are an important source of energy substrates to fuel metabolism, yielding ~4 kcal/g. Carbohydrates come in many forms (e.g., starch, dietary fiber), but all must be broken down into monosaccharides (glucose, galactose, and fructose) before they can be absorbed by the intestinal epithelium.

1. **Starch digestion:** Starch comprises a chain of glucose molecules. **Amylose** is a linear glucose chain linked by α1,4 bonds. **Amylopectin** is starch containing branches linked by α1,6 bonds. Starch is digested by the pancreatic enzyme **pancreatic amylase**, but it can only break internal α1,4 bonds to yield maltose, maltotriose, glucose oligomers, and α-limit dextrin (Fig. 31.3). Oligo- and disaccharidases then reduce these digestion products to monosaccharides.

2. **Oligosaccharide and disaccharide digestion:** Oligosaccharides and disaccharides are reduced to monosaccharides by four brush-border **disaccharidases** (Table 31.1). Their membrane association means that their digestion products are released close to the transporters responsible for absorbing them across the apical membrane. For example, lactose breakdown into glucose and galactose is facilitated by lactase, which is located close to the transporters responsible for glucose and galactose absorption (Fig. 31.4).

3. **Dietary fiber:** Dietary fiber includes **soluble** (e.g., pectin) and **insoluble** (e.g., cellulose) forms (Table 31.2). Dietary fiber contains bonds that human digestive enzymes are unable to break. For example, cellulose contains linear β1,4 glucose bonds that neither salivary nor pancreatic amylase can disrupt. Because dietary fiber cannot be digested and absorbed, it ends up in the colon, where it either fuels colonic bacterial fermentation or increases fecal bulk. Bulking can be beneficial in that it increases intestinal motility and defecation frequency.

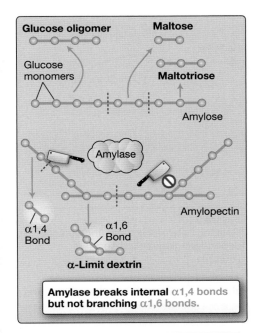

Figure 31.3.
Carbohydrate digestion.

Table 31.1: Membrane-Bound Disaccharidases

Enzyme	Substrate(s)	Product(s)
Gluco-amylase	Maltose and maltotriose	Glucose
Isomaltase	α-Limit dextrins, maltose, and maltotriose	Glucose
Lactase	Lactose	Glucose and galactose
Sucrase	Maltose, maltotriose, and sucrose	Glucose and fructose

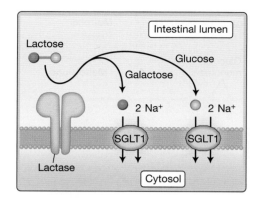

Figure 31.4.
Relation between lactase and sodium-dependent glucose transporters (SGLT1).

Table 31.2: Dietary Fiber Classifications

Types	Solubility
Cellulose	Insoluble
Hemicellulose	Insoluble
Lignin	Insoluble
Gums	Soluble
Pectins	Soluble

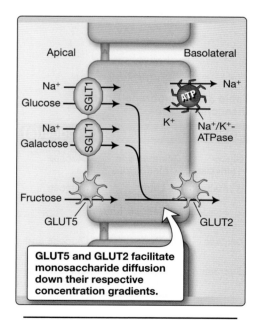

GLUT5 and GLUT2 facilitate monosaccharide diffusion down their respective concentration gradients.

Figure 31.5.
Apical and basolateral monosaccharide transport. GLUT2 and 5 = members of the glucose transporter family, SGLT1 = sodium-dependent glucose transporter 1.

Clinical Application 31.1: Lactose Intolerance

Lactose is an important source of carbohydrates in infants, but lactase expression decreases after weaning in most ethnic groups (Caucasians of northern European descent being the main exception), which can lead to lactose intolerance. Lactose intolerance is a malabsorption syndrome. Symptoms are caused by lactose passing undigested through the small intestine and entering the large intestine, where intestinal bacteria break it down to short-chain fatty acids and H_2. If the ingested lactose load is excessive, the hydrogen gas causes bloating, flatulence, and abdominal pain. Undigested lactose causes an osmotic diarrhea. Individuals with conditions that impair lactose digestion (e.g., inflammatory bowel disease) are especially prone to lactose intolerance. Intolerance can be assessed using a 50-g oral lactose challenge, followed by monitoring for symptoms of intolerance, an H_2 breath test, or by testing for the appearance of lactose digestion products (glucose and galactose) in blood. Avoidance of dietary lactose or using a commercially available lactase preparation is curative.

4. **Monosaccharide absorption:** Glucose, galactose, and fructose are hydrophilic, so transporters are required to move them across the intestinal epithelium.

   a. **Apical membrane transport:** Glucose and galactose are transported across the apical membrane by **SGLT1** (*SLC5A1* gene), a Na⁺-dependent glucose cotransporter (Fig. 31.5). The basolateral Na⁺/K⁺-ATPase provides the driving force for secondary active glucose transport against a glucose concentration gradient. Fructose uptake occurs by facilitated diffusion via GLUT5 (*SLC2A5* gene), a glucose transporter family member.

   b. **Basolateral membrane transport:** Transport across the basolateral membrane is facilitated by GLUT2 (*SLC2A2* gene). GLUT2 transports glucose, galactose, and fructose. The monosaccharides then enter the portal circulation.

D. Protein digestion and absorption

Proteins are primarily used as building blocks for reassembly into other proteins, although they also can be used for energy production (~4 kcal/g) during fasting. Small amounts of proteins and peptides are absorbed via phagocytosis across the apical membranes of enterocytes and specialized mucosal immune cells (M cells). Most proteins are broken down into amino acids and oligopeptides to facilitate absorption. Protein digestion begins with pepsin in the stomach and is continued in the small intestine by several pancreatic and brush-border proteases.

1. **Digestion:** The pancreas secretes five proteases as proenzymes that activate in the small intestine. Three endopeptidases (**trypsin**, **chymotrypsin**, and **elastase**) disassemble proteins into oligopeptides containing six or fewer amino acids. Two exopeptidases (**carboxypeptidase A** and **B**) cleave off single amino acids from oligopeptides (Fig. 31.6). About 70% of total protein

is reduced to oligopeptides and ~30% to single amino acids. The proteases also digest each other, so luminal activity rapidly decreases once the enzymes become active. Brush-border peptidases (e.g., enterokinase and aminopeptidase) may also break down small peptides and oligopeptides into individual amino acids.

2. **Absorption:** Amino acids are transported across the apical membrane via numerous classes of amino acid cotransporters with varying substrate specificities. Di- and tripeptides are transported across the apical membrane by a **H⁺/oligopeptide cotransporter** ([**PepT1**] *SLC15A1* gene) that uses a transmembrane H^+ gradient to power uptake. The peptides are then broken down into individual amino acids by cytosolic peptidases (Fig. 31.7). Transport across the basolateral membrane occurs by facilitated diffusion via three or more different amino acid transporters.

E. Lipid digestion and absorption

Lipids are calorically denser (~9 kcal/g) than carbohydrates or proteins and therefore represent a significant source of energy substrates. The diet includes several different types of lipid, but most is in the form of triacylglycerols ([TAGs] >90%). TAGs comprise three fatty acid "tails" esterified to a glycerol backbone (Fig. 31.8A). Other dietary fats include phospholipids, cholesterol, and free fatty acids (FFAs). TAGs are water insoluble, so extracting them from chyme (which is aqueous) requires that they be broken down into smaller FFAs and 2-monoacylglycerol (2-MAG). The first step in lipid disassembly is emulsification.

1. **Emulsification:** Emulsification is a process whereby oils are mechanically broken up into small droplets. Just as olive oil is whisked with vinegar to form an emulsion familiar as a vinaigrette salad dressing, chewing (mouth) and churning motions (stomach and small intestines) emulsify dietary lipids. The purpose of emulsification is to increase net lipid surface area available to lipases. During emulsification, lipid droplets inevitably become coated with other food components and bile salts, which prevents the droplets from recoalescing.

2. **Digestion:** Lipid digestion begins in the mouth with **lingual lipase** but not in physiologically significant amounts. In the stomach, **gastric lipase** cleaves SCFAs (<6 carbons long), medium-chain fatty acids ([MCFAs] 6–12 carbons long), and some long-chain fatty acids (LCFAs) from the TAG backbone, leaving diacylglycerol

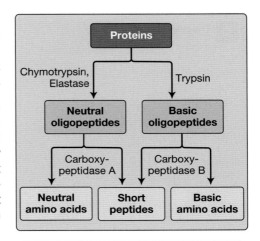

Figure 31.6.
Protein and peptide digestion.

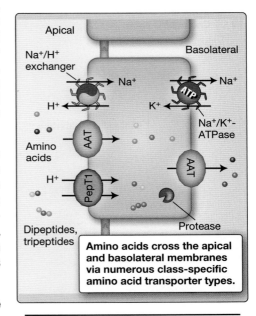

Figure 31.7.
Amino acid (AAT) and di- and tripeptide transporters. PepT1 = H⁺/oligopeptide cotransporter.

Clinical Application 31.2: Hartnup Disease

Hartnup disease is an autosomal-recessive disorder caused by mutations in *SLC6A19*, a gene that encodes a Na⁺-dependent neutral amino acid cotransporter found in the small intestine and renal proximal tubule. Affected individuals may develop pellagra-like symptoms (photosensitivity, dementia) due to tryptophan deficiency. Tryptophan is a precursor in niacin (vitamin B_3) synthesis, so the mutation can result in niacin deficiency. Alternate routes for tryptophan absorption (e.g., PepT1 and phagocytosis) can partially compensate for the transport defect, but patients usually require dietary niacin supplementation.

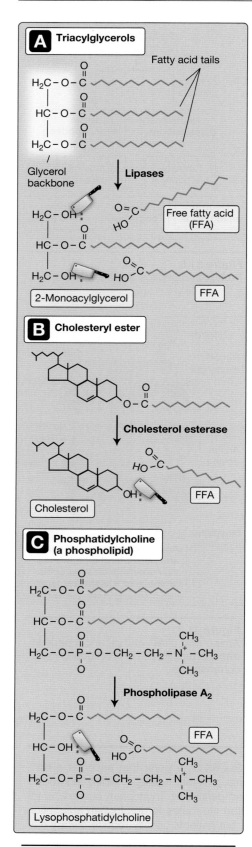

Figure 31.8.
Lipid digestion.

(DAG). The majority of lipid digestion occurs in the small intestine and is carried out by pancreatic enzymes.

a. Pancreatic lipase: The pancreas secretes **pancreatic lipase** in amounts that far exceed requirements, thereby maximizing TAG digestion. Pancreatic lipase cleaves the fatty acids (FAs) at carbons 1 and 3 from TAGs to yield 2-MAG and two FFAs (see Fig. 31.8A). Pancreatic lipase is tethered to the lipid droplet by **colipase**, which anchors itself to the droplet surface by binding to bile salts. Colipase is a cofactor secreted by the pancreas as procolipase and activated in the intestinal lumen by trypsin.

b. Other lipases: Gastric and pancreatic lipases only digest TAGs. Cholesteryl esters are digested into cholesterol and FAs by **cholesterol esterase** (see Fig. 31.8B). **Phospholipase A$_2$** targets glycerophospholipids (e.g., phosphatidylcholine) to release the FA at carbon 2, leaving lysophospholipids (see Fig. 31.8C).

3. Absorption: Principal lipid digestion products include FFAs, 2-MAG, cholesterol, and lysophospholipids. SCFAs and MCFAs are water soluble and dissolve in chyme following their release from a lipid droplet (Fig. 31.9). LCFAs and other digestion products are less soluble, forming coin-shaped disks called **mixed micelles**. The micelles have a hydrophobic core and a hydrophilic rim containing bile salts and polar lipid head groups. Micelles carry lipid digestion products to the enterocyte apical membrane.

a. Apical absorption: SCFAs and MCFAs readily cross the apical membrane by diffusion. The apical surface layers are acidic compared with chyme, which causes the micelles to disperse. Freed LCFAs may either diffuse directly across the apical membrane or be carried across by fatty acid transporters or in association with **fatty acid–binding proteins**. 2-MAGs and cholesterol are absorbed via similar pathways.

> Niemann-Pick C1-like 1 protein (NPC1L1, a product of the *SLC65A2* gene) is a cholesterol-specific carrier found in hepatocytes and small intestinal enterocytes. Pharmacologic inhibition of NPC1L1 may be used to decrease cholesterol uptake and lower circulating cholesterol levels in patients with dyslipidemia.[1]

b. Basolateral absorption: SCFAs and MCFAs traverse the basolateral membrane by diffusion and enter the portal circulation, where they bind to albumin and are transported to the liver (see Fig. 31.9). The other products of lipid disassembly

[1]For more information on drugs used to treat hyperlipidemia, see *LIR Pharmacology*, 8e, Chapter 14·III.

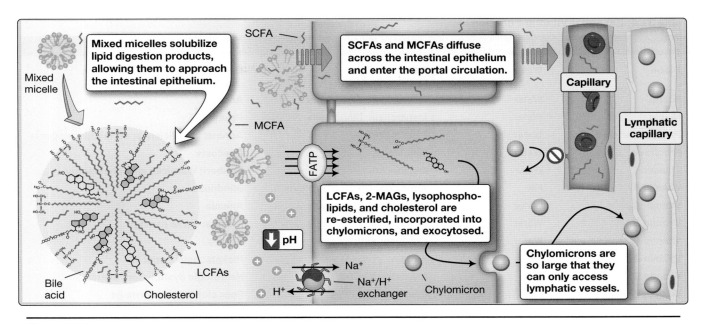

Figure 31.9.
Lipid absorption mechanisms. 2-MAG = monoacylglycerol; FATP = fatty acid transport protein; LCFA, MCFA, and SCFA = long-, medium-, and short-chain fatty acid, respectively.

are reassembled (esterified) into TAGs, cholesteryl ester, or phospholipids to form lipid droplets. A protein coat (mainly apolipoprotein B-48 and other apolipoproteins) is added as an emulsifier that prevents the droplets coalescing, and the resulting package (**chylomicron**) is exocytosed across the basolateral membrane. Chylomicrons are too large (75–450 μm) to fit through capillary fenestrations, so, instead, they enter the lymphatic system for transportation (see Fig. 31.9).

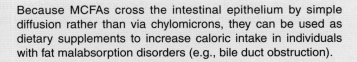

Because MCFAs cross the intestinal epithelium by simple diffusion rather than via chylomicrons, they can be used as dietary supplements to increase caloric intake in individuals with fat malabsorption disorders (e.g., bile duct obstruction).

Table 31.3: Mineral Function

Mineral	Functions
Calcium	Bone and teeth; cell excitability; blood clotting
Chloride	Cell excitability
Copper	Enzyme cofactor; collagen
Iron	Metabolism; oxygen binding; collagen
Iodine	Hormone synthesis
Magnesium	Metabolism
Phosphorus	Bone and teeth; energy storage; cell signaling
Potassium	Cell excitability
Sodium	Cell excitability; establish extracellular fluid osmolality
Zinc	Enzyme cofactor

F. Vitamin and mineral absorption

The small intestine absorbs numerous essential vitamins and minerals, summarized in Tables 31.3 and 31.4.

1. **Vitamins:** Fat-soluble vitamins (vitamins A, D, E, and K) are absorbed in a similar manner to LCFAs and are distributed in chylomicrons.[1] Water-soluble vitamins, with the exception of

 [1]For more information on vitamins, see *LIR Biochemistry*, 8e, Chapter 28.

Table 31.4: Essential Vitamin Function

Vitamin	Solubility	Function	Deficiency
Biotin	Water	Metabolism	Dermatitis; alopecia; neurologic symptoms
Folate	Water	Metabolism; blood cells	Anemia
Niacin	Water	Metabolism; blood cells	Pellagra
Pantothenic acid	Water	Metabolism	Adrenal insufficiency; dermatitis, alopecia
Riboflavin	Water	Metabolism	Cheilosis
Thiamin	Water	Metabolism	Beriberi; Wernicke encephalopathy
Vitamin A	Fat	Antioxidant; vision; proteins	Blindness
Vitamin B_6	Water	Metabolism; blood cells	Anemia
Vitamin B_{12}	Water	Metabolism; blood cells	Anemia; nerve damage
Vitamin C	Water	Collagen; antioxidant	Scurvy
Vitamin D	Fat	Proteins	Rickets; osteomalacia
Vitamin E	Fat	Antioxidant	Neurologic dysfunction; anemia
Vitamin K	Fat	Blood cells	Poor clotting; hemorrhage

vitamin B_{12} (cobalamin), are absorbed via Na^+ cotransport. Vitamin B_{12} is absorbed in a five-step process:

Step 1: Vitamin B_{12} is ingested attached to dietary proteins and is liberated by low stomach pH and pepsin.

Step 2: The vitamin immediately binds to haptocorrin, a glycoprotein produced by salivary glands, which protects it from denaturation by gastric acid.

Step 3: On entering the duodenum, pancreatic proteases degrade haptocorrin, so vitamin B_{12} now binds to intrinsic factor (IF). IF is secreted by gastric parietal cells but does not interact with the vitamin until it reaches the small intestine. IF protects vitamin B_{12} from degradation by proteases.

Step 4: The IF–vitamin B_{12} complex binds to a specific receptor expressed by ileal enterocytes, and it is absorbed by endocytosis.

Step 5: Once inside the enterocyte, the complex dissociates, and vitamin B_{12} binds to cobalamin II for absorption across the basolateral membrane and transportation through the portal circulation.

2. **Minerals:** Ion uptake by the small intestine occurs paracellularly and by various transcellular pathways.[1] Ca^{2+} is absorbed throughout the small intestine, but the duodenum is the site of regulated absorption and participates in Ca^{2+} homeostasis (see Chapter 35·IV·D·3). The strategies used for Ca^{2+} absorption are very similar to those involved in regulated Ca^{2+} reabsorption by the renal distal convoluted tubule (see Chapter 27·III·C). Absorption across the apical membrane occurs passively via a TRPV6 Ca^{2+} channel (Fig. 31.10). Ca^{2+} then binds to **calbindin**, which acts as a buffer to keep intracellular levels low and the transapical gradient for Ca^{2+}

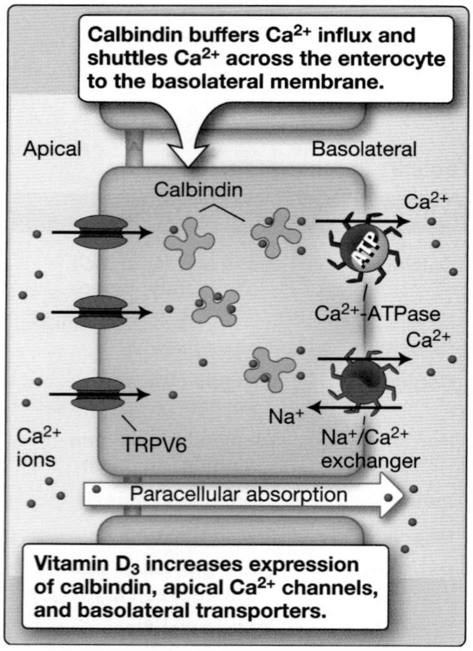

Calbindin buffers Ca^{2+} influx and shuttles Ca^{2+} across the enterocyte to the basolateral membrane.

Vitamin D_3 increases expression of calbindin, apical Ca^{2+} channels, and basolateral transporters.

Figure 31.10.
Calcium absorption pathways.

[1]For more information on minerals, see *LIR Biochemistry*, 8e, Chapter 29.

diffusion high. Calbindin also shuttles Ca^{2+} across to the baso-lateral membrane, where it is transported to the interstitium by a Ca^{2+}-ATPase and a Ca^{2+}/Na^+ exchanger. Vitamin D_3 increases expression levels of all pathway components.

G. Water absorption

We ingest an average of 2.0 to 2.5 L of water a day (beverages and water contained in food; Fig. 31.11). Salivary, gastric, pancreatic, biliary, and GI secretions pour another 7 L of fluid into the GI lumen. Most of this (~80%) is reabsorbed in the small intestine. Reabsorption largely occurs paracellularly in response to osmotic gradients created by solute absorption.

III. LARGE INTESTINE

The large intestine comprises the cecum; ascending, transverse, descending, and sigmoid colon; rectum; and anus (Fig. 31.12). Food digestion and absorption is largely complete by the time intestinal contents are moved into the colon, so the main function of the colon is to recover remaining water and store feces until eliminated from the GI tract (**defecation**).

A. Motility

Waste enters the colon via the **ileocecal sphincter**, which acts as a one-way valve between the ileum and cecum (see Fig. 31.12). Waste is propelled through the sphincter by peristaltic contractions and facilitated by sphincter relaxation, triggered by ileal distension and irritation (stimulation of chemical afferents). The ileocecal sphincter also relaxes, and the ileum contracts peristaltically immediately following a meal, a response known as the **gastroileal reflex**. The reflex is likely controlled by gastrin and CCK. Distension of the cecum by entering waste inhibits peristalsis and contracts the sphincter to prevent backflow into the ileum.

1. **Haustration and mass movements:** The task of the colon is to recover most of the remaining ~2 L of water from feces, turning it into a semisolid. Recovery is facilitated by periodic segmentation contractions lasting 12 to 60 seconds that churn the waste and increase contact between intestinal contents and the epithelium. The segmentation contractions divide the intestine into a series of small pouches (**haustra**), giving the large intestine a characteristic bead-like appearance in imaging studies (Fig. 31.13). Approximately one to three times a day, the segmentation contractions cease, and a strong peristaltic wave drives intestinal contents toward the anus by ~20 cm (**mass movements**).

2. **Rectum and anal canal:** The rectum is usually empty, in part because segmentation contractions in the distal colon tend to retard forward movement, but also because retrograde rectal contractions force fecal matter back into the colon. The anal canal is normally closed by tonic contraction of the internal anal sphincter (IAS) and external anal sphincter (EAS). The IAS is composed

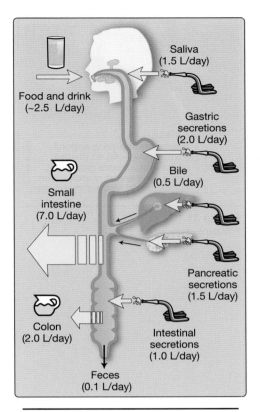

Figure 31.11.
Fluid intake, secretion, and absorption by the GI tract.

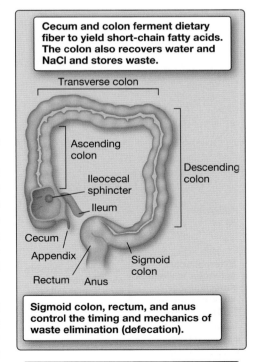

Figure 31.12.
Large intestine.

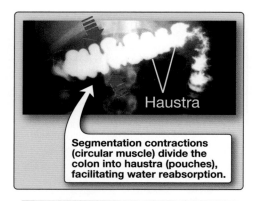

Segmentation contractions (circular muscle) divide the colon into haustra (pouches), facilitating water reabsorption.

Figure 31.13.
Haustra.

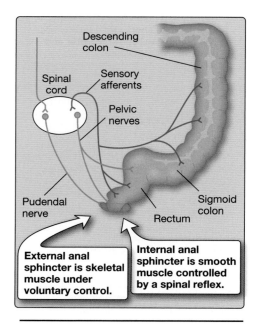

Figure 31.14.
Innervation of the colon, rectum, and anus.

External anal sphincter is skeletal muscle under voluntary control.

Internal anal sphincter is smooth muscle controlled by a spinal reflex.

of smooth muscle. The EAS is composed of skeletal muscle that is under somatic control via the pudendal nerve (Fig. 31.14). A colonic mass movement forces feces into the rectum and triggers the urge to defecate (a "call to stool").

a. **Rectosphincteric reflex:** The **rectosphincteric reflex** is a spinal reflex triggered by rectal distension. Rectal sensory afferents travel in the pelvic splanchnic nerve. The rectum contracts and the IAS relaxes reflexively, mediated by pelvic splanchnic parasympathetic efferents. The EAS reflexively increases tone to prevent elimination. Opening of the anal canal allows rectal contents to be sampled by an anal sensory area to determine composition (e.g., solid vs. gas) and liquidity. If defecation is inconvenient, the EAS remains voluntarily contracted, and elimination may usually be postponed.

b. **Defecation:** If defecation is desired, the EAS is relaxed voluntarily via the pudendal nerve. The puborectalis muscle relaxes reflexively also (via pelvic nerve branches), which straightens the anorectal angle to facilitate elimination. The puborectalis muscle is a part of the pelvic floor that forms a sling around the rectum and crimps the wall to form a ~90° bend that helps prevent accidental defecation. Squatting or sitting also straightens the anorectal angle. Voluntary straining raises intraabdominal pressure to assist colorectal peristaltic activity, and rectal contents are expelled.

B. Absorption

The large intestine does not actively participate in digestion, but it does have important transport functions.

1. **Short-chain fatty acids:** The cecum and colon evolved to create a fermenter that makes use of colonic bacteria to break down indigestible plant fibers (e.g., cellulose) to release energy-rich SCFAs (principally butyrate, acetate, and propionate). Colonic enterocytes avidly absorb 95% of released SCFAs using an apical anion exchanger (DRA, a product of the *SLC26A3* gene) and a monocarboxylate transporter ([MCT1] *SLC16A1* gene) (Fig. 31.15). DRA exchanges an SCFA for HCO_3^-, the latter helping maintain a neutral colonic pH during fermentation reactions. Butyrate is retained by the colonocyte and used to fuel metabolism. Acetate and propionate enter the portal circulation and are transported to the liver.

2. **Electrolytes and water:** The proximal colon reabsorbs Na^+ and Cl^- via an apical NHE3 Na^+/H^+ exchanger and DRA (see Fig. 31.15). The colon receives ~2 L of water per day, most of which is reabsorbed paracellularly, driven by the osmotic gradient created by NaCl reabsorption. Bacterial enterotoxins (e.g., cholera toxin) inhibit this pathway, resulting in a profound secretory diarrhea. The distal colon reabsorbs Na^+ via an epithelial sodium channel (ENaC). This pathway is regulated by aldosterone. When there is a need to conserve Na^+ or water, aldosterone increases the

activity and expression levels of ENaC and the basolateral Na⁺/K⁺-ATPase in a similar manner to that described for the renal tubule (see Chapter 27·IV·B). K⁺ is passively secreted throughout the colon via apical K⁺ channels (see Fig. 31.15). The distal colon also actively secretes K⁺ in an aldosterone-dependent manner.

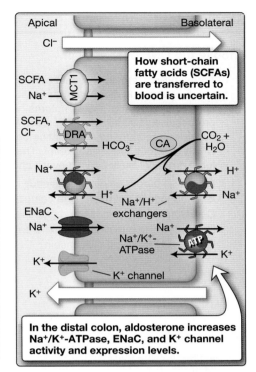

Figure 31.15.
Colonic transport pathways. CA = carbonic anhydrase, DRA = anion exchanger, MCT1 = monocarboxylate transporter 1.

Clinical Application 31.3: Diarrhea and Metabolic Acidosis

Chronic (>4 weeks), persistent (2–4 weeks), or acute (<2 weeks) diarrhea, if severe, can result in the excretion of large amounts of HCO_3^- and other ions. Diarrhea refers to the passage of semisolid or fluid stool several times daily. Most cases of diarrhea are due to viral, bacterial, or protozoan infection. Diarrhea results when osmotic pressure develops within the intestinal lumen, which prevents fluid reabsorption and, in severe cases, may cause net water secretion. This can lead to hypovolemia, and to a lower plasma pH as a result of HCO_3^- loss. Because Na⁺ and Cl⁻ are excreted along with HCO_3^-, the anion gap does not appreciably change. Thus, this type of acid–base disturbance can be classified as **metabolic acidosis** with a normal anion gap.

Chapter Summary

- Small intestinal **motility** involves **segmentation** contractions, which mix and churn food, and peristaltic contractions that propel chyme toward the anus. **Migrating motor complexes** sweep the intestinal lumen free of residual particles between meals.

- Starch is a glucose polymer digested by **pancreatic amylase**, which cleaves internal **α1,4 bonds**. Apical membrane-bound **disaccharidases** convert starch digestion products to **monosaccharides** (glucose, galactose, and fructose) for absorption.

- Glucose and galactose are absorbed across the apical membrane by a Na⁺/glucose cotransporter. Fructose is absorbed by facilitated diffusion. Basolateral monosaccharide transport also occurs by facilitated diffusion.

- Pancreatic proteases (**trypsin**, **chymotrypsin**, **elastase**, and **carboxypeptidases**) digest proteins to yield smaller peptides. Membrane-bound **peptidases** digest these protein fragments to yield amino acids, dipeptides, and tripeptides.

- Amino acids are absorbed across the apical membrane by Na⁺-dependent cotransporters. Small peptides are absorbed by H⁺-oligopeptide cotransport and then broken down to release amino acids within the enterocyte. Basolateral amino acid transport occurs by facilitated diffusion via various amino acid transporters.

- **Bile acids** emulsify lipids, allowing pancreatic lipase to cleave fatty acids (FAs) from triacylglycerols (TAGs). Dietary cholesterol esters are digested into cholesterol and FAs by **cholesterol esterase**. Lipid digestion products collect in **mixed micelles**.

- FAs and cholesterol diffuse across the apical membrane or may be assisted by transporters. FAs are reassembled to form TAGs and repackaged into **chylomicrons** within the enterocyte. Chylomicrons are exocytosed across the basolateral membrane and transported by the lymphatic circulation.

- The **ileocecal sphincter** regulates movement of waste between small and large intestines. Internal and external anal sphincters regulate feces elimination. Large intestine motility includes **segmentation** contractions and **peristaltic** contractions that result in **mass movements**. Defecation is coordinated by reflex pathways and voluntary relaxation of the external anal sphincter.

- The large intestine absorbs Na⁺, Cl⁻, and water and secretes K⁺ and HCO_3^-.

Study Questions

Choose the ONE best answer.

31.1. A 25-year-old female presents with abdominal discomfort, bloating, and diarrhea after consuming dairy products. She notes that the symptoms have been getting progressively worse over the past few months. The physician recommends eliminating dairy from her diet, which provides significant relief. The patient's symptoms are most likely caused by which of the following?

A. Brush-border enzyme deficiency
B. Changes in gut microbiota composition
C. Decreased galactose and glucose uptake
D. Decreased intestinal surface area
E. Inability to digest milk casein

Best answer = A. The patient is lactose intolerant. Lactase is a brush-border enzyme present in newborns but whose expression decreases after weaning in most ethnic groups. Lactase deficiency causes lactose (a milk sugar), to pass through the small intestine to the colon. Here, it is digested by gut microbiota to release short-chain fatty acids and hydrogen gas. The gas is the cause of the patient's discomfort and bloating (see Clinical Application 31.1). Changes in gut microbiota composition (B) can cause bloating and discomfort, but this patient's symptoms are due to lactase deficiency. Lactase converts lactose to glucose and galactose, which are then absorbed by the small intestine. Decreased glucose and galactose uptake (C) is not responsible for the patient's symptoms. Decreased intestinal surface area (D) describes celiac disease caused by a reaction to gluten. Casein (E) is a milk protein and would not be affected by lactase deficiency.

31.2. Genetic studies have suggested that mutations in a gene encoding an H^+-dependent transporter gene that localizes to the small intestine may be involved in the etiology of irritable bowel disease. This transporter is most likely required for which of the following small intestinal actions?

A. Apical dipeptide uptake
B. Apical fructose uptake
C. Apical glycerol uptake
D. Basolateral amino acid transport
E. Basolateral glucose transport

Best answer = A. Apical absorption of dipeptides occurs via PepT1, which is a cotransporter powered by an inward H^+ gradient (see Section II·D·2). Apical fructose uptake (B) occurs via GLUT5, which is a facilitated transporter that does not require a H^+ or Na^+ gradient to power transport. Apical glycerol uptake (C) does not require the assistance of any ion or specialized transport protein. Glycerol uptake occurs by diffusion across the epithelial cell membrane. Basolateral amino acid transport (D) occurs via either individual or group transporters independently of ion gradients. Similarly, basolateral glucose transport (E) occurs via a GLUT2 transporter which facilitates diffusional uptake of glucose independently of ions.

31.3. A 6-year-old male presents to the office with chronic diarrhea, foul-smelling greasy stools, and failure to thrive. Laboratory studies reveal pancreatic insufficiency, leading to a deficiency in pancreatic enzymes and a cofactor. Which of the following most likely describes the function of this cofactor normally?

A. Anchoring lipase to a lipid droplet
B. Cholesterol transport
C. Converting trypsinogen to trypsin
D. Digesting DNA
E. Vitamin B_{12} transport

Best answer = A. Colipase is cofactor synthesized by the pancreas that binds to both pancreatic lipase and bile salts on the surface of lipid droplets, thereby anchoring the lipase to the droplet (see Section II·E) and allowing it to digest the lipids within. Pancreatic deficiency impairs lipid digestion. Undigested lipids are carried to the colon, where they are acted on by gut microbes. Digestion products result in diarrhea and foul-smelling, greasy stools. Cholesterol absorption is facilitated by NPC1L1, a cholesterol-specific carrier (B). Trypsinogen is converted to trypsin (C) by proteases within the small intestine lumen. DNA digestion (D) is carried out by pancreatic deoxyribonuclease. Vitamin B_{12} is transported (E) through the small intestine by intrinsic factor, which is a product of gastric parietal cells.

Exocrine Pancreas and Liver

32

I. OVERVIEW

The **pancreas**, **gallbladder**, and **liver** are intestinal accessory organs (Fig. 32.1) that produce specialized secretions to help the intestines digest carbohydrates, proteins, and lipids. The pancreas produces HCO_3^- and digestive enzymes. The liver produces **bile**, which is stored in the gallbladder until needed. Bile has a special place in the history in medicine, comprising two of second-century Greek physician Galen's four humors: yellow bile, black bile, blood, and phlegm. The notion that disturbances in these humors are the root cause of all disease has long since been discarded, but bile does serve a critical role in fat digestion and absorption. Food digestion products are taken up from the intestinal lumen, transferred to the **portal circulation**, and then transported to the liver for processing (see Fig. 20.10). The liver has numerous essential functions, including metabolism and storage of absorbed nutrients; bile production; and the detoxification, biotransformation, and excretion of hormones, drugs, and other bioactive compounds.

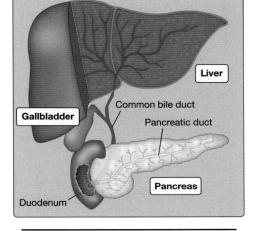

Figure 32.1.
Intestinal accessory organs.

II. EXOCRINE PANCREAS

The pancreas is composed of numerous lobules containing secretory acini (Fig. 32.2). Acini are lined with a secretory epithelium composed of **acinar cells**, which synthesize and secrete digestive enzymes. Enzymes are released into an **intercalated duct**, the lining of which secretes HCO_3^--rich fluid. Secretions are channeled through an extensive ductal system (intralobular ducts, interlobular ducts, and the pancreatic duct) that empties into the duodenum.

A. Secretions

Acinar and ductal cells are specialized to produce two different secretory products and are regulated independently.

1. **Acinar cells:** Acinar cells synthesize large quantities of protein, as evidenced by an extensive rough endoplasmic reticulum. The cells produce a variety of digestive enzymes in either active or precursor form (**zymogens**) that are activated in the small intestine (Table 32.1). Newly synthesized proteins are ferried to the Golgi apparatus for packaging in large **condensing vacuoles**, which slowly decrease in diameter to

Table 32.1: Zymogen Granule Enzymes and Enzyme Precursors

Zymogen granule protein	Description
Amylase	Carbohydrase
Chymotrypsinogen	Protease precursor
Deoxyribonuclease	Nuclease
Lipase	Lipase
Procarboxypeptidase A and B	Protease precursors
Proelastase	Protease precursor
Prophospholipase A₂	Lipase precursor
Procolipase	Colipase precursor (lipase cofactor)
Ribonuclease	Nuclease
Trypsinogen	Protease precursor

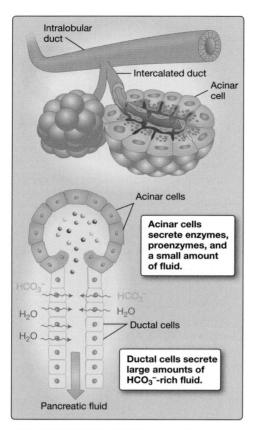

Figure 32.2.
Pancreatic secretions.

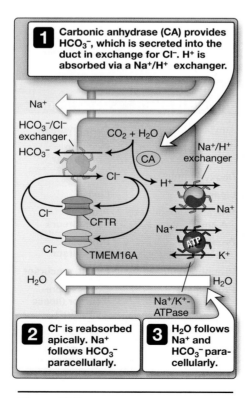

Figure 32.3.
Ductal cell secretion mechanisms.

yield secretory **zymogen granules**. These granules localize to the acinar cell apical membrane, where they dock and await release. Acinar cells also secrete an isotonic NaCl-rich fluid to hydrate and wash the proteins out of the acinus (see Fig. 32.2). Acinar cells take up Cl^- from blood using a basolateral $Na^+/K^+/2Cl^-$ cotransporter and then secrete it into the acinus via apical Cl^- channels. Na^+ and water follow paracellularly, driven, respectively, by electrochemical and osmotic gradients (not shown).

2. **Ductal cells:** Ductal cells secrete an isotonic sodium bicarbonate ($NaHCO_3$) solution. HCO_3^- is generated from CO_2 and H_2O, facilitated by carbonic anhydrase (CA), then crosses the apical membrane via a HCO_3^-/Cl^- exchanger (Fig. 32.3). HCO_3^- may also be taken up from blood by a Na^+/HCO_3^- cotransporter (not shown). Apical HCO_3^-/Cl^- exchange cannot occur unless there is Cl^- in the ductal lumen, so the membrane also contains Cl^- channels (cystic fibrosis transmembrane conductance regulator [CFTR] and TMEM16A) that recycle the incoming Cl^-. Defects in CFTR lead to the pancreatic symptoms associated with cystic fibrosis (see Clinical Application 32.1). The H^+ that is coproduced with HCO_3^- leaves the cell with the assistance of a basolateral Na^+/H^+ exchanger (see Fig. 32.3) and a H^+-ATPase (not shown).

B. Regulation

Between meals, pancreatic secretion rate is low, with periodic bursts of output coinciding with cleansing migrating motor complexes that sweep sloughed dead enterocytes and other debris out of the intestines (see Chapter 31·II·A). Secretion rate increases up to 20-fold to ~1.5 L/day in response to feeding. The ionic composition of the secreted fluid changes with flow rate, with HCO_3^- levels rising and Cl^- levels falling in parallel (Fig. 32.4).

1. **Pathways:** Acinar cells and ductal cells both express muscarinic (M_3) acetylcholine (ACh) receptors that facilitate increased enzyme and serous fluid secretion in response to vagal stimulation. Ductal cells additionally express **secretin** receptors, which mediate HCO_3^--rich fluid secretion in response to low duodenal pH. HCO_3^- is secreted into the duodenum to neutralize gastric acid entering via the pylorus, thereby maintaining an optimal pH for pancreatic enzyme activity and protecting the duodenal lining from acid damage.

2. **Phases:** There are three phases to pancreatic secretion (see also Chapter 30·IV·D·3). An initial, brief **cephalic phase** (~25% of total response) is mediated by the vagal nerve. Stomach distension stimulates a weak **gastric phase** (10% of total), again mediated by the vagal nerve. The **intestinal phase** is long-lasting (hours) and accounts for up to 75% of the total response. It is mediated hormonally through secretin released from duodenal S cells and **cholecystokinin (CCK)** released from duodenal I cells. The primary stimulus for CCK secretion is the presence of lipids in the duodenum, although protein also stimulates release. CCK binds to CCK_1 receptors on vagal afferents, thus modifying pancreatic function indirectly (CCK does not directly affect human acinar output).

III. HEPATOBILIARY SYSTEM

The hepatobiliary system produces, stores, and secretes bile. The liver is composed of ~100,000 lobules separated by connective tissue. The lobules are functional units common to exocrine glands, but their shape is roughly hexagonal rather than globular (Fig. 32.5). Within each lobule, numerous sheets of **hepatocytes** are arranged radially. The basolateral membrane backs up onto a **hepatic sinusoid**. The interstitial space between the basolateral membrane and sinusoid is known as the **space of Disse**. Hepatic sinusoids are perfused by blood from the portal vein, which is relatively O_2 poor, and also by the hepatic artery, which is a systemic artery carrying O_2-rich blood (see Fig. 20.10). The sinusoid drains into a hepatic vein located at the center of the hexagon. Hepatocytes take up material from sinusoidal blood across the basolateral membrane via the space of Disse. Bile is secreted across their apical membrane into a ~1-μm diameter **bile canaliculus**. Both membranes sprout numerous microvilli that greatly amplify membrane surface area available for absorption or secretion. The bile canaliculus drains into a bile duct, which, together with branches of the portal vein and hepatic artery, forms a **portal triad** (see Fig. 32.5).

A. Bile

Hepatic bile is an isotonic fluid containing electrolytes, bile acids, and salts (67% of organic components), phospholipids (22%), cholesterol (5%), protein (5%), and bile pigments (Table 32.2). Bile has two main physiologic functions. The first is to aid lipid digestion and absorption by the small intestine. Second, bile serves as the principal excretory pathway for many substances that are lipophilic (e.g., cholesterol, bile pigments, drugs) and cannot be filtered out and removed from the circulation by the kidney.

1. **Bile acids and bile salts:** Hepatocytes synthesize **primary bile acids** (principally cholic acid and chenodeoxycholic acid) from cholesterol using 7α-hydroxylase (*CYP7A1* gene). Hepatocytes also take up **secondary bile acids** (deoxycholic acid and lithocholic acid) that form by bacterial action on primary bile acids in the terminal ileum and colon. Hepatocytes may then conjugate bile acids to glycine or taurine to form **bile salts** (e.g., taurocholic acid and glycocholic acid). Deprotonated, unconjugated bile acids are also bile salts. Bile acids and salts are actively secreted across the apical membrane into canaliculi by multidrug resistance protein 2 ([MRP2] *ABCC2* gene) and a bile salt export pump ([BSEP]

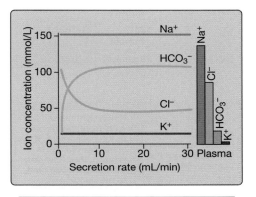

Figure 32.4.
Effect of pancreatic secretion rate.

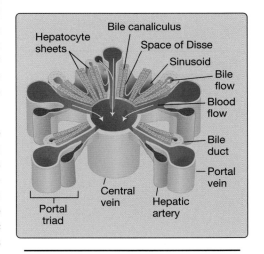

Figure 32.5.
Hepatic lobule organization. A sixth portal triad has been omitted for clarity.

Table 32.2: Hepatic and Gallbladder Bile Composition

Substance	Hepatic Bile	Gallbladder Bile
Bile salts	1 g/dL	5-fold ↑
Bilirubin	0.04 g/dL	10-fold ↑
Cholesterol	0.1 g/dL	5-fold ↑
Fatty acids	0.12 g/dL	6-fold ↑
Lecithin	0.04 g/dL	10-fold ↑
Na^+	145 mmol/L	Slight ↓
K^+	5 mmol/L	3-fold ↑
Ca^{2+}	2.5 mmol/L	5-fold ↑
Cl^-	100 mmol/L	10-fold ↓
HCO_3^-	28 mmol/L	3-fold ↓

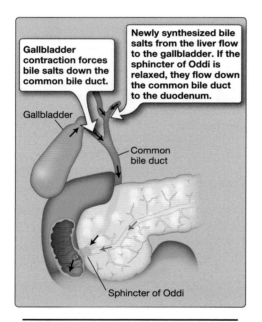

Figure 32.6.
Bile storage and secretion.

ABCB11 gene). MRP2 and BSEP are ATP-binding cassette transporters similar to those found in the renal proximal tubule (see Chapter 26·IV·B).

2. **Cholesterol and phospholipids:** Bile contains cholesterol and lecithins (phospholipids). Lecithins help keep cholesterol and other bile constituents in solution.

3. **Pigments and organic molecules:** The principal bile pigment is bilirubin, which is a byproduct of heme catabolism. Free (unconjugated) bilirubin is poorly water soluble, but it rapidly complexes with albumin and is then extracted from blood by hepatocytes during transit through the liver. Hepatocytes conjugate bilirubin to glucuronic acid to make it water soluble and then secrete it into bile via MRP2. Hepatocytes also take up a wide variety of organic anions (e.g., steroid and thyroid hormones, many drugs) and organic cations (e.g., choline, thiamine, many drugs) and actively excrete them into bile using MRP2 and other apical transporters.

B. Bile flow and storage

Hepatocytes secrete bile continuously. After entering a canaliculus, bile is channeled through ~2 km of bile ductules and ducts whose diameter increases and number decreases as they progressively merge. The ductal system is lined with **cholangiocytes**, specialized epithelial cells that monitor flow and modify bile by removing valuable components (e.g., glucose) and adding water and HCO_3^-. The biliary tree converges on the **common hepatic duct**. Bile may then flow either through the **common bile duct** into the duodenum or through the **cystic duct** to the **gallbladder** (Fig. 32.6). Access to the duodenum is controlled by the **sphincter of Oddi**. When the sphincter is relaxed, bile flows from the common hepatic duct into the duodenum. When the sphincter is contracted, the common bile duct has a high resistance to flow, so bile travels to the gallbladder. Bile may be stored in the gallbladder for several hours between meals, during which time it is concentrated by isosmotic NaCl and water reabsorption to form **gallbladder bile** (see Table 32.2). Na^+ is reabsorbed in exchange for H^+, whereas Cl^- is reabsorbed in exchange for HCO_3^- (Fig. 32.7). HCO_3^- secretion does not quite balance H^+ secretion, so pH drops from ~7.5 to ~6.0, which helps prevent insoluble calcium salt formation. Water follows ions isosmotically by paracellular and transcellular routes. Transcellular reabsorption occurs via aquaporins (AQP1 and AQP8).

C. Bile release

Bile release occurs when lipids and protein entering the duodenum trigger CCK secretion from I cells. CCK stimulates vagal afferents, causing sphincter of Oddi relaxation and rhythmic gallbladder contraction, which expels stored bile (Fig. 32.8). CCK also stimulates gallbladder smooth muscle contraction directly. Somatostatin and norepinephrine inhibit bile acid release.

D. Bile acid recycling

Hepatocytes synthesize less than a gram of bile acids per day and an equivalent amount is excreted in feces, yet the amount secreted

Clinical Application 32.2: Cholelithiasis

Cholelithiasis refers to the presence of gallstones in the gallbladder. Gallstones can be classified as cholesterol stones or pigment stones, the latter containing significant amounts of bilirubin. Gallstones form when cholesterol monohydrate crystals precipitate out of bile and are trapped by mucus. In time, they grow to form concretions of cholesterol, calcium salts, and variable amounts of bilirubin. Cholesterol stones are more common. Several processes contribute to the pathophysiology of stone formation, including genetics, bile stasis, and supersaturation of bile with cholesterol. Gallstones can obstruct the bile duct, limiting bile secretion and leading to fat malabsorption.

Gallstones.

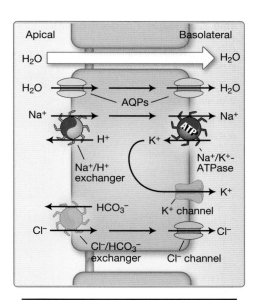

Figure 32.7.
Gallbladder bile concentration pathways.

each day is up to 60 times higher. This is made possible by bile acid recycling via the **enterohepatic circulation**. Unconjugated bile acids are reabsorbed passively in the terminal small intestine (95%) and in the large intestine (5%). Conjugated bile acids require a special transport system involving an apical sodium–bile salt transporter ([ASBT] *SLC10A2* gene) and a basolateral organic anion transporter (OAT). Bile acids travel back to the liver via the portal vein, are taken up by hepatocytes, and secreted back into bile.

IV. NONBILIARY LIVER FUNCTIONS

In addition to producing bile, the liver generates energy substrates, stores them, and makes them available to tissues as needed. The liver also synthesizes numerous proteins and stores fat-soluble vitamins and trace elements. Finally, the liver detoxifies various chemicals and removes bacteria.

A. Metabolism

The liver metabolizes carbohydrates, fats, and proteins and stores vitamins and trace elements.

1. **Carbohydrate:** The liver is one of three key organs involved in glucose homeostasis (see Chapter 33·II). The liver can generate glucose from triacylglycerols, amino acids, or lactate (**gluconeogenesis**). The liver stores ~100 g of glucose in the form of **glycogen** and can break it down through **glycogenolysis** to liberate glucose for release into the systemic circulation.

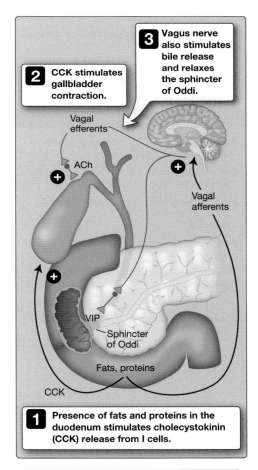

Figure 32.8.
Neural and endocrine control of bile secretion. VIP = vasoactive intestinal peptide.

Table 32.3: Major Plasma Proteins

Albumin
α_1-Globulins
α_1-Antitrypsin
α_1-Acid glycoprotein
α-Lipoprotein
α_2-Globulins
α_2-Macroglobulin
Ceruloplasmin
Haptoglobin
β_1–β_2-Globulins
β-lipoprotein
Complement C3
Hemopexin
Transferrin
γ-Globulins
Fibrinogen
C-reactive protein
Immunoglobins*

*Produced by lymphocytes and other plasma cells.

Table 32.4: Hormone-Binding and Transport Proteins

Plasma Protein	Section
Corticosteroid-binding globulin	Chapter 34·II·B
Growth hormone–binding protein	Chapters 33·IV·B and 34·IV·A
Insulin-like growth factor 1–binding proteins	Chapter 33·IV·C
Sex hormone–binding globulin	Chapters 36·III·B and 36·IV·B
Thyroxine-binding globulin	Chapter 35·II·C
Transthyretin	Chapter 35·II·C
Vitamin D–binding protein	Chapter 35·IV·D

2. **Lipid:** The liver is a primary site of lipid metabolism. The liver produces lipoproteins, phospholipids, ketone bodies, and cholesterol and can convert amino acids and carbohydrates to new lipids. The liver mobilizes fatty acids during fasting through **lipolysis** and then releases these energy substrates into the sinusoids and systemic circulation.

3. **Protein:** The liver takes up amino acids from blood and uses them to synthesize and export numerous proteins. When blood amino acids levels exceed need, the liver metabolizes the excess via the urea cycle. Liver export proteins include most major plasma proteins (Table 32.3), which can be separated by electrophoresis to yield five major peaks corresponding to albumin, α_1 globulins, α_2 globulins, a β_1–β_2 globulin double peak, and γ globulins. Albumin and, to a lesser extent, the globulins, are responsible for maintaining plasma colloid osmotic pressure (see Chapter 18·VII). The liver also produces several hormone-binding and transport proteins (Table 32.4), prohormones (e.g., angiotensinogen), and most components of the clotting cascade.

4. **Vitamins and minerals:** Many vitamins and minerals are delivered to the liver by the portal circulation. The liver stores lipid-soluble vitamins (vitamins A, D, E, and K) and releases them to the circulation during fasting. The liver also stores certain minerals, such as iron and copper.

B. Detoxification

The liver metabolizes various harmful compounds, such as ammonia and ethanol. It also carries out biotransformation reactions to detoxify numerous endogenous and exogenous noxious chemicals, including pharmaceuticals.

1. **Metabolism:** The intestines (primarily the large intestine) are responsible for about ~50% of daily ammonia production. Most of this arrives at the liver via the portal circulation and is metabolized using the urea cycle. Urea is released into the systemic circulation and is excreted by the kidney. Alcohol is metabolized using alcohol dehydrogenase, which facilitates the conversion of ethanol into acetaldehyde and reduced nicotinamide adenine dinucleotide. These products are converted into acetyl coenzyme A by peripheral tissues (e.g., skeletal muscle).

2. **Drug biotransformations:** Biotransformation occurs in two phases: **phase I** (**oxidation**) and **phase II** (**conjugation** and **elimination**).

   a. **Phase I:** During phase I, P450 cytochromes (CYPs) are used to introduce or expose polar groups that increase a drug's water solubility and thereby decrease its bioactivity. Phase I reactions include oxidation, reduction, and hydrolysis. Most drug classes are deactivated by phase I reactions, the exception being amine-containing compounds. Phase I reactions may also be exploited by the pharmaceutical industry to activate drugs (e.g., angiotensin-converting enzyme inhibitors).

b. Phase II: Phase II reactions involve a group of transferases that conjugate phase I products to gluconurate, acetate, amino acids, or other substrates to further increase water solubility. This allows the biotransformation end products to be released into blood for excretion by the kidney or to be secreted into the small intestine with bile for excretion in feces.

Some oral pharmaceuticals are almost completely metabolized during first pass through the liver. This is why certain drugs must be delivered in topical, inhaled, or injectable forms.

C. Immune functions

Portal blood often contains intestinal bacteria that have traversed the intestinal epithelium. These are removed from blood by **Kupffer cells**, which are specialized phagocytic macrophages incorporated into the sinusoidal epithelium (Fig. 32.9). Kupffer cells also engulf and remove particulates, including red blood cell fragments arriving from the spleen via the portal circulation. The liver is also a major source of lymph and immunoglobulin A, the latter being released into bile to help counter microbial growth.

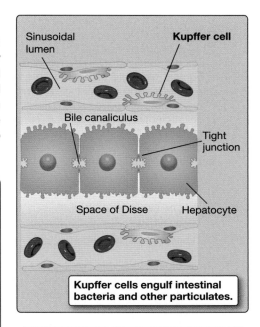

Kupffer cells engulf intestinal bacteria and other particulates.

Figure 32.9.
Kupffer cells.

Biologic Sex and Aging 32:1: Liver Function

Liver volume decreases by 20%–40% with age, and liver blood flow is reduced to a similar extent. These changes have minimal effect on liver enzymes or their functions, however, with the exception of cytochrome (CYP) P450. CYP content drops by ~30% by age 70 years, which impedes clearance of many pharmaceuticals by 20%–40%. This can result in dosing errors. Overall decreases in the regenerative capacity of the aging liver reduce functional reserve and lower the ability to withstand hepatic insults. Consequently, the incidence of hepatitis, fatty liver disease (both alcoholic and nonalcoholic), cancer, and jaundice are increased in older individuals.

Chapter Summary

- The exocrine pancreas secretes enzymes, ions, and serous solutions. Enzymes are typically secreted in inactive precursor form. HCO_3^- is secreted to neutralize stomach acid. Exocrine pancreatic secretion is stimulated by **secretin** and **cholecystokinin (CCK)**.

- Bile contains **primary** and **secondary bile acids**, electrolytes, phospholipids, cholesterol, and bilirubin. Bile acids are recycled via the **enterohepatic circulation**. They are reabsorbed from the intestinal lumen, transported back to the liver, taken up by hepatocytes, and secreted back into the biliary tree.

- Bile is stored in the **gallbladder** between meals. The gallbladder removes salt and water during storage, which concentrates bile. Bile release is stimulated by CCK, which initiates gallbladder contraction.

- The **sphincter of Oddi** regulates bile release into the duodenum. CCK causes sphincter relaxation, allowing bile to enter the small intestine.

- The liver is a key site of carbohydrate, lipid, and protein metabolism. The liver can either store or release energy substrates depending on feeding state. It also stores lipid-soluble vitamins and some minerals.

- The liver **detoxifies** and clears drugs, hormones, and ammonia from the circulation. It also has immune functions, including removal of intestinal bacteria from the circulation.

Study Questions

Choose the ONE best answer.

32.1. A 17-year-old male presents to the office with chronic diarrhea, abdominal pain, and bloating. He notes that his stools are often pale, greasy, and difficult to flush. A fecal elastase test is suggestive of pancreatic insufficiency, and a sweat chloride test confirms the diagnosis of cystic fibrosis. Pancreatic insufficiency is most likely due to which of the following changes in pancreatic duct function?

A. Decreased Cl^- secretion
B. Decreased K^+ reabsorption
C. Increased HCO_3^- secretion
D. Increased Na^+ secretion
E. Increased luminal pH

Best answer = A. Cystic fibrosis symptoms are caused by a defect in CFTR, which is a Cl^- channel (see Clinical Application 32.1). When the ability to secrete Cl^- into the ductal lumen is impaired, Cl^-/HCO_3^- exchange cannot occur which, in turn, prevents Na^+ and water secretion. Resulting pancreatic secretions are thick and viscous and may block the duct. Pancreatic lipase is required for lipid digestion, so pancreatic insufficiency leads to greasy stools and bloating. K^+ reabsorption (B) is not required for ductal secretion. HCO_3^- (C) and Na^+ secretion (D) are both decreased, not increased, by a defective CFTR, and luminal pH (E) would be expected to be lower, not higher in the absence of HCO_3^- secretion.

32.2. A 58-year-old male presents with a history of alcohol use disorder and increasing abdominal pain. Imaging reveals a large hepatic mass suspicious for hepatocellular carcinoma. During hepatic surgery, bile is sampled from the liver and then from the gallbladder. What is the most likely difference between gallbladder bile and hepatic bile composition?

A. Higher bilirubin concentration
B. Higher Cl^- concentration
C. Lower bile salt concentration
D. Lower cholesterol concentration
E. Lower fatty acid concentration

Best answer = A. Bile is produced by the liver and stored by the gallbladder until needed to aid fat digestion (see Section III·A). The gallbladder concentrates and modifies bile composition during storage, causing bilirubin levels to rise 10-fold. Cl^- (B) is reabsorbed along with some other ions during biliary concentration, its levels falling 10-fold, not increasing. Bile salt (C), cholesterol (D), and fatty acid (E) concentrations similarly increase.

32.3. To evaluate possible cholecystitis, cholecystokinin (CCK) is given during a cholescintigraphy procedure in which bile constituents are radioactively labeled and biliary secretions tracked. What is the most likely purpose of administering CCK during this procedure?

A. To contract the gallbladder
B. To decrease primary bile salt formation
C. To decrease secondary bile salt formation
D. To inhibit bicarbonate secretion
E. To stimulate local sympathetic efferents

Best answer = A. CCK has several roles in GI function, including releasing bile into the intestinal lumen. Release is accomplished by relaxing the sphincter of Oddi and contracting the gallbladder (see Section III·B). Bile release is also stimulated by parasympathetic nervous system ACh release. CCK does not regulate primary (B) or secondary bile salt formation (C). CCK stimulates rather than inhibits HCO_3^- secretion (D). The sympathetic nervous system (E) does not contribute to bile release, and norepinephrine is classified as an inhibitor of bile secretion.

32.4. A 42-year-old female presents with diarrhea, steatorrhea, and weight loss. Her history is significant for recurrent inflammation of the terminal ileum (Crohn disease). A stool sample is positive for fats and a random urinary bile acid test shows decreased levels. What is the most likely explanation for the urinary bile acid findings?

A. Cholelithiasis
B. Decreased bile acid reabsorption
C. Impaired bile acid synthesis
D. Increased fecal excretion of bile acids
E. Pancreatic insufficiency

Best answer = B. Bile acids are released into the small intestine and then recycled by the enterohepatic circulation 4–12 times a day (see Section III·D). Reabsorption occurs primarily in the terminal ileum, so inflammation of this region can interfere with recycling and cause bile acids to be excreted in feces. Decreased bile acid reabsorption results in less circulating bile acids and less bile acids appearing in urine. Cholelithiasis (A) and obstruction of bile flow would increase circulating and urinary bile acid levels. Impaired bile acid synthesis (C) would decrease urinary bile acid levels, but interrupted recycling is more likely given the patient's Crohn disease. Interruption of bile acid recycling would lead to increased fecal excretion (D) but is not the primary cause. Pancreatic insufficiency (E) would lead to steatorrhea but is a less likely cause of decreased urinary bile acids.

Endocrine Pancreas and Liver

33

I. OVERVIEW

All tissues require a steady supply of energy substrates to fuel metabolism and to grow. Key energy substrates include **glucose**, **free fatty acids (FFAs)**, **amino acids**, and **ketone bodies**. Energy substrates can be obtained from food, but food availability can be variable and unreliable, with wide swings between feast and famine. Therefore, the body must be able to store energy substrates during times of feasting to ensure continued supply to tissues when food supply is limited. Key organs involved in energy substrate storage are the **liver**, **muscle**, and **adipose tissue**, whose activity is coordinated by the endocrine system and autonomic nervous system (ANS). The endocrine pancreas plays a pivotal role in regulating energy substrate availability through secretion of **insulin**, which promotes energy storage (in the form of **glycogen**, **triacylglycerols [TAGs]**, and **protein**) when feasting, and **glucagon**, which mobilizes glucose in times of famine. The sympathetic nervous system (SNS) is also able to mobilize energy substrates when the body is stressed to help prepare the body for "fight or flight" responses. SNS actions are mediated largely by **epinephrine** release from **adrenal glands**.

II. GLUCOSE HOMEOSTASIS

Glucose is arguably the most important energy substrate because it forms the backbone of cellular energetics (i.e., glycolysis, citric acid cycle, and oxidative phosphorylation). Although levels spike during a meal, fasting glucose is strictly maintained in the range of 3.9 to 5.5 mmol/L (Fig. 33.1).

A. Hyperglycemia

Chronic elevation of blood glucose (**hyperglycemia**) is highly detrimental to tissues because it promotes protein glycation. Hyperglycemia can also cause insulin resistance and diabetes mellitus (see Clinical Application 33.1). During a meal, rising blood glucose stimulates insulin

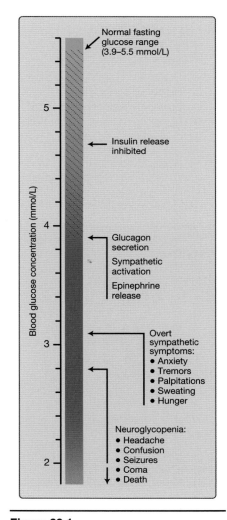

Figure 33.1.
Thresholds for responses to hypoglycemia.

439

release from the pancreas. Insulin signals the liver, muscle, and adipose tissue that energy substrates are available. They respond by increasing glucose uptake and storing it as glycogen or TAGs.

B. Hypoglycemia

The brain is both a principal consumer of glucose (60%–70% of total) and uniquely dependent on glucose for continued function. Most other tissues can use alternate energy substrates (e.g., FFAs and amino acids) to fuel metabolism, but FFAs do not cross the blood–brain barrier in significant amounts, and amino acids are rapidly consumed in neurotransmitter synthesis. Therefore, the brain must ensure that blood glucose does not fall below a certain critical level, or neuronal activity will falter and then cease. There are four principal lines of defense against hypoglycemia (see Fig. 33.1).

1. **Decreased insulin secretion:** Insulin is released in response to high blood glucose. When glucose levels fall into the low-normal range (4.4–4.7 mmol/L), insulin secretion is inhibited, and energy substrate uptake by insulin-sensitive tissues declines.

2. **Glucagon secretion:** Glucagon is a pancreatic hormone released at a blood glucose of 3.6 to 3.9 mmol/L. Glucagon stimulates glucose production by the liver (see Section III·C).

3. **Sympathetic activation:** Glucose-sensitive neurons in the hypothalamus and other brain regions cause SNS activation when glucose falls to 3.6 to 3.9 mmol/L. SNS stimulation of the pancreas inhibits insulin secretion via synaptic α-adrenergic receptors. Epinephrine release from the adrenals (see Chapter 34·V) increases hepatic glucose production, acting through β-adrenergic receptors. When glucose levels fall below 3.1 mmol/L, the symptoms of SNS activation (anxiety, tremors, heart palpitations, sweating, and hunger) become overt, serving as a warning of the need to seek out and ingest food.

4. **Growth hormone and cortisol secretion:** Chronic hypoglycemia stimulates growth hormone (GH) and cortisol secretion (see Section IV·B; see also Chapter 34·II), hormones that potentiate hepatic glucose production and inhibit glucose uptake by tissues. If blood glucose levels fall to below 2.8 mmol/L, normal brain function is impaired, manifesting as headaches and confusion (**neuroglycopenia**). If blood glucose declines further, coma, seizures, and death can result.

III. ENDOCRINE PANCREAS

The pancreas secretes several hormones that are synthesized by clusters of endocrine cells called **islets of Langerhans**. The most important of these hormones for glucose homeostasis are insulin and glucagon.

A. Islet structure

Islets are small (50–300 μm), numerous (>1,000,000), and scattered throughout the pancreas. They contain four principal endocrine cell types, each of which produces one or more hormones. α **Cells** secrete glucagon, β **cells** secrete insulin and **amylin** (see Chapter 30·V·C), δ **cells** secrete **somatostatin (SS)**, and **PP cells** secrete

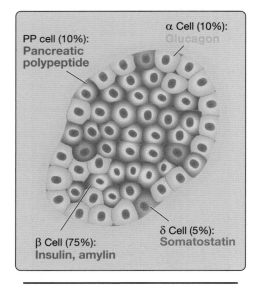

PP cell (10%):
Pancreatic polypeptide

α Cell (10%):
Glucagon

δ Cell (5%):
Somatostatin

β Cell (75%):
Insulin, amylin

Figure 33.2.
Pancreatic islet cellular composition. Percentages in parentheses indicate prevalence of the cell types indicated.

pancreatic polypeptide (Fig. 33.2). β Cells are in the majority (70–75% of total) and are located centrally, whereas α cells (10–20% of total) and δ cells (7–10% of total) are located peripherally. Adjacent cells within the islet are connected via gap junctions, allowing for direct cell-to-cell communication (see Chapter 4·II·F). Islet cells also communicate with each other through the extracellular space and via their blood supply.

1. **Blood supply:** Islets are extensively vascularized and have high flow rates relative to their weight. The flow rate facilitates assessment of prevailing blood glucose availability. Arterial blood enters islets at their center and then flows toward the edge (Fig. 33.3), much as the water in a fountain bubbles up from the center and flows outward. This flow pattern allows for local (paracrine) hormonal signaling within an islet. Thus, insulin released from centrally located β cells inhibits glucagon release from peripherally located α cells. Secreted hormones enter the portal circulation and are carried to the liver (see Fig. 20.10). The liver has a key role in energy substrate storage and metabolism, so this arrangement allows islet hormones (which coordinate energy substrate availability) to communicate with hepatocytes directly.

2. **Innervation:** Islets are innervated both by the parasympathetic nervous system (PSNS) and the SNS. PSNS activation stimulates hormone release via postsynaptic cholinergic (muscarinic) receptors. β Cells express both α- and β-adrenergic receptors, the latter responding to circulating epinephrine rather than norepinephrine (from SNS nerve terminals). SNS activation decreases insulin secretion.

B. Insulin

Insulin is a "feast" hormone, released from β cells when circulating energy substrate levels are high. Insulin is a product of the *INS* gene, which encodes preproinsulin. A 24–amino acid signal peptide is cleaved during translation in the rough endoplasmic reticulum to yield proinsulin (Fig. 33.4). Proinsulin is packed into secretory granules, along with proteases, in the Golgi apparatus. During insulin maturation, the proteases release C-peptide (31 amino acids) and two chains (A and B) linked together by two disulfide bonds. The linked chains comprise insulin, a 51–amino acid peptide hormone. Insulin's half-life in the circulation is about 3 to 8 minutes. More than 50% of insulin secreted from islets into portal blood is removed during its first pass through the liver. The kidneys and other peripheral tissues remove much of the remainder.

C-peptide is cosecreted with insulin in an equimolar ratio. It is not removed from the circulation during its first pass through the liver and has a longer half-life than insulin (~35 min). Thus, it can be used clinically to monitor pancreatic β-cell output, which may be helpful in assessing β-cell functionality in a patient with diabetes receiving exogenous insulin, for example.

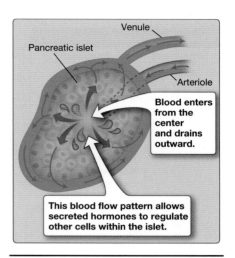

Figure 33.3.
Islet blood flow.

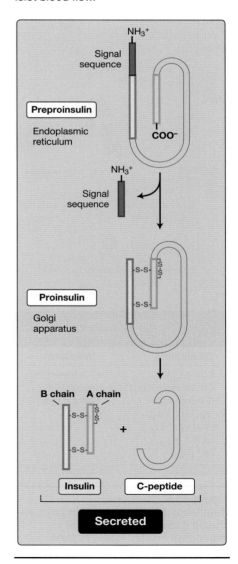

Figure 33.4.
Insulin processing.

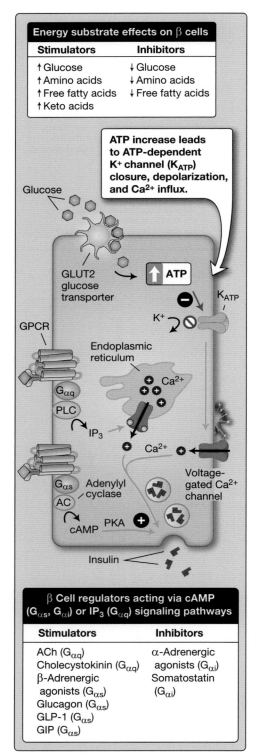

Energy substrate effects on β cells

Stimulators	Inhibitors
↑ Glucose	↓ Glucose
↑ Amino acids	↓ Amino acids
↑ Free fatty acids	↓ Free fatty acids
↑ Keto acids	

ATP increase leads to ATP-dependent K+ channel (K_{ATP}) closure, depolarization, and Ca^{2+} influx.

Glucose

GLUT2 glucose transporter

↑ ATP

K_{ATP}

K^+

GPCR

Endoplasmic reticulum

Ca^{2+}

$G_{\alpha q}$
PLC

IP_3

Ca^{2+}

Voltage-gated Ca^{2+} channel

$G_{\alpha s}$ Adenylyl cyclase
AC

cAMP   PKA

Insulin

β Cell regulators acting via cAMP ($G_{\alpha s}$, $G_{\alpha i}$) or IP_3 ($G_{\alpha q}$) signaling pathways

Stimulators	Inhibitors
ACh ($G_{\alpha q}$)	α-Adrenergic
Cholecystokinin ($G_{\alpha q}$)	agonists ($G_{\alpha i}$)
β-Adrenergic	Somatostatin
agonists ($G_{\alpha s}$)	($G_{\alpha i}$)
Glucagon ($G_{\alpha s}$)	
GLP-1 ($G_{\alpha s}$)	
GIP ($G_{\alpha s}$)	

Figure 33.5.
Regulation of insulin secretion by pancreatic β cells. GIP = glucose-dependent insulinotropic peptide, GLP-1 = glucagon-like peptide 1, PKA = protein kinase A, PLC = phospholipase C.

1. **Secretion:** Insulin secretion by β cells is stimulated primarily by circulating glucose and the PSNS. Insulin secretion is pulsatile with a periodicity of 2–5 min. Pulsatile release prevents target cell desensitization. Several hormones modulate glucose responsiveness (Fig. 33.5).

   a. **Sensory mechanism:** Glucose is taken up by GLUT2 glucose transporters (products of the *SLC2A2* gene), phosphorylated by glucokinase to yield glucose 6-phosphate (G6P) and then metabolized to yield ATP. Rising intracellular ATP levels inhibit an **ATP-sensitive K+ channel** (K_{ATP}), which terminates K_{ATP}-mediated K^+ efflux. The membrane depolarizes as a result, causing voltage-gated Ca^{2+} channels to open. The resulting Ca^{2+} influx prompts docking and fusion of insulin-laden vesicles with the cell membrane, followed by insulin release. β Cells also respond to circulating amino acids, FFAs, and keto acid levels through similar mechanisms. PSNS activation can also stimulate insulin release through acetylcholine (ACh) binding to a muscarinic receptor and intracellular Ca^{2+} release.

   b. **Regulation:** Glucagon, **glucose-dependent insulinotropic peptide** (**GIP**), glucagon-like peptide 1 (GLP-1), and cholecystokinin (CCK) potentiate glucose-induced insulin secretion (see Fig. 33.5). GIP and GLP-1 are both known as **incretins**. They are secreted by the intestinal mucosa in response to increased glucose levels in the gut. Think of this as a news flash alerting the pancreas that glucose will soon be appearing in the circulation. Incretins account for up to half of the insulin response to a carbohydrate meal. Inhibition of insulin secretion occurs primarily through SNS synaptic α-adrenergic-receptor activation and the cAMP signaling pathway. SS can also inhibit insulin release via this signaling pathway.

2. **Function:** Insulin signals increased plasma glucose availability to receptive tissues, including the liver, skeletal muscle, and adipose tissue. These tissues respond by increasing uptake of glucose and other energy substrates and using them to replenish energy reserves in the form of glycogen, fats, and proteins (Table 33.1). Insulin binds to a **tyrosine kinase receptor** (see Chapter 1·VII·C), which phosphorylates several adapter proteins, including a family of insulin receptor substrate proteins (IRS-1–IRS-4). Adapter proteins activate signaling cascades that mediate insulin's myriad cellular effects.

Table 33.1: Changes in Energy Substrate Availability and Storage During Feeding (High Insulin, Low Glucagon)

Glucose	Free Fatty Acids	Amino Acids
↑ Glucose uptake (M, A), retention (L)	↑ Fatty acid synthesis (L)	↑ Amino acid uptake (L, M)
↑ Glucose use (L, M, A)	↓ Lipolysis (A)	↑ Protein synthesis (M)
↑ Glycogenesis (L, M)	↓ Ketogenesis (L)	↓ Proteolysis (M)
↑ Glycolysis (L, M, A)		↓ Urea cycle activity (L)
↓ Glycogenolysis (L, M)		
↓ Gluconeogenesis (L)		

A = adipose tissue, L = liver, M = muscle.

a. **Glucose uptake:** Insulin increases glucose uptake by muscle and adipose tissue by causing GLUT4 glucose transporters (products of the *SLC2A4* gene) to be inserted into the membrane (Fig. 33.6). Insulin-dependent upregulation of transport capacity means that muscle can dramatically impact blood glucose levels when insulin levels are high (e.g., when feasting) but does not appreciably affect blood glucose when insulin levels are low (e.g., when fasting). The liver handles glucose using GLUT2, which is not insulin sensitive. Insulin also facilitates glucose uptake by stimulating glucose conversion to G6P. Phosphorylation lowers cytosolic glucose levels and helps maintain the concentration gradient driving glucose uptake from blood (GLUTs operate by facilitated diffusion rather than active transport, so they are very sensitive to concentration gradients). Phosphorylation also prevents glucose back-diffusion via GLUTs, effectively trapping the substrate within the cell for use in generating ATP (via glycolysis) or building energy reserves (glycogen, lipids, amino acids, and ketone bodies).

b. **Glycolysis:** Insulin activates phosphofructokinase and pyruvate dehydrogenase in muscle and the liver (Fig. 33.7). Insulin

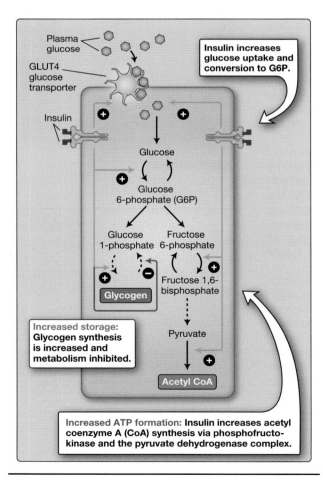

Figure 33.6.
Insulin effects on skeletal muscle.

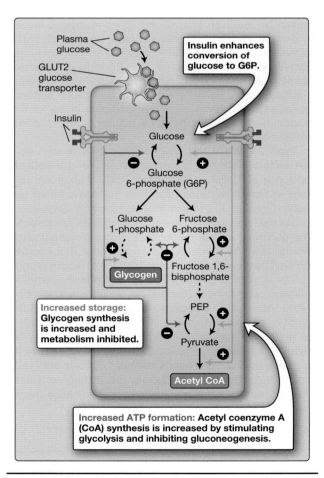

Figure 33.7.
Insulin effects on hepatocytes. PEP = phosphoenolpyruvate.

Clinical Application 33.1: Diabetes Mellitus

Diabetes mellitus (DM) is a serious metabolic disease affecting an estimated 537 million people worldwide. According to the World Health Organization, 1,600,000 deaths are directly attributable to diabetes annually.[1] DM is characterized by glucose homeostatic dysfunction and hyperglycemia, which is responsible for an accelerated rate of advanced glycation end product (AGE) formation. AGEs bind to AGE receptors (RAGEs) on macrophages and T cells and on vascular endothelium and smooth muscle, triggering reactions that cause widespread damage to the vasculature and, to a lesser degree, other organs. RAGE-related changes promote atherosclerosis, hypertension, and heart disease. Thickening of the basement membrane and increased leakiness of smaller vessels leads to diabetic nephropathy, retinopathy, and peripheral neuropathy. Damage to individual tissues can have additive effects, such as when peripheral vascular damage and peripheral neuropathy lead to foot ulcers. Hemoglobin (Hb) is also a target for glycation, yielding HbA_{1c}. HbA_{1c} accumulates during the ~120-day life of a red blood cell, such that average blood HbA_{1c} content approximates 6% of total Hb in a normal, healthy individual. Glycation rate is proportional to blood glucose levels, so HbA_{1c} can be used to monitor average blood glucose levels during the prior 2–3 months, which can be helpful in monitoring efficacy of glycemic control measures in patients with diabetes. Hyperglycemia in diabetes is defined as fasting plasma glucose ≥7.0 mmol/L or ≥11.1 mmol/L after an oral glucose tolerance test (rapid ingestion of 75 g glucose in 300 mL water and glucose monitoring over the following 120 minutes). There are two broad classifications of the disease.

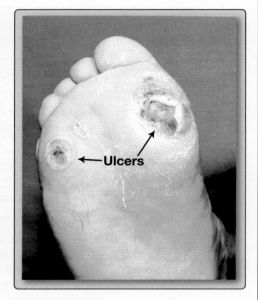

Diabetes ulcer.

Type 1 DM (T1DM) is caused by autoantibodies against islet β-cell components, which leads to β-cell destruction. T1DM has a genetic component, whereby susceptible individuals develop overt disease after exposure to an environmental trigger (e.g., a virus). The lack of insulin removes the brakes on glucagon-induced glucose and ketone body formation by the liver. The plasma glucose excess leads to polyuria (due to osmotic diuresis), glucosuria (see Chapter 26·III·B), and polydipsia. In the absence of medical intervention, hyperglycemia can progress to diabetic ketoacidosis (see Chapter 28·VI·E) or a hyperosmolar hyperglycemic state and diabetic coma. Glucosuria can be easily detected with a urine dipstick. Patients with T1DM are insulin dependent, and disease management involves insulin replacement in the form of daily injections or administered via a portable insulin pump, depending on personal preference.

Type 2 DM (T2DM) is a complex disease with several independent risk factors, including genetics, obesity, and a sedentary lifestyle. The incidence of T2DM in the United States has grown to epidemic proportions in recent decades, mirroring the concomitant rise in obesity. The early stages of T2DM are characterized by decreased β-cell sensitivity to glucose and by insulin-sensitive tissues becoming insulin resistant. The pancreas initially responds to rising blood glucose levels with β-cell hyperplasia and increased insulin secretion, but compensation is usually followed by β-cell failure and inadequacy of insulin production.

The increased risk of diabetes with obesity correlates with the accumulation of abdominal (central) fat, which is a functional endocrine organ producing various adipokines (e.g., adiponectin, leptin, tumor necrosis factor-α, and resistin), some of which promote hyperglycemia. Abdominal fat generates excessive amounts of free fatty acids, resulting in dyslipidemia that can adversely affect glucose homeostasis and contribute to cardiovascular disease. Treatment options involve diet modification, weight loss, exercise, metformin (an insulin sensitizer), sulfonylureas (ATP-sensitive K^+ channel inhibitors that enhance insulin secretion), and insulin injections.

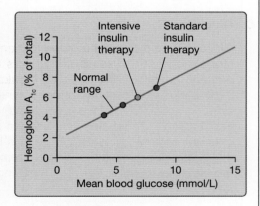

Hemoglobin A_{1c}.

[1]For more information on diabetes mellitus and its treatment options, see *LIR Pharmacology*, 8e, Chapter 24.

additionally activates pyruvate kinase in the liver, which, when coupled with the aforementioned increase in glucokinase activity, facilitates glucose consumption. In hepatocytes, gluconeogenesis (i.e., the reverse pathway) is repressed simultaneously to prevent these pathways competing for their respective substrates and products.

c. **Glycogenesis:** Insulin increases glycogen formation in both muscle and the liver. Insulin first stimulates conversion of glucose to G6P by glucokinase (see Fig. 33.7). Glycogen synthase then converts G6P to glycogen. Insulin simultaneously inhibits glycogen phosphorylase to facilitate glycogen accumulation (see Fig. 33.7).

d. **Lipogenesis:** Insulin stimulation of adipocytes increases **lipoprotein lipase** (**LPL**) and decreases **hormone-sensitive lipase** (**HSL**) activity. LPL facilitates the breakdown of chylomicrons and other low-density lipoproteins into FFAs, which can then be absorbed and stored as TAGs. HSL, which is activated by epinephrine and other catecholamines, normally breaks down TAGs during SNS activation.

e. **Ketone bodies:** In hepatocytes, ketone body formation and secretion are inhibited in the presence of insulin because this hormone inhibits the rate-limiting carnitine shuttle. The carnitine shuttle consists of transferase and translocase enzymes that move long-chain fatty acyl coenzyme A into the mitochondria for processing.[1]

f. **Protein synthesis:** In skeletal muscle and hepatocytes, insulin promotes amino acid uptake and protein synthesis and inhibits protein catabolism. The anabolic effect of insulin involves both the mTOR (mammalian target of rapamycin) pathway and an increase in amino acid uptake by cells. The mTOR pathway decreases proteolysis and increases ribosomal production and assembly.

C. Glucagon

Glucagon is a "famine" hormone that ensures continued energy substrate availability to tissues when food supplies are limited. It does this primarily by stimulating glycogen breakdown (Fig. 33.8). Glucagon is a 29–amino acid peptide hormone synthesized by pancreatic islet α cells. It is a product of the *GCG* gene, which encodes preproglucagon. Proteolysis liberates proglucagon and then glucagon, plus two inactive protein fragments. Glucagon's half-life in the circulation is 5 to 10 minutes. Glucagon is largely removed from the circulation on the first pass through the liver, meaning that the liver is the primary functional target.

1. **Secretion:** Glucagon is secreted when plasma glucose levels fall. α Cells express a complement of channels and transporters similar

[1]For more information on the carnitine shuttle, see *LIR Biochemistry*, 8e, Chapter 8·IV·B·1.

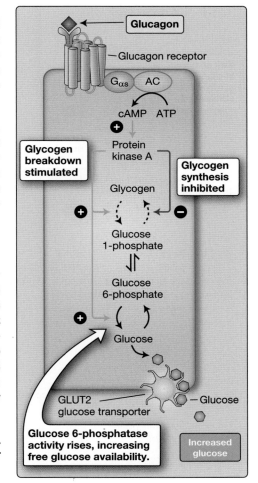

Figure 33.8.
Glucagon stimulation of glycogenolysis in hepatocytes. AC = adenylyl cyclase.

Table 33.2: Changes in Energy Substrate Availability and Storage During Fasting (High Glucagon, Low Insulin)

Glucose	Free Fatty Acids	Amino Acids
↑ Glycogenolysis (L, M)	↑ Ketogenesis (L)	↑ Proteolysis (M)
↑ Gluconeogenesis (L)	↑ Lipolysis (A)	↑ Urea cycle activity (L)
↓ Glucose uptake (M, A), retention (L)	↓ Fatty acid synthesis (L)	↓ Amino acid uptake (L, M)
↓ Glucose use (L, M, A)		↓ Protein synthesis (M)
↓ Glycogenesis (L, M)		
↓ Glycolysis (L, M, A)		

A = adipose tissue, L = liver, M = muscle.

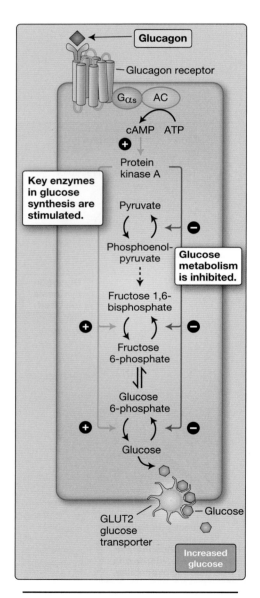

to that of β cells, and it is possible that they sense glucose directly, or they may be regulated through paracrine signaling by β cells. Glucagon secretion is inhibited by insulin, amylin, and SS. Glucagon secretion is stimulated by epinephrine released from the adrenal glands and by rising plasma amino acid levels during a meal.

2. **Function:** Glucagon opposes insulin actions when blood glucose levels fall, mobilizing reserves and making energy substrates available for use by tissues between meals or during stress (Table 33.2). Glucagon's primary target is the liver, although muscle and adipocytes are also glucagon sensitive. Glucagon receptors are G protein–coupled receptors (GPCRs) that exert intracellular effects via the cAMP signaling pathway. When activated, the receptors stimulate increased blood glucose, fatty acid, and ketone body availability by glycogenolysis, gluconeogenesis, lipolysis, and ketogenesis.

   a. **Glycogenolysis:** Glucagon increases glycogen breakdown to liberate glucose. This is accomplished through increased **glycogen phosphorylase** and glucose 6-phosphatase activity (see Fig. 33.8). Glycogen phosphorylase is activated by protein kinase A (PKA)-dependent phosphorylation following glycogen receptor binding. PKA simultaneously phosphorylates and inhibits glycogen synthesis by glycogen synthase, thereby facilitating glucose mobilization.[1] Glucose then exits the cell by facilitated transport via GLUT2. In muscle, glucose from glucagon-stimulated glycogenolysis supports increased ATP production and contractile activity.

   b. **Gluconeogenesis:** Glucagon also stimulates glucose synthesis from noncarbohydrate sources, such as lipids and proteins. Gluconeogenesis is mediated by pathways that include glucose 6-phosphatase and fructose 2,6-bisphosphatase (Fig. 33.9).[2] Glucagon simultaneously inhibits enzymes that break down glucose, including glucose kinase, phosphofructokinase, and pyruvate kinase.

Figure 33.9.
Glucagon stimulation of gluconeogenesis in hepatocytes. AC = adenylyl cyclase.

[1]For more information on glycogen breakdown, see *LIR Biochemistry*, 8e, Chapter 11.

[2]For more information on gluconeogenesis, see *LIR Biochemistry*, 8e, Chapter 10.

c. **Lipolysis:** Adipocytes respond to hypoglycemia or glucagon by breaking down TAGs into glycerol and FFAs. Lipolysis is mediated by HSL.

d. **Ketogenesis:** The FFAs produced by lipolysis are absorbed by hepatocytes. These are then transported into mitochondria via a carnitine shuttle for processing. Incomplete fatty acid oxidation results in ketone body formation (acetoacetate and β-hydroxybutyrate), which are then released into the circulation. Ketone bodies are water soluble and easily absorbed by extrahepatic tissues, where they are converted back into acetyl coenzyme A for use in aerobic metabolism.

> Ketone bodies such as acetone (familiar as an ingredient of nail-polish remover) are organic volatiles with a characteristic fruity aroma that can be detected on the breath of individuals metabolizing them. Excess ketone production can cause ketoacidosis, a high anion gap metabolic acidosis (see Chapter 28·VI·E).

IV. LIVER

The liver is a key energy substrate source and storehouse, and it also helps regulate energy substrate availability through **insulin-like growth factor-1** (**IGF-1**) production. IGF-1 release is controlled through the **hypothalamic–pituitary–liver** (**HPL**) **axis** (Fig. 33.10).

A. Hypothalamic–pituitary–liver axis

The HPL axis is unique in that both the pituitary hormone (i.e., **growth hormone**) and target tissue hormone (i.e., IGF-1) have widespread biologic effects. In other endocrine axes (discussed in Chapters 34–36), only the target hormone has effects on tissues outside the axis.

1. **Insulin-like growth factor-1 secretion:** The hypothalamic arcuate nucleus contains small diameter neurons that secrete **GH-releasing hormone** (**GHRH**) into the hypophyseal portal circulation. These hormones target anterior pituitary **somatotropes** (see Table 7.2 and Fig. 33.10). GHRH binds to a GPCR linked to the cAMP signaling pathway, which increases GH synthesis and stimulates GH release when active. GH has wide-ranging effects, including stimulating IGF-1 release from the liver. IGF-1 is produced by many tissues, but hepatocytes contain 100-fold more IGF-1 mRNA than do cells in other tissues, meaning that most circulating IGF-1 has hepatic origins. Hepatic **GH receptors** couple to the Janus kinase/signal transducer and activator of transcription (JAK/STAT) signaling pathway. Receptor activation increases IGF-1 production and release into the circulation.

2. **Role of somatostatin:** Somatotropes also express SS receptors, which are GPCRs linked to the cAMP signaling pathway (see Fig. 33.10). SS is produced by the hypothalamic paraventricular

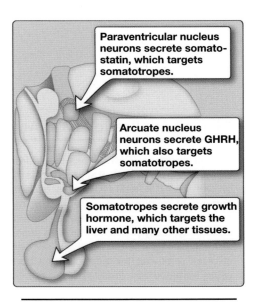

Figure 33.10.
Hypothalamic nuclei controlling growth-hormone release. GHRH = growth hormone–releasing hormone.

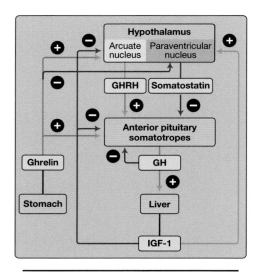

Figure 33.11.
Hypothalamic–pituitary–liver hormone axis. GH = growth hormone, GHRH = GH-releasing hormone, IGF-1 = insulin-like growth factor 1.

nucleus and is a potent inhibitor of GH release, dominating basal somatotrope secretory activity.

3. **Feedback inhibition:** IGF-1 inhibits the HPL axis through a direct effect on somatotropes (Fig. 33.11). It also inhibits GHRH secretion and promotes SS release from the hypothalamus. GH also feeds back on the somatotropes to inhibit its own release.

4. **Ghrelin:** Ghrelin is a peptide hormone produced by the stomach (see Chapter 30·V·B). Familiarly known as the "hunger hormone," ghrelin stimulates appetite, but it also stimulates GHRH secretion by the hypothalamus and GH secretion by the anterior pituitary, while simultaneously suppressing SS secretion.

B. Growth hormone

GH is a peptide hormone. Alternate splicing yields a 20-kDa form and a more abundant 22-kDa form. The preprohormone is a product of the *GH1* gene. Once secreted, a portion of GH binds weakly to **GH-binding protein** and other plasma proteins before ultimately being broken down by the liver. The half-life of GH in the circulation is ~20 minutes.

1. **Secretion:** GH is released in response to hypoglycemia, increasing amino acid levels, and by exercise and stress. GH is released in a pulsatile manner throughout the day and night, with surges occurring at 3- to 5-hour intervals, with larger surges occurring during sleep. Pulsatile secretion is advantageous because target tissues tend to "tune out" a constant signal (as do students listening to a humdrum faculty lecture). Increasing blood glucose and fatty acid levels inhibit GH release.

2. **Function:** GH has several targets, including the liver, cartilage, bone, muscle, and adipose tissue. In cartilage and muscle, GH stimulates amino acid uptake and protein synthesis. Collagen formation and chondrocyte size and number increase in the presence of GH. In adipose tissue, GH increases the breakdown of TAGs and decreases glucose uptake. This decrease in glucose uptake is sometimes referred to as the "anti-insulin effect."

Clinical Application 33.2: Acromegaly

Growth hormone (GH) excess in children prior to epiphyseal growth plate closure results in **pituitary gigantism**. In adults, GH (and insulin-like growth factor 1 [IGF-1]) excess causes **acromegaly**. Acromegaly most commonly results from a GH-producing pituitary adenoma and is associated with overgrowth of bone, cartilage, connective tissue, and skin. Patients typically present in the fifth decade with a prominent brow and jaw; skin thickening; and enlarged and swollen hands, feet, and nose. Treatment involves removing the adenoma via transsphenoidal surgical resection or using somatostatin analogs (e.g., octreotide) to suppress GH and IGF-1 secretion.

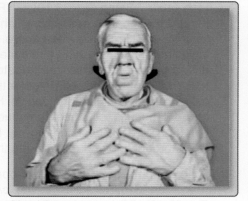

Gigantism.

C. Insulin-like growth factor 1

IGF-1 is a peptide hormone with some structural similarity to insulin (hence the "insulin-like" moniker). It is produced and secreted by hepatocytes in response to GH binding. Unlike GH, IGF-1 binds tightly to plasma proteins, which protects the hormone from degradation and imparts a half-life of ~20 hours. IGF-1 mediates many long-term actions of GH. Increased IGF-1 correlates with growth spurts during adolescence. In adults, IGF-1 primarily targets the musculoskeletal system, where it increases amino acid and glucose uptake and stimulates protein synthesis.

 Because of the pulsatile nature and short half-life of GH, measurement of the more stable IGF-1 in the plasma can provide a better way of assessing HPL axis status.

Clinical Application 33.3: Insulin-like Growth Factor 1 Deficiencies

Insulin-like growth factor 1 (IGF-1) deficiencies are seen in high prevalence in certain ethnic groups such as the Bayaka of Central Africa (one of the traditional pygmy peoples). In this ethnic group, many persons are proportioned normally but have very short stature. Adult men are often <5 ft (150 cm) tall. In these individuals, growth hormone levels are normal to high, whereas IGF-1 concentration is very low.

Chapter Summary

- **Glucose homeostasis** is essential for sustaining organ function and avoiding detrimental effects associated with advanced glycation end product accumulation. Glucose homeostasis is a primary endocrine pancreatic function.

- **Insulin** is secreted by **pancreatic β cells**. Insulin signals energy substrate availability and facilitates energy substrate uptake by tissues.

- Insulin causes **GLUT4** glucose transporters to be inserted into the plasma membrane of skeletal muscle cells and adipocytes. **GLUT2** transporters are insulin insensitive and are constitutively active in tissues such as the liver.

- **Insulin receptor** activation stimulates glycogen, fat, and protein production within target tissues.

- **Diabetes mellitus** is a disease associated with glucose homeostasis failure. **Type 1** diabetes mellitus results from autoimmune destruction of pancreatic β cells and an inability to secrete insulin. **Type 2** diabetes mellitus results from glucose or insulin insensitivity.

- **Glucagon** is secreted by **pancreatic α cells**. Glucagon opposes insulin actions and increases circulating energy substrate levels. This is accomplished by breaking down glycogen, triacylglycerols, and proteins and by forming new glucose from noncarbohydrate sources.

- **Growth hormone** (GH) is secreted from the anterior pituitary in response to hypothalamic **GH-releasing hormone**. GH can acutely affect growth and uptake of glucose and amino acids, but most GH functions are mediated by insulin-like growth factor 1 (IGF-1).

- **IGF-1** is secreted by the liver in response to GH. IGF-1 is responsible for many of the long-term effects of GH, including increased amino acid uptake and protein synthesis in cartilage and muscle and breakdown of triacylglycerols in adipocytes.

Study Questions

Choose the ONE best answer.

33.1. Which of the following hormones most likely binds to a cytosolic receptor that then translocates to the nucleus to induce transcriptional changes?

A. Aldosterone
B. Epinephrine
C. Growth hormone
D. Insulin
E. Insulin-like growth factor

Best answer = A. Aldosterone binds to mineralocorticoid receptors within target cells, and the complex is then translocated to the nucleus, where it binds to a mineralocorticoid response element on DNA (see Chapter 27·IV·B·2; see also Chapter 34). Epinephrine (B) binds to α- or β-adrenergic receptors in the plasma membrane (see Table 5.2). Growth hormone (C), insulin (D; see Section III·B), and insulin-like growth factor (E) bind to cell surface receptors that contain or are associated with a tyrosine kinase. All of the cell membrane receptors mentioned can have genomic effects by stimulating various pathways but not via direct binding to DNA.

33.2. A 22-year-old female is participating in a drug study affecting pancreatic hormones. If the test drug greatly elevates glucagon levels while having no effect on insulin release, which of the following processes are most likely to increase?

A. Gluconeogenesis in neurons
B. Glucose uptake by hepatocytes
C. Glycolysis in skeletal and cardiac muscle
D. Ketone uptake by neurons
E. Lipolysis in adipocytes

Best answer = E. Glucagon's principal function is to mobilize energy substrates and release them to the circulation for use by cells (see Section III·C·2). Its actions include stimulating hormone-sensitive lipase in adipocytes. The lipase breaks down triacylglycerols into free fatty acids and glycerol (lipolysis), which are then released into the circulation. Glucagon increases gluconeogenesis (glucose synthesis) and ketone release (not uptake, D) in hepatocytes, but not neurons (A). Glucagon simulates glucose release from hepatocytes, not uptake (B). Glycolysis (C) is inhibited by glucagon.

33.3. A 10-year-old male with an autoimmune disease that has destroyed his pancreatic β cells is most likely to exhibit which of the following signs and symptoms?

A. Decreased circulating fatty acid levels
B. Enhanced glucose uptake by adipocytes
C. Enhanced protein storage in muscle
D. Hyperglycemia and diuresis
E. Hyperkalemia

Best answer = D. Selective loss of pancreatic β cells results in type 1 diabetes mellitus (see Clinical Application 33.1). Symptoms include high blood glucose (hyperglycemia) that can spill over into the urine and cause diuresis (see Clinical Application 26.1). Patients with type 1 diabetes have increased, not decreased, levels of circulating fatty acids (A) and triglycerides. Insulin normally stimulates glucose uptake by adipocytes and protein storage in muscle, so both actions would be decreased in type 1 diabetes, not increased (B, C). Insulin does normally help regulate K^+ balance through its effects on the Na^+/K^+-ATPase, but hyperkalemia (E) is associated with kidney disease, not pancreatic disease.

33.4. A 20-year-old male has sudden onset of weakness, dizziness, altered vision, and rapid breathing. He reports frequent urination and insatiable thirst. He has no significant medical history. On physical examination, his breath is notable for a fruity aroma. A urinalysis is positive for glucose. Blood values of which of the following are most likely to be elevated in this patient?

A. Fructose
B. Hemoglobin A_{1c}
C. High-density lipoprotein
D. Hormone-sensitive lipase
E. Ketones

Best answer = E. The patient is exhibiting signs and symptoms of acute hyperglycemia (see Clinical Application 33.1). In patients with diabetes, ketone bodies are elevated because of the inability to rely on glucose as the primary energy substrate. Acetone imparts a fruity aroma to the breath, which is consistent with such a diagnosis. Rapid breathing also indicates elevated ketones, as this is a respiratory response to a metabolic acidosis (ketoacidosis; see Chapter 28·VI·E). Fructose (A) and high-density lipoproteins (C) are not factors in hyperglycemia reactions. Hemoglobin A_{1c} (B) is elevated during prolonged, not acute, hyperglycemia. Hormone-sensitive lipase (D) is stimulated by glucagon and epinephrine rather than insulin. Moreover, the lipase is membrane bound and therefore not likely circulating in plasma.

Adrenal Glands

34

I. OVERVIEW

Adrenal glands (also known as suprarenal glands) sit atop the superior pole of the kidneys. They produce two distinct classes of hormones, reflecting the fact that the cortex and medulla have different embryologic origins. The medulla, which accounts for ~10% of gland weight, is derived from neural crest and produces epinephrine. The cortex (~90% of total weight) is derived from mesodermal mesenchyme and produces steroid hormones. Three histologically distinct zones can be distinguished within the cortex: the **zona glomerulosa**, **zona fasciculata**, and **zona reticularis** (Fig. 34.1). The three zones express subsets of enzymes involved in steroid synthesis, which biases their output toward a single steroid hormone. The zona glomerulosa secretes the mineralocorticoid aldosterone. The zona fasciculata produces the glucocorticoid cortisol. The zona reticularis secretes adrenal androgens.

Epinephrine and **cortisol** are both stress hormones. The sounding of the body's alarms and mobilization of defenses helps an individual survive physical threats, endure pain, and tap the body's physical and metabolic reserves. Epinephrine release from the medulla is under control of the sympathetic nervous system (SNS), which allows for very rapid responses (i.e., within seconds) to psychologic or physiologic stress. Responses include increasing blood pressure and mobilizing energy substrate reserves, for example. Cortisol is a longer-acting stress hormone (hours to days) that regulates metabolism, inflammation, and immune responses. **Aldosterone** regulates plasma volume through modulation of salt and water retention by the kidneys. The **adrenal androgens** participate in secondary sex characteristics (e.g., hair growth) during puberty and adolescence. Stress, salt, and sex management is a heavy responsibility for glands measuring only ~1.5 by 7.5 cm and weighing ~8 to 10 g.

II. ADRENAL CORTEX: CORTISOL

Cortisol, like all adrenal cortical hormones, is derived from cholesterol (Fig. 34.2). Cholesterol is readily obtained from blood in the form of low-density lipoprotein (LDL), but it can also be synthesized *de novo* as needed. Cholesterol is stored in cytoplasmic lipid droplets until used in steroid production, a process that begins in mitochondria. Cholesterol movement from the outer to the inner mitochondrial membrane is facilitated by **steroidogenic acute regulatory protein** (**StAR**). This transfer is a rate-limiting and regulated step in cholesterol

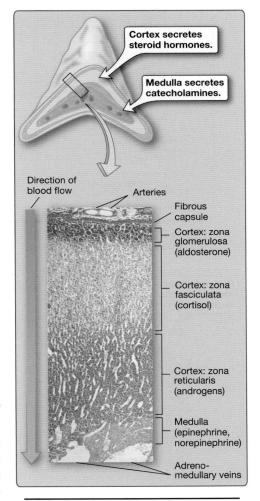

Figure 34.1.
Adrenal gland structure and secretory products.

451

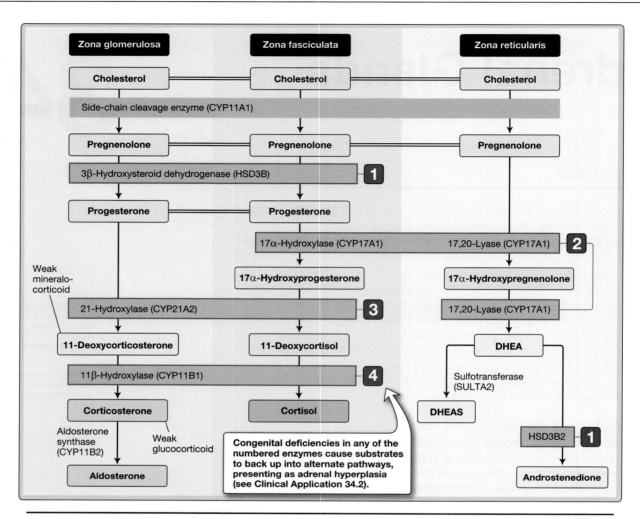

Figure 34.2.
Adrenocortical hormone synthesis.

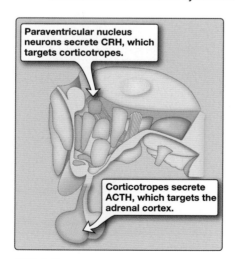

Paraventricular nucleus neurons secrete CRH, which targets corticotropes.

Corticotropes secrete ACTH, which targets the adrenal cortex.

Figure 34.3.
Hypothalamus and pituitary gland.
ACTH = adrenocorticotropic hormone,
CRH = corticotropin-releasing hormone.

production. Regulation is accomplished through the **hypothalamic–pituitary–adrenal (HPA) axis**.

A. Hypothalamic–pituitary–adrenal axis

The HPA axis includes **corticotropin-releasing hormone (CRH)**, **adrenocorticotropic hormone (ACTH)**, and cortisol.

1. **Hypothalamus:** CRH is a 41–amino acid peptide hormone synthesized by small-bodied neurons in the hypothalamic **paraventricular nucleus** and released into the **hypophyseal portal circulation** for transport to the anterior pituitary (Fig. 34.3).

2. **Pituitary gland:** CRH binds to corticotropes in the anterior pituitary. Corticotropes express a type 1 CRH receptor, which is a member of the G protein–coupled receptor (GPCR) superfamily. CRH binding activates the cAMP signaling system

and protein kinase A (PKA), which initiates ACTH release. CRH binding also activates transcription factors that upregulate **proopiomelanocortin** gene (**POMC**) expression. *POMC* encodes ACTH.

> *POMC* encodes a 241–amino acid precursor peptide (pre-POMC) whose sequence includes several hormones, including ACTH, α-melanocyte-stimulating hormone (MSH), β-endorphin, and enkephalin. The precursor is processed by tissue-specific proteases to yield the different secretory products. Corticotropes process the peptide to yield ACTH, whereas melanocytes secrete MSH.

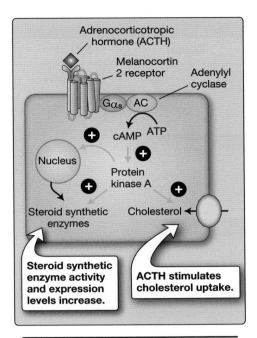

Figure 34.4.
Melanocortin 2–receptor signaling.

3. **Adrenal cortex:** ACTH is a 39–amino acid peptide that acts through **melanocortin 2 receptors** (**MC2Rs**), which are members of the GPCR superfamily. MC2R binding activates the cAMP signaling pathway and PKA (Fig. 34.4) to acutely increase StAR and side-chain cleavage enzyme (SCC) activity (see Fig. 34.2). In the longer term, ACTH increases expression of enzymes involved in cortisol synthesis and also of the LDL receptor that facilitates cholesterol uptake from blood.

B. Synthesis and secretion

ACTH stimulates cholesterol conversion to pregnenolone by SCC in all three cortical layers. Pregnenolone is then converted to progesterone. CYP17A1 hydroxylates progesterone to form 17α-hydroxyprogesterone, which is then converted to 11-deoxycortisol and then cortisol. Alternate routes and intermediaries for cortisol synthesis are shown in Figure 34.2. CYP17A1 is not present in the glomerulosa, whereas cortisol substrates are directed toward androgen synthesis in the reticularis, so cortisol production is largely restricted to the fasciculata. Cortisol diffuses from the cortex and enters the circulation, where most of it (~90%) binds **corticosteroid-binding globulin** with high affinity. Albumin binds another 5% to 7%. Cortisol has a half-life of ~60 minutes and is removed from the circulation by the liver.

C. Regulation

The pituitary releases ACTH with a diurnal rhythm influenced by the hypothalamic suprachiasmatic nucleus (see Chapter 7·VII·E), with surges occurring throughout the day. ACTH release is also stimulated by higher CNS control centers in response to physical, emotional, and biochemical stress (e.g., hypoglycemia). The HPA axis and cortisol release is subject to negative feedback control by both cortisol and ACTH (Fig. 34.5).

D. Function

Cortisol prepares the body for stress. Cortisol diffuses across the cell membrane and binds to an intracellular **glucocorticoid receptor** (**GR**). The hormone–receptor complex translocates to the nucleus

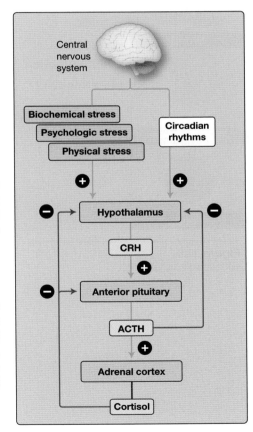

Figure 34.5.
Hypothalamic–pituitary–adrenal axis.
ACTH = adrenocorticotropic hormone,
CRH = corticotropin-releasing hormone.

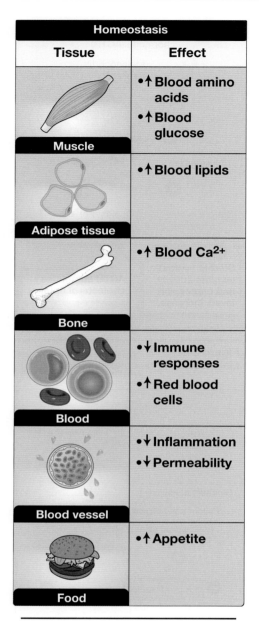

Homeostasis	
Tissue	**Effect**
Muscle	• ↑ Blood amino acids • ↑ Blood glucose
Adipose tissue	• ↑ Blood lipids
Bone	• ↑ Blood Ca^{2+}
Blood	• ↓ Immune responses • ↑ Red blood cells
Blood vessel	• ↓ Inflammation • ↓ Permeability
Food	• ↑ Appetite

Figure 34.6.
Glucocorticoid effects.

and binds a **glucocorticoid-response element** on DNA, modifying gene transcription. Cortisol has numerous physiologic effects involving all tissues (Fig. 34.6).

1. **Metabolic:** Cortisol increases plasma glucose (see Chapter 33·II), free fatty acid (FFA), and amino acid levels to provide energy substrates to fuel stress responses. In skeletal muscle, cortisol decreases protein synthesis and promotes proteolysis to liberate amino acids that are then converted to glucose via hepatic gluconeogenesis (see Chapter 32·IV·A and Chapter 33·II). Cortisol stimulates lipolysis in white adipose tissue to liberate FFAs. Cortisol also stimulates appetite to increase energy substrate intake. Ingesting food helps fuel stress responses, but increased appetite can lead to weight gain if the stress response does not involve physical activity.

2. **Immune:** Cortisol suppresses immune responses and inflammation.[1] Although this response may seem counterproductive in stressful conditions, when the life of the organism is in danger, directing resources to fight illness is less important than addressing an immediate physical threat. Immunosuppression involves decreasing the number of circulating T cells and inhibiting the production of various cytokines. The anti-inflammatory effects of cortisol are due to decreased synthesis of prostaglandins and leukotrienes that normally mediate vasodilation, increased blood flow, and increased capillary permeability, thereby decreasing accumulation of inflammatory mediators.

3. **Musculoskeletal:** Cortisol increases bone resorption through inhibition of anabolic remodeling (see Chapter 14·IV) and reduces Ca^{2+} absorption from the gastrointestinal tract and Ca^{2+} reabsorption from the renal tubule. Chronic hypercortisolemia can thus cause osteoporosis. Cortisol decreases collagen formation throughout the body. Cortisol-induced skeletal muscle protein catabolism can eventually lead to muscle weakness and early onset of fatigue during physical activity.

4. **Cardiovascular:** Cortisol increases erythropoietin release from kidneys, which stimulates red blood cell production (see Chapter 23·II·B). Cortisol potentiates vasoconstrictor responses by blocking local vasodilators, such as nitric oxide and prostaglandins. Cortisol also binds to GRs on vascular smooth muscle and exerts direct constrictor effects by raising intracellular Ca^{2+} levels. Cortisol also potentiates catecholamine actions on the cardiovascular system (e.g., increased cardiac inotropy, vasoconstriction) through adrenergic receptor upregulation.

[1]Immune suppressive and anti-inflammatory effects of glucocorticoids can be exploited pharmacologically. Drugs like prednisone, which is structurally similar to cortisol, can be used as immune suppressants for autoimmune diseases. For more information, see *LIR Pharmacology*, 8e, Chapter 26·II·C.

Clinical Application 34.1: Cushing Syndrome

Patients with **Cushing syndrome** may present with muscle weakness, osteoporosis, hypertension, type 2 diabetes mellitus (T2DM), increased hair growth (hirsutism), and weight gain with fat redistribution. These symptoms reflect chronic elevations in glucocorticoid levels. Muscle weakness results from skeletal muscle protein catabolism. Osteoporosis results from bone Ca^{2+} resorption. T2DM is a result of chronic hyperglycemia. Hirsutism is not a cortisol effect but a response to adrenal androgens (see Section III), whose levels are typically also increased in Cushing syndrome. The weight gain is caused by cortisol-induced increases in appetite and fat redistribution. For reasons that are not apparent, unused fatty acids are redeposited in the face and upper back, causing a "moon face" appearance and development of "buffalo hump." Cushing syndrome is commonly caused by adrenocorticotropic hormone (ACTH) excess, either from an ACTH-secreting pituitary adenoma or small-cell lung cancer.

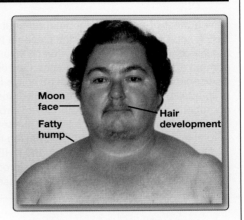

Female with Cushing syndrome.

III. ADRENAL CORTEX: ANDROGENS

The adrenal gland produces numerous androgens, including dehydroepiandrosterone (DHEA), DHEA sulfate (DHEAS), androstenedione, androstenediol, and 11β-hydroxyandrostenedione. DHEAS is the most abundant circulating adrenal androgen. Adrenal androgens have very weak androgenic activity compared with testosterone and estrogens. Limited information is available about how adrenal androgen production is regulated. Release follows a circadian rhythm like that of ACTH, suggesting HPA axis involvement, and increases during various stages of growth and development.

A. Synthesis and secretion

Adrenal androgens are synthesized and secreted primarily by the zona reticularis (see Figs. 34.1 and 34.2) for three main reasons. First, HSD3B2 is expressed at relatively low levels in the reticularis compared with the rest of the cortex, which favors androgen production. Second, the reticularis expresses CYB5A, a CYP17A1 cofactor that favors conversion of 17-hydroxypregnenolone to DHEA rather than entering the cortisol production pathway. Finally, the reticularis is the primary location of the sulfotransferase (SULTA2) that converts DHEA to DHEAS. In the circulation, DHEA and androstenedione bind with low affinity to albumin and other globulins, so they have a half-life of 15 to 30 minutes. DHEAS has a higher affinity for albumin and has a half-life of 8 to 10 hours, demonstrating the ability of carrier proteins to extend the half-lives of hormones and serve as temporary storage sites.

B. Function

Adrenal androgens have such weak androgenic activity that their physiologic role is largely limited to providing substrates for testosterone and estrogen synthesis. Androgens are required for secondary sex characteristic development during childhood and adolescence (see Biologic Sex and Aging 34.1). Androgens are also notable for the significant virilization and hirsutism they can cause in females when in pathologic excess (see Clinical Application 34.2).

Biologic Sex and Aging 34.1: Adrenal Androgens

During most of gestation, the fetal adrenal gland is dominated by a single "fetal zone" that secretes massive amounts of DHEA and DHEAS for use by the placenta in estrogen formation. At term, gland androgen output is greater than in an adult, but the zone rapidly regresses after birth, and DHEAS levels fall. In the run up to puberty, beginning at around age 6 years, adrenal androgen output again rises ("adrenarche") and stimulates axillary and pubic hair growth. DHEAS also stimulates skin sebaceous gland development, which is associated with oily skin, acne, and body odor (see Chapter 15·VI·C).

In adults, adrenal androgens supply substrates for testosterone and estrogen synthesis by peripheral tissues. In males, androgens are used to generate <5% of circulating testosterone. In young healthy females, the role of DHEA and DHEAS is more significant. During the follicular phase of the menstrual cycle, adrenal androgens contribute substrates for 65%–75% of total testosterone synthesis, decreasing to ~40% of total during later phases when ovarian hormones dominate.

After about age 30 years, DHEA and DHEAS levels decline at a rate of about 2%–5% per year in both males and females. By the time individuals are in their 80s, production of DHEA and DHEAS has dropped to <30% of earlier peak production. This decline has been termed "adrenopause," which is independent of menopause.

Clinical Application 34.2: Congenital Adrenal Hyperplasia

Congenital adrenal hyperplasia (CAH) is an autosomal-recessive disorder characterized by a defect in any one of several enzymes involved in cortisol synthesis. Cortisol deficiency frees the hypothalamic–pituitary–adrenal axis (see Fig. 34.5) from negative feedback and promotes adrenocorticotropic hormone (ACTH) secretion. ACTH stimulates the adrenal cortex, causing adrenal hyperplasia and an excess of adrenal steroid substrates. CAH presents with characteristic symptoms that vary with the severity and location of the enzyme defect in the steroid synthetic pathway (bracketed numerals correspond to enzymes shown in Fig. 34.2).

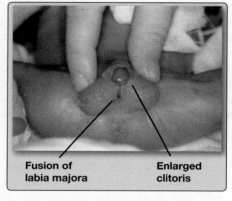

Fusion of labia majora — Enlarged clitoris

Female with ambiguous genitalia.

3β-Hydroxysteroid dehydrogenase [1]: *HSD3B2* mutation is a rare cause of CAH (<1% of cases), but it blocks formation of all adrenal steroid hormones, leading to hyponatremia, hypotension, and hyperkalemia, all because of aldosterone deficiency (see Section IV). ACTH drives increased pregnenolone formation, which is channeled into DHEA synthesis. DHEA can be converted to testosterone by peripheral tissues, and the resulting testosterone excess causes labial fusion and clitoromegaly in females during *in utero* development (ambiguous genitalia; see figure).

17α-Hydroxylase [2]: *CYP17A1* gene mutations are extremely rare, but they block formation of cortisol and all androgens. CYP17A1 substrates instead back up into the mineralocorticoid pathway, leading to an excess of 11-deoxycorticosterone (11-DOC), which is a weak mineralocorticoid. Hypernatremia, hypertension, and hypokalemia result, along with feedback suppression of aldosterone synthesis by high blood pressure. CYP17A1 is required for estrogen formation, so females present with primary amenorrhea, whereas the lack of androgens in males results in ambiguous genitalia.

21-Hydroxylase [3]: *CYP21A2* mutations account for >90% of CAH cases. Symptoms vary depending on the severity of the enzyme deficit, but the "classic" severe form is characterized by hyponatremia, hypotension, and hyperkalemia from mineralocorticoid deficiency (see Section IV). Androgen excess causes genital ambiguity and menstrual irregularities in females, whereas males show early virilization (pubic and body hair growth and development of body odor).

11β-Hydroxylase [4]: *CYP11B1* gene mutations cause 5%–8% of CAH cases. They block aldosterone and cortisol synthesis, causing substrates to be channeled into androgen production, with associated genital ambiguity and menstrual irregularities in females, and early virilization in males. However, whereas aldosterone deficiency usually leads to salt wasting and hypotension, ACTH-driven adrenal hyperactivity creates an 11-DOC excess that causes hypernatremia, hypertension, and hypokalemia.

IV. ADRENAL CORTEX: ALDOSTERONE

Aldosterone is a mineralocorticoid produced by the zona glomerulosa. It is an important determinant of extracellular fluid (ECF) volume and blood pressure.

A. Synthesis and secretion

The glomerulosa does not express CYP17A1, which prevents it from synthesizing cortisol or androgens. Instead, available substrates are directed toward aldosterone synthesis. The glomerulosa is also the only cortical region to express **aldosterone synthase** ([**AS**] *CYP11B2* gene), which limits aldosterone synthesis to this region. AS produces aldosterone from corticosterone. Once released into the cortical capillary sinusoids, aldosterone binds with low affinity to corticosteroid-binding protein and albumin. The hormone has a half-life of ~20 minutes in the circulation.

B. Regulation

Aldosterone secretion is not regulated by the HPA axis, although the HPA axis can indirectly impact aldosterone production through changes in substrate availability. Aldosterone release is regulated by angiotensin II (Ang-II) and plasma K^+ levels (Fig. 34.7). Ang-II is a hormonal component of the **renin–angiotensin–aldosterone system** (**RAAS**). The RAAS activates in response to decreased arterial and renal perfusion pressure and increased SNS activity

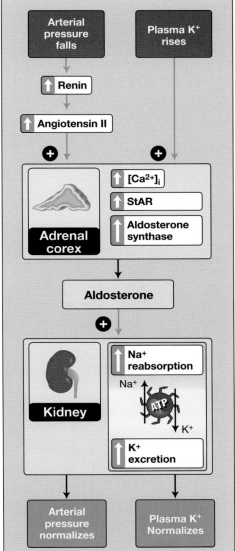

Figure 34.7.
Aldosterone regulation. StAR = steroidogenic acute regulatory protein.

Clinical Application 34.3: Primary Adrenal Insufficiency

Primary adrenal insufficiency commonly results from destruction of the adrenal cortex by autoantibodies. Symptoms include fatigue, dehydration, hyponatremia, and hypotension due to loss of glucocorticoids and mineralocorticoids. Adrenal hormone deficiency releases the brakes on corticotropin-releasing hormone release and proopiomelanocortin gene (*POMC*) expression through relief from feedback inhibition, which increases circulating levels of adrenocorticotropic hormone (ACTH). The ACTH peptide sequence includes melanocyte-stimulating hormone (MSH), which allows it to

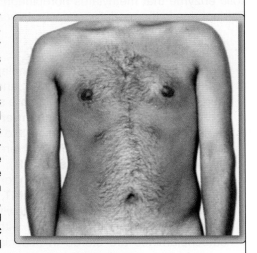

Bronze skin and nipple hyperpigmentation.

interact with MSH receptors on melanocytes and stimulate melanin production. Hyperpigmentation of hands, feet, nipples, axillae, and the oral cavity results. Treatment involves fluid replacement and exogenous glucocorticoids, such as hydrocortisone. Mineralocorticoid replacement therapy is usually also needed to correct the effects of Na^+ depletion and hyperkalemia.

(see Chapter 28·III·C). Ang-II binds to an AT1 angiotensin receptor, which couples via G_q to the IP_3 signaling pathway, increasing intracellular Ca^{2+} concentration when activated. Hyperkalemia similarly raises intracellular Ca^{2+} by depolarizing the glomerulosa cells and activating voltage-dependent Ca^{2+} channels. The Ca^{2+} flux then increases StAR and AS expression levels.

C. Function

The effect of aldosterone on ions (minerals) is reflected in its class name, mineralocorticoid. Aldosterone binds to intracellular **mineralocorticoid receptors** (**MRs**) that modify gene expression levels in target organs (see Fig. 27.12). Target organs include the kidney, where aldosterone increases Na^+ and dependent water reabsorption, thereby correcting hypovolemia and hypotension. It simultaneously increases urinary K^+ excretion to correct hyperkalemia (see Chapter 27·IV). Aldosterone also increases Na^+ absorption and reabsorption by large intestinal enterocytes, which increases body Na^+ stores and promotes water uptake (see Chapter 31·III·B).

V. ADRENAL MEDULLA: CATECHOLAMINES

The adrenal medulla derives from neural crest, which migrates into and becomes encapsulated within the cortex during development. The medulla is composed of **chromaffin cells**, which are the structural and functional equivalent of undifferentiated SNS postganglionic neurons. They are kept undifferentiated by high local cortisol concentrations released from the neighboring reticularis. Cortisol also induces expression of phenylethanolamine-*N*-methyltransferase (PNMT), a cytosolic enzyme that methylates norepinephrine (NE) to yield epinephrine. Epinephrine is the medulla's principal secretory product.

A. Synthesis and secretion

Chromaffin cells synthesize catecholamines (L-3,4-dihydroxyphenylalanine [L-dopa], dopamine, NE, and epinephrine) from tyrosine (Fig. 34.8). L-Dopa and dopamine are synthesized in the cytosol, and then a **catecholamine/H+ exchanger** (vesicular monoamine transporter 1 [**VMAT1**]) moves dopamine into secretory **chromaffin granules** (VMAT1 is encoded by *SLC18A1*). Here, dopamine is converted to NE by **dopamine β-hydroxylase** (**DBH**), as occurs in SNS postganglionic nerve terminals in preparation for synaptic release. Conversion of NE to epinephrine relies on PNMT, which is a cytosolic enzyme, so NE returns to the cytoplasm. Newly formed epinephrine is taken up by chromaffin granules, along with NE, by VMAT1, where both catecholamines are stored in association with **chromogranin** (an acidic binding protein) until released (see Fig. 34.8). Chromaffin cells secrete NE and epinephrine in an approximate 1:4 ratio into the medullary capillary sinusoids for delivery to tissues. Medullary capillaries are highly fenestrated to facilitate secretion. Catecholamine half-lives range from 10 to 90 seconds, which allows for a more sustained SNS response than that yielded by synaptic signaling. Catecholamines are metabolized in the liver and kidney primarily by catechol-*O*-methyltransferase to yield vanillylmandelic acid and metanephrine.

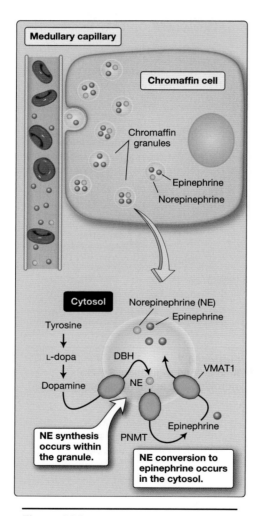

Figure 34.8.
Chromaffin cell. DBH = dopamine β-hydroxylase; L-dopa = L-3,4-dihydroxyphenylalanine; PNMT = phenylethanolamine-*N*-methyltransferase, VMAT1 = vesicular monoamine transporter 1.

All circulating epinephrine is derived from the adrenal medulla. Only 30% of circulating NE is secreted by chromaffin cells, however. The remaining 70% represents spillover from SNS postganglionic synapses.

B. Regulation

Catecholamine release is regulated by the SNS rather than the HPA axis. SNS preganglionic neurons release acetylcholine (ACh) onto chromaffin cells, which express **nicotinic type 2 ACh receptors** (**[nAChRs]** see Fig. 7.5). nAChR activation stimulates NE and epinephrine synthesis by enhancing tyrosine hydroxylase and DBH activity and by triggering chromaffin granule exocytosis.

C. Function

Epinephrine produces classic fight-or-flight responses, or an "adrenaline rush." Epinephrine and NE actions are similar to those of the SNS (see Fig. 7.4). Delivery via the circulation means that responses to hormones, although typically slower, are wider ranging because they can reach receptor populations not specifically located within a SNS synaptic cleft. Adrenergic receptor types differ in their affinities for catecholamines. β-Adrenergic receptors have a higher affinity for epinephrine compared with NE. α-Adrenergic receptors have a higher affinity for NE. The physiologic effects of catecholamines are related to the amount secreted and target tissue adrenergic receptor type and density.

Clinical Application 34.4: Pheochromocytoma

Pheochromocytomas are rare catecholamine-producing neoplasms arising from chromaffin cells. Symptoms include a classic triad of headache, tachycardia, and profuse sweating. Other symptoms may include chronic or paroxysmal hypertension, palpitations (forceful cardiac contractions), tremor, pallor, dyspnea, and anxiety (a sense of impending doom). Symptoms can be episodic or sustained, depending on the nature of the catecholamine release. Testing for pheochromocytomas involves measuring 24-hour urinary catecholamine and metanephrine levels. Treatment involves adrenergic receptor antagonists to control symptoms and surgical resection of the tumor.

Chapter Summary

- The **hypothalamic–pituitary–adrenal axis** involves secretion of **corticotropin-releasing hormone** from the hypothalamus, which stimulates **adrenocorticotropic hormone** (**ACTH**) release from the anterior pituitary. ACTH then promotes **glucocorticoid** and **adrenal androgen** secretion by the adrenal cortex.

- Glucocorticoids (**cortisol** and **corticosterone**) increase blood glucose and suppress immunity and inflammation, among many other responses.

- Adrenal androgens (**DHEA, DHEAS,** and **androstenedione**) participate in the development of secondary sex characteristics and serve as substrates in peripheral conversion of androgens to testosterone and estrogens.

- Secretion of the **mineralocorticoid aldosterone** is regulated by **angiotensin II** and plasma K^+. Aldosterone increases renal Na^+ and water reabsorption to increase circulating fluid volume and promotes excretion of excess K^+.

- **Catecholamines (epinephrine and norepinephrine)** are produced and secreted by adrenal medullary **chromaffin cells**, which are regulated by the sympathetic nervous system. Catecholamines prepare the body to face stressful events by increasing heart rate and inotropy to raise blood pressure and by mobilizing stored energy reserves.

Study Questions

Choose the ONE best answer.

34.1. Blood tests on a 34-year-old male identified high levels of circulating adrenocorticotropic hormone (ACTH). Levels of which of the following adrenocortical hormones would most likely be normal in this patient?

A. Aldosterone
B. Androstenedione
C. Corticosterone
D. Cortisol
E. Dehydroepiandrosterone sulfate

Best answer = A. The primary regulators of aldosterone release are angiotensin II and hyperkalemia, not ACTH (see Section IV·B). ACTH is released by the anterior pituitary, and its principal actions are to regulate adrenal corticosteroid production and release (see Section II·A). Production of adrenal androgens (androstenedione [B] and dehydroepiandrosterone sulfate [E]; see Section III·A) and glucocorticoids (corticosterone [C] and cortisol [D]; see Section II·B) are all directly controlled by ACTH, and levels of all four hormones would most likely be elevated in this patient. ACTH stimulates side-chain cleavage enzyme, which is one of the key rate-limiting steps for adrenocortical hormone production.

34.2. A 32-year-old male with suspected adrenocortical insufficiency is being treated with a synthetic cortisol (hydrocortisone). High doses improve his symptoms. If this dosing regimen is continued for a prolonged period, what is the most likely result?

A. Adrenal gland hypertrophy
B. β-Adrenergic receptor desensitization
C. Bone deposition and collagen formation
D. Muscle weakness
E. Virilization

Best answer = D. Corticosteroids normally prepare the body for stress by mobilizing energy substrates, such as glucose and free fatty acids (see Section II·D). This is accomplished in part through skeletal muscle protein catabolism, so prolonged cortisol administration can cause muscle weakness. Cortisol suppresses, not stimulates, adrenal gland growth (A) due to feedback inhibition of adrenocorticotropic hormone. β-Adrenergic receptor desensitization (B) occurs in response to chronically high catecholamine levels, not glucocorticoids. Cortisol stimulates Ca^{2+} resorption from bone, not deposition (C). Glucocorticoids can stimulate mineralocorticoid receptors at high levels, but the androgen receptors involved in virilization (E) are relatively insensitive.

34.3. A 10-day-old infant is brought to the emergency department because of vomiting and severe dehydration. The mother is a recent immigrant and had not received prenatal care. Physical examination of the infant reveals hypotension and ambiguous genitalia. Laboratory studies show hyponatremia and hyperkalemia. The patient's condition is most likely due to a deficiency in which of the following enzymes?

A. Aldosterone synthase
B. 3β-Hydroxysteroid dehydrogenase
C. 11β-Hydroxylase
D. 17α-Hydroxylase
E. 21-Hydroxylase

Best answer = E. The patient most likely has congenital adrenal hyperplasia (CAH) resulting from a defect in the gene that encodes 21-hydroxylase. *CYP21A2* mutations are the most common cause of CAH (see Clinical application 34.2). The condition causes hyponatremia, hypotension, hyperkalemia, and genital ambiguity. Aldosterone synthase deficiency (A) would not cause genital ambiguity. 3β-Hydroxysteroid dehydrogenase deficiency (B) causes similar symptoms to the patient's but is extremely rare. 11β-Hydroxylase deficiency (C) and 17α-hydroxylase deficiency (D) cause hypernatremia, hypertension, and hypokalemia.

34.4. A 55-year-old female comes to the emergency department because her heart is racing and she's sweating profusely. She says she has had similar episodes previously. A CT scan of the abdomen reveals a right adrenal mass. Which of the following additional findings is most likely in this patient?

A. Decreased blood pressure
B. Decreased plasma glucose
C. Headache
D. Increased urinary cortisol levels
E. Increased urinary potassium levels

Best answer = C. The patient presents with pheochromocytoma, which is a catecholamine-producing tumor. The classic symptoms are headache, tachycardia, and profuse sweating (see Clinical Application 34.4). Catecholamines increase blood pressure and circulating glucose levels, not decrease them (A, B). Urinary vanillylmandelic acid and metanephrine levels would be increased (catecholamine breakdown products; see Section V·A) in this patient, not cortisol (D). Urinary potassium levels would be increased (E) in response to hyperaldosteronism.

Thyroid and Parathyroid Glands

35

I. OVERVIEW

The cells that make up the human body vary widely with respect to their metabolic and development rates. Thyroid hormones provide the brain with a global method to govern these processes in the long term and outside of conditions of acute stress. The **thyroid gland** is located in the neck just below the larynx (Fig. 35.1). It produces two thyroid hormones: **triiodothyronine** (T_3) and **tetraiodothyronine** (T_4, or **thyroxine**). T_3 and T_4 induce transcription, translation, and synthesis of transporters, enzymes, and cellular scaffolding. The thyroid gland also has a minor role in Ca^{2+} balance via **parafollicular C cells**, which release **calcitonin**. The principal overseers of Ca^{2+} balance are the **parathyroid glands**, located in the inferior and superior margins of the thyroid gland (see Fig. 35.1) and **vitamin D**. **Parathyroid hormone** (**PTH**) and vitamin D together control plasma Ca^{2+} and PO_4^{3-} levels. Ca^{2+} balance is a critical homeostatic function. Ca^{2+} is required for numerous cellular events, including cell signaling (e.g., initiating muscle contraction and vesicle secretion), sustaining action potentials (e.g., in cardiac myocytes), and maintaining bone mineral density. Ca^{2+} balance disturbances can cause muscle spasms, seizures, confusion, and potentially fatal cardiac rhythm disturbances.

II. THYROID GLAND: THYROID HORMONES

The thyroid gland is a highly vascularized assemblage of numerous, 200- to 300-μm diameter hollow spheres (**follicles**) filled with a viscous, protein-rich fluid called **colloid** (Fig. 35.2). Follicles are lined with a specialized epithelium composed of follicular cells resting on a basement membrane. Follicular cells synthesize and secrete thyroid hormones (Fig. 35.3). The oxidative chemistry involved in thyroid hormone synthesis can be very harmful to cells, so it is performed extracellularly within the follicle lumen. This is a similar concept to walling off hydrogen peroxide reactions within cytosolic peroxisomes.

A. Synthesis

Thyroid hormone synthesis and secretion is a multistep process involving iodination and conjugation of adjacent tyrosyl residues on **thyroglobulin** (**Tg**), a 660,000-MW glycoprotein homodimer. Thyroid hormone synthesis can be broken into eight steps that correspond to the following sections (Fig. 35.4).

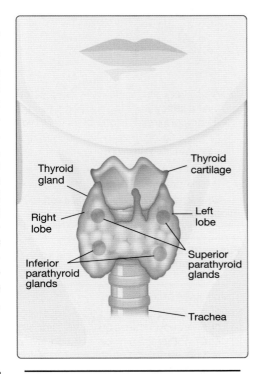

Figure 35.1.
Thyroid and parathyroid glands.

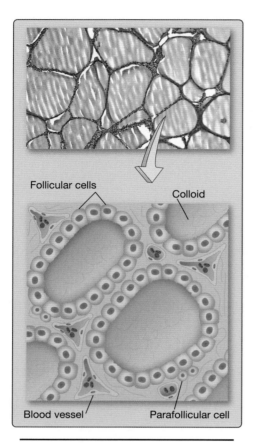

Figure 35.2.
Thyroid gland organization.

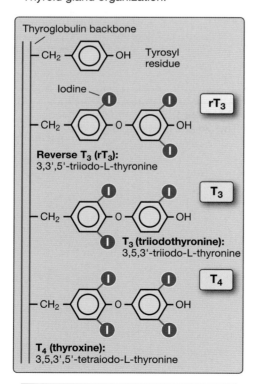

Figure 35.3.
Thyroid hormones.

1. **Iodide uptake:** Thyroid hormone synthesis begins when follicular cells take up I^- from blood, a process called "I^- **trapping**." I^- is moved across the basolateral membrane by a Na^+/I^- cotransporter (or, sodium/iodide symporter [**NIS**]), powered by the Na^+ gradient established by the basolateral Na^+/K^+-ATPase. NIS is a member of the solute carrier 5 family and is encoded by the *SLC5A5* gene. Follicular cells take up ~20% of the I^- made available to them by their blood supply on each pass through the thyroid gland, raising thyroid I^- concentration to ~20 to 50 times plasma levels.

> Thyroid hormones cannot be produced without I^-, which must be obtained from dietary sources (U.S. recommended dietary allowance is 150 µg/d). Iodine deficiency results in hypothyroidism and presents as goiter (see Clinical Application 35.1).

2. **Apical secretion:** I^- then moves across the apical membrane and enters colloid via a Ca^{2+}-**dependent anion channel** (TMEM16A, also known as anoctamin 1) and **pendrin**. Pendrin is a Cl^-/I^- exchanger. Mutations in the pendrin gene (*SLC26A4*) can cause Pendred syndrome, which is associated with goiter, hypothyroidism, and deafness (pendrin is involved in establishing the endocochlear potential used to transduce auditory stimuli; see Chapter 9·IV·C). Tg is synthesized in follicular cells and then secreted across the apical membrane into colloid by exocytosis.

3. **Oxidation:** The Tg-laden secretory vesicles express **thyroid peroxidase** (**TPO**), a heme-containing enzyme, on their inner surfaces. When the vesicles fuse with the apical membrane, TPO is presented to the colloid lumen and immediately catalyzes an oxidation reaction in which iodide is combined with H_2O_2 to form iodine (I^0) and H_2O. H_2O_2 is generated by **dual oxidase 2** (**DUOX2**), which is a member of the nicotinamide adenine dinucleotide phosphate (NADPH) oxidase family. DUOX2 complexes with TPO in the apical membrane (not shown in Fig. 35.4), where it combines intermediates from the pentose phosphate pathway with O_2 to yield H_2O_2.

4. **Iodination:** TPO also facilitates iodination (or **organification**) of Tg tyrosyl residues to form **monoiodotyrosine** (**MIT**) and **diiodotyrosine** (**DIT**). The reason why some tyrosyl residues bind only a single I^0, whereas others bind two is not clearly understood. The Tg peptide sequence contains ~100 tyrosyl residues, but, ultimately, only ~20 are involved in thyroid hormone synthesis.

5. **Conjugation:** MIT and DIT combine to form T_3 or **reverse T_3** (**rT_3**), whereas two DIT residues combine to form T_4 (see Fig. 35.3). rT_3 has no biologic activity. The two hormones and rT_3 remain attached to Tg until internalized by the follicular cells. Conjugation is also facilitated by TPO.

6. **Endocytosis:** The iodinated and conjugated Tg binds to a megalin receptor on the follicular cell membrane, causing it to be endocytosed back into the cell, along with a portion of the colloid.

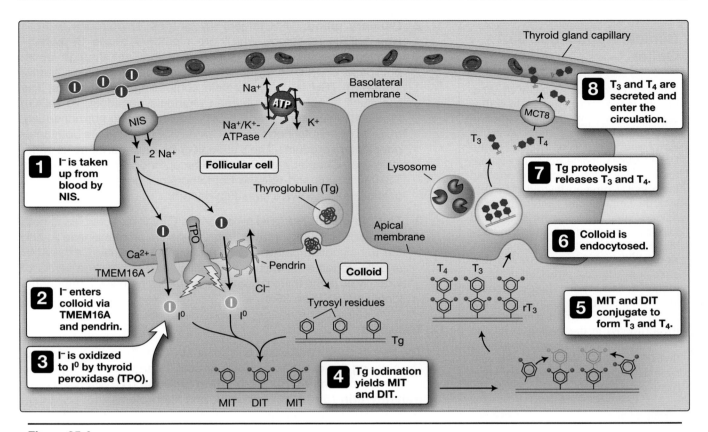

Figure 35.4.
Thyroid hormone biosynthesis. DIT = diiodotyrosine, MCT8 = monocarboxylate transporter 8, MIT = monoiodotyrosine, NIS = sodium/iodide symporter, rT_3 = reverse T_3, T_3 = triiodothyronine, T_4 = tetraiodothyronine.

7. **Proteolysis:** The endocytosed vesicle fuses with a lysosome, and iodinated Tg is degraded to liberate T_3 and T_4, which are then released into the follicular cytosol. Tg degradation fragments and other endocytosed colloid materials are recycled.

8. **Secretion:** Following release, T_3 and T_4 cross the basolateral cell membrane via a monocarboxylate transporter (MCT8; *SLC16A2* gene) and enter the extensive thyroid vascular network for distribution.

B. Hypothalamic–pituitary–thyroid axis

Thyroid hormone synthesis and secretion is controlled through the **hypothalamic–pituitary–thyroid (HPT) axis**, which involves **thyrotropin-releasing hormone (TRH)**, **thyroid-stimulating hormone (TSH)**, and the two thyroid hormones (Fig. 35.5).

1. **Hypothalamus:** TRH is a tripeptide hormone produced by parvocellular (i.e., small-bodied) neurons in the hypothalamic **paraventricular nucleus** ([PVN] see Fig. 35.5). The neurons project to the **median eminence** (a circumventricular organ; see Chapter 7·VII·C), where TRH is stored in PVN nerve terminals awaiting release. When stimulated appropriately, the neurons secrete TRH into the hypophyseal portal circulation for transport to the anterior pituitary.

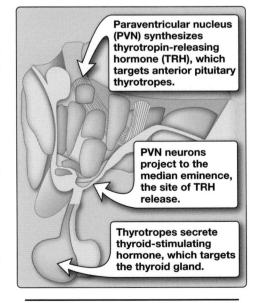

Paraventricular nucleus (PVN) synthesizes thyrotropin-releasing hormone (TRH), which targets anterior pituitary thyrotropes.

PVN neurons project to the median eminence, the site of TRH release.

Thyrotropes secrete thyroid-stimulating hormone, which targets the thyroid gland.

Figure 35.5.
Hypothalamic nuclei involved in thyroid gland regulation.

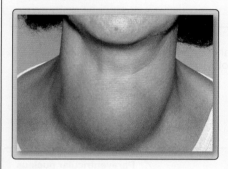
2. **Pituitary gland:** TRH targets thyrotropes. Thyrotropes express type 1 TRH receptors, which are members of the G protein–coupled receptor (GPCR) superfamily that act through phospholipase C (PLC) and the IP_3 signaling pathway (see Chapter 1·VII·B·3). TSH is a glycoprotein comprising two subunits. The α subunit is common to human chorionic gonadotropin (hCG), follicle-stimulating hormone (FSH), and luteinizing hormone (LH). hCG is a placental hormone, whereas FSH and LH are both anterior pituitary products. The β subunit gives each hormone its specificity (see Chapter 7·VII·D·3). TRH receptor occupancy stimulates TSH synthesis and release from secretory granules. Release is pulsatile and follows a circadian rhythm.

3. **Thyroid gland:** The TSH receptor is a GPCR that couples to both the cAMP signaling pathway, which acts via protein kinase A, and the IP_3 signaling pathway, which raises intracellular Ca^{2+}. cAMP and IP_3 together stimulate the activity and expression levels of virtually all of the thyroid hormone synthesis pathway components identified in Figure 35.4.

C. Transport

The thyroid gland synthesizes and secretes T_3 and T_4 in a ratio of ~1:10. On entering the circulation, >99% of both T_3 and T_4 bind to plasma proteins, including **thyroxine-binding globulin** ([**TBG**] 75%–80%), **transthyretin** (5%–10%), albumin, and lipoproteins (15% of total). Less than 0.5% of either hormone remains free. Binding proteins protect thyroid hormones from degradation and increase their half-life to as long as a week. Binding proteins also create a hormone reserve that buffers changes in hormone production. Only the free hormones have biologic activity.

> Maternal thyroid hormone levels rise twofold during the first two trimesters of pregnancy to meet the increased metabolic needs of supporting fetal growth. TBG production rises to a similar degree to keep free thyroid hormone levels relatively constant.

D. Regulation

The HPT axis is regulated by circulating free thyroid hormone levels (Fig. 35.6). The axis is also regulated by nutritional state and body temperature.

1. **Feedback control:** Even very small changes in circulating free T_3 and T_4 have a significant effect on the output of both the hypothalamus and pituitary. Increased thyroid hormone levels decrease TRH expression, processing, and release from hypothalamic neurons. HPT axis regulation involves **tanycytes**, specialized glia-like cells embedded in the ependymal lining of the third ventricle that project to several hypothalamic areas, including the median eminence. Here, they respond to T_3 and T_4 by secretion of an endopeptidase that degrades the TRH being released from PVN terminals. A rise in free thyroid hormone levels also decreases TRH1 receptor expression levels and TSH release from the anterior pituitary.

2. **Nutritional state:** Thyroid hormones increase energy substrate use, which can be counterproductive when food is limited. TRH-producing neurons in the PVN receive inputs from the hypothalamic arcuate nucleus that either suppress or increase TRH release according to feeding state. The arcuate nucleus senses and responds to circulating feeding-related cues, such as glucose, insulin, ghrelin, and leptin (see Chapter 30·V).

3. **Temperature:** Cold exposure stimulates increased TRH secretion by the PVN, which raises thyroid hormone levels. T_3 and T_4 stimulate metabolism and contribute to nonshivering thermogenesis (see Chapter 38·II·D·4). This cold-induced effect is mediated in part by catecholamines.

E. Functions

Thyroid hormones affect the function of virtually all cells in the body. T_4 and T_3 diffuse across the target cell membrane, and then much of the T_4 is converted to T_3 (the more active form) by one or more monodeiodinases. T_3 and T_4 bind to a nuclear thyroid hormone receptor (TR) that complexes with a retinoid X receptor (RXR). This TR–RXR receptor complex binds to a thyroid-response element within DNA, which, through release of a corepressor and addition of a coactivator, initiates transcription (Fig. 35.7). Proteins synthesized in response to thyroid hormones mediate a wide range of cellular responses, including increases in growth and development, energy substrate availability, and basal metabolic rate (BMR).

1. **Growth and development:** Normal growth and development are dependent on thyroid hormones. In nervous tissue, T_3 and T_4 affect the timing and rate of development, including, for example, stretch reflex development (see Chapter 11·III·B). Thyroid hormones increase bone ossification and linear growth in children and adolescents. Thyroid hormone deficiencies can, thus, result in mental impairments (**congenital hypothyroidism**) and short stature (**achondroplasia**).

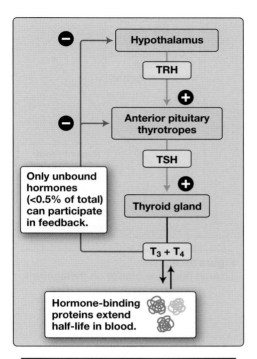

Figure 35.6.
Hypothalamic–pituitary–thyroid axis. T_3 = triiodothyronine, T_4 = tetraiodothyronine, TRH = thyrotropin-releasing hormone, TSH = thyroid-stimulating hormone.

Table 35.1: Thyroid Hormone Effects

	↓ T_3 + T_4	↑ T_3 + T_4
Basal metabolic rate	↓	↑
O₂ consumption	↓	↑
Carbohydrates	↓ Glycogenolysis	↑ Glycogenolysis
	↓ Gluconeogenesis	↑ Gluconeogenesis
	↔ Glucose	↔ Glucose
Lipids	↓ Lipolysis	↑ Lipolysis
	↓ Lipogenesis	↑ Lipogenesis
Proteins	↓ Proteolysis	↑ Proteolysis
	↓ Protein synthesis	↑ Protein synthesis
Sympathetic nervous system	None	↑ β-Adrenergic sensitivity

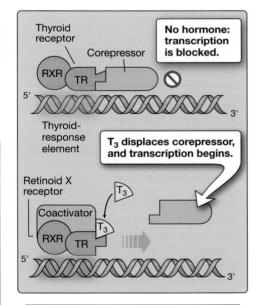

Figure 35.7.
Thyroid hormone–receptor binding. RXR = retinoid X receptor, T_3 = triiodothyronine, TR = thyroid receptor.

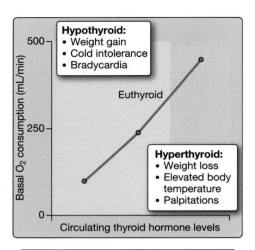

Figure 35.8.
Thyroid hormone effects on metabolism.

2. **Energy substrates:** Thyroid hormones increase energy substrate availability to support increased metabolism (Table 35.1). T_3 and T_4 both increase glycogen breakdown (glycogenolysis) to liberate glucose and stimulate glucose formation (gluconeogenesis), although plasma glucose levels remain stable. T_3 and T_4 also increase the breakdown of stored triacylglycerols into free fatty acids to be used as energy substrates for gluconeogenesis, but the hormones simultaneously promote lipid formation (lipogenesis) to sustain energy reserves. In muscle (primarily), thyroid hormones stimulate proteolysis to release amino acids to be used in gluconeogenesis, but they also stimulate protein synthesis to offset protein breakdown effects.

3. **Basal metabolic rate:** Simultaneous stimulation of catabolic and anabolic pathways significantly increases BMR, generates heat, and increases basal O_2 consumption (Fig. 35.8). A significant proportion of O_2 usage is attributable to increased Na^+/K^+-ATPase activity. Thyroid hormones increase Na^+/K^+-ATPase expression. Because the ATPase is always active, O_2 consumption increases in proportion to expression level.

4. **Thyroid hormone and catecholamine synergy:** When T_3 and T_4 and norepinephrine from sympathetic nervous system activation are released in concert (e.g., during severe cold stress), the physiologic functions of both are heightened. Thyroid hormones also upregulate β-adrenergic receptors, which potentiates synergism, resulting in increased heart rate, myocardial inotropy, and cardiac output, for example.

III. THYROID GLAND: CALCITONIN

Calcitonin is a 32–amino acid peptide synthesized by parafollicular C (clear) cells, which are randomly distributed between follicles throughout the thyroid gland. Calcitonin is released in response to very high plasma Ca^{2+} concentrations, and its primary effect is to block bone resorption and Ca^{2+} mobilization by osteoclasts. The physiologic role of calcitonin in humans, if any, may be limited. Calcitonin is used clinically as a biomarker for medullary thyroid cancer and to suppress osteoclast activity in patients with Paget disease of bone. It can also be used in the treatment of osteoporosis.[1] Plasma Ca^{2+} and PO_4^{3-} levels are regulated primarily by PTH and vitamin D, as discussed in the next section.

IV. PARATHYROID GLAND: PARATHYROID HORMONE

The parathyroid glands synthesize PTH, which, in conjunction with vitamin D, is responsible for maintaining Ca^{2+} balance. Ca^{2+} homeostasis necessarily also involves regulating plasma PO_4^{3-} levels because increases in either precipitate hydroxyapatite crystal formation. PTH is

---

Clinical Application 35.2: Hashimoto Thyroiditis

Hashimoto thyroiditis is the most common form of **hypothyroidism**, characterized by an autoimmune response to various thyroid gland antigens, including thyroglobulin, thyroid peroxidase, and the thyroid-stimulating hormone receptor. The gland becomes inflamed, and gland follicles are destroyed, which decreases thyroid hormone production. Extensive inflammatory infiltrates, together with deposition of fibrous tissue to replace lost follicles, may cause diffuse gland enlargement and a goiter. Patients typically lack energy, are easily fatigued, are cold intolerant, and gain weight due to metabolic rate decline. Hashimoto disease is more common in females than males (~7:1), and the incidence increases with age.

---

 [1]For a discussion of the use of calcitonin to treat osteoporosis, see *LIR Pharmacology*, 8e, Chapter 27·IV·F.

Clinical Application 35.3: Graves Disease

Graves disease causes **hyperthyroidism** via an antibody (**thyroid-stimulating immunoglobulin**) that binds to and activates the thyroid-stimulating hormone (TSH) receptor. TSH receptor stimulation causes glandular hypertrophy, which, in turn, causes goiter. Symptoms of hyperthyroidism include agitation, tremor, muscle weakness, heat intolerance, and increased appetite. Graves disease is also associated with myxedema and ophthalmopathy. Myxedema is a dermal thickening caused by overproduction of mucopolysaccharides (modified carbohydrates) and resultant edema, lymphocytic infiltration, and deposition of connective tissue. Ophthalmopathy is characterized by inflammation and mucopolysaccharide deposition in tissues surrounding the eye. Swelling of the extraocular muscles caused by fibrosis and protein deposition produces a characteristic orbit protrusion (exophthalmos).

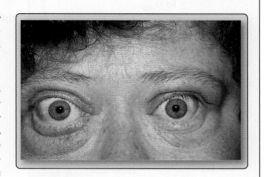

Exophthalmos.

synthesized by chief cells as a 115–amino acid preprohormone encoded by the *PTH* gene. Posttranslational processing yields an 84–amino acid hormone ("intact PTH") that has a half-life of <4 minutes in the circulation. Degradation begins even during PTH storage in secretory granules. The parathyroid thus secretes intact PTH; an inactive C terminal fragment; and shorter, active N-terminal fragments.

A. Tally sheet

Total plasma Ca^{2+} is maintained at 2.1 to 2.8 mmol/L, but Ca^{2+} balance focuses on free Ca^{2+}, which is held at 1.0 to 1.3 mmol/L. The difference between total and free Ca^{2+} reflects the fact that >55% is bound to anionic sites on proteins or is complexed with small organic anions. Total body Ca^{2+} is much higher (~1,000 g), the bulk of it being deposited in association with PO_4^{3-} as hydroxyapatite in bone (Fig. 35.9). About 1.0 g of Ca^{2+} is ingested daily in food. A small proportion of total (~0.175 g) is absorbed by the gastrointestinal (GI) tract, and the remainder is excreted in feces. The kidneys excrete ~0.175 g of Ca^{2+} daily to hold plasma Ca^{2+} in balance, but this tally sheet hides the fact that a significant amount of Ca^{2+} is constantly being released from and then redeposited in bone during routine bone remodeling (see Chapter 14).

B. Sensory mechanism

Free Ca^{2+} levels are sensed by a calcium-sensing receptor (CaSR) expressed by parathyroid chief cells, which synthesize PTH. CaSR is a GPCR encoded by the *CASR* gene. When CaSR binds Ca^{2+}, it initiates Ca^{2+} release from intracellular stores and activates protein kinase C via $G_{\alpha q}$ and the IP_3 signaling system (Fig. 35.10).

C. Regulation

PTH regulation is unusual in that release is tonically inhibited by the CaSR at normal free plasma Ca^{2+} levels. When free Ca^{2+} falls, the CaSR stops signaling, and PTH is released into the circulation (see Fig. 35.10). The relationship between free Ca^{2+} and PTH release is sigmoidal, with the steepest portion of the curve being in the

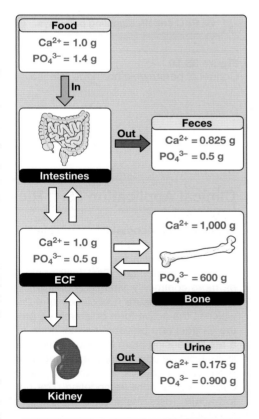

Figure 35.9.
Calcium and phosphate balance.

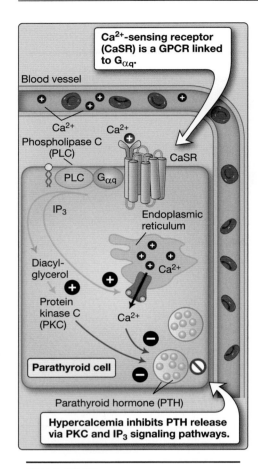

Figure 35.10.
Control of parathyroid hormone release.

physiologic range of plasma Ca^{2+}. The CaSR also binds Mg^{2+}, meaning that PTH release is similarly dependent on plasma free Mg^{2+}. The PTH receptor (PTH1R) is a GPCR that acts through both the cAMP and IP_3 signaling systems. PTH1R activation raises plasma Ca^{2+} by increasing Ca^{2+} resorption from bone and increasing reabsorption from the kidney tubule (see Fig. 35.10).

1. **Bone:** PTH1R is found on osteoblasts. PTH binding initiates release of cytokines that recruit osteoclast precursors to a bone remodeling site (see Chapter 14·IV·C). It also increases osteoclast activity. Osteoclasts digest bone, thereby releasing Ca^{2+} and PO_4^{3-} to the circulation.

2. **Kidney:** PTH1R is expressed by the renal distal convoluted tubule ([DCT] see Chapter 27·III·C). PTH binding increases Ca^{2+} recovery from tubule fluid by enhanced Ca^{2+}-channel and Ca^{2+}-pump activity. The proximal tubule (PT) also expresses PTH1R, but, here, PTH binding promotes PO_4^{3-} excretion (see Chapter 26·VI) by decreasing expression of transporters that mediate PO_4^{3-} reabsorption. PO_4^{3-} wasting ensures that the PTH-induced rise in plasma Ca^{2+} concentrations does not trigger hydroxyapatite formation in tissues.

D. **Vitamin D**

Vitamin D refers to a group of related sterols. One of these (1,25-dihydroxyvitamin D_3 [**1,25-diOH-D$_3$**], also known as **calcitriol**) is both a dietary requirement and a hormone. Vitamin D is hydrophobic and is carried in blood in association with vitamin D–binding protein.

1. **Synthesis:** Vitamin D is synthesized from 7-dehydrocholesterol, a cholesterol derivative. 7-Dehydrocholesterol photolyzes when exposed to ultraviolet light and spontaneously forms cholecalciferol (vitamin D_3). The skin is a primary source of vitamin D_3 for this reason. Vitamins D_3 and D_2 (ergocalciferol) may also be obtained from dietary sources. Vitamins D_2 and D_3 are converted to 25-hydroxyvitamin D_3 (25-OH-D_3) in the liver. The renal PT then

---

Clinical Application 35.4: Rickets and Osteomalacia

Rickets and **osteomalacia** are bone mineralization defects. Rickets describes deficient mineralization and deformation of the growth plate in children. Osteomalacia is a bone cortical mineralization deficiency. Rickets can be caused by inadequate Ca^{2+} (calcipenic rickets); PO_4^{3-} (phosphopenic rickets); or, more commonly, vitamin D deficiency due to lack of dietary intake or ultraviolet light exposure. Deficiency of 1,25-dihydroxyvitamin D_3 (1,25-diOH-D_3), which is the active form of vitamin D, prevents Ca^{2+} and PO_4^{3-} absorption from the GI tract and prevents normal bone mineralization. Bones are very weak as a result, causing them to distort when stressed physically. In children, the lower limbs typically bow. Epiphyses widen, and the metaphyses flare due to cartilage overgrowth. In adults, 1,25-diOH-D_3 deficiency prevents osteoid mineralization during normal bone remodeling. The bone weakens and becomes vulnerable to fracture (osteomalacia), most commonly affecting vertebral bodies and the femoral heads.

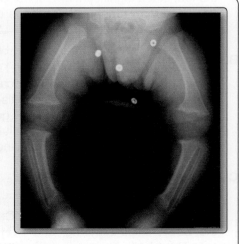

Rickets.

synthesizes 1,25-diOH-D$_3$ from 25-OH-D$_3$ using 1α-hydroxylase (*CYP27B1* gene). The PT also expresses 24α-hydroxylase (*CYP24A1*) that inactivates 1,25-diOH-D$_3$, so 1,25-diOH-D$_3$ availability always reflects a balance between the activities of these two enzymes.

2. **Regulation:** PTH targets the kidney and the final step in 1,25-diOH-D$_3$ synthesis. PTH increases 1α-hydroxylase activity and expression levels and simultaneously decreases 24α-hydroxylase activity and expression levels. Low levels of Ca^{2+} and PO$_4^{3-}$, as well as hormones, such as growth hormone, prolactin, and estrogen, also increase 1,25-diOH-D$_3$ levels. Conversely, 1,25-diOH-D$_3$ production is decreased by bone-derived fibroblast factor 23 and by 1,25-diOH-D$_3$ itself.

3. **Function:** 1,25-diOH-D$_3$ helps maintain Ca^{2+} balance primarily through effects on the small intestine with additional lesser effects on the kidneys and bone.

   a. **Small intestine:** 1,25-diOH-D$_3$ is lipid soluble and readily diffuses across enterocyte surface membranes. Once inside the cell, it binds to a nuclear vitamin D receptor (VDR). VDR complexes with RXR (see Section II·E) and increases the expression of vitamin D–sensitive genes. These include apical membrane Ca^{2+} channels and basolateral Ca^{2+} transporters, thereby enhancing Ca^{2+} absorption from the intestinal lumen. Importantly, 1,25-diOH-D$_3$ also increases the expression of calbindin to maintain low intracellular Ca^{2+} levels and facilitate transcellular Ca^{2+} movement (see Fig. 31.10).

   b. **Kidneys and bone:** In the kidney, 1,25-diOH-D$_3$ targets the PT, where it increases the expression of apical Na$^+$/P$_i$ cotransporters and potentiates PTH-enhanced Ca^{2+} reabsorption. In bone, 1,25-diOH-D$_3$ can increase mature osteoclast numbers, which allows for increased bone resorption. Finally, 1,25-diOH-D$_3$ decreases PTH synthesis and release by the parathyroid glands through a negative feedback pathway (Fig. 35.11).

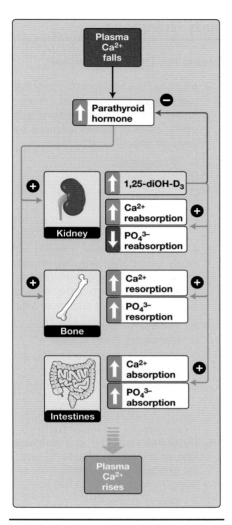

Figure 35.11.
Regulation of plasma Ca^{2+}. 1,25-diOH-D$_3$ = 1,25-dihydroxyvitamin D$_3$.

Chapter Summary

- The thyroid gland secretes **triiodothyronine (T$_3$)** and **thyroxine (T$_4$)** in a 1:10 ratio. Thyroid intermediates include **monoiodotyrosine (MIT)** and **diiodotyrosine (DIT)**. Target tissues convert T$_4$ to T$_3$, which is the more active form.

- **Thyroid-stimulating hormone (TSH)** regulates T$_3$ and T$_4$ secretion by stimulating most synthetic and release pathway components. TSH also causes thyroid gland hypertrophy.

- T$_3$ and T$_4$ bind to **thyroxine-binding globulin** for blood transport. Relatively little circulates in the free form.

- T$_3$ and T$_4$ diffuse across the cell membrane and bind to nuclear thyroid receptors (TRs). T$_3$ has a greater affinity for TRs than T$_4$, which gives it greater biologic activity. TRs modulate gene transcription and translation, thereby affecting protein expression levels.

- Thyroid hormones increase glycogen and fat breakdown and utilization and increase metabolic rate and heat production. Thyroid hormones are also imperative for normal growth and development.

- **Parathyroid hormone (PTH)** increases circulating Ca^{2+} levels by increasing Ca^{2+} reabsorption in the kidney and by increasing bone resorption. PTH also stimulates **vitamin D** synthesis.

- The active form of vitamin D (**1,25-dihydroxyvitamin D$_3$**) is synthesized in the renal proximal tubule. Its primary function is to increase circulating Ca^{2+} levels via absorption from the intestines.

Study Questions

Choose the ONE best answer.

35.1. A 55-year-old female with hyperthyroidism is given propylthiouracil (PTU), which inhibits thyroid peroxidase. Based on PTU's mechanism of action, which of the following steps in thyroid hormone synthesis would most likely be affected first?

A. Conjugation
B. Iodination
C. Oxidation
D. Proteolysis
E. Trapping

Best answer = C. The enzyme thyroid peroxidase (TPO) oxidizes iodide as it crosses the follicular cell's apical membrane into colloid (see Section II·A). Conjugation ([A] combining iodinated tyrosyl residues) and iodination ([B] adding iodine to tyrosyl residues) also require TPO, but these events occur after oxidation. Neither proteolysis (D) nor trapping (E) involve TPO and would not be affected by their inhibition.

35.2. A 45-year-old female presents with symptoms commonly associated with hypothyroidism. A blood sample reveals that thyroid-stimulating hormone levels are above normal. Which of the following results most likely describes the effect on thyroid follicular cells?

A. Increased blood supply
B. Increased iodide uptake
C. Increased thyroxine-binding globulin synthesis
D. Inhibition of growth
E. Inhibition of pendrin insertion

Best answer = B. Thyroid follicular cells are epithelial cells specialized for thyroid hormone synthesis and release. Thyroid-stimulating hormone (TSH) regulates many steps in the synthetic pathway (see Section II·B), including iodide uptake by a Na^+/I^- symporter (NIS). TSH upregulates NIS. Blood supply to follicular cells (A) is not affected by TSH. Thyroxine-binding globulin (C) is synthesized by the liver, not follicular cells. TSH stimulates thyroid tissue growth, rather than inhibiting it (D). Pendrin is an apical Cl^-/I^- cotransporter required to move I^- across the apical membrane and into the follicular lumen. Its expression levels are not regulated by TSH (E).

35.3. Triiodothyronine (T_3) and thyroxine (T_4) have a multitude of peripheral effects. T_3 and T_4 most likely have the greatest biologic activity in which of the following states?

A. Bound to albumin
B. Bound to thyroglobulin
C. Bound to thyroxine-binding globulin
D. Bound to transthyretin
E. Unbound

Best answer = E. Plasma proteins, such as albumin (A) and thyroxine-binding globulin (C), are important in maintaining a circulating pool of T_3 and T_4 (see Section II·C), but these hormones are not biologically active when bound to transport proteins. Only free hormones can exert peripheral effects (see Section II·E), and this is one reason why both the free and bound states of thyroid hormones are measured in the blood during a thyroid panel. Thyroglobulin (B) is a protein involved in thyroid hormone synthesis by the thyroid gland. Transthyretin (D), also known as prealbumin, is an albumin precursor protein that binds T_3 and T_4 in the circulation.

35.4. A researcher interested in Ca^{2+} homeostatic mechanisms is studying a hormone that increases Ca^{2+} resorption from bone and Ca^{2+} reabsorption in the renal distal convoluted tubule. The hormone also decreases PO_4^{3-} reabsorption in the renal proximal tubule. The hormone under study is most likely which of the following?

A. 1,25-Dihydroxyvitamin D_3
B. 7-Dehydrocholesterol
C. Calcitonin
D. Calsequestrin
E. Parathyroid hormone

Best answer = E. Parathyroid hormone (PTH) increases plasma Ca^{2+} availability by increasing reabsorption of Ca^{2+} from the renal tubule while simultaneously decreasing PO_4^{3-} reabsorption. PTH also increases Ca^{2+} resorption from bone (see Section IV·C). The vitamin D derivatives 1,25-dihydroxyvitamin D_3 (A) and, to a lesser extent, 7-dehydrocholesterol (B) increase renal reabsorption of both Ca^{2+} and PO_4^{3-} (see Section IV·D). Calcitonin's (C) mechanism of action is via its effect on osteoclasts, not the kidney. Calsequestrin (D) is a Ca^{2+}-binding protein, not a hormone that regulates blood Ca^{2+} levels.

Male and Female Gonads

36

I. OVERVIEW

"Gonad" is derived from the Greek *gonos*, which translates to "seed" or "family." In a very basic sense, the primary purpose of any species is to pass along a unique set of DNA to the next generation. Gonads, together with various accessory structures, make reproduction possible. Gonads are endocrine glands that, like other such glands, are regulated by a **hypothalamic–pituitary axis**, but they have an additional exocrine function. Gonads produce seed cells (gametes) through gametogenesis. Male gonads (**testes**; Fig. 36.1) produce sperm (**spermatogenesis**), whereas female gonads (**ovaries**) produce oocytes (**oogenesis**). Gonadal endocrine hormones have wide-ranging actions, including coordinating gametogenesis, guiding sexual development and maturation, facilitating conception, and supporting pregnancy.

II. SEX AND GENDER

Sex is determined by chromosomal complement, or karyotype. A pair of X chromosomes (46,XX karyotype) identifies a female, whereas a 46,XY karyotype identifies a male. **Gonadal sex** is defined by gonadal type. Gonadal females have ovaries, whereas gonadal males have testes. **Phenotypic sex** is determined by the characteristics of the genital tract and external genitalia. Finally, **gender** is a psychosocial term used to identify a person as a man or woman based on a set of characteristics, attributes, or social norms. These definitions become blurred in genetic variants such as Klinefelter syndrome, which is associated with an extra X chromosome: 47,XXY ("47" denotes an extra chromosome because a normal karyotype comprises 46 chromosomes arranged in 23 pairs). Hereditary androgen resistance can yield a genotypic male (46,XY) with a female phenotype. Phenotypic sex can be changed by surgery. In this chapter, we use gonadal sex denotations.

III. MALE GONADS

The testes (or testicles) lie outside the abdominal cavity within a scrotal sac. Their location reduces local temperature to 2 to 3°C below internal temperature, which is optimal for spermatogenesis. Testes produce testosterone under the control of the **hypothalamic–pituitary–testicular (HPT) axis**.

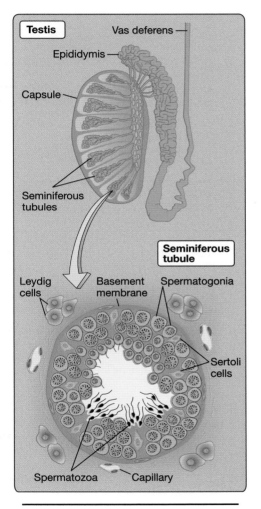

Figure 36.1.
Testicular anatomy.

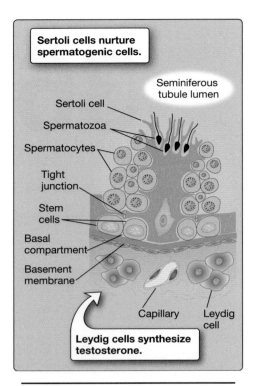

Figure 36.2.
Sertoli cell structure.

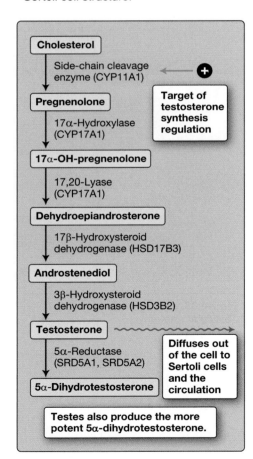

Figure 36.3.
Testosterone synthesis by Leydig cells.

A. Anatomy

The testes are encased within a thick, connective tissue capsule (tunica albuginea). Each testis is densely packed with finely coiled **seminiferous tubules** arranged in ~250 lobules (see Fig. 36.1). Lobules are separated by connective tissue septa that extend from the capsule. Seminiferous tubules produce sperm. The interstitial spaces between adjacent tubules contain **Leydig cells**, which produce testosterone.

1. **Seminiferous tubules:** Seminiferous tubules are ~50 cm long and 150 to 250 μm in diameter. They are lined with a seminiferous epithelium supported by a basement membrane. The epithelium contains two principal cell types, **spermatogenic cells** and **Sertoli cells** (see Fig. 36.1).

   a. **Spermatogenic cells:** Sperm cells arise from spermatogonia, which are stem cells tightly adhered to the basement membrane. Serial mitotic and meiotic divisions yield primary and secondary spermatocytes; spermatids; and, finally, spermatozoa. Spermatids and spermatozoa are haploid (23 chromosomes). During the final transition from spermatid to spermatozoa (**spermiogenesis**), cells elongate, cytoplasm is removed (forming a second structure, a residual body), and organelles are reoriented. Notably, a flagellum forms during this stage, which facilitates sperm motility. Spermatogenetic intermediaries migrate toward the tubule lumen during sperm formation, so mature spermatozoa come to reside on the luminal (apical) epithelial surface. Spermiogenesis takes ~72 days.

   b. **Sertoli cells:** Sertoli cells (supporting or sustentacular cells) are nurse cells whose apical and lateral surfaces are modified to surround and nurture spermatogenic cells in their various stages of development (Fig. 36.2). Sertoli cells and their charges are connected via adherens junctions and gap junctions (see Chapter 4·II), the latter allowing for intercellular communication. Release of mature spermatozoa into the tubule lumen (**spermiation**) ultimately requires that these junctional structures be broken down. Adjacent Sertoli cells are also coupled by tight junctions to create a blood–testis barrier that prevents sperm and other luminal contents from entering blood. The barrier also forms the roof of a basal compartment optimized for germ-cell activity. The tight junctions must be disassembled to allow passage of spermatocytes and then immediately reassembled.

2. **Leydig cells:** Large, polygonal Leydig cells are specialized for hormone synthesis (see Fig. 36.2). They are notable for prominent lipid droplets, an extensive smooth endoplasmic reticulum, and numerous mitochondria, all of which are required to support steroidogenesis.

B. Testosterone synthesis and secretion

Leydig cells synthesize testosterone from cholesterol via the steroid synthetic pathway common to the adrenal cortex (Fig. 36.3; see also Fig. 34.2). Low-density lipoprotein (LDL) cholesterol is taken up via LDL receptors and clathrin-mediated endocytosis from blood or formed *de novo* and then ferried to mitochondria. **Steroidogenic** acute regulatory protein (StAR) helps move cholesterol across the outer mitochondrial membrane. Cholesterol is then converted to pregnenolone by side-chain cleavage enzyme ([SCC] CYP11A1). 17α-Hydroxylase (CYP17A1) converts pregnenolone to 17α-hydroxypregnenolone, and then 17,20-lyase (CYP17A1) forms dehydroepiandrosterone (DHEA). Leydig cells differ from the adrenal zona reticularis in that they express 17α-hydroxysteroid dehydrogenase (HSD17B3), which allows them to convert DHEA to androstenediol, which then is converted to testosterone by 3β-hydroxysteroid hydrogenase (HSD3B2). Testosterone is the principal secretory product, but the testes produce other steroid synthesis pathway intermediaries such as androstenedione. The testes are also able to convert testosterone to **5α-dihydrotestosterone** (DIT) but not in significant quantities. Secreted testosterone enters the circulation and binds to sex hormone–binding globulin ([SHBG] ~44%) and to corticosteroid-binding globulin (CBG) and albumin (~54%), with ~2% remaining in free form. Testosterone has a circulatory half-life of 2 to 4 hours. The liver inactivates and processes testosterone. Testosterone and byproducts are excreted in urine and feces.

C. Hypothalamic–pituitary–testicular axis

The HPT axis involves **gonadotropin-releasing hormone (GnRH)**, **luteinizing hormone (LH)**, **follicle-stimulating hormone (FSH)**, and **testosterone** (Fig. 36.4).

1. **Hypothalamus:** GnRH is a decapeptide hormone produced by parvocellular neurons concentrated in preoptic and arcuate nuclei (Fig. 36.5). The *GNRH1* gene encodes a 92–amino acid preprohormone. Removal of the signal sequence yields a 69–amino acid prehormone, which is then processed to release GnRH and a 56–amino acid GnRH-associated protein (GAP). GnRH neurons transport GnRH (and GAP) to the **median eminence** for release into the hypophyseal portal system for transport to the anterior pituitary. GnRH has a half-life in the circulation of 2 to 4 minutes. GnRH release is pulsatile, which has the advantage of using less resources and avoids target tissue receptor desensitization. GnRH neurons receive inputs from many other brain regions, which modify their output in response to changing day length and stressors, for example. GnRH targets **gonadotropes** in the anterior pituitary.

2. **Pituitary:** GnRH binds to a GnRH receptor, which is a G protein–coupled receptor (GPCR) that acts via phospholipase C and the IP_3 signaling pathway (see Chapter 1·VII·B·3). Gonadotropes respond by secreting LH and FSH, which are related glycoprotein hormones comprising two peptide chains. They share an identical α subunit common to human chorionic gonadotropin (hCG) and thyroid-stimulating hormone (see Chapter 35·II·B·2). A 115–amino acid β subunit gives LH and FSH their specificity.

3. **Testis:** LH and FSH target Leydig and Sertoli cells, respectively (see Fig. 36.4).

   a. **Leydig cells:** Leydig cells express an LH receptor, which is a GPCR linked to adenylyl cyclase (AC) and the cAMP signaling pathway, which activates protein kinase A (PKA). PKA increases testosterone production.

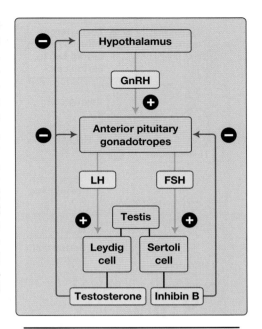

Figure 36.4.
Hypothalamic–pituitary–testicular axis regulation. FSH = follicle-stimulating hormone, GnRH = gonadotropin-releasing hormone, LH = luteinizing hormone.

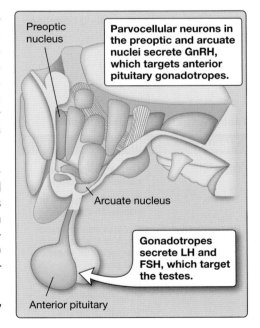

Figure 36.5.
Hypothalamic nuclei involved in testicular control. FSH = follicle-stimulating hormone, GnRH = gonadotropin-releasing hormone, LH = luteinizing hormone.

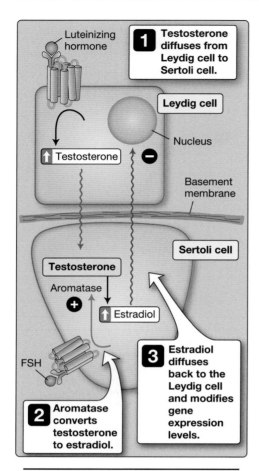

Figure 36.6.
Interactions between Leydig and Sertoli cells. FSH = follicle-stimulating hormone.

b. Sertoli cells: Sertoli cells express FSH receptors, which are also GPCRs that act via the cAMP signaling pathway. FSH stimulates several Sertoli cell functions essential for spermatogenesis, including secretion of growth factors, **androgen-binding protein (ABP)**, and **inhibin B**.

    i. Growth factors: Sertoli cell growth factors increase stem cell activity, thus increasing spermatozoa production. Growth factors are also required to maintain Leydig cell numbers. They also regulate peritubular myoid cells that maintain the basement membrane.

    ii. Androgen-binding protein: ABP is closely related to SHBG, which carries testosterone and 17β-estradiol (estradiol) through the circulation. ABG maintains luminal testosterone at the high levels (~100 times that of plasma) required to support spermatogenesis.

    iii. Inhibin B: Inhibins are glycoprotein hormones comprising an αβ-subunit dimer linked by a disulfide bond. Sertoli cells produce inhibin B ($\alpha\beta_B$), which exerts negative feedback control on the HPT axis (see Fig. 36.4). It also stimulates Leydig cell growth.

D. Testosterone regulation

LH is secreted in a pulsatile fashion that follows the pattern of GnRH release. LH increases testosterone secretion by increasing SCC activity and increasing the expression of StAR and most of the enzymes involved in steroid synthesis (see Fig. 36.3). Testosterone exerts negative feedback control of the HPT axis at the level of both the hypothalamus and pituitary (see Fig. 36.4). An additional layer of control is exerted through Sertoli cells. Sertoli cells express aromatase (*CYP19A1* gene), which converts testosterone diffusing from Leydig cells to estradiol (Fig. 36.6). Estradiol then decreases LH-stimulated testosterone production. Aromatase activity is stimulated by FSH. Sertoli cells also produce inhibin B, which inhibits FSH secretion by gonadotropes (see Fig. 36.4).

E. Testosterone functions

Testosterone is lipophilic, which allows it to enter cells passively by diffusion. Once inside a cell, it has several potential fates and actions (Table 36.1).

1. Receptor binding: Testosterone binds to a high-affinity androgen receptor (AR), which is also a transcription factor. AR receptor dimers and bound hormone interact with androgen-response elements on DNA and modify expression of several proteins. In addition to stimulating spermatogenesis and regulating sex glands, testosterone also increases fat deposition and has anabolic effects on bone and skeletal muscle.

2. Conversion to 5α-dihydrotestosterone: Some tissues, most notably the testes and urogenital tract, sebaceous glands, hair follicles, and liver, express 5α-reductase (*SRD5A2* gene), which converts testosterone to DIT. DIT acts through AR and is ~50 times more potent than testosterone.

Table 36.1: Effect of Testosterone

Tissue	Effect
Bone	↑ Growth of bone and connective tissue
Muscle	↑ Growth of muscle and connective tissue
Reproductive organs	↑ Growth and development of testes, prostate, seminal vesicles, and penis
	↑ Growth of facial, axillary, and pubic hair
	↑ Growth of larynx
	↑ Spermatogenesis
Skin	↑ Sebaceous gland size and secretions

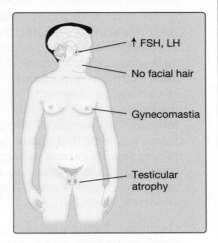
3. **Conversion to estradiol:** Some tissues, including Sertoli cells and the brain and liver, convert testosterone to estradiol, which exerts its own effects on many tissues.

IV. FEMALE GONADS

The two ovaries are located within the pelvic cavity close to the oviducts. Ovaries produce estrogens and progestins under the control of the **hypothalamic–pituitary–ovarian (HPO) axis**. In females of reproductive age, the HPO axis cycles with an average periodicity of ~28 days. The **ovarian cycle** comprises a **follicular phase** and a **luteal phase**, with phase transition occurring midcycle and being marked by **ovulation**.

A. Anatomy

An ovary comprises an outer cortex and inner medulla enclosed within a connective tissue capsule (tunica albuginea). The medulla contains blood vessels, lymphatic vessels, nerves, and loose connective tissue. The outer cortex contains numerous **ovarian follicles** embedded in a cellular connective tissue matrix (Fig. 36.7). Follicles are the functional units of an ovary. The majority of follicles (primordial follicles) comprise an oocyte surrounded by a single layer of flattened pregranulosa cells resting on a basal lamina and surrounded by stroma. Oocytes are formed by oogenesis during fetal development. None are produced after birth. Each month, a cohort of 10 to 30 primordial follicles undergo growth and maturation (**folliculogenesis**).

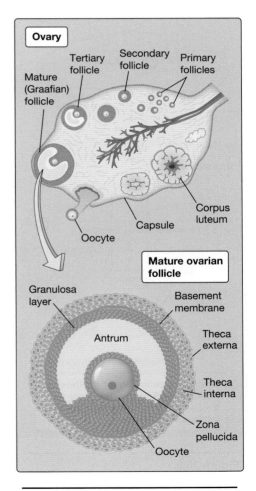

Figure 36.7.
Ovarian anatomy.

1. **Folliculogenesis:** Folliculogenesis successively produces primary, secondary, tertiary, and mature (Graafian) follicles. A primordial follicle requires several months to mature fully, so the cortex normally contains follicles in various stages of development. Maturation prepares the follicle for ovulation. During primordial follicle growth, the oocyte enlarges, and the pregranulosa cells undergo rapid division to form a stratified layer of **granulosa cells** (**stratum granulosum**), which are extensively interconnected via gap junctions. Stromal cells immediately surrounding the follicle also grow and develop into a sheath called the **theca folliculi.** Two functional layers can be distinguished within the theca: the outer **theca externa** is a connective tissue layer containing smooth muscle cells and collagen, and the **theca interna** comprises steroidogenic secretory cells (**theca cells**) supported by a rich vascular network. Theca cells are responsible for ovarian endocrine functions.

2. **Ovulation:** Each month, a single follicle becomes dominant, completes maturation, and releases its oocyte (ovulation). The other cohort members undergo atresia. The empty follicle persists as the **corpus luteum** and serves an important endocrine function in preparing the uterine lining for implantation (see Section V). The follicle undergoes a phenotypic transformation following ovulation, such that theca cells give rise to **theca-lutein cells**, and granulosa cells give rise to **granulosa-lutein cells**. A rich vascular supply supports the endocrine functions of both cell types.

B. Estrogen synthesis and secretion

Estrogen synthesis by theca cells follows a pathway like that described for testicular Leydig cells (see Fig. 36.3). Theca cells take up cholesterol from blood, and they may also synthesize cholesterol as needed. Theca cells do not express the aromatase (CYP19A1) needed to convert androstenedione or testosterone to estrogens, but neighboring granulosa cells do (Fig. 36.8). Granulosa cells, in turn, do not have a ready supply of cholesterol (they are too far removed from their blood supply), and they lack the CYP17A1 theca cells use to synthesize androgens from progesterone, so estrogen formation requires both cell types. Androgens diffuse from theca cells and enter granulosa cells. Here, androstenedione is converted to estrone. Estrone and testosterone are then converted to the more potent estradiol. Estradiol reaches blood by diffusion, where it binds to SHBG (38%) and albumin (60%). About 2% circulates in free form. The liver processes estrogens, and degradation products are excreted in urine.

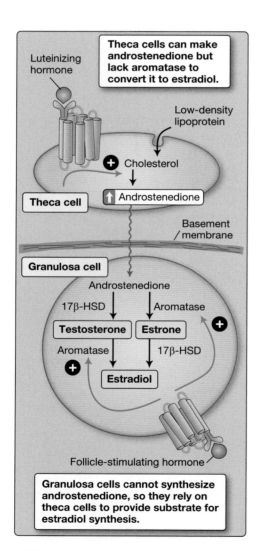

Figure 36.8.
Cooperation between theca and granulosa cells. 17β-HSD = 17β-Hydroxysteroid dehydrogenase.

SHBG has a high affinity for both testosterone and estrogen, but it is a low-capacity carrier protein. Albumin binds androgens with low affinity, but its high concentration in blood confers a very high capacity. In practice, this means that the majority of both hormones are transported in association with albumin.

C. Progestin synthesis and secretion

Relatively small amounts of progesterone and 17α-hydroxyprogesterone are synthesized by and diffuse from theca cells as described earlier. Synthesis is sensitive to the same factors regulating estradiol production. After ovulation, granulosa cells gain a blood and cholesterol supply, at which point they become a significant and biologically important source of progestins. Of the two progestins, progesterone is the more potent form and circulates at higher levels. Progestins are carried in blood in association with CBG (18%) and albumin (80%), with 2% remaining free. Progestins have a half-life of ~5 minutes. The liver processes progesterone similarly to other steroid hormones, and degradation products are excreted in urine.

D. Hypothalamic–pituitary–ovarian axis

The hypothalamic and pituitary components of the HPO axis are identical to that of the HPT axis (see Fig. 36.4). Hypothalamic neurons secrete GnRH pulses, which target anterior pituitary gonadotropes. Gonadotropes secrete LH and FSH, which target ovarian theca cells and granulosa cells. As in the testis, LH and FSH receptors are GPCRs that couple to the cAMP signaling pathway and PKA. Theca cells express LH receptors, whereas granulosa cells express both LH and FSH receptors. As discussed earlier, theca and granulosa cells act cooperatively to produce progestins and estrogens. LH and FSH binding increases production of both steroid hormones. Granulosa cells also produce inhibin A ($\alpha\beta_A$), inhibin B ($\alpha\beta_B$), and activins in response to FSH. Activins are related hormones comprising inhibin β subunit homodimers ($\beta_A\beta_A$, $\beta_A\beta_B$, and $\beta_B\beta_B$).

E. Ovarian cycle regulation

Ovarian hormone production is subject to interrelated feedback loops acting at each level of the HPO axis, involving estrogens, progestins, inhibins, and activins (Fig. 36.9). The HPO axis is unusual in that the feedback relationships between the participants change during the course of the ovarian cycle.

1. **Follicular phase:** The follicular phase begins with the demise of the corpus luteum (Fig. 36.10A). FSH causes a new cohort of large follicles to begin growing rapidly in an LH- and FSH-dependent manner. The gonadotropins also stimulate secretion of estrogen and progesterone, both of which feed back negatively on the hypothalamus and anterior pituitary (see Fig. 36.9A). FSH also stimulates granulosa cells to produce activins and inhibins that either stimulate or inhibit FSH release, respectively.

2. **Ovulation:** The growing follicles release increasingly larger amounts of steroids as they develop. Once estrogen levels reach a certain threshold value, they sensitize gonadotropes to GnRH, causing an increase in the amplitude of pulsatile LH and FSH release. Estrogen also increases the frequency and amplitude of GnRH release from the hypothalamus. In practice, this means that estrogen and progesterone now feed back positively on the HPO axis (see Fig. 36.9B), culminating in a dramatic surge in LH and FSH at around day 13 of the cycle (see Fig. 36.10A). A day later, the follicle ruptures and expels its oocyte (see Fig. 36.10B).

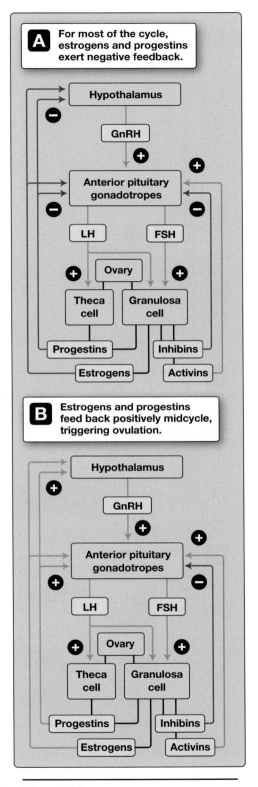

Figure 36.9.
Changes in hypothalamic–pituitary–ovarian axis regulation during the ovarian cycle. FSH = follicle-stimulating hormone, GnRH = gonadotropin-releasing hormone, LH = luteinizing hormone.

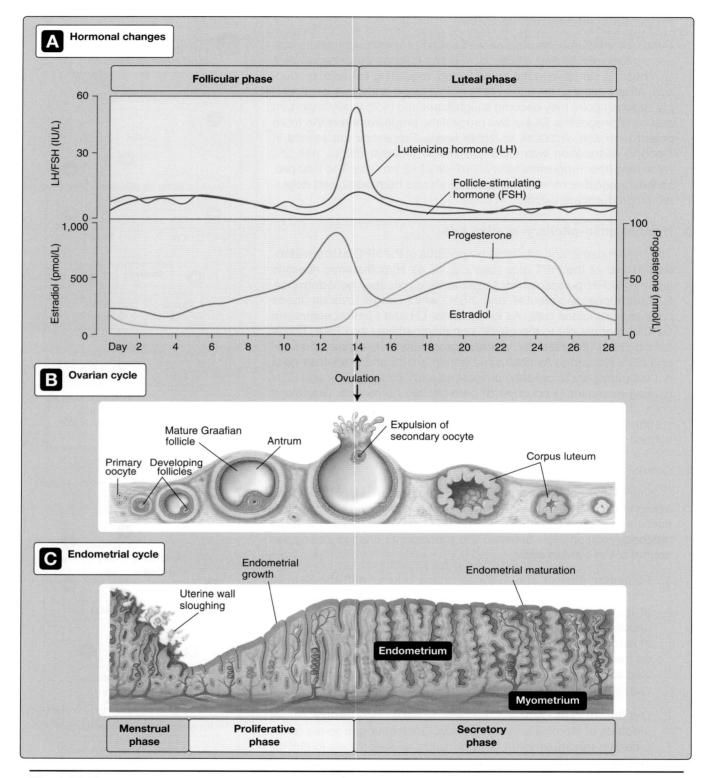

Figure 36.10.
Ovarian and endometrial cycle. FSH = follicle-stimulating hormone, LH = luteinizing hormone.

Clinical Application 36.2: Polycystic Ovary Syndrome

Polycystic ovary syndrome (PCOS) affects 5%–10% of females in their reproductive years. PCOS is a disease of unknown etiology that presents as androgen excess and menstrual irregularities. Most patients with PCOS either have abnormal menstrual cycles or are amenorrheic and infertile. Ovaries may become enlarged with numerous subcapsular cystic follicles. Hormonal alterations in PCOS include an increase in the ratio of luteinizing hormone (LH) to follicle-stimulating hormone (FSH). LH stimulates theca cells, which produce excessive amounts of androstenedione and testosterone. Because FSH levels are not increased simultaneously (and are often lower than normal), granulosa cells are not stimulated to convert the androgen excess to estradiol. Hyperandrogenism presents as hirsutism, acne, and virilization. Other common PCOS symptoms include obesity, insulin resistance, dyslipidemia, coronary heart disease, sleep apnea, and depression.

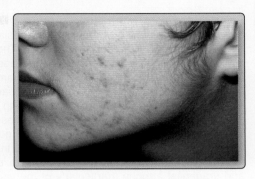

Severe cystic acne and hirsutism.

3. **Luteal phase:** Following ovulation, the HPO axis is again subject to feedback inhibition, and gonadotropin levels fall (see Fig. 36.9A). The corpus luteum secretes estrogen and progesterone in increasing amounts, which decreases the amplitude and frequency of the LH pulses. LH (or a related hormone, such as hCG; see following text) is required for luteal survival, so, as LH levels fall, the corpus luteum declines and regresses (see Fig. 36.10B), forming a nonfunctional scarlike structure (corpus albicans). The corpus albicans slowly migrates deeper into the ovary and is degraded. Corpus luteum regression occurs 10 to 12 days after ovulation. Thus, duration of the luteal phase is fairly constant at ~14 days.

F. Functions

Estrogens have several functions, both genomic and nongenomic. The nongenomic effects are mediated by cell membrane receptors and do not directly affect gene transcription, translation, and protein synthesis. Most of estrogens' effects are genomic. The two classes of **estrogen receptors (ERs)**, ERα and ERβ, function as dimers. ERα is primarily expressed in the reproductive organs, whereas ERβ is primarily expressed in granulosa cells and in nonreproductive organs. ERα and ERβ are cytosolic and nuclear. Once estrogen binds to its receptor, the receptor homodimerizes and binds to an estrogen-response element on DNA, inducing gene expression. Estrogen effects are summarized in Table 36.2. Progesterone effects are mediated by homodimeric progesterone receptors (PRs), which bind to a progesterone-response element on target genes. Progesterone actions are largely restricted to reproductive tissues (Table 36.3; see also Chapter 37·III·C).

V. ENDOMETRIAL CYCLE

The hormonal changes that accompany the ovarian cycle induce structural changes in the uterine lining (endometrium) designed to support implantation and pregnancy. The endometrial cycle is characterized by a proliferative phase, a secretory phase, and menstruation (see Fig. 36.10C).

Table 36.2: Effect of Estrogens

Tissue	Effect
Bone	↑ Growth via osteoblasts
Endocrine	↑ Progesterone responses
Liver	↑ Clotting factors
	↑ Steroid-binding proteins
	↓ Total and low-density lipoprotein
	↑ High-density lipoprotein
Reproductive organs	↑ Uterine growth
	↑ Vaginal and fallopian tube growth
	↑ Breast growth
	↑ Cervical mucus secretion
	↑ Luteinizing hormone receptors on granulosa cells

LDL = low-density lipoprotein, LH = luteinizing hormone, HDL = high-density lipoprotein.

Table 36.3: Effect of Progestins

Tissue	Effect
Breast	↑ Lobular development
	↓ Milk production
Reproductive organs	↓ Endometrial growth
	↑ Endometrial secretions
	Mucosal secretions thicken
Temperature	↑ Internal temperature

Biologic Sex and Aging 36.1: Menopause

Females are born with a fixed number of follicles that decline steadily throughout life because of progressive atresia. Follicle depletion is associated with ovarian failure and **menopause**. The mean age of menopause is 51 years, although the normal menopausal age range is wide (42–60 years). Menopause can occur prematurely, if, for example, the ovaries are removed or develop functional abnormalities. Menopause is typically preceded by a transition period (**perimenopause**) lasting several years, during which time the menstrual cycle slows and becomes irregular. Follicular depletion and loss of estrogens lead to increases in gonadotropin levels (especially follicle-stimulating hormone) and a decrease in inhibins. Typical symptoms of menopause include hot flashes, night sweats, decreased vaginal secretions, and urogenital atrophy (particularly of the vaginal epithelium and ovaries) because of the lower female gonadal hormone levels. Some women develop areas of skin hyperpigmentation (melasma). Later changes include a net decrease in bone mineral density and an increase in cholesterol, which both

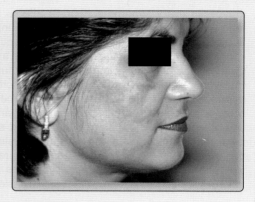

Melasma.

increase the risk of osteoporosis and bone fractures (see Clinical Application 14.3). Hormone replacement therapy can alleviate many of these risks but also carries with it a marginally increased risk of venous thrombosis and certain cancers that increases significantly with age postmenopause.[1]

A. Proliferative phase

Rising estrogen levels associated with follicular growth stimulate hyperplasia of all components of the endometrial stratum functionalis, causing it to increase in thickness from an initial 1 to 2 mm to 8 to 10 mm by phase end. Estrogen also stimulates expression of PRs, which primes the endometrium to respond to progesterone during the secretory phase.

B. Secretory phase

After ovulation, the corpus luteum secretes high levels of progesterone, which has several physiologic effects. First, it downregulates endometrial ERs and inactivates estradiol, thereby halting endometrial proliferation. Progesterone also markedly stimulates endometrial glands, increasing their activity and causing them to become tortuous and engorged (see Fig. 36.10C). Progesterone also promotes spiral artery development to supply endometrial activity. Finally, the thermogenic effects of progesterone cause body temperature to rise by between 0.22° and 0.5°C.

C. Menstruation

If conception does not occur, and the corpus luteum is not rescued by hCG produced by the developing embryo, the stratum functionalis is shed. Menstruation begins with a pronounced prostaglandin-mediated vasoconstriction of spiral arteries, which causes local ischemic injury. Inflammatory cells infiltrate the area and cause further breakdown of the lining (see Fig. 36.10C). Fibrinolysins prevent clot formation and maintain bleeding until the lining is sloughed off the uterine wall.

[1]For additional information on hormone replacement therapy, see *LIR Pharmacology*, 8e, Chapter 25·II·B·2.

VI. MAMMARY GLANDS

The human breast consists of a mammary gland, adipose tissue, and a ductal system that channels gland secretions to a nipple. A mammary gland comprises ~20 secretory lobules, each containing clusters of secretory epithelium-lined alveoli. The alveoli secrete colostrum and milk, which provides optimal nourishment for neonates. Milk's principal components are protein (casein, lactalbumin), fats, lactose, Ca^{2+}, PO_4^{3-}, other electrolytes, and water. Breast growth and development occurs during puberty in response to female gonadal hormones coordinating the acquisition of secondary sex characteristics. During pregnancy, a breast develops more fully in response to high levels of estrogens, progestins, hCG from the fetal placenta, and **prolactin** (**PRL**). PRL also controls milk production. Suckling requires milk "let-down" and ejection, a process mediated by **oxytocin** (Fig. 36.11).

A. Prolactin

PRL is a peptide hormone produced and secreted by anterior pituitary lactotropes. Unlike other anterior pituitary hormones, it is not associated with a hormone axis. It is secreted in both males and females. During pregnancy, estrogen promotes lactotrope proliferation and hypertrophy, which increases PRL secretion.

1. **Secretion:** PRL secretion is tonically suppressed by **dopamine** emanating from the hypothalamic paraventricular and arcuate nuclei. Suckling is a primary stimulus for PRL release, mediated by touch-sensitive neurons in the nipple and inhibition of hypothalamic dopamine release (see Fig. 36.11). Numerous factors enhance PRL secretion rates, including estrogen, oxytocin, thyrotropin-releasing hormone, sleep, and stress. Somatostatin and growth hormone decrease PRL secretion. These changes in PRL secretion occur either through direct actions on lactotropes or via inhibition of the hypothalamic dopaminergic neurons. PRL is not associated with a blood hormone-binding protein and has a half-life of ~20 minutes.

2. **Function:** In females, PRL stimulates mammary development and ductal proliferation, and it initiates and sustains milk production. PRL effects are mediated by a cytokine cell membrane receptor that stimulates the Janus kinase/signal transducer and activator of transcription (JAK/STAT) signaling pathway. The role of PRL in males is not well understood.

B. Oxytocin

Oxytocin is a small peptide hormone produced in magnocellular neurons of the paraventricular and supraoptic nuclei of the hypothalamus and secreted from the posterior pituitary (see Chapter 7·VII·D·4). Once in the blood, oxytocin has a very short half-life (3–5 minutes).

1. **Secretion:** Cervical stretch and breast suckling stimulate oxytocin secretion through neuroendocrine reflexes (see Fig. 36.11). Strong or intense emotional stimuli (e.g., fear and pain) can decrease blood oxytocin levels. Oxytocin actions may also be modulated through up- or downregulation of oxytocin receptors.

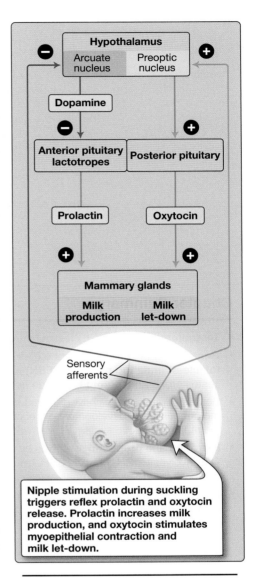

Nipple stimulation during suckling triggers reflex prolactin and oxytocin release. Prolactin increases milk production, and oxytocin stimulates myoepithelial contraction and milk let-down.

Figure 36.11.
Hormonal regulation of lactation.

Estrogen, for example, dramatically increases oxytocin receptor expression during pregnancy, which, in turn, potentiates oxytocin efficacy.

2. **Function:** Oxytocin has two primary functions in females. First, it causes myoepithelial cells surrounding milk-producing alveoli to contract, thereby forcing milk flow through the ductal system (milk let-down). Secondly, it stimulates uterine muscle contraction during parturition (see Chapter 37·VI·B·3). Secondary functions in females include reinforcing the need to care for the infant, stimulating PRL release, and decreasing nociception. The effects of oxytocin on the breast, uterus, and central nervous system are mediated by cell membrane oxytocin receptors (GPCR superfamily members) and work primarily through the IP_3 signaling pathway.

Chapter Summary

- **Testosterone** synthesis by testicular Leydig cells is regulated by the hypothalamic–pituitary–testicular axis.
- Gonadotropin-releasing hormone (GnRH) released from the hypothalamus stimulates **luteinizing hormone** (**LH**) and **follicle-stimulating hormone** (**FSH**) released from anterior pituitary **gonadotropes**. LH and FSH target the testes. Testosterone and **inhibin B** provide negative feedback at the level of the anterior pituitary.
- Some peripheral cells convert testosterone to **5α-dihydrotestosterone** (**DHT**). Testosterone and DHT regulate spermatogenesis and development of secondary sex characteristics, increase fat deposition, and have anabolic effects on the musculoskeletal system.
- **Estrogen** and **progestin** synthesis by the ovaries is regulated by the **hypothalamic–pituitary–ovarian axis**, which also involves GnRH, LH, and FSH. LH and FSH target the ovaries. Estrogens, progestins, and inhibins provide negative feedback for the axis, whereas activins provide positive feedback at the level of the anterior pituitary.
- FSH and LH control folliculogenesis. The follicular phase of the monthly **ovarian cycle** is associated with follicular maturation and culminates in an LH and FSH surge that triggers **ovulation**. Progesterone secretion by the corpus luteum during the luteal phase facilitates implantation of a fertilized ovum.
- Estrogen production involves cooperation between **granulosa** and **theca cells**. Both produce progesterone, but only theca cells can process this into **androstenedione**. Androstenedione and testosterone migrate to granulosa cells, where they are converted into estradiol and secreted.
- Estrogens stimulate growth of the female reproductive organs and associated structures.
- Progestins aid in the conversion of endometrial cycling from a proliferative phase to a secretory phase. During pregnancy, progestins are involved in preparing mammary glands for lactation.

Study Questions

Choose the ONE best answer.

36.1. A 21-year-old female comes to the physician concerned that she has not yet started menstruating. Physical examination reveals a short, blind-ended vagina, an absent uterus, and testes in the inguinal canal. Genetic analysis reveals a 46,XY karyotype and a loss-of-function mutation in the androgen receptor gene. Laboratory studies would most likely reveal which of the following sets of testosterone, luteinizing hormone (LH), and gonadotropin-releasing hormone (GnRH) findings?

A. ↓ Testosterone, ↑ LH, ↑ GnRH
B. ↓ Testosterone, ↓ LH, ↑ GnRH
C. ↓ Testosterone, ↓ LH, ↓ GnRH
D. ↑ Testosterone, ↓ LH, ↓ GnRH
E. ↑ Testosterone, ↑ LH, ↑ GnRH

Best answer = E. The patient is a genotypic male who has developed as a phenotypic female due to congenital androgen-insensitivity syndrome (CAIS). An inability to sense and respond to testosterone leads to loss of normal feedback inhibition of the hypothalamic–pituitary–testis (HPT) endocrine axis (see Section III·C), so testosterone levels rise, along with LH and GnRH levels. Testosterone levels decrease (A, B, C) due to primary testicular failure or to pituitary or hypothalamic dysfunction, respectively. A rise in testosterone levels combined with decreased LH and GnRH (D) suggests an exogenous source of testosterone is suppressing the HPT axis.

36.2. A 16-year-old male has a 5α-reductase deficiency. This individual was raised as a female, but, at puberty, male secondary sex characteristics emerged, and male genital growth occurred. Levels of which of the following steroids are most likely to be reduced because of this deficiency until overwhelmed by testosterone during puberty?

A. Androstenedione
B. Dihydrotestosterone
C. Estradiol
D. Estrone
E. Progesterone

Best answer = B. 5α-Reductase is an enzyme normally found in several tissues that converts testosterone to dihydrotestosterone ([DHT] see Section III·B). Therefore, a 5α-reductase deficiency would result in reduced DHT levels. DHT mediates many androgen effects, so individuals with a 5α-reductase deficiency do not express many male secondary sex characteristics until puberty, when testosterone levels dramatically increase. Androstenedione (A) is a substrate for HSD17B3, an enzyme involved in testosterone synthesis (see Fig. 36.3). This reaction, however, is located upstream of 5α-reductase, which means that androstenedione levels would not be actively reduced. Estradiol (C), estrone (D), and progesterone (E) are all gonadal hormones, but none are directly associated with male secondary sex characteristics.

36.3. A 32-year-old female comes to the physician because of headaches and worsening visual deficits. She also complains of a milky discharge from her nipples. A pregnancy test is negative. A head MRI reveals a pituitary tumor pressing on the optic chiasm. The tumor is most likely secreting which of the following hormones?

A. Follicle-stimulating hormone
B. Gonadotropin-releasing hormone
C. Luteinizing hormone
D. Oxytocin
E. Prolactin

Best answer = E. The patient has a pituitary adenoma that is secreting prolactin (PRL) in an unregulated manner. Such tumors can cause visual deficits when they compress the optic nerves. PRL is synthesized and secreted by lactotropes that are normally under tonic inhibition by dopamine from the hypothalamus (see Section VI·A). PRL stimulates mammary gland development and secretory activity. Follicle-stimulating hormone (A) and luteinizing hormone (C) regulate the ovarian cycle and ovulation (see Section IV·E). Gonadotropin-releasing hormone (B) is a hypothalamic hormone that targets pituitary gonadotropes. Oxytocin (D) is a posterior pituitary hormone that stimulates milk let-down during breast suckling and uterine contractions during parturition (see Section VI·B).

36.4. A 20-year-old female who was administered a gonadotrope-stimulating drug responded with an increase in plasma luteinizing hormone (LH) levels, but follicle-stimulating hormone (FSH) levels remained low. Levels of which of the following hormones would most likely be unaffected by such a drug?

A. Androstenedione
B. DHEA
C. Estradiol
D. Progesterone
E. Testosterone

Best answer = C. Gonadotropes, which are located in the anterior pituitary, normally respond to a stimulating hormone by releasing both FSH and LH (see Section IV·D). FSH stimulates aromatase in granulosa cells to convert androgens from surrounding theca cells into estrogens, including estradiol (see Section IV·B). If FSH release does not occur, estradiol levels would remain low also. LH stimulates side-chain cleavage-enzyme complex activity and increases production of androstenedione (A), progesterone (D), and testosterone (E). DHEA (B) is primarily an adrenal androgen that is not directly stimulated by LH or FSH, although small increases could occur in the gonads.

36.5. A 32-year-old female with a history of type 1 diabetes presents to the emergency room with confusion, irritability, and diaphoresis. She reports having skipped a meal after taking insulin and then engaging in strenuous physical activity. Laboratory studies reveal a blood glucose of 50 mg/dL (2.8 mmol/L). Which of the following processes would most likely be increased in this patient?

A. Glycogenesis
B. Glycolysis
C. Ketogenesis
D. Lipogenesis
E. Protein synthesis

Best answer = C. The patient presents with signs of sympathetic activation and neuroglycopenia caused by low blood glucose levels (see Chapter 33·II·B). Hypoglycemia stimulates glucagon secretion from pancreatic α cells (see Chapter 33·III·C). Adipocytes respond to glucagon by releasing fatty acids, which hepatocytes then take up and use in ketogenesis. Lipogenesis (D) is simultaneously suppressed by glucagon. Glucagon also suppresses glycogenesis (A) in the liver and muscle and glucose breakdown (B) in the liver, muscle, and adipose tissue. Muscle also responds to glucagon with reduced protein synthesis (E).

36.6. A 48-year-old male presents to the office with muscle weakness and frequent headaches for several weeks. He has a history of persistent hypertension. His temperature is 37°C (98.6°F), pulse is 75 beats/min, respirations are 18 breaths/min, and his blood pressure is 160/100 mm Hg. Laboratory studies reveal elevated aldosterone and decreased plasma renin activity. Which pair of plasma sodium and potassium levels is most likely to be noted in this patient?

A. Sodium: 125 mEq/L, potassium: 4.5 mEq/L
B. Sodium: 132 mEq/L, potassium: 5.3 mEq/L
C. Sodium: 140 mEq/L, potassium: 3.5 mEq/L
D. Sodium: 142 mEq/L, potassium: 4.5 mEq/L
E. Sodium: 146 mEq/L, potassium: 3.2 mEq/L

Best answer = E. The patient has hyperaldosteronism. Aldosterone is a mineralocorticoid that promotes renal Na^+ retention and K^+ excretion (see Chapter 34·IV). The patient's Na^+ is high, causing hypertension. Hypokalemia leads to muscle weakness. Low Na^+ and normal K^+ values (A) are consistent with excessive water retention and hypervolemia. Low Na^+ and high K^+ values (B) are seen in patients with primary adrenal insufficiency. Low plasma K^+ in a setting of normal Na^+ (C) would be consistent with metabolic alkalosis. Values of 142 mEq/L Na^+ and 4.5 mEq/L K^+ (D) are both within a normal range.

36.7. A 42-year-old male comes to the office with numbness and tingling around his lips and in his fingers, muscle cramps, and twitching. A history reveals recent thyroid surgery. Laboratory studies are notable for low serum calcium and parathyroid hormone (PTH) levels. Which of the following would most likely be increased in this patient?

A. Osteoclast activity
B. Plasma phosphate
C. Plasma vitamin D_3
D. PTH secretion
E. Renal Ca^{2+} retention

Best answer = B. The patient presents with symptoms of hypocalcemia (paresthesias and increased muscle excitability) and decreased PTH levels following thyroid surgery, suggesting a diagnosis of hypoparathyroidism. Damage to or excision of the parathyroid glands is a common complication of thyroid surgery. PTH normally promotes renal phosphate wasting (see Chapter 26·VI), so plasma phosphate would be increased in this patient. PTH also increases osteoclast activity (A) and renal Ca^{2+} retention (E), so both would be reduced. Vitamin D_3 levels (C) are increased by skin exposure to the sun. The patient has damaged or absent parathyroid glands, so PTH secretion (D) would be decreased.

Pregnancy and Birth

37

I. OVERVIEW

Pregnancy and birth are exceptional phenomena that place extreme demands on both mother and fetus. Although the likelihood of success seems improbable once the complexity of the underlying physiology is appreciated, the current global population of more than 8 billion shows it to be a highly reliable way of perpetuating the species. A successful pregnancy requires that several challenges be met. After fertilization, the developing embryo must implant in the uterine endometrium. The placenta must assume hormonal control of uterine growth to create an environment that allows the fetus to develop undisturbed for the next several months. An interface must be established between maternal and fetal circulations that allows for exchange of nutrients and waste products. The mother's body must adapt to meet the needs of the growing fetus. Finally, at term, the link between mother and fetus must be broken in a manner that allows both individuals to survive and thrive. Pregnancy begins with fertilization of the ovum and ends at **parturition** (childbirth). Pinpointing the moment of fertilization is typically difficult, so the progression of pregnancy is usually measured with reference to the first day of a woman's last menstrual period. By this measure, pregnancy lasts approximately 40 weeks.

II. IMPLANTATION

Fertilization typically occurs within the **fallopian tube** ampulla (Fig. 37.1). The fallopian tube is lined with motile cilia that sweep the newly fertilized ovum downward toward the uterine cavity. The embryo remains free in the mother's reproductive tract for 6 or 7 days, during which time it undergoes a series of rapid divisions to form a **blastocyst** with a fluid-filled cavity (**blastocoele**) at its center. A thin layer of **trophoblast** cells around the central cavity's outer edge ultimately becomes the **placenta** and the membranes that enclose and protect the developing embryo. The embryo develops from an inner cell mass. By the time the blastocyst is ready to attach to and invade the uterine wall, the endometrium has been readied for implantation (**predecidualization**) under the influence of progesterone from the **corpus luteum**. During implantation,

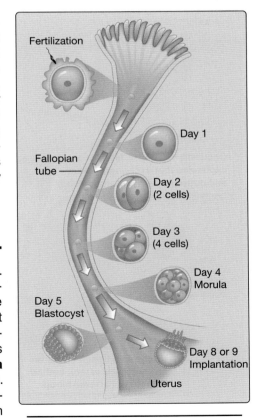

Figure 37.1.
Embryo development and implantation.

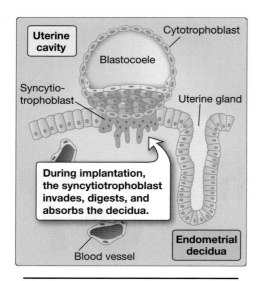

Figure 37.2.
Implantation.

the syncytiotrophoblast enzymatically digests and invades the maternal uterine endometrium (Fig. 37.2). **Decidual cells** within the endometrium sustain the embryo with glycogen and other nutrients until the placenta is formed and functional. Erosion of capillaries by the invading syncytiotrophoblast allows blood to escape the maternal vasculature. Small pools (**lacunae**) ultimately coalesce to form a lake of maternal blood that fills the space between maternal and fetal placenta.

III. PLACENTA

The placenta is a disk-shaped organ that forms an interface between fetus and mother (Fig. 37.3). The placenta has several important functions. It anchors the fetus to the uterus. It brings fetal and maternal blood into close apposition to facilitate exchange of materials between the two circulations and to allow for hormonal crosstalk between mother and fetus. It is an endocrine organ that manipulates maternal reproductive physiology to sustain the pregnancy. Finally, it functions as a barrier that prevents maternal and fetal blood from mixing.

A. Structure

The placenta comprises a **fetal placenta** and a **maternal placenta**.

1. **Fetal:** The fetal placenta is attached to the fetus by the **umbilical cord**, a ropelike, muscular tether containing two **umbilical arteries** and an umbilical vein (see Fig. 37.3). O_2-poor blood is carried

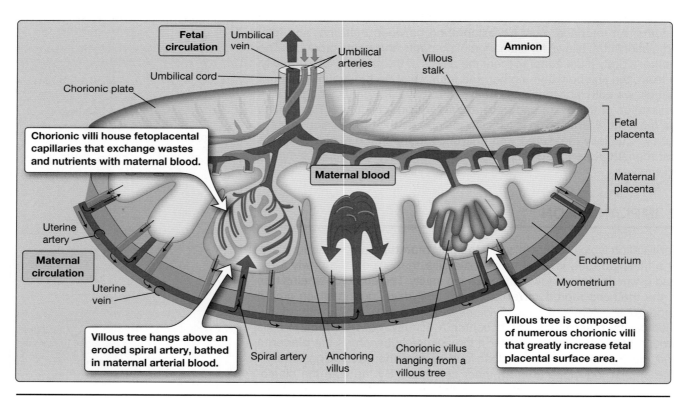

Figure 37.3.
Placental anatomy and blood flow.

from the fetus to the placenta by umbilical arteries. They penetrate the placental **chorionic plate** and then branch and distribute arterial blood to 60 to 70 **villous trees** assembled in groups called **cotyledons** (15–20 per placenta). Villous trees are branching structures covered with **chorionic villi** (see Fig. 37.3). Villi contain fetoplacental capillaries and are the primary site of exchange between maternal and fetal circulations. O_2^- and nutrient-rich blood is carried from the villi to the fetus by the umbilical vein. Some villi are structural. They are physically attached to the maternal placenta and serve as placental anchors (see Fig. 37.3).

2. **Maternal:** The maternal **placental bed** (area below the fetal placenta) resembles an egg carton. The endometrium is hollowed out to form an array of blood-filled sinuses in which the villous trees hang (see Fig. 37.3). The intervillous space between the fetal placenta and endometrium is filled with ~500 mL of maternal blood. Blood enters this space via **uteroplacental** vessels, which are the remnants of spiral arteries that have been eroded by the fetal syncytiotrophoblast layer during implantation and placental development. Blood drains from the space via uterine veins located in the floor of the maternal placental bed (see Fig. 37.3).

B. Exchange

Everything the fetus needs to develop and grow must cross the fetoplacental barrier separating maternal and fetal circulations. Most materials cross by simple or facilitated diffusion, driven by concentration gradients. Small amounts of material may cross by transcytosis (Fig. 37.4). Three features of placental design optimize transfer: the minimal nature of the barrier, its large surface area, and the positioning of villous trees above maternal blood vessels.

1. **Barrier:** Blood flowing through the maternal placenta is outside the usual confines of the maternal vasculature. In practice, this means that the barrier between maternal and fetal blood comprises a single endothelial cell layer (fetal capillary wall), a thin layer of connective tissue (basal lamina), and a thin layer of **syncytiotrophoblast**. The fetal capillary wall and basal lamina are relatively leaky, so the syncytiotrophoblast forms the barrier proper. By the end of gestation ("**term**"), the fetoplacental barrier has thinned to <5 μm to maximize diffusional exchange between mother and fetus.

2. **Surface area:** Fetal villi hang from the villous trees like bunches of bananas (see Fig. 37.3). The syncytiotrophoblast's apical (maternal) surface is densely packed with **microvilli**, which amplify the surface area available for diffusion. By term, total villous surface area amounts to ~10 to 12 m².

3. **Villi:** Villi develop directly above and around plumes of blood spewing from an eroded maternal artery (see Fig. 37.3), such that they are constantly doused with arterial blood. Constancy of flow maintains the steep concentration gradients that drive diffusional exchange of nutrients and waste products between maternal and fetal blood.

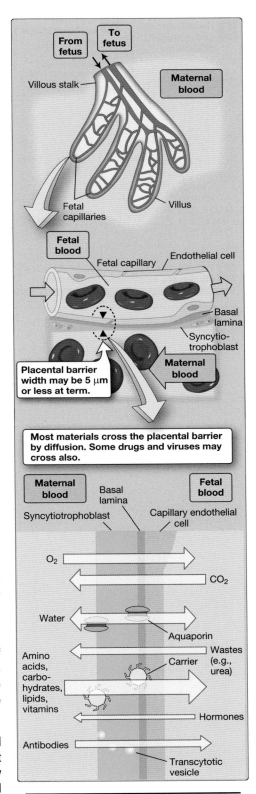

Figure 37.4.
Exchange across the placental barrier.

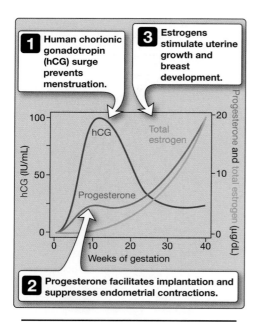

Figure 37.5.
Placental hormones.

C. Endocrine functions

A nonpregnant female's uterus sheds its lining (outer endometrial layer) every 4 weeks and then begins the **menstrual cycle** anew, the cycle's timing being controlled by female reproductive hormones (see Chapter 36·V). A successful pregnancy requires that the menstrual cycle be interrupted and the fetus left undisturbed for ~9 months. Cycle interruption is accomplished by the fetal placenta, which secretes several key hormones that manipulate maternal reproductive physiology, including **human chorionic gonadotropin** (**hCG**), **progesterone**, and **estrogens** (Fig. 37.5).

1. **Human chorionic gonadotropin:** The syncytiotrophoblast (fetoplacental precursor) begins secreting hCG within days of fertilization. hCG signals the corpus luteum that fertilization has occurred and impels it to sustain progesterone and estrogen production. These hormones prevent the uterus from shedding its lining and prepare it for implantation. The corpus luteum continues releasing progesterone and estrogen at ever-increasing levels in response to placental hCG, until the placenta takes over hormonal control around week 10 (see Fig. 37.5).

> The rapid rise in hCG that follows fertilization provides the basis for home pregnancy tests, which detect the presence of hCG in maternal urine. Test strips typically have a detection threshold of 25–50 mIU/mL, levels that are not achieved until after implantation has occurred (up to 10 days after ovulation).

2. **Progesterone:** The placental syncytiotrophoblast produces large amounts of progesterone (see Fig. 37.5). The hormone initially helps prepare the endometrium for implantation. Progesterone also reduces myometrial excitability, thereby preventing contractions that might expel the developing embryo. Progesterone also stimulates breast development.

3. **Estrogens:** The placenta produces several estrogens (see Fig. 37.5), the main one being **estriol**. Estrogens stimulate growth and development of the mother's uterus and breasts. The placenta does not have all the necessary substrates (e.g., cholesterol) and enzymes (e.g., 17α-hydroxylase) to synthesize steroids (see Fig. 36.3), relying on both the fetus and mother to provide pathway intermediates.

IV. MATERNAL PHYSIOLOGY

The fetus relies on the mother to supply it with O_2 and nutrients and to dispose of CO_2, heat, and other metabolic waste products. Meeting fetal demands taxes all maternal organ systems and puts the maternal cardiovascular (CV) system under considerable stress. By term, maternal cardiac output (CO) and circulating blood volume have risen by 40% to 50%

(2 L). Much of this capacity increase is needed to perfuse the maternal placenta, but flow to the skin, kidneys, liver, and gastrointestinal (GI) tract increases substantially also.

A. Uterine blood flow

A nonpregnant female's uterus receives <5% of total CO. The main source of uterine vascular resistance is the highly muscular **spiral arteries** (Fig. 37.6), which are resistance vessels that modulate uterine blood flow in response to changing metabolic needs (i.e., **autoregulation**; see Chapter 19·II·B·1). The developing fetal placenta erodes and invades the spiral arteries (see Fig. 37.6B). Arterial walls are remodeled, the smooth muscle layers being replaced with fibrous material to create wide tortuous vessels with very high flow rates. The functional advantages of this remodeling to the fetus are obvious. Blood now pulses directly from the uterine supply arteries at a pressure of >70 mm Hg and washes over the fetoplacental villi, bringing with it needed O_2 and nutrients (see Fig. 37.6C). The CV consequences for the mother are profound, both in terms of flow increase and inability to control flow through these vessels.

1. **Flow:** Resistance vessels are flow regulators that limit the amount of blood a tissue receives to match its prevailing metabolic needs (see Chapter 19·II). Spiral artery erosion and widening allows blood to flow unhindered into the placental lake, so overall uterine blood flow increases dramatically during pregnancy. Flow rises in direct proportion to falling resistance, increasing from ~200 mL/min at 10 weeks' gestation to 600–700 mL/min at term.

2. **Regulation:** Eradication of maternal resistance vessels maximizes flow to the placental site but simultaneously limits the ability of the uterus to control blood flow. Thus, flow becomes a direct function of uterine arterial pressure, as predicted by the hemodynamic

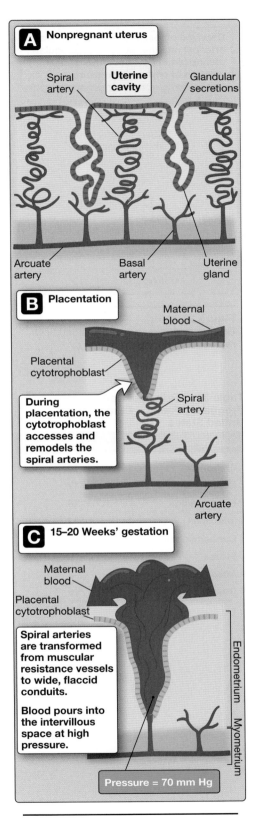

Figure 37.6.
Erosion and invasion of spiral arteries during placentation.

A Nonpregnant uterus

Spiral artery · Uterine cavity · Glandular secretions

Arcuate artery · Basal artery · Uterine gland

B Placentation

Maternal blood

Placental cytotrophoblast

During placentation, the cytotrophoblast accesses and remodels the spiral arteries.

Spiral artery

Arcuate artery

C 15–20 Weeks' gestation

Maternal blood

Placental cytotrophoblast

Spiral arteries are transformed from muscular resistance vessels to wide, flaccid conduits.

Blood pours into the intervillous space at high pressure.

Endometrium

Myometrium

Pressure = 70 mm Hg

Clinical Application 37.1: Preeclampsia

Preeclampsia is a syndrome characterized by new-onset hypertension (systolic blood pressure [SBP] of ≥140 mm Hg or diastolic blood pressure of ≥90 mm Hg) and proteinuria (≥0.3 g/24 hr) that develops after 20 weeks' gestation. Other symptoms may include severe headaches, visual disturbances, epigastric pain, and abnormal liver function. These symptoms reflect a generalized endothelial dysfunction that increases vascular tone and permeability and is associated with a coagulopathy affecting all organs, including the brain, kidneys, liver, and placenta. Although the underlying molecular mechanisms are not yet well defined, preeclampsia is believed to result from incomplete remodeling of the spiral arteries during placental development. In practice, this means that blood flow to the placenta is suboptimal. The placenta responds to hypoperfusion by releasing factors that inhibit angiogenesis and disrupt normal maternal endothelial function. **Severe preeclampsia** (SBP ≥160 mm Hg and/or DBP ≥110 mm Hg) carries a significant risk of maternal seizure, stroke, and death. Immediate delivery regardless of gestational age is generally indicated and normally resolves pressure issues.

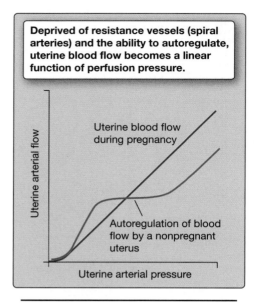

Deprived of resistance vessels (spiral arteries) and the ability to autoregulate, uterine blood flow becomes a linear function of perfusion pressure.

Uterine blood flow during pregnancy

Autoregulation of blood flow by a nonpregnant uterus

Uterine arterial flow

Uterine arterial pressure

Figure 37.7.
Uterine perfusion pressure–blood flow relationship during pregnancy.

Table 37.1: Principal Causes of Pregnancy-Related Deaths in the United States

Cause	% of Total
Embolism	20
Hemorrhage	17
Hypertension	16
Infection	13
Cardiomyopathy	8
Stroke	5
Anesthesia	2
Other	19

equivalent of the Ohm law (Flow = Pressure ÷ Resistance) as shown in Figure 37.7 (also see Chapter 18·IV and Fig. 19.4).

Loss of uterine blood flow control mechanisms puts the mother in grave risk of massive blood loss in the event of premature placental detachment. Hemorrhage is a leading cause of pregnancy-related death in the United States (Table 37.1).

B. Hemodynamic profile

The uterus is a systemic vascular bed, so when uterine vascular resistance falls, systemic vascular resistance (SVR) falls along with it (Fig. 37.8). A fall in SVR causes CO to rise to maintain mean arterial pressure (MAP): MAP = CO × SVR (see Chapter 17·IV·A).

1. **Systemic vascular resistance:** SVR falls steadily over the first 20 weeks of gestation due, in part, to widespread peripheral vasodilation caused by a rise in circulating prostaglandin and relaxin levels. Relaxin stimulates nitric oxide production by the vascular endothelium. Continuing erosion of maternal resistance vessels by the fetal placenta contributes to the decline in SVR, as does the growing need to dissipate heat and eliminate fetal waste products that is facilitated by a decrease in the resistance of the cutaneous and renal vascular beds.

2. **Cardiac output:** The growing need for increased CO is met by increasing stroke volume (SV) and heart rate (HR; see Fig. 37.8). HR rises slowly during pregnancy, averaging 15 to 20 beats/min higher compared with nonpregnant values by 32 weeks. SV begins to rise very early in pregnancy, mediated by an increase in preload and contractility.

   a. **Preload:** The body responds to a sustained need for increased CO by increasing circulating blood volume through Na^+ and water retention. Placental hormones potentiate this effect by stimulating thirst and activating the renin–angiotensin–aldosterone system ([RAAS] see Chapter 19·IV).

   b. **Contractility:** Sustained increases in CO also stimulate eccentric ventricular hypertrophy. The heart enlarges to accommodate increased end-diastolic volumes (preload), ventricular mass rises, and overall contractility increases.

3. **Mean arterial pressure:** MAP must be maintained at prepregnancy levels to ensure adequate perfusion of all vascular beds (see Fig. 37.8), but the introduction of a low-resistance pathway into the maternal vascular circuit (i.e., the placenta) means that blood escapes the arterial system more easily during diastole (increased **diastolic runoff**; see Chapter 18·V·C·2) compared with the nonpregnant state. Thus, diastolic blood pressure falls during pregnancy, and **pulse pressure** widens.

C. Physiologic anemia

Increased Na^+ and water retention during pregnancy causes maternal plasma volume to increase by 40% to 50% (Fig. 37.9). Red blood cell (RBC) production does not keep pace with the rapid expansion of blood volume, increasing by only 25% to 35%. The gap between volume expansion and RBC production results in a **physiologic anemia of pregnancy**, which is a dilutional anemia. Although anemia reduces total O_2-carrying capacity, there are clear physiologic benefits because it reduces blood viscosity, which, in turn, reduces shear stress. It can also cause benign murmurs.

1. **Shear stress:** Blood moves through the maternal arteries and veins at high velocity to support the sustained increases in CO that accompany pregnancy. High-velocity flow increases **shear stress** on the vascular lining, to the point where it could become damaging. Shear stress is proportional to both blood velocity and viscosity (**Reynolds equation**; see Chapter 18·V·A). Because hematocrit is the primary determinant of blood viscosity, anemia reduces stress levels and lessens the risk of vascular endothelial damage.

2. **Murmurs:** One benign consequence of decreased blood viscosity is an increased tendency for turbulent blood flow. The Reynolds equation predicts that turbulence is most likely to occur in regions of the CV system where flow velocities are highest. In practice, this means that mothers often develop **functional** (i.e., innocent) **murmurs** associated with blood ejection through the aortic and pulmonary valves. They can also develop a **venous hum**, a sound associated with high-velocity, turbulent blood flow through the larger veins.

D. Edema

The combined weight of the uterus and its contents (fetus, placenta, and amniotic fluid = ~8–10 kg total at term) compresses and retards flow through the inferior vena cava and other smaller veins returning blood from the lower extremities. Compression causes venous pressures in the lower extremities to rise, which increases both mean capillary hydrostatic pressure and net fluid filtration from blood to the interstitium (see Chapter 18·VII). The resulting edema and swelling of the feet (**pedal edema**) and ankles is common in pregnant women. The tendency for edema formation is increased by a fall in **plasma colloid osmotic pressure** (i.e., plasma protein concentration) by 30% to 40% during pregnancy (from ~25 mm Hg prior to pregnancy to ~15 mm Hg postpartum).

E. Respiratory system

The O_2 demands of the mother and growing fetus increase rapidly during pregnancy; O_2 consumption at term is increased ~30% over nonpregnant values. These increased needs are met by a progressive increase in minute ventilation to ~50% over nonpregnant values during the second trimester. The ventilation increase is accomplished largely by an increase in tidal volume and only a small rise in respiratory rate (2–3 breaths/min). The net effect is that arterial

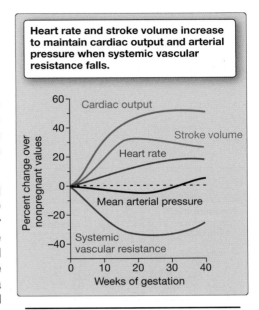

Heart rate and stroke volume increase to maintain cardiac output and arterial pressure when systemic vascular resistance falls.

Figure 37.8.
Changes in maternal hemodynamic profile during pregnancy.

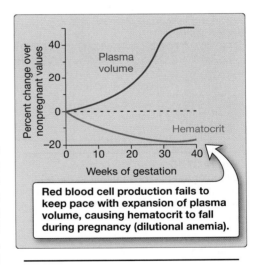

Red blood cell production fails to keep pace with expansion of plasma volume, causing hematocrit to fall during pregnancy (dilutional anemia).

Figure 37.9.
Physiologic anemia of pregnancy.

P_{O_2} (P_aO_2) rises by ~10 mm Hg, and arterial P_{CO_2} (P_aCO_2) falls by ~8 mm Hg, causing a slight respiratory alkalosis (<0.1 pH units). Other significant respiratory changes include a 20% decrease in functional residual capacity, expiratory reserve volume, and residual volume (see Chapter 21·IX) caused by a rise in the diaphragm, which may limit the mother's ability to compensate for increased O_2 demand during exercise, for example.

F. Renal

Glomerular filtration rate rises steadily to ~50% above normal values at 16 weeks' gestation and remains elevated until parturition. The increase reflects a mother's need to excrete fetal wastes, including urea and nonvolatile acid.

V. FETAL PHYSIOLOGY

Because a fetus receives everything it needs for successful development from the maternal circulation, few fetal organ systems are *required* to support normal growth, although most gain a degree of functionality before birth. The principal exception is the CV system, which becomes functional very early in pregnancy.

A. Vasculature

During initial development, the embryo relies on simple diffusion to obtain nutrients from fallopian and other maternal secretions. Once the embryo attains a size that exceeds the ability of O_2 and other nutrients to reach the innermost cells by diffusion alone, a functional CV system is required to sustain further growth. A rudimentary single-chambered heart begins pumping blood resembling interstitial fluid during the fourth week after conception. In the adult circulation, the path that blood follows is dictated by a need to pick up O_2 from the lungs and nutrients from the GI tract. The placenta provides all the fetus' nutritional requirements, and the vascular circuitry is modified accordingly. Four adaptations to the adult vascular circuit occur in the fetus: the **placenta**, **ductus venosus**, **foramen ovale**, and **ductus arteriosus** (Fig. 37.10).

1. **Placenta:** The fetal placenta functions as fetal lungs, kidneys, GI tract, and liver, thus forming a major low-resistance circuit that receives ~40% of fetal CO (Table 37.2).

2. **Ductus venosus:** Blood coursing from the fetal placenta is shunted past the liver by the ductus venosus (see Fig. 37.10). In the adult, the liver filters and processes nutrient-rich blood from the GI system. In the fetus, the GI tract is largely nonfunctional; therefore, both GI organs are bypassed. Both receive sufficient blood to meet their nutritional needs via lesser vascular circuits.

3. **Foramen ovale:** Blood entering the fetal right heart from the inferior vena cava is O_2-rich after passing through the placenta (75% saturation). The fetal lungs do not participate in gas exchange, so passage through the pulmonary circulation would serve no purpose. Pulmonary vascular resistance (PVR) is high also, which

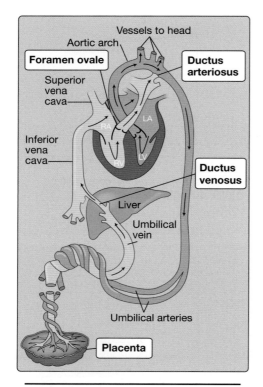

Figure 37.10.
Vascular adaptations unique to the fetal circulation. LA and LV = left atrium and ventricle, RA and RV = right atrium and ventricle.

Table 37.2: Fetal and Adult Cardiac Output Distribution*

Organ System	Fetus	Adult
Lungs	6	100
Heart	5	5
Kidneys	2	20
Brain	20	20
Musculoskeletal	20	20
Splanchnic	7	30
Placenta	40	—

*All values are approximate and given as a percentage of total cardiac output.

makes the lungs difficult to perfuse (see the next section). Thus, O_2-rich blood is shunted past the lungs from the right atrium directly into the left atrium via the foramen ovale (see Fig. 37.10).

4. **Ductus arteriosus:** Blood entering the right heart via the superior vena cava is O_2 poor (25% saturation; see Section C) after having traversed fetal systemic vascular beds. It is pumped through the right heart and then through the ductus arteriosus to the descending aorta, thereby bypassing the lungs (see Fig. 37.10). The foramen ovale and ductus arteriosus together create a vascular circuit in which left and right hearts are arranged in parallel.

B. Vascular resistance

In the adult circulation, SVR > PVR. The adult circulation is dominated by the left heart. In the fetal circulation, PVR > SVR. Fetal lungs are fluid filled, and the air spaces are collapsed. The pulmonary vasculature is tonically constricted as a response to low O_2 levels (**hypoxic vasoconstriction**; see Chapter 22·III·E), so fetal PVR is high. By contrast, the fetal systemic circulation includes the placenta, which is a very low-resistance pathway for blood flow, so fetal SVR is low.

C. Oxygen transfer

Maternal uterine blood has an O_2 saturation of ~80% to 100%. Although the barrier separating fetal and maternal circulations is minimal, the placental route is a relatively inefficient means of gas exchange compared with the lungs, and fetal blood can only achieve P_{O_2} levels of 30 to 35 mm Hg at best (compare to a P_{aO_2} of 98–100 mm Hg in an adult pulmonary vein). Despite the inherent limitations of the transfer route, fetal blood carries amounts of O_2 similar to the adult circulation. This is made possible by hemoglobin F (HbF), a fetal Hb isoform that has a leftward-shifted O_2 dissociation curve (see Chapter 23·IV·C·2). HbF's high O_2 affinity is well adapted to take up O_2 at partial pressures common to the maternal placenta, meaning that blood traveling from the placenta to the fetus in the umbilical veins typically has an O_2 saturation of 75% (Fig. 37.11). Fetal blood also contains ~20% more Hb than adult blood, which increases overall O_2-carrying capacity.

D. Oxygen distribution

Blood traveling from the placenta to the fetus via the umbilical vein is O_2 rich. It streams around the liver via the ductus venosus but then encounters O_2-poor blood returning from the lower regions of the body in the inferior vena cava ([IVC] see Fig. 37.11). A filmy membrane ensures that little mixing occurs at the point where the two bloodstreams merge, and the O_2-rich stream is preserved in the instant it takes to reach the right atrium (streamline flow; see Chapter 18·V·A). A valve at the entrance to the right atrium (**eustachian valve**) directs blood entering from the IVC toward the foramen ovale where the O_2-rich and -poor streams are cleaved by the interatrial septum (**crista dividens**). The O_2-rich stream passes preferentially into the left heart and then into the aorta. The first arteries to branch off the aorta feed the myocardium and the brain, so flow streamlining ensures that these two critical circulations receive highly oxygenated blood.

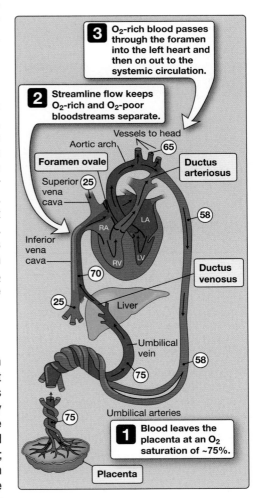

Figure 37.11.
O_2 distribution by the fetal circulatory system. Circled numbers represent O_2 saturations. LA and LV = left atrium and ventricle, RA and RV = right atrium and ventricle.

E. Renal function

Fetal kidneys start producing urine within 9 to 10 weeks of conception. The ability to concentrate urine is gained around 4 weeks later, but the fetus remains dependent on the placenta for fluid and electrolyte balance throughout gestation. By 18 weeks, the kidneys are producing more than 10 mL of urine per hour, and fetal urine has become the primary source of amniotic fluid.

VI. PARTURITION

Human gestation lasts 40 weeks on average. At the end of this time, the fetus is forcibly expelled from the uterus, and the physical link with the mother is broken (**parturition**). The birthing process requires careful coordination if both mother and newborn are to survive.

A. Staging

Parturition can be divided into three stages of variable duration: **dilation**, **fetal expulsion**, and **placental**.

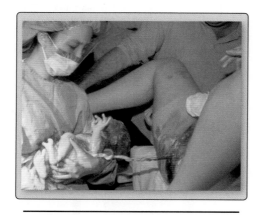

Figure 37.12.
Newly delivered infant.

1. **Dilation:** Stage 1 begins with labor and ends when the cervix is dilated fully. The fetus is enclosed within the amniotic sac, but the main barrier preventing it from exiting the uterus is the cervix. During the first stage of parturition, the myometrium begins contracting rhythmically and with increasing intensity. Contraction begins in the uterine fundus and spreads caudally, which pushes the fetus against the cervix and causes the latter to thin and dilate. Stage 1 usually lasts for ~8 to 15 hours.

2. **Fetal expulsion:** The fetus is forcibly expelled from the uterus through the cervix and vaginal canal by frequent and intense waves of contraction. Stage 2 is complete within 45 to 100 minutes (Fig. 37.12). The umbilical cord is traditionally clamped shortly after birth, although preterm infants may benefit from delayed clamping and gently milking blood from the cord toward the newborn to increase the infant's hematocrit.

3. **Placental:** The uterus continues contracting after the fetus has been expelled, which causes it to shrink in size (**involution**). Shrinkage shears the placenta from the uterine wall. The placenta and associated membranes are later expelled as afterbirth. Stage 3 is usually complete within minutes of fetal expulsion.

B. Hormones

Irregular waves of weak uterine contraction occur throughout pregnancy (**Braxton Hicks contractions**). The reasons why contractions abruptly transition to the forceful contraction of parturition are not understood, although several hormones have been implicated. The fetus stretches the myometrium and increases its overall excitability as it grows and may be a contributing factor. The major hormones driving parturition include estrogen and progesterone, prostaglandins, oxytocin, and cortisol.

1. **Estrogen/progesterone ratio:** Progesterone suppresses uterine contraction during pregnancy. Estrogens promote excitability by increased expression of Na^+ and Ca^{2+} channels and gap junctions

between adjacent smooth muscle cells within the myometrium. The gap junctions allow developing waves of excitation to sweep through the uterine wall, manifesting as a wave of contraction. At parturition, the estrogen/progesterone ratio increases, and the uterus becomes excitable.

2. **Prostaglandins:** The uterus, placenta, and fetus all produce prostaglandins (PGE_2 and $PGF_{2\alpha}$), which stimulate uterine contractions. Increasing estrogen levels likewise increase prostaglandin production.

3. **Oxytocin:** Oxytocin is a powerful stimulant of uterine contractions. It is released from the posterior pituitary in response to cervical distension (see Chapter 36·VI·B), providing a positive feedback mechanism that couples fetal expulsion with the motive force required for expulsion.

4. **Cortisol:** The fetal hypothalamic–pituitary–adrenal axis is activated to release cortisol (see Chapter 34·II). Cortisol increases the estrogen/progesterone ratio.

C. Circulatory transition from fetal to adult

Parturition breaks the link between mother and fetus and forces the fetal vasculature to adopt the serial circulatory pattern common to the adult. Transition follows a rapid sequence of events that are initiated by lung inflation and a decrease in PVR; an increase in SVR; closure of the ductus arteriosus, foramen ovale, and ductus venosus; and culminating in a shift from right- to left-sided circulatory dominance (Fig. 37.13).

> Remnants of the umbilical arteries and vein can be observed in the adult as the **medial umbilical ligaments** and **ligamentum teres**, respectively.

1. **Lung inflation:** The umbilicus is a muscular structure that contracts spontaneously in response to the trauma of birth. Compression and occlusion of umbilical vessels halts blood flow and deprives the fetus of O_2, causing asphyxia. This, together with the sudden cooling experienced by the infant at birth, stimulates respiratory control centers in the fetal brainstem, causing the neonate to gasp and take several breaths. Intraalveolar pressure drops below atmospheric pressure, creating a pressure gradient that drives air inflow, and the lungs inflate.

2. **Pulmonary vascular resistance:** During development, PVR is high because the lungs are collapsed, and pulmonary arteries are compressed and constricted in response to low Po_2. The first breaths cause alveolar and pulmonary arterial Po_2 levels to rise dramatically, promoting vasodilation. Lung inflation also increases the internal diameter of pulmonary blood vessels through radial traction (see Fig. 22.3). PVR drops dramatically as a result (see Fig. 37.13), and there is a concomitant increase in pulmonary blood flow.

3. **Systemic vascular resistance:** Contraction of the umbilicus during parturition occludes the umbilical arteries and vein and terminates

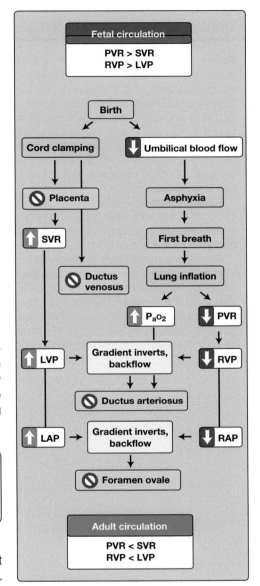

Figure 37.13.
Changes in fetal vascular circuitry during parturition. LAP and LVP = left atrial and ventricular pressure, respectively; P_aO_2 = partial pressure of oxygen (arterial); PVR and SVR = pulmonary and systemic vascular resistance, respectively; RAP and RVP = right atrial and ventricular pressure, respectively.

Clinical Application 37.2: Circulatory Transition Difficulties

Patent ductus arteriosus (PDA) is a common congenital heart defect, particularly in premature and very low–birthweight infants among whom the incidence may be as high as 30%. If the ductus arteriosus remains open, high-pressure blood from the systemic circulation shunts into the pulmonary circulation. Depending on severity, the shunt can cause pulmonary hypertension and may result in right-sided heart failure if not addressed. Ductus arteriosus patency is maintained during development in part by high circulating levels of prostaglandin E_2, so administering a cyclooxygenase inhibitor, such as indomethacin, is often sufficient to prompt complete closure.[1] Surgical ligation or occlusion may be required if pharmaceutical intervention is unsuccessful.

Patent foramen ovale is a congenital heart lesion affecting 25%–30% of the general population. Although the pathway between right and left atria remains intact, left atrial pressure is usually higher than right atrial pressure, and the foramen remains occluded by the one-way valve. Healthy individuals usually remain asymptomatic.

Persistent pulmonary hypertension of the newborn results from failure of pulmonary vascular resistance (PVR) to decrease normally at birth. Causes may include underdevelopment of the pulmonary vasculature and vascular anomalies, or bacterial infections that increase PVR of a normally developed lung. A high PVR causes right-to-left shunting via the PDA. Neonates present with symptoms of respiratory distress, including tachypnea, grunting, and cyanosis. Management includes supportive measures such as O_2 supplementation, mechanical ventilation, and inhaled nitric oxide to promote pulmonary vasodilation.

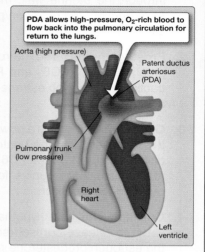

PDA allows high-pressure, O_2-rich blood to flow back into the pulmonary circulation for return to the lungs.

Aorta (high pressure)

Patent ductus arteriosus (PDA)

Pulmonary trunk (low pressure)

Right heart

Left ventricle

Patent ductus arteriosus.

flow to the placenta. Fetal SVR increases when this low-resistance pathway is removed from the systemic vascular circuit.

4. **Ductus arteriosus:** The fall in PVR and loss of flow from the umbilical vein causes right atrial pressure to fall. SVR simultaneously increases due to loss of the placental circuit, so left ventricular and aortic pressure rises. The sudden inversion of the fetal right-to-left pressure gradient (see Fig. 37.13) causes a reversal of blood flow in the ductus arteriosus, and it constricts, probably in response to rising P_aO_2 and falling circulating prostaglandin levels (see Clinical Application 37.2). Complete anatomic closure takes several months, and vestiges of the fetal shunt persist in the adult as the **ligamentum arteriosum**.

5. **Foramen ovale:** The right-to-left blood pressure inversion pushes on a valve-like flap that then covers the foramen ovale. Gradually escalating left atrial pressures hold the flap closed to isolate the left and right sides of the heart. In time, the flap fuses with the interatrial septum to seal off the foramen permanently (seen as the **fossa ovale** in an adult heart).

6. **Ductus venosus:** The ductus venosus closes by a sphincter-like mechanism, persisting as the **ligamentum venosum** in the adult. The mechanism of closure is unknown.

[1]For more information on the actions and uses of cyclooxygenase inhibitors, see *LIR Pharmacology*, 8e, Chapter 40·III·A·2.

7. **Circulatory dominance:** Over ensuing weeks, the left ventricle slowly hypertrophies as a response to a rising SVR. Meanwhile, the right heart pumps against a lower PVR than it did during gestation, so its muscle mass slowly decreases relative to the left heart.

D. Maternal blood loss

A mother typically loses ~500 mL of blood from the placental site during a normal delivery. Although this represents a substantial hemorrhage, the mother has been well prepared for the loss by the expansion of blood volume that occurs during the first weeks of pregnancy. Further loss is prevented by intense uterine contractions, which compress the uterine vasculature and allow hemostasis. The contractions are stimulated by oxytocin during the third stage of parturition.

Chapter Summary

- Pregnancy begins with fertilization. The embryo divides rapidly over subsequent days to form a **blastocyst** and then **implants** in the maternal uterine wall. Implantation is accomplished by the **syncytiotrophoblast**, which invades and digests the maternal **decidua** and creates an interface between fetal and maternal circulations (the **placenta**).
- The placenta exchanges nutrients, hormones, and waste products between fetal and maternal circulations.
- The fetal placenta comprises 60 to 70 **villous trees** that increase interface surface area. The maternal placenta comprises 15 to 20 blood-filled, eroded endometrial sinuses sculpted by the fetal syncytiotrophoblast during placentation. The space between fetal and maternal placenta is filled with ~500 mL of maternal blood. Blood flows in an unregulated manner at relatively high pressure (~70 mm Hg) from eroded **spiral arteries** and washes over the fetal villous trees.
- The placenta is also an endocrine organ that secretes **human chorionic gonadotropin**, **progesterone**, and **estrogens**.
- Supporting the needs of a developing fetus involves most of a mother's organ systems, including the **cardiovascular system**, **kidneys** (increased disposal of waste products), **lungs** (~30% increase in O_2 demand), **GI tract**, **liver**, and **skin** (thermoregulation).
- Maternal **cardiac output** increases ~50% during gestation, accomplished through increases in **heart rate** and **stroke volume (SV)**. SV increases as a result of fluid retention and increased **circulating blood volume**.
- Blood volume increases faster than red blood cell production, causing a **physiologic anemia of pregnancy**. The resulting decrease in blood **viscosity** reduces **shear stress** on the heart and vascular lining.
- The uterus and its contents gain significant weight during pregnancy, which, depending on posture, compresses and impairs blood flow from the mother's lower extremities. The result is **pedal edema**.
- The fetal circulation includes three **shunts** that allow blood from the umbilical vein to bypass the liver (**ductus venosus**) and lungs (**foramen ovale** and **ductus arteriosus**) for distribution to the developing organs.
- Fetal blood contains a **fetal hemoglobin (HbF)** isoform that has a high O_2 affinity. HbF helps compensate for the fact that the placenta is a less efficient route for O_2 transfer than the lungs and allows fetal blood to carry near-adult levels of O_2.
- **Parturition** is initiated and sustained by changing levels of hormones produced by both the mother and the fetal placenta. Uterine contractions expel the fetus and placenta then compress and collapse the uterine vasculature. Compression limits maternal blood loss during delivery.
- At birth, the neonate's **pulmonary vascular resistance** decreases, and **systemic vascular resistance** increases, thereby establishing the left side–dominated circulatory system common to an adult.

Study Questions

Choose the ONE best answer.

37.1. A 23-year-old pregnant female in her third trimester mentions to her friend that her feet and ankles are frequently swollen. The swelling is most likely caused by which of the following?

A. Decreased blood viscosity
B. Excessive fluid retention
C. High pedal venous pressures
D. Hypertension (preeclampsia)
E. Increased left ventricular preload

Best answer = C. The gravid uterus compresses veins returning blood from the lower extremities, causing pedal venous pressures to rise (see Section IV·D). Pedal capillary hydrostatic pressure rises as a result, promoting fluid filtration and edema. Blood viscosity changes (A) and hypertension (D) do not significantly affect the Starling forces governing fluid filtration from capillaries under physiologic conditions. Maternal cardiac output increases through fluid retention (B) to increase left ventricular preload (E), but the increased output is needed to supply the placenta and other organs and does not raise arterial pressure or contribute significantly to a rise in pedal venous pressure.

37.2. A pregnant 30-year-old female in her third trimester (32 weeks' gestation) comes to the office for a routine prenatal examination. Her systolic blood pressure was noted to be elevated during two previous visits. Physical examination reveals her blood pressure to be increased above normal (but is still well below 160 mm Hg) and a systolic heart murmur is recorded. What is the most likely cause of this murmur?

A. Endothelial dysfunction
B. Increased cardiac output
C. Low plasma colloid osmotic pressure
D. Physiologic anemia
E. Preeclampsia

Best answer = D. Maternal cardiac output (CO) increases to support fetal needs during pregnancy in part through fluid retention. Red blood cell production lags behind plasma volume expansion, causing a physiologic anemia. Anemia decreases blood viscosity and increases turbulent blood flow, manifesting as functional murmurs (see Section IV·C). Cardiac output (B) does increase during pregnancy, but it would not be expected to cause a murmur in the absence of viscosity changes. Colloid osmotic pressure does decrease (C) during pregnancy but does not significantly impact blood viscosity. Endothelial dysfunction (A) can lead to preeclampsia (E), a BP >160 mm Hg, (see Clinical Application 37.1), but neither condition would be expected to cause a heart murmur.

37.3. A researcher is comparing physiologic changes induced by pregnancy and aerobic exercise training in healthy female volunteers. Which of the following changes would most likely be observed in both groups of volunteers?

A. Afterload-induced ventricular hypertrophy
B. Decreased hemoglobin concentration
C. Increased minute ventilation
D. Increased resting diastolic blood pressure
E. Increased resting heart rate

Best answer = C. Both pregnancy and aerobic training increase minute ventilation to maximize tissue O_2 delivery (see Section IV·E and Chapter 39·IX·C). Ventricular hypertrophy (A) during pregnancy and training occurs in response to a chronically increased preload and stroke volume (SV). Hb levels fall during pregnancy, whereas training increases Hb (B). Resting diastolic blood pressure (D) falls during pregnancy due to peripheral vasodilation but decreases minimally with aerobic exercise training. Resting heart rate (E) rises during pregnancy to help meet the increased demands for cardiac output (CO). Resting CO is not changed by training, however, so the SV increase decreases resting heart rate.

37.4. The answer choices below compare pairs of cardiovascular system variables. Which of these pairings most likely describes the fetal cardiovascular system?

A. Atrial pressure: left > right
B. Flow: pulmonary vein > descending aorta
C. Hemoglobin levels: adult > fetal
D. O_2 saturation: inferior vena cava > aorta
E. Vascular resistance: systemic > pulmonary

Best answer = D. Fetal blood is oxygenated in the placenta and then flows via the umbilical vein at ~85% saturation into the inferior vena cava (~70%), through the heart, and into the aorta (see Section V·C). Saturation has fallen to ~65% by venous admixture during passage. Pulmonary vascular resistance is much higher than systemic vascular resistance in the fetus (E), so right atrial pressure is higher than left (A). Shunts (foramen ovale and ductus arteriosus) direct blood past the high-resistance pulmonary circuit; therefore, pulmonary blood flow is lower than aortic flow (B). Fetal blood is enriched with hemoglobin (C) to enhance its O_2-carrying capacity.

Thermal Stress and Fever

38

I. OVERVIEW

The biochemical pathways that sustain the body's organ systems function optimally within a narrow temperature range (36.5°–37.5°C; Fig. 38.1). Temperature extremes outside this range can cause severe organ dysfunction. Severe hyperthermia (>42°C), for example, causes oxidative uncoupling, enzymatic failure, and a systemic inflammatory response that may result in shock and death. Severe hypothermia (<28°C) causes unconsciousness. Cooling to below 24°C results in irreversible organ damage and death. Thus, thermoregulation is one of the body's primary homeostatic functions. An ability to thermoregulate requires a feedback control system that minimally consists of a sensor capable of detecting changes in body temperature, a temperature control center, and a way of changing body temperature when deviations from optimum are detected. The skin is very sensitive to temperature, but, as we know from experience, skin temperature varies widely with external temperature, so it is not a reliable instrument for monitoring internal (core) temperature. Internal temperature is instead monitored primarily by sensors buried deep in the brain (hypothalamus). The hypothalamus also houses a thermoregulatory control center. Thermoregulatory responses are produced in two ways. First, behavior can be modified to avoid external temperature extremes (e.g., seeking shade or putting on warm clothes). Second, internal temperature can be adjusted. When internal temperature is too high, heat can be transferred to the environment by directing blood to the skin and sweating. When internal temperature is low, heat can be generated by shivering or using metabolic futile cycles. This combination of behavior and physiologic response has allowed humans to occupy most regions of the Earth's surface, including Dallol, Ethiopia (average annual temperature = 35°C), and Plateau Station, Antarctica (average annual temperature = –55°C).

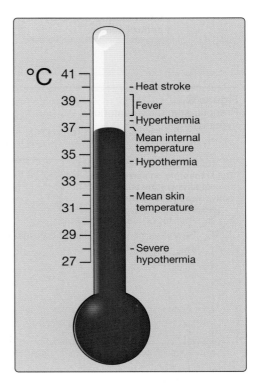

Figure 38.1.
Body temperatures.

II. THERMOREGULATION

Thermoregulation is accomplished by means of a feedback system employing thermosensors to monitor internal and external temperatures, a thermoregulatory control center, and effector pathways to adjust internal temperature as needed.

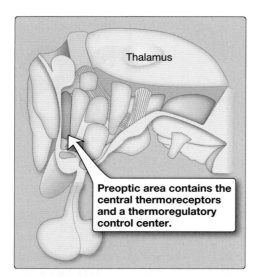

Figure 38.2.
Preoptic area of the anterior hypothalamus.

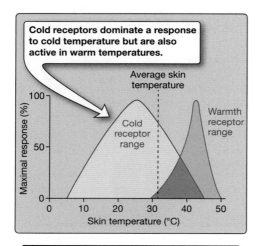

Figure 38.3.
Skin thermoreceptor sensitivity.

Internal temperature can be accurately assessed using temperature probes placed in the esophagus or rectum. Oral temperature measurements can also yield good estimates of internal temperature, provided that the patient is breathing nasally, and ventilation is low. Oral temperature measurements are 0.25°–0.5°C lower compared with rectal temperature.

A. Thermosensors

The body possesses two groups of thermosensors. Central thermoreceptors monitor internal body temperature, whereas skin thermoreceptors provide information about the external thermal environment.

1. **Central:** Internal body temperature is sensed primarily by warmth-sensitive neurons located in the **preoptic area** (**POA**) of the anterior hypothalamus (Fig. 38.2), although spinal and visceral thermosensors also contribute temperature data. Warmth-sensitive POA neurons are tonically active at normal body temperature. A rise in internal temperature (as reflected in the temperature of blood bathing the POA) increases their firing rate, whereas cooling decreases firing rate.

2. **Skin:** Skin contains two distinct populations of thermoreceptors involved in thermoregulation (see Chapter 15·VII·B). **Cold receptors** mediate neutral, cool, and cold temperature sensation (5°–45°C). **Warmth receptors** sense neutral and warm temperatures (30°–50°C). The overlapping temperature ranges of the two receptor populations mean that both contribute thermosensory data at normal average skin temperature (Fig. 38.3). Skin temperature is usually lower than normal internal temperature due to heat loss to the environment.

B. Control center

The thermoregulatory integration and control center resides in the POA. The control center normally holds internal temperature at 37°C, although there is daily variance of 1°C (i.e., 36.5°–37.5°C). The temperature nadir occurs at 6 AM, with the peak occurring between 4 PM and 6 PM. Cooling the POA elicits heating responses and behaviors, whereas heating this area activates cooling responses and behaviors. If the POA is damaged (e.g., by trauma or ischemia), internal temperature fluctuates over an exaggerated range, and responses to thermal stress are impaired. POA output is based primarily on input from central thermoreceptors, but the control center integrates signals from many other areas also. These include skin thermoreceptors; the immune system (see Section IV); and areas of the central nervous system that regulate other systemic variables, such as blood pressure, plasma glucose concentration, and plasma osmolality.

C. Effector pathways

The POA initiates most thermoregulatory responses via the sympathetic nervous system (SNS). Sympathetic signals travel in T1–L3 spinal nerves (see Chapter 7·III·B).

D. Response

The primary effectors of cooling and heating responses are skin blood vessels, sweat glands, skeletal muscle, and brown adipose tissue ([BAT]

Fig. 38.4). Heat stress reduces tissue insulation and initiates sweating. Conversely, cold stress increases tissue insulation and increases metabolic rate via shivering and nonshivering thermogenesis (heat production).

1. **Skin blood vessels:** Heat loss or gain is regulated most effectively at the level of the skin, which presents ~2 m² of body surface area to the external environment. Glabrous skin contains deep arteriovenous anastomoses (AVAs) that allow blood to bypass surface capillary beds when conserving heat is necessary. Hairy skin does not have AVAs, but it does have superficial and deep capillaries.

   a. **Heat stress:** Initial responses to heat stress involve suppression of tonic SNS vasoconstrictor influence. Next, active vasodilation may increase cutaneous blood flow by as much as 8 L/min (Fig. 38.5). Vasodilation is accomplished by nitric oxide and ACh release from SNS cholinergic nerves acting in concert with one or more coreleased neurotransmitters. The pathways involved are not yet fully delineated. Flow through the superficial capillaries brings blood to within 1–2 mm of the body surface for efficient heat transfer to the environment. Vasodilation also increases venous capacity, which enhances the temperature gradient driving heat transfer. Supporting cutaneous blood flow during heat stress involves increased cardiac output and redirection of flow from other vascular beds (e.g., renal and gastrointestinal).

   b. **Cold stress:** During cold stress, SNS adrenergic nerves constrict cutaneous arteries, veins, and AVAs, which shunts blood away from the skin surface to reduce heat loss. Cutaneous blood flow may drop to <6 mL/min total in efforts to conserve heat, causing skin surface temperature to approach that of the external environment. Because such extreme flow diversion risks tissue damage and frostbite if skin temperatures fall too far, a safety mechanism is built in. The AVAs dilate reflexively when skin temperatures drop below a certain level. This allows blood to surge from deeper venous plexuses into superficial veins to rewarm surrounding tissues. So, for example, the temperature of fingers immersed in water at 5°C remains at around 9°C due to cold-induced AVA vasodilation.

2. **Sweat glands:** Evaporative cooling during heat stress is mediated by **eccrine sweat glands** (see Chapter 15·VI·D). Sweating is initiated by SNS cholinergic nerves, but the glands are also stimulated by epinephrine and norepinephrine. Sweating can dehydrate the body and cause a hypertonic volume contraction. Even minor fluid losses (2% body weight) can decrease work performance and impair heat-loss mechanisms.

3. **Muscles:** Skeletal muscle produces large amounts of heat during contraction because force production is only ~20% efficient. The remaining 80% of expended energy is liberated as heat. Shivering is a rapid, cyclical contraction that produces minimal force and maximal heat. Shivering is mediated by somatic motor pathways rather than the SNS, but the response is initiated by the POA.

4. **Nonshivering thermogenesis:** Nonshivering thermogenesis is an SNS-mediated increase in metabolic rate in muscle and other tissues designed to liberate heat. In BAT, SNS stimulation activates uncoupling protein ([UCP1] thermogenin) in the inner

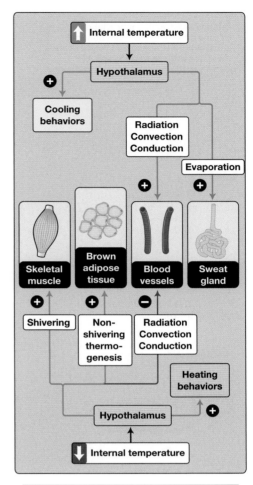

Figure 38.4.
Thermoregulatory response mediators.

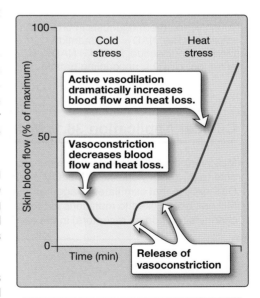

Figure 38.5.
Skin blood flow during heat stress.

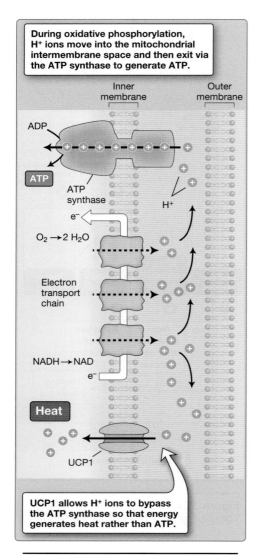

During oxidative phosphorylation, H$^+$ ions move into the mitochondrial intermembrane space and then exit via the ATP synthase to generate ATP.

ADP

ATP

ATP synthase

Inner membrane

Outer membrane

e$^-$

O$_2$ → 2 H$_2$O

Electron transport chain

NADH → NAD

e$^-$

Heat

H$^+$

UCP1

UCP1 allows H$^+$ ions to bypass the ATP synthase so that energy generates heat rather than ATP.

Figure 38.6.
Nonshivering thermogenesis in brown adipose tissue. NAD = nicotinamide adenine dinucleotide, NADH = NAD hydrogen, UCP1 = uncoupling protein.

mitochondrial membrane (Fig. 38.6). UCP1 forms a H$^+$ channel that allows H$^+$ to reenter the mitochondrial interior and bypass ATP synthase, thereby liberating heat rather than generating ATP. UCP1 is inhibited at normal body temperature, but β$_3$-adrenergic–receptor binding following SNS activation generates free fatty acids, which activate the channel. Thyroid hormones also increase heat production by increasing UCP1 expression levels.

5. **Behavior:** Skin temperature sensors forewarn the POA about changes in external temperature and promote behaviors that can decrease or even eliminate thermal stress. Although driven by the POA, these behaviors can be overridden or modified by other brain areas. Conscious behavioral responses to heat stress include drinking fluids to facilitate sweating, removing clothing, seeking shade, and turning on a fan. Cold stress–related behaviors include increasing insulation (e.g., putting on a coat), increasing physical activity to increase metabolic rate, and seeking an external heat source.

III. HEAT PRODUCTION AND TRANSFER

The amount of heat stored within the body reflects a balance between heat production and the amount transferred to the external environment. Heat storage can be quantified theoretically using the heat balance equation:

$$S = (M - Wk) \pm (R + K + C) - E$$

where S is heat storage; M is metabolism; Wk is external work; and R, K, C, and E describe radiative, conductive, convective, and evaporative heat transfer, respectively.

A. Production

Heat is a metabolic byproduct reflecting the inefficiency of the chemical pathways involved. The amount of heat produced at rest is related to **basal metabolic rate** (**BMR**), which, in turn, is related to body mass (e.g., a person with a mass of 50 kg has a BMR of ~1,315 kcal/d; a 90-kg person has a ~2,045 kcal/d BMR). Any increase in tissue metabolism increases heat production. Food digestion and assimilation increase energy expenditure (amount of energy used is known as the **thermic effect of food**), as do body motions and exercise.

Clinical Application 38.1: Malignant Hyperthermia

Malignant hyperthermia (MH) is a syndrome triggered by some inhalation anesthetics (e.g., isoflurane) and muscle relaxants (e.g., succinylcholine).[1] Metabolic rate increases so rapidly that it far outpaces heat dissipation due to a sarcoplasmic Ca^{2+} excess that stimulates exaggerated and prolonged excitation–contraction coupling in skeletal muscle. Susceptibility to MH has been linked to mutations in the genes encoding L-type Ca^{2+} channels (dihydropyridine receptors) and Ca^{2+}-release channels (ryanodine receptors) in the sarcoplasmic reticulum (see Chapter 12·III·A).

 [1]For more information on the use of anesthetics, see *LIR Pharmacology*, 8e, Chapter 20·IV.

B. Heat transfer

Heat produced by metabolism must be transferred to the external environment, principally via the skin, although a small amount of heat is transferred via the respiratory tract. Human skin temperature is ~32°C (~90°F) in normothermic (i.e., temperatures that support a normal body temperature) environments. Because outside temperatures are usually lower, body heat can be transferred to air or surrounding objects. When the external temperature is higher than skin temperature, the body gains heat. The four primary mechanisms by which heat is transferred to the environment are **radiation**, **conduction**, **convection**, and **evaporation** (Fig. 38.7).

1. **Radiation:** Radiation transfers thermal energy to external objects via electromagnetic waves. Heat energy is carried in the infrared spectrum, and the amount transferred depends on the temperature and the emissivity (ability to absorb energy) of the object. At rest, radiation is a primary body heat dissipation mechanism (Table 38.1).

2. **Conduction:** Conduction is the transfer of thermal energy between two objects in close physical contact. The high kinetic energy of molecules in a warm region dissipates by collisions with adjacent molecules in a cool region. Solids vary in their ability to conduct heat. Substances with low thermal conductivity are called thermal insulators.

3. **Convection:** Convection transfers heat to a moving fluid (e.g., air or water). Generally, heating reduces the density of air and water, and gravity creates a "natural" fluid convection current near the skin as the warmer, low-density fluid rises. Forced convection results when an alternative energy source propels the fluid past the skin (e.g., fan, wind, water current). Convective heat loss or gain is proportional to the specific heat of the fluid, the temperature gradient, and the square root of fluid velocity.

4. **Evaporation:** Evaporation dissipates heat by using thermal energy to convert water from a liquid to a vapor. The skin and respiratory tract are primary sites of evaporative heat loss. Evaporation is a highly effective means of dissipating heat: one liter of sweat transfers ~580 kcal of body heat to the environment. Evaporation is dependent on the relative humidity of ambient air: Humid air attenuates, whereas dry air facilitates sweating. During exercise or when ambient air temperature is above skin temperature, sweat evaporation provides the primary and, often, only mode of heat dissipation (see Table 38.1).

IV. CLINICAL ASPECTS

Clinically, it has been useful to define optimal body temperature as a hypothalamic set point against which prevailing body temperature is compared. If any deviation from optimum is detected, effector pathways are activated to restore the optimum. Internal temperature is normally maintained at 37°C (98.6°F), but the hypothalamus may adjust the set point and allow internal temperature to increase above normal in attempts

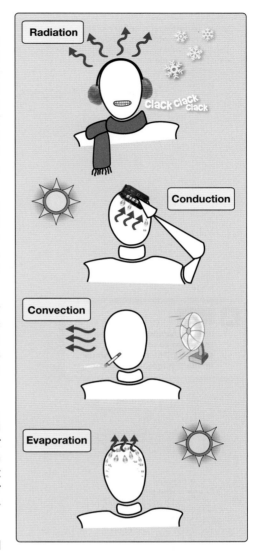

Figure 38.7.
Heat-transfer mechanisms.

Table 38.1: Heat Dissipation Pathways

Heat Transfer	Sitting Indoors in 25°C	Walking Outside in 30°C
Radiation	60%	Minimal*
Convection	15%	10%
Conduction	5%	Minimal
Evaporation	20%	90%

*Radiation involves both heat gain and loss, but the net change is minimal.

to thwart a pathogen. This temperature increase manifests as **fever**. Deviation from normal may also occur when the body's thermoregulatory systems are overwhelmed, resulting in **hypothermia** or **hyperthermia**.

A. Fever

Fever, which has long been recognized as a symptom of illness, is caused by **exogenous** and **endogenous pyrogens**. Exogenous pyrogens include microorganisms, such as *Staphylococcus aureus*, and their byproducts or toxins. Fever is usually caused by endogenous pyrogens released by macrophages and monocytes as a response to infection.[1] Endogenous pyrogens are interferons and cytokines, including interleukins (e.g., IL-1 and IL-6) and tumor necrosis factor. Circulating pyrogens are sensed by the organum vasculosum of the lamina terminalis (OVLT), which is a circumventricular organ (see Chapter 7·VII·C). The OVLT signals pyrogen detection to the POA via prostaglandin E_2 release. POA neurons express an EP_3 prostaglandin receptor that mediates the fever response. Receptor activation resets the hypothalamic set point toward a higher value (Fig. 38.8A), causing thermoregulatory effector organs to raise body temperature until the new, elevated temperature optimum has been attained. Higher temperatures are thought to both enhance immune system performance and decrease pathogen growth and proliferation. The febrile state is distinct from the body temperature increase associated with muscle contraction and exercise or ambient heat exposure. In both cases, the body attempts to dissipate heat with the goal of returning internal temperature to 37°C (see Fig. 38.8B).

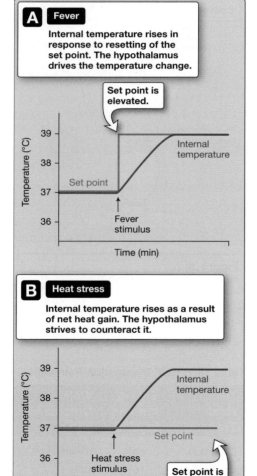

The hypothalamic set point can be returned toward 37°C and the symptoms of fever reduced by administering nonsteroidal anti-inflammatory drugs (NSAIDs), such as aspirin, ibuprofen, and acetaminophen. NSAIDs are cyclooxygenase inhibitors that block prostaglandin synthesis and, thus, have antipyretic effects.[2]

B. Hypothermia and hyperthermia

Hypothermia and hyperthermia result from thermoregulatory system failure. Failure occurs most commonly during weather extremes and in patients with an impaired ability to respond to thermal stress due a congenital inability to secrete sweat, for example (**congenital insensitivity to pain with anhidrosis**).

1. **Hypothermia:** Hypothermia (internal temperature of <35°C) commonly results from cold water immersion because water transfers

Figure 38.8.
Fever vs. heat stress.

[1]For cellular defense mechanisms and associated cytokines, see *LIR Immunology*, 3e, Chapter 5·IV.

[2]For more on aspirin and other nonsteroidal anti-inflammatory drugs, see *LIR Pharm*

heat 25 times faster than does air. Heat production is insufficient to compensate for lost heat, resulting in hypothermia. Hypothermia causes symptoms associated with cold-induced decreases in neuronal metabolic rate, including drowsiness, slurred speech, bradycardia, and hypoventilation. Severe hypothermia (internal temperature of <28°C, 82.4°F) can cause coma, hypotension, oliguria, and fatal cardiac arrhythmias (ventricular fibrillation). Peripheral tissues can also be injured by cold extremes. In frostbite, the water in skin and subcutaneous tissues crystallizes (freezes), disrupting cell membranes and causing tissue necrosis (Fig. 38.9). Necrosed areas commonly require amputation.

2. **Hyperthermia:** Precise definition of hyperthermia is not possible without assessing the cause. For example, internal temperatures of >39°C (102.2°F) can be achieved during exercise without developing a heat illness. Heat illnesses form a continuum, from a milder **heat exhaustion** to a more severe **heat stroke**.

   a. **Heat exhaustion:** Heat exhaustion is characterized by an inability to maintain cardiac output, which compromises skin perfusion and heat transfer. Heat exhaustion is often accompanied by dehydration and decreased blood volume from exercise or heat stress-induced sweating.

   b. **Heat stroke:** Heat stroke refers to failure of the heat dissipation mechanisms during thermal stress, allowing internal temperature to exceed 40°C (104°F). This can occur because of passive heat exposure (nonexertional heat stroke) or sustained heavy exercise (exertional heat stroke). Some of the first symptoms are cognitive: collapse, confusion, and even seizures. Heat stroke can lead to uncoupling of oxidative phosphorylation, disseminated intravascular coagulation, systemic inflammatory response syndrome, and distributive shock (see Chapter 40·III·C). Multiple organ dysfunction syndrome and death may result, even with cooling and medical intervention.

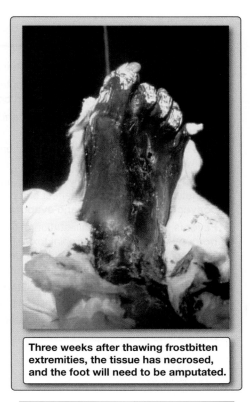

Three weeks after thawing frostbitten extremities, the tissue has necrosed, and the foot will need to be amputated.

Figure 38.9.
Frostbite.

Biologic Sex and Aging 38.1: Thermoregulation

Infants and neonates rely on brown adipose tissue (BAT) for heat production. BAT is found on the neck and in the supraclavicular, interscapular, paraaortic, paravertebral, and suprarenal regions. BAT deposits keep the surrounding tissue warm during cold exposure through nonshivering thermogenesis. BAT was previously considered important only in neonates and infants. However, recent technologic advances in radionucleotide imagery have allowed investigators to identify BAT in adults and relate it to individual metabolic rate and body mass differences.

Basal body temperature does not change with age, but older individuals are at significantly increased risk of morbidity and mortality from hypothermia and heat stroke following ambient temperature changes. Aging decreases skin thermoreceptor sensitivity to warmth and cold, which makes it less likely that individuals will respond behaviorally. Responsiveness to thermal challenge is impaired. Cutaneous vascular responsiveness to active vasodilator (nitric oxide, reactive oxygen species) and vasoconstrictor (adrenergic) influences is decreased. Sweating is impaired, and sweat glands are activated at higher temperatures than in young adults. Decreased muscle mass and shivering thermogenesis is a significant contributor to hypothermia in older adults, as is reduced fat mass and nonshivering thermogenesis capability.

Chapter Summary

- Internal temperature is normally maintained at 37.0°C ± 0.5°C.

- Internal temperature is sensed and controlled by the **preoptic area of the anterior hypothalamus**. Environmental temperature is sensed by **warmth** and **cold skin thermoreceptors**.

- Heat stress induces skin vasodilation and sweating to help offload body heat. Cold-seeking behaviors are also stimulated.

- Cold stress induces skin vasoconstriction to decrease heat loss and shivering to increase heat production. Heat-seeking behaviors are stimulated.

- Heat balance is achieved by matching heat production with heat loss. Heat production includes the amount generated by metabolism, the thermic effect of food, spontaneous movements, and exercise. Heat transfer occurs via **radiation**, **convection**, **conduction**, and **evaporation**.

- **Fever** is the external manifestation of the hypothalamic temperature set point resetting to a higher value. Both **exogenous pyrogens** (e.g., microbial toxins) and **endogenous pyrogens** (e.g., interferons, interleukins, and tumor necrosis factor) can increase the set point via **prostaglandin E_2** production. The body then defends this higher value by normal means, such as shivering and cutaneous vasoconstriction to raise temperature.

- **Hypothermia** is an abnormally low internal temperature clinically associated with processes that slow body metabolism. **Frostbite** results in tissue necrosis from fluid crystallizing within and between cells.

- **Hyperthermia** is an abnormally elevated internal temperature. The most serious form of heat illness is **heat stroke**.

Study Questions

Choose the ONE best answer.

38.1. A 52-year-old female with multiple sclerosis has been admitted to the emergency department on three occasions with mild hypothermia (rectal temperature 34°–35°C) during fall camping trips. What is the most likely reason she might be experiencing low temperatures during prolonged cold exposure?

   A. Caudal cerebellar lesions
   B. Decreased cutaneous pain sensation
   C. Decreased cutaneous warmth sensation
   D. Increased sweat secretion
   E. Preoptic hypothalamic lesions

Best answer = E. Multiple sclerosis causes demyelination and decreased axonal conduction. If a sclerotic lesion occurs in the preoptic area, signals from skin temperature receptors may not be processed appropriately (see Section II·B). This can effectively blunt responses to thermal stress and allow body temperature to fluctuate more with ambient temperatures. Lesions occurring in the cerebellum (A) would affect movement and coordination rather than temperature regulation. Individuals with multiple sclerosis can develop peripheral neuropathy, but skin warmth and pain receptors (B, C) are not directly involved in cold sensation. Sweating (D) occurs during heat exposure, not cold exposure (see Section II·D·2).

38.2. A 6-month-old female is inadvertently exposed to a cold environment after she and her parents were in a rainstorm on a cold day. What thermogenic tissue has a mitochondrial uncoupling protein that can most likely aid her thermoregulation?

   A. Brown adipose tissue
   B. Cardiac muscle
   C. Skeletal muscle
   D. Smooth muscle
   E. White adipose tissue

Best answer = A. Nonshivering thermogenesis is a process of increasing metabolic rate to generate heat without shivering (see Section II·D·4). Brown adipose tissue, which is proportionally higher in infants than in adults, has a specialized mitochondrial uncoupling protein (thermogenin) that generates heat without producing useful work. White adipose tissue (E) does not have this capability. Muscle, primarily skeletal (C), can participate in nonshivering thermogenesis but does not contain a thermogenin-like protein. Cardiac (B) and smooth muscle (D) do not participate directly in cold responses.

38.3. A pharmaceutical company is interested in developing a product that modifies central thermoregulatory control. During preliminary studies, a researcher notes that a cyclooxygenase inhibitor brought the internal temperature of a test subject back to within the 36.5°–37.5°C range. The test subject was most likely suffering from which of the following conditions?

   A. Ambient cold stress
   B. Ambient heat stress
   C. Fever
   D. Heat exhaustion
   E. Severe hypothermia

Best answer = C. Nonsteroidal anti-inflammatory medications, such as aspirin, are cyclooxygenase inhibitors that block prostaglandin synthesis. Their actions include inhibiting prostaglandin EP_3-receptor activation in the preoptic area of the anterior hypothalamus, thereby lowering a thermoregulatory set point that has been elevated by pyrogens during fever (see Section IV·A). Ambient cold (A) and heat stress (B) challenge thermoregulation, but the internal set point is maintained within a normal range. Heat exhaustion (D) is a hyperthermic condition, and severe hypothermia (E) is defined as an internal temperature <28°C, but, again, the body attempts to regulate internal temperature within normal range (see Section IV·B).

38.4. Three males ages 6 to 11 years fall into a lake after walking out onto thin ice during a recent prolonged cold spell. By the time emergency services are able to rescue them, two have arrested and cannot be resuscitated. The third is transferred to an intensive care unit. Which of the following is most likely the principal mechanism helping maintain body temperature in this child prior to being extracted from the lake?

   A. Basal metabolic rate increase
   B. Cutaneous vasodilation
   C. Resetting of the hypothalamic set point
   D. Shivering
   E. Uncoupling protein activation

Best answer = D. Cold water immersion is a common cause of hypothermia because water transfers heat away from the body 25 times faster than does air (see Section IV·B). Shivering describes rapid, cyclical skeletal muscle contractions that generate minimal force but liberate heat through cross-bridge cycling and ATP hydrolysis (see Section II·D·3). Basal metabolic rate (A) is under hormonal control and would increase too slowly to prevent acute hypothermia. Cutaneous vasodilation (B) would warm the skin but increase the rate of heat loss from the body. The hypothalamic set point resets (C) during fever, not cold water immersion. Uncoupling protein (E) generates heat in brown adipose tissue (BAT) by uncoupling electron transport from ATP production. BAT is abundant in neonates but decreases with age and is not as effective as shivering in generating heat acutely.

39 Exercise and Bed Rest

I. OVERVIEW

Exercise, like pregnancy (discussed in Chapter 37), presents one of the greatest challenges a human body must ever face. During intense aerobic exercise, working skeletal muscle can demand blood flow in excess of 20 L/min, or approximately four times basal cardiac output ([CO] Fig. 39.1). Working muscle generates heat, which must be transferred to the environment by increasing blood flow to the skin, which places further demands on CO. A working muscle also requires increased O_2 supply, which necessitates a 25-fold increase in minute ventilation. ATP synthesis must increase to supply energy for contractions. The demands of intense physical exercise are so great that they stress the physiologic systems that support it to the maximum and force the central nervous system (CNS) to decide whether to permit body heat to rise to sustain blood flow to muscle rather than to skin.

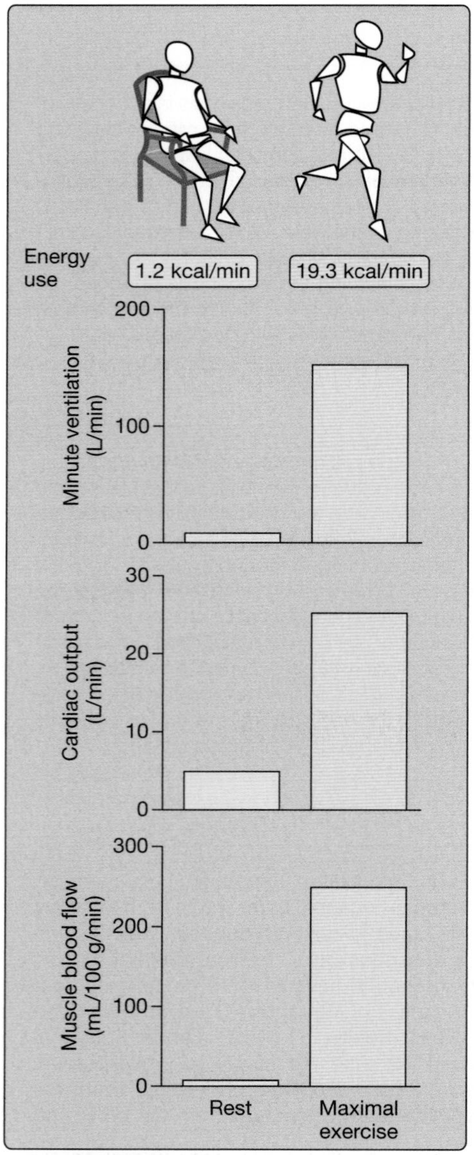

Historically, humans have engaged in physical activity to fetch water, forage for food, and journey to distant hunting sites. In modern society, water is delivered to our homes, food is readily available from local stores, and multiple transportation options facilitate travel. Most individuals are now sedentary, spending long hours sitting at workstations or in front of the television. This is despite the fact that exercise and exercise training have numerous physiologic and psychologic benefits, whereas inactivity significantly increases morbidity and risk of premature death. Regular moderate exercise decreases the risk of coronary heart disease; hypertension; stroke; type 2 diabetes; and colon, breast, and other cancers and can help avoid the numerous deleterious consequences of weight gain. The recognition that inactivity is so detrimental to cardiac, pulmonary, and muscular function has forced the medical community to rethink its traditional recommendations of prolonged bed rest for patients convalescing from severe illnesses, such as myocardial infarction.

II. DEFINITIONS

Physical activity can take many forms and employs different types of muscle contractions; it can thus be classified in different ways based on type of movement (**isotonic vs. isometric** contractions; see Chapter 12·IV·B) or the predominant energy system used to support contractions (**aerobic vs. anaerobic**; Table 39.1). "Exercise" is a planned, structured, repetitive form of physical activity designed to improve physical fitness and health. Activities such as running, cycling, and swimming that involve brief, repeated

Figure 39.1.
Rest vs. maximal exercise.

contractions are considered **dynamic exercise**. Weightlifting involves sustained contractions and is considered **static exercise**. "Exercise training" involves repeated bouts of exercise that stimulate physiologic adaptations to improve responsiveness to subsequent physical stress. Adaptation begins as soon as the exercise training program is initiated but may require months to years to manifest fully. Many of these adaptations can have prophylactic health benefits and can be used in physical rehabilitation to improve work capacity after injury or disease.

III. OXYGEN UPTAKE AND CONSUMPTION

An individual's physiologic capacity for exercise can be expressed in terms of maximal O_2 uptake and consumption ($\dot{V}O_{2max}$). $\dot{V}O_{2max}$ averages 2.1–2.8 L O_2/min (30–40 mL O_2/kg/min) in moderately active adults. $\dot{V}O_{2max}$ is dependent on an individual's genetic history, sex, age, body size and composition (percent muscle vs. fat), and training state.

A. $\dot{V}O_{2max}$ determinants

$\dot{V}O_{2max}$ is determined by the rate at which O_2 is delivered to and extracted by muscle. O_2 delivery to muscle ($\dot{V}_aO_2$) is a product of cardiac output (CO) and arterial O_2 content (C_aO_2):

$$\dot{V}_aO_2 = CO \times C_aO_2$$

in mL O_2/min.

O_2 extraction by muscle ($\dot{V}O_2$) is a product of muscle blood flow (Q) and the arteriovenous O_2 difference:

$$\dot{V}O_2 = Q \times \left(C_aO_2 - C_vO_2\right)$$

where C_vO_2 is the O_2 content of venous blood.

B. Steps in O_2 delivery and extraction

There are multiple steps between O_2 being taken up from the atmosphere and being delivered to muscle mitochondria for use by the enzymes involved in aerobic respiration. We can assess potential limitations of each of these steps in terms of pulmonary, cardiovascular, and muscle performance (Fig. 39.2).

1. **Lungs:** The amount of O_2 taken up by the lungs depends on alveolar ventilation, integrity of the blood–gas interface, and pulmonary perfusion. In healthy individuals, O_2 uptake by the lungs is not rate limiting, even during maximal exercise (see Chapter 22·V·B). Obstruction of airflow or compromise of blood–gas interface integrity, as seen in patients with chronic obstructive pulmonary disease or pulmonary fibrosis, can significantly decrease $\dot{V}O_{2max}$, however,

2. **Cardiovascular system:** Delivery of O_2 to muscle depends on blood O_2-carrying capacity and CO. C_aO_2 is determined by hematocrit (Hct) and blood hemoglobin (Hb) concentration. Hct and Hb concentration is normally lower in females compared with males, which contributes to a lower $\dot{V}O_{2max}$ (see Biologic Sex and Aging 39.1). $\dot{V}O_{2max}$ is also decreased in anemic patients. CO is

Table 39.1: Exercise Classification

Exercise	Type
400-m sprint	Anaerobic
10-km run	Aerobic
Track cycling (1 km)	Anaerobic
Road cycling (40 km)	Aerobic
100-m freestyle swim	Anaerobic
1,500-m freestyle swim	Aerobic

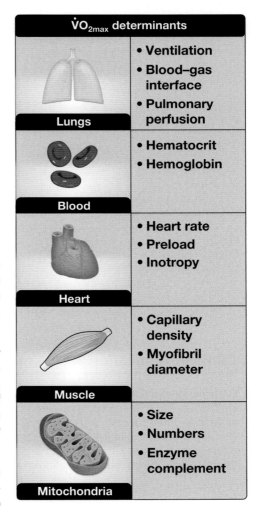

Figure 39.2.
Determinants of oxygen uptake and consumption.

a product of heart rate (HR) and stroke volume ([SV] see Chapter 17·IV). Training-induced increases in SV significantly increase $\dot{V}O_{2max}$ (see Section VI), but the ability to increase CO during exercise remains a principal limiting factor, even in elite athletes.

3. **Muscle:** The ability of muscle to extract O_2 from blood is dependent on muscle capillary density and muscle's ability to generate ATP via aerobic respiration. Exercise training induces angiogenesis, an increase in the size and numbers of mitochondria, and increased expression of enzymes involved in aerobic respiration (see Section IX), suggesting that O_2 extraction may also limit $\dot{V}O_{2max}$.

IV. SKELETAL MUSCLE

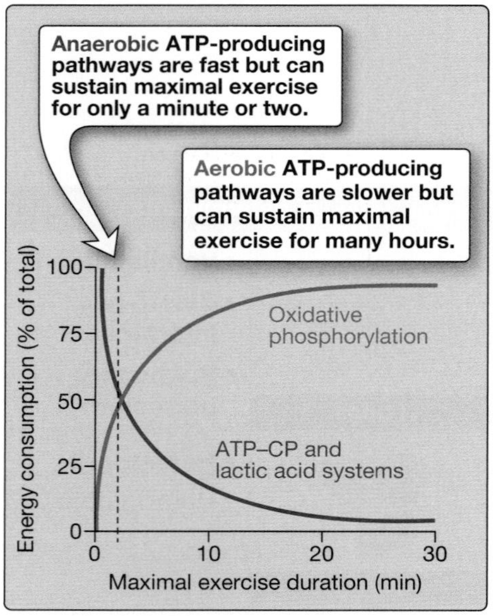

Exercise uses skeletal muscles to generate force that is transmitted through tendons to bone. Bones then move along a force vector within a joint-specific range of motion to transfer that force to a pedal to move a bicycle or to launch a basketball toward a hoop, for example. A contracting muscle consumes energy in the form of ATP. There are three principal pathways for generating ATP that differ in their respective capacities and time courses: the ATP–creatine phosphate (CP) system, the lactic acid system, and oxidative phosphorylation (Fig. 39.3).

A. ATP–creatine phosphate system

CP (also known as phosphocreatine) contains a high-energy phosphate bond that can be used to rapidly regenerate ATP from ADP. The conversion is catalyzed by creatine kinase, a sarcoplasmic enzyme. Muscles contain sufficient CP reserves to support contractions lasting 8 to 10 seconds.

B. Lactic acid system

The lactic acid system generates ATP at about half the rate of the ATP–CP system and uses glucose as a substrate. Glycolysis uses two ATP molecules to produce four more, for a net gain of two ATP per glucose molecule, with the final product being pyruvic acid. Pyruvate-to-lactate conversion does not produce additional ATP but rather regenerates reducing equivalents.[1] Muscles continue to shuttle pyruvate to lactate to extend maximal contraction time to 0.5 to 2.5 minutes (see Fig. 39.3).

Anaerobic **ATP-producing** pathways are fast but can sustain maximal exercise for only a minute or two.

Aerobic **ATP-producing** pathways are slower but can sustain maximal exercise for many hours.

Figure 39.3.
Energy system time course.
CP = creatine phosphate.

Clinical Application 39.1: Creatine Kinase

Creatine kinase (CK) can be used as an index of muscle damage because circulating levels increase when the sarcolemma has been breached. There are three different CK isoforms. Skeletal muscle contains a CK-MM isoform, CK-MB is specific for heart muscle, and the CK-BB isoform is found in nervous tissue. The CK-MM isoform may be released into the circulation of healthy individuals after long-duration aerobic exercise (e.g., running a marathon) that causes minor muscle damage. Individuals with Duchenne and Becker muscular dystrophies may also show 25- to 200-fold increases in blood CK-MM levels during muscle breakdown associated with atrophy (see Clinical Application 12.1).

---

[1]For a discussion of the pyruvate-to-lactate reaction, see *LIR Biochemistry*, 8e, Chapter 8·V·K and L.

C. Oxidative phosphorylation

Oxidative phosphorylation generates ATP at about half the rate of the lactic acid system. Oxidative phosphorylation also involves glycolysis, but pyruvate then enters the citric acid cycle via the pyruvate dehydrogenase complex. The citric acid cycle produces ATP and CO_2. Its main potential energy products are reducing equivalents that enter the electron transport chain, a process that uses O_2 as the final electron acceptor and regenerates ATP from ADP.[1] Oxidative phosphorylation nets ~30 ATP molecules per glucose molecule and can continue for hours, depending on exercise intensity and energy substrate availability (i.e., glucose, fatty acids, ketone bodies, and amino acids; see Fig. 39.3).

V. MOTOR AND AUTONOMIC CONTROL

Physical activity is planned and initiated by the motor cortex and then monitored constantly. Needed adjustments to motor and visceral functions involve all nervous system divisions. The way these systems are coordinated during exercise can be illustrated by considering the pathways required to take a mountain bike ride along a winding single track.

A. Peripheral nervous system

Peripheral nerves convey motor commands via motor neurons from the CNS to the various muscles required to ride a bike. These nerves also convey sensory information from muscle sensors to the CNS.

1. **Motor neurons:** Motor neurons are stimulated at the level of the spinal cord from the descending corticospinal tract. α-Motor neurons initiate force production by extrafusal muscle fibers. γ-Motor neurons, which innervate intrafusal fibers, contract simultaneously to maintain the stretch sensitivity of sensory muscle spindles within the working muscle (see Chapter 11·II·A).

2. **Motor units:** Cycling on level terrain typically requires that submaximal force be applied to pedals. Skeletal muscle fibers produce force in an all-or-none manner, but force production can be graded by activating subsets of motor units (see Chapter 12·IV·D). Alternating active motor units prevents individual motor unit fatigue. If pedal resistance increases (e.g., when ascending a hill), more motor units are recruited to supply the necessary force to maintain forward motion (Fig. 39.4).

3. **Muscle sensors:** Three types of muscle sensor provide the CNS with feedback regarding muscle and joint position during exercise. **Muscle spindles** convey information about limb position through changes in muscle stretch. **Golgi tendon organs (GTOs)** located in the musculotendinous junction sense muscle tension (see Chapter 11·II·B). **Muscle afferents** monitor the local mechanical and chemical environment and relay information back to

 [1]For a discussion of the electron transport chain and oxidative phosphorylation, see *LIR Biochemistry*, 8e, Chapter 6·V.

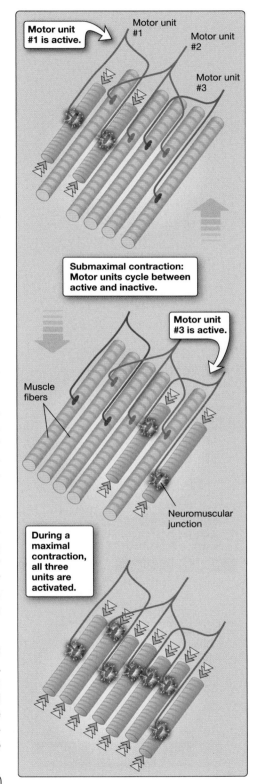

Motor unit #1 is active.

Motor unit #1

Motor unit #2

Motor unit #3

Submaximal contraction: Motor units cycle between active and inactive.

Motor unit #3 is active.

Muscle fibers

Neuromuscular junction

During a maximal contraction, all three units are activated.

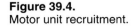

Figure 39.4.
Motor unit recruitment.

autonomic nervous system (ANS) control centers that coordinate cardiopulmonary responses to exercise. Muscle afferents mediate an **exercise pressor reflex**, or a reflex increase in blood pressure observed during exercise. Although some sensory overlap exists between them, there are two main classes of muscle afferents: group III and group IV.

   a. **Group III:** Group III afferents are thin, myelinated nerve endings associated with the collagen sheaths of muscle, nerves, and blood vessels. They respond primarily to mechanical stimuli, such as stretch, compression, and pressure. They activate as soon as contraction begins.

   b. **Group IV:** Group IV afferents are unmyelinated fibers located near muscle blood and lymphatic vessels. They activate shortly after exercise begins in response to building metabolite levels (e.g., lactate, H^+, K^+, adenosine), arachidonate, and bradykinin.

4. **Cardiovascular receptors:** Arterial and cardiopulmonary baroreceptors monitor blood pressure and allow ANS control centers to maintain arterial pressure at levels sufficient to ensure flow to active muscles and other vital systems during exercise (see Chapter 19·III). Peripheral chemoreceptors located in carotid bodies monitor arterial P_{CO_2} and H^+ levels and allow ANS respiratory centers in the medulla to adjust ventilation (see Chapter 24·III·B).

B. Central nervous system

Higher brain centers provide the motivation to go on the bike ride, and they also regulate the muscle contractions required to pedal. The ANS ensures adequacy of blood flow and O_2 delivery to leg muscles and those involved in maintaining stability and posture (e.g., back, arms, shoulders). Information flow from the primary senses helps one balance the bike, stay on the path, and avoid being unsaddled by a tree limb.

1. **Somatic:** The premotor cortex, supplemental motor cortex, and basal ganglia aid in motor program development (see Chapter 11·IV). The motor program coordinates basic motor patterns with reference to sensory information about where the pedals are located and whether the foot is firmly on the pedal, for example. The motor program is executed by the primary motor cortex and signaled via the corticospinal tract. The cerebellum coordinates leg and foot movements during exercise by integrating sensory feedback with motor input.

2. **Autonomic:** Autonomic systems are required to redistribute flow among the various vascular beds to maintain arterial pressure at levels that ensure adequate flow and O_2 delivery to the active muscles. This is accomplished via sympathetic feedforward and feedback pathways (Fig. 39.5).

   a. **Feedforward:** Feedforward control increases CO, arterial pressure, and ventilation in anticipation of or at the beginning of exercise (e.g., when donning a bike helmet or beginning to pedal). Feedforward control is mediated by **central command** acting through the ANS and prepares the cardiopulmonary

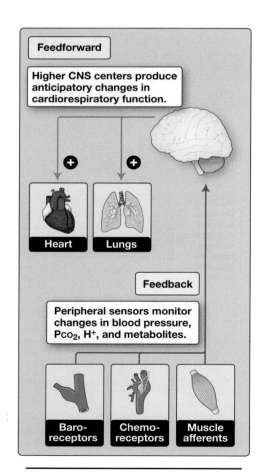

Figure 39.5.
Feedforward and feedback signals.

system for needed increases in O_2 uptake and delivery to skeletal muscles.

   b. Feedback: Feedback occurs via group III and IV muscle afferents and through other autonomic afferents (baroreceptors and chemoreceptors) discussed earlier.

3. Senses: Vision plays a major role in the mountain bike ride, providing information about potential obstacles and the nature of the terrain. Hearing plays a lesser role, but it does help provide clues about the location of other riders, gearing, and under-the-tire terrain. The vestibular system provides information regarding linear acceleration (otolith organs), head rotation, and head position (semicircular canals) when scanning the path ahead and taking note of passing scenery (see Chapter 9·V·A).

VI. CARDIOVASCULAR SYSTEM

A contracting skeletal muscle requires increased blood flow to deliver nutrients and remove metabolic byproducts, including heat. Skeletal muscle receives ~1 L/min at rest, but strenuous exercise can increase demand to >21 L/min (see Fig. 20.1). Such dramatic increases in flow cannot occur without changes in both cardiac and vascular function.

A. Arterial pressure

Sympathetic nervous system (SNS) activation in anticipation of exercise causes mean arterial pressure (MAP) to rise, mediated by increases in heart rate (HR), myocardial inotropy, and systemic vascular resistance (SVR) and by venoconstriction. Aerobic and anaerobic activities differ in their effects on pulse pressure (SBP – DBP; Fig. 39.6).

1. Aerobic exercise: Central command increases CO and systolic blood pressure (SBP) in anticipation of exercise. Once exercise begins, metabolite levels within active muscles build, causing local vasodilation. Dilation facilitates increased flow and O_2 delivery to the active muscles but creates a low-resistance circuit within the systemic vasculature that decreases SVR. Diastolic blood pressure (DBP) decreases as a result, and pulse pressure widens (see Fig. 39.6).

2. Anaerobic exercise: Weightlifting employs isometric contractions that cause SBP, DBP, and MAP to rise dramatically because SVR increases rather than decreases (see Fig. 39.6). MAP values of >275 mm Hg have been recorded during two-leg presses (knee and hip extension), for example. The reason is that sustained contractions compress and occlude arterial supply vessels for prolonged periods, which greatly increases muscle vascular resistance and raises SVR. SVR also increases as a result of high sympathetic drive affecting all systemic vascular beds. An increase in SVR in the setting of increased cardiac contractility causes the excessive increase in MAP (MAP = CO × SVR). Shoveling heavy snow similarly causes SVR and MAP to increase, which is why older adults and patients with hypertension and coronary artery disease are advised against clearing their own walks and driveways.

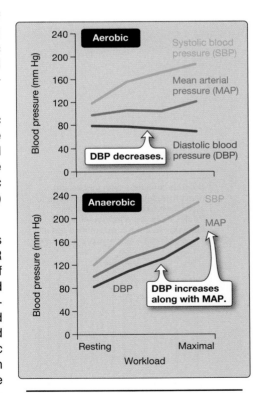

Figure 39.6.
Blood pressure responses during aerobic exercise and weightlifting (anaerobic).

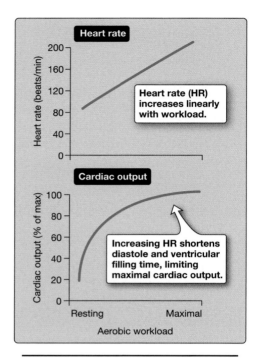

Figure 39.7.
Effects of aerobic exercise on heart rate and stroke volume (SV).

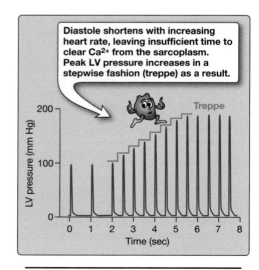

Figure 39.8.
Staircase effect. LV = left ventricular.

B. Cardiac output

Increased flow through the skeletal musculature requires a corresponding increase in CO. CO during aerobic exercise is workload dependent, rising from 5 L/min at rest to 25 L/min during maximal aerobic exercise in a moderately active individual. CO increases are accomplished through increases in HR and contractility (Fig. 39.7).

1. **Heart rate:** HR increases linearly with workload during aerobic exercise (see Fig. 39.7), which is why HR can be used as a rough estimate of how hard the body is working. The HR increase results from concomitant withdrawal of parasympathetic outflow and increase in sympathetic outflow to cardiac nodal cells. HR rises from ~65 beats/min at rest to a maximum of ~195 beats/min, depending on age (see Chapter 16·IV·A and Biologic Sex and Aging 16.1). The rise in HR also reflects direct effects of exercise-induced internal temperature rises on nodal cell automaticity (~8 beats/min/°C). Increasing HR occurs at the expense of diastolic filling time. For example, the duration of diastole drops by ~450 msec when HR increases from 60/min to 160/min. SNS-mediated decreases in atrioventricular nodal conduction time and increased rate of myocardial relaxation help offset the reduced filling time, but stroke volume (SV) still falls linearly with increasing HR in the absence of any other change in cardiac performance. The effect of ejecting more stroke volumes per minute ultimately outweighs the effect of declining SV on CO at low-to-moderate levels of exercise, but SV can become limiting at maximal exercise levels (see Fig. 39.7). In practice, SV does not fall precipitously with rising HR, at least at low-to-moderate levels of exercise, because myocardial contractility increases simultaneously.

2. **Contractility:** Myocardial contractility increases with exercise intensity, in part because of SNS activation, and in part because of a staircase effect.

   a. **SNS activation:** SNS stimulation of the myocardium increases ejection fraction (EF), which decreases end-systolic volume (ESV) but increases SV. The SNS simultaneously stimulates epinephrine release from the adrenal medulla. Epinephrine also increases contractility through binding to β-adrenergic receptors on the myocardium.

   b. **Staircase effect:** The **Bowditch staircase phenomenon**, also known as the **treppe effect** (*treppe* is German for staircase), refers to the observation that myocardial contractile force increases with HR independently of external factors (Fig. 39.8). In addition to reducing filling time, a rise in HR limits the time available to renormalize intracellular Ca^{2+} concentration between contractions (see Chapter 17·IV·F). Each successive contraction begins at a slightly higher intracellular Ca^{2+} level as a result, producing a stepwise increase in contractility (see Fig. 39.8).

C. Venous return

When CO rises to 25 L/min to support intense exercise, venous return (VR) must necessarily increase to 25 L/min to sustain LV preloading and CO. Increased VR is facilitated by the SNS, which decreases venous capacity through venoconstriction. Venoconstriction both increases effective circulating blood volume and speeds the rate at which blood traverses the system (see Chapter 19·V·B). Increased ventilation also assists VR by increasing the pressure gradient driving flow between skeletal muscle veins and the right atrium. The gradient is enhanced during the deep inspirations that typically accompany aerobic exercise. However, the main force driving VR is a venous (or muscle) pump (see Fig. 19.19). Muscle contractions compress the veins within, forcing blood back to the heart. The venous pump causes central venous pressure to rise slightly during exercise, which assists ventricular preloading.

D. Flow redistribution

SNS-mediated vasoconstriction in the vascular beds supplying inactive muscles and other organs that are not directly involved in exercise (e.g., gastrointestinal system, kidneys) diverts blood flow temporarily to support active muscle, the myocardium, and skin (Fig. 39.9). The vasoconstrictor signal is transmitted to both active and inactive muscle groups, but the command is overridden in active muscle by local metabolic and mechanical factors responsible for maximizing flow. Blood flow to the skin increases during exercise to dissipate heat to the environment. Cutaneous blood flow can rise to ~8 L/min when vasodilated fully, which forces the CNS to prioritize flow to support active muscle or the needs of thermoregulation (see Chapter 38) during intense dynamic exercise. In practice, cutaneous flow is reduced and internal temperature allowed to rise by ~2°C in order to support flow to muscle. Excess heat is stored primarily in active and inactive muscle (muscle temperature may rise by ~5°C) and shed over a period of ~90 mins once exercise ends.

VII. RESPIRATORY SYSTEM

Aerobic exercise increases the body's O_2 requirements, from ~0.25 L/min at rest to >4.0 L/min during maximal aerobic exercise in an aerobically trained person. These O_2 needs are met through increases in pulmonary minute ventilation (V_E) and O_2 extraction by tissues.

A. Ventilation

V_E increases from ~6 L/min at rest to ~150 L/min during maximal aerobic exercise through increases in both respiratory rate and tidal volume. There is an anticipatory increase in V_E at the beginning of exercise, mediated primarily by central command acting on medullary respiratory control centers. Peripheral feedback from muscles and chemoreceptors (via P_aCO_2) causes a linear V_E increase throughout low-to-moderate exercise. During heavy exercise, V_E increases sharply because of H^+ accumulation, which is sensed by peripheral chemoreceptors (Fig. 39.10). This transition point is referred to as the **ventilatory threshold**.

B. Oxygen extraction

Muscles consume O_2 faster when exercised, which decreases P_{O_2} locally and enhances the magnitude of the gradient driving O_2

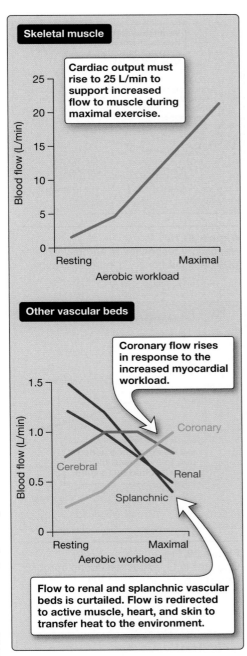

Figure 39.9.
Blood flow redistribution during exercise.

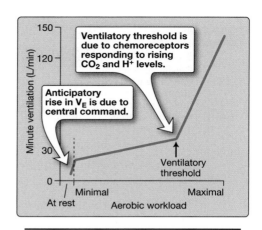

Figure 39.10.
Ventilatory responses to aerobic exercise. V_E = pulmonary minute ventilation.

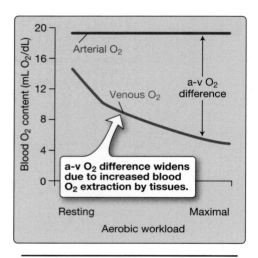

Figure 39.11.
Changes in O_2 extraction during aerobic exercise. a-v = arteriovenous.

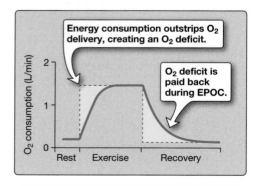

Figure 39.12.
Excess postexercise O_2 consumption (EPOC).

diffusion from the atmosphere to the musculature. This manifests as a widening of the arteriovenous (a-v) O_2 difference, from ~5 mL O_2/dL of blood at rest to ~15 mL O_2/dL during maximal aerobic exercise (Fig. 39.11). O_2 delivery to the active tissues is facilitated by a decrease in Hb–O_2 binding affinity, which increases off-loading. The rightward shift in the Hb–O_2 dissociation curve occurs due to increased CO_2 and H^+ production and rising local temperatures (see Chapter 23·IV·C).

C. Excess postexercise oxygen consumption

Excess postexercise oxygen consumption (**EPOC**) describes the concept of repaying an O_2 debt incurred during initial exercise-induced increases in O_2 consumption. When exercise begins, there is a brief period during which O_2 consumption outstrips O_2 delivery, forcing the muscle to rely on high-energy phosphate groups (CP) and glycogen to generate ATP and causing accumulation of metabolic byproducts (e.g., H^+ and lactate). When activity ceases, the energy stores must be regenerated and byproducts cleared from the sarcoplasm, contributing to EPOC (Fig. 39.12).

VIII. ENDOCRINE SYSTEM

Exercise initiates a stress response that engages several endocrine glands to liberate stored energy substrates for use by muscle. These include the adrenal medulla, which releases **catecholamines** such as epinephrine and norepinephrine (see Chapter 34·V·A) and the adrenal cortex, which releases **cortisol** (see Chapter 34·II·C). **Insulin** release from the pancreas is suppressed during aerobic exercise, whereas **glucagon** secretion increases. The stress hormones and glucagon together increase blood glucose, free fatty acids, and amino acids by increasing glycogenolysis, gluconeogenesis, lipolysis, and proteolysis. **Antidiuretic hormone** (posterior pituitary) and **aldosterone** (adrenal cortex) levels both increase during exercise. These hormones help conserve fluid by increasing water and sodium reabsorption by the kidney (see Chapter 19·IV). Limiting extracellular fluid (ECF) losses helps maintain blood volume. Thyroid hormones (**triiodothyronine [T_3]** and **thyroxine [T_4]**) increase during exercise. T_3 and T_4 regulate metabolic rate and may participate in recovery after exercise. **Growth hormone** and **insulin-like growth factor 1** increase with exercise and contribute to recovery by stimulating tissue growth and repair by their effects on protein synthesis (see Chapter 33·IV·B and C).

IX. TRAINING

Exercise training increases the body's ability to take up and use oxygen, manifesting as an increase in $\dot{V}O_{2max}$. All the body's organ systems are affected to some degree, but the changes in muscular and cardiovascular performance are the most evident.

A. Skeletal muscle

Aerobic exercise training and anaerobic exercise training affect muscle performance in different ways.

1. **Aerobic exercise training:** Aerobic training (**endurance training**, e.g., cross-country skiing, running, cycling, swimming) promotes cellular adaptations that increase muscles' ability to store and then process energy substrates aerobically.

    a. **Energy stores:** Training increases myocyte glycogen stores. Glycogen provides a readily available carbohydrate energy source to supplement plasma glucose uptake during exercise. Depletion of muscle glycogen reserves results in fatigue, and work rate must slow, corresponding to "hitting the wall" in marathon running.

    b. **Metabolism:** Exercise training increases aerobic capacity in several ways. Training increases mitochondrial size and numbers. It also upregulates oxidative enzymes involved in the citric acid cycle and in oxidative phosphorylation as well as enzymes that break down glycogen and those involved in β-oxidation (Table 39.2). Muscle myoglobin content increases, thereby enhancing O_2 storage.

2. **Anaerobic exercise training:** Anaerobic exercise training (**resistance training**, e.g., weightlifting and team sports, such as ice hockey, that involve bursts of intense activity) increases force production and fatigue resistance.

    a. **Force production:** Training promotes neural adaptations that increase the efficiency of motor unit activation, which increases force production. Parallel replication of myofibrils within types I and II muscle fibers increases their cross-sectional area and force–production capacity (Fig. 39.13). Type IIx fibers are specialized for speed and force production, but they rely primarily on glycolytic pathways that make them prone to fatigue (see Chapter 12·V·B).

    b. **Fatigue resistance:** Fatigue resistance increases by upregulating enzymes associated with glycolysis to enhance ATP synthesis via the lactic acid system and creatine kinase in the ATP–CP system. Training also increases intramuscular glycogen stores, as seen during aerobic exercise training.

B. Cardiovascular system

Cardiovascular changes occurring during endurance training include an increase in blood volume, cardiac hypertrophy, and increased vascular compliance.

1. **Blood volume:** As noted previously, aerobic exercise activates the renin–angiotensin–aldosterone system (RAAS), leading to salt and water retention and expansion of ECF volume. Although it takes 48 hours for RAAS to become maximally effective, increased lymph flow during exercise flushes albumin and other proteins out of the interstitium, causing plasma oncotic pressure to increase. Plasma volume increases as a result, which raises LV preload. A single intense training session results in a 10% increase in blood volume that persists for days post-exercise. Repeated training sessions increase plasma volume further. Red blood cell production lags behind expansion of plasma volume by as much as a

Table 39.2: Enzymes Upregulated as a Result of Aerobic Exercise Training

Pathway	Enzyme
Glycolysis	Glucokinase
Glycolysis	Phosphofructokinase
Citric acid cycle	Citrate synthase
Citric acid cycle	Succinate dehydrogenase
Electron transport chain	Cytochrome c
β Oxidation	Carnitine palmitoyltransferase

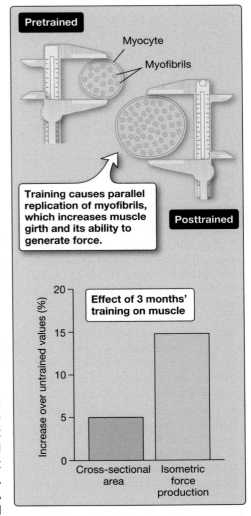

Figure 39.13.
Effects of 3 months of anaerobic weight training on skeletal muscle.

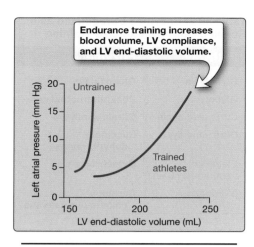

Figure 39.14.
Effects of endurance training on end-diastolic volume. LV = left ventricular.

month, resulting in a physiologic anemia resembling that seen during pregnancy (see Chapter 37·IV·C).

2. **Cardiac hypertrophy:** Expansion of blood volume increases end-diastolic volume (EDV) and results in **eccentric cardiac hypertrophy** (Fig. 39.14). Serial addition of myofibrils within myocytes increases left-ventricular (LV) chamber size and LV mass, which increases inotropy but has minimal effect on myocyte cross-sectional area or LV wall thickness. LV compliance increases to facilitate preloading. Hypertrophy increases resting EDV and SV, which is why trained athletes typically have a slower resting HR than untrained individuals. Resting demand for CO is ~5 L/min in both cases, and CO = HR × SV. Maximal HR decreases by ~3%–7%, most likely to limit the deleterious effects of increasing HR on SV (see Section VI·B).

> Anaerobic exercise training, which involves repeatedly forcing the LV to eject against an elevated MAP, stimulates LV hypertrophy reminiscent of that seen in patients with aortic stenosis and untreated hypertension (see Clinical Application 17.4). **Concentric hypertrophy** is characterized by parallel addition of myofibrils within the myocytes, which increases LV wall thickness at the expense of lumen diameter.

3. **Vascular compliance:** Training increases systemic arterial compliance to facilitate the increased flow during exercise. Training also increases the vasodilator capability of skeletal and cardiac muscle, both effects involving increased production of nitric oxide and other vasodilators. Over time, angiogenesis also increases the density of coronary arterioles, collaterals, and capillaries, thereby decreasing the distance for diffusional exchange of O_2 and nutrients between blood and myocytes.

C. Respiratory system

Aerobic exercise training has no significant impact on lung volumes or capacities, but it does increase ventilation and tissues' ability to extract O_2 from blood.

1. **Ventilation:** Maximal alveolar ventilation and V_E both increase with aerobic exercise training. This likely occurs via aerobic training adaptations in the respiratory muscles to increase fatigue resistance.

2. **Arteriovenous oxygen difference:** Aerobic exercise training increases the a-v O_2 difference, reflecting an increased ability of working muscle to extract O_2 from blood. Adaptations in skeletal muscle O_2 processing, increased capillary density with concomitant decreased diffusional distance between blood and myocytes, and increased blood flow likely account for this effect. Total Hb also increases with aerobic training, which increases blood O_2-carrying capacity.

Biologic Sex and Aging 39.1: Exercise

Exercise capability, as reflected in maximal O_2 uptake ($\dot{V}O_{2max}$), is significantly higher in males compared with females (42–46 mL O_2/kg/min vs. 38–41 mL O_2/kg/min, respectively). This reflects, in part, higher hemoglobin levels (13.5–17.5 g/dL vs. 12.0–16.0 g/dL in females) and higher muscle to fat ratio in males. $\dot{V}O_{2max}$ decreases by ~10% per decade. The effects of aging on exercise capability have not been studied extensively, but the decline in $\dot{V}O_{2max}$ can be explained, in part, by a steady decline in maximal heart rate with age (see Biologic Sex and Aging 16.1) and a 50% decrease in muscle mass (see Biologic Sex and Aging 12.1). A lifelong exercise training program blunts the impact of age significantly: $\dot{V}O_{2max}$ decreases by only 5% per decade in healthy exercising individuals.

3. **Oxygen consumption:** The increase in CO, alveolar ventilation, and a-v O_2 difference combine to increase $\dot{V}O_{2max}$ during training. During periods of physical inactivity, such as bed rest, $\dot{V}O_{2max}$ decreases (Fig. 39.15).

X. BED REST

Sedentary behaviors such as prolonged bed rest or sitting at a workstation or video game console reduce $\dot{V}O_{2max}$ and reverse most of the chronic adaptations to exercise. Reversal occurs more rapidly than many of the exercise adaptations. For example, $\dot{V}O_{2max}$ may decrease 25% to 30% during 2 weeks of bed rest (see Fig. 39.15), whereas it takes 30 to 45 days of aerobic training to reobtain pre–bed rest values. Bed rest decreases skeletal muscle cross-sectional area and strength. For example, 6 weeks of bed rest cause knee extensor cross-sectional area to fall by 15%, and strength decreases by 20%. Skeletal muscle metabolic enzyme activity (e.g., citrate synthase, succinate dehydrogenase, malate dehydrogenase, and β-hydroxyacyl-CoA dehydrogenase) decreases 30% to 40% during ~2 months of inactivity. Plasma volume contracts with inactivity. Cardiac muscle is affected similarly. For example, 21 days of inactivity decreases LV EDV and SV by ~10% each. The decrease in SV causes a corresponding increase in HR to maintain resting CO.

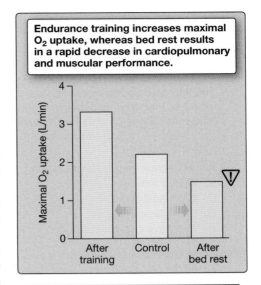

Figure 39.15.
Changes in maximal O_2 uptake with 6 months of endurance exercise (cycling) and 1 month of bed rest.

Chapter Summary

- Exercise involves somatic control of voluntary movement and autonomic nervous system control of the cardiopulmonary system to ensure that working muscles are adequately supplied with oxygenated blood.
- An individual's physiologic capacity for exercise is a function of maximal O_2 uptake and consumption ($\dot{V}O_{2max}$), which is determined by the ability to deliver oxygenated blood to muscle and muscle's ability to extract O_2.
- The **citric acid cycle** and **oxidative phosphorylation** provide most of the energy needed for skeletal muscle during aerobic exercise.
- The **ATP–creatine phosphate** and **lactic acid systems** are used to generate ATP during anaerobic exercise.
- Aerobic activities such as running, swimming, and cycling are characterized by decreased **systemic vascular resistance (SVR)** and a compensatory increase in **cardiac output (CO)** to maintain **mean arterial blood pressure (MAP)** at a level that to ensures adequate perfusion of working muscle and other vascular beds. **Systolic blood pressure (SBP)** increases and diastolic blood pressure (**DBP**) falls due to increased blood flow to active vascular beds during diastole.
- Anaerobic activities such as weightlifting increase MAP, SBP, and DBP due to the effects of sustained contractions on SVR.
- **Endurance training** is associated with increased blood volume to chronically increase **cardiac preload**. End-diastolic volume increases and the myocardium undergoes **eccentric hypertrophy**. Resting stroke volume increases and resting heart rate decreases.
- Simultaneous increases in muscle capillary density and mitochondrial size and numbers increase O_2 extraction.
- **Resistance training** increases muscle size, strength, stored energy substrates, and anaerobic enzyme levels. The myocardium undergoes **concentric hypertrophy** in response to the high arterial pressures associated with such resistance training.
- Bed rest and other sedentary behaviors reverse training adaptations and decrease an individual's physiologic capacity for exercise.

Study Questions

Choose the ONE best answer.

39.1. A 25-year-old female undergoes surgery for a strained anterior cruciate ligament. Postsurgical rehabilitation includes isometric quadriceps contractions that are held until fatigue (~60 s). These exercises most likely use which of the following as the primary energy system?

A. ATP–creatine phosphate system
B. ATP stores
C. Citric acid cycle
D. Lactic acid system
E. Oxidative phosphorylation

Best answer = D. The lactic acid system predominates during maximal exercise with durations between 30 seconds and 2.5 minutes that causes fatigue (see Section III·A). Glycolysis produces pyruvic acid, which is then shuttled to lactic acid. ATP stores (B) can support exercise for a few seconds, and the ATP–creatine phosphate system (A) extends this time to 8–10 seconds. Aerobic metabolism (citric acid cycle [C] and oxidative phosphorylation [E]) is the primary energy system used to synthesize ATP during maximal exercise lasting >2.5 min.

39.2. At the beginning of exercise, a feedforward mechanism increases heart and respiratory rates. What is the term or receptor most likely responsible for this mechanism?

A. Arterial baroreceptors
B. Central command
C. Class III muscle afferents
D. Class IV muscle afferents
E. Peripheral chemoreceptors

Best answer = B. Central command is the feedforward signal that increases cardiovascular and respiratory function at the beginning of or in anticipation of exercise (see Section IV·B·2). Baroreceptors (A) provide feedback regarding arterial pressures at the aortic arch and carotid arteries. Class III and IV muscle afferents (C, D) provide feedback regarding muscle stretch, compression and metabolic status. Chemoreceptors (E) provide feedback regarding arterial partial pressure of CO_2 and O_2 and H^+ levels (see Section IV·A).

39.3. An exercise-induced rightward shift in the O_2–hemoglobin (Hb) dissociation curve is most likely responsible for increasing which of the following respiratory parameters during aerobic exercise?

A. Alveolar ventilation
B. Arteriovenous O_2 difference
C. Excess postexercise oxygen consumption
D. O_2-carrying capacity
E. Work of breathing

Best answer = B. The arteriovenous (a-v) O_2 difference widens with aerobic exercise through increased O_2 offloading by Hb (see Section VI·B). This manifests as a decrease in venous O_2 content, so that even though arterial levels are unchanged, the a-v difference increases. Increased offloading occurs due to a rightward shift in the Hb–O_2 dissociation curve (see Chapter 23·IV·C). Alveolar ventilation (A) and work of breathing (E) both increase during exercise but are unrelated to the Hb–O_2 dissociation curve. Postexercise O_2 consumption (C) does not impact O_2 usage during exercise. Blood's O_2-carrying capacity (D) is determined primarily by Hb concentration, not by Hb's O_2 affinity.

39.4. A 65-year-old male has a history of chronic obstructive pulmonary disease (COPD). He is on the standard pharmacology treatment regimen for mild COPD but is also encouraged to begin an outpatient pulmonary rehabilitation program that involves aerobic exercise. Which of the following lung parameters will most likely increase because of this program?

A. Forced expiratory volume in 1 second
B. Forced vital capacity
C. Maximal minute ventilation
D. Resting physiologic dead space
E. Total lung capacity

Best answer = C. Maximal minute and alveolar ventilation both increase with aerobic exercise training, due to improvements in respiratory muscle function. Increased ventilation enhances alveolar gas exchange (see Section VI·D). Pulmonary function test volumes (A) and capacities (B, E) do not change with aerobic exercise training. Physiologic dead space (D) remains the same or may decrease slightly with aerobic exercise training.

Systems Failure

40

I. OVERVIEW

I like a look of Agony,
Because I know it's true –
Men do not sham Convulsion,
Nor simulate, a Throe –

The Eyes glaze once – and that is Death –
Impossible to feign
The Beads upon the Forehead
By homely Anguish strung.

Emily Dickinson

We are all destined to die.

Living depends on a delicate homeostatic balancing act. During life, we rely on our organ systems to compensate for changes in innumerable internal parameters, including P_{O_2} and P_{CO_2}, pH, electrolyte levels, and body temperature. Ultimately, however, all these compensatory systems slowly falter and then fail. At the cellular level, this process is known as senescence and apoptosis. At the organismal level, we know it as aging and death.

The Centers for Disease Control and Prevention periodically publishes a list of the leading causes of death in the United States (Table 40.1). The list does not include "old age" because it is based on death certificates, which require physicians to identify a specific causal event (e.g., heart failure [HF]). From a physiologic perspective, however, corporeal death usually reflects a long series of individual, aging-related cell deaths. Cell by cell, all organs age and, eventually, fail. Which organ falls off the homeostatic tightrope first may be a matter of chance or may be determined by an underlying disease or lifestyle choice. In this final chapter, we consider various causes and consequences of individual organ failure. There are many other causes of death (e.g., accidents and trauma), as shown in Table 40.1, but, regardless, death of the individual occurs when the cerebral hemispheres are O_2 deprived and the cortex dies, either because of cardiovascular, respiratory, renal, or multiple organ system failure. *The Eyes glaze once.*

II. AGING AND DEATH

Gerontology is a relatively new discipline dealing with old issues (and issues encountered by older adults). Although researchers have forwarded many ideas as to why cells and organ systems inevitably lose functionality and fail, there are no solutions to the age-old problem of why we die. Preprogrammed cell death (**apoptosis**) is probably just one of many contributing factors. Regardless, the human lifespan is limited to ~120 years. Medical advances over the past 100 years may have

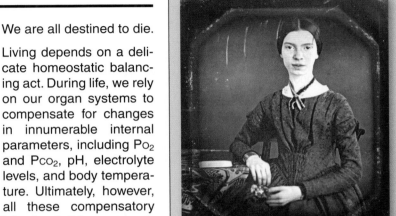

Figure 40.1.
Emily Dickinson (American poet, 1830–1886).

Table 40.1: Leading Causes of Death in the United States in 2021

Rank	Cause of Death
1	Heart disease
2	Cancer
3	COVID-19
4	Accidents
5	Stroke
6	Chronic lower respiratory disease
7	Alzheimer disease
8	Diabetes mellitus
9	Chronic liver disease and cirrhosis
10	Kidney disease

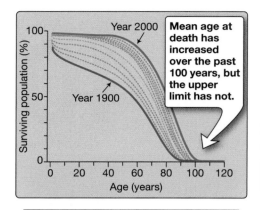

Figure 40.2.
Mean life expectancy in the United States, 1900–2000.

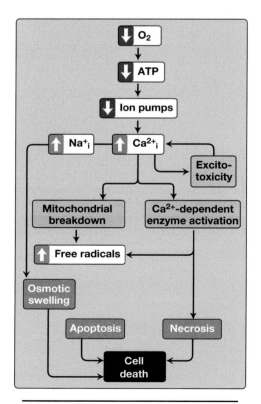

Figure 40.3.
Ischemic cascade. Ca^{2+}_i and Na^+_i = intracellular Ca^{2+} and Na^+ concentration.

increased mean life expectancy but not the upper limit (Fig. 40.2), suggesting that failure and death may be genetically predetermined.

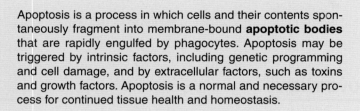

Apoptosis is a process in which cells and their contents spontaneously fragment into membrane-bound **apoptotic bodies** that are rapidly engulfed by phagocytes. Apoptosis may be triggered by intrinsic factors, including genetic programming and cell damage, and by extracellular factors, such as toxins and growth factors. Apoptosis is a normal and necessary process for continued tissue health and homeostasis.

A. Homeostenosis

Individual physiologic responses to aging vary widely and can be significantly impacted by physical fitness and underlying disease, but aging is generally accompanied by progressive decreases in cell numbers and reduced functionality of most organ systems. Although basic homeostatic responsibilities are typically not affected, aging does reduce the ability to respond to challenge ("**homeostenosis**"). Homeostenosis manifests as a gradual loss of physiologic reserve and homeostatic agility. For example, although aging has minimal effect on basal heart rate (HR), stroke volume, ejection fraction (EF), or cardiac output (CO), a progressive decline in maximal HR limits an individual's ability to increase CO during exercise. Eventually, even responding to routine daily challenges may become limiting (e.g., climbing stairs). Similar changes occur throughout the body's various tissues. Decreased reserves can limit responses to environmental challenges, increasing susceptibility to heatstroke and hypothermia, for example. Although physiologic aging is not a disease, homeostenosis greatly increases vulnerability to disease. Decreased reserves and slowed responsiveness can allow a disease process to rapidly intensify and accelerate once established.

B. Cell death

Death has many causes, but the final common pathway for most diseases and failing organ systems is O_2 deprivation resulting from inadequacy of perfusion (**shock**; see Sections III and IV). All organs are dependent on O_2 for continued survival. O_2 restriction caused by interruption of local blood supply (**ischemia**) or reduced arterial O_2 levels (**hypoxemia**) initiates a sequence of biochemical events known as an **ischemic cascade**. Significant events include switchover to anaerobic metabolism, dissipation of ion gradients, Ca^{2+}-induced toxicity, mitochondrial breakdown, apoptosis, and necrosis (Fig. 40.3).

1. **Anaerobic metabolism:** O_2 deprivation forces cells to convert from primarily aerobic to exclusively anaerobic metabolism to generate ATP. The transition is the metabolic equivalent of switching over to an emergency gas-powered generator during a domestic power outage. The generator keeps a few vital systems running, but output is limited by the size of the generator and the capacity of the gas tank (glycogen stores). Also, the exhaust fumes can be deadly in the absence of adequate ventilation. Anaerobic "exhaust"

comes in the form of lactic acid, which causes acidosis. Lactic acid is produced even in healthy individuals during intense muscle activity (see Chapter 39·IV·B), but local levels remain relatively low because the circulation limits build up. If the biologic power outage reflects perfusion failure, however, lactic acid levels build rapidly, and intracellular pH falls, which further compromises cell function.

2. **Ion gradients:** Falling ATP levels limit the ability of ion pumps (e.g., Na^+/K^+-ATPase and Ca^{2+}-ATPase) to maintain transmembranous ion gradients (Fig. 40.4). Membrane potential depolarizes, and intracellular Ca^{2+} concentration rises as a result. In excitable cells, depolarization triggers cationic influx via voltage-dependent Na^+ and Ca^{2+} channels and K^+ efflux via K^+ channels, which effectively collapses the ion gradients within seconds. Intracellular fluid (ICF) osmolality rises as a result, causing water to enter by osmosis.

3. **Calcium toxicity:** Ca^{2+} influx and release from intracellular stores activates several signaling pathways that ultimately destroy the cell. These include ATPases, lipases, endonucleases, and Ca^{2+}-activated proteases such as calpains. Calpains are regulatory proteases under normal circumstances. When activated by ischemia-induced rises in intracellular Ca^{2+} concentration, calpains destroy the cytoskeleton and, with help from Ca^{2+}-dependent lipases, digest the plasma and intracellular membranes (Fig. 40.5). The cell swells, lyses, and dies (**necrosis**).

> Necrosis is pathologic cell death caused by trauma, culminating in lysis and release of cellular contents. These materials trigger an inflammatory response that typically causes extensive cellular damage. This contrasts with apoptosis, in which damaged and dying cells stimulate phagocytosis, and their contents remain contained within membranes.

4. **Excitotoxicity: Excitotoxicity** is an aggressive positive feedback pathway that makes the brain highly vulnerable to O_2 deprivation (see Fig. 40.3). Ischemia-induced increases in intracellular Ca^{2+} concentration cause synaptic vesicles to fuse with the synaptic membrane, releasing their contents into the synaptic cleft (see Chapter 5·IV·C). These vesicles often contain glutamate, which is the brain's principal excitatory neurotransmitter. Postsynaptic glutamate receptors (e.g., *N*-methyl-D-aspartate receptors) are Ca^{2+} permeable, so intracellular Ca^{2+} levels rise even faster in neurons than they do in nonexcitable tissues (see Table 5.2). Thus, the ischemic cascade is accelerated in brain tissue.

5. **Mitochondrial breakdown:** Reduced O_2 availability impairs mitochondrial function and increases **reactive oxygen species (ROS)** accumulation. ROS include the superoxide anion ($O_2^{-}\cdot$), hydrogen peroxide (H_2O_2), and the hydroxyl radical ($OH\cdot$), all produced by the mitochondrial electron transport chain (Fig. 40.6). ROS are extremely damaging to cells because they react with and break molecular bonds in lipids, proteins, and DNA. Cells normally aggressively defend themselves against ROS using enzymes

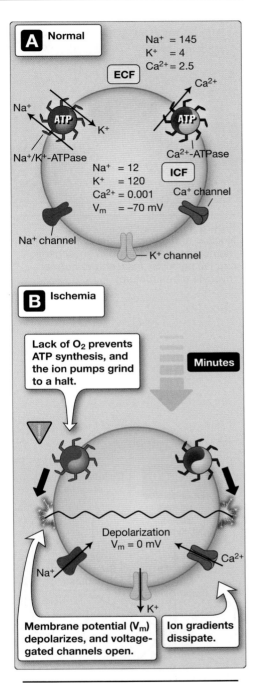

Figure 40.4.
Dissipation of transmembrane ion gradient during ischemia. All ion concentrations are given in mmol/L.

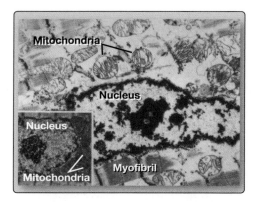

Figure 40.5.
Nuclear and mitochondrial swelling and membrane deterioration in an ischemic cardiac myocyte. *Inset* shows normal myocyte ultrastructure.

Clinical Application 40.1: Therapeutic Hypothermia

Most patients (95%) who suffer cardiac arrest outside of a hospital do not survive, even with attempted resuscitation. Death occurs largely due to neurologic damage sustained during ischemic cascade progression and exacerbated by distribution of inflammatory mediators when the circulation is restored (**reperfusion injury**). The chances of surviving myocardial infarction and avoiding neurologic damage have improved significantly in recent years through use of **targeted temperature management**, or **therapeutic hypothermia (TH)**, during which body temperature is reduced to 33°–36°C (91.4°–96.8°F) for 12–24 hours following the ischemic event. Target temperatures are achieved by infusing a patient with chilled intravenous fluids, often combined with surface cooling. TH is beneficial because it reduces the extent of mitochondrial breakdown and limits inflammatory mediator release during an ischemic cascade.

(e.g., superoxide dismutase and peroxidase) and ROS scavengers (e.g., vitamins C and E). During ischemia, however, rising ROS levels increase mitochondrial membrane permeability, causing organellar swelling and release of electron-chain constituents that initiate apoptosis. If cell necrosis does not occur within the first few minutes, apoptotic pathways impel cellular suicide over a prolonged timescale, but the end result is nevertheless the same.

C. Brain death

Brain death *is* death. Although our bodily tissues can be sustained artificially following brain death, every trait we associate with being human, including personality, intellect, and awareness of self and others, is a function of the brain. Therefore, when the brain dies, we die. Verifying brain death clinically requires performing a set of neurologic tests designed to establish a complete and irreversible loss of critical brain functions and reflexes, even though spinal reflexes may persist. Assessing brain function includes testing for the absence of a pupillary light reflex (see Chapter 8·II·C) or caloric reflex (response to irrigating the ear canal with warm or cold water; see Clinical Application 9.3). Both assess brainstem function. Establishing brain death also requires that a patient be provided with 100% O_2 and then disconnected from a ventilator and observed for 8 to 10 minutes to confirm the complete absence of spontaneous respiration even as arterial P_{CO_2} climbs >60 mm Hg (**apnea test**). Reflex increases in respiratory effort induced by hypercapnia are one of the most basic and essential brain functions (see Chapter 24·III·C). Some patients may survive a severe ischemic event and progress to a **persistent vegetative state (PVS)**. PVS patients retain sufficient autonomic brainstem function to preserve basic cardiovascular and pulmonary reflexes yet show no signs of awareness or comprehension. PVS patients typically die from multiorgan failure, infection, or other causes within 2 to 5 years.

III. SHOCK CLASSIFICATIONS

All tissues in the body, including the heart and vasculature, are dependent on the cardiovascular system to deliver O_2 in amounts sufficient to meet their metabolic needs. The brain has a high dependence on O_2, and loss of consciousness occurs within seconds of interrupting blood flow.

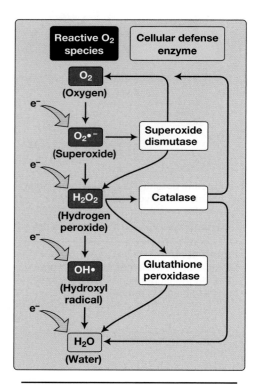

Figure 40.6.
Reactive O_2 species produced by the mitochondrial electron (e⁻) chain.

Tissues with low O_2 demands can tolerate ischemia for longer periods, but ultimately all tissues die when O_2 deprived. Inadequacy of flow and O_2 delivery results in **shock**. There are four main types of shock: **hypovolemic**, **cardiogenic**, **obstructive**, and **distributive**.

A. Hypovolemic

Hypovolemic shock is caused by a decrease in circulating blood volume. When blood volume falls, the extent to which the left ventricle (LV) fills during diastole (i.e., LV preload; see Chapter 17·IV·D) decreases also, which compromises CO as shown in Figure 40.7. When CO falls, mean arterial pressure (MAP) falls also, which reduces the amount of oxygenated blood reaching tissues. Hypovolemic shock can be further divided into two categories: hemorrhagic shock and shock caused by loss of extracellular fluid (ECF).

1. **Hemorrhagic:** Hemorrhagic shock results from loss of whole blood from the vasculature (**extravasation**). Blood loss to the external environment typically occurs as a result of trauma (see Fig. 40.7B) but can also occur internally upon rupture of esophageal or stomach varices, for example. A bone fracture or ruptured abdominal aortic aneurism can also cause significant blood loss to internal compartments.

2. **Fluid loss:** Hypovolemic shock can also result from ECF volume contraction, due either to fluid loss to the external environment or to the interstitium and abdominal cavities ("**third spacing**"). Fluid is lost to the environment via the kidneys; through sweating, vomiting, and diarrhea; and following significant skin burns (see Chapter 15·III·B). Third spacing occurs when plasma protein concentrations fall, either because of liver failure and impaired ability to synthesize plasma proteins or increased capillary permeability to proteins.

> Plasma proteins create an osmotic potential (plasma colloid osmotic pressure [π_c]) that is the main force holding fluid in the vasculature, as defined by the **Starling law of the capillary**:
>
> $$Q = K_f \left[\left(P_C - P_{if} \right) - \left(\pi_C - \pi_{if} \right) \right]$$
>
> where Q is net fluid flow across the capillary wall, K_f is a filtration coefficient, P_c is capillary hydrostatic pressure, P_{if} is interstitial fluid pressure, and π_{if} is interstitial colloid osmotic pressure (see Chapter 18·VII·D).

B. Cardiogenic

Cardiogenic shock is caused by cardiac pump failure. Common causes include **dysrhythmia**, **mechanical issues**, and **cardiomyopathies**.

1. **Dysrhythmia:** Cardiogenic shock can result from atrial or ventricular dysrhythmias. Dysrhythmias prevent or impair coordinated contraction of one or more cardiac chambers, which reduces CO. Ventricular tachycardia and fibrillation cause complete loss of CO and prove rapidly fatal unless the arrhythmia is corrected by cardioversion using an external electrical defibrillator (see Chapter 16·VI·D).

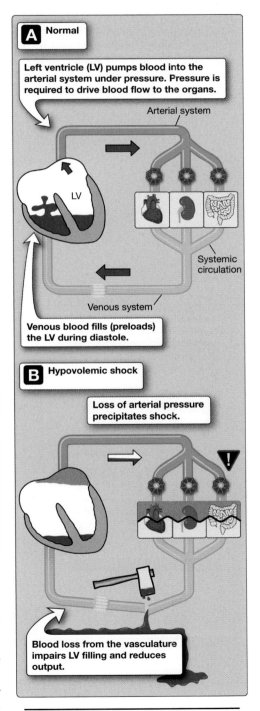

A Normal

Left ventricle (LV) pumps blood into the arterial system under pressure. Pressure is required to drive blood flow to the organs.

Arterial system

LV

Systemic circulation

Venous system

Venous blood fills (preloads) the LV during diastole.

B Hypovolemic shock

Loss of arterial pressure precipitates shock.

Blood loss from the vasculature impairs LV filling and reduces output.

Figure 40.7.
Hypovolemic shock.

Clinical Application 40.2: Sepsis

Sepsis is a clinical syndrome reflecting a systemic inflammatory response to infection. It is characterized by bacteremia, fever, tachycardia, and increased respiratory rate. Although sepsis can be caused by a variety of organisms, it is often seen in association with gram-negative infections, in which a bacterial cell wall component (lipopolysaccharide [LPS]) triggers an inflammatory process leading to **systemic inflammatory response syndrome**.[1] LPS binds to and is recognized by a receptor on the surface of phagocytes, which respond by releasing cytokines and initiating an inflammatory response and fever. Vascular endothelial cells respond to bloodborne cytokines by releasing more cytokines and chemokines, thereby amplifying the inflammatory response. They also initiate blood coagulation. This inflammatory cascade also includes neutrophil activation and release of reactive O_2 species, causing extensive and widespread vascular damage. Resistance vessels and veins lose their resting tone, thereby increasing vascular capacity. Capillary permeability may be increased also, allowing plasma proteins and fluids to leak into the interstitium. Sepsis mortality rates can be as high as 50%, increasing to 90% when shock develops. Treatment options include antibiotics to address the underlying infection, intravenous fluids to help maintain effective circulating blood volume, and vasopressors to increase vascular tone.

2. **Mechanical:** Incompetent and stenotic heart valves both challenge the ability of the myocardium to maintain basal CO due either to backflow or impaired forward flow, respectively. Septal defects that allow left-to-right ventricular backflow can also precipitate cardiogenic shock.

3. **Cardiomyopathy:** The causes and consequences of heart disease are considered in more detail in Section V. Myocardial infarction (MI) that damages >40% of the LV wall is the most common cause of cardiogenic shock and death.

C. Obstructive

Obstructive shock is due to **obstruction of ventricular outflow** or **venous return** (VR) or **impaired diastolic filling**.

1. **Ventricular outflow:** Obstruction of ventricular outflow may occur as a result of **pulmonary embolism** (**PE**) (Fig. 40.8). It may also occur because of systemic emboli, **pulmonary hypertension**, or **aortic stenosis**. PE and pulmonary hypertension limit right ventricular (RV) output, which limits LV preload.

2. **Venous return:** Common causes of decreased VR include **intrathoracic tumors** and **tension pneumothorax**. Tension pneumothorax is a condition in which pressure within the chest increases with every breath caused by air entrapment within the pleural space.

3. **Diastolic filling:** Conditions that impair diastolic filling include **tamponade**, **constrictive pericarditis**, and **restrictive cardiomyopathy**. Tamponade is caused by fluid accumulation (e.g., blood or a pericardial effusion) between the pericardium and heart wall. The presence of fluid prevents normal ventricular filling (Fig. 40.9). Inflammation-induced pericardial thickening or a grossly hypertrophied and stiffened ventricular wall can similarly limit ventricular filling.

D. Distributive

Most arterial and venous vessels have a resting tone controlled by the sympathetic nervous system (SNS) as a way of limiting cardiovascular

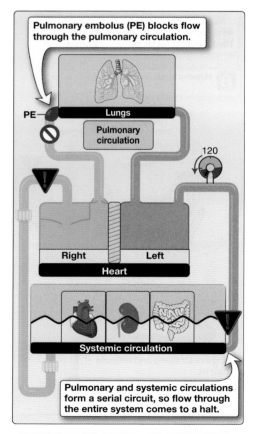

Pulmonary embolus (PE) blocks flow through the pulmonary circulation.

Pulmonary and systemic circulations form a serial circuit, so flow through the entire system comes to a halt.

Figure 40.8.
Consequences of pulmonary embolism.

[1]For more information on differences between gram-positive and gram-negative bacteria, see *LIR Microbiology*, 4e, Chapter 6·I·C.

capacity to ~5 L (see Chapter 19·V). Distributive, or **vasodilatory**, shock occurs when SNS vascular control is compromised, resulting in vasodilation and an exponential increase in vascular capacity. MAP dissipates rapidly as blood surges through dilated resistance vessels and becomes trapped in capillary beds and veins (Fig. 40.10). Distributive shock has many causes. The most common include **sepsis** (see Clinical Application 40.2), **systemic inflammatory response syndrome**, and **anaphylaxis**.

IV. SHOCK STAGES

We can divide shock progression into three stages, beginning with the initial causal event and then progressing in a sequential manner through **preshock** with compensation, progressive or non-compensated **shock**, and end-organ dysfunction and **failure** (irreversible shock). The following discussion uses **hemorrhagic shock** as an example to illustrate how the body responds to the initial causal event and how the systems that attempt to compensate for loss of MAP can create positive feedback spirals that may ultimately hasten failure and lead to death.

A. Preshock

Hemorrhage depletes blood volume and drains the venous reservoir. Hemorrhage depletes veins preferentially because the heart continues transferring blood from the venous compartment to arteries and their dependent capillaries until the venous compartment is depleted. Loss of preload causes MAP to fall, triggering an SNS-mediated baroreceptor reflex (see Fig. 19.12; also see Chapter 19·III). The SNS redirects blood flow away from less essential organs, increases cardiac inotropy and HR (Fig. 40.11), and mobilizes blood reservoirs. These pathways are summarized in Figure 40.12.

1. **Redirection of flow:** Flow to less essential organs is reduced by selective SNS-mediated constriction of resistance vessels. Systemic vascular resistance (SVR) rises as blood flow is directed away from splanchnic, cutaneous, muscle, and renal vascular beds. Deprived tissues increase their O_2 extraction from the residual supply, causing mixed venous O_2 saturation (SvO_2) to fall. Reduced flow to the kidney triggers renin release from the juxtaglomerular apparatus (JGA) and activates the long-term fluid retention pathways. Two key components (**angiotensin II** and **antidiuretic hormone**) are vasoactive and potentiate SNS-mediated vasoconstriction (see Chapter 28·III).

2. **Cardiac efficiency:** SNS stimulation of the myocardium increases HR and contractility to help compensate for loss of preload (see Fig. 40.11). Epinephrine release from adrenal glands during SNS activation contributes to tachycardia and increased inotropy during preshock.

3. **Venoconstriction:** SNS stimulation of veins increases their tone and decreases their capacity, forcing blood back to the heart. VR is aided by a steepening of the pressure gradient between capillary beds and the right atrium.

4. **Transcapillary refill:** The baroreceptor reflex helps preserve flow to critical organs during the first few minutes after a hemorrhage. It also buys time that allows fluid to migrate from the interstitium to the

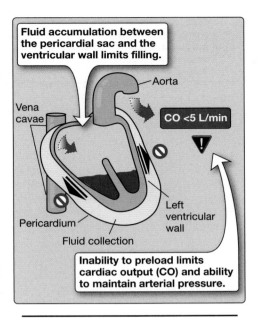

Figure 40.9.
Tamponade effects on ventricular filling.

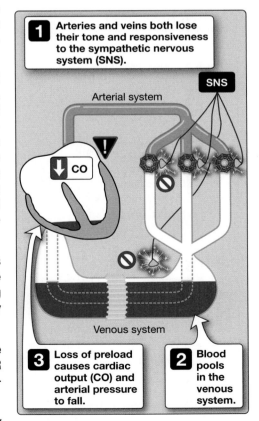

Figure 40.10.
Distributive shock.

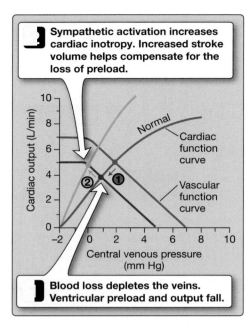

Sympathetic activation increases cardiac inotropy. Increased stroke volume helps compensate for the loss of preload.

Blood loss depletes the veins. Ventricular preload and output fall.

Figure 40.11.
Reflex increases in cardiac inotropy during hypovolemia.

vasculature, a process called **transcapillary refill**. Transcapillary refill is a primary survival mechanism that relies on Starling forces to recruit interstitial fluid (see earlier discussion). Resistance vessel constriction reduces capillary hydrostatic pressure and allows π_c to drive fluid movement from the interstitium into the vasculature (see Fig. 18.28). Fluid influx dilutes the plasma proteins and reduces π_c, but transcapillary refill can still replace ~75% of lost blood volume during hemorrhage (see Fig. 40.12).

B. Shock

The baroreceptor reflex effectively compensates for decreases in circulating blood volume of ~10%, so the only early sign of imminent shock may be mild tachycardia. Once blood volume drops by more than ~10%, however, compensatory mechanisms may no longer be adequate to sustain perfusion in critical circulations, and signs of shock may become overt. These signs include **hypotension**; cold, clammy skin; decreased urine output; and a rise in plasma lactate levels.

> Most organ systems have **functional reserves** that permit homeostasis even as system capacity is reduced. The efficacy of cardiovascular reserves explains why an individual can donate a unit of blood with little or no detrimental effect on MAP. Reserve capacity dwindles with age, as discussed previously.

1. **Hypotension:** When blood volume drops below ~60%, increases in HR and inotropy alone are unable to compensate for loss of preload, and systolic blood pressure drops to 90 mm Hg or below. Intense SNS-mediated constriction of the vasculature limits blood outflow from the arterial system and keeps diastolic blood pressure high; thus, MAP is maintained at a level that sustains blood flow to the critical cerebral and coronary circulations. These two circulations are regulated primarily through autoregulatory mechanisms (e.g., CO_2, K^+, and lactate release) and therefore are not directly influenced by SNS activity.

2. **Skin:** Intense SNS activation raises SVR by effectively shutting off flow to the vascular beds that occupy the lowest positions on the circulatory hierarchy (see Chapter 19·II·F), including the gastrointestinal (GI) and cutaneous circulations. The intensity of SNS activation is clearly apparent in the skin, which becomes cold and clammy. Cooling is due to intense cutaneous vasoconstriction, which reduces flow to <6 mL/min. Blood drains from the skin, and its temperature cools accordingly. SNS activation also stimulates sweat glands. Because sweat is a modified blood filtrate, when blood flow is curtailed, output is minimal. The skin becomes slightly clammy to the touch.

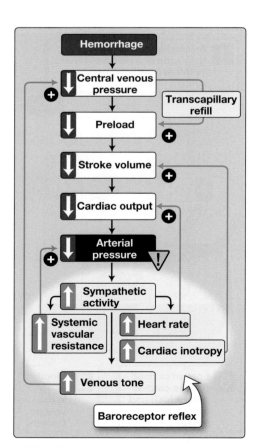

Figure 40.12.
Pathways that preserve arterial pressure during preshock.

> Peripheral perfusion can be quickly assessed using **capillary refill time**. Pressure is applied to the nailbed of a finger or toe for 10 sec to cause blanching. Normally, it should take 2 sec or less for reperfusion to restore a normal color once pressure is released. Delayed capillary refill is a reliable indicator of hypovolemia.

3. **Urinary output:** The renal glomerular afferent and efferent arterioles are both resistance vessels. During intense SNS activation, flow through both is severely curtailed, and glomerular hydrostatic pressure (P_{GC}) falls dramatically (Fig. 40.13; see Chapter 25·IV·F). P_{GC} determines glomerular filtration rate (GFR), so flow through the tubule slows, and urine output can drop to <30 mL per hour (**oliguria**).

4. **Metabolic acidosis:** Plasma lactate levels are normally 0.5 to 1.5 mmol/L, but hypoxia forces many tissues to rely on anaerobic metabolism, causing lactate levels to rise. A plasma lactate of >4 mmol/L is consistent with shock, although levels can rise under other circumstances also (e.g., ketoacidosis and anaerobic exercise).

C. System failure

The actions described earlier may be insufficient to ensure patient survival, even though arterial pressure may renormalize for an hour or two. Blood pressure alone may not reliably reflect adequacy of perfusion in early shock because the central nervous system (CNS) cardiovascular control centers have the ability and determination to maintain MAP at levels that ensure continued flow to the cerebral circulation to the very last. In cases of severe hemorrhage, this is accomplished by holding SVR at levels that compromise organs occupying lower positions on the circulatory hierarchy (see Chapter 19·II·F), including the kidneys and GI system. Once the invisible line demarcating compensated and irreversible shock has been crossed, a positive feedback spiral begins that leads inevitably to organ failure and death (Fig. 40.14).

The need to restore circulating blood volume as soon as possible after trauma is a major reason for the widespread use of mobile trauma teams and Medevac helicopters. Rapid-response units allow medical personnel to reach the scene of an accident and administer intravenous fluids to a patient within the critical window before irreversible tissue damage occurs (a period of variable duration often referred to in emergency medicine as the "**golden hour**").

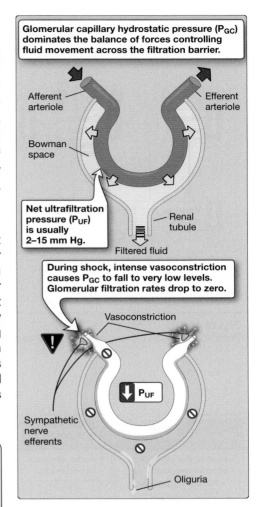

Glomerular capillary hydrostatic pressure (P_{GC}) dominates the balance of forces controlling fluid movement across the filtration barrier.

Afferent arteriole. Efferent arteriole. Bowman space. Net ultrafiltration pressure (P_{UF}) is usually 2–15 mm Hg. Filtered fluid. Renal tubule.

During shock, intense vasoconstriction causes P_{GC} to fall to very low levels. Glomerular filtration rates drop to zero.

Vasoconstriction. Sympathetic nerve efferents. P_{UF}. Oliguria.

Figure 40.13.
Effects of intense sympathetic activation on glomerular blood flow.

1. **Cardiac depression:** If MAP drops below 60 mm Hg, the myocardium becomes ischemic through inadequacy of coronary perfusion. Ischemia impairs myocardial contractility and pressure falls further. So begins a positive feedback cycle that results in acute HF. During hemorrhage, one or more **myocardial-depressant factors** may be released from ischemic tissues that further challenge cardiac function (see Fig. 40.14).

2. **Sympathetic escape:** The SNS cannot maintain intense vasoconstriction for prolonged periods, so SVR eventually falls. Resistance vessel dilation ("**sympathetic escape**") may be due to SNS neurotransmitter depletion, α-adrenergic receptor desensitization, or chronically elevated metabolite concentrations overriding central control. Venoconstrictor influence ultimately fails also, thereby impairing VR and preload (see Fig. 40.10).

3. **Acidemia:** Lactic acid and high P_aCO_2 (due to inadequacy of tissue perfusion and pulmonary and renal dysfunction) together cause significant acidemia. Acidemia impairs myocyte function and further

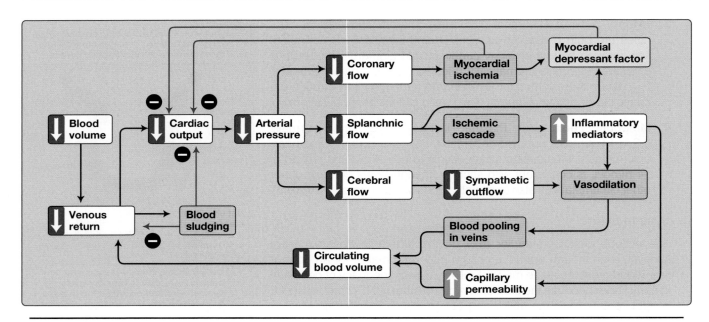

Figure 40.14.
Positive feedback pathways that cause cardiovascular system failure.

challenges the ability of the myocardium and vasculature to sustain CO and SVR, respectively. MAP falls yet further as a result.

4. **Increased blood viscosity:** When blood is moving slowly, red blood cells (RBCs) and other blood components adhere to each other, which raises blood viscosity (see Chapter 18·III·C). The process is exacerbated by acidemia and involves not only RBCs but also leukocytes and platelets, causing blood "**sludging**" (see Fig. 40.14). Sludging increases resistance to flow through the vasculature, and, because MAP cannot rise to compensate, tissue perfusion falls. In time, the microvessels (i.e., capillaries, arterioles) become plugged with clots.

5. **Cellular deterioration:** With prolonged hypoxemia, cell integrity breaks down, triggering an inflammatory response. Inflammatory mediators increase vascular permeability, and plasma exudes into the interstitial space at the expense of blood volume. Deterioration of the GI epithelial lining breaches the barrier separating the gut contents from the vasculature, allowing microorganisms to gain access to the circulation. The likelihood of septic shock when (and if) circulation is restored is increased greatly as a result.

6. **Cerebral depression:** Prolonged hypoxemia ultimately impacts the brain. Neural activity is depressed, and the cardiovascular and respiratory control centers fail. As sympathetic output wanes, MAP declines. Cerebral perfusion pressure decreases also, and brain death soon follows.

V. HEART FAILURE

HF can be the final common pathway for virtually all forms of cardiac disease, and, accordingly, there are many underlying causes (Fig. 40.15). Although the time course of failure can vary widely, it may ultimately result in cardiogenic shock.

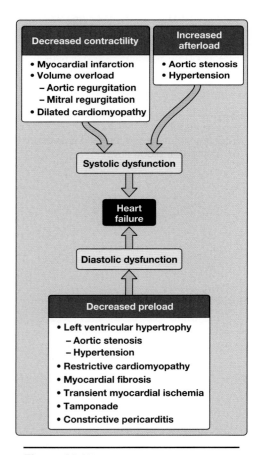

Figure 40.15.
Common causes of heart failure.

A. Causes

There can be considerable overlap in how the various underlying causes of HF impact cardiac performance. The RV and LV face different challenges and can fail independently of each other, but the left heart is so dependent on the right (and vice versa) that failure of either side independently elicits similar compensatory mechanisms.

1. **Right heart:** The right heart is a thin-walled chamber that generates low peak systolic pressures against a low pulmonary vascular resistance (PVR). If flow resistance increases as a result of PE or pulmonary hypertension, for example, it has limited ability to compensate; hence, failure develops.

> || The most common cause of right HF is left HF.

2. **Left heart:** The left heart is a thick-walled chamber well adapted to stress associated with generating high peak systolic pressures against a high SVR. Causes of left HF can be grouped according to whether they impair filling (**diastolic HF**, or HF with preserved EF [**HFpEF**]) or ejection (**systolic HF**, or HF with reduced EF [**HFrEF**]).

   a. **Diastolic (EF ≥50%):** One common cause of diastolic failure is LV hypertrophy, due either to a chronically increased afterload or cardiomyopathy. Untreated hypertension and aortic stenosis both impair CO by increasing LV afterload. The myocardium hypertrophies to repeatedly generate the high pressures required to maintain a normal CO (see Clinical Application 17.4). New myofibrils are added in parallel with old myofibrils, causing individual myocytes to increase their girth, thickening the ventricular wall, a process known as **concentric hypertrophy** (Fig. 40.16). The advantage to a thicker wall is that it helps offset the effects of high intraventricular pressure on wall stress, as described by the law of Laplace (Fig. 40.17; see Chapter 17·VI·B). The disadvantages to concentric hypertrophy are twofold. First, myocyte diameter may exceed the diffusional limits for O_2, which increases the likelihood of ischemia (see Fig. 40.16) and arrhythmias. Second, the ventricle stiffens and becomes increasingly difficult to fill, requiring higher RV ejection pressures. Ultimately, both ventricles fail under such circumstances.

   b. **Systolic (EF ≤40%):** Systolic failure occurs when the LV fails to maintain adequate output. This can be due to impaired contractility or an excessive afterload, but the most common cause of systolic failure is MI, as discussed in the next section.

B. Myocardial infarction

MI, or a "heart attack," is one of the most common causes of HF. An MI typically occurs when an atherosclerotic plaque ruptures and forms a blood clot that occludes a coronary supply vessel. Myocytes that had previously been served by the occluded vessel become ischemic and die, which impairs myocardial contractility. The chances of surviving such an event depend on many factors. If the infarcted area is relatively

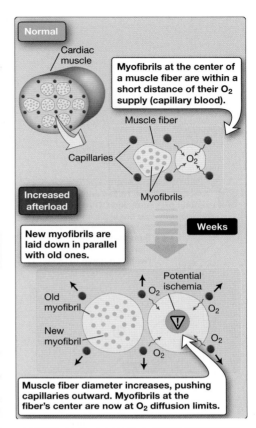

Figure 40.16.
Effects of concentric myocardial hypertrophy on O_2 delivery to myofibrils.

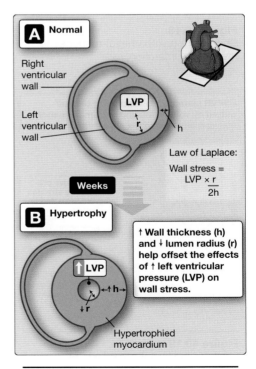

Figure 40.17.
Increases in ventricular wall thickness during cardiac hypertrophy.

small, short-term responses may allow the patient to survive the initial insult until long-term compensatory mechanisms become active.

C. Compensation

A small infarct triggers short- and long-term compensatory mechanisms simultaneously. Short-term events help maintain output until the long-term pathways have had time to activate fully.

1. **Short term:** The short-term response to myocardial ischemia includes both local and central reflexes.

   a. **Local:** Interruption of blood flow to myocytes causes interstitial metabolite levels (e.g., adenosine, K^+, CO_2, lactate) to rise. All resistance vessels in the immediate vicinity dilate reflexively through local vascular control mechanisms. Collaterals are tonically constricted normally, but they also participate in the vasodilatory response to rising metabolite levels. Blood flow through collaterals may allow areas peripheral to a focal infarct to survive the initial ischemic event (see Chapter 20·III·D).

   b. **Central:** Death of myocytes impairs myocardial contractility, which reduces LV stroke volume and CO (Fig. 40.18). MAP falls as a result, triggering a baroreceptor reflex that involves all the same effector mechanisms described in Section IV·A. If the infarct is small, these pathways may be sufficient to restore MAP.

2. **Long term:** A decrease in MAP also activates the renin–angiotensin–aldosterone system, regardless of cause (see Chapter 19·IV). It takes 24 to 48 hours for Na^+ and water retention mechanisms to expand ECF volume and support the failing myocardium with increased preload and increased end-diastolic volume (EDV). In the days and weeks following an ischemic event, the body begins repairing some of the tissue damage wrought by infarction. Collateral vessels enlarge, and the myocardium hypertrophies in response to the chronic increases in EDV through the serial addition of sarcomeres (**eccentric hypertrophy**), which helps compensate for loss of contractility.

D. Preload penalty

The Frank-Starling mechanism is highly effective in compensating for minor decreases in cardiac inotropy (see Chapter 17·IV·D). Some individuals may suffer a series of these insults and remain unaware for years until compensation causes symptoms (e.g., dyspnea associated with pulmonary congestion). Dyspnea is only one of several penalties associated with preloading, however. Others include limits to the benefits of length-dependent sarcomeric activation, excessive LV wall stress, dysrhythmias, valve incompetency, and edema.

1. **Length-dependent activation limits:** Preloading a healthy heart increases CO through length-dependent activation of sarcomeres (see Chapter 17·IV·D). The end-systolic pressure–volume relationship has a plateau region, however: once myocytes have been stretched to a length that optimizes force generation, further increases in preload are ineffective in generating additional force (Fig. 40.19).

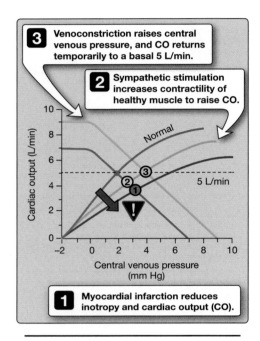

Figure 40.18.
Short-term (sympathetic) response to myocardial infarction.

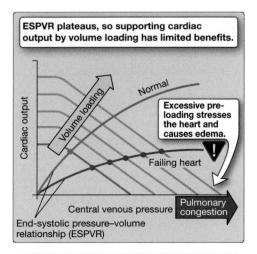

Figure 40.19.
Limits to the beneficial effects of preloading.

2. **Wall stress:** Preloading dilates the ventricle and increases wall tension, as predicted by the law of Laplace. Wall tension contributes to afterload, so, although preloading does help support output within normal physiologic ranges, high preloads also increase cardiac workload and reduce its efficacy.

3. **Dysrhythmias:** Excessive preloading stretches the ventricular wall and distorts conduction pathways, which predisposes the myocardium to potentially fatal dysrhythmias and arrhythmias.

4. **Valve incompetency:** Excessive preloading also stretches and distorts the cartilaginous valve rings and prevents the valve leaflets from coming into close apposition upon closure, which allows retrograde blood flow. Regurgitation further increases the workload required of a failing heart.

5. **Edema:** Central venous pressure (CVP) elevation raises mean capillary hydrostatic pressure and favors fluid filtration from blood into the interstitium. In the systemic vasculature, interstitial fluid excess manifests as swollen ankles and feet. The lower extremities are particularly prone to edema in an erect individual because vascular pressures in these regions are enhanced by gravity. In the lungs, fluid filters from the pulmonary capillaries and collects in the alveolar sacs, where it interferes with gas exchange (pulmonary congestion) as shown in Figure 40.20. Pulmonary edema may cause **orthopnea** (shortness of breath when lying flat), forcing patients to sleep sitting upright. Gravity helps reduce pulmonary perfusion pressures, thus decreasing the likelihood of fluid accumulation in the air spaces.

E. System failure

A failing heart becomes locked in a decompensatory spiral in which preload supports output yet ultimately limits efficiency through its effects on wall tension. Unless this cycle is interrupted and managed, it can prove fatal. Thus, the goal of medical intervention is to decrease preload using diuretics while simultaneously supporting the myocardium with inotropes that help it work more efficiently at a lower filling volume.[1] In end-stage HF, myocytes continue to die one by one, slowly chipping away at contractility and the ability to sustain MAP. Even mild physical exertion causes severe dyspnea because cardiac reserve has dropped to the point where even minimal muscular activity places demands on output that exceed myocardial capabilities, so patients become bedridden (see Chapter 20·III·B). Bed rest exacerbates frailty by causing muscle atrophy and decreased bone density. Excessive volume loading causes pulmonary edema and hypoxic respiratory failure. The liver fails due to passive congestion and restricted O_2 delivery caused by systemic edema. Loss of glomerular pressure precipitates renal failure. Each additional organ loss raises mortality risk by ~20%.

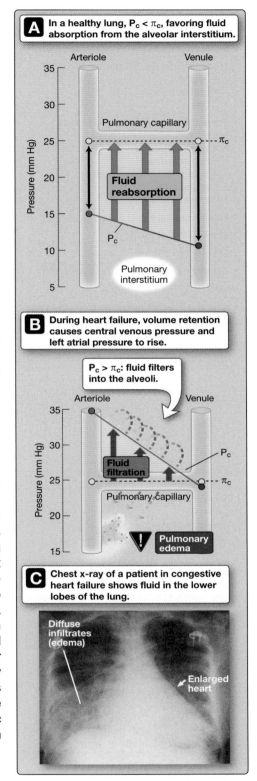

A In a healthy lung, $P_c < \pi_c$, favoring fluid absorption from the alveolar interstitium.

B During heart failure, volume retention causes central venous pressure and left atrial pressure to rise.

$P_c > \pi_c$: fluid filters into the alveoli.

C Chest x-ray of a patient in congestive heart failure shows fluid in the lower lobes of the lung.

Figure 40.20.
Pulmonary edema resulting from heart failure. P_c = capillary hydrostatic pressure; π_c = plasma colloid osmotic pressure.

[1]For more information on pharmaceutical approaches to treatment of heart failure, see *LIR Pharmacology*, 8e, Chapter 10.

Table 40.2: Common Causes of Respiratory Failure

Impaired Ventilation
• Upper airway obstruction
• Infection
• Foreign body
• Tumor
• Weakness or paralysis of respiratory muscles
• Cerebral trauma
• Drug overdose
• Guillain-Barré syndrome
• Muscular dystrophy
• Spinal cord injury
• Chest wall injury
Impaired Diffusion
• Pulmonary edema
• Acute respiratory distress syndrome
Impaired $\dot{V}_A/\dot{Q}$ Matching
• Chronic obstructive pulmonary disease
• Restrictive lung disease
• Pneumonia
• Atelectasis

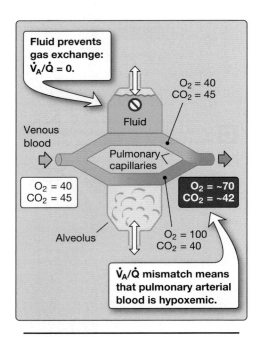

Figure 40.21.
A $\dot{V}_A/\dot{Q}$ mismatch, one common cause of hypoxemia. All partial-pressure values are given in mm Hg. Q and $\dot{V}_A$ = alveolar perfusion and ventilation, respectively.

VI. RESPIRATORY FAILURE

Respiratory failure occurs when the respiratory system is unable to fulfill one or both of its gas exchange functions, namely, O_2 uptake or CO_2 elimination. Clinically, this manifests as **hypoxemic respiratory failure** (**type 1**) or **hypercapnic respiratory failure** (**type 2**), respectively. The two failure types represent **syndromes** (sets of related symptoms) rather than the end result of any specific disease.

A. Causes

Respiratory failure may develop chronically or acutely, usually from trauma (see Section VI·D for a discussion of **acute respiratory distress syndrome [ARDS]**). Conditions causing respiratory failure can be grouped according to whether they impair ventilation (air-pump function and control), diffusion (integrity of the blood–gas interface), or ventilation–perfusion $\left(\dot{V}_A/\dot{Q}\right)$ matching (Table 40.2).

B. Hypoxemic respiratory failure (type 1)

Hypoxemic respiratory failure is characterized by a P_aO_2 <60 mm. Hypoxemia can be caused by hypoventilation, but, because attaining a normal P_aO_2 (100 mm Hg) requires that the full area of the blood–gas interface be functional, processes that decrease this area and allow venous blood to pass through the lungs without being arterialized cause some degree of hypoxemia. Thus, hypoxemic respiratory failure usually occurs when air or blood is unable to access the interface (i.e., $\dot{V}_A/\dot{Q}$ mismatch).

1. **Ventilation/perfusion mismatch:** Regional $\dot{V}_A/\dot{Q}$ mismatch is common in a healthy lung but has minimal impact on overall respiratory function (see Chapter 22·IV·B). In disease states, large numbers of alveoli may collapse and seal (**atelectasis**) or fill with fluid (pulmonary edema), pus (**pneumonia**), or blood (hemorrhage), all of which effectively prohibit O_2 uptake and cause hypoxemia (Fig. 40.21).

2. **Compensation:** Hypoxemia is detected primarily by the carotid and aortic chemoreceptors (see Chapter 24·III·C). CNS respiratory control centers respond by increasing minute ventilation, and CNS cardiovascular control centers simultaneously increase CO to help maximize the O_2 diffusion gradient across the exchange barrier.

3. **Consequences:** The physiologic consequences of hypoxemia were discussed in relation to the effects of ascent to altitude (see Chapter 24·V·A). Mild hypoxemia causes slight impairment of mental function and visual acuity. When P_aO_2 drops below ~40 to 50 mm Hg, patients become confused and prone to personality changes and irritability. Hypoxemia also initiates a positive feedback spiral in which the pulmonary vasculature constricts reflexively and further reduces O_2 uptake. Vascular constriction also increases RV afterload and induces pulmonary hypertension, which stresses the RV. These symptoms can usually be reversed clinically by administering O_2 to maximize $\dot{V}_A/\dot{Q}$ ratios and, at least temporarily, restore P_aO_2 until the underlying cause of hypoxemia can be evaluated and addressed.

C. Hypercapnic respiratory failure (type 2)

Hypercapnic respiratory failure is indicated by an acute rise in P_{aCO_2} to >50 mm Hg. Hypercapnia that develops over a course of months is tolerated well, however, so failure may not occur until P_{aCO_2} reaches ~70 to 90 mm Hg. Unlike hypoxemia, hypercapnia can be corrected relatively easily by adjusting alveolar ventilation. Thus, hypercapnic respiratory failure usually only occurs when ventilatory control is impaired. Hypercapnia is usually associated with varying degrees of hypoxemia.

1. **Ventilation:** Ventilatory failure occurs if the respiratory center or its neural pathways are damaged by stroke, drug overdose, or neuromuscular diseases (e.g., myasthenia gravis), but the most common causes of hypercapnic respiratory failure are impairment of air-pump function (chest wall and respiratory muscles) and chronic airway disorders.

    a. **Chest wall:** Movement of the chest wall can become severely limited by obesity and by abnormal spine curvature. Kyphosis (forward flexion curvature), as shown in Figure 40.22A, and scoliosis (lateral curvature), as shown in Figure 40.22B, are congenital disorders, but the former is also seen in association with arthritis and osteoporosis (see Fig. 40.22A). Both can develop into debilitating curvatures that severely limit chest wall excursions and hasten failure in a compromised lung.

    b. **Muscles:** The respiratory muscles (diaphragm and intercostals) increase intrathoracic volume and expand the lungs during inspiration. They are skeletal muscles and, therefore, susceptible to wasting diseases such as muscular dystrophy. They are also subject to **fatigue**, which is a principal concern when addressing problems underlying respiratory failure. Chronic conditions that reduce chest wall or lung compliance (restrictive lung diseases) increase the work of breathing, and fatigue ultimately reduces contractility and causes hypoventilation and hypercapnia.

    c. **Airways:** Chronic obstructive pulmonary disease and asthma increase airway resistance and can reduce alveolar ventilation and raise P_{aCO_2}.

2. **Compensation:** P_{aCO_2} is monitored by central and peripheral chemoreceptors. The respiratory center responds to acute hypercapnia by increasing ventilation rate, even if such an action causes respiratory muscle fatigue and precipitates a respiratory crisis. During chronic hypercapnia, the chemoreceptors adapt to persistent elevation of P_{aCO_2}, so ventilation rates remain normal. CO_2 retention decreases plasma pH (respiratory acidosis), but the kidneys compensate by retaining HCO_3^-, allowing pH to remain within a normal range even as P_{aCO_2} climbs above ~70 to 90 mm Hg (see Chapter 28·VI·C). Such patients typically have a limited pulmonary reserve, however, and may decompensate quickly if illness creates additional demands on an already compromised system.

3. **Consequences:** The cerebral vasculature is highly sensitive to P_{aCO_2}. Acute CO_2 retention causes cerebral vasodilation, which causes headaches and intracranial hypertension. The latter may

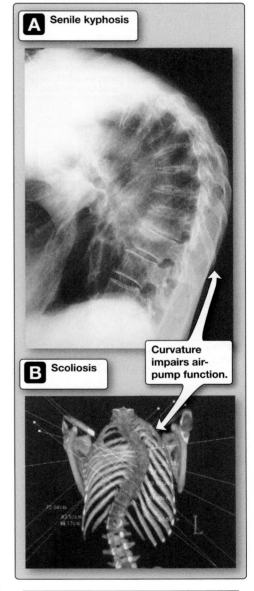

A Senile kyphosis

B Scoliosis

Curvature impairs air-pump function.

Figure 40.22.
Kyphosis and scoliosis.

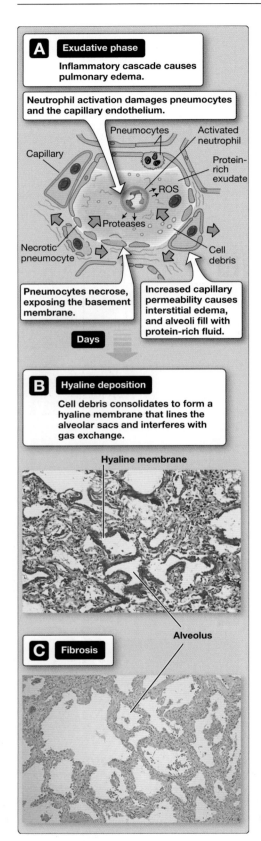

Figure 40.23.
Acute respiratory distress syndrome.
ROS = reactive oxygen species.

manifest as a swelling of the optic disc and can cause blindness. High CO_2 levels also cause dyspnea and neurologic symptoms, such as involuntary wrist movements and hand tremors.

D. Acute respiratory distress syndrome

ARDS is the leading cause of respiratory failure in young adults, with mortality rates currently as high as 46%. In the military, ARDS was known originally as "shock lung," reflecting the similarities between the onset of ARDS and septic shock.

1. **Causes:** ARDS may develop in association with a wide range of conditions, the most common being sepsis, aspiration of stomach contents, non-fatal drowning, multiple blood transfusions, trauma, bone fractures, and pneumonia.

2. **Stages:** ARDS is precipitated by local or circulating inflammatory mediators (e.g., histamine), which trigger an inflammatory cascade that severely damages the alveolar endothelial cells (pneumocytes) and capillary endothelial cells that comprise the blood–gas interface (Fig. 40.23; see Chapter 22·II·C). ARDS has three discrete stages, characterized by exudate formation, hyaline membrane deposition, and fibrosis.

    a. **Exudates:** The initial stages of ARDS (24–48 hours) are characterized by an inflammatory reaction within the lung parenchyma that damages the alveolar epithelium and causes a profound increase in capillary permeability. The lungs fill with a bloody exudate containing plasma proteins and cellular debris (see Fig. 40.23A). Chest x-rays typically reveal diffuse bilateral infiltrates reminiscent of pulmonary edema but that occur when CVP and left atrial pressure is normal.

    b. **Hyaline membrane:** In the following 2 to 8 days, hyaline membranes begin to settle from the exudate and cover the alveolar lining (see Fig. 40.23B). Hyaline is a fibrous matrix of plasma proteins and disrupted cell debris.

    c. **Fibrosis:** After ~8 days, the interstitium is infiltrated by fibroblasts, which deposit collagen and other fibrous materials.

3. **Consequences:** The alveolar infiltrates and hyaline membranes prevent gas exchange, causing hypoxemia. The infiltrates also inactivate surfactant and suppress surfactant production by type II pneumocytes, causing alveolar collapse. Surfactant loss and atelectasis make the lung extremely stiff and difficult to expand, which is one of the hallmarks of ARDS (Fig. 40.24). Atelectasis not only impairs ventilation but also reduces the area of the blood–gas interface, thereby exacerbating hypoxemia. Support from a mechanical ventilator is lifesaving while the underlying process is addressed.

E. System failure

Hypoxemia associated with acute respiratory failure elicits the same SNS responses as shock but in the context of an intact and functional vasculature. SNS-stimulated increases in CO and SVR can cause MAP to rise to levels that rupture cerebral blood vessels. Chronic hypoxemia causes a gradual decline in neural function that inhibits the pathways controlling ventilation and blood pressure. The most obvious external sign of hypoxemia is cyanosis, a blue discoloration of the skin and mucous membranes that reflects the color of deoxyhemoglobin (cyanosis occurs when deoxyhemoglobin levels rise to ~5 g/dL). Acute hypercapnia produces respiratory acidosis through CO_2 retention, but the body's response is dominated by reflex responses to the hypoxemia that accompanies hypercapnia. Acute hypercapnia causes anesthetic-like effects on the CNS (**CO_2 narcosis**). Narcosis appears at a P_{CO_2} of ~90 mm Hg, causing confusion and lethargy. High P_{CO_2} depresses respiratory center function and suppresses ventilatory drive, thus creating a positive feedback cycle that potentiates CO_2 retention and, ultimately, results in coma and death (P_{CO_2} ~130 mm Hg).

VII. KIDNEY FAILURE

Two forms of kidney failure are recognized. **Acute kidney injury** (**AKI**) develops abruptly (within 48 hours) but usually can be treated if patients have no other complicating medical issues. Kidney function and risk of failure can be assessed using RIFLE criteria as shown in Table 40.3. **Chronic kidney disease** (**CKD**) develops over the course of many years due to chronic decreases in renal perfusion (renal artery stenosis, heart disease), renal vascular disease, glomerular disease, or intrinsic tubule or interstitial disease. CKD is characterized by progressive and irreversible loss of nephrons, but the development of dialysis and transplant technologies means that CKD is not necessarily fatal. The mortality rate of patients on dialysis is very high (dialysis increases lifespan by only 4.5 years in 60–64-year-olds), but death usually occurs from cardiovascular disease, infection, or **cachexia** (a wasting syndrome). AKI is a primary cause of death (~75%) in urgent care facilities, however, where patients may be elderly and have other underlying pathologies.

A. Causes

AKI can be precipitated by numerous factors, which are usually grouped according to etiology: **prerenal**, **intrarenal**, and **postrenal** (Table 40.4).

1. **Prerenal:** Prerenal failure is the most common cause of AKI and is characterized by a profound drop in GFR due to decreased renal blood flow and glomerular perfusion pressure (see Figure 40.13). Prerenal failure usually occurs secondarily to hypovolemia, hypotension, and shock.

2. **Intrarenal:** Intrarenal failure occurs when the renal tubule or surrounding interstitium is injured. The most common cause of intrarenal failure is **acute tubular necrosis** (**ATN**). ATN usually results from ischemia caused by prolonged hypotension and reduced renal blood flow, but the tubule may also be injured by drugs and other toxins, infections, and inflammation.

   a. **Ischemia:** The renal epithelium's transport functions create a high ATP dependency and coincident susceptibility to ischemia. The transport epithelium receives O_2 via the peritubular capillary network, whose flow is governed by glomerular arterioles

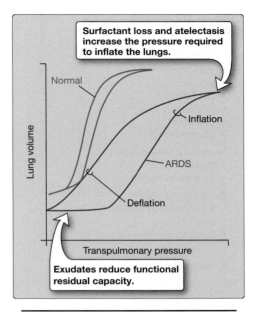

Figure 40.24.
Change in lung function accompanying acute respiratory distress syndrome (ARDS).

Table 40.3: Assessing Kidney Injury and Failure

	GFR (Serum Creatinine)*	Urine Output
Risk	↓25% (↑1.5×)	<0.5 mL/kg for 6 hr
Injury	↓50% (↑2×)	<0.5 mL/kg for 12 hr
Failure	↓75% (↑3×)	<0.3 mL/kg for 24 hr or anuria for 12 hr
Loss	Loss of renal function for >4 wk	
ESRD	End-stage renal disease (ESRD)	

GFR = glomerular filtration rate.
*Deviation from baseline values.

Table 40.4: Common Causes of Acute Kidney Injury

Prerenal
• **Hypovolemia**   • **Hemorrhage**   • **Dehydration**   • **Prolonged vomiting**   • **Diarrhea**   • **Severe burns**
• **Peripheral vasodilation**   • **Sepsis**   • **Anaphylactic shock**
• **Cardiogenic shock**   • **Myocardial infarction**   • **Heart failure**   • **Cardiac tamponade**
• **Renal vasoconstriction**   • **Vasoactive drugs**
Intrarenal
• **Vascular occlusion**   • **Renal artery occlusion**   • **Renal vein thrombosis**   • **Vasculitis**
• **Glomerulonephritis**   • **Acute tubular necrosis**   • **Ischemia**   • **Nephrotoxic drugs, heavy metals, organic solvents**   • **Intratubular deposits (uric acid casts, muscle proteins)**
Postrenal
• **Urolithiasis**
• **Ureterocele**
• **Prostatic hyperplasia**
• **Malignancies**

(see Chapter 26·II·C). Tubule ischemia usually occurs during a hypotensive crisis when both arterioles are constricted, filtration has ceased, and peritubular flow no longer meets the epithelium's basal O_2 needs. Ischemic cells may respond by sloughing their apical villi into the tubule lumen, which reduces overall surface area, transporter density, and O_2 requirements.

  b. **Toxins:** The renal tubule's ability to concentrate drugs and toxins makes it very vulnerable to nephrotoxicity. Although all tubule regions are at risk, necrosis most commonly occurs in the proximal tubule, which is responsible for secreting many drugs (see Chapter 26·IV).

3. **Postrenal:** Postrenal failure is caused by urinary outflow obstruction, which can occur at any point within the tubule, collecting duct system, ureters, bladder, or urethra. Common causes include kidney stones (**calculi**; see Clinical Application 4.2) and an enlarged prostate (**benign prostatic hyperplasia**). Obstruction causes pressure in more proximal tubule segments to build to the point where they negate ultrafiltration pressure (P_{UF}), and glomerular filtration stops. In time, the affected tubule segments may dilate and atrophy.

B. Consequences

Renal failure is considered to have occurred when GFR is reduced to 25% of normal values. Kidneys lose their ability to control water and electrolyte levels when GFR is so low, which manifests as hypervolemia, hyperkalemia, metabolic acidosis, and accumulation of nitrogenous wastes (azotemia).

C. System failure

An inability to excrete nitrogenous wastes and maintain electrolyte balance negatively impacts all organs, but the neurologic effects dominate. Patients become lethargic, drowsy, and delirious and eventually slip into a coma. Death typically occurs due to cardiac arrhythmia caused by hyperkalemia.

VIII. MULTIPLE ORGAN DYSFUNCTION SYNDROME

Emily Dickinson's generation was well acquainted with the signs of organ system failure and impending death (Fig. 40.1). People usually died at home in the care of family and loved ones. The poem that opened this final chapter accurately notes the effects of acute hypoxemia on the CNS (convulsion) and the consequences of intense SNS-mediated stimulation of sweat glands (*The Beads upon the Forehead/By homely Anguish strung*). In more recent times, the final throes occur in medical facilities, witnessed mainly by health care professionals. Patients arriving at emergency departments and intensive care units often already have long-standing diseases and have coped with progressive failure of one or more organs for many months or years. They present when an infection or some other seminal event has precipitated **multiple organ dysfunction syndrome** (a medical crisis involving two or more organ systems), at which point medical intervention is required for continued survival. It is the job of care providers to help restore homeostasis and create an environment in which the body may recover from an acute disease state. In the final reckoning, however, the physiologic systems that maintain homeostasis and that have been described in the preceding chapters are robust and have remarkable recuperative capabilities. They readily reassert control if given a chance, either on their own or by timely medical intervention as needed. Whether or not they take the chance is the medical mystery that is Life.

THE END

Chapter Summary

- **Aging** is accompanied by a progressive reduction in total cell number, organ functionality, and the ability to respond to homeostatic challenge. Aging ultimately results in organ failure and death.

- The final common pathway for most instances of cell death is **ischemia**. Ischemia initiates a series of events that comprise an **ischemic cascade**. Significant events include acidosis, dissipation of ion gradients, Ca^{2+} activation of proteases and other degradative enzymes, and mitochondrial lysis.

- Ischemia usually results from lack of perfusion due to **circulatory shock**. **Hypovolemic shock** results from hemorrhage or reduced ECF volume. **Cardiogenic shock** is caused by loss of cardiac pump function. **Obstructive shock** occurs when blood flow to, through, and out of the heart is impeded. **Distributive shock** occurs when the CNS loses vascular control, and the resulting systemic vasodilation allows blood to become trapped in capillaries and veins.

- Shock can be divided into three stages: **preshock**, **shock**, and **organ failure**. During preshock, sympathetic nervous system (SNS)-mediated increases in cardiac and vascular function help compensate for falling mean arterial pressure.

- During shock, perfusion of critical circulations (cerebral, coronary) becomes suboptimal, and signs of intense SNS activity become overt (hypotension, decreased urinary output, acidosis).

- **System failure** occurs when shock becomes irreversible. The cardiovascular system becomes locked in a positive feedback spiral that results in loss of myocardial and vascular contractility; acidemia; blood clotting; cellular deterioration; and, ultimately, loss of cerebral perfusion pressure and brain death.

- **Heart failure** is the leading cause of death in the United States and is the final common pathway for many cardiac diseases. The right heart typically fails as a result of increased pulmonary vascular resistance. Causes of **left heart failure** include filling impairment (**diastolic failure**), loss of contractility, and excessive afterload that impairs output (**systolic failure**).

- **Myocardial infarction** is a common cause of heart failure. Long-term compensatory mechanisms that support cardiac output through volume retention and invocation of the **Frank-Starling mechanism** ultimately become counterproductive and precipitate failure. Symptoms of congestive heart failure include **edema** and **shortness of breath** upon exertion.

- Respiratory failure occurs when the pulmonary system is unable to take up O_2 or eliminate CO_2 from the body. **Hypoxemic respiratory failure (type 1)** usually occurs as a result of impaired alveolar ventilation due to alveolar collapse or accumulation of fluid, pus, or blood in alveoli.

- The central respiratory control centers readily adapt to increases in $P_{a}CO_2$, so **hypercapnic respiratory failure (type 2)** usually reflects impaired air-pump function (respiratory muscles and chest wall).

- **Acute respiratory distress syndrome (ARDS)** is associated with inflammatory reactions that damage lung parenchyma and increase pulmonary capillary permeability. Lungs fill with infiltrates and become stiff and difficult to expand.

- **Renal failure** can be precipitated by inadequacy of perfusion, kidney tubule deterioration, or obstruction of urinary outflow. The kidney's resulting inability to control water and electrolyte levels allows K^+ levels to rise (**hyperkalemia**), and death usually results from cardiac arrhythmia.

Study Questions

Choose the ONE best answer.

40.1. A 21-year-old male experiences extensive blood loss from deep wounds inflicted by an improvised explosive device. Which of the following most likely identifies the primary mechanism responsible for maintaining blood volume until fluids can be administered?

A. Aldosterone release
B. Decreased renal blood flow
C. Recruitment of interstitial fluid
D. Resistance vessel constriction
E. Venoconstriction

Best answer = C. Hemorrhage causes central venous pressure to fall, which reduces mean capillary hydrostatic pressure in all circulations (see Section IV·A). This causes fluid to move into the vasculature from the interstitium ("transcapillary refill") under the influence of plasma colloid osmotic pressure, thereby helping support blood volume. Aldosterone release (A) increases fluid retention by the kidney but only after several hours. Renal blood flow decreases (B) as a result of resistance vessel constriction (D) that directs blood away from less essential organs but does not increase blood volume. Venoconstriction (E) forces blood out of veins but does not affect total blood volume.

40.2. A 67-year-old female is brought to the emergency department in shock. Assessment of her cardiovascular function shows that heart rate (HR) is high and cardiac output (CO) is increased, whereas left atrial pressure (LAP), mean arterial pressure (MAP), and systemic vascular resistance (SVR) are all low. What is the most likely cause?

A. Cardiac tamponade
B. Cardiogenic shock
C. Hypovolemic shock
D. Pulmonary embolism
E. Septic shock

Best answer = E. Septic (distributive) shock is associated with inflammatory reactions that decrease SVR and cause MAP to fall. A decrease in MAP triggers an increase in HR and CO in attempts to restore MAP and maintain O_2 delivery to the brain (see Section IV·B). In tamponade (A), a cause of obstructive shock, and cardiogenic shock (B), LAP (preload) would be increased in attempts to support CO. Hypovolemic shock (C) reduces CO. Pulmonary embolism (D), which is another cause of obstructive shock, would cause LAP and CO to fall, but SVR would be very high.

40.3. A 55-year-old female is having difficulty sleeping because of shortness of breath. She also shows signs of peripheral edema secondary to heart failure. She is being treated with bumetanide (loop diuretic) and a combination of hydralazine and isosorbide dinitrate (vasodilator and venodilator, respectively). The combination of hydralazine and isosorbide dinitrate is most likely to increase which of the following in this patient?

A. Coronary perfusion
B. Left-ventricular contractility
C. Left-ventricular ejection fraction
D. Left-ventricular preload
E. Pulmonary vascular resistance

Best answer = C. The left ventricle (LV) in patients with congestive heart failure is usually operating at end-diastolic volumes (EDVs) in the plateau region of the Frank-Starling curve, which reduces cardiac efficiency (see Section V·D). Venodilators are beneficial because they decrease, not increase, central venous pressure and LV preload (D), reducing EDV and increasing ejection fraction (EF). Vasodilators decrease afterload, which further increases EF. Coronary perfusion (A) should be reduced in parallel with cardiac workload by the drugs. LV contractility (B) is regulated by inotropes, not combination vasodilator and venodilator therapy. Pulmonary vascular resistance (E) would be decreased, not increased, by a vasodilator.

40.4. A 48-year-old male with pneumonia is hospitalized when he develops acute respiratory distress syndrome (ARDS) and requires a mechanical ventilator to support breathing. Why is a mechanical ventilator most likely to be helpful in this patient?

A. Fluid within alveoli decreases lung compliance.
B. Inflammatory exudates impair surfactant function.
C. Ventilators increase cardiac output.
D. Ventilators prevent hyaline membrane formation.
E. Ventilators prevent pulmonary fibrosis.

Best answer = B. The lungs of patients with ARDS are highly noncompliant and require the assistance of a mechanical ventilator to expand, mainly because the inflammatory exudates inactivate surfactant and inhibit its production (see Section VI·D). Fluid in the pulmonary interstitium reduces lung compliance but not fluid within alveoli (A) because fluid-filled lungs are easier to expand than normal because surface tension effects are negated (see Chapter 21·IV·B). Ventilation may decrease, not increase left ventricular preload and cardiac output (C) and has no effect on hyaline membrane formation (D), which interferes with gas exchange. Fibrosis, which may further reduce compliance over time, is not prevented by ventilation (E).

APPENDIX A
Clinical Cases

I. CASE VIGNETTES AND QUESTIONS (Answers to questions begin on p. 553)

CASE 1: RECURRENT RESPIRATORY INFECTIONS AND PNEUMONIA

A 22-year-old female presents to the clinic with recurrent respiratory infections, persistent cough, and difficulty breathing. Her medical history is significant for cystic fibrosis (CF) and frequent hospitalizations for pneumonia and sinusitis.

On physical examination, she demonstrates mild respiratory distress with a persistent cough. Chest examination reveals diffuse wheezing and crackles. Her oxygen saturation is 92% on room air. Her body mass index (BMI) is 17.5 kg/m^2 (normal = 18.5–25 kg/m^2). Arterial blood gas (ABG) analysis shows mild respiratory acidosis with a pH of 7.33, partial pressure of carbon dioxide (Pco_2) of 54 mm Hg, and partial pressure of oxygen (Po_2) of 75 mm Hg. Chest x-ray reveals bilateral bronchiectasis, which is consistent with her history of CF. A sweat test records a chloride level of 84 mmol/L (normal = <30 mmol/L).

Genetic analysis reveals a homozygous ΔF508 mutation in the *CFTR* gene. A sputum culture is obtained, and the patient is started on antibiotics to treat her for *Pseudomonas aeruginosa* infection.

THOUGHT QUESTIONS

TQ1.1 What does the *CFTR* gene encode?

TQ1.2 How does ΔF508 mutation affect CFTR function?

TQ1.3 How is *CFTR* gene mutation related to recurrent chest infections?

TQ1.4 Why does the patient have respiratory acidosis?

TQ1.5 What is the significance of the patient's low BMI?

TQ1.6 What is the significance of the sweat test result?

REVIEW QUESTIONS: Choose the ONE best answer

RQ1.1 Which of the following hormones is most likely to increase CFTR activity?
- A. Cholecystokinin
- B. Epinephrine
- C. Insulin
- D. Testosterone
- E. Thyroxine

RQ1.2 Neonates with CF may have stools that are pale, foul-smelling, and float in the toilet bowl. These findings are most likely the result of deficiency in which of the following enzymes?
- A. Elastase
- B. Lingual lipase
- C. Pancreatic lipase
- D. Salivary amylase
- E. Trypsin

CASE 2: ASCENDING WEAKNESS FOLLOWING A SEVERE DIARRHEAL ILLNESS

A 56-year-old female presents to the emergency department with progressive weakness and difficulty walking. She notes that her symptoms began a week ago as mild tingling and numbness in her toes, which then spread rapidly over a matter of days to involve both legs and then her arms. She is now unable to walk without assistance. Her history reveals diarrhea 2 weeks ago accompanied by severe abdominal cramps and fever lasting several days, which she attributes to eating undercooked chicken at a family picnic.

Physical examination shows the patient to be weak and fatigued. Neurologic examination reveals generalized hyporeflexia, symmetric muscle weakness, and reduced sensation in the distal extremities. Deep tendon reflexes are absent.

The patient's history and clinical findings lead to a preliminary diagnosis of Guillain-Barré syndrome (GBS), an acute inflammatory demyelinating polyneuropathy affecting peripheral nerves. Electrophysiologic studies, including nerve conduction studies (NCS) and electromyography, confirm the likely diagnosis. The patient is started on intravenous immunoglobulin (IVIG) therapy to halt progression of the disease.

THOUGHT QUESTIONS

TQ2.1 What did the nerve conduction studies most likely reveal?

TQ2.2 What are deep tendon reflexes, and why are they absent in this patient?

TQ2.3 What is myelin?

TQ2.4 What is the role of myelin in sensory and motor nerve signaling?

TQ2.5 Is the patient's muscle weakness due to loss of α motor neurons?

TQ2.6 What is the relationship between GBS and multiple sclerosis (MS)?

TQ2.7 What is the likely relationship, if any, between the prior diarrheal illness and development of GBS?

TQ2.8 What is the likely mechanism of action of IVIG treatment?

REVIEW QUESTIONS: Choose the ONE best answer

RQ2.1 In the absence of myelin, electrical signals most likely decay due to the presence of which of the following axonal membrane conductance pathways?
A. Ca^{2+} channels
B. Cl^- channels
C. Cl^-/HCO_3^- exchangers
D. K^+ channels
E. Na^+/Ca^{2+} exchangers

RQ2.2 Which of the following changes is most likely to increase neuronal conduction velocity?
A. Decreased axonal diameter
B. Decreased neuronal stimulation frequency
C. Decreased rate of recovery from Na^+-channel inactivation
D. Increased axonal length
E. Increased Na^+ channel opening rate

CASE 3: MUSCLE WEAKNESS AND FATIGUE

A 21-year-old male presents to the office with progressive muscle weakness and fatigue over the past 6 months. He reports difficulty with daily activities such as walking and climbing stairs.

Physical examination reveals proximal muscle weakness with a waddling gait. He has difficulty rising from a seated position without assistance. Deep tendon reflexes are diminished.

Laboratory studies are notable for a creatine kinase (CK) of 950 U/L (normal = 25–90 U/L). Electromyography (EMG) and nerve conduction studies indicate short-duration, low-amplitude motor unit potentials and early recruitment. A muscle biopsy shows scattered nemaline rods within the muscle fibers (Fig. C3.1). Genetic analysis reveals a mutation in the actin *ACTA1* gene linked to nemaline myopathy (NM).

NM is a rare hereditary disease (incidence of 1:50,000 live births) most commonly caused by mutations in the *NEB* gene (50% of total), which encodes nebulin, and *ACTA1* (15%–25%). The patient begins physical therapy to improve muscle strength and manage his symptoms.

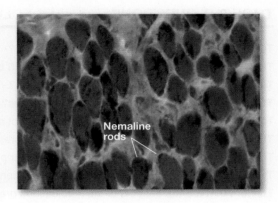

Figure C3.1.
Nemaline myopathy showing muscle fibers filled with nemaline rods.

THOUGHT QUESTIONS

TQ3.1 What is the role of actin and nebulin in muscle function?

TQ3.2 What is the significance of the CK finding?

TQ3.3 What is the significance of the EMG findings?

TQ3.4 What does early recruitment indicate?

TQ3.5 Patients with NM usually have a shortened life expectancy. What is the likely cause of increased mortality?

REVIEW QUESTIONS: Choose the ONE best answer

RQ3.1 Which of the following proteins most likely tethers the ends of actin thin filaments to Z disks?
   A. Actinin
   B. Dystrophin
   C. Titin
   D. Tropomyosin
   E. Troponin

RQ3.2 Which of the following most likely activates ryanodine receptors during skeletal muscle contraction?
   A. Dihydropyridine receptor activation
   B. IP_3 formation
   C. Phosphorylation by protein kinase A
   D. Rise in sarcoplasmic Ca^{2+} levels
   E. Sarcolemmal Na^+-channel activation

CASE 4: FATIGUE AND SHORTNESS OF BREATH DURING PREGNANCY

A 26-year-old female, gravida 1, para 1, comes to the clinic for her first prenatal visit at 24 weeks' gestation. She reports feeling increasingly fatigued and short of breath, especially when she exerts herself. She also notes occasional chest discomfort and palpitations.

She is a recent immigrant from Southeast Asia. She states that she has previously been healthy with no significant medical history. Her vital signs are within normal limits. Physical examination reveals a loud first heart sound (S_1) with an opening snap and a low-pitched murmur. An echocardiogram confirms mitral stenosis (MS) with a valve area of 1.3 cm^2 (normal = 4.0–6.0 cm^2). The left atrium is dilated, and there is evidence of pulmonary hypertension. Left-ventricular ejection fraction is within a normal range.

The patient is started on diuretics, β-blockers, and low-molecular-weight heparin (LMWH) and advised to limit physical activity for the duration of her pregnancy.

THOUGHT QUESTIONS

TQ4.1 How does MS affect cardiac performance?

TQ4.2 When would a murmur associated with MS be expected to occur?

TQ4.3 How would MS affect left-heart pressures?

TQ4.4 How would MS affect right-heart pressures?

TQ4.5 Why has this patient's valvular disorder not caused symptoms previously?

TQ4.6 Is this patient's fatigue and shortness of breath related to pulmonary or cardiac changes?

TQ4.7 What drives the changes in cardiovascular performance that accompany pregnancy?

TQ4.8 What is the most likely etiology of this patient's MS?

TQ4.9 What are the benefits of the medications prescribed for this patient to address her symptoms?

REVIEW QUESTIONS: Choose the ONE best answer

RQ4.1 Mitral valve regurgitation would most likely be associated with which of the following?
   A. Continuous murmur
   B. Diastolic murmur
   C. Reduced S_2
   D. S_3 heart sound
   E. S_4 heart sound

RQ4.2 Pregnancy is most likely associated with a decrease in which of the following cardiovascular parameters?
   A. End-diastolic volume
   B. Heart rate
   C. Hematocrit
   D. Pulse pressure
   E. Stroke volume

CASE 5: DIZZINESS ON STANDING

A 76-year-old female is brought to the emergency department after falling at home. She reports having felt dizzy and weak for several days. On the day of the fall, she had been eating breakfast and stood up very quickly to answer the phone, causing her to lose her balance. She has a history of hypertension and osteoarthritis, for which she is taking a diuretic, a β-blocker, and a bisphosphonate.

Physical examination reveals a pulse of 78 beats/min and a blood pressure of 138/86 mm Hg when supine. On standing, her pulse rises to 102 beats/min, and her blood pressure drops to 88/58 mm Hg. A 12-lead ECG is unremarkable. These findings are consistent with a diagnosis of orthostatic hypotension (OH).

The patient is advised to increase her fluid intake and to stand up slowly to allow her body time to adjust to the change in position. She is given an outpatient cardiology referral and discharged.

THOUGHT QUESTIONS

TQ5.1 What is the underlying cause of dizziness and syncope?

TQ5.2 What happens to the distribution of blood volume on standing?

TQ5.3 How are changes in arterial pressure sensed?

TQ5.4 What normally prevents syncope on standing?

TQ5.5 What is the role of cardiopulmonary receptors in this response?

TQ5.6 Which of the patient's medications might predispose her to OH?

TQ5.7 Why is OH more common in elderly populations?

REVIEW QUESTIONS: Choose the ONE best answer

RQ5.1 Which of the following is most likely to increase on standing in this patient?
   A. Left ventricular end-diastolic volume
   B. Left ventricular end-systolic volume
   C. Left ventricular preload
   D. Systemic vascular resistance
   E. Venous return

RQ5.2 Which of the following is most likely to decrease on standing in this patient?
   A. Blood volume
   B. Circulating norepinephrine
   C. Heart rate
   D. Sympathetic nervous system activity
   E. Vagal nerve afferent firing rate

CASE 6: NON-FATAL DROWNING

A 25-year-old male is brought to the emergency department following a near-drowning incident at a local lake. He was fully submerged for at least 3 minutes. On arrival, the patient appears cyanotic and is using accessory muscles to breathe. His temperature is 36.5°C (97.7°F), pulse is 116 beats/min, respirations are 30 breaths/min, and his blood pressure is 108/72 mm Hg. His oxygen saturation (SaO_2) is 82% on room air.

Arterial blood (reference ranges shown in parentheses):

pH	7.22 (7.35–7.45)
P_{CO_2}	56 mm Hg (33–45 mm Hg)
P_{O_2}	60 mm Hg (75–105 mm Hg)
HCO_3^-	22 mEq/L (22–28 mEq/L)

Supplemental O_2 is provided via a non-rebreather mask and a chest x-ray ordered to evaluate for signs of pulmonary edema or aspiration.

THOUGHT QUESTIONS

TQ6.1 Why is the patient hypoxic and hypercapnic?

TQ6.2 Why is the patient cyanotic?

TQ6.3 Why is the patient using respiratory accessory muscles?

TQ6.4 What is the patient's acid–base status?

TQ6.5 Considering that water aspiration damages the alveolar wall and causes inflammation, what condition is a possible clinical concern in this patient?

REVIEW QUESTIONS: Choose the ONE best answer

RQ6.1 In addition to preventing atelectasis, surfactant most likely has which of the following functions?
   A. Enhancing immune responses
   B. Increasing lung-diffusing capacity
   C. Keeping the alveolar lumen dry
   D. Moistening the alveolar epithelium
   E. Promoting a cough reflex

RQ6.2 Which of the following conditions is most likely associated with a high $\dot{V}_A/\dot{Q}$ ratio?
   A. Asthma
   B. Bronchitis
   C. Emphysema
   D. Pulmonary arteriovenous formation
   E. Pulmonary edema

CASE 7: HEADACHE, NAUSEA, AND SHORTNESS OF BREATH

A 22-year-old male presents to the emergency department with severe headache, nausea, and shortness of breath. He reports that he lives in Los Angeles, CA (sea level), but is visiting a college friend in Provo, UT (elevation = 1,387 m; 4,551 ft), with whom he has spent the past 2 days hiking in the Wasatch Mountain range at elevations averaging 3,000 m (9,843 ft).

Physical examination shows the patient to be fatigued and short of breath. His pulse is 106 beats/min, respirations are 20 breaths/min, and blood pressure is 128/82 mm Hg. Oxygen saturation (SaO_2) is 92% on room air. Laboratory studies show a hemoglobin (Hb) of 16.3 g/dL (normal = 13.5–17.5 g/dL) and hematocrit of 49% (normal = 41%–53%).

The clinical presentation is consistent with acute mountain sickness (AMS). The patient is advised to rest, ensure adequate hydration, and use an over-the-counter pain medication for the headaches.

THOUGHT QUESTIONS

TQ7.1 Why does ascent to high altitude cause symptoms associated with AMS?

TQ7.2 What are F_IO_2, P_IO_2, and P_AO_2, and what are their expected values for this patient during his hike?

TQ7.3 What is the patient's blood O_2-carrying capacity?

TQ7.4 How is SaO_2 measured, and what determines SaO_2?

TQ7.5 How is P_aO_2 sensed, and how would P_aCO_2 and blood pH be expected to be affected by ascent to high altitude?

TQ7.6 How would cerebral blood flow be affected by ascent to high altitude?

TQ7.7 Why might prophylactic use of a carbonic anhydrase (CA) inhibitor be helpful in preventing AMS?

REVIEW QUESTIONS: Choose the ONE best answer

RQ7.1 Which of the following is most likely to decrease during high-altitude adaptation and acclimation?
  A. 2,3-Bisphosphoglycerate levels
  B. Blood volume
  C. Hematocrit
  D. Hemoglobin O_2 affinity
  E. Right-ventricular afterload

RQ7.2 Rapid ascent to high altitude can result in fatal high-altitude cerebral edema (HACE). An increase in which of the following is most likely to contribute to HACE?
  A. Cerebral blood flow
  B. Cerebral capillary hydrostatic pressure
  C. Intracranial pressure
  D. Right-ventricular afterload
  E. Ventilation

CASE 8: CHILD WITH FACIAL AND LEG SWELLING

A 5-year-old male is brought to the office with concerns over sudden onset of swelling in his face and legs. The patient has no significant medical history and no recent illnesses or infections.

Physical examination reveals periorbital and pedal edema but is otherwise unremarkable (Fig. C8.1). Vital signs are within normal range for his age.

Laboratory studies (reference ranges shown in parentheses):

Serum

Na^+	138 mEq/L (136–145 mEq/L)
K^+	4.4 mEq/L (3.5–5.5 mEq/L)
Ca^{2+}	7.6 mg/dL (8.8–10.1 mg/dL)
Urea nitrogen	18 mg/dL (5–25 mg/dL)
Creatinine	0.15 mg/dL (0.12–1.06 mg/dL)
Albumin	2.2 g/dL (3.7–5.5 g/dL)
Cholesterol	310 mg/dL (135–200 mg/dL)
Glucose	95 mg/dL (70–115 mg/dL)

Urinalysis reveals 3+ protein and fatty casts. The presentation and laboratory findings are consistent with minimal change disease (idiopathic nephrotic syndrome).

Figure C8.1.
Periorbital edema.

THOUGHT QUESTIONS

TQ8.1 What are the most common causes of interstitial edema?

TQ8.2 Considering the laboratory studies, what is the most likely cause of edema in this patient?

TQ8.3 Why is his serum albumin low?

TQ8.4 How are proteins normally prevented from entering the renal tubule?

TQ8.5 Why are Na^+ and glucose levels not reduced in this patient?

REVIEW QUESTIONS: Choose the ONE best answer

RQ8.1 Plasma proteins become concentrated during blood's passage through renal glomerular capillaries. If glomerular capillary hydrostatic pressure (P_{GC}) is 50 mm Hg, hydrostatic pressure in the Bowman space (P_{BS}) is 15 mm Hg, Bowman space oncotic pressure (π_{BS}) is 0 mm Hg, and plasma oncotic pressure (π_{GC}) is 25 mm Hg and 33 mm Hg at the afferent and efferent arteriolar ends, respectively, which of the following most likely corresponds to average glomerular capillary ultrafiltration pressure (P_{UF})?

    A. 21 mm Hg

    B. 15 mm Hg

    C. 10 mm Hg

    D. 6 mm Hg

    E. 2 mm Hg

RQ8.2 Which of the following most likely has the highest renal clearance under normal conditions?

    A. Albumin

    B. Glucose

    C. Na^+

    D. Uric acid

    E. Water

CASE 9: HEADACHES, DIZZINESS, AND SHORTNESS OF BREATH

A 56-year-old male presents to the office with a 3-month history of worsening headaches, dizziness, and shortness of breath. His medical history is significant for hypertension that has been difficult to manage, currently necessitating three antihypertensive agents.

Physical examination reveals a bruit over the left flank. Pulse is 92 beats/min, and blood pressure is 175/102 mm Hg. Laboratory studies are notable for a serum creatinine of 1.6 mg/dL (normal = 0.6–1.2 mg/dL), a blood urea nitrogen (BUN) of 32 mg/dL (normal = 7–18 mg/dL), and a potassium of 3.3 mEq/L (normal = 3.5–5.0 mEq/L).

Renal artery Doppler ultrasound and computed tomography (CT) angiography reveal that the left kidney appears to be smaller than the right with significant stenosis in the left renal artery. A diagnosis of renal artery stenosis (RAS) is made, and the patient is referred to a vascular surgeon for further management.

The patient undergoes renal artery angioplasty with stenting. Following the procedure, his blood pressure and laboratory values improve. His headaches, dizziness, and shortness of breath resolve. He is instructed to continue taking his antihypertensive medications and to follow a low-sodium diet to help control his blood pressure and prevent further kidney damage.

THOUGHT QUESTIONS

TQ9.1 Why are the patient's creatinine and BUN elevated?

TQ9.2 What is a bruit, and what is the likely cause in this patient?

TQ9.3 Why does vessel narrowing cause turbulence?

TQ9.4 What do the imaging studies suggest regarding left kidney function?

TQ9.5 What impact does RAS have on glomerular function?

TQ9.6 How does RAS lead to hypertension?

TQ9.7 Why is the patient hypokalemic?

REVIEW QUESTIONS: Choose the ONE best answer

RQ9.1 NaCl sensing by the macula densa is most likely inhibited by which of the following diuretics?
  A. Acetazolamide
  B. Amiloride
  C. Furosemide
  D. Hydrochlorothiazide
  E. Mannitol

RQ9.2 Which of the following most likely decrease(s) glomerular arteriolar resistance and increase(s) renal blood flow and glomerular filtration rate?
  A. Atrial natriuretic peptide
  B. Endothelins
  C. Leukotrienes
  D. Norepinephrine
  E. Parasympathetic stimulation

CASE 10: CRUSH INJURY

A 28-year-old male is brought to the emergency department from a construction site after a reinforced concrete column fell on his right leg. The patient reports severe leg pain and weakness. On physical examination, his right leg appears severely swollen with a deep laceration that is bleeding heavily.

His pulse is 112 beats/min, and his blood pressure is 92/58 mm Hg. An electrocardiogram (ECG) shows characteristic changes associated with hyperkalemia. X-rays show multiple fractures of the tibia and fibula, with significant soft tissue swelling and displacement.

Laboratory studies reveal a potassium level of 6.7 mEq/L (normal = 3.5–5.0 mEq/L) and a creatinine of 1.6 mg/dL (normal = 0.6–1.2 mg/dL). The patient is given intravenous fluids to improve his blood pressure, calcium gluconate to stabilize cardiac cell membranes, and insulin and glucose to lower potassium levels. He is taken for emergency surgery to explore the extent of his injuries.

In the postoperative period, the patient's urine darkens, and his serum creatine kinase (CK) increases to 90,000 U/L (normal = 20–200 U/L). The clinical and laboratory findings are consistent with rhabdomyolysis.

THOUGHT QUESTIONS

TQ10.1   What is the cause of the hyperkalemia?

TQ10.2   How would hyperkalemia be expected to affect the ECG?

TQ10.3   How does hyperkalemia affect muscle function?

TQ10.4   How does calcium gluconate stabilize cardiac cell membranes?

TQ10.5   How do insulin and glucose lower potassium levels?

TQ10.6   Why is the patient's urine dark and serum CK elevated?

REVIEW QUESTIONS: Choose the ONE best answer

RQ10.1 Which of the following hormones is most likely to increase renal K^+ excretion in response to hyperkalemia?
  A. Aldosterone
  B. Angiotensin II
  C. Antidiuretic hormone
  D. Epinephrine
  E. Insulin

RQ10.2 Which of the following is most likely to cause hyperkalemia?
  A. Alkalosis
  B. Chemotherapy
  C. Liddle syndrome
  D. Loop diuretic use
  E. Prolonged vomiting

CASE 11: EXCESSIVE THIRST AND FREQUENT URINATION

A 38-year-old male presents to the clinic with excessive thirst and frequent urination for the past 3 months. He reports drinking several liters of water daily, but he still feels thirsty, and his mouth is always dry. He feels excessively tired during the daytime because he awakens several times during the night to urinate. His past medical history is significant for bipolar disorder for which he is taking lithium, and he mentions that his father has recently been diagnosed with chronic kidney disease.

Physical examination is unremarkable, except for mild dehydration of the oral mucosa. Vital signs are normal. Laboratory studies reveal serum sodium levels of 152 mEq/L (normal = 136–145 mEq/L), serum osmolality of 320 mOsm/kg H_2O (normal = 275–295 mOsm/kg H_2O), and a low urine osmolality of 75 mOsm/kg H_2O (normal = 50–1,400 mOsm/kg H_2O). Serum creatinine is within normal limits.

The clinical presentation and laboratory studies are consistent with arginine vasopressin resistance (AVP-R), which is confirmed with further testing. The patient is advised to stop taking lithium and is started on a thiazide diuretic to reduce urine output.

THOUGHT QUESTIONS

TQ11.1 Why is the patient thirsty if he is drinking so much water?

TQ11.2 Why is the patient's serum sodium and osmolality high if he is drinking several liters of water a day?

TQ11.3 How does antidiuretic hormone (ADH) regulate renal water retention?

TQ11.4 How might AVP-R be distinguished from other potential causes of excessive thirst and urination?

TQ11.5 Why is the patient advised to stop taking lithium?

TQ11.6 How would a diuretic help a patient presenting with excessive urination?

REVIEW QUESTIONS: Choose the ONE best answer

RQ11.1 ADH functions most likely include which of the following?
   A. Decreasing renal Na^+ reabsorption
   B. Increasing systemic vascular resistance
   C. Stimulating angiotensin II secretion
   D. Stimulating thirst
   E. Venoconstriction

RQ11.2 In addition to ADH, the posterior pituitary most likely secretes which of the following hormones?
   A. Dopamine
   B. α-Melanocyte-stimulating hormone
   C. Oxytocin
   D. Prolactin
   E. Somatostatin

CASE 12: ACCIDENTAL TOXIN INGESTION

A 4-year-old female is brought to the emergency department by her parents with severe flank pain and vomiting. They report that she was found in the family garage and is believed to have ingested a bright green liquid from an unlabeled container approximately 2 hours ago.

Physical examination finds the patient to be agitated, disoriented, and diaphoretic. Her temperature is 37.7°C (99.9°F), pulse is 132 beats/min, respirations are 32 breaths/min, and her blood pressure is 102/68 mm Hg. Her abdomen is tender and distended.

Laboratory studies reveal (reference ranges shown in parentheses):

Serum

Na$^+$	140 mEq/L	(136–145 mEq/L)
K$^+$	4.2 mEq/L	(3.5–5.0 mEq/L)
Cl$^-$	110 mEq/L	(95–105 mEq/L)
HCO$_3^-$	8 mEq/L	(22–28 mEq/L)
Glucose	125 mg/dL	(70–115 mg/dL)
Urea nitrogen (BUN)	30 mg/dL	(7–18 mg/dL)

Arterial blood

pH	7.20	(7.35–7.45)
Pco$_2$	20 mm Hg	(33–45 mm Hg)
Po$_2$	98 mm Hg	(75–105 mm Hg)

Based on the presentation and laboratory findings, ethylene glycol poisoning is suspected. An alcohol dehydrogenase (ADH) inhibitor is administered.

THOUGHT QUESTIONS

TQ12.1 What is the patient's acid–base status?

TQ12.2 Why is the patient's respiratory rate so high?

TQ12.3 How would the kidneys be expected to respond to the acid–base disturbance?

TQ12.4 What is this patient's anion gap, and what does this tell us about the possible cause of the acid–base disorder?

TQ12.5 How does ethylene glycol cause an acid–base disturbance?

TQ12.6 What does the elevated BUN suggest?

TQ12.7 Why is an ADH inhibitor helpful?

REVIEW QUESTIONS: Choose the ONE best answer

RQ12.1 Which of the following conditions is most likely to cause metabolic alkalosis?
- A. Hysteria
- B. Myasthenia gravis
- C. Prolonged vomiting
- D. Pulmonary edema
- E. Severe diarrhea

RQ12.2 Which of the following findings would most likely be consistent with respiratory acidosis?
- A. Decreased Pco$_2$
- B. Decreased plasma Cl$^-$
- C. Decreased plasma HCO$_3^-$
- D. Increased plasma K$^+$
- E. Increased plasma pH

CASE 13: EPIGASTRIC PAIN, BLOATING, AND DIARRHEA

A 52-year-old male presents to the office with epigastric pain, bloating, and diarrhea. He has a 6-year history of peptic ulcer disease that has been treated with proton-pump inhibitors (PPIs). He reports that in recent months he has experienced intermittent acid reflux symptoms and has lost about 4 kg (9 lb) in weight despite eating well.

Physical examination reveals a soft, nontender abdomen with no palpable masses or hepatosplenomegaly. His vital signs are normal. Laboratory studies are notable for a serum gastrin level of 680 pg/mL (normal =

0–100 pg/mL) and gastric acid production rate of 30 mmol/h (normal = 0–10 mmol/h). Pancreatic enzymes are normal.

Upper endoscopy is significant for a duodenal ulcer that is negative for *Helicobacter pylori*. Abdominal CT reveals a 2.5-cm mass in the head of the pancreas. A secretin stimulation test markedly increases serum gastrin to 1,150 pg/mL, confirming the diagnosis of Zollinger-Ellison syndrome (ZES).

The patient is started on high-dose proton pump inhibitor (PPI) therapy and referred to a surgeon for resection of the gastrinoma.

THOUGHT QUESTIONS

TQ13.1 Why does the patient have an elevated gastric acid production rate?

TQ13.2 How is gastrin secretion and acid production limited normally?

TQ13.3 Why does secretin increase serum gastrin?

TQ13.4 Why is the ulcer duodenal rather than gastric?

TQ13.5 What is the most common cause of peptic ulcers?

TQ13.6 Why does the patient have bloating and diarrhea?

TQ13.7 Why are PPIs so effective at controlling acid reflux symptoms?

REVIEW QUESTIONS: Choose the ONE best answer

RQ13.1 Parietal cell secretory products are most likely essential for absorption of which of the following nutrients?
- A. Vitamin A
- B. Vitamin B_{12}
- C. Vitamin C
- D. Vitamin D
- E. Vitamin K

RQ13.2 Gastric chief cells most likely produce which of the following?
- A. Cholecystokinin
- B. Gastric lipase
- C. Lactoferrin
- D. Mucus
- E. Trypsinogen

CASE 14: FATIGUE AND ITCHY, DRY SKIN

A 35-year-old female presents to the clinic with fatigue and itchy, dry skin for the past 6 months. She feels increasingly sluggish and lacking in energy. She also reports a weight gain of 9 kg (20 lb). She has a family history of autoimmune disease, including systemic lupus erythematosus and rheumatoid arthritis.

Physical examination reveals a rubbery, diffusely enlarged thyroid gland that is nontender. Her skin appears dry and pale. Laboratory testing reveals an elevated antithyroid peroxidase antibody level of 320 IU/mL (normal = <9 IU/mL).

The patient is started on thyroid hormone replacement therapy and referred to an endocrinologist for further management.

THOUGHT QUESTIONS

TQ14.1 What is the likely cause of the patient's hypothyroidism?

TQ14.2 How would the hypothalamic–pituitary–thyroid (HPT) endocrine axis be affected?

TQ14.3 Would this be considered primary, secondary, or tertiary hypothyroidism?

TQ14.4 What might a radioactive iodine uptake (RAIU) test reveal?

TQ14.5 Thyroiditis often presents initially as hyperthyroidism: how would the HPT axis be affected during this phase?

REVIEW QUESTIONS: Choose the ONE best answer

RQ14.1 Which of the following most likely describes the primary means by which thyroid hormones are carried in blood?
A. Bound to albumin
B. Bound to lipoproteins
C. Bound to thyroxine-binding globulin
D. Bound to transthyretin
E. In free form

RQ14.2 Thyroid hormone control of thyrotropin-releasing hormone (TRH) release from the hypothalamus most likely involves which of the following?
A. Activation of the cyclic adenosine monophosphate (cAMP) signaling pathway
B. Activation of the inositol trisphosphate (IP_3) signaling pathway
C. Increased TRH proteolysis
D. Increased TRH uptake from the median eminence
E. Thyroid receptor phosphorylation

CASE 15: MENSTRUAL IRREGULARITIES, EXCESSIVE HAIR GROWTH, AND ACNE

A 16-year-old female comes to the clinic concerned about her irregular menstrual periods, excessive hair growth, and acne, symptoms that have been getting progressively worse over the past year. She has no significant medical history. She is not taking any medications and denies using any supplements or hormonal contraceptives.

Physical examination reveals a blood pressure of 152/94 mm Hg and hirsutism affecting her face, chest, and back. Acne lesions are evident on her face and upper back.

Laboratory studies reveal elevated levels of 17-hydroxyprogesterone and 11-deoxycorticosterone (DOC), whereas aldosterone levels are decreased. Based on these findings, 11β-hydroxylase (11β-OH) deficiency is suspected. Genetic analysis confirms a mutation in *CYP11B1*, the gene that encodes 11β-OH.

THOUGHT QUESTIONS

TQ15.1 What is the function of 11β-OH normally?

TQ15.2 What are the likely consequences of 11β-OH deficiency for adrenal function?

TQ15.3 What is causing the patient's menstrual irregularities, hirsutism, and acne?

TQ15.4 What is the function of aldosterone normally?

TQ15.5 Why is the patient hypertensive if aldosterone levels are reduced?

TQ15.6 How might 11β-OH deficiency be expected to affect the renin–angiotensin–aldosterone system (RAAS)?

TQ15.7 How might the hypothalamic–pituitary–adrenal (HPA) endocrine axis be affected by 11β-OH deficiency?

TQ15.8 What are likely treatment options for this patient?

REVIEW QUESTIONS: Choose the ONE best answer

RQ15.1 ACTH is synthesized from the proopiomelanocortin (POMC) gene (*POMC*) that most likely also encodes which of the following?
A. Angiotensinogen
B. β-Endorphin
C. Melanin
D. Melatonin
E. Testosterone

RQ15.2 The mechanism by which both angiotensin II (Ang-II) and K^+ stimulate aldosterone secretion most likely involves which of the following?
A. Activating a receptor tyrosine kinase
B. Activating an Ang-II type 1 (AT1) receptor
C. Activating G_s
D. Increasing intracellular Ca^{2+}
E. Raising intracellular cGMP levels

II. CLINICAL CASE ANSWERS

CASE 1: Answers to Thought Questions

TQ1.1 The *CFTR* gene encodes the cystic fibrosis transmembrane conductance regulator (CFTR). CFTR is an adenosine triphosphate (ATP)-binding cassette transporter that has lost its transport capabilities and functions instead as an anion channel that supports transmembrane Cl^- and HCO_3^- fluxes. Other family members include multidrug-resistance proteins found in the renal proximal tubule (see Chapter 26·IV·B). CFTR requires ATP binding and phosphorylation by cyclic adenosine monophosphate (cAMP)-dependent protein kinase A (PKA) to open.

TQ1.2 ΔF508 mutation causes deletion of phenylalanine at position 508 of the CFTR peptide sequence. This interferes with normal protein folding and causes CFTR to be degraded. CFTR is normally expressed in the apical membrane of many secretory epithelia, where it mediates Cl^- fluxes (see Chapter 15·VI·D, Chapter 30·II·C and IV·D, and Chapter 32·II·A). Decreased CFTR expression reduces apical Cl^- movements. CFTR mutations are recessive, but the mutation is so prevalent within populations of Northern European descent that CF is a relatively common disease in the United States.

TQ1.3 Patients with CF produce an unusually thick and viscous airway mucus, which makes it difficult to clear from the lung. Airway mucus normally functions to trap inhaled particulates, including bacteria (such as *Pseudomonas aeruginosa*), which are then eliminated via the mucociliary escalator (see Chapter 21·II·A). In patients with CF, these bacteria can become entrenched, causing recurrent cough and respiratory infections that result in progressive decline in lung function and bronchiectasis (airways enlargement). Although the precise role of CFTR in airway mucus formation is unclear, it is likely that normal Cl^- channel function is required for mucus hydration and thinning.

TQ1.4 The patient's respiratory infection has interfered with normal ventilation and gas exchange, as evidenced by the abnormally low P_{O_2} and high P_{CO_2}. CO_2 dissolves in plasma to form carbonic acid, which lowers blood pH (see Chapter 23·VI).

TQ1.5 The patient is underweight, most likely because of pancreatic enzyme insufficiency affecting nutrient absorption. CFTR normally facilitates pancreatic secretion of enzymes and HCO_3^--rich fluid. In patients with CF, pancreatic secretions are thick and viscous and may remain trapped within the pancreas. Pancreatic enzymes are required for normal digestion of lipids, carbohydrates, and proteins, so patients with CF are typically malnourished and require pancreatic enzyme replacement therapy to aid food digestion (see Clinical Application 32.1).

TQ1.6 Sweat gland secretory coils produce a sweat precursor fluid that resembles plasma. Cells lining the sweat gland duct then reabsorb Na^+ and Cl^- via ENaC (epithelial Na^+ channel) and CFTR, respectively, leaving sweat hypotonic (Fig. C1.1; see also Chapter 15·VI·D). Sweat normally contains <30 mmol/L Cl^- for this reason. CFTR mutation impairs Cl^- reabsorption, so sweat Cl^- levels are elevated. A sweat test Cl^- level of 40–59 mmol/L is considered "borderline" and ≥60 mmol/L is diagnostic of CF. Sweat tests are routinely used to screen newborns for CF in the United States.

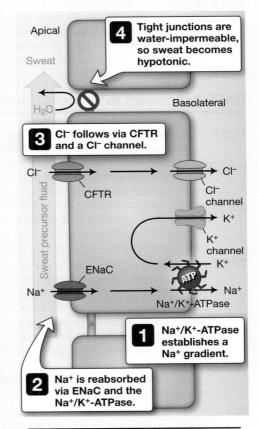

Figure C1.1.
Ductal modification of sweat gland precursor fluid. CFTR = cystic fibrosis transmembrane conductance regulator, ENaC = epithelial Na^+ channel.

Summary

CF is caused by *CFTR* gene mutation, most commonly ΔF508. CFTR is expressed by most epithelia, where it supports apical Cl⁻ fluxes. Na⁺ follows Cl⁻ to create an osmotic gradient that leads to fluid secretion (occurs in lungs, pancreas, intestines, and reproductive tracts). CF symptoms reflect impaired transepithelial fluid movement leading to secretions that are thick and viscous. In sweat glands, CFTR allows Cl⁻ (and Na⁺) to be reclaimed from the duct lumen, rendering sweat hypotonic.

CASE 1: Answers to Review Questions

RQ1.1 Answer = B. CFTR activity is increased by protein kinase A (PKA)-dependent phosphorylation of intracellular regulatory domains. Epinephrine binds to adrenergic receptors, which are GPCRs coupled to the cAMP-signaling pathway and PKA (see Chapter 1·VII·B). This pathway is believed to be responsible for increasing airway fluid secretion during exercise and is reduced in patients with exercise-induced asthma. Cholecystokinin (A) acts through the IP_3 signaling pathway. Insulin (C) binds to a receptor tyrosine kinase. Testosterone (D) and thyroxine (E) bind to intracellular receptors.

RQ1.2 Answer = C. Stools that are pale, foul-smelling, and float in the toilet bowl are an indication of undigested fat. Although lingual lipase (B) does digest some fat in the mouth and stomach, its contribution to total fat digestion is minimal compared with pancreatic lipase (see Chapter 31·II·E). Pancreatic insufficiency in CF patients means that fats travel through the gastrointestinal tract and are excreted without being digested and absorbed. Elastase (A) and trypsin (E) are pancreatic proteases. Salivary amylase (D) digests carbohydrates.

CASE 2: Answers to Thought Questions

TQ2.1 NCS assess the functional integrity of peripheral nerves (see Fig. C2.1). Electrodes are commonly placed on the body surface at three points along a nerve's pathway. A brief electrical stimulus is applied at all three sites. A recording electrode placed over the belly of the target muscle records a compound muscle action potential (CMAP), the height of which is proportional to the number of fibers stimulated. Delay times between CMAPs evoked by the stimulating electrodes are used to calculate nerve conduction velocity. Similar techniques are used to assess sensory nerve function. This patient has reduced sensation in the extremities and muscle weakness, suggesting that sensory- and motor-nerve signaling is impaired. NCS would most likely show reduced nerve signal amplitude and, given that GBS is a demyelinating disease, slowed conduction velocity (e.g., compare Fig. C2.1B and C).

TQ2.2 A deep tendon reflex, also known as a stretch reflex or myotatic reflex, is an involuntary muscle contraction triggered by stretching a homonymous muscle (see Chapter 11·III·B). It is mediated by a spinal cord reflex arc. For example, the patellar reflex is a reflex contraction of the quadriceps muscle caused by tapping the patellar ligament. Tapping the ligament stretches the quadriceps muscle, activating muscle spindles within. Sensory signals from spindles are carried by type 1a sensory nerve afferents to the spinal cord, where they synapse with interneurons that excite α motor neurons innervating the same muscle. The leg jerks forward as a result (see Fig. 11.5). Other commonly tested deep tendon reflexes include the ankle (Achilles), biceps, triceps, and brachioradialis reflexes. Deep tendon reflexes rely on normal functioning of sensory and motor nerves innervating a muscle, so the lack of deep tendon reflexes in this patient is consistent with impaired nerve conduction.

TQ2.3 Myelin is a lipid-rich, multilayered, insulating sheath that envelops nerve axons in both the peripheral and central nervous systems (see Chapter 5·V·A). Myelin is formed by glial cells that extend processes that rotate around an axon more than 100 times (see Fig. 5.9). Cytoplasm is squeezed out from between the membrane layers as they build, creating a highly compacted insulating structure. Peripheral neurons are myelinated by Schwann cells, whereas oligodendrocytes myelinate central neurons.

TQ2.4 Sensory and motor neurons both signal using Na⁺-based action potentials (APs). Stimulating a nerve opens rapidly activating, voltage-dependent Na⁺ channels that support a Na⁺ flux that depolarizes the membrane and yields the upstroke of an AP (see Chapter 5·III·B). Na⁺ channels then rapidly inactivate, and voltage-dependent K⁺ channels open to facilitate membrane repolarization to yield the AP downstroke. Myelin acts an electrical insulator that reduces signal decay and thereby increases the rate at which APs propagate down the length of a nerve. For example, myelination allows α motor nerves to conduct APs at 80–120 m/s (see Table 11.1), whereas unmyelinated pain fibers conduct at around 1 m/s (see Fig. 5.4). Myelin loss significantly slows nerve conduction velocity and may block signaling, leading to motor and sensory nerve deficits.

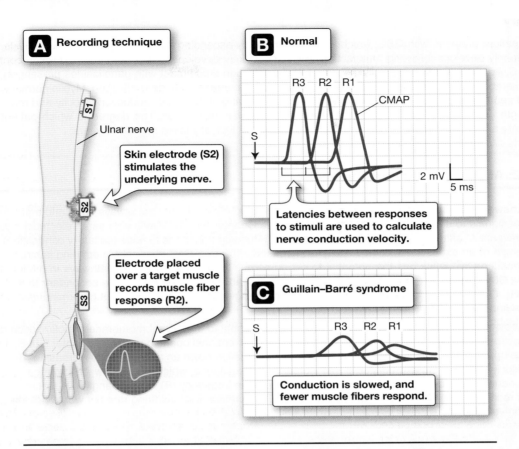

Figure C2.1.
Nerve conduction studies. The three overlapping traces in B and C were triggered at the time indicated by the arrows labeled S. Knowing the distance between the three points of stimulation allows nerve conduction time to be calculated from the time delay between the muscle responses. CMAP = compound muscle action potential.

TQ2.5 GBS symptoms are due to the production of antibodies that attack Schwann cells, leading to local inflammation and activation of macrophages that then proceed to strip the myelin layers from axons. Loss of myelin impairs motor neuron signaling and is sufficient to explain the observed muscle weakness, although axonal inflammation can also result in loss of motor neurons and muscle paralysis. The latter is of particular concern because the diaphragm and other respiratory muscles (see Table 21.1) are skeletal muscles innervated by α motor neurons.

TQ2.6 GBS and MS are both demyelinating diseases, with GBS primarily affecting peripheral neurons, and MS affecting central neurons. The cause of MS is unknown (see Clinical Application 5.2). Patients present with a variety of neurologic symptoms, including fatigue and headaches, vision impairment, muscle weakness, numbness, gait disturbances, and autonomic dysfunction.

TQ2.7 GBS often follows a respiratory or gastrointestinal infection caused by Epstein-Barr virus; herpes simplex virus; or, most commonly, *Campylobacter jejuni*. *C. jejuni* causes acute gastroenteritis and diarrhea and is the most likely cause of GBS in this patient. *C. jejuni* inhabits the digestive tract of a variety of animals, including poultry. *C. jejuni* infections lead to the development of antibodies that cross-react with specific gangliosides expressed by Schwann cells and components of the axonal membrane, leading to demyelination.

TQ2.8 IVIG therapy involves IV infusion of immunoglobulins from donor blood. The mechanism of action is poorly defined, but IVIG may inhibit complement or immune cell activation, or it may inhibit binding of antiganglioside antibodies to their targets.

Summary

The patient presents with GBS, which is characterized by ascending muscle weakness and paralysis that most commonly develops following an infection. The patient likely developed GBS from eating chicken contaminated with *C. jejuni*. Antibodies formed in response to the infection cross-react with gangliosides present on Schwann cell membranes (i.e., myelin) and the axon, leading to inflammation and demyelination. Myelin normally facilitates rapid neuronal signaling, so demyelination leads to a slowing of neuronal conduction velocity and reduced signal strength. Sensory deficits, muscle weakness, and paralysis may result. The degree to which patients recover sensory and motor functions following GBS is variable but typically takes months to years.

CASE 2: Answers to Review Questions

RQ2.1 Answer = D. All excitable membranes contain K^+ "leak" channels that are always open and that allow electrical signals to decay with distance from origin (see Chapter 5·III·B). Myelin acts as an electrical insulator that prevents K^+ efflux and signal decay, allowing an AP to jump from node to node (saltatory conduction) down the length of an axon. Conduction velocity is increased >100-fold. Ca^{2+}-channel (A) opening is strictly regulated because Ca^{2+} influx has the potential to set numerous intracellular signaling pathways in motion. Cl^- fluxes via Cl^- channels (B) tend to dampen membrane excitability, but their effects are secondary to K^+ leak channels. Cl^-/HCO_3^- exchangers (C) and Na^+/Ca^{2+} exchangers (E) are not significant determinants of membrane excitability.

RQ2.2 Answer = E. Neuronal conduction velocity is proportional to the rate of membrane depolarization during the upstroke of the AP, which, in turn, is proportional to Na^+-channel opening rate (see Chapter 5·III·B). Increasing the opening rate would increase the rate of AP propagation down an axon, thereby increasing AP conduction velocity. Decreasing axon diameter (A) increases its resistance, which would slow AP propagation. Neuronal conduction velocity is largely independent of stimulation frequency (B) or axon length (D). Decreasing the rate of recovery from Na^+-channel activation (C) would decrease maximal firing rate but would not affect conduction velocity.

CASE 3: Answers to Thought Questions

TQ3.1 Actin is the principal component of thin filaments that, together with myosin, form sarcomeric contractile arrays in striated muscle (see Chapter 12·II). Nebulin is a filamentous cytoskeletal protein that runs the length of a thin filament and is believed to act as a molecular ruler during thin-filament assembly. Actin and nebulin are both essential for normal muscle function.

TQ3.2 CK is a muscle enzyme used to transfer a high-energy phosphate bond from phosphocreatine to ADP to regenerate ATP (see Chapter 39·IV). The presence of CK in blood indicates muscle damage and myocyte lysis (see Clinical Application 39.1). Mutations in *ACTA1* and *NEB* not only impair contractility, they also make myocytes vulnerable to damage during contraction.

TQ3.3 Motor units are groups of muscle fibers that may be spatially separate but are innervated by the same α motor neuron (see Chapter 12·IV·D). Motor unit potentials (MUPs) are compound potentials caused by electrical excitation of the fibers that make up a motor unit, recorded by electrodes implanted in the body of a muscle (Fig. C3.2). MUP amplitude and duration are proportional to the number and size of the generating muscle fibers. MUPs in healthy muscle are typically <2 mV in amplitude and 10–15 ms in duration. MUPs in patients with muscle disease are typically small (0.5 mV) with a duration of only 5–10 ms, indicating decreased numbers of fibers within the motor unit.

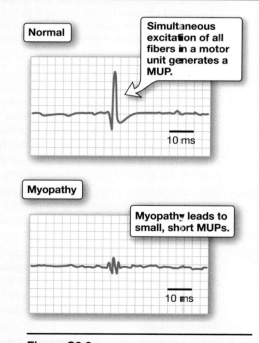

Normal

Simultaneous excitation of all fibers in a motor unit generates a MUP.

10 ms

Myopathy

Myopathy leads to small, short MUPs.

10 ms

Figure C3.2.
Motor unit potentials (MUPs) recorded from normal, healthy muscle and a muscle from a patient with myopathy. MUPs were recorded by electrodes implanted in muscle following α motor neuron stimulation.

TQ3.4 Skeletal muscle contractile strength is increased by recruiting additional motor units (see Chapter 12·IV·D). Early recruitment suggests that individual motor units have low contractile strength, so increasing the force of contraction requires recruiting additional motor units earlier and in greater numbers than in healthy muscle.

TQ3.5 Mutations affecting skeletal muscle structural proteins impair contractility and increase vulnerability to damage during contraction. Over time, muscle wasting occurs, leading to progressive muscle weakness. Loss of proximal muscle (muscles closest to the trunk) makes climbing stairs, standing after having been sitting, walking, and other daily activities challenging but not fatal. Respiratory muscles such as the diaphragm and intercostals are also composed of skeletal muscle, so NM may cause respiratory muscle weakness and hypoventilation, ultimately leading to respiratory failure.

Summary

NM is a hereditary muscle wasting disorder most commonly caused by actin and nebulin gene mutations. Muscle wasting occurs due to increased vulnerability to damage during contraction. Nemaline rods (thread-like inclusions composed of Z-disk proteins) are a common but nonspecific histologic finding. Muscle wasting leads to weakness, decreased deep tendon reflexes, swallowing difficulties, and hypoventilation.

CASE 3: Answers to Review Questions

RQ3.1 Answer = A. α-Actinin is a sarcomeric structural protein that tethers the ends of actin thin filaments to Z disks (see Chapter 12·II·C). Dystrophin (B) is also a sarcomeric structural protein that associates with Z disks, but it anchors the contractile array to the cytoskeleton. Titin (C) attaches myosin thick filaments to Z disks. It acts as a molecular spring that limits sarcomeric stretching and centers thick filaments within the sarcomere. Tropomyosin (D) filaments lie along the surface of actin filaments and conceal myosin-binding sites until contraction begins. Troponin (E) is a Ca^{2+} sensor that initiates contraction once sarcoplasmic calcium levels rise.

RQ3.2 Answer = A. Ryanodine receptors (RyRs) are Ca^{2+}-release channels located in the sarcoplasmic reticulum (SR). They activate when depolarization of T tubules activates dihydropyridine (DHP) receptors (see Chapter 12·III·A). DHP receptors are L-type Ca^{2+} channels. IP_3 (B) activates Ca^{2+}-release channels in the SR of smooth muscle. Protein kinase A (C) phosphorylates Ca^{2+}-release channels in cardiac muscle as a way of modulating Ca^{2+} release and contractility. These channels are activated by a rise in sarcoplasmic Ca^{2+} levels (D). Sarcolemmal Na^+-channel activation (E) leads to T-tubule depolarization, but RyR activation requires the conformational change associated with DHP receptor activation.

CASE 4: Answers to Thought Questions

TQ4.1 The mitral valve (MV) is located between the left atrium (LA) and left ventricle ([LV] see Fig. 16.2). When open, it allows blood to flow from the LA into the LV. The MV orifice is normally sufficiently wide that it offers minimal resistance to flow during LV preloading. When the valve orifice narrows (i.e., becomes stenotic), flow is impeded, which can limit preloading and cardiac output (CO). Symptoms typically do not manifest until valve orifice cross-sectional area drops below 1.5 cm^2.

TQ4.2 Heart murmurs are caused by turbulent blood flow (see Clinical Application 17.1 and Chapter 18·V·A). Flow through the MV occurs during diastole, so the murmur associated with MS is diastolic. The pressure driving flow through the MV from LA to LV is usually relatively low in the early stages of MS (<20 mm Hg), so murmur intensity is minimal.

TQ4.3 In the early stages, MS has minimal effect on left heart pressures. When the valve orifice narrows to the point of interfering with LV preloading, the body begins retaining fluid (through activation of the renin–angiotensin–aldosterone system [RAAS]) to increase LA pressure (LAP). The pressure increase is required to offset the effect of stenosis on flow by increasing flow velocity (see Chapter 18·V·A·3). Normally, the pressure difference between LA and LV is ~2 mm Hg or less, but MS causes a significant pressure gradient to develop across the valve that increases with orifice narrowing (Fig. C4.1). This gradient is a hallmark of MS. LV function is generally not affected by MS, and the patient's ejection fraction is indeed noted to be normal.

TQ4.4 RAAS activation causes the kidney to retain salt and water to increase extracellular fluid (ECF) and blood volume (plasma is an ECF component; see Chapter 19·IV). Central venous pressure (CVP) rises as a result. The goal is to raise LAP to maintain LV preloading, but the right heart and lungs lie between the central veins and the LA. A rise in CVP increases right-ventricular preload, pulmonary arterial pressure, and pulmonary venous pressure. The patient presents with right-atrial (RA) enlargement (due to the increased preloading) and with pulmonary hypertension, which is consistent with an increase in CVP. Despite an inbuilt pressure safety margin (see Fig. 40.20), the danger of raising pulmonary vascular pressures is that pulmonary capillary hydrostatic pressure can rise to the point where the Starling forces holding fluid in the vasculature are overwhelmed, resulting in pulmonary edema. Pulmonary edema decreases lung compliance and increases the work of breathing. Fluid within alveoli interferes with gas exchange and results in hypoxemia and shortness of breath. Also, RA enlargement distorts conduction pathways and may lead to atrial fibrillation (AF), which may be felt as palpitations.

TQ4.5 MS is an insidious condition that may remain undetected until the cardiovascular system is stressed—by pregnancy, for example. Pregnancy places extraordinary demands on all the body's organ systems. CO increases by ~50% (see Fig. 37.8) to support the fetus plus the increased needs of maternal organs during pregnancy. Although MS may not produce symptoms when flow through the valve is ~5 L/min, increasing flow to ~7.5 L/min may reveal a flow limitation that presents as fatigue and shortness of breath.

TQ4.6 The patient presents with fatigue and shortness of breath (dyspnea). Fatigue is most likely caused by CO limitations, which become more evident with physical exertion. Dyspnea can be caused by inadequacy of CO or pulmonary hypertension and edema.

TQ4.7 The first ~20 weeks of gestation are accompanied by a significant drop in systemic vascular resistance ([SVR] see Fig. 37.8). The decrease is due in part to placentation, during which the uterus becomes a low-resistance pathway within the systemic circulation. The corpus luteum and placenta also release relaxin to promote systemic vasodilation and facilitate increased flow through the vasculature, which further decreases SVR. Finally, cutaneous and renal vascular resistance decrease to facilitate the flow increases needed to dissipate heat and filter blood to remove additional wastes produced by the pregnancy. A fall in SVR activates the RAAS to increase CO, as discussed earlier, the goal being to maintain MAP at levels that ensure adequate perfusion of all maternal organs (MAP = CO × SVR).

TQ4.8 The most common cause of MS is rheumatic fever resulting from *Streptococcus pyogenes* infection. Rheumatic fever is usually treated promptly with antibiotics in developed countries, so valve involvement and cardiac complications are uncommon. In developing countries, infection of heart valves may lead to leaflet damage and thickening, resulting in insufficiency and/or stenosis.

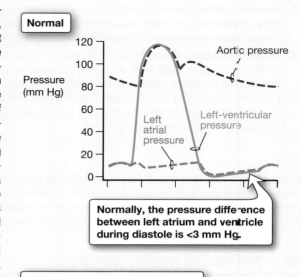

Normal

Normally, the pressure difference between left atrium and ventricle during diastole is <3 mm Hg.

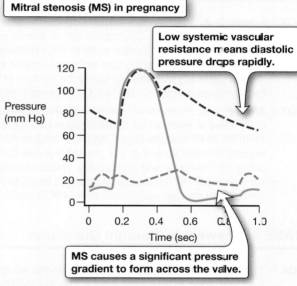

Mitral stenosis (MS) in pregnancy

Low systemic vascular resistance means diastolic pressure drops rapidly.

MS causes a significant pressure gradient to form across the valve.

Figure C4.1.
Effects of mitral stenosis on left-heart pressures.

TQ4.9 Treatment options for MS during pregnancy are limited because of potential harm to the fetus. β-Blockers reduce heart rate, which prolongs diastolic filling and lowers LAP. Diuretics reduce symptoms associated with volume overload (pulmonary edema). LMWH is an anticoagulant that reduces the risk of stroke. Pregnancy is a hypercoagulable state, plus atrial enlargement and AF both increase the likelihood of clot formation due to incomplete atrial emptying and blood stagnation (see Clinical Application 18.2).

Summary

The patient presents with newly diagnosed MS, most likely caused by untreated *S. pyogenes* infection during childhood. The narrowed valve orifice limits LV preloading and CO, the effects being felt during physical exertion. The body responds to CO limitations by expanding blood volume and CVP to increase LAP, causing a significant pressure gradient to develop across the valve between LA and LV. The rise in CVP leads to increased right-heart pressures and pulmonary hypertension, potentially causing pulmonary edema. Right-heart failure may result if the condition remains untreated. The patient's condition revealed itself during pregnancy, when the defective valve prevented the rise in CO needed to support fetal development.

CASE 4: Answers to Review Questions

RQ4.1 Answer = D. Mitral regurgitation (MR) allows blood to leak backward from the LV to the LA during systole (see Clinical Application 17.1). The regurgitant blood volume adds to blood accumulating in the LA from the pulmonary veins, which increases LV preload. The enhanced blood volume can cause an S_3 heart sound associated with increased LV filling. MR is associated with a systolic murmur, not a continuous (A) or diastolic murmur (B). The former is associated with patent ductus arteriosus, the latter with aortic or pulmonary valve regurgitation. MR can cause softening of S_1, which is caused by closure of the atrioventricular valves, but would not affect S_2 (C). An S_4 heart sound (E) is associated with atrial hypertrophy to assist filling of a stiffened LV.

RQ4.2 Answer = C. Blood volume increases during pregnancy from RAAS activation. The expansion of blood volume outpaces red blood cell production, so hematocrit falls, resulting in a physiologic anemia of pregnancy (see Chapter 37·IV·C). Anemia increases the likelihood of turbulence, so murmurs associated with high-volume, high-velocity flow are common in pregnant patients. End-diastolic volume (A) increases during pregnancy due to the enhanced preloading caused by expansion of blood volume. Heart rate (B) increases to help support a need for increased CO. Pulse pressure (D) increases due to the drop in diastolic blood pressure caused by a decrease in SVR. Stroke volume (E) increases during pregnancy because of the enhanced cardiac preloading.

CASE 5: Answers to Thought Questions

TQ5.1 Dizziness and fainting are most commonly due to cerebral hypoperfusion. The tissues of the brain are highly dependent on oxidative phosphorylation for energy production, which creates a heavy reliance on uninterrupted cerebral blood flow (see Chapter 20·II). A decrease in cerebral perfusion pressure and resulting drop in cerebral flow quickly leads to dizziness and lightheadedness and can result in loss of consciousness (syncope).

TQ5.2 The distribution of blood within the body is heavily influenced by gravity (see Chapter 19·III·D). Standing after having been supine or sitting causes up to 800 mL of blood to be pulled downward into the lower extremities (see Fig. 19.11). Veins in the feet and ankles swell passively to accommodate the additional blood volume. Meanwhile, central veins have become volume depleted, which negatively impacts left-ventricular (LV) preloading and cardiac output (CO). Mean arterial pressure (MAP) may fall by several mm Hg as a result. Cerebral perfusion pressure is dependent on MAP, so changes in blood volume distribution potentially may adversely affect cerebral blood flow and brain function.

TQ5.3 Arterial pressure is monitored by baroreceptors located in the carotid sinus and aortic arch. Baroreceptors comprise mechanosensitive neurons embedded in the vessel walls whose firing rate is a direct reflection of wall stretch (see Chapter 19·III·A). Increases in arterial pressure stretch the vessel walls to a greater degree, and the baroreceptors respond with a volley of action potentials. Sensory information is relayed to a cardiovascular control center located in the brainstem medulla via cranial nerves (CNs) IX and X.

TQ5.4 MAP is a product of CO and systemic vascular resistance (SVR). Standing causes central blood volume and cardiac preload to fall, and CO and MAP soon follow. The drop in MAP decreases baroreceptor afferent firing rate, and the cardiovascular control center responds by decreasing parasympathetic (PSNS) and increasing sympathetic nervous system (SNS) output to the heart and vasculature ("orthostatic reflex"; see Fig. 19.11). Heart rate and myocardial contractility increase to enhance stroke volume (SV) and CO. SV is further increased when venoconstriction forces blood back to the heart. SNS-induced constriction of resistance vessels raises SVR. The resistance increase is key to helping limit a drop in MAP on standing (Fig. C5.1).

TQ5.5 Carotid baroreceptors and, to a lesser degree, aortic baroreceptors mediate short-term responses to changes in arterial pressure. Cardiopulmonary receptors located in low-pressure parts of the cardiovascular system (vena cavae, atria, pulmonary vasculature) relay information about the fullness of the vascular system and are more important for long-term control of arterial pressure (see Chapter 19·III·A·2). Because vascular fullness

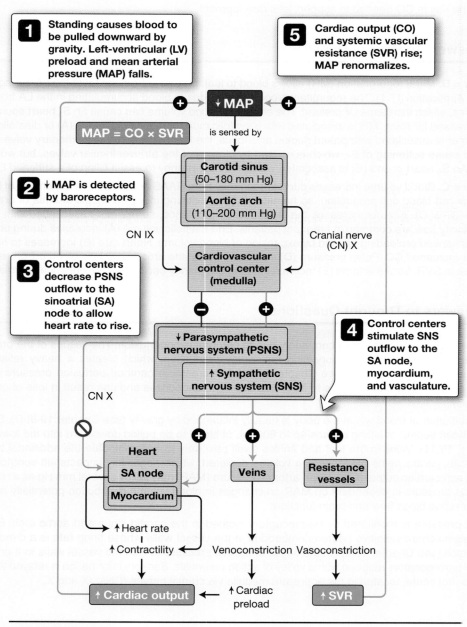

Figure C5.1.
Baroreflex response to standing.

determines cardiac preload, information received by the medullary cardiovascular control center from cardio-pulmonary receptors gets integrated and used to formulate an appropriate response.

TQ5.6 The patient is currently being treated with a β-blocker and a diuretic to address her hypertension. β-Blockers lower heart rate, thus limiting compensation for the decrease in MAP. Diuretics reduce blood volume and can cause hypovolemia, which reduces cardiac preloading. Both drugs likely contributed to the patient's impaired ability to respond appropriately to standing with increased CO.

TQ5.7 Aging negatively impacts many aspects of cardiovascular function and control center responsiveness to changes in MAP, but the main cause of the increased incidence of OH in elderly populations is arteriosclerosis (see Clinical Application 19.1). Arterial hardening reduces wall compliance and the degree of stretching in response to luminal pressure changes. This reduces baroreceptor nerve ending responsiveness to arterial pressure changes and delays compensatory changes in cardiovascular control center output.

Summary

The patient's medications and advanced age have impaired timely cardiovascular responsiveness to the shift in blood volume that occurs on standing from a sitting position. Blood tends to accumulate in the lower extremities when erect, and central venous blood volume falls. The resulting drop in cardiac preload and output causes arterial pressure and cerebral perfusion pressure to fall, leading to dizziness. Normally, the cardiovascular control centers would quickly respond to postural changes in arterial pressure with increased cardiac efficiency and SVR to ensure that cerebral blood flow is impacted minimally. Patients are advised to ensure adequate hydration and stand slowly after periods of sitting or lying supine.

CASE 5: Answers to Review Questions

RQ5.1 Answer = D. The decrease in MAP that occurs on standing is sensed by arterial baroreceptors, triggering a reflex increase in SNS output to the heart and vasculature (see Chapter 19·III·D). Resistance vessels in vascular beds throughout the body constrict in response to SNS activation, which increases their resistance to flow. SVR increases as a result. Venous return (E) decreases on standing because gravity causes blood to draw downward out of the central veins and into the lower extremities. LV end-diastolic volume (A) is established by LV preload (C), both of which decrease in parallel with venous return. SNS activation increases LV contractility, which helps maintain SV even as preloading decreases. Consequently, LV end-systolic volume (B) decreases.

RQ5.2 Answer = E. Baroreceptor output is directly proportional to the degree the carotid and aortic walls are stretched by luminal pressure (see Fig. 19.8). The decrease in MAP that occurs on standing causes baroreceptor output to fall. Output from baroreceptors in the aortic arch travels via the aortic nerve and vagus nerve (CN X) to the cardiovascular control center in the medulla, so vagal nerve afferent firing rate would be decreased. Blood redistributes on standing, but blood volume (A) does not change. Both circulating epinephrine and norepinephrine (NE) levels (B) and heart rate (C) increase during an orthostatic reflex due to increased sympathetic nerve activity (D). NE is released in response to SNS stimulation of the adrenal glands and spills over into blood from active sympathetic nerve terminals.

CASE 6: Answers to Thought Questions

TQ6.1 The patient's respiration rate is 30 breaths/min yet he is hypoxemic, suggesting that gas exchange is impaired. The difference between alveolar and arterial P_{O_2} (A-a gradient) can be used to assess the efficacy of oxygenation (see Chapter 22·IV·D). The A-a gradient for a healthy individual is age-dependent, but can be estimated from:

$$2.5 + (0.21 \times \text{age in years})$$

For this patient, the predicted A-a gradient is:

$$2.5 + (0.21 \times 25) = 7.75 \text{ mm Hg}.$$

This patient's arterial P_{O_2} is 60 mm Hg. His alveolar P_{O_2} can be calculated from the alveolar gas equation (see equation 22.1) as:

$$150 - (56/0.8) = 80 \text{ mm Hg}$$

The measured A-a gradient is, therefore, 80 − 60 mm Hg = 20 mm Hg, which is 12.25 mm Hg wider than predicted, indicating impaired O_2 uptake across the blood–gas interface.

TQ6.2 Cyanosis is a bluish discoloration of the skin, lips, nails, and fingertips caused by severe hypoxemia. In this instance, water aspiration has damaged alveolar functioning, producing a $\dot{V}_A/\dot{Q}$ mismatch (see Chapter 22·IV·C) and the widened A-a gradient.

TQ6.3 Use of accessory muscles (see Table 21.1) indicates increased work of breathing. Drowning victims may aspirate very little water initially because reflex laryngospasm occludes the upper airway (Fig. C6.1), but the ensuing hypoxia terminates the laryngeal response, and aspiration continues. Aspiration of as little as 100 mL of water washes out surfactant. Surfactant is essential for countering the effects of surface tension on lung function (see Chapter 21·IV·B). Surfactant washout makes the lungs difficult to inflate and leads to atelectasis (alveolar collapse), amplifying $\dot{V}_A/\dot{Q}$ mismatching. Aspiration also damages and increases the permeability of the alveolar wall, leading to fluid shifts that further reduce surfactant levels. Surfactant washout during near drowning reduces lung compliance by 10%–40%.

TQ6.4 Arterial blood gas values are consistent with respiratory acidosis (pH < 7.35) caused by impaired ability to transfer CO_2 to the atmosphere (P_{CO_2} > 45 mm Hg; see Chapter 28·VI·C). The kidneys have not yet had time to compensate by excreting H^+ and generating HCO_3^-, so HCO_3^- levels remain on the low side of normal.

TQ6.5 Near drowning leads to acute lung injury with damage to the alveolar wall, underlying parenchyma, and pulmonary capillaries. Acute respiratory distress syndrome ([ARDS] see Chapter 40·VI·D) may develop as a result. In ARDS, the alveoli fill with inflammatory exudate that suppresses surfactant production, thereby compounding lung stiffness (decreased lung compliance and expansion). Mechanical ventilation is typically required for support as the lung recovers.

Summary

This patient aspirated water during a non-fatal (near) drowning event. Fresh water washes out surfactant, promoting alveolar collapse and atelectasis and creating a right-to-left shunt due to $\dot{V}_A/\dot{Q}$ mismatch. The result is hypoxemia, hypercapnia, and respiratory acidosis. Non-fatal downing is a potential cause of ARDS, which requires ventilatory support via mechanical ventilation.

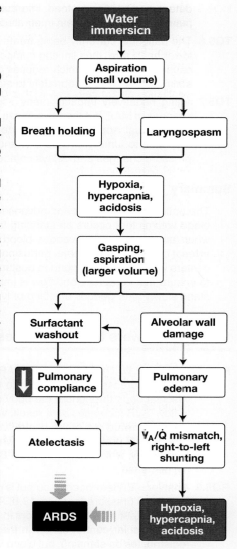

Figure C6.1.
Effects of near drowning on lung function. ARDS = acute respiratory distress syndrome.

CASE 6: Answers to Review Questions

RQ6.1 Answer = C. Alveolar lining fluid creates surface tension that both favors alveolar collapse and creates a driving force for fluid movement from the pulmonary interstitium (see Chapter 21·IV·B). Surfactant decreases surface tension and thereby helps keep the alveolar lumen dry. Surfactant does not participate in immune responses (A). Lung-diffusing capacity (B) is a measure of the rate at which gases are exchanged across the blood–gas interface, which is not affected by surfactant. The alveolar epithelium remains moist (D) by virtue of alveolar lining fluid, not surfactant. The cough reflex (E) is triggered by activation of specialized irritant receptors, not surfactant.

RQ6.2 Answer = C. Emphysema is characterized by destruction of alveolar walls and their small vessel blood supply, leaving behind large, cystic, air-filled spaces (see Chapter 21·VII·A). Loss of alveolar capillaries reduces pulmonary perfusion ($\dot{Q}$), while ventilation ($\dot{V}_A$) is preserved, leading to a high $\dot{V}_A/\dot{Q}$ ratio. Asthma exacerbations (A) and bronchitis (B) obstruct airflow, which reduces ventilation and decreases the $\dot{V}_A/\dot{Q}$ ratio. Pulmonary arteriovenous malformations (D) allow venous blood to bypass the blood–gas interface without participating in gas exchange (right-to-left shunt). Pulmonary edema (E) is associated with fluid in the alveoli, which reduces ventilation and impairs gas exchange across the blood–gas interface.

CASE 7: Answers to Thought Questions

TQ7.1 Atmospheric pressure (P_{atm}) decreases with height above sea level (see Chapter 24·V). For example, P_{atm} decreases from 760 mm Hg at sea level to around 526 mm Hg at 3,000 m (the average elevation during this patient's hikes). The partial pressure of O_2 (P_{O_2}) falls in parallel, which reduces the pressure gradient driving O_2 uptake by the lungs and causes hypoxemia. Although the likelihood of developing AMS following ascent to high altitude varies with the individual and extent of prior adaptation, the symptoms of AMS are associated with hypoxemia.

TQ7.2 F_{IO_2}, is the fraction of O_2 in inspired air. Atmospheric air contains N_2 (78%), O_2 (21%), Ar (1%), CO_2 (0.03%), and trace amounts of other gases (see Chapter 24·V·A). Fractional composition does not change with altitude, so $F_{IO_2} = 0.21$.

P_{IO_2} is the partial pressure of O_2 in inspired air once delivered to the alveoli. The airways add water to air during inspiration, which reduces the fractional composition of O_2 to 0.197. P_{IO_2} equals P_{atm} at 3,000 m (526 mm Hg) times fractional composition, or 104 mm Hg. This is considerably less than P_{IO_2} at sea level (150 mm Hg).

P_{AO_2} is the partial pressure of O_2 in alveolar air after equilibration with pulmonary blood. Blood adds CO_2 to alveolar air during equilibration, which affects P_{O_2}, so we would need to know P_{ACO_2} to calculate P_{AO_2}. For example, if P_{ACO_2} is 35 mm Hg at 3,000 m, the alveolar gas equation (see Equation 22.1) indicates that P_{AO_2} would be 60 mm Hg.

TQ7.3 Blood O_2-carrying capacity is determined by the amount of O_2 that dissolves in blood plus blood Hb content. The amount of O_2 that dissolves in blood approximates 3 mL/L (see Chapter 23·I). Hb binds 1.39 mL O_2/g (see Chapter 23·IV·A), so, with a blood Hb content of 16.3 g/dL (= 163 g/L), this patient's O_2-carrying capacity is 3 mL/L + 227 mL/L = 230 mL/L.

TQ7.4 Blood O_2 saturation (SaO_2) is a measure of the number of Hb's four O_2-binding sites that are occupied. SaO_2 is typically measured with pulse oximetry, which uses a light-emitting probe to estimate Hb O_2 saturation based on the extent of the protein's color shift from saturated (bright red) to unsaturated (dark blue; see Chapter 23·IV). The probe is usually attached to an earlobe, finger, or toe, so a reading may often be cited as SpO_2, denoting peripheral O_2 saturation.

TQ7.5 The partial pressure of arterial oxygen, or P_{aO_2}, is sensed by peripheral chemoreceptors concentrated within carotid bodies located at the bifurcation of the common carotid arteries (see Chapter 24·III·B). Chemoreceptors show the strongest response once P_{aO_2} drops below ~60 mm Hg, which coincides with the steepest part of the Hb–O_2 dissociation curve. Respiratory centers in the brainstem respond to chemoreceptor activation by increasing respiration rate and alveolar ventilation. Note that increasing ventilation blows off CO_2 to the atmosphere at a faster rate, causing P_{aCO_2} to fall below normal. P_{CO_2} is a principal determinant of plasma pH, as defined by the Henderson-Hasselbalch equation (see Chapter 23·VI):

$$pH = pK + \log\frac{[HCO_3^-]}{[CO_2]}$$

When P_{aCO_2} falls, plasma pH rises to cause respiratory alkalosis that may be partly responsible for the symptoms associated with AMS.

TQ7.6 Cerebral blood flow is very sensitive to changes in local P_{CO_2}: when CO_2 levels rise, cerebral resistance vessels dilate, and flow increases (see Chapter 20·II·C). Conversely, when CO_2 levels fall, cerebral vascular resistance increases, and flow decreases. Normally (i.e., when O_2 availability is not limited), a drop in P_{aCO_2} caused by hyperventilation would be expected to decrease cerebral blood flow. At high altitude when O_2 availability is limited, hypoxemia becomes the primary determinant of cerebral flow. Cerebral resistance vessels dilate, and flow increases to compensate for the reduced O_2 availability. The resulting increase in cerebral perfusion pressure is believed to be responsible, at least in part, for the headaches and nausea that characterize AMS.

TQ7.7 Individuals with a history of AMS may be prescribed a CA inhibitor (e.g., acetazolamide) prophylactically to reduce the likelihood of symptom development during ascent to high altitude. CA inhibitors impair HCO_3^- reabsorption in the renal proximal tubule (see Chapter 26·IX·A), causing metabolic acidosis and mild diuresis. The fall in blood pH is sensed by central chemoreceptors and communicated to the brainstem respiratory control center, which responds by increasing ventilation. The ventilation increase helps compensate for the reduced O_2 availability upon ascent. CA inhibitors also help prevent or compensate for the bicarbonate excess that develops from hyperventilation at high altitude. Alkalosis is believed to contribute to AMS symptoms.

Summary

O_2 availability falls in parallel with decreasing atmospheric pressure during ascent to high altitude, leading to hypoxia. Ventilation increases reflexively to compensate, resulting in hypocapnia and respiratory alkalosis. Cerebral resistance vessels dilate in response to hypoxemia, which increases cerebral blood flow and pressure. The combination of increased pressure and alkalosis may result in symptoms associated with AMS, including headaches and nausea.

CASE 7: Answers to Review Questions

RQ7.1 Answer = D. The O_2 affinity of Hb decreases during high-altitude adaptation and acclimation to enhance O_2 offloading to tissues (see Chapter 24·V·A). The rightward shift in the Hb–O_2 dissociation curve is due to a rise in 2,3-bisphosphoglycerate levels (A) stimulated by alkalosis (see Fig. 23.11). Blood volume (B) and hematocrit (C) both increase during acclimation to enhance blood's O_2-carrying capacity. Declining blood O_2 levels during ascent cause reflex hypoxic vasoconstriction of the pulmonary vasculature, which increases resistance to flow. Right-ventricular afterload (E) increases as a result.

RQ7.2 Answer = B. Cerebral resistance vessels respond to the hypoxemia that results from rapid ascent to high altitude by vasodilating, which raises cerebral capillary hydrostatic pressure (P_c). P_c is a force that favors fluid filtration from the vasculature into surrounding tissues (see Chapter 18·VII·D). If P_c becomes excessive, the ability of the meningeal lymphatic system to drain fluid may be overwhelmed, resulting in edema. Cerebral blood flow (A) does increase with ascent to altitude, but this is secondary to vasodilation. An increase in intracranial pressure (C) would tend to oppose fluid filtration from the cerebral vasculature. The increase in right-ventricular afterload (D) caused by hypoxic pulmonary vasoconstriction does not affect cerebral blood flow. Ventilation increases (E) help compensate for reduced O_2 availability at high altitude and would thus tend to counter the development of HACE.

CASE 8: Answers to Thought Questions

TQ8.1 The likelihood of edema formation is determined by the balance of Starling forces across the capillary wall (see Chapter 18·VII·D). The principal forces that favor fluid filtration from blood are capillary hydrostatic pressure (P_c) and interstitial oncotic pressure (π_i). The principal forces opposing fluid filtration and likelihood of edema are plasma oncotic pressure (π_c) and interstitial hydrostatic pressure (P_i). The most common causes of edema are a rise in P_c (due to plasma volume expansion, for example), a decrease in π_c due to liver disease that impairs the ability to synthesize plasma proteins, and lymphatic obstruction that allows proteins to accumulate in the interstitium, thereby raising π_i.

TQ8.2 Plasma oncotic pressure is dependent on plasma proteins such as albumin and globulins (see Chapter 18·VII·A). Albumin is the principal determinant of π_c (80%), so the fact that this patient's serum albumin level is well below normal would explain the periorbital and leg edema.

TQ8.3 In a child, serum albumin levels are most likely to fall either because the liver is unable to synthesize proteins due to starvation with resulting lack of available substrates (kwashiorkor), or because albumin is being excreted in urine. Urinalysis reveals a 3+ protein level, which corresponds to severe proteinuria, so the patient's presenting symptoms are most likely due to renal dysfunction.

TQ8.4 The glomerulus is designed to filter copious amounts of cell- and protein-free fluid from blood into the renal tubule (see Chapter 25·III). The filtration barrier is a multilayered structure comprising glomerular capillary endothelial cells, a basement membrane, and a filtration slit diaphragm (see Fig. 25.6). The capillary endothelial cells are fenestrated to facilitate fluid movement, while preventing cells from leaving the vasculature. The basement membrane is negatively charged to reflect proteins back into blood. The filtration slit diaphragm is a proteinaceous sieve formed between two podocyte foot processes that ultimately determines filtrate composition. The filtration slit diaphragm also carries a negative charge to repulse proteins, so the fluid entering the renal tubule is normally protein-free. The fact that large amounts of protein are appearing in the patient's urine suggests that the barrier has been severely damaged.

TQ8.5 Na^+ and glucose are freely filtered by the glomerulus (see Table 25.1) and then reabsorbed by the renal tubule. Glucose is reabsorbed in the proximal tubule (PT). Na^+ is reabsorbed along the length of the tubule in amounts needed to optimize plasma Na^+ levels. Damage to the glomerular filtration barrier would not affect the amount of either solute being filtered or reabsorbed. The patient's serum Ca^{2+} levels are low because laboratory tests usually measure total Ca^{2+}, a significant proportion of which is bound to albumin and being lost to urine in this patient. Ionized Ca^{2+} levels would not be expected to be affected.

Summary

The cause of glomerular damage in minimal change disease (MCD) is poorly understood, but the condition often follows an infection and is believed to be due to circulating cytokines. It is the most common form of nephrotic syndrome in children. Biopsy samples usually show podocyte effacement (loss of foot processes). The result is massive plasma protein loss in urine, causing edema. MCD is also associated with hyperlipidemia, perhaps to compensate for reduced oncotic pressure. Indicators of renal function, such as blood urea nitrogen and creatinine, are usually normal. MCD usually responds well to treatment with steroids.

CASE 8: Answers to Review Questions

RQ8.1 Answer = D. π_{GC} rises from 25 to 33 mm Hg as blood travels the length of the glomerular capillary and the proteins become concentrated by fluid filtration. Average π_{GC} is $(25 + 33)/2 = 29$ mm Hg. Using the Starling equation to calculate P_{UF}:

$$P_{UF} = (P_{GC} - P_{BS}) - (\pi_{GC} - \pi_{BS})$$

$$P_{UF} = (50 - 15) - (29 - 0) = 6 \text{ mm Hg.}$$

A value of 21 mm Hg (A) ignores the contribution of the Bowman space to net filtration pressure. A value of 15 mm Hg (B) is hydrostatic pressure in the Bowman space. Values of 10 mm Hg (C) and 2 mm Hg (E) are the net filtration pressures at the afferent arteriole and efferent arteriole, respectively.

RQ8.2 Answer = D. Renal clearance is calculated as the volume of blood that is completely cleared of a substance during passage through the renal vasculature (see Chapter 25·V·A). Uric acid is a waste product that filters freely across the glomerular barrier, but it is also actively secreted by organic acid transporters in the PT, unlike the other substances listed. Albumin (A) is trapped in the vasculature, so its clearance rate is normally zero. Glucose (B) filters freely at the glomerulus but is immediately recovered in the PT, so clearance is normally close to zero. Na^+ (D) and water (E) filter freely, and all but 1% of the respective filtered loads are reclaimed during passage through the tubule.

CASE 9: Answers to Thought Questions

TQ9.1 Creatinine and urea (the principal determinant of BUN) are both waste products excreted by the kidneys. Both are mildly elevated, indicating impairment of renal function.

TQ9.2 A bruit [broot] is an abnormal sound heard over an artery that is caused by turbulent blood flow. Given that the sound is heard over the patient's left flank, and imaging studies reveal narrowing of the left renal artery, the bruit is most likely caused by blood jetting at high velocity through the narrowed vessel and causing turbulence distal to the site of obstruction. Velocity is proportional to flow divided by cross-sectional area of the vessel (see Chapter 18·V·A). Stenosis reduces cross-sectional area, so velocity increases proportionally.

TQ9.3 Flow through the cardiovascular system is normally layered, orderly, and silent (see Chapter 18·V·A). Laminar flow, also known as streamline flow, is highly efficient, with the different layers slipping over each other down the length of the vessel. Under some conditions, the layers can break apart to form vortices and eddy currents that waste energy (turbulent flow). Turbulence causes vessel walls to reverberate, and the resulting pressure waves travel up to the body surface to be detected as an audible bruit or murmur. The likelihood of turbulence is predicted by the Reynolds number (Re), which is proportional to flow velocity, vessel diameter, and blood density and inversely proportional to blood viscosity. In practice, this means that turbulence in the cardiovascular system occurs in areas where flow velocity is high or when blood viscosity is reduced. Blood viscosity is proportional to hematocrit, so anemia may lead to physiologic murmurs (e.g., pregnant patients may exhibit a venous hum associated with high-velocity, low-viscosity flow through the veins; see Chapter 37·IV·C).

TQ9.4 In addition to revealing significant stenosis of the left renal artery, the left kidney is smaller than the right. Stenosis impairs blood supply to tissues downstream and may cause ischemia and infarction. RAS may lead to ischemia-induced nephron loss. Damaged nephrons are replaced with fibrous material, and the kidney atrophies. The decrease in the size of the left kidney is consistent with stenosis and impaired renal function, as reflected in the elevated creatinine and BUN levels.

TQ9.5 RAS is usually caused by atherosclerotic plaque formation. Plaque builds slowly over time, producing small, incremental decreases in renal blood flow (RBF) and glomerular perfusion pressure. The kidneys require high glomerular pressure to produce plasma ultrafiltrate, so two local feedback pathways monitor pressure and glomerular filtration rate ([GFR] Fig. C9.1). Both involve the juxtaglomerular apparatus (JGA), which includes the afferent and efferent arterioles, mesangial cells, and the macula densa (see Chapter 25·IV). The macula

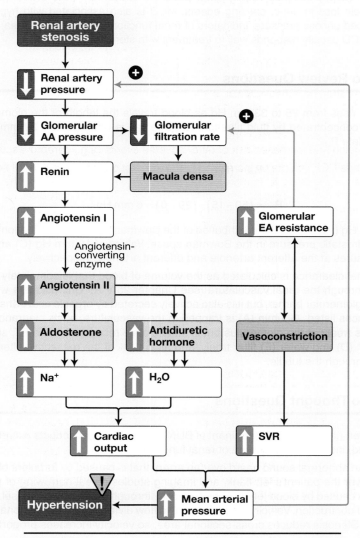

Figure C9.1.
Concept map detailing how renal artery stenosis leads to hypertension. AA = afferent arteriole, EA = efferent arteriole, SVR = systemic vascular resistance.

densa is a specialized region of the thick ascending limb of the loop of Henle located at the site of contact with the glomerular arterioles.

- The glomerular afferent arteriole (AA) is a baroreceptor that responds to decreased pressure with renin release from granular cells (juxtaglomerular cells). Renin converts angiotensinogen, a plasma protein, to angiotensin I (Ang-I). Ang-I is converted to Ang-II by angiotensin-converting enzyme (ACE), which is expressed by most tissues, including the kidney.

- Flow through the tubule is sensed through the amount of NaCl presented to the macula densa. When GFR decreases, the amount of NaCl reaching the macula densa falls, initiating a signal that increases renin release from AA granular cells.

Acting as a glomerular autoregulator, Ang-II constricts the efferent arteriole (EA) and increases its resistance, which makes it more difficult for blood to leave the glomerulus. Glomerular ultrafiltration pressure (P_{UF}) and GFR rise as a result.

TQ9.6 In addition to being a local renal autoregulatory agent, Ang-II has a key role in helping maintain mean arterial pressure (MAP) through its effects on systemic vascular resistance (SVR) and extracellular fluid (ECF) volume (see Fig. C9.1).

- Ang-II constricts resistance vessels throughout the body to raise SVR. MAP is a product of cardiac output (CO) and SVR (MAP = CO × SVR), and increases in SVR are highly effective in raising MAP.

- ECF volume is regulated through the amount of Na^+ and H_2O retained by the kidneys. Ang-II increases renal Na^+ retention directly (see Table 28.2), and it also stimulates aldosterone release from the adrenal cortex. Aldosterone promotes renal Na^+ retention. Ang-II also stimulates thirst to increase water intake and antidiuretic hormone (ADH) release from the posterior pituitary to promote renal water retention.

Na^+ and H_2O retention increases ECF volume, which increases ventricular preload and CO, which further increases MAP. In a healthy cardiovascular system, such changes are essential for long-term blood pressure control. RAS creates a situation in which increases in MAP are being driven by the kidney regardless of how such changes may affect other systemic organs. The goal of renin release is to increase renal arterial pressure until flow through the obstructed renal vessel is sufficient to renormalize P_{UF}, even though this may cause systemic hypertension and increased risk of other adverse events (e.g., cerebrovascular accident).

TQ9.7 Aldosterone expands ECF volume by increasing Na^+ reabsorption in the renal distal segments. The pathways involved simultaneously secrete K^+ into the tubule lumen, leading to K^+ wasting. The observed mild hypokalemia is consistent with renin–angiotensin–aldosterone system (RAAS) activation, although hypokalemia is a rare presentation in cases of RAS. The patient is currently taking three antihypertensives, and it is very likely that one of these (e.g., a thiazide diuretic) is responsible for the low K^+ reading.

Summary

The patient has RAS affecting the left renal artery. Obstruction of blood flow impairs kidney function, leading to atrophy. Increased creatinine and BUN levels are indicative of renal impairment. The left kidney responds to decreased renal blood flow by activating the RAAS, causing systemic hypertension. The rise in systemic blood pressure forces blood through the left renal artery at increased velocity, leading to turbulence that is detected at the body surface as a bruit.

CASE 9: Answers to Review Questions

RQ9.1 Answer = C. The apical membrane of macula densa cells contains an NKCC2 $Na^+/K^+/2Cl^-$ cotransporter that facilitates Na^+ and Cl^- entry. Cl^- then exits via a basolateral Cl^- channel, causing depolarization (see Chapter 25·IV·C). The magnitude of the response is proportional to the amount of NaCl in tubule fluid and constitutes a signal that is used to adjust glomerular arteriolar tone to optimize GFR. Loop diuretics such as furosemide block this cotransporter (see Clinical Application 27.1), thereby rendering the macula densa blind to changes in tubule fluid composition. Acetazolamide (A) is a carbonic anhydrase inhibitor that targets the proximal tubule. Amiloride (B) is a potassium-sparing diuretic that inhibits ENaC in the distal tubule. Hydrochlorothiazide (D) blocks the NCC Na^+/Cl^- cotransporter in the distal convoluted tubule. Mannitol (E) is an osmotic diuretic that has no effect on NKCC2.

RQ9.2 Answer = A. Atrial natriuretic peptide is secreted by atrial myocytes when stressed by excessive blood volume (see Chapter 28·III·C). It limits further increases in ECF volume by antagonizing Ang-II and ADH effects. At the glomerular level, it stimulates arteriolar dilation, which increases RBF and GFR. Endothelins (B) are potent

---

Clinical Cases Answers: Cases 9, 10

vasoconstrictors released in response to trauma. They reduce RBF and GFR, as do leukotrienes (C), whose levels rise during inflammation. Norepinephrine (D) released during sympathetic activation constricts both glomerular arterioles. Glomerular arterioles are not regulated by the parasympathetic nervous system (E).

CASE 10: Answers to Thought Questions

TQ10.1 The patient has a significant soft-tissue crush injury. Most (98%) of total body potassium is concentrated within cells by the ubiquitous Na⁺/K⁺-ATPase (see Chapter 28·IV). Roughly half of body mass is skeletal muscle and half of that is in the lower extremities. Crushing muscle causes rhabdomyolysis, allowing the contents of the myocytes to spill into the interstitial space and vasculature. Crush injuries are a common cause of hyperkalemia.

TQ10.2 All cells rely on a strong outwardly directed K⁺ gradient to establish membrane potential (see Chapter 2·II). Hyperkalemia blunts the concentration gradient and causes membrane depolarization (see Fig. 2.7). Neurons, myocytes, and other excitable cells are especially susceptible to membrane depolarization because it may either initiate or inhibit action potential (AP) formation. An ECG records waves of electrical activity moving through the myocardium generated by thousands of cardiac myocytes (see Chapter 16·VI). Hyperkalemia typically manifests as a tall, peaked T wave, prolongation of the PR interval, flattening of the P wave, and widening of the QRS complex (Fig. C10.1).

TQ10.3 Hyperkalemia depolarizes myocytes and brings membrane potential closer to the threshold for AP formation and contraction. The voltage-dependent Na⁺ channel that yields the upstroke of an AP in skeletal muscle and in atrial and ventricular myocytes may not be able to recover from inactivation following an AP, potentially leading to muscle weakness and paralysis (see Clinical Application 2.1).

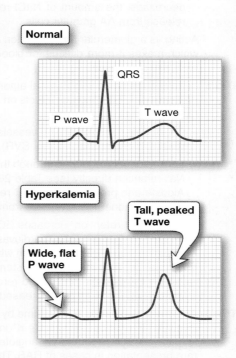

Figure C10.1.
Hypokalemia effects on the electrocardiogram (ECG).

TQ10.4 Calcium gluconate is a calcium salt of gluconic acid, which is formed by oxidation of glucose. Administration of calcium gluconate raises intracellular Ca²⁺ and activates Ca²⁺/calmodulin-dependent protein kinase II (CAMKII). CAMKII shifts the voltage sensitivity of the Na⁺ channel and threshold for AP formation toward more positive voltages, which helps restore excitability despite the hyperkalemia-induced membrane depolarization.

TQ10.5 Insulin stimulates Na⁺/K⁺-ATPase activity, which causes K⁺ to shift from extracellular fluid (ECF) into cells (see Chapter 28·IV). Storing K⁺ temporarily within cells allows more time for the kidneys to restore K⁺ homeostasis. Glucose stimulates insulin secretion from pancreatic β cells, thereby enhancing the effects of exogenous insulin.

TQ10.6 The sarcoplasm contains high levels of myoglobin (an O₂-binding protein) and CK, which is used to transfer a high-energy bond from creatine to ADP to regenerate ATP (see Chapter 39·IV). Rhabdomyolysis causes both proteins to spill into the vasculature. Myoglobin, like hemoglobin, is pigmented, and causes urine to become dark when filtered out of blood by the kidneys.

Summary

The patient has experienced a significant crush injury and rhabdomyolysis that has caused K⁺, myoglobin, and muscle enzymes (e.g., CK) to spill into blood. The resulting hyperkalemia has caused muscle weakness, abdominal discomfort, and ECG changes. Myoglobin darkens the urine and may damage the glomerular filtration barrier, causing renal dysfunction (as evidenced by elevated serum creatinine).

CASE 10: Answers to Review Questions

RQ10.1 Answer = A. Hyperkalemia stimulates aldosterone release from the adrenal cortex (see Chapter 28·IV). Aldosterone targets the renal distal segments, where it increases the expression of the Na^+/K^+-ATPase and an apical ROMK K^+ channel, causing K^+ to be excreted in urine. Angiotensin II (B) stimulates aldosterone release. Angiotensin II is involved in blood pressure regulation, not K^+ homeostasis. Antidiuretic hormone (C) is released from the posterior pituitary in response to increased ECF osmolality. It promotes renal water retention. Epinephrine (D) and insulin (E) both stimulate Na^+/K^+-ATPase activity, causing K^+ to move from ECF into cells.

RQ10.2 Answer = B. Chemotherapy is used to destroy tumor cells which, like crush injuries, causes the contents of lysed cells to spill into blood. If the tumor burden is high, chemotherapy may potentially cause "tumor lysis syndrome," characterized by hyperkalemia, hyperphosphatemia, hypocalcemia, and hyperuricemia (the latter due to catabolism of released nucleic acids). All the other answer choices are associated with hypokalemia rather than hyperkalemia. Alkalosis (A) causes K^+ to enter cells to compensate for the charge imbalance caused by reduced intracellular H^+ levels. The internal K^+ balance shift results in hypokalemia. Liddle syndrome (C) is a genetic disorder that increases renal Na^+ reabsorption in exchange for K^+, resulting in hypernatremia, hypertension, and hypokalemia. Loop diuretic use (D) leads to renal K^+ wasting and hypokalemia. Gastric juice is K^+-rich, so prolonged vomiting (E) can cause excessive K^+ losses and hypokalemia.

CASE 11: Answers to Thought Questions

TQ11.1 The sensation of thirst originates in cortical areas (anterior and posterior cingulate cortex and insular cortex) in response to increased extracellular fluid (ECF) osmolality. Osmolality is monitored by the lamina terminalis, an area bordering the third ventricle containing three nuclei: the subfornical organ, the organum vasculosum of the lamina terminalis, and the median preoptic nucleus. Together, these three nuclei comprise a thirst center (see Chapter 28·II). The patient is thirsty because his serum osmolality is abnormally high (320 mOsm/kg H_2O), and he is dehydrated, as noted during physical examination of his oral mucosa.

TQ11.2 Water balance and ECF osmolality are controlled through regulation of water intake (i.e., thirst) and output (see Chapter 28·II). Output is controlled at the level of the kidneys. The osmoreceptors in the lamina terminalis control ADH release from the posterior pituitary. ADH is also known as arginine vasopressin (AVP). ADH regulates how much water is retained by the kidney. Thus, the patient's condition could be caused by a malfunction in the pathway leading to ADH secretion (AVP deficiency [AVP-D], formerly known as central diabetes insipidus), or an inability of the kidney to respond to ADH by increasing water retention (AVP resistance [AVP-R], formerly known as nephrogenic diabetes insipidus).

TQ11.3 ADH binds to vasopressin V_2 receptors (*AVPR2* gene) on principal cells in the renal collecting ducts (CDs). These receptors promote the insertion of vesicles laden with aquaporins (AQP2) into the apical membrane (see Chapter 27·V·C). The basolateral membrane contains two constitutive AQPs (AQP3 and AQP4), so inserting AQP2 into the apical membrane completes a pathway for water to leave the CD lumen and return to the vasculature. Urine becomes concentrated as a result. In the absence of circulating ADH, the CDs are water-impermeable, and urine becomes dilute.

TQ11.4 The potential causes of polydipsia (excessive thirst with increased fluid intake) and polyuria (increased urine production: > 3L/day) can be narrowed by a water deprivation test and by administering desmopressin, which is a synthetic ADH analog (Fig. C11.1). If urine osmolality increases after restricting water intake, the likely cause is typically psychogenic (i.e., primary polydipsia). If urine osmolality increases after administering desmopressin, the likely cause is AVP-D because the kidney has demonstrated its ability to concentrate urine when stimulated appropriately. If desmopressin has no effect on urine osmolality, then the patient most likely has AVP-R.

TQ11.5 The most common causes of AVP-R are hereditary defects (*AVPR2* gene mutations), lithium toxicity, and hypocalcemia. *AVPR2* gene mutations usually present during childhood. Lithium is used clinically for treatment of bipolar and unipolar (major depressive) disorders. Lithium is a monovalent cation (Li^+) that can substitute for Na^+ and permeate apical Na^+ channels (ENaCs) in CD principal cells. Here, it interferes with ADH receptor signaling and AQP2 expression, leading to AVP-R. Symptoms usually resolve with cessation of lithium treatment. However, prolonged lithium use (decades) causes the CDs to undergo remodeling, and the AVP-R becomes irreversible.

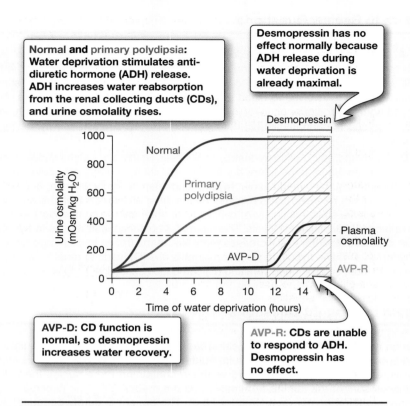

Figure C11.1.
Water deprivation and desmopressin stimulation test. AVP-D
and AVP-R = arginine vasopressin deficiency and resistance,
respectively.

TQ11.6 Diuretics increase urinary volume, so recommending a thiazide diuretic to treat polyuria might seem question-able, at best. Patients with AVP-R do seem to benefit, however. Thiazide diuretics inhibit Na^+ uptake by the NCC Na^+/Cl^- cotransporter in the distal convoluted tubule (see Clinical Application 27.2). The amount of Na^+ deliv-ered to the distal segments and CDs increases as a result. Here, it is reabsorbed via an aldosterone-regulated pathway, and water apparently follows by osmosis, meaning that daily urinary output decreases.

Summary

The patient has AVP-R caused by the lithium being used to treat his bipolar disorder. Lithium permeates the renal tubule epithelium and interferes with normal AQP2 expression in the distal segments and collecting ducts. The resulting inability to regulate water output leads to an increase in serum Na^+ levels and ECF osmolality which, in turn, stimulates polydipsia and polyuria.

CASE 11: Answers to Review Questions

RQ11.1 Answer = B. ADH was originally known as arginine vasopressin, reflecting its ability to constrict resistance vessels and raise systemic vascular resistance and blood pressure when circulating at high levels (see Chapter 28·III·C). Resistance vessels express a vasopressin V_1 receptor that mediates the response. The name was changed in the 1950s to acknowledge ADH's primary role in maintaining water balance. ADH increases (not decreases) Na^+ reabsorption (A) by the thick ascending limb and cortical collecting duct. Angiotensin II, or Ang-II (C), is synthesized from angiotensin I by angiotensin-converting enzyme. Circulating Ang-II levels reflect plasma renin activity. Thirst (D) is under the control of plasma osmolality and Ang-II. Venoconstriction (E) occurs in response to sympathetic activation.

RQ11.2 Answer = C. The posterior pituitary secretes ADH and oxytocin, which are related nonapeptide hormones that differ by only two amino acids (see Chapter 7·VII·D). Oxytocin stimulates milk let-down in mammary glands and uterine contractions during parturition (see Chapter 36·VI·B). Dopamine (A) is secreted continuously into the hypophyseal portal system by the hypothalamus. It targets the anterior pituitary, where it tonically inhibits prolactin (D) secretion. α-Melanocyte-stimulating hormone (B) is produced in the skin, where it regulates melanin production. Somatostatin (E) is secreted by the hypothalamus and enteroendocrine cells in the gastrointestinal system.

CASE 12: Answers to Thought Questions

TQ12.1 The patient's arterial blood gas (ABG) reveals a pH of 7.20, which is below normal range. The patient is acidemic.

TQ12.2 The patient's respiratory rate is high (30 breaths/min; normal = 12–20 breaths/min) because the lungs are trying to correct for the acidosis by blowing off CO_2 to the atmosphere (note the low arterial P_{CO_2}). Acid–base homeostasis is controlled by the lungs (immediate, short-term compensation) and kidneys (prolonged, long-term compensation). Acidemia is sensed by central chemoreceptors, which prompt medullary respiratory centers to increase ventilation. P_{CO_2} is a principal determinant of plasma pH (see Chapter 23·VI·A), so increasing the rate at which CO_2 is ventilated to the atmosphere causes plasma pH to rise, as predicted by the Henderson-Hasselbalch equation.

TQ12.3 The kidneys' response to acidemia is slower than that of the lungs and involves increasing the amount of H^+ excreted in urine and generating HCO_3^-. The proximal tubule is responsible for excreting the greatest amount of acid normally (see Chapter 26·IX). It does this using titratable acids (e.g., hydrogen phosphate, urate, creatinine, lactate, and pyruvate) and by generating NH_3 from glutamine. NH_3 diffuses into the tubule lumen and traps H^+ in the form of NH_4^+. Generating NH_3 simultaneously creates new HCO_3^- that is transferred to the vasculature. Acidosis increases H^+ secretion and NH_3 synthesis. pH balance is the responsibility of the distal segments and collecting ducts (see Chapter 27·IV·D and 27·V·E). Here, α intercalated cells secrete H^+ into the tubule lumen using a V-ATPase and an H^+/K^+-ATPase. Acidosis increases the number of α intercalated cells to handle the acid excess.

TQ12.4 The anion gap is calculated from the difference between plasma Na^+ concentration (the principal cation) and the sum of plasma Cl^- and HCO_3^- concentrations (the principal anions; see Chapter 28·VI·B). This patient's anion gap is calculated as 140 mEq/L − (110 mEq/L + 8 mEq/L) = 22 mEq/L. This is above the normal range for an anion gap (8–16 mEq/L). An anion gap represents contributions from all the minor anions, including proteins, phosphates, citrate, and lactate. Abnormal widening of the anion gap indicates that the patient has an organic acid excess that has caused an anion gap metabolic acidosis.

TQ12.5 Ethylene glycol is commonly used as an automotive coolant and antifreeze. Although it is colorless, commercial coolant formulations of ethylene glycol are usually bright green. It has a sweet taste. Although ethylene glycol itself is relatively nontoxic (it is a central nervous system depressant, like other alcohols), it is metabolized by the liver to yield glycolate and oxalate (Fig. C12.1), which are highly toxic (see Chapter 28·VI·E·1·b). Accumulation of these anions is responsible for the anion gap metabolic acidosis.

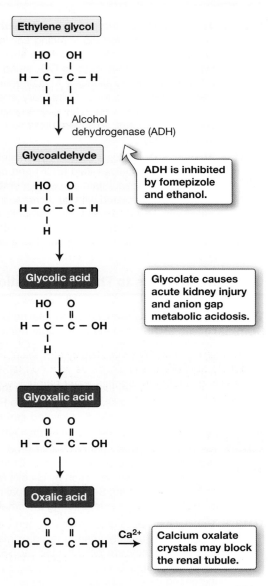

Figure C12.1.
Ethylene glycol metabolism.

TQ12.6 The elevated BUN suggests that kidney function is impaired. Glycolate and oxalate both target the kidney and can cause acute kidney injury. Glycolate damages the renal tubule, whereas oxalate may complex with Ca^{2+} to form calcium oxalate crystals that block the tubule (see Fig. C12.1).

TQ12.7 The first step in ethylene glycol metabolism is conversion to glycoaldehyde, catalyzed by ADH. Competitive inhibition of ADH with fomepizole or ethanol blocks ethylene glycol metabolism and allows the poison to be excreted via the kidneys within around 24 hours.

Summary

This patient accidently ingested ethylene glycol, which is used as an engine coolant and radiator antifreeze. Ethylene glycol is rapidly metabolized in the liver via a pathway that includes ADH to yield glycolate and oxalate. These and other byproducts of metabolism lead to an anion gap metabolic acidosis. Glycolate is highly toxic to the renal tubule and causes acute kidney injury.

CASE 12: Answers to Review Questions

RQ12.1 Answer = C. Gastric parietal cells secrete H^+ into the stomach lumen to maintain a low pH between and during meals. Prolonged vomiting can result in significant H^+ loss from the body, leaving behind a base excess that manifests as metabolic alkalosis (see Chapter 28·VI·F). Hysteria (A) is usually accompanied by hyperventilation that leads to CO_2 being transferred to the atmosphere at an inappropriately high rate. The result is respiratory alkalosis. Myasthenia gravis (B) can cause respiratory muscle paralysis, hypoventilation, and respiratory acidosis. Pulmonary edema (D) also causes respiratory acidosis by impairing transfer of CO_2 to the atmosphere. The large intestine secretes HCO_3^- to buffer acid produced by microbial action on undigested wastes, so severe diarrhea (E) can cause metabolic acidosis because of excessive HCO_3^- loss from the body.

RQ12.2 Answer = D. Acidosis causes hyperkalemia (see Chapter 28·IV·D). H^+ is a small cation that readily enters cells from extracellular fluid (ECF) and displaces K^+ from intracellular binding sites. The displaced K^+ leaks out of the cell to maintain a normal membrane potential. The result is an increase in plasma K^+. Respiratory acidosis is associated with an increase in Pco_2, not a decrease (A). Decreases in plasma Cl^- (B) are associated with HCO_3^- excess and metabolic alkalosis. The rise in Pco_2 that leads to respiratory acidosis causes plasma HCO_3^- levels to increase simultaneously, not decrease (C), and HCO_3^- rises further during renal compensation. Acidosis is defined as a decrease in plasma pH, not an increase (E).

CASE 13: Answers to Thought Questions

TQ13.1 Gastric acid is produced by parietal cells (see Chapter 30·IV·D). Parietal cell activity is stimulated by acetylcholine (ACh) released from vagal nerve terminals, histamine from enterochromaffin-like (ECL) cells, and gastrin from G cells (Fig. C13.1). ECL cells are enteroendocrine cells located below the gastric epithelium close to parietal cells. Histamine, acting as a paracrine, is the primary acid secretagogue. During a meal, parasympathetic nervous system (PSNS) activation causes ACh release from vagal nerve terminals, which increases acid production directly, but the vagus nerve also stimulates histamine and gastrin release from ECL cells and G cells, respectively (see Fig. 30.10). Gastrin also promotes histamine release from ECL cells. The patient has markedly elevated serum gastrin levels, so the increased rate of acid production is most likely a response to abnormal stimulation by excessive gastrin and histamine secretion.

TQ13.2 Negative feedback control of acid production is provided by somatostatin (SST) released from D cells in response to low stomach pH (see Chapter 30·IV·D). SST inhibits ECL and G-cell activity. Secretin also suppresses gastrin secretion by G cells (see Fig. C13.1). Secretin is produced by S cells located in the duodenum and jejunum. Secretin's principal function is to stimulate HCO_3^--rich secretions from the pancreas to neutralize acidic chyme entering the small intestine via the pylorus and thereby maintain a neutral pH.

TQ13.3 The patient has a gastrinoma, a neuroendocrine tumor that secretes gastrin in an unregulated manner. Gastrinomas usually arise in the proximal duodenum, but 25%–30% are pancreatic. Their remote location means that they are not affected by SST released by gastric D cells (some gastrinomas are insensitive to SST). Gastric G cells are inhibited by secretin, but gastrinomas paradoxically respond to secretin with increased gastrin release. Secretin is infused intravenously. A rise in circulating levels of >120 pg/mL over basal rates after 10 minutes is highly suggestive of a gastrinoma.

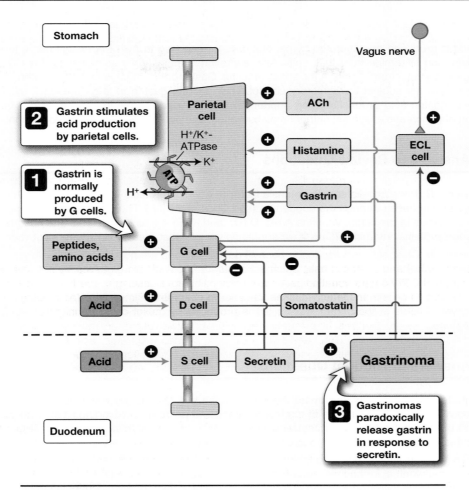

Figure C13.1.
Parietal cell regulation in Zollinger-Ellison syndrome. ACh = acetylcholine, ECL = enterochromaffin-like.

TQ13.4 The gastric epithelium is well protected against acid and gastric enzymes by the gastric mucosal barrier (see Chapter 30·IV·G). The barrier consists of a thick mucus coat that acts as a physical barrier, and surface epithelial cells secrete HCO_3^-, which remains trapped below the mucus coat and acts as a chemical barrier to acid and pepsin. Gastric epithelial tight junctions are also highly acid-impermeable. The duodenum does not have such protection, and its only defense against acid is HCO_3^--rich pancreatic secretions. In ZES, the stomach produces acid in amounts that overwhelm the duodenal defenses and may lead to erosion and ulceration of the duodenal mucosa.

TQ13.5 The most common cause of peptic ulcers (breaks in the intestinal epithelium lining the esophagus, stomach, or duodenum that exposes underlying tissue blood vessels) is *Helicobacter pylori* infection and nonsteroidal anti-inflammatory drug (NSAID) use (see Clinical Application 30.3). *H. pylori* establishes itself below the protective mucus layer and invades the underlying epithelium. The resultant inflammatory reaction can cause ulceration. NSAIDs inhibit the synthesis of prostaglandins that promote mucus secretion and support the intestinal epithelium through increased blood flow, which can leave the mucosa vulnerable to acid and gastric enzymes.

TQ13.6 Patients with ZES produce gastric acid in amounts that overwhelm the duodenum's ability to maintain a neutral pH. Duodenal enzymes function optimally near pH 7.0 and are inactivated by gastric acid. These include pancreatic enzymes, such as pancreatic lipase. Undigested nutrients travel through the small intestine to the large intestine, where they are acted on by colonic bacteria. Digestion products include gas that causes bloating, colonic irritants, and osmotically active solutes that cause diarrhea.

TQ13.7 PPIs bind to and irreversibly inactivate the parietal cell H^+/K^+-ATPase, which halts gastric acid production until new proton pumps can be synthesized and inserted into the parietal cell apical membrane.

Summary

Gastric acid is produced and secreted by parietal cells in response to ACh from vagal nerve terminals, gastrin from G cells, and histamine from ECL cells. Acid secretion is inhibited by somatostatin from D cells and, to a lesser degree, secretin from S cells in the small intestine. Gastrinomas, most commonly located in the duodenum or pancreas, are not subject to regulatory control. They secrete gastrin in large amounts, stimulating chronic increases in gastric acid production. The duodenum is ill-equipped to neutralize this acid excess, leading to inactivation of pancreatic enzymes; nutrient malabsorption; and potentially, peptic ulcer formation.

CASE 13: Answers to Review Questions

RQ13.1 **Answer = B.** Parietal cells secrete intrinsic factor (IF), which protects vitamin B_{12} from degradation by proteases during transit through the duodenum (see Chapter 31·II·F). The IF–vitamin B_{12} complex is absorbed by endocytosis in the ileum. Vitamins A (A), D (D), and K (E) are all lipid soluble and absorbed by fatty acid transport proteins. Vitamin C (C) is a water-soluble vitamin absorbed by Na^+-dependent vitamin C transport.

RQ13.2 **Answer = B.** Chief cells localize to the base of gastric pits and are notable for producing pepsinogen (inactive proenzyme) and gastric lipase, which accounts for ~30% of lipid digestion by the gastrointestinal tract (see Chapter 30·IV·F). Pancreatic lipases are responsible for digesting the remainder. Cholecystokinin (A) is secreted by I cells in the small intestine. Lactoferrin (C) is an antimicrobial iron-binding protein produced by salivary glands in the mouth. Mucus (D) is produced by goblet cells. Trypsinogen (E) is a pancreatic proenzyme.

CASE 14: Answers to Thought Questions

TQ14.1 The patient presents with Hashimoto thyroiditis, which is a chronic autoimmune disease characterized by extensive lymphocytic infiltration of the thyroid gland and destruction of follicles. Laboratory studies usually (>90% of patients) reveal high circulating levels of antibodies to thyroid peroxidase (TPO), as seen in the patient, and to thyroglobulin ([Tg] see Clinical Application 35.2).

TQ14.2 Thyroid gland activity is regulated by the HPT axis (see Chapter 35·II·B). Parvocellular neurons in the hypothalamic paraventricular nucleus secrete thyrotropin-releasing hormone (TRH) into the hypophyseal circulation. TRH targets thyrotropes in the anterior pituitary, which secrete thyroid-stimulating hormone (TSH) into the systemic circulation. TSH binds to a TSH receptor on thyroid follicles, which increases the activity and expression levels of most components of the thyroid-hormone synthetic pathway. Thyroid hormones exert negative feedback control of the hypothalamus and posterior pituitary. When circulating thyroid hormones levels fall, feedback inhibition is released, so circulating levels of both TRH and TSH would be expected to be high in this patient (Fig. C14.1).

TQ14.3 A dysfunctional thyroid gland that is unable to sustain thyroid hormone synthesis is considered primary hypothyroidism. Hypothyroidism caused by disorders of the pituitary (secondary) and hypothalamus (tertiary) are known as central hypothyroidism and are rare (<1% of cases). Secondary hypothyroidism is usually a result of pituitary disorders affecting all pituitary endocrine cell types (panhypopituitarism, or "panhypopit").

TQ14.4 An RAIU test involves having a patient ingest radioactive iodine in liquid or capsule form and then using a gamma probe to detect how much radioactivity has been taken up by the thyroid gland after 6–24 hours. Thyroid follicular cells avidly take up iodine from blood for use in hormone synthesis using a Na^+/iodide symporter (see Chapter 35·II·A). Destruction of thyroid follicles by autoimmune antibodies would be expected to reduce radiolabel uptake.

TQ14.5 Autoimmune destruction of thyroid follicles releases pre-formed thyroid hormones into the vasculature, causing circulating levels to rise. The result may be hyperthyroidism, characterized by increased metabolic rate, hyperactivity, heat intolerance and sweating, an irregular heart rate, and palpitations. The relationships between thyroid hormone levels and TRH and TSH secretion mean that circulating levels of both would decrease.

Summary

The patient has Hashimoto thyroiditis caused by autoimmune destruction of thyroid follicles. This is the most common cause of hypothyroidism. The disease is characterized by high levels of circulating antibodies against TPO and Tg and by low thyroid hormone levels. TRH and TSH levels both increase due to lack of feedback inhibition of the HPT axis. Increased TSH stimulates cell hyperplasia, resulting in enlargement of the thyroid gland.

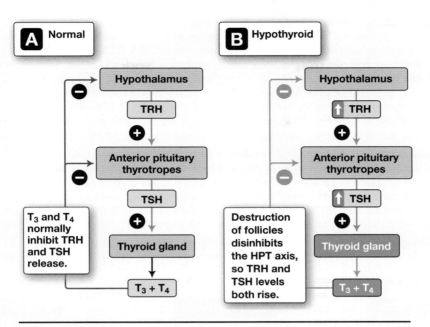

Figure C14.1.
Effects of hypothyroidism on the hypothalamic–pituitary–thyroid (HPT) axis. T_3 = triiodothyronine, T_4 = tetraiodothyronine, TRH = thyrotropin-releasing hormone, TSH = thyroid-stimulating hormone.

CASE 14: Answers to Review Questions

RQ14.1 Answer = C. After having been released from thyroid follicular cells, 99% of thyroid hormones bind to and are carried through the vasculature in association with plasma proteins (see Chapter 35·II·C). Thyroxine-binding globulin binds 75%–80%, albumin (A), and lipoproteins (B) carry <15% combined. Transthyretin binds 5%–10% (D), and <1% travels in free form (E).

RQ14.2 Answer = C. TRH is released into the hypophyseal portal circulation from hypothalamic paraventricular nucleus nerve terminals. Tanycytes, glia-like cells embedded in the lining of the third ventricle, control TRH levels through secretion of an endopeptidase that degrades TRH following release (see Chapter 35·II·D). Thyroid hormones stimulate endopeptidase secretion by tanycytes, thereby inhibiting the HPT axis. Thyroid hormones signal through binding to an intracellular receptor and modification of gene expression, not by activating the cAMP (A) or IP₃ signaling pathway (B). TRH is either carried to the anterior pituitary or degraded once released; it is not taken up from the median eminence and recycled (D). Tyrosine receptor kinases include the insulin receptor, not thyroid hormone receptors (E).

CASE 15: Answers to Thought Questions

TQ15.1 11β-OH is a key enzyme in the adrenal hormone synthetic pathway (see Fig. 34.2). In the zona glomerulosa, it converts DOC to corticosterone, which then becomes a substrate used by aldosterone synthase (AS) to generate aldosterone. In the zona fasciculata, 11β-OH converts 11-deoxycortisol to cortisol.

TQ15.2 Depending on the severity of the defect caused by the *CYP11B1* mutation and the degree of residual 11β-OH function, aldosterone and cortisol levels would be expected to be reduced or absent. Aldosterone is a mineralocorticoid involved in Na⁺ and K⁺ balance (see Chapter 28·III and IV). Aldosterone deficiency causes hyponatremia, hyperkalemia, and hypotension. Cortisol is an essential adrenal hormone involved in stress responses (see Chapter 34·II·D). Cortisol deficiency causes fatigue, weight loss, anxiety, and depression.

TQ15.3 When an enzyme deficiency interrupts a synthetic pathway, substrates of the affected enzyme accumulate behind the blockage, much as water builds up behind a beaver dam that has obstructed a stream. Because the adrenal cortex contains three steroid hormone–synthetic pathways that involve common substrates (e.g., cholesterol and pregnenolone), such blockages can cause an excess of substrate from one pathway to spill over into one of the other pathways, leading to a hormone excess (Fig. C15.1). In the case of 11β-OH deficiency, 11β-OH substrates spill over into the zona reticularis, where they feed excess adrenal androgen production. The principal adrenal androgens include dehydroepiandrosterone (DHEA), dehydroepiandrosterone sulfate (DHEAS), and androstenedione. Although these steroid hormones have little androgenic activity, they provide a pool of circulating precursors for peripheral conversion to more potent androgens, such as testosterone and estrogens, causing hirsutism and menstrual irregularities. Adrenal androgens also stimulate sebum production by sebaceous glands in the skin, leading to clogged pores, infection, inflammation, and acne.

TQ15.4 Aldosterone is synthesized from corticosterone, the reaction being catalyzed by AS (see Chapter 34·IV). AS activity is regulated by angiotensin II (Ang-II) and hyperkalemia. Ang-II circulates when blood pressure is low and when the sympathetic nervous system (SNS) is active. Once released into the circulation, aldosterone targets the renal distal segments, where it promotes simultaneous Na$^+$ retention and K$^+$ secretion. Na$^+$ retention expands extracellular fluid (ECF) and blood volume, which raises cardiac output and returns arterial pressure toward normal. K$^+$ secretion causes any K$^+$ excess to be excreted in urine, thereby rectifying the hyperkalemia.

TQ15.5 Because long-term control of blood pressure relies on aldosterone and its regulation of Na$^+$ reabsorption by the renal tubule, hypoaldosteronism would be expected to cause hyponatremia and hypotension. A

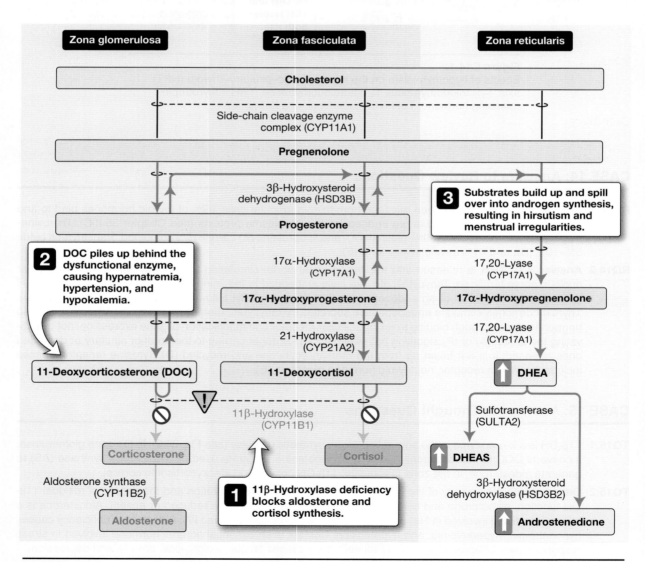

Figure C15.1.
Effect of 11β-hydroxylase deficiency on adrenal hormone production.

peculiarity of 11β-OH deficiency is that it instead causes hypertension due to DOC accumulation. DOC is the principal 11β-OH substrate in the zona glomerulosa (see Fig. C15.1). DOC has weak mineralocorticoid activity, but 11β-OH deficiency can cause DOC levels to build to the point that they adversely affect renal function, causing hypernatremia and hypertension.

TQ15.6 The RAAS is engaged when blood pressure falls and the SNS activates. SNS targets include juxtaglomerular cells in the wall of the glomerular afferent arteriole, which secrete renin when stimulated appropriately. Renin is also released in response to a decrease in glomerular perfusion pressure and glomerular filtration rate (GFR). Hypertension suppresses the SNS and increases glomerular perfusion pressure and GFR. Circulating levels of renin and RAAS intermediaries should all be low in this patient.

TQ15.7 Cortisol is controlled by the HPA axis (see Chapter 34·II·A). The hypothalamus secretes corticotropin-releasing hormone (CRH), which stimulates adrenocorticotropic hormone (ACTH) release from anterior pituitary corticotropes. ACTH stimulates side-chain cleavage (SCC) enzyme activity in the adrenals, thereby increasing substrate availability for cortisol synthesis. Cortisol exerts negative feedback control on the HPA axis at the level of both the hypothalamus and anterior pituitary. 11β-OH deficiency blocks cortisol synthesis, disinhibits the HPA, and leads to a chronic increase in circulating ACTH levels. The patient is not showing overt signs of hypocortisolism or ACTH excess (e.g., hypoglycemia and skin pigmentation), however, so it might be inferred that the 11β-OH deficiency is incomplete, and cortisol production is reduced rather than being inhibited 100%. The HPA axis would still be expected to be disinhibited, however, which would lead to increased circulating ACTH levels and adrenal hyperplasia to increase cortisol production capacity. 11β-OH deficiency is the second most common form of congenital adrenal hyperplasia (CAH), accounting for 5%–8% of total cases. The most common form (~95%) results from 21-hydroxylase deficiency.

TQ15.8 CAH treatment options include administration of corticosteroids to suppress the HPA axis, thereby reducing circulating ACTH levels as well as overproduction of DOC and adrenal androgens. Symptoms of androgen excess (hirsutism, acne) can be suppressed with androgen-receptor antagonists.

Summary

This patient has CAH caused by 11β-OH deficiency. 11β-OH is a key enzyme used in the synthesis of two adrenal steroid hormones: aldosterone and cortisol. In 11β-OH deficiency, aldosterone and cortisol levels fall, whereas the two 11β-OH substrates DOC and 11-deoxycortisol back up into the adrenal androgen synthetic pathway. 11-Deoxycortisol and 17-hydroxyprogesterone (a substrate for 11-deoxycortisol synthesis) levels both rise and can be detected in blood. DOC is a mineralocorticoid that stimulates Na^+ retention and K^+ excretion by the kidney, leading to hypernatremia, hypertension, and hypokalemia. Accumulating 11β-OH substrates fuel adrenal androgen excess, presenting as hirsutism, menstrual irregularities, and acne.

CASE 15: Answers to **Review Questions**

RQ15.1 Answer = B. *POMC* encodes several biologically active peptides in addition to ACTH. These include β-endorphin, enkephalin, melanocyte-stimulating hormone (MSH), and lipotropins (see Chapter 34·II·A). Anterior pituitary corticotropes process POMC to yield ACTH, but other tissues express enzymes necessary to release alternate gene products. β-Endorphin is an endogenous opioid neuropeptide produced by various neurons in the central and peripheral nervous systems. Angiotensinogen (A) is a plasma protein synthesized by the liver that is a renin substrate. Melanin (C) is a dark pigment produced by skin melanocytes. Melatonin (D) is a hormone produced by the pineal gland that helps coordinate the sleep–wake cycle. Testosterone (E) is a steroid hormone, not a peptide hormone.

RQ15.2 Answer = D. Ang-II and K^+ both raise intracellular Ca^{2+} levels, which increases steroidogenic acute regulatory protein (StAR) and AS expression levels (see Chapter 34·IV·B). The mechanism by which the two stimuli raise Ca^{2+} is different, however. Ang-II binds to an AT1 receptor (B), which couples to G_q, not G_s, (C) and the IP_3 signaling pathway. K^+ depolarizes the glomerulosa cell and activates voltage-dependent Ca^{2+} channels. Ang-II does not activate a receptor tyrosine kinase (A). Nitric oxide uses cGMP (E) as a second messenger, not Ang-II.

APPENDIX B
Normal Physiologic Values

Note that normal values typically vary with age, weight, and sex.

	Normal Range	
	Conventional Units	**S.I. Units**
Blood (serum or plasma)		
Anion gap	8–16 mEq/L	8–16 mmol/L
Bicarbonate (HCO_3^-)	22–28 mEq/L	22–28 mmol/L
Calcium, ionized (Ca^{2+})	8.4–10.2 mg/dL	2.1–2.8 mmol/L
Chloride (Cl^-)	95–105 mg/dL	95–105 mmol/L
Creatine kinase, male	25–90 U/L	25–90 U/L
Creatine kinase, female	10–70 U/L	10–70 U/L
Creatinine, male	0.7–1.3 mg/dL	62–115 µmol/L
Creatinine, female	0.6–1.1 mg/dL	53–97 µmol/L
Glucose, fasting	70–110 mg/dL	3.8–6.1 mmol/L
Glucose, 2-h postprandial	<120 mg/dL	<6.6 mmol/L
Hydrogen ion (H^+)	3.6–4.4 ng/dL	36–44 nmol/L
Magnesium, ionized (Mg^{2+})	1.5–2.0 mEq/L	0.75–1.0 mmol/L
O_2 saturation (SaO_2)	96%–100%	96%–100%
Osmolality	275–295 mOsmol/kg H_2O	275–295 mOsmol/kg H_2O
P_aO_2	75–105 mm Hg	10–14 kPa
P_aCO_2	33–45 mm Hg	4.4–5.9 kPa
pH	7.35–7.45	7.35–7.45
Phosphorus (inorganic)	3.0–4.5 mg/dL	1.0–1.5 mmol/L
Potassium (K^+)	3.5–5.0 mEq/L	3.5–5.0 mmol/L
Protein, total	6.0–7.8 g/dL	60–78 g/L
Protein, albumin	3.5–5.5 g/dL	35–55 g/L
Protein, globulin	2.3–3.5 g/dL	23–35 g/L
Sodium (Na^+)	136–145 mEq/L	136–145 mmol/L
Urea nitrogen (BUN)	7–18 mg/dL	1.2–3.0 mmol/L
Uric acid, male	4.5–8.0 mg/dL	0.27–0.47 mmol/L
Uric acid, female	2.5–6.2 mg/dL	0.15–0.37 mmol/L
Hematologic		
Erythrocyte count, male	4.3–5.9 million/mm^3	4.3–5.9 × 10^{12}/L
Erythrocyte count, female	3.5–5.5 million/mm^3	3.5–5.5 × 10^{12}/L
Hematocrit, male	41%–53%	0.41–0.53
Hematocrit, female	36%–46%	0.36–0.46
Hemoglobin A1c	≤6%	≤0.06
Hemoglobin, male	13.5–17.5 g/dL	2.09–2.71 mmol/L
Hemoglobin, female	12.0–16.0 g/dL	1.86–2.48 mmol/L
Mean corpuscular volume, male	80–94 µm³	80–94 fL
Mean corpuscular volume, female	81–100 µm³	81–100 fL

	Normal Range	
	Conventional Units	**S.I. Units**
Red blood cell distribution width	29–46 fL	11.6%–14.6%
Reticulocyte count	0.5%–1.5%	0.005–0.015
Cardiovascular		
Cardiac output, resting	5–7 L/min	5–7 L/min
Cardiac output, maximal	15–21 L/min	15–21 L/min
Central venous pressure	3–8 mm Hg	5–12 cm H_2O
Circulating blood volume	~5 L	~5 L
Ejection fraction	55%–75%	55%–75%
Heart rate, resting	60–100 beats/min	60–100 beats/min
Heart rate, maximal	220–age (in years) beats/min	220–age (in years) beats/min
Left atrial pressure	4–12 mm Hg	4–12 mm Hg
Mean arterial pressure	70–105 mm Hg	70–105 mm Hg
Mean pulmonary arterial pressure	10–20 mm Hg	10–20 mm Hg
Pulmonary vascular resistance	2–3.1 mm Hg·L·min^{-1}	2–3.1 mm Hg·L·min^{-1}
Right atrial pressure	1–8 mm Hg	1–8 mm Hg
Stroke volume	60–100 mL	60–100 mL
Systemic vascular resistance	11–15 mm Hg·L·min^{-1}	11–15 mm Hg·L·min^{-1}
Respiratory		
a-v O_2 difference, resting	4–5 mL/dL	4–5 mL/dL
a-v O_2 difference, maximal	~15 mL/dL	~15 mL/dL
Alveolar–arterial O_2 difference	(age/4) + 4	(age/4) + 4
Anatomic dead space	~150 mL	~150 mL
O_2 consumption, resting	~0.25 L/min	~0.25 L/min
O_2 consumption, maximal	2.5–4.0 L/min	2.5–4.0 L/min
Respiration rate	12–20 breaths/min	12–20 breaths/min
Tidal volume	~0.5 L	~0.5 L
Total lung capacity, male	~6.0 L	~6.0 L
Total lung capacity, female	~4.2 L	~4.2 L
Vital capacity, male	~4.8 L	~4.8 L
Vital capacity, female	~3.1 L	~3.1 L
Renal		
Glomerular filtration rate	90–160 mL/min/1.73 m^2	90–160 mL/min/1.73 m^2
Interstitial fluid	~10.5 L	~10.5 L
Intracellular fluid	~28 L	~28 L
Plasma volume	~3.5 L	~3.5 L
Renal blood flow	~1,200 mL/min	~1,200 mL/min
Renal plasma flow	~625 mL/min	~625 mL/min
Total body water	~42 L (60% of total body weight)	~42 L (60% of total body weight)
Endocrine		
Aldosterone (supine serum)	2–5 ng/dL	55–138 pmol/L

	Normal Range	
	Conventional Units	**S.I. Units**
Antidiuretic hormone (plasma)		
<290 mOsmol/kg H_2O	<1.7 pg/mL	<2 pmol/L
>290 mOsmol/kg H_2O	1–10 pg/mL	2–12 pmol/L
Cortisol (free 24-h urine)	20–70 µg/dL	11–135 nmol
Estradiol (serum), male	10–60 pg/mL	<105 pmol/L
Estradiol (serum), female		
follicular phase	25–75 pg/mL	145–1,375 pmol/L
midcycle	200–600 pg/mL	350–2,800 pmol/L
luteal phase	100–300 pg/mL	175–1,615 pmol/L
postmenopausal	5–25 pg/mL	<40 pmol/L
Glucagon (plasma)	<60 pg/mL	<17 pmol/L
Growth hormone (serum)	<10 ng/mL	<465 pmol/L
Insulin (serum), fasting	<0.8 ng/mL	<135 pmol/L
Parathyroid hormone (serum)	14–72 pg/mL	1.0–6.5 pmol/L
Testosterone (serum), male	280–1,100 ng/dL	8.65–38.15 nmol/L
Testosterone (serum), female premenopausal	15–70 ng/dL	0.52–2.43 nmol/L
Testosterone (serum), female postmenopausal	8–35 ng/dL	0.28–1.22 nmol/L
Thyroxine (free serum)	5–12 µg/dL	10.5–35 pmol/L

APPENDIX C
Reference Equations

Diffusion (Fick law)	$J = P \times A(C_1 - C_2)$
Equilibrium potential (Nernst equation)	$E_x = \dfrac{-RT}{zF} \ln \dfrac{[X]_i}{[X]_o}$
Hemodynamic Ohm law	$P = Q \times R$
Systemic vascular resistance	$SVR = \dfrac{MAP - CVP}{CO}$
Mean arterial pressure	$MAP = DBP + \dfrac{(SBP - DBP)}{3}$
Resistance (Poiseuille law)	$R = \dfrac{8L\eta}{\pi r^4}$
Turbulence (Reynolds equation)	$Re = \dfrac{vd\rho}{\eta}$
Wall stress (Laplace law)	$\sigma = P \times \dfrac{r}{2h}$
Compliance	$C = \dfrac{\Delta V}{\Delta P}$
Cardiac output	$CO = HR \times SV$
Ejection fraction	$EF = \dfrac{SV}{EDV}$
Starling law of the capillary	$Q = K_f \left[(P_c - P_{if}) - (\pi_c - \pi_{if}) \right]$
Alveolar ventilation	$V_A = (TV - V_D) \times breaths/min$
Alveolar gas equation	$P_A o_2 = P_I o_2 - \dfrac{P_A co_2}{RER}$
Dead space	$V_D = TV \times \dfrac{P_a co_2 - P_E co_2}{P_A co_2}$
Vital capacity	$VC = TV + IRV + ERV$
Respiratory exchange ratio	$RER = \dfrac{\dot{V}co_2}{\dot{V}o_2}$
Renal clearance	$C = \dfrac{UV}{P}$
Glomerular filtration rate	$GFR = \dfrac{[U]_{Creatinine} \times V}{[P]_{Creatinine}}$
Free water clearance	$C_{H2O} = V - C_{Osm}$
Henderson-Hasselbalch equation	$pH = pK + \log \dfrac{[HCO_3^-]}{[CO_2]}$
Serum anion gap	$Anion\ gap = [Na^+] - ([Cl^-] + [HCO_3^-])$
Plasma osmolality	$Osmolality = 2[Na^+] + \dfrac{[glucose]}{18} + \dfrac{[BUN]}{2.8}$

Figure Credits

Figure 4.4 (lower): Stedman's Medical Dictionary.

Figure 5.1: Modified from Jennings HS. *Behavior of the Lower Organisms*. The Columbia University Press; 1906.

Figure 5.12: Modified from Moore KL, Dalley AF. *Clinical Oriented Anatomy*. 4th ed. Lippincott Williams & Wilkins; 1999.

Figure 12.3: Modified from Seifter J, Ratner A, Sloane D. *Concepts in Medical Physiology*. Lippincott Williams & Wilkins; 2005.

Figure 12.4B: From Dudek RW. *High-Yield Histopathology*. 2nd ed. Wolters Kluwer; 2010.

Figure 12.8A: Photograph from Cohen BJ, Taylor JJ. *Memmler's the Human Body in Health and Disease*. 11th ed. Lippincott Williams & Wilkins; 2009.

Figure 13.1 (upper): Micrograph from Moore KL, Agur A. *Essential Clinical Anatomy*. 7th ed. Lippincott Williams & Wilkins; 2024.

Figure 13.1 (lower panel): Micrograph from Eroschenko VP. *Atlas of Histology with Functional Correlations*. 13th ed. Lippincott Williams & Wilkins; 2017.

Figure 13.2: Data from Kuo KH, Seow CY. *J Cell Sci*. 2003;117:1503-1511.

Figure 14.3 (upper): Crystal data from Robinson RA. *J Bone Jt Surg Am*. 1952;34:389-476.

Figure 14.4: Model (lower) from Thurner PJ. *Nanomed Nanobiotechnol*. 2009;1:624-629.

Figure 16.4: Cui D, Daley W. *Histology from a Clinical Perspective*. 2nd ed. Wolters Kluwer; 2023.

Figure 20.4: Data from Harper AM. *Acta Neurol Scand*. 1965;14:94.

Figure 20.5: Data from Ingvar DH. *Brain Res*. 1976;107:181-197.

Figure 21.9: From Kahn GP, Lynch JP. *Pulmonary Disease Diagnosis and Therapy: A Practical Approach*. Lippincott Williams & Wilkins; 1997.

Figure 21.17 (panels A, B, and D): Cagle PT. *Color Atlas and Text of Pulmonary Pathology*. Lippincott Williams & Wilkins; 2005.

Figure 23.2: From the Centers for Disease Control and Prevention Public Health Images Library. No. 7315 (photo courtesy of Janice Carr).

Figure 24.17: From National Oceanic and Atmospheric Administration (photo credit Doug Kesling).

Figure 25.6 (lower two micrographs): From Schrier RW. *Diseases of the Kidney and Urinary Tract*. 8th ed. Lippincott Williams & Wilkins; 2006.

Figure 26.1B: Reprinted from Clapp WL, Park CH, Madsen KM, Tisher CC. Axial heterogeneity in the handling of albumin by the rabbit proximal tubule. *Lab Invest*. 1988;58(5), 549–558. Copyright © 1988 United States & Canadian Academy of Pathology. With permission.

Figure 26.6: Data from Rector FC. *Am J Physiol*. 1983;244:F461-F471.

Figure 27.5: Based on data from Pannabecker TL, Dantzler WH, Layton HE, et al. *Am J Physiol*. 2008;295:F1271-F1285.

Figure 27.7: © Stephen Barnett, www. OzAnimals.com; https://creativecommons.org/licenses/by/2.0/

Figure 27.15 (lower): Reprinted from Clapp WL, Madsen KM, Verlander JW, Tisher CC. Morphologic heterogeneity along the rat inner medullary collecting duct. *Lab Invest*. 1989;60(2), 219-230. Copyright © 1989 United States & Canadian Academy of Pathology. With permission.

Figure 31.13: Radiograph from Dean D, Herbener TE. *Cross-Sectional Human Anatomy*. Lippincott Williams & Wilkins; 2000.

Figure 35.7: Modified from Golan DE, Tashjian AH, Armstrong EJ. *Principles of Pharmacology: The Pathophysiologic Basis of Drug Therapy*. 4th ed. Wolters Kluwer Health; 2017.

Figure 36.10B,C: ACC Anatomy I, 2008-05-08 0841, OB/GYN Anatomy, Female Reproductive System, Anatomical Chart Company.

Figure 38.9: Mills WJ Jr. *Papers, Archives and Special Collections, Consortium Library*. University of Alaska Anchorage; 1993.

Figure 40.1: Amherst College Archives & Special Collections. http://en.wikipedia.org/wiki/Emily_Dickinson

Figure 40.20C: Radiograph from Topol EJ, Califf RM, Isner J. *Textbook of Cardiovascular Medicine*. 3rd ed. Lippincott Williams & Wilkins; 2006.

Figure 40.22B: Radiograph from Frymoyer JW, Wiesel SW. *The Adult and Pediatric Spine*. 3rd ed. Lippincott Williams & Wilkins; 2004.

Figure 40.23 (lower), Clinical Application 14.3 (upper): Radiograph from Strayer DS, Saffitz JE, Rubin E. *Rubin's Pathology: Mechanisms of Human Disease*. 8th ed. Lippincott Williams & Wilkins; 2020.

Clinical Application 3.1: Photograph from Eisenberg RL. *Clinical Imaging: An Atlas of Differential Diagnosis*. 4th ed. Lippincott Williams & Wilkins; 2003.

Clinical Application 4.1: Photograph from Berg D, Worzala K. *Atlas of Adult Physical Diagnosis*. Lippincott Williams & Wilkins; 2006.

Clinical Application 4.3: Photograph reprinted with permission from Doshi DC, Limdi PK, Parekh NV, Gohil NR. Oculodentodigital dysplasia. *Ind J Ophthalmol*. 2016;64(3):227-230. Copyright © 2016 Indian Journal of Ophthalmology.

Clinical Application 5.2, Clinical Application 33.1 (upper): Photograph from Smeltzer SC, Bare BG, Hinkle JL, et al. *Brunner and Suddarth's Textbook of Medical—Surgical Nursing*. 12th ed. Lippincott Williams & Wilkins; 2011.

Clinical Application 11.1: Magnetic resonance image courtesy of Nokes SR, Little Rock, Arkansas.

Clinical Application 14.1: Radiograph courtesy of Wei T, Portland, Oregon.

Clinical Application 14.2: Bone scan from Koopman WJ, Moreland LW. *Arthritis and Allied Conditions—A Textbook of Rheumatology*. 15th ed. Lippincott Williams & Wilkins; 2005.

Clinical Application 15.4: Photograph from Elder DE. *Lever's Histopathology of the Skin*. 12th ed. Lippincott Williams & Wilkins; 2023.

Clinical Application 18.2 (upper): Clinical Cardiopulmonary LifeART Collection, OVID - CAD.

Clinical Application 20.2. Photograph courtesy of Medtronics, Peripheral Division, Santa Rosa, California.

Clinical Application 23.1: Micrographs from Pereira I, George TI, Arber DA. *Atlas of Peripheral Blood: The Primary Diagnostic Tool*. Lippincott Williams & Wilkins; 2011.

Biologic Sex and Aging 23.2 (right figure): From Anderson SC. *Anderson's Atlas of Hematology*. 3rd ed. Lippincott Williams & Wilkins; 2020.

Biologic Sex and Aging 25.1: Micrographs from Jennette JC, D'Agati VD, Olson JL, Silva FG. *Heptinstall's Pathology of the Kidney*. 8th ed. Lippincott Williams & Wilkins; 2024.

Clinical Application 26.2: From https://commons.wikimedia.org/wiki/File:Fluorescent_uric_acid.JPG. CC BY-SA 4.0

Clinical Application 30.1: Photograph from Gold DH, Weingeist TA. *Color Atlas of the Eye in Systemic Disease*. Lippincott Williams & Wilkins; 2000.

Clinical Application 33.2: Photograph from Willis MC. *Medical Terminology: A Programmed Learning Approach to the Language of Health Care*. 2nd ed. Lippincott Williams & Wilkins; 2008.

Clinical Application 34.1: Photograph from Reisner H, Rubin E. *Essentials of Rubin's Pathology*. 6th ed. Lippincott Williams & Wilkins; 2014.

Clinical Application 34.2: Photograph from Schaaf CP, Zschocke J, Potocki L. *Human Genetics*. Lippincott Williams & Wilkins; 2011.

Clinical Application 35.1: Photograph from Weber JR, Kelley JH. *Health Assessment in Nursing*. 7th ed. Lippincott Williams & Wilkins; 2022.

Clinical Application 35.4: Photograph from Flynn JM, Skaggs DL, Waters PM. *Rockwood and Wilkins' Fractures in Children*. 9th ed. Lippincott Williams & Wilkins; 2020. Used with permission of the Children's Orthopaedic Center, Children's Hospital Los Angeles.

From Gartner LP. *Gartner & Hiatt's Atlas and Text of Histology*. 8th ed. Wolters Kluwer; 2023. Micrographs appearing in Figures 4.15 (upper panel) and 16.4.

Modified from Bear MF, Connors BW, Paradiso MA. *Neuroscience—Exploring the Brain*. 4th ed. Lippincott Williams & Wilkins; 2016. Figures 7.4, 7.11, 7.12, 9.3, 9.4, 9.6, 9.7, 9.13, 12.7, 15.12A, 36.10A, and Clinical Application 8.2.

From the Centers for Disease Control and Prevention, Public Health Image Library: Photographs appearing in Clinical Applications 5.1 (Photo Credit Dr. Fred Murphy, Sylvia Whitfield, 1975), 12.4, 18.3, and 19.1 (Photo Credit Dr. Edwin P. Ewing, Jr, 1972). Data appearing in Tables 37.1 and 40.1.

Modified from Chandar N, Viselli S. *Lippincott® Illustrated Reviews: Cell and Molecular Biology*. 3rd ed. Lippincott Williams & Wilkins; 2023. Figures 1.2, 1.3, 1.4, 1.5, 1.6, 1.12, 1.13, 1.18, 1.21, 1.22, 1.23, 4.5, and 4.10.

Modified from Clarke MA, Finkel R, Rey JA, et al. *Lippincott® Illustrated Reviews: Pharmacology*. 5th ed. Lippincott Williams & Wilkins; 2012. Figures 1.17 and 1.20.

From Daffner RH. *Clinical Radiology—The Essentials*. 4th ed. Lippincott Williams & Wilkins; 2014, Photographs appearing in Figure 21.12 and 22.12 and Clinical Application 4.2.

From Feigenbaum H., Armstrong WF, Ryan T. *Feigenbaum's Echocardiography*. 6th ed. Lippincott Williams & Wilkins; 2004. Photographs appearing in Clinical Applications 17.1 (upper panel) and 18.2 (lower panel).

From Fleisher GR, Ludwig W, Baskin MN. *Atlas of Pediatric Emergency Medicine*. Lippincott Williams & Wilkins; 2004. Photographs appearing in Clinical Applications 1.2, 6.1, and 25.2 (lower).

From Goodheart HP. *Goodheart's Photoguide of Common Skin Disorders*. 3rd ed. Lippincott Williams & Wilkins; 2009. Photographs appearing in Figure 15.2 and Clinical Applications 35.3 and 36.2 and Biologic Sex and Aging 36.1.

From Gorbach SL, Bartlett JG, Blacklow NR. *Infectious Diseases*. 3rd ed. Lippincott Williams & Wilkins; 2004. Photographs appearing in Clinical Applications 4.5 and 29.2 (right).

Modified from Harvey RA, Ferrier DR. *Lippincott® Illustrated Reviews: Biochemistry*. 5th ed. Lippincott Williams & Wilkins; 2011. Figures 23.7, 23.11, 23.12, and 40.6, Clinical Application 33.1 (lower panel).

Modified from Ferrier DR. *Lippincott® Illustrated Reviews: Biochemistry*. 7th ed. Lippincott Williams & Wilkins; 2017. Figures 23.5, 23.10, and Clinical Application 23.2.

Modified from Klabunde RE. *Cardiovascular Physiology Concepts*. 3rd ed. Lippincott Williams & Wilkins; 2022. Figures 16.18, 17.1, 17.5, 19.5, and 20.13.

From Klossner NJ, Hatfield N. *Introductory Maternity and Pediatric Nursing*. 5th ed. Lippincott Williams & Wilkins; 2022. Photographs appearing in Figure 37.12 and Clinical Application 15.1.

Modified from Krebs C, Weinberg J, Akesson E. *Lippincott® Illustrated Reviews: Neuroscience*. 2nd ed. Lippincott Williams & Wilkins; 2018. Figures 6.5, 6.7, 6.11, 6.12, 7.9, 8.1, 8.9, 8.10, 8.14, 9.2, 9.15, 10.5, 10.6, 11.4, 15.13, and 20.2.

Modified from McArdle WD, Katch FI, Katch VL. *Exercise Physiology*. 8th ed. Lippincott Williams & Wilkins; 2015. Figures 12.6, 12.11, and 23.4.

Modified from McConnell TH. *The Nature of Disease Pathology for the Health Professions*. 2nd ed. Lippincott Williams & Wilkins; 2014. Clinical Application 15.2 and photographs appearing in Figure 23.14 and Clinical Applications 12.1, 17.4, 22.1, 25.1, 30.4, and 34.3.

From Mills SE. *Histology for Pathologists*. 5th ed. Lippincott Williams & Wilkins; 2020. Photographs appearing in Figures 1.9, 4.4A, 10.5B, 13.1 (middle panel), 14.7, 25.12 (lower panel), 31.2, 34.1 (lower panel), and 35.2 (upper panel).

Modified from Porth CM. *Essentials of Pathophysiology*. 2nd ed. Lippincott Williams & Wilkins; 2007. Figures 14.6 and 25.14.

Modified from Premkumar K. *The Massage Connection: Anatomy and Physiology*. Lippincott Williams & Wilkins; 2004. Figures 15.3 and 36.11 (lower panel).

Modified from Rhoades RA, Bell DR. *Medical Physiology*. 6th ed. Lippincott Williams & Wilkins; 2022. Figures 12.14, 19.13, 20.8, 21.23, 25.5, 25.9, and 37.5.

Modified from Rubin E, Farber JL. *Pathology*. 3rd ed. Lippincott Williams & Wilkins; 1999. Clinical Application 36.1, and Photographs appearing in Figures 14.5, 20.9A, 21.17C, and 40.5; and Clinical Applications 16.1, 25.2 (upper panel), and 32.2.

Modified from Siegel A, Sapru HN. *Essential Neuroscience*. 4th ed. Lippincott Williams & Wilkins; 2019. Figures 6.6, 6.8, 11.11, 11.12, and 11.13.

From Tasman W, Jaeger E. *The Wills Eye Hospital Atlas of Clinical Ophthalmology*. 2nd ed. Lippincott Williams & Wilkins; 2007. Photographs appearing in Figure 8.4 and Clinical Applications 8.1 and 12.2.

Modified from Taylor CR, Lynn P, Bartlett J. *Fundamentals of Nursing, The Art and Science of Nursing Care*. 10th ed. Lippincott Williams & Wilkins; 2022. Figure 38.7 and Clinical Application 24.1.

From Uflacker R. *Feigenbaum's Atlas of Vascular Anatomy: An Angiographic Approach*. 2nd ed. Lippincott Williams & Wilkins; 2006. Photographs appearing in Figures 25.4 and 25.6 (upper panel). Reprinted with permission from Sampaio, F.J.B.

From Yochum TR, Rowe LJ. *Yochum and Rowe's Essentials of Skeletal Radiology*. 3rd ed. Lippincott Williams & Wilkins; 2004. Photographs appearing in Figure 40.22A and Clinical Application 14.3 (lower).

Modified from West JB. *Best and Taylor's Physiological Basis of Medical Practice*. 12th ed. Williams & Wilkins; 1991. Figures 22.9, 37.10, and 37.11.

Modified from West JB, Luks AM. *Respiratory Physiology—The Essentials*. 11th ed. Lippincott Williams & Wilkins; 2022. Figures 21.1, 21.2, 21.3, 21.14, 21.19, 21.22, 22.10, 22.11, 23.8, 24.10, 24.12, 28.12, and 28.14-28.17.

Case 3: Photograph from https://commons.wikimedia.org/wiki/File:Congenital_nema-line_myopathy.jpg. CC BY-SA 3.0

Case 8: Photograph from https://en.wiki-pedia.org/wiki/Minimal_change_disease#/media/File:Nephrotic_eyes.JPG. CC BY-SA 3.0

Index

O

P